T0134396

Springer Series in Computational Mathematics

Volume 52

Editorial Board

R.E. Bank
R.L. Graham
W. Hackbusch
J. Stoer
R.S. Varga
H. Yserentant

More information about this series at http://www.springer.com/series/797

Joachim Gwinner • Ernst Peter Stephan

Advanced Boundary Element Methods

Treatment of Boundary Value, Transmission and Contact Problems

Springer

Joachim Gwinner
Fakultät für Luft- und Raumfahrttechnik
Universität der Bundeswehr München
Neubiberg/München
Germany

Ernst Peter Stephan
Institut für Angewandte Mathematik
Leibniz Universität Hannover
Hannover, Germany

ISSN 0179-3632 ISSN 2198-3712 (electronic)
Springer Series in Computational Mathematics
ISBN 978-3-030-06346-7 ISBN 978-3-319-92001-6 (eBook)
https://doi.org/10.1007/978-3-319-92001-6

Mathematics Subject Classification (2010): 35-XX, 45-XX, 49-XX, 65-XX

© Springer International Publishing AG, part of Springer Nature 2018
Softcover re-print of the Hardcover 1st edition 2018
This work is subject to copyright. All rights are reserved by the Publisher, whether the whole or part of the material is concerned, specifically the rights of translation, reprinting, reuse of illustrations, recitation, broadcasting, reproduction on microfilms or in any other physical way, and transmission or information storage and retrieval, electronic adaptation, computer software, or by similar or dissimilar methodology now known or hereafter developed.
The use of general descriptive names, registered names, trademarks, service marks, etc. in this publication does not imply, even in the absence of a specific statement, that such names are exempt from the relevant protective laws and regulations and therefore free for general use.
The publisher, the authors and the editors are safe to assume that the advice and information in this book are believed to be true and accurate at the date of publication. Neither the publisher nor the authors or the editors give a warranty, express or implied, with respect to the material contained herein or for any errors or omissions that may have been made. The publisher remains neutral with regard to jurisdictional claims in published maps and institutional affiliations.

Printed on acid-free paper

This Springer imprint is published by the registered company Springer International Publishing AG part of Springer Nature.
The registered company address is: Gewerbestrasse 11, 6330 Cham, Switzerland

To our wives Hannelore and Karin Sabine for their love and understanding.

Preface

The boundary element method (BEM) has become an important tool to provide approximate solutions for boundary integral equations covering a rich area of applications in engineering and physics. Today, there exist many books and survey articles on boundary integral equations and on boundary element methods [98, 112, 225, 259, 260, 276, 304, 359, 362, 391]. However, we believe that modern topics like adaptive methods; treatment of general transmission, screen, crack, and contact problems; and the *hp*-version of the BEM are dealt with in special research papers only. In this book, we collect some of the key results of these topics, prove them in detail, and describe the most important approaches. We elaborate on the mathematical analysis of both the boundary integral equations and the BEM and demonstrate the power of the BEM with numerical results for representative problems from various applications in acoustics, electromagnetics, and solid mechanics covering Laplace, Helmholtz, Navier–Lame, and Maxwell partial differential equations.

Our book introduces the reader into the classical setting of boundary integral equations and standard boundary element methods in Chaps. 1–3 and Chap. 6. The book covers advanced boundary element methods in recent research areas as mentioned above in Chaps. 4, 5, 7–13.

In Chap. 4, we apply the modern tool of pseudodifferential operators to mixed boundary value problems and transmission problems.

In Chap. 5, we focus on the Signorini problem and more nonsmooth BVPs, dealing with unilateral contact without and with friction and nonmonotone contact in delamination.

In Chap. 6, we collect basic issues of BEM, covering Galerkin and collocation methods with modifications and extensions (augmented boundary elements, duality estimates, and qualocation).

In Chap. 7, we turn to boundary value problems (BVPs) in nonsmooth domains and present improved BEM with graded meshes and higher polynomial approximation.

In Chap. 8, we investigate in detail the exponential convergence of the *hp*-version BEM on geometrically graded meshes.

In Chap. 9, we employ the Mellin transform and analyze the boundary integral operators on polygonal domains in depth.

In Chap. 10, we study the adaptive BEM using error estimators of residual type and of hierarchical type, and also we give results on the convergence of adaptive boundary element schemes.

In Chap. 11, we extend the BEM to unilateral contact problems without and with friction and nonmonotone contact problems from delamination.

In Chap. 12, we analyze the symmetric FEM–BEM coupling for various transmission problems in applications.

The final Chap. 13 is devoted to the time-dependent BEM (TD-BEM). We treat the scattering of waves at screens and time-dependent contact problems using retarded potentials.

In the Appendix, we collect some fundamental concepts of linear operator theory and also provide some supplementary material on Fourier transform and pseudodifferential operators. Further, we present a short course on convex and nonsmooth analysis leading to linear and nonlinear variational inequalities and their approximation. Also, some aspects of implementations of BEM are given.

For the ease of the reader, the chapters are self-contained; hence, it is unavoidable that the text has some repetitions.

Different from standard textbooks and monographs on BEM, we stress on first kind integral equations, adaptive methods, the *hp*-version of BEM, and the application of BEM to contact problems with recent developments for the dynamic case. Our book is addressed to mathematicians and engineers as well as to graduate students. Therefore, we provide the necessary foundations of BEM and demonstrate the applicability of BEM via prototype problems. We put specific emphasis on numerical approaches underlined by representative numerical simulations.

One of the main concerns of the book is the abstract setting of the convergence of the boundary element method. This is dealt with by the key theorems on the convergence of the projection method (Theorems 1.1, 1.2, 6.1, 6.11). Another prime topic of the book is the regularity of solutions of elliptic boundary value problems in polygonal and polyhedral domains and hence of solutions of the corresponding boundary integral equations on polygonal curves and polyhedral surfaces. Here, the reduced regularity of the solution near corners and edges requires special boundary element methods like enrichment by singularity functions or the use of graded meshes or *hp*-techniques. The latter are investigated in detail, and especially exponentially fast *hp*-methods are described. Another way to tackle the loss of regularity of the solutions is to use adaptive boundary element methods, also described in detail for *h*, and *p*-versions. A further prime topic is the use of BEM for unilateral contact problems and thus the analysis of boundary variational inequalities. Furthermore, the symmetric FEM/BEM coupling is analyzed and various applications are given. Also, the time-domain boundary element method is investigated for the time-dependent acoustic scattering. Important mathematical tools for the analysis, presented here, are Fourier and Mellin transform together with pseudodifferential operator techniques.

Chapter 1 introduces to the theory of approximation methods for the solution of operator equations and for the solution of related variational problems. Chapter 2 is of introductory character and gives the standard approach from potential theory to boundary integral equations. Chapter 3 introduces the concept of periodic Sobolev spaces with the help of Fourier series and constructs the solution to the interior and exterior Dirichlet problems for the Laplacian on the unit sphere by a Fourier series approach. The mapping properties of weakly singular and hypersingular boundary integral operators are analyzed by Fourier series; with this tool, a Gårding inequality is derived for the first time. A perturbation argument allows us to go from the unit circle to smooth curves. Chapter 4 deals first with smooth surfaces and uses the concept of Fourier transformation and pseudodifferential operators to treat the mixed Dirichlet–Neumann BVPs for the Laplacian and the acoustic interface problem with the Helmholtz equation as well as crack/screen problems and time-harmonic exterior Maxwell problems. Generalizations to Lipschitz curves and surfaces are done for interface problems in linear elasticity in Sect. 4.4. In Chap. 5, we present the boundary integral approach for the scalar Signorini problem with the Laplacian and for unilateral contact problems without and with friction; in addition, we treat nonmonotone contact problems from delamination by a combination of boundary integral methods and regularization techniques from nondifferentiable optimization. Chapter 6 starts with an abstract setting for the Galerkin method for strongly elliptic operator equations. The h-version BEM (Galerkin and collocation) is presented in the frame of general projection methods. Sections 6.1–6.5 provide some fundamental facts of BEM, including BEM on quasiuniform meshes, Aubin–Nitsche duality estimate, superapproximation, and local/L^∞ error estimates. Sections 6.6–6.10 cover special topics like discrete collocation, augmentation of the boundary element space by special singular functions, and modified collocation and qualocation. In Sect. 6.11, a meshless method with radial basis functions is presented for integral equations of the first kind; herewith scattered satellite data can be accounted for. Chapter 7 is devoted to the hp-version BEM on polygonal and polyhedral domains using first uniform meshes. Then the results for the h-version are extended to graded meshes. Chapter 8 presents the hp-version of the BEM on geometrically refined meshes and shows its exponentially fast convergence. In Chap. 9, the notion of Mellin symbols for the boundary integral operators is introduced and their mapping properties in countably normed spaces are derived. Chapter 10 is devoted to adaptive boundary element methods. The results on residual error estimators for integral equations on curves are given in Sects. 10.1–10.2 and on surfaces in Sects. 10.3–10.4. Special emphasis is given to the two-level approach with hierarchical error estimator in Sect. 10.5 for the h-version and Sect. 10.6 for the p-version. The convergence of adaptive BEM schemes is investigated in Sect. 10.7. Chapter 11 extends the BEM for contact problems with special emphasis on the use of Gauss–Lobatto–Lagrange basis functions for the hp-version in Sects. 11.2,11.3 and of biorthogonal basis functions in the mixed scheme in Sect. 11.4. Sections 11.5, 11.5 combine regularization techniques of nondifferentiable optimization with h-BEM or hp-BEM for delamination problems. Chapter 12 overviews the symmetric FEM/BEM coupling method. Interface problems together with contact conditions and strongly nonlinear operators in the FEM domain are analyzed. Also, different mixed formulations (primal/dual) are considered. Moreover, least squares coupling methods are studied. Further, the symmetric coupling for the time-harmonic eddy current problem from electromagnetics is addressed. Also, for a parabolic-elliptic interface problem a FEM/BEM coupling is given. The final Chap. 13 considers dynamic scattering and contact problems and uses the tool of retarded potentials to obtain Galerkin approximations with the TD-BEM based on marching-on-in-time (MOT) schemes. The Appendix supports reading of the book and has 4 parts: In Appendix A, we give the fundamentals of linear operator theory. In Appendix B, we present a short introduction into pseudodifferential operators. In Appendix C, we collect some aspects on variational inequalities and convex and nonsmooth analysis. Finally in Appendix D, we describe the implementation of the BEM for some representative examples on curves and surfaces.

The introductory part of this monograph (Sects. 1.1–6.5 and Appendix) grew out of lecture notes from courses given by the authors at the Universität der Bundeswehr München and at the Leibniz Universität Hannover, whereas the other sections deal with research topics.

First of all, we want to thank our wives Hannelore Raith and Karin Sabine Stephan for their great understanding and support during the work-intensive time, it took us to write our book.

The authors thank their colleagues L. Banz, C. Carstensen, A. Chernov, M. Costabel, J. Elschner, G. Gatica, H. Gimperlein, N. Heuer,F. Leydecker, M. Maischak, P. Mund, N. Ovcharova, D. Praetorius, T. von Petersdorff, T. Tran, and W.L. Wendland for their cooperation which has highly influenced the contents of the book. Especially we thank C. Özdemir for his continuous, generous, and very pleasant support in producing the manuscript.

München, Germany — Joachim Gwinner
Hannover, Germany — Ernst Peter Stephan
2018

first kind BIE's for
BVP's, transmission, contact
Chapter 2,4,5

general projection methods
Chapter 1,
Section 6.1,6.2,6.8,10.1,12.1

standard BEM
h-version
Chapter 6

advanced BEM
h,p,hp-versions
Chapter 7,8

Tools:
Fourier series (Chapter 3)
Fourier transformation (Chapter 4)
Mellin transformation (Chapter 9)
linear operator theory (Appendix A)
pseudodifferential operators
(Appendix B)
variational inequalities, convex and
nonsmooth analysis (Appendix C)
some implementations
for BEM (Appendix D)

A-BEM
Chapter 10

BEM for contact
Chapter 11

FE/BE coupling
Chapter 12

TD-BEM
Chapter 13

Contents

Chapter 1
Introduction

This chapter gives an introduction to the theory of approximation methods for the solution of operator equations and for the solution of related variational problems. In the first section we formulate the basic approximation problems and their setting. Then in the following section we present a collection of various examples and model applications in a simplified way. In the following chapters we shall elaborate at these examples at the more deeper level of boundary value problems that arise from diverse fields of mathematical physics. Then we shall reformulate these boundary value problems as first kind integral equations and focus to boundary element methods for their numerical treatment.

In here a heuristic approach is given in order to show briefly the fundamental questions in the theory of approximation methods. This chapter should motivate the reader to go into the next chapters with at least some knowledge about what is going on.

Furthermore one can read this chapter a second time after the development of the boundary element methods to learn how these methods are related to general methods, as for example with Galerkin‘s method for the solution of operator equations.

We write $f \lesssim g$ provided there exists a constant C such that $f \leq Cg$.

1.1 The Basic Approximation Problems

Problem 1.1 Let $A : X \to Y$ be a continuous linear operator between two separable Banach spaces X and Y. The question is how to find a $u \in X$ such that, for given $f \in Y$, $Au = f$ holds ? The idea of projection methods is to solve the

© Springer International Publishing AG, part of Springer Nature 2018

J. Gwinner, E. P. Stephan, *Advanced Boundary Element Methods*,
Springer Series in Computational Mathematics 52,
https://doi.org/10.1007/978-3-319-92001-6_1

above equation in certain subspaces $X_N \subset X$, $Y_N \subset Y$:

$$A_N u_N = f_N \in Y_N \quad \text{for } N \in \mathbb{N},$$

where A_N is an approximation to A and to hope that $u_N \in X_N$ is a "reasonable" approximation to $u \in X$.

Remark 1.1 Obvious requirements for a "successful" approximation method such that $u_N \in X_N$ converges to u for $N \to \infty$ are

1. The perturbed right hand sides f_N should converge to the given right hand side $f \in Y$.
2. The subspaces X_N should exhaust the entire space X in the sense that

$$\overline{\bigcup_{N \in \mathbb{N}} X_N} = X.$$

A more delicate question is how the operators A_N are defined on the subspaces X_N and how these operators should approximate the given operator A. Let us note that the simple choice of $A_N := A|X_N$, the restriction of A to the subspace X_N seldom works in the applications.

Linear operator equations abound in applied mathematics. They result from a linear problem modelling or – more often – within a linearization procedure, as e.g. Newton's method, for genuine nonlinear problems. Here we concentrate on well-posed operator equations that arise from the classical problems of mathematical physics. Thus as an introductory example we have Symm's integral equation:

$$V\psi(x) = -\frac{1}{\pi} \int_\Gamma \ln|x - y| \psi(y) ds_y = f(x), \tag{1.1}$$

where Γ is the boundary of a bounded domain in $\mathbb{R}^2$, such that (1.1) corresponds to $A = V$, $X = H^{-1/2}(\Gamma)$, $Y = H^{1/2}(\Gamma)$ (see Chapter 2). An equally basic problem is that of a variational problem, that is the minimization of a functional without or under some constraints. We start with the most simple variational problem, namely with the unconstrained one.

Problem 1.2 Assume that $A : X \to X'$ is a continuous linear operator from a Banach space X into its dual X'. Assume that the quadratic form $\langle Ax, x \rangle$ is nonnegative. Further let some $l \in X'$ be given. The question now is how to find a $x \in X$ that minimizes the functional

$$F(x) = \frac{1}{2} \langle Ax, x \rangle - l(x)$$

in X. The idea of approximation methods (in particular the so-called Ritz method) is to solve the above variational problem in certain subspaces $X_N \subset X$ and likewise to "hope" that these (approximate) solutions u_N converge to u.

Note for Symm's integral equation the quadratical functional becomes

$$F(\psi) = \frac{1}{2} \langle V\psi, \psi\rangle - l(\psi), \quad \psi \in H^{-1/2}(\Gamma),$$

where $\langle\cdot,\cdot\rangle$ denotes the duality between $H^{-1/2}(\Gamma)$ and $H^{1/2}(\Gamma)$ and $l(\psi) = \langle f, \psi\rangle$. Considering a polygon Γ we can take X_N as space of piecewise constants on quasi-uniform mesh where the vertices of Γ belong to the mesh points.

A variational constrained problem in general terms is the following.

Problem 1.3 Assume that $F : X \to \mathbb{R}$ is a continuous functional on a Banach space X. Moreover, let C be a closed subset of X. The question now is how to find a $u \in C$ that minimizes F in C. The idea of projection methods is to solve the above variational problem in certain subsets $C_N \subset X_N$, where again $X_N \subset X$ are subspaces (of finite dimension), and likewise to hope that the associated minimizers u_N of F in C_N converge to u.

Here F may be of the form of Problem 1.2. In the most simple case, the set C is an affine subspace of X, but may be more generally a convex cone or a convex subset. The approximation problem becomes more delicate if $C_N \subset C$ does not hold (so-called nonconforming approximation).

A typical example is:

$$\text{Minimize } F(u) = \frac{1}{2}\langle u, Su\rangle - l(u) = \frac{1}{2}\int_\Gamma u\frac{\partial u}{\partial n} ds - \langle f, u\rangle \tag{1.2}$$

$$\text{subject to } u \leq g \text{ on } \Gamma_c. \tag{1.3}$$

Here Γ_c is a part of Γ and $S : H^{1/2}(\Gamma) \to H^{-1/2}(\Gamma)$ is the Dirichlet-to-Neumann map (Poincaré-Steklov operator), see Chap. 5.

When the functional F is convex and differentiable, the minimization of F on C is equivalent to the variational inequality: Find $x \in C$ such that

$$\langle F'(x), y - x\rangle \geq 0, \quad \forall y \in C.$$

Here $\langle F'(x), z\rangle$ denotes the directional derivative of F at x in direction z.

In the model application above we obtain the variational inequality (VI):

Find $u \in C$ such that

$$\langle Su, v - u\rangle \geq \langle f, v - u\rangle \qquad \forall v \in C,$$

where $C := \{v \in H^{1/2}(\Gamma) | v|_{\Gamma_c} \leq g\}$. Now we may choose X_N as continuous piecewise linear functions on a quasi-uniform mesh and C_N is given as those functions $u_N \in X_N$ that satisfy $u_N \leq g$ in the mesh points, see Chap. 11.

1.2 Convergence of Projection Methods

Now we consider the situation of the general Galerkin method, i.e. for Hilbert spaces X, Y and a linear, continuous and bijective mapping, $A : X \to Y$ we want to find an approximation u_N of the solution $u \in X$ of

$$Au = f \tag{1.4}$$

for $f \in Y$ given. Thus let $X_N \subset X$ and $T_N \subset Y'$ be the spaces of trial and test functions, respectively, with $\dim T_N = \dim X_N = N < \infty$. Then we want to find $u_N \in X_N$ such that

$$\langle t, Au_N \rangle = \langle t, f \rangle \quad \forall t \in T_N\,. \tag{1.5}$$

Theorem 1.1 (Galerkin Method) *Let $X = Y'$, $Y = X'$ and $T_N = X_N$ for the above situation. If A is positive definite, i.e. $\exists \alpha > 0 : \langle x, Ax \rangle \geq \alpha \|x\|_X^2 \quad \forall x \in X$, the following holds:*

1. ***Existence of a unique solution***
   $\forall N\ \exists!\, u_N \in X_N : \langle t, Au_N \rangle = \langle t, f \rangle = \langle t, Au \rangle \quad \forall t \in T_N$
2. ***Stability of the method***
   $\exists\, M_{(independent\ of\ N)} : \|u_N\|_X \leq M\|u\|_X$
3. ***Quasioptimal error estimate***
   $\exists\, C_{(independent\ of\ N)} : \|u - u_N\|_X \leq C \inf_{v \in X_N} \|u - v\|_X =: C\, d(u, X_N)$
4. ***Convergence of the method***
   $d(v, X_N) \overset{N\to\infty}{\longrightarrow} 0\ \forall v \in X \implies \|u - u_N\|_X \overset{N\to\infty}{\longrightarrow} 0$

Proof As already suggested above, let $\{b_1, \ldots, b_N\}$ and $\{t_1, \ldots, t_N\}$ be a basis of X_N and T_N respectively, leading to

$$\sum_{k=1}^{N} \alpha_k \langle t_j, Ab_k \rangle = \langle t_j, f \rangle\,, \quad j = 1, \ldots, N\,. \tag{1.6}$$

1. With A being positive definite, the matrix $\left(\langle b_j, Ab_k \rangle\right)_{j,k=1\ldots N}$ is positive definite, too. Hence it is invertible, implying the existence of a unique solution u_N.
2. The assumption of A being positive definite further yields

$$\begin{aligned} \|u_N\|^2 &\leq \frac{1}{\alpha}\langle u_N, Au_N \rangle = \frac{1}{\alpha}\langle u_N, f \rangle = \frac{1}{\alpha}\langle u_N, Au \rangle \\ &\leq \frac{1}{\alpha}\|u_N\|_X\, \|Au\|_{X'} \leq \frac{1}{\alpha}\|u_N\|_X\, \|A\|_{X\to X'}\|u\|_X \end{aligned}$$

$$\implies \|u_N\|_X \leq \frac{1}{\alpha}\|A\|_{X\to X'}\|\, u\|_X =: M\|u\|_X$$

3. First of all, the triangle inequality yields for all $v \in X_N$

$$\|u - u_N\|_X \le \|u - v\|_X + \|v - u_N\|_X,$$

where u_N is defined by $\langle t, Au_N\rangle = \langle t, Au\rangle \ \forall\, t \in X_N$,

Let $G_N : \begin{cases} X \longrightarrow X_N \\ u \mapsto u_N \end{cases}$ be the so-called *Galerkin projector.*

This linear operator is continuous by 2., further a projector onto X_N, since $G_N(v) = v$, for all $v \in X_N$ (because with $\langle t, Av_N\rangle = \langle t, Av\rangle \ \forall t \in X_N$ and $G_N v := v_N$ for $v_N \in X_N$ unique, it follows $v = v_N$).

For the Galerkin projector we have

$$\|u_N\|_X \;=\; \|G_N u\|_X \;\le\; \frac{\|A\|_{X\to X'}}{\alpha}\|u\|_X \;\Longrightarrow\; |\!|\!|G_N|\!|\!| \;\le\; M = \frac{\|A\|_{X\to X'}}{\alpha}$$

Hence for all $v \in X_N$:

$$\begin{aligned} \|u - u_N\|_X &= \|u - v + v - u_N\|_X = \|u - v + G_N v - G_N u\|_X \\ &= \|(1 - G_N)(u - v)\|_X \le (1 + |\!|\!|G_N|\!|\!|)\|u - v\|_X \end{aligned}$$

$$\Longrightarrow \|u - u_N\|_X \le \left(1 + \frac{\|A\|_{X\to X'}}{\alpha}\right)\|u - v\|_X$$

Thus 3.,

$$\|u - u_N\|_X \le C \inf_{v\in X_N} \|u - v\|_X \ , \ C := \left(1 + \frac{\|A\|_{X\to X'}}{\alpha}\right).$$

4. By 3., $\lim\limits_{N\to\infty} \inf\limits_{v\in X_N} \|u - v\|_X = 0 \implies \lim\limits_{N\to\infty} \|u - u_N\|_X = 0.$ □

For the situation of the general Petrov-Galerkin method, i.e. with $X \neq Y'$ and $X_N \neq T_N$ we shall consider the following stability criteria:

a)

$$\begin{aligned} &\exists \alpha > 0_{(\text{independent of } N)}\ \forall v \in X_N\ \exists t \in T_N\backslash\{0\} : \\ &|\langle t, Av\rangle| \ \ge\ \alpha\|v\|_X \cdot \|t\|_{Y'} \end{aligned} \tag{1.7}$$

b)

$$\exists \alpha > 0 : \quad \inf_{v\in X_N\backslash\{0\}} \sup_{t\in T_N\backslash\{0\}} \frac{|\langle t, Av\rangle|}{\|v\|_X\|t\|_{Y'}} \ \ge\ \alpha \tag{1.8}$$

$$\textit{Babuška} - \textit{Brezzi} - \textit{condition}$$

c)

$$\begin{aligned} &\exists\, Q_N : X_N \to T_N,\ \exists\, M > 0\, \exists\, \alpha > 0_{\text{(independent of } N)} : \\ &|\langle Q_N v, Aw\rangle| \le M \|v\|_X \|w\|_X \ \ \forall\, v \in X_N, w \in X \\ &|\langle Q_N v, Av\rangle| \ge \alpha \|v\|_X^2 \ \ \forall\, v \in X_N \end{aligned} \tag{1.9}$$

Remark 1.2 It is easily verified that the above three conditions are equivalent. For the *Babuška – Brezzi – condition* (1.8) see also e.g. [56, 60, 65].

Theorem 1.2 (A More General Projection Method) *Let one (and thus all) of the above stability criteria* (1.7)–(1.9) *be satisfied, then the statements of Theorem 1.1 also hold for the projection method (1.5).*

Proof

1. We will show that the kernel of the matrix defined in (1.6) only consists of 0: Let $\sum\limits_k \alpha_k \langle t_j, Ab_k\rangle = 0, \quad$ for $j = 1, \ldots, N$, what is

$$\langle t, Au_N\rangle \;=\; 0 \ \forall\, t \in T_N \ \ (\text{with } u_N = \sum_k \alpha_k b_k).$$

By (1.7) $\exists\, t \in T_N \backslash \{0\} :\ 0 = \langle t, Au_N\rangle \ \ge\ \alpha \|u_N\|_X\, \|t\|_{Y'}$. Hence $u_N = 0$. Thus the matrix is injective. Since it is also quadratic it must be bijective, assuring the existence of a unique solution u_N .

2. $\|u_N\|_X \le \frac{1}{\alpha}\frac{1}{\|t\|_{Y'}} |\langle t, Au_N\rangle| \quad$ (using (1.7) with $v = u_N$)
$= \frac{1}{\alpha}\frac{1}{\|t\|_{Y'}} |\langle t, Au\rangle| \le \frac{1}{\alpha}\frac{1}{\|t\|_{Y'}} \|t\|_{Y'}\, \|Au\|_Y$
$= \frac{1}{\alpha} \|Au\|_Y \le \frac{1}{\alpha} |||A|||\, \|u\|_X$

3. The Galerkin projector G_N is well-defined by 1. Furthermore, 2. yields $|||G_N||| \le \frac{1}{\alpha} |||A|||$. Thus, for all $v \in X_N$

$$\begin{aligned} \|u - u_N\|_X &\le \|u - v\|_X + \|G_N(u - v)\|_X \\ &\le (1 + |||G_N|||) \|u - v\|_X \\ &\le \left(1 + \frac{|||A|||}{\alpha}\right) \|u - v\|_X \\ &=: C \|u - v\|_X \quad \forall\, v \in X_N \end{aligned}$$

4. Follows directly from 3. □

The following result by Hildebrandt and Wienholtz [244] generalizes Theorem 1.1 to strongly elliptic operators.

Theorem 1.3 *Let X be a Hilbert space with dual X' and $A, D : X \to X'$ isomorphisms such that $T = A - D : X \to X'$ compact. Let $\{S_h\}_{h>0}$ be a family of subspaces of X such that the equations*

$$\langle Dw_h, v\rangle = \langle Dw, v\rangle \quad \textit{for all } v \in S_h \tag{1.10}$$

define an operator $G_D^h : w \in X \mapsto w_h \in S_h$ *with the property*

$$\|G_D^h w - w\| \to 0 \quad as \quad h \to 0 \quad for\ all\ w \in X. \tag{1.11}$$

Then for small $h > 0$ *the equations*

$$\langle Au_h, v\rangle = \langle Au, v\rangle \quad for\ all\ v \in S_h \tag{1.12}$$

define an operator $G_A^h : u \in X \mapsto u_h \in S_h$ *such that*

$$\|G_A^h\| \leq C$$

with C *independent of* h.

Proof From (1.11) and the compactness of T follows $\|A^{-1}T(1 - G_D^h)\| \to 0$ $(h \to 0)$. Therefore for small h $\tilde{G}_A^h := G_D^h[1 - A^{-1}T(1 - G_D^h)]^{-1}$ exists and $\|\tilde{G}_A^h\|$ is uniformly bounded. From equations (1.10) and (1.12) it is easily verified that $\tilde{G}_A^h = G_A^h$. □

Chapter 2
Some Elements of Potential Theory

In this chapter we collect well-known concepts and results of classical potential theory that are necessary for the understanding of BEM. We focus to the elementary level of the Laplace equation in $\mathbb{R}^2$, respectively in $\mathbb{R}^3$ in Sect. 2.4.

First by the Gauss divergence theorem and classical limit arguments we derive the representation formula. This leads to the definition of the single- and double-layer potential. Then based on a distribution approach we provide the jump relations of the associated boundary integral (BI) operators and study their mapping properties, first in classical spaces of smooth functions, then by extension in the relevant Sobolev spaces of fractional order. Here we are concerned with the capacity of a Lipschitz curve what provides a sufficient condition for positive definiteness of the single-layer potential operator. Also we discuss in detail the bijectivity of the hypersingular operator. All 4 singular BI operators (single-layer, double-layer, adjoint of double-layer, hypersingular) enter the Calderon projector what expresses the jump relations in a compact form.

We conclude this chapter with another elementary approach based on complex function theory. Here again we derive the representation formula and are concerned with the single-layer and the hypersingular operator; for the latter we give an applicable representation.

2.1 Representation Formulas

Let Ω be a bounded domain in $\mathbb{R}^2$ with smooth boundary Γ. We can later relax this assumption on Γ.

© Springer International Publishing AG, part of Springer Nature 2018
J. Gwinner, E. P. Stephan, *Advanced Boundary Element Methods*,
Springer Series in Computational Mathematics 52,
https://doi.org/10.1007/978-3-319-92001-6_2

Let us first recall the basic Green formula for the Laplacian Δ:

$$\int_{\Omega} (\Delta u\, v - u\, \Delta v)\, dx = \int_{\Gamma} \left(\frac{\partial u}{\partial n} v - u \frac{\partial v}{\partial n} \right) ds\,, \tag{2.1}$$

where $\frac{\partial}{\partial n}$ denotes the outer normal derivative, that is, the directional derivative in the direction of the outer normal unit vector n that points towards the exterior of Ω. In view of $\Delta_x \log(|x-y|) = 0$, for $x \neq y$, $|x| \mapsto \log|x|$ is a fundamental solution of the Laplacian in $\mathbb{R}^2$.

For later use we note

$$\nabla_x \log(|x-y|) = \frac{x-y}{|x-y|^2}\,, \quad \frac{\partial}{\partial n(x)} \log(|x-y|) = \frac{\langle x-y, n(x)\rangle}{|x-y|^2}\,.$$

We now want to prove the representation formula for the Laplace equation.

Theorem 2.1 (Representation Formula) *Let Ω be a bounded simply connected region with a smooth boundary $\Gamma = \partial\Omega$, and let $u \in C^2(\bar{\Omega})$. Then there holds the following representation for u in Ω :*

$$\begin{aligned} u(x) = {} & \frac{1}{2\pi} \int\limits_{\Omega} \ln|x-y| \Delta u(y)\, dy + \\ & + \frac{1}{2\pi} \int\limits_{\Gamma} \big(u(y) \partial_{n_y} \ln|x-y| - \partial_{n_y} u(y) \ln|x-y| \big)\, ds_y, \quad x \in \Omega \end{aligned} \tag{2.2}$$

Proof Consider the following Fig. 2.1.

Let $J_\epsilon(x) := \int\limits_{\Omega \setminus B_\epsilon(x)} \ln|x-y|\, \Delta u(y)\, dy$. Then using (2.1)

$$J_\epsilon(x) = \int\limits_{\Omega \setminus B_\epsilon(x)} \left(\ln|x-y|\, \Delta u(y) \; - \; u(y)\, \underbrace{\Delta_y \ln|x-y|}_{=0} \right) dy$$

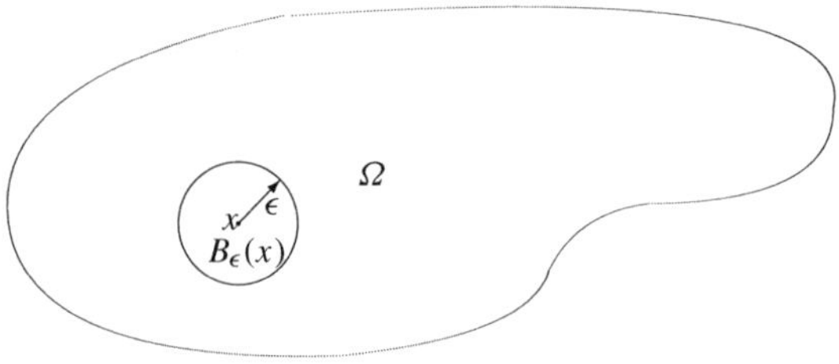

Fig. 2.1 Geometrical setting

$$= \int_{\partial(\Omega\setminus B_\epsilon(x))} \big(\ln|x-y|\,\partial_n u(y) \;-\; u(y)\,\partial_{n_y}\ln|x-y|\big)\;ds_y$$

$$= \int_{\partial\Omega} \big(\ln|x-y|\,\partial_n u(y) \;-\; u(y)\,\partial_{n_y}\ln|x-y|\big)\;ds_y \;+\; J_{\epsilon^*}\,,$$

where $J_{\epsilon^*}(x) := -\int_{\partial B_\epsilon(x)} \big(\ln|x-y|\,\partial_n u(y) \;-\; u(y)\,\partial_{n_y}\ln|x-y|\big)\;ds_y.$

With $u \in C^2(\Omega)$, we further have

$$\begin{aligned}\int_{\partial B_\epsilon(x)} \ln|x-y|\partial_n u(y)\,ds_y &= \int_0^{2\pi} \ln\epsilon\cdot \tfrac{d}{dr}u(x+re^{i\varphi})|_{r=\epsilon}\cdot\epsilon\,d\varphi\\ &= \epsilon\cdot\ln\epsilon\cdot\int_0^{2\pi}\partial_r u(x+re^{i\varphi})|_{r=\epsilon}\,d\varphi\\ &\le \epsilon\cdot\ln\epsilon\cdot\sup_{x\in\overline{B_\epsilon}}\|\nabla u(x)\|\cdot 2\pi \overset{\epsilon\to 0}{\longrightarrow} 0\,,\end{aligned}$$

$$\begin{aligned}\int_{\partial B_\epsilon(x)} u(y)\partial_{n_y}\ln|x-y|\,ds_y &= \int_0^{2\pi} u(x+\epsilon\cdot e^{i\varphi})\tfrac{d}{dr}\ln\, r|_{r=\epsilon}\cdot\epsilon\,d\varphi\\ &= \int_0^{2\pi} u(x+\epsilon\cdot e^{i\varphi})d\varphi \overset{\epsilon\to 0}{\longrightarrow} 2\pi u(x)\,.\end{aligned}$$

Thus, the assertion follows with $\int_\Omega \ln|x-y|\,\Delta u(y)\,dy = \lim_{\epsilon\to 0} J_\epsilon\,.$ □

Remark 2.1 Setting $G(x,y) = \frac{1}{2\pi}\ln|x-y|$, we leave it as an exercise to the reader to show that for $f \in C_0^0(\mathbb{R}^2)$ (space of continuous functions with compact support supp (f)) the function

$$u(x) = \int_{\mathbb{R}^2} G(x,y)f(y)dy$$

satisfies the Poisson equation $\Delta u = f$ in $\mathbb{R}^2$.

Corollary 2.1 *Let G define the integral operator* $f \in C_0^0(\mathbb{R}^2) \mapsto Gf$ *by*

$$Gf(x) := \int_{\mathbb{R}^2} G(x,y)f(y)\,dy \;=\; \frac{1}{2\pi}\int_{\mathbb{R}^2}\ln|x-y|f(y)\,dy$$

Then we have for $u \in C_0^2(\mathbb{R}^2)$

$$u = G\Delta u = \Delta G u. \tag{2.3}$$

Proof The equation $u = \Delta G u$ follows from Remark 2.1. Take Ω such that supp $(u) \subset\subset \Omega$, then the representation formula (2.2) further yields $u(x) = G\Delta u(x), \forall x \in \Omega$. □

Let $\Omega' := \mathbb{R}^2 \setminus \overline{\Omega}$ denote the "exterior" domain. Then the basic "Green's representation formula" for a harmonic function (see (2.5) below) can be stated as follows.

Theorem 2.2 *Let $u \in C^2(\Omega) \cap C^2(\Omega')$. Assume there exist the limits*

$$u|_{int\,\Gamma}(x) = \lim_{\substack{z \to x \in \Gamma \\ z \in \Omega}} u(z), \qquad u|_{ext\,\Gamma}(x) := \lim_{\substack{z \to x \in \Gamma \\ z \in \Omega'}} u(z),$$

and the analogously defined limits $\frac{\partial u}{\partial n}\Big|_{int\,\Gamma}, \Big|\frac{\partial u}{\partial n}\Big|_{ext\,\Gamma}$. *Let*

$$[u(x)] := u(x)|_{int\,\Gamma} - u(x)|_{ext\,\Gamma}$$

$$\left[\frac{\partial u(x)}{\partial n}\right] := \frac{\partial u(x)}{\partial n}|_{int\,\Gamma} - \frac{\partial u(x)}{\partial n}|_{ext\,\Gamma}$$

be the jump of the trace and of the normal derivative of u, respectively.

Moreover let u satisfy

$$\begin{cases} \Delta u = 0 \text{ in } \Omega \cup \Omega' \\ \left.\begin{aligned} u(y) &= O\left(\frac{1}{|y|}\right) \\ |\nabla u(y)| &= O\left(\frac{1}{|y|^2}\right) \end{aligned}\right\} \text{for } |y| \to +\infty \end{cases} \tag{2.4}$$

Then there holds for $y \in \Omega \cup \Omega'$,

$$u(y) = -\frac{1}{2\pi}\left\{\int_\Gamma \left[\frac{\partial u(x)}{\partial n}\right] \log(|x-y|)\, ds(x) - \int_\Gamma [u(x)]\, \frac{\partial}{\partial n(x)} (\log(|x-y|)\, ds(x)\right\}, \tag{2.5}$$

and for $y \in \Gamma$,

$$\frac{u(y)|_{int\,\Gamma} + u(y)|_{ext\,\Gamma}}{2} = -\frac{1}{2\pi}\left\{\int_\Gamma \left[\frac{\partial u(x)}{\partial n}\right] \log(|x-y|) ds(x) - \int_\Gamma [u(x)]\, \frac{\partial}{\partial n(x)} \log(|x-y|)\, ds(x)\right\}. \tag{2.6}$$

Proof Let $y \in \Omega \cup \Omega'$. Choose $\varepsilon > 0$ such that the ε-ball $B(y, \varepsilon)$ with boundary S_ε is contained in Ω, respectively in Ω', and moreover $\Omega \cup B(y, \varepsilon)$ is contained in the R–ball $B(0, R)$ with boundary S_R for large enough $R > 0$.

Since $\Delta u(x) = 0$, $\Delta \log(|x - y|) = 0$ for $x \neq y$, we can apply Green's formula both in $\Omega \setminus B(y, \varepsilon)$ and in $\Omega' \cap B(0, R) \setminus B(y, \varepsilon)$. Thus we obtain for $y \in \Omega \cup \Omega'$ the following equations

$$\int_\Gamma \left\{ u(x)|_{\text{int }\Gamma} \frac{\partial}{\partial n(x)} \log(|x - y|) - \frac{\partial u(x)}{\partial n(x)}|_{\text{int }\Gamma} \log(|x - y|) \right\} ds(x)$$
$$+ \int_{S_\varepsilon} \left\{ u(x) \frac{\partial}{\partial n(x)} \log(|x - y|) - \frac{\partial u(x)}{\partial n(x)} \log(|x - y|) \right\} ds(x) = 0$$

$$\int_\Gamma \left\{ -u(x)|_{\text{ext }\Gamma} \frac{\partial}{\partial n(x)} \log(|x - y|) + \frac{\partial u(x)}{\partial n(x)}|_{\text{ext }\Gamma} \log|x - y| \right\} ds(x)$$
$$+ \int_{S_R(\cup S_\varepsilon)} \left\{ u(x) \frac{\partial}{\partial n(x)} \log(|x - y|) - \frac{\partial u(x)}{\partial n(x)} \log(|x - y|) \right\} ds(x) = 0$$

By addition

$$\int_\Gamma [u(x)] \frac{\partial}{\partial n(x)} \log(|x - y|)\, ds(x) - \int_\Gamma \left[\frac{\partial u(x)}{\partial n(x)} \right] \log(|x - y|)\, ds(x)$$
$$+ \int_{S_R} \left\{ u(x) \frac{\partial}{\partial n(x)} \log(|x - y|) - \frac{\partial u(x)}{\partial n(x)} \log(|x - y|) \right\} ds(x)$$
$$+ \int_{S_\varepsilon} \left\{ u(x) \frac{\partial}{\partial n(x)} \log(|x - y|) - \frac{\partial u(x)}{\partial n(x)} \log(|x - y|) \right\} ds(x) = 0$$

Let $\varepsilon \to 0$. Then

$$\left| \int_{S_\varepsilon} \frac{\partial u(x)}{\partial n} \log(|x - y|)\, ds(x) \right| \leq C \log \varepsilon \cdot 2\pi\varepsilon \to 0,$$

$$\int_{S_\varepsilon} u(x) \frac{\partial}{\partial n(x)} \log(|x - y|)\, ds(x) = -\varepsilon^{-1} \int_{S_\varepsilon} u(x)\, ds(x) \to -2\pi u(y),$$

where the minus sign results from the normal $\vec{n}$ pointing to the interior of S_ε and where we can apply the integral mean value theorem. Further let $R \to \infty$. Then

$$\int_{S_R} \left\{ u(x) \frac{\partial}{\partial n(x)} \log(|x - y|) - \frac{\partial u(x)}{\partial n} \log(|x - y|) \right\} ds(x) \to 0,$$

since by (2.4), $\left|u(x)\frac{x-y}{|x-y|^2}\right| \le C\frac{1}{R}\cdot\frac{1}{R}$, $\int \left|u(x)\dfrac{x-y}{|x-y|^2}\right| ds(x) \le \dfrac{2\pi C}{R} \to 0$ and

$$\left|\frac{\partial u(x)}{\partial n}\log(|x-y|)\right| \le C\frac{1}{R^2}\log R, \int |\ldots| ds(x) \le \frac{2\pi C}{R}\log R \to 0.$$

Thus we arrive at (2.5). To obtain (2.6) note that for $y \in \Gamma$, Γ smooth we have

$$\int_{S_\varepsilon} u(x)\frac{\partial}{\partial n(x)}\log(|x-y|)\, ds(x) \to -2\pi\frac{u(y)|_{\text{int}\,\Gamma} + u(y)|_{\text{ext}\Gamma}}{2}.$$

This can be seen as follows. Approximate Γ by the tangent in y that separates S_ε in two half-balls. Introduce polar coordinates. Then for $\varrho = \varepsilon, 0 < \varphi < \pi$ we have $u(\varrho,\varphi) \approx u(y)|_{\text{int}\,\Gamma}$, whereas for $\pi < \varphi < 2\pi$ we have $u(\varrho,\varphi) \approx u(y)|_{\text{ext}\,\Gamma}$. □

Remark 2.2 1. Formula (2.5) holds also for a Lipschitz domain with corners. If $y \in \Gamma$ is a vertex with inner angle δ (= angle of the two tangents in x along Γ) and outer angle $2\pi - \delta$, then the left hand side of (2.6) is to be changed to

$$\frac{\delta}{2\pi}u(y)|_{\text{int}\,\Gamma} + \frac{2\pi-\delta}{2\pi}u(y)|_{\text{ext}\,\Gamma}$$

2. Consider the special case $u = 0$ on Ω'. Then there holds for $y \in \Omega$

$$u(y) = -\frac{1}{2\pi}\int_\Gamma \left\{\frac{\partial u(x)}{\partial n}\log(|x-y|) - u(x)\frac{\partial}{\partial n(x)}\log(|x-y|)\right\} ds(x).$$

This means that $u|\Gamma$ and $\frac{\partial u}{\partial n}|\Gamma$ determine u on Ω; however $u|\Gamma$ and $\frac{\partial u}{\partial n}|\Gamma$ cannot be prescribed independently (remind the Dirichlet–Problem and the Neumann problem).

From Theorem 2.2 we derive the following formulas in a "weak" sense.

Theorem 2.3 *Let u satisfy the assumptions of Theorem 2.2. In addition, suppose $u|_{int\,\Gamma} - u|_{ext\,\Gamma} = 0$. Then using $q = \left[\frac{\partial u}{\partial n}\right] = \frac{\partial u}{\partial n}|_{int\,\Gamma} - \frac{\partial u}{\partial n}|_{ext\,\Gamma}$ there holds for all $y \in \Gamma$*

$$\frac{\partial u(y)}{\partial n}|_{ext\,\Gamma} = -\frac{1}{2}q(y) - \frac{1}{2\pi}\int_\Gamma q(x)\frac{\partial}{\partial n(y)}\log(|x-y|)\, ds(x), \tag{2.7}$$

$$\frac{\partial u(y)}{\partial n}|_{int\,\Gamma} = \frac{1}{2}q(y) - \frac{1}{2\pi}\int_\Gamma q(x)\frac{\partial}{\partial n(y)}\log(|x-y|)\, ds(x), \tag{2.8}$$

Remark 2.3 Obviously the formulas (2.7) and (2.8) are equivalent to

$$\frac{1}{2}\left\{\frac{\partial u(y)}{\partial n}|_{\text{ext}\,\Gamma} + \frac{\partial u(y)}{\partial n}|_{\text{int}\,\Gamma}\right\} = -\frac{1}{2\pi}\int_\Gamma q(x)\frac{\partial}{\partial n(y)}\log(|x-y|)\,ds(x) \tag{2.9}$$

Proof (of Theorem 2.3) Let $\varphi \in C_0^\infty(\mathbb{R}^n)$, i.e. infinitely differentiable with compact support supp φ. Then by the Gauss divergence theorem on $\Omega' \cap \operatorname{supp}\varphi$ (note the orientation of $\vec{n}$) using (2.5) in Theorem 2.2

$$\begin{aligned}-\int_\Gamma \frac{\partial u(y)}{\partial n}|_{\text{ext}\,\Gamma}\,\varphi(y)\,ds(y) &= \int_{\Omega'}\langle\nabla u(z), \nabla\varphi(z)\rangle\,d(z_1,z_2)\\ &= \frac{1}{2\pi}\int_{\Omega'}\int_\Gamma q(x)\frac{\langle x-z, \nabla\varphi(z)\rangle}{|x-z|^2}\,ds(x)\,dz,\end{aligned}$$

where the latter integral exists on $\Omega' \times \Gamma$, since q is continuous as a difference of continuous functions, $(\Omega' \cap \operatorname{supp}\varphi) \times \Gamma$ compact, and

$$\int_{\Omega'\cap\operatorname{supp}\varphi}\int_\Gamma \frac{1}{|x-z|}\,ds(x)\,dz < +\infty,$$

as seen as follows: Introduce polar coordinates $z - x = \rho(\cos\theta, \sin\theta)^T$, $(z_1, z_2)^T = (x_1 + \rho\cos\theta, x_2 + \rho\sin\theta)^T$, $\left|\frac{\partial(z_1,z_2)}{\partial(\varrho,\theta)}\right| = \varrho$ and obtain

$$\int_{\Omega'\cap\operatorname{supp}\varphi}\int_\Gamma \frac{1}{\varrho}\,ds(x)\,\varrho\,d\theta < +\infty.$$

Interchanging the integrations according to the theorem of Fubini leads to

$$-\int_\Gamma \frac{\partial u(y)}{\partial n}|_{\text{ext}\,\Gamma}\,\varphi(y)\,ds(y) = \frac{1}{2\pi}\int_\Gamma q(x)\left(\int_{\Omega'}\frac{\langle x-z, \nabla\varphi(z)\rangle}{|x-z|^2}dz\right)ds(x).$$

With $B_\varepsilon := B(x;\varepsilon)$ and $S_\varepsilon := \partial B_\varepsilon$ we have as an improper integral

$$\int_{\Omega'}\frac{\langle x-z, \nabla\varphi(z)\rangle}{|x-z|^2}dz = \lim_{\varepsilon\to 0}\int_{\Omega'\setminus(\Omega'\cap B_\varepsilon)}\frac{\langle x-z, \nabla\varphi(z)\rangle}{|x-z|^2}\,dz.$$

Once more we apply the divergence theorem to $-\log(|x-z|)$ $(z \neq x, x$ fixed on $\Omega' \setminus B_\varepsilon \cap \operatorname{supp}\varphi)$. By $\nabla_z(-\log|x-z|) = \frac{x-z}{|x-z|^2}$, by the orientation of $\vec{n}(y)$, and by $\vec{n}(y) = \frac{x-y}{|x-y|}$ on S_ε we have

$$\int_{\Omega'\setminus B_\varepsilon} \frac{\langle x-z, \nabla\varphi(z)\rangle}{|x-z|^2} dz = -\int_{\Gamma\setminus\Gamma\cap B_\varepsilon} \frac{\langle x-y, \vec{n}(y)\rangle}{|x-y|^2} \varphi(y)\, ds(y) + \int_{S_\varepsilon\cap\Omega'} \frac{1}{|x-y|} \varphi(y)\, ds(y)\,.$$

Taking $\varepsilon \to 0$ we arrive at

$$\int_{\Omega'} \frac{\langle x-z, \nabla\varphi(z)\rangle}{|x-z|^2} dz = -\int_\Gamma \frac{\langle x-y, \vec{n}(y)\rangle}{|x-y|^2} \varphi(y)\, ds(y) + \pi\varphi(x)$$

and hence (y instead of x)

$$-\int_\Gamma \frac{\partial u(y)}{\partial n}|_{\mathrm{ext}\,\Gamma}\, \varphi(y)\, ds(y) = -\frac{1}{2\pi}\int_\Gamma\int_\Gamma \frac{\langle x-y, \vec{n}(y)\rangle}{|x-y|^2} q(x)\varphi(y)\, ds(x) ds(y) + \frac{1}{2}\int_\Gamma q(y)\varphi(y)\, ds(y)$$

Thus (2.7) holds true in a weak sense. The proof of (2.8) follows similar lines. □

2.2 Single- and Double-Layer Potential

Definition 2.1 Let $\Gamma \in Lip$,i.e.Γ is locally the graph of a Lipschitz function, and $\varphi \in C(\Gamma)$. Then we define for $x \notin \Gamma$

i) the *single-layer potential $S\varphi$ with density φ* by

$$(S\varphi)(x) := S\varphi(x) := -\frac{1}{\pi}\int_\Gamma \varphi(y) \ln|x-y|\, ds_y, \quad x \notin \Gamma \tag{2.10}$$

ii) the *double-layer potential $D\varphi$ with density φ* by

$$(D\varphi)(x) := D\varphi(x) := -\frac{1}{\pi}\int_\Gamma \varphi(y)\partial_{n_y} \ln|x-y|\, ds_y, \quad x \notin \Gamma \tag{2.11}$$

Corollary 2.2

i) Let Ω be bounded, $\Gamma \in Lip$, $u \in C^2(\overline{\Omega})$ and $\Delta u = 0$ in Ω. Defining $\gamma_0 u := u|_\Gamma$, $\gamma_1 u := \partial_n u|_\Gamma$ and Ω^-, Ω^+ (corresponding to Fig. 2.2) denoting the

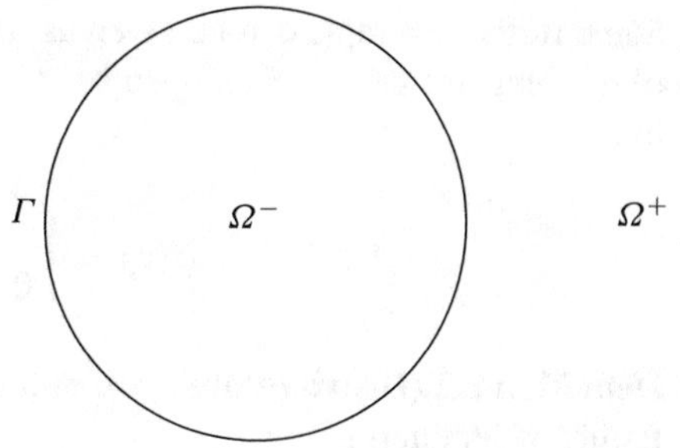

Fig. 2.2 Exterior domain Ω^+ and interior domain Ω^-

interior domain, exterior domain, respectively, there holds the representation formula

$$u = -\frac{1}{2}D(\gamma_0 u) + \frac{1}{2}S(\gamma_1 u), \quad in\ \Omega = \Omega^-. \tag{2.12}$$

ii) Let $\Gamma \in Lip$ and $u \in C_0^2(\mathbb{R}^2\backslash\Gamma)$, i.e. $u|_\Omega \in C^2(\Omega)$ and $u|_{\mathbb{R}^2\backslash\overline{\Omega}} \in C_0^2(\mathbb{R}^2\backslash\overline{\Omega})$. Setting $u^{+/-} := u|_{\Omega^{+/-}}$, we define:

a) $[\gamma_0 u] := \gamma_0 u^+ - \gamma_0 u^- = u^+|_\Gamma - u^-|_\Gamma$
b) $[\gamma_1 u] := \gamma_1 u^+ - \gamma_1 u^- = \partial_n u^+|_\Gamma - \partial_n u^-|_\Gamma$

This yields with $f := \Delta u$ the representation

$$u = Gf + \frac{1}{2}D[\gamma_0 u] - \frac{1}{2}S[\gamma_1 u] \qquad in\ \mathbb{R}^2\backslash\overline{\Gamma}. \tag{2.13}$$

This is just the statement of Theorem 2.2, since in the definition of the jump $[\cdot,\cdot]$ the role of $\pm$ are interchanged.

Remark 2.4 The notation *potential* in Definition 2.1 is justified, since for $\varphi \in C(\Gamma)$ the identity $\Delta_x \ln|x - y| = 0$ for $x \neq y$ yields in $\mathbb{R}^2\backslash\Gamma$ with Δ and $\int$ interchanged

$$\Delta S\varphi = 0 = \Delta D\varphi\,. \tag{2.14}$$

2.2.1 *Some Remarks on Distributions*

We now want to introduce some basic definitions and results on distributions, in order to derive the classical theorems on boundary integral equations with modern methods. For this, we first need the notion of a *test function*, i.e. a function $\varphi \in C_0^\infty(\mathbb{R}^2)$, as for example

$$\varphi(x) = \begin{cases} \exp\left(\frac{1}{|x|^2 - R^2}\right) & : \ |x| < R, \\ 0 & : \ |x| \geq R. \end{cases}$$

Some further example can be given as follows. Let $A \subset O$ be a closed, bounded (thus compact) subset of an open set O. Then we may define a function $\varphi \in C^\infty$ by:

$$\varphi(x) = \begin{cases} 1 & : \quad x \in A, \\ 0 & : \quad x \notin O. \end{cases}$$

Definition 2.2 (Distributions) We define $\mathscr{D}(\Omega) := C_0^\infty(\Omega)$, endowed with the family of seminorms

$$\sup_{x \in K} |D^\alpha \varphi(x)| < \infty\,, \qquad \forall \alpha \in \mathbb{N}_0^2 \, \forall K \subset\subset \Omega.$$

Then we denote with $\mathscr{D}'(\Omega)$ the *space of distributions on* Ω, i.e. the set of linear continuous functionals f on $\mathscr{D}(\Omega)$.

Any function $f \in C(\mathbb{R}^2)$ defines a regular distribution

$$\varphi \longmapsto \langle f, \varphi \rangle := \int_{\mathbb{R}^2} f(x)\varphi(x)dx \quad \forall \varphi \in C_0^\infty(\mathbb{R}^2)\,.$$

Example 2.1 Let $0 \in \Omega \subset \mathbb{R}^2$. The Dirac Delta-function $\delta_0 \in \mathscr{D}'(\Omega)$ is defined by

$$\langle \delta_0, \varphi \rangle \ = \ \varphi(0), \quad \varphi \in \mathscr{D}(\Omega)\,.$$

Definition 2.3 The support of a distribution $\varphi \in \mathscr{D}'(\Omega)$ is defined as the set of all points x in $\overline{\Omega}$ for which for any $\eta > 0$ the restriction of φ to the domain $\Omega \cap \{y||y - x| < \eta\}$ differs from the zero distribution.

Thus, for the Dirac Delta-function we have that supp $(\delta_0) = \{0\}$.

Example 2.2 A further example is given as follows:
For $\psi \in C(\Gamma)$ we define $\gamma_0^* \psi$ by

$$\langle \gamma_0^* \psi, \varphi \rangle \ := \ \int_\Gamma \psi(x)\varphi(x)ds_x\,, \qquad \forall \varphi \in C_0^\infty(\mathbb{R}^2)$$

and $\gamma_1^* \psi$ by

$$\langle \gamma_1^* \psi, \varphi \rangle \ := \ \int_\Gamma \psi(x)\partial_n\varphi(x)ds_x\,, \qquad \forall \varphi \in C_0^\infty(\mathbb{R}^2)\,.$$

Definition 2.4 (Derivatives of Distributions) Let $t \in \mathscr{D}'(\mathbb{R}^2)$. Then we define the partial derivative $\partial_j t$ by

$$\langle \partial_j t, \varphi \rangle \ := \ - \langle t, \partial_j \varphi \rangle\,, \qquad \forall \varphi \in C_0^\infty(\mathbb{R}^2).$$

We may also define the operator G as given in Corollary 2.1 for distributions $t \in \mathscr{D}'(\mathbb{R}^2)$ with compact support as follows:

First let $f \in C_0^0(\mathbb{R}^2)$ and $\varphi \in C_0^\infty(\mathbb{R}^2)$. Then we may define for

$$\langle Gf, \varphi\rangle = \int_{\mathbb{R}^2} (Gf)(x)\varphi(x)dx = \int_{\mathbb{R}^2}\int_{\mathbb{R}^2} G(x,y) f(y)\,dy\,\varphi(x)\,dx$$

$$= \int_{\mathbb{R}^2} f(y) \int_{\mathbb{R}^2} G^*(y,x)\varphi(x)\,dxdy = \langle f, G^*\varphi\rangle = \langle f, \chi G^*\varphi\rangle\,,$$

where $\chi \in C_0^\infty$ is any cut-off function with $\chi \equiv 1$ on supp (f). Thus, for $\varphi \in C_0^\infty$ we have $\chi G^*\varphi \in C_0^\infty$ since $\partial_j G^*\varphi = -G^*\partial_j\varphi$, implying that $G^*\varphi \in C^\infty$ and thus $\chi G^*\varphi \in C_0^\infty$. This leads to the following

Definition 2.5 For $t \in \mathscr{D}'(\mathbb{R}^2)$ with supp $(t) \subseteq \{\chi \equiv 1\}$ and $\varphi \in C_0^\infty$ we define

$$\langle Gt, \varphi\rangle := \langle t, \chi G^*\varphi\rangle$$

Application of Definition 2.5

1. Let $\chi \in C_0^\infty$ with $\chi \equiv 1$ in some neighbourhood of the origin. Then we may show that $G\delta$ is regular:

$$\langle G\delta, \varphi\rangle = \langle \delta, \chi G^*\varphi\rangle = (\chi G^*\varphi)(0) = \chi(0)\cdot(G^*\varphi)(0)$$

$$= \int_{\mathbb{R}^2} G(y,0)\varphi(y)\,dy = \langle G(\cdot,0), \varphi\rangle$$

$$\Longrightarrow \quad (G\delta)(y) = G(y,0) = \frac{1}{2\pi}\ln|y|$$

2. We want to consider $G(\gamma_0^*\psi)$. Therefore, we now assume that $\chi \equiv 1$ in some neighbourhood of Γ. With $\varphi \in C_0^\infty$ we have:

$$\langle G\gamma_0^*\psi, \varphi\rangle = \langle \gamma_0^*\psi, \chi G^*\varphi\rangle = \int_\Gamma \psi(x)(\chi G^*\varphi)(x)\,ds_x$$

$$= \int_\Gamma \psi(x)(G^*\varphi)(x)\,ds_x = \int_\Gamma \psi(x)\int_{\mathbb{R}^2} G(y,x)\varphi(y)dy\,ds_x$$

Hence by Fubini with single-layer potential $S\psi$, see (2.10)

$$\langle G\gamma_0^*\psi, \varphi\rangle = \int_{\mathbb{R}^2}\varphi(y)\int_\Gamma G(y,x)\psi(x)\,ds_x\,dy = \frac{1}{2}\int_{\mathbb{R}^2}\varphi(y)(-S\psi)(y)\,dy$$

Lemma 2.1 *Let $\psi \in \mathscr{D}'(\mathbb{R}^2)$ be a distribution with compact support, then*

$$\Delta G\psi = G\Delta\psi = \psi .$$

Proof From Definition 2.4 and the definition of G we have, since $G^* = G$ for all $\varphi \in C_0^\infty$, using (2.3)

$$\langle \Delta G\psi, \varphi\rangle = \langle G\psi, \Delta\varphi\rangle = \langle \psi, G^*\Delta\varphi\rangle = \langle \psi, G\Delta\varphi\rangle = \langle \psi, \varphi\rangle ,$$
$$\langle G\Delta\psi, \varphi\rangle = \langle \Delta\psi, G^*\varphi\rangle = \langle \psi, \Delta G\varphi\rangle = \langle \psi, \varphi\rangle . \qquad \square$$

As an example let us now assume that $u \in C_0^2(\Omega^-) \cup C_0^2(\Omega^+)$, i.e. $u \in C^2(\mathbb{R}^2 \backslash \overline{\Gamma})$ with supp (u) compact. Then we want to find a representation for the Laplacian Δu of u. Using the same notations as in Corollary 2.2, the second Green formula implies for all $\varphi \in C_0^\infty$,

$$\begin{aligned}\langle \Delta u, \varphi\rangle = \langle u, \Delta\varphi\rangle &= \int_{\mathbb{R}^2} u\Delta\varphi\, dx = \int_{\Omega^-} u^-\Delta\varphi\, dx + \int_{\Omega^+} u^+\Delta\varphi\, dx \\ &= \int_{\Omega^-} \Delta u^-\varphi\, dx + \int_{\Omega^+} \Delta u^+\varphi\, dx + \int_{\Gamma} (u^-\partial_n\varphi - \partial_n u^-\varphi - u^+\partial_n\varphi + \partial_n u^+\varphi)\, ds .\end{aligned}$$

Setting $f := \Delta u^- + \Delta u^+$, this yields

$$\begin{aligned}\langle \Delta u, \varphi\rangle &= \langle f, \varphi\rangle + \int_{\Gamma} [\partial_n u]\varphi\, ds - \int_{\Gamma} [u]\partial_n\varphi\, ds \\ &= \langle f, \varphi\rangle + \langle \gamma_0^*[\partial_n u], \varphi\rangle - \langle \gamma_1^*[u], \varphi\rangle ,\end{aligned}$$

using $\langle \gamma_1^*\psi, \varphi\rangle = \int_{\Gamma} \psi\partial_n\varphi\, ds$ and $\langle \gamma_0^*\psi, \varphi\rangle = \int_{\Gamma} \psi\varphi\, ds$. Thus, we finally obtain the result

$$\Delta u = f + \gamma_0^*[\gamma_1 u] - \gamma_1^*[\gamma_0 u] . \tag{2.15}$$

Therefore, Lemma 2.1 implies that

$$u = G\Delta u = Gf + G\gamma_0^*[\partial_n u] - G\gamma_1^*[u] .$$

Comparing this result with the representation of u in (2.13), we find a new relation for the double- and single-layer potential,

$$S\psi = -2G\gamma_0^*\psi \tag{2.16}$$
$$D\psi = -2G\gamma_1^*\psi \quad . \tag{2.17}$$

2.2.2 Jump Relations

In this subsection we want to derive the jump relations for the single- and double-layer potentials.

In the following we will make extensive use of the following boundary integral operators

Definition 2.6 Let $x \in \Gamma$, then we denote by

$$V\varphi(x) := -\frac{1}{\pi}\int\limits_{\Gamma} \varphi(y) \ln|x-y|\, ds_y \tag{2.18}$$

the **single layer potential operator** and the **double layer potential operator** is given by

$$K\varphi(x) := -\frac{1}{\pi}\int\limits_{\Gamma} \varphi(y)\partial_{n_y} \ln|x-y|\, ds_y. \tag{2.19}$$

Moreover, we define the **adjoint double layer potential operator** as

$$K'\varphi(x) := -\frac{1}{\pi}\int\limits_{\Gamma} \varphi(y)\partial_{n_x} \ln|x-y|\, ds_y \tag{2.20}$$

and introduce the so called **hypersingular operator** by

$$W\varphi(x) := -\partial_{n_x} K\varphi(x) \tag{2.21}$$

From (2.16) and (2.17) it follows using Lemma 2.1 that

$$\Delta(S\psi) = -2\gamma_0^*\psi \qquad \Delta(D\psi) = -2\gamma_1^*\psi\,. \tag{2.22}$$

If now we apply (2.15) for $u = \frac{1}{2}S\psi$ and $u = \frac{1}{2}D\psi$, respectively, a comparison with (2.22) will prove the following jump relations.

Lemma 2.2 (Jump Relations) *Suppose* $\Gamma = \partial\Omega \in C^3$ *and* $\psi \in C^2(\Gamma)$, *such that we have* $S\psi,\ D\psi \in C^2(\Omega^+ \cup \Omega^-)$. *Then there holds*

i) $[\gamma_0(S\psi)] = 0$
$[\gamma_1(S\psi)] = -2\psi$

ii) $[\gamma_0(D\psi)] = 2\psi$
$[\gamma_1(D\psi)] = 0$

Proof From (2.22), (2.15) and (2.14) we deduce

$$-\gamma_0^*\psi = \Delta(\frac{1}{2}S\psi) = 0 + \gamma_0^*[\gamma_1(\frac{1}{2}S\psi)] - \gamma_1^*[\gamma_0(\frac{1}{2}S\psi)]\,,$$

hence by uniqueness of representation

$$-2\psi = [\gamma_1(S\psi)], \quad 0 = [\gamma_0(S\psi)] .$$

Similarly from

$$-\gamma_1^*\psi = \Delta(\frac{1}{2}D\psi) = 0 + \gamma_0^*[\gamma_1(\frac{1}{2}D\psi)] - \gamma_1^*[\gamma_0(\frac{1}{2}D\psi)]$$

it follows

$$2\psi = [\gamma_0(D\psi)], \quad 0 = [\gamma_1(D\psi)] . \qquad \square$$

Remark 2.5 Using Lemma 2.1 one observes that the proof of Lemma 2.2 remains valid for $\Gamma \in Lip$ and $\psi \in \mathscr{D}'(\mathbb{R}^2)$, since Green's formulas hold for Lipschitz domains, i.e. domains Ω with $\Gamma = \partial\Omega \in Lip$. Note that for $\Omega \subset \mathbb{R}^d$, $\Gamma = \partial\Omega \in Lip$ if every point on Γ has a neighborhood $N \subset \mathbb{R}^d$ such that, after an affine change of coordinates (translation and rotation), $\Gamma \cap N$ is described by the equation $x_d = \varphi(x_1, \ldots, x_{d-1})$, where φ is uniformly Lipschitz continuous. Moreover, $\Omega \cap N$ is on one side of $\partial\Omega \cap N$, e.g. $\Omega \cap N = \{x \in N : x_d < \varphi(x_1, \ldots, x_{d-1})\}$ (see [327]).

Lemma 2.3 *Under the assumptions of the above lemma there holds,*

i) $\gamma_1(S\psi)^+ = K'\psi - \psi$
 $\gamma_1(S\psi)^- = K'\psi + \psi$
ii) $\gamma_0(D\psi)^+ = K\psi + \psi$
 $\gamma_0(D\psi)^- = K\psi - \psi$

Proof We consider $\psi \in C_0^2(\mathbb{R}^2)$ and u satisfying the equation $\Delta u = 0$ in Ω. Then for $\varphi \in C^1(\Omega) \cap C(\overline{\Omega})$,

$$\int_\Omega \nabla u \nabla \varphi \, dx = \int_\Gamma \frac{\partial u}{\partial n} \varphi \, ds$$

and setting $u = (\frac{1}{2}S\psi)^-$ we have

$$\begin{aligned}\int_\Gamma \gamma_1 u \, \gamma_0 \varphi &= \int_\Omega \nabla(\frac{1}{2}S\psi)\nabla\varphi \\ &= \int_\Omega \nabla_x \lim_{\epsilon\to 0}\Big(-\frac{1}{2\pi}\int_{\substack{y\in\Gamma \\ |y-x|\geq\epsilon}} \ln|x-y|\psi(y)\, ds_y\Big) \cdot \nabla_x\varphi(x)dx \\ &= -\frac{1}{2\pi}\int_\Gamma \psi(y)\Big(\lim_{\epsilon\to 0}\int_{\substack{x\in\Omega \\ |y-x|\geq\epsilon}} \nabla_x \ln|x-y| \cdot \nabla_x\varphi(x)dx\Big)ds_y\end{aligned}$$

where the last identity is obtained with the Fubini's theorem. Now using Green's first formula, we obtain

$$\int_{\substack{x\in\Omega\\|y-x|\geq\epsilon}} \nabla_x \ln|x-y| \cdot \nabla_x\varphi(x)dx$$

$$= \int_{\partial(\Omega\setminus B_\epsilon)} \gamma_{1,x} \ln|x-y|\gamma_0\varphi(x)ds_x - \int_{\Omega\setminus B_\epsilon} (\Delta_x \ln|x-y|)\varphi(x)dx$$

$$= \int_{\substack{x\in\Omega\\|x-y|=\epsilon}} \gamma_{1,x} \ln|x-y|\varphi(x)\,ds_x + \int_{\substack{x\in\Gamma\\|x-y|\geq\epsilon}} \gamma_{1,x} \ln|x-y|\,\gamma_0\varphi(x)\,ds_x$$

$$= \int_{\substack{x\in\Omega\\|x-y|=\epsilon}} \gamma_{1,x} \ln|x-y|(\varphi(x)-\varphi(y))\,ds_x + \varphi(y)\int_{\substack{x\in\Omega\\|x-y|=\epsilon}} \gamma_{1,x} \ln|x-y|\,ds_x$$

$$+ \int_{\substack{x\in\Gamma\\|x-y|\geq\epsilon}} \gamma_{1,x} \ln|x-y|\,\gamma_0\varphi(x)\,ds_x$$

We use polar coordinates

$$\gamma_{1,x} \ln|x-y|\Big|_{r=\epsilon} = -\frac{\partial}{\partial r}\ln r\Big|_{r=\epsilon} = -\frac{1}{\epsilon}$$

such that

$$\int_{\substack{x\in\Omega\\|x-y|=\epsilon}} \gamma_{1,x} \ln|x-y|\,ds_x = -\frac{1}{\epsilon}\int_{\substack{x\in\Omega\\|x-y|=\epsilon}} ds_x = -\pi \quad \text{for } \epsilon\to 0.$$

On the other hand

$$\int_{\substack{x\in\Omega\\|x-y|=\epsilon}} \gamma_{1,x} \ln|x-y|(\varphi(x)-\varphi(y))ds_x \leq \max_{\substack{x\in\Omega\\|x-y|=\epsilon}} |\varphi(x)-\varphi(y)| \int_{\substack{x\in\Omega\\|x-y|=\epsilon}} \big|\gamma_{1,x} \ln|x-y|\big|\,ds_x$$

Thus, when $\epsilon \to 0$, the above term tends to zero. Therefore

$$\int_\Gamma \gamma_1 u\, \gamma_0\varphi ds = -\frac{1}{2\pi}\int_\Gamma \psi(y)(-\pi\varphi(y))ds_y$$

$$-\frac{1}{2\pi}\int_\Gamma \psi(y) \lim_{\epsilon\to 0}\int_{\substack{x\in\Gamma\\|x-y|\geq\epsilon}} \gamma_{1,x} \ln|x-y|\gamma_0\varphi(x)ds_x\,ds_y$$

$$= \frac{1}{2}\int_\Gamma \psi(y)\varphi(y)ds_y + \int_\Gamma \psi(y)\Big(-\frac{1}{2\pi}\int_\Gamma \gamma_{1,x} \ln|x-y|\gamma_0\varphi(x)\,ds_x\Big)\,ds_y$$

$$= \frac{1}{2}\int_\Gamma \psi(y)\varphi(y)ds_y + \int_\Gamma \varphi(x)\Big(-\frac{1}{2\pi}\int_\Gamma \gamma_{1,x} \ln|x-y|\psi(y)ds_y\Big)\,ds_x$$

or shortly,

$$\gamma_1(\frac{1}{2}S\psi)^- = \frac{1}{2}\psi + \frac{1}{2}K'\psi.$$

In the same way we can also prove

$$\gamma_0(D\psi)^\pm = K\psi \pm \psi$$
$$\gamma_1(S\psi)^+ = K'\psi - \psi \qquad \square$$

Let us look again at the homogeneous Laplace problem. The representation formula (2.2) yields for all $x \in \Omega$ and for u with $\Delta u = 0$:

$$u(x) = \frac{1}{2\pi}\int_\Gamma u(y)\partial_{n_y} \ln|x-y|\,ds_y - \frac{1}{2\pi}\int_\Gamma \partial_n u(y)\ln|x-y|\,ds_y$$
$$= -\frac{1}{2}Du(x) + \frac{1}{2}S\frac{\partial u(x)}{\partial n}.$$

Making use of the jump relations, letting $x \to \Gamma$ we have for $x \in \partial\Omega = \Gamma$:

$$u(x) = \frac{1}{2\pi}\int_\Gamma u(y)\partial_{n_y} \ln|x-y|\,ds_y + \frac{u(x)}{2} - \frac{1}{2\pi}\int_\Gamma \partial_n u(y)\ln|x-y|\,ds_y$$
$$= -\frac{1}{2}Ku(x) + u(x)/2 + \frac{1}{2}V\frac{\partial u(x)}{\partial n}$$

leading with $u = g$ for the Dirichlet problem to

$$f(x) := g(x) - \frac{1}{\pi}\int_\Gamma g(y)\partial_{n_y}\ln|x-y|\,ds_y = -\frac{1}{\pi}\int_\Gamma \partial_n u(y)\ln|x-y|\,ds_y\,.$$

Thus, for the inhomogeneous Dirichlet problem we finally obtain Symm's integral equation

$$Vq = f$$

for the unknown $q = \frac{\partial u}{\partial n}$.

The above topic was extended to a Lipschitzian boundary curve in the works by Verchota, Mitrea and others (see [307, 421]) on one hand and by Costabel on the other hand (see [114]).

2.3 Mapping Properties of Boundary Integral Operators

In this section we follow [116].

Lemma 2.4 *Let $\Gamma = \partial\Omega$, $\Omega \subset \mathbb{R}^2$, and $\varphi \in C(\Gamma)$ with $\varphi \neq 0$ such that*

$$\int_\Gamma \varphi(x)\, ds_x = 0. \tag{2.23}$$

Then we have

$$\langle \varphi, V\varphi \rangle := \int_\Gamma \varphi(x)(V\varphi)(x)\, ds_x > 0.$$

Proof A simple calculation yields for large $|y|$:

$$\begin{aligned}\ln|x-y| &= \ln|y| + \frac{1}{2}\ln(1 - 2\frac{(x,y)}{|y|^2} + \frac{|x|^2}{|y|^2})\\ &= \ln|y| - \frac{(x,y)}{|y|^2} + \frac{1}{2}\frac{|x|^2}{|y|^2} + O(\frac{1}{|y|^2}).\end{aligned}$$

Hence for $x \notin \Gamma$,

$$-\pi S\varphi(y) = \ln|y| \int_\Gamma \varphi(x) ds_x - \frac{1}{|y|^2}\{y_1 \int_\Gamma x_1\varphi(x) ds_x + y_2 \int_\Gamma x_2\varphi(x) ds_x\} + O(\frac{1}{|y|^2})$$

implies by (2.23)

$$S\varphi(y) = O\left(|y|^{-1}\right).$$

Also we have

$$\nabla S\varphi(y) = c\ln|y| + O\left(|y|^{-1}\right) \quad \text{for } |y| \longrightarrow \infty,$$

since

$$\left|\frac{x-y}{|x-y|^2}\right| = \frac{1}{|y|}\frac{|y|}{|x-y|} \leq \frac{1}{|y|}\left(1 + \frac{|x|}{|y-x|}\right) = O\left(|y|^{-1}\right).$$

Using the jump relations given in Lemma 2.2 and Lemma 2.3 and setting $u := S\varphi$, we obtain using Green's first formula

$$2\langle \varphi, V\varphi \rangle = \langle \gamma_1 u^-, \gamma_0 u^- \rangle - \langle \gamma_1 u^+, \gamma_0 u^+ \rangle$$
$$= \int_{\Omega^-} |\nabla u|^2 + \int_{\Omega^+} |\nabla u|^2 = \int_{\mathbb{R}^2} |\nabla S\varphi|^2 .$$

Suppose $\int_{\mathbb{R}^2} |\nabla S\varphi|^2 = 0$, hence $S\varphi = const.$ and thus $2\varphi = -[\gamma_1 S\varphi] = 0$, a contradiction. Hence it follows $\langle \varphi, V\varphi \rangle > 0$ for $\varphi \neq 0$. □

Definition 2.7 For $m(\phi) := \frac{1}{L} \int_\Gamma \varphi \, ds$ with $L := \int_\Gamma ds$ and $\varphi_0 := \varphi - m(\varphi)$ we define for $\varphi \in C(\Gamma)$ a norm by

$$\|\varphi\|_V^2 := \langle \varphi_0, V\varphi_0 \rangle + m^2(\varphi) .$$

Definition 2.8 We define the space H_V to be the completion or closure of $L^2(\Gamma)$ (or of $C(\Gamma)$ or of $C^\infty(\Gamma)$) with respect to the norm $\|.\|_V$, i.e. H_V is a Hilbert space with inner product $(\varphi, \psi)_V := \langle \varphi_0, V\psi_0 \rangle + m(\varphi)m(\psi)$.

Remark 2.6 The spaces $C(\Gamma)$, $C^\infty(\Gamma)$ and $L^2(\Gamma)$ are dense in H_V. - The mapping $V : L^2(\Gamma) \to L^2(\Gamma)$ can be extended continuously to $V : H_V \to H_V'$, H_V' denoting the dual space which lies dense in $L^2(\Gamma)$. Thus, $L^2(\Gamma)$ is self-dual, $H_V' \subset L^2(\Gamma) \subset H_V$.

Lemma 2.5 *For $\Gamma = \partial\Omega \in Lip$ the following statements are equivalent:*

(i) $V : H_V \to H_V'$ is bijective.
(ii) The equation $V\psi = 1$ has a solution in H_V.
(iii) The equation $V\psi = 0$ has only the trivial solution.

Proof (i) $\Rightarrow$ (ii): clear.
(ii) $\Rightarrow$ (i): Let $Ve \equiv 1$. Then the operator $V : H_V \to H_V'$ is

- injective, since: Let $V\psi \equiv 0$, then for all $\varphi \in H_V$

$$\langle \varphi, V\psi \rangle = \langle V\varphi, \psi \rangle = 0.$$

Thus

$$0 = \langle Ve, \psi \rangle = \langle 1, \psi \rangle = \int_\Gamma \psi \, ds ,$$

hence by Lemma 2.4 $\psi = 0$, since $\langle V\psi, \psi \rangle = 0$

- surjective, since:

  For $t \in H'_V$ the *Riesz* representation theorem yields: $\exists \psi \in H_V$ such that $\forall \varphi \in H_V$

$$\begin{aligned}\langle t, \varphi\rangle &= (\psi, \varphi)_V = \langle \psi_0, V\varphi_0\rangle + m(\varphi)m(\psi) \\ &= \langle V\psi_0, \varphi\rangle + m(\psi)m(\varphi) - \langle V\psi_0, m(\varphi)\rangle \\ &= \int\limits_\Gamma \left(V\psi_0 + \frac{m(\psi)}{L} - m(V\psi_0)\right)\varphi(x)\, ds_x\,.\end{aligned}$$

Hence, $t = V\left(\psi_0 + \left[\frac{m(\psi)}{L} - m(V\psi_0)\right]e\right)$, $with Ve \equiv 1$.

(i) ⇒ (iii): clear.

(iii) ⇒ (i): The boundary integral operator V is a Fredholm operator (see Definition A.8 in Appendix A) with zero index, see Sections 4.2 and 4.3, see also [259, Section 10.3]. □

As we shall see below, the boundary integral operator V is bijective and even positive definite, if Ω is contained in a disk with radius < 1, what can always be arranged by scaling.

More precisely, due to Gaier [185, Satz 11], Sloan and Spence [380, Section 2] based on analytic function theory [245, Chapter 16] the equation $V\psi = 1$ admits a (unique) solution $\psi = e$, provided cap$(\Gamma) \neq 1$, where cap(Γ) denotes the *logarithmic capacity* or *transfinite diameter* of a Lipschitz curve Γ. Note, that if e exists, we always have that $m(e) \neq 0$, since: $m(e) = 0 \Rightarrow \langle e, Ve\rangle > 0$ by Lemma 2.4, but $\langle e, Ve\rangle = \langle e, 1\rangle = Lm(e) = 0$, which is a contradiction. Further, cap(Γ) scales linearly, i.e. cap$(\Gamma_r) = r$ cap(Γ), where for some scalar $r > 0$,

$$\Gamma_r := r \cdot \Gamma = \left\{x \in \mathbb{R}^2 : \frac{x}{r} \in \Gamma\right\}.$$

In what follows, the single-layer operator V on Γ will be denoted by V_Γ.

Theorem 2.4 *Let $\Gamma \in Lip$ contained in $\mathbb{R}^2$. Then*

(i) If cap$(\Gamma) \neq 1$, then cap$(\Gamma) = \frac{1}{r_0}$ with $r_0 = exp(\frac{\pi}{Lm(e)})$
(ii) V_{Γ_r} is positive definite if and only if cap$(\Gamma) < 1$.

Proof

(i) We first transform V_{Γ_r} acting on Γ_r onto the boundary Γ as follows:

For u defined on Γ let $u_r : \Gamma_r \longrightarrow \mathbb{R}$ be given by $u_r(x) := u\left(\frac{x}{r}\right)$. Further, for $x \in \Gamma$ let $(V^r_\Gamma u)(x) := (V_{\Gamma_r} u_r)(rx)$. Then, V^r_Γ is bijective if and only if V_{Γ_r} is bijective. One calculates:

$$\begin{aligned}(V_\Gamma^r)u(x) &= -\frac{1}{\pi}\int_{\Gamma_r} \ln|rx-y|u_r(y)\,ds(y)\\ &\overset{y=rz}{=} -\frac{1}{\pi}\int_{\Gamma} \ln|rx-rz|u_r(rz)r\,ds(z)\\ &= -\frac{1}{\pi}\int_{\Gamma} \ln(r|x-z|)u(z)r\,ds(z)\\ &= r\left((V_\Gamma u)(x) - \frac{\ln r}{\pi}\int_\Gamma u(z)\,ds(z)\right)\\ &= r\,(V_\Gamma u(x) - \ln r\cdot c(u))\end{aligned}$$

with $c(u) = \frac{1}{\pi}\int_\Gamma u(z)\,ds(z)$. Let e solve $V_\Gamma\psi \equiv 1$. Then $V_\Gamma^r e = r(1 - c(e)\ln r)$. Hence in particular $V_\Gamma^{r_0}e = 0$ for $r_0 = \exp(\frac{1}{c(e)})$. Thus, we have that for V_Γ bijective, the operator $V_{\Gamma_{r_0}}$ becomes not bijective and thus

$$\text{cap}\,\Gamma_{r_0} = r_0\text{cap}\,\Gamma = 1\,.$$

With r_0 as given above, $c(e) = \frac{1}{\pi}\int_\Gamma e\,ds$, and $m(e) = \frac{\pi}{L\ln r_0}$ the first part of the theorem is proved.

(ii): To show that V is positive definite, let $\varphi\in H_V$ and $\varphi_0 := \varphi - \frac{m(\varphi)}{m(e)}e$. By Lemma 2.4, without loss of generality, $m(\varphi)\neq 0$. Hence

$$m(\varphi_0) = m(\varphi) - \frac{m(\varphi)}{m(e)}m(e) = 0\,,$$

$$\langle\varphi, V\varphi\rangle = \langle\varphi_0, V\varphi_0\rangle + \langle\varphi_0, V\frac{m(\varphi)}{m(e)}e\rangle + \langle\frac{m(\varphi)}{m(e)}e, V\varphi_0\rangle + \left(\frac{m(\varphi)}{m(e)}\right)^2\langle e, Ve\rangle\,.$$

Since $\langle e, V\varphi_0\rangle = \langle\varphi_0, Ve\rangle = \langle\varphi_0, 1\rangle = 0$, the second summand and third summand vanish. Further with $\mu := \frac{m(\varphi)}{m(e)} \neq 0$ and again by Lemma 2.4, $\langle\varphi, V\varphi\rangle > L\mu^2 m(e)$. Now, $Lm(e) = \frac{\pi}{\ln r_0} > 0 \Leftrightarrow r_0 > 1 \Leftrightarrow \text{cap}(\Gamma) < 1$.
Conversely let V be positive definite. Then V is injective. By Lemma 2.5, V is bijective and there exists e with $Ve = 1$. Since V is positive definite, $\langle e, Ve\varphi\rangle = Lm(e) > 0$. Hence by definition $r_0 > 1 \Leftrightarrow \text{cap}(\Gamma) < 1$. The assertion of (ii) follows.

□

Now from

$$\text{cap}\,(\Gamma) \leq \text{cap}\,\partial B(0;a) = a\,,$$

whenever Γ is contained in a disk $B(0;a)$ with radius a, diam $\Gamma < 1$ is a sufficient condition for V to be bijective and positive definite.

Theorem 2.5 *Let Γ be a Lipschitz curve. Then for the operators V, K, K' and W the following holds:*

(i) $V : H^{-\frac{1}{2}}(\Gamma) \longrightarrow H^{\frac{1}{2}}(\Gamma)$ *is continuous, positive definite* $\Longleftrightarrow \mathrm{cap}(\Gamma) < 1$

(ii)

$$1 + K : \begin{cases} H^{\frac{1}{2}}(\Gamma) \longrightarrow H^{\frac{1}{2}}(\Gamma) \\ L^2(\Gamma) \longrightarrow L^2(\Gamma) \end{cases}$$

is continuous with $\ker(1 + K) = \mathbb{R} =$ *constants*

(iii)

$$1 + K' : \begin{cases} H^{-\frac{1}{2}}(\Gamma) \longrightarrow H^{-\frac{1}{2}}(\Gamma) \\ L^2(\Gamma) \longrightarrow L^2(\Gamma) \end{cases} \text{ is continuous and injective with}$$

$(1 + K')u = g$ *solvable* $\Leftrightarrow m(g) = \frac{1}{L} \int_{\Gamma} g = 0$

(iv)

$$1 - K : \begin{cases} H^{\frac{1}{2}}(\Gamma) \longrightarrow H^{\frac{1}{2}}(\Gamma) \\ L^2(\Gamma) \longrightarrow L^2(\Gamma) \end{cases} \text{ is continuous and bijective}$$

(v)

$$1 - K' : \begin{cases} H^{-\frac{1}{2}}(\Gamma) \longrightarrow H^{-\frac{1}{2}}(\Gamma) \\ L^2(\Gamma) \longrightarrow L^2(\Gamma) \end{cases} \text{ is continuous and bijective}$$

(vi)

$$W : H^{\frac{1}{2}}(\Gamma) \longrightarrow H^{-\frac{1}{2}}(\Gamma) \text{ is continuous with either}$$

$$\ker W = \{\text{constants}\} \text{ or } \operatorname{coker} W = \{\text{constants}\}$$

$$\Longrightarrow W : H_0^{\frac{1}{2}}(\Gamma) := \{\varphi \in H^{\frac{1}{2}}(\Gamma) : m(\varphi) = 0\} \longrightarrow H^{-\frac{1}{2}}(\Gamma)/\mathbb{R}$$

is continuous and bijective with

$$\langle W\varphi, \varphi\rangle \geq 0 \,\forall\, \varphi \in H^{\frac{1}{2}}(\Gamma),\ \langle W\varphi, \varphi\rangle > 0 \text{ if } m(\varphi) = 0 \text{ and } \varphi \neq 0$$

Proof We only prove the last assertion (vi) and leave the other items as an exercise. Let $u(x) := \frac{1}{2} D\varphi(x)$ for $x \notin \Gamma$. Then by the jump relations given in Lemma 2.2 (ii) we have

$$[\gamma_0 u] = \varphi,\ \ [\gamma_1 u] = 0 \text{ and } \gamma_1^+ u = \gamma_1^- u = -W\varphi$$

Thus,

$$\langle \varphi, W\varphi \rangle = -\langle [\gamma_0 u], \gamma_1^- u \rangle = \langle \gamma_0 u^-, \gamma_1 u^- \rangle - \langle \gamma_0 u^+, \gamma_1 u^+ \rangle = \int\limits_{\Omega^-} |\nabla u|^2 + \int\limits_{\Omega^+} |\nabla u|^2$$

using Green's formulas, since $u = O\left(\frac{1}{|x|}\right)$ as $x \to \infty$. It follows $\langle \varphi, W\varphi \rangle \geq 0$.

Let now $u \in H^1_{loc}(\mathbb{R}^2)$, i.e. $u \in H^1(K)\ \forall K \supset\supset \Omega^-,\ K compact$. With $u = O\left(\frac{1}{|x|}\right)$ as $x \to \infty$ it follows that

$$\int\limits_{\Omega^-} |\nabla u|^2 + \int\limits_{\Omega^+} |\nabla u|^2 < \infty .$$

By the trace theorem we further have $u^{+/-}|_\Gamma =: \gamma_0^{+/-} u \in H^{\frac{1}{2}}(\Gamma)$ and thus

$$\langle\, [\gamma_0 u], \gamma_1 u^{+/-}\, \rangle < \infty .$$

Furthermore, W is continuous, since

$$\|W\varphi\|_{H^{-\frac{1}{2}}(\Gamma)} := \sup_{\varphi \in H^{\frac{1}{2}}(\Gamma)} \frac{\langle \varphi, W\varphi \rangle}{\|\varphi\|_{H^{\frac{1}{2}}(\Gamma)}} < \infty .$$

And W is bijective, since

$$\begin{aligned} \langle \varphi, W\varphi \rangle = 0 &\Longleftrightarrow u^{+/-} = c(constant) \\ &\Longrightarrow u^+ = 0 \ \ (since\, u = O\left(\tfrac{1}{|x|}\right) \text{ as } x \to \infty) \\ &\Longrightarrow u^- = c \Longrightarrow \gamma_1 u^- = -W\varphi = 0 \end{aligned}$$

With $\varphi = [\gamma_0 u] = -c$ and $m(\varphi) = 0$ it follows $\varphi \equiv 0$. □

2.4 Laplace's Equation in $\mathbb{R}^3$

In this chapter we want to consider the situation that $\Omega \subseteq \mathbb{R}^3$. We may assume that for a neighbourhood U of any point $x_0 \in \Gamma = \partial\Omega$ there exists a local parameter representation (Fig. 2.3).

Thus, $\mathbb{R}^3 \supseteq O \ni (u, v, w) \mapsto x(u, v, w) \in U$ with $x_0 \in U$, such that

$$\det \frac{\partial x}{\partial(u, v, w)} \neq 0\ \forall (u, v, w) \in O .$$

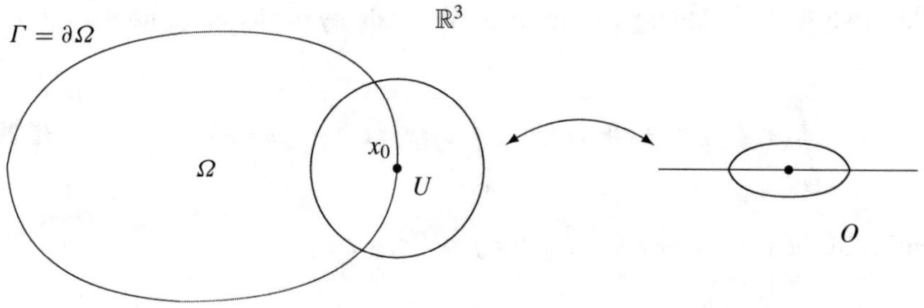

Fig. 2.3 Geometrical setting

We can arrange the local parameter representation such that

$$\begin{aligned} w = 0 &\Leftrightarrow x(u, v, w)(= x(u, v)) \in \Gamma \\ w > 0 &\Leftrightarrow x(u, v, w) \qquad \in \Omega^- = \Omega \\ w < 0 &\Leftrightarrow x(u, v, w) \qquad \in \Omega^+ = \mathbb{R}^3 \backslash \overline{\Omega} \end{aligned}$$

Hence, we have that $\Gamma \in C^m$ if and only if $\forall\, x_0 \in \Gamma$ there exists a local parameter representation in $C^m(O)$.

Let the tangential and normal vectors to the curve Γ be denoted by $(\frac{\partial x}{\partial u}, \frac{\partial x}{\partial v})^T$ and $\overrightarrow{n} = +/- \frac{\frac{\partial x}{\partial u} \times \frac{\partial x}{\partial v}}{\| \frac{\partial x}{\partial u} \times \frac{\partial x}{\partial v} \|}$, such that $\overrightarrow{n}$ points from Ω^- to Ω^+. Then, the normal derivative is given by $\partial_n u := \overrightarrow{n} \cdot grad(u)$. For the surface measure we have

$$ds := (\det g)^{\frac{1}{2}}\, du\, dv \quad \text{for } g_{ij} = \partial_i x \times \partial_j x$$

(with $\partial_1 x = \frac{\partial x}{\partial u}, \partial_2 x = \frac{\partial x}{\partial v}$), which is defined on $\tilde{U} \subset \Gamma$ independent of the parameter representation. Thus we have

$$\int_\Gamma \varphi(x)\, ds(x) := \sum_{j=1}^{J} \int_{O_j \cap \mathbb{R}^2} \varphi_j(u, v)(det(g(u, v)))^{\frac{1}{2}}\, du\, dv \tag{2.24}$$

with $\Gamma \subseteq \bigcup_{j=1}^{J} O_j, \quad \text{supp}\,(\varphi_j) \subseteq O_j \cap \Gamma, \quad \sum \varphi_j = \varphi\,.$

For $f \in C_0^1(\mathbb{R}^3)$ integration by parts yields

$$\int_\Omega \partial_j f\, dx = \int_\Gamma n_j(x) f(x)\, ds(x)\,. \tag{2.25}$$

This leads to the following conclusions. We obtain by partial integration for $f, g \in C_0^1$

$$\int_{\Omega} (\partial_j f \cdot g + f \cdot \partial_j g)\, dx = \int_{\Gamma} n_j(x) f(x) g(x)\, ds(x)\,, \tag{2.26}$$

further by the first Green's formula for $f \in C_0^2$, $g \in C_0^1$

$$\int_{\Omega} (\nabla f \cdot \nabla g + \Delta f \cdot g)\, dx = \int_{\Gamma} \partial_n f \cdot g\, ds(x)\,, \tag{2.27}$$

and by the second Green's formula for $f, g \in C_0^2$

$$\int_{\Omega} (\Delta f \cdot g - f \cdot \Delta g)\, dx = \int_{\Gamma} (\partial_n f(x) \cdot g(x) - f \cdot \partial_n g(x))\, ds(x)\,. \tag{2.28}$$

For a formulation of Green's formulas in a distributional form we shall consider $(u^-, u^+) \in C_0^2(\overline{\Omega^-}) \times C_0^2(\overline{\Omega^+})$, where shortly $\overline{\Omega^-} = \Omega^- \cup \Gamma$, $\overline{\Omega^+} = \Omega^+ \cup \Gamma$, and let

$$u := \begin{cases} u^- & \text{on } \Omega^- \\ u^+ & \text{on } \Omega^+ \end{cases}, \qquad f := \Delta u|_{I\!R^3 \setminus \Gamma} \ .$$

Then for any $\chi \in C_0^\infty(\mathbb{R}^3)$, $\gamma_0 \chi = \chi|_\Gamma$ and $\gamma_1 \chi = \partial_n \chi|_\Gamma$.

With γ_0^*, γ_1^* defined by

$$\langle \gamma_0^* \varphi, \chi \rangle := \int_{\mathbb{R}^3} \varphi \cdot \gamma_0 \chi\, dx := \int_{\Gamma} \varphi \cdot \chi\, ds\ ,$$
$$\langle \gamma_1^* \varphi, \chi \rangle := \int_{\mathbb{R}^3} \varphi \cdot \gamma_1 \chi\, dx := \int_{\Gamma} \varphi \cdot \partial_n \chi\, ds$$

respectively, we obtain as in Sect. 2.2.1 for the Laplacian Δu the representation analogous to (2.15)

$$\Delta u = f + \gamma_0^*[\gamma_1 u] - \gamma_1^*[\gamma_0 u] \tag{2.29}$$

2.4.1 Representation Formula

Analogous to Sect. 2.1 we now want to derive a representation formula for $u \in C_0^2(\overline{\Omega})$. Let

$$G(x, y) = -\frac{1}{4\pi} \frac{1}{|x - y|}\,,$$

Fig. 2.4 Geometrical setting

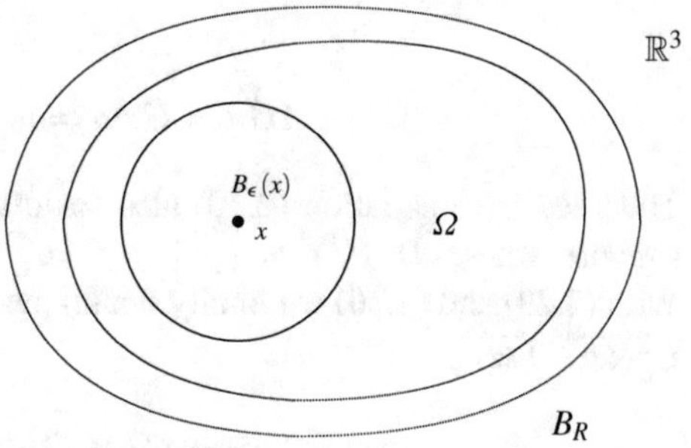

which is a fundamental solution, since it is easy to show (Exercise) that $\Delta_x G(x, y) = 0, \;\; \forall x \neq y$. Consider now the following Fig. 2.4:
Let $u \in C_0^2(\mathbb{R}^3)$, $f = \Delta u$ and

$$\int_{\mathbb{R}^3} G(x, y) f(y)\, dy := \lim_{\epsilon \to 0} \underbrace{\int_{\epsilon < |x-y| < R} G(x, y) f(y)\, dy}_{:= J_\epsilon} .$$

Then by 2. Green formula (2.28) in $\Omega \backslash B_\epsilon(x)$,

$$J_\epsilon = \int_{\partial B_\epsilon} (\partial_n u \cdot G - u \cdot \partial_{n_y} G)\, ds_y \; + \; 0$$

and

$$\left| \int_{\partial B_\epsilon(x)} \partial_n u(y) \cdot G(x, y)\, ds_y \right| \leq \sup |\nabla u| \cdot \frac{1}{4\pi\epsilon} \cdot 4\pi\epsilon^2 \stackrel{\epsilon \to 0}{\longrightarrow} 0 .$$

We further have

$$\begin{aligned} \int_{\partial B_\epsilon(x)} u(y) \partial_{n_y} G(x, y)\, ds_y &= \int_{\partial B_\epsilon(0)} u(x+y)[\partial_r(-\tfrac{1}{4\pi r})]_{r=\epsilon}\, ds_y \\ &= \frac{1}{4\pi\epsilon^2} \int_{\partial B_\epsilon(0)} u(x+y)\, ds_y \\ &= \frac{1}{4\pi} \int_{\partial B_1(0)} u(x+\epsilon y)\, ds_y \stackrel{\epsilon \to 0}{\longrightarrow} u(x) . \end{aligned}$$

This gives

$$\int_{\mathbb{R}^3} G(x, y) f(y)\, dy = u(x)$$

and

$$\Delta G u = G \Delta u = u, \; \forall\, u \in C_0^2(\mathbb{R}^3) \tag{2.30}$$

Thus, we have equation (2.30) also for distributions $u \in D'(\mathbb{R}^3)$ with compact support, where $D'(\mathbb{R}^3) = \{\varphi \mid \varphi : C_0^\infty(\mathbb{R}^3) \longrightarrow \mathbb{R}$, linear and continuous$\}$. With (2.29) and (2.30) we finally obtain the desired representation formula for $u \in C_0^2(\overline{\Omega}^+ \cup \overline{\Omega}^-)$:

$$u = G\Delta u = Gf + G\gamma_0^*[\gamma_1 u] - G\gamma_1^*[\gamma_0 u] \,.$$

In correspondence with Sect. 2.2 we define the potential operators as follows.

Definition 2.9 For $x \notin \Gamma$ the operator

$$S\varphi(x) := -2G\gamma_0^*\varphi(x) = \frac{1}{2\pi}\int_\Gamma \frac{\varphi(y)}{|x-y|}\,ds(y) \tag{2.31}$$

is called the *single-layer potential* with density φ;

$$D\varphi(x) := -2G\gamma_1^*\varphi(x) = \frac{1}{2\pi}\int_\Gamma \varphi(y)\partial_{n_y}\frac{1}{|x-y|}\,ds(y) \tag{2.32}$$

is called the *double-layer potential* with density φ.

With these definitions the above representation formula reads

$$u = Gf + \frac{1}{2}D[\gamma_0 u] - \frac{1}{2}S[\gamma_1 u].$$

Remark 2.7 We may again prove the jump relations: (i) $[\gamma_0 S\psi] = 0$ and $[\gamma_1 S\psi] = -2\psi$ (ii) $[\gamma_0 D\psi] = 2\psi$ and $[\gamma_1 D\psi] = 0$.

Remark 2.8 The operators $(1 \pm K)$ in $H^{1/2}(\Gamma)$ and $(1 \pm K')$ in $H^{-1/2}(\Gamma)$ are contractions and they give rise to coercive bilinear forms. This allows to solve the corresponding boundary integral equations by the use of Banach's fixed point theorem; and Neumann's series always converges [118, 394].

2.5 Calderon Projector

As seen above the solution $u \in C^2(\overline{\Omega})$ of $\Delta u = 0$ in $\Omega = \Omega^-$ has a representation $u = -\frac{1}{2}Dv + \frac{1}{2}S\phi$ with $v = \gamma_0 u$, $\phi = \gamma_1 u$. Then the jump relations yield

$$\begin{aligned} 2\gamma_0 u &= -\gamma_0 Dv + \gamma_0 S\phi = v - Kv + V\phi \\ 2\gamma_1 u &= -\gamma_1 Dv + \gamma_1 S\phi = Wv + \phi + K'\phi. \end{aligned}$$

or more compactly written

$$\begin{pmatrix} \gamma_0 u \\ \gamma_1 u \end{pmatrix} \equiv \begin{pmatrix} v \\ \varphi \end{pmatrix} = \frac{1}{2}(Id + \mathscr{A}) \begin{pmatrix} v \\ \phi \end{pmatrix} =: \mathscr{C} \begin{pmatrix} v \\ \phi \end{pmatrix} \tag{2.33}$$

with

$$\mathscr{A} := \begin{pmatrix} -K & V \\ W & K' \end{pmatrix}$$

Definition 2.10 The operator $\frac{1}{2}(Id + \mathscr{A}) =: \mathscr{C}(= \mathscr{C}^-)$ is called **Calderon projector** (for the domain Ω^- and the Laplacian).

Theorem 2.6 *Suppose $\Gamma \in C^\infty$.*

1. *Let $v, \phi \in C^\infty(\Gamma)$. Then there exists $u \in C^\infty(\overline{\Omega})$ with $\Delta u = 0$ in Ω, $v = \gamma_0 u$ and $\phi = \gamma_1 u$ if and only if $\begin{pmatrix} v \\ \phi \end{pmatrix} = \mathscr{C} \begin{pmatrix} v \\ \phi \end{pmatrix}$.*
2. $\mathscr{C}^2 = \mathscr{C}$.
3. $\mathscr{C}^- + \mathscr{C}^+ = Id$.

Proof

1. Let $v, \phi \in C^\infty(\Gamma)$ be given and set $u := -\frac{1}{2}Dv + \frac{1}{2}S\phi$. Then $u \in C^\infty(\overline{\Omega})$, $\Delta u = 0$ and $\begin{pmatrix} \gamma_0 u \\ \gamma_1 u \end{pmatrix} = \mathscr{C} \begin{pmatrix} v \\ \phi \end{pmatrix}$ as shown above. If also there holds $\begin{pmatrix} v \\ \phi \end{pmatrix} = \mathscr{C} \begin{pmatrix} v \\ \phi \end{pmatrix}$ then it follows that $\begin{pmatrix} v \\ \phi \end{pmatrix} = \begin{pmatrix} \gamma_0 u \\ \gamma_1 u \end{pmatrix}$.
   The inverse assertion is just (2.33).
2. Let ϕ, v both in $C^\infty(\Gamma)$ and set $u := -\frac{1}{2}Dv + \frac{1}{2}S\phi$. Then there holds $\begin{pmatrix} \gamma_0 u \\ \gamma_1 u \end{pmatrix} = \mathscr{C} \begin{pmatrix} v \\ \phi \end{pmatrix}$ and $\begin{pmatrix} \gamma_0 u \\ \gamma_1 u \end{pmatrix} = \mathscr{C} \begin{pmatrix} \gamma_0 u \\ \gamma_1 u \end{pmatrix} = \mathscr{C}^2 \begin{pmatrix} v \\ \phi \end{pmatrix}$

$$\text{Hence } \mathscr{C} \begin{pmatrix} v \\ \phi \end{pmatrix} = \mathscr{C}^2 \begin{pmatrix} v \\ \phi \end{pmatrix} \text{ implying } \mathscr{C}^2 = \mathscr{C}$$

3. The jump relations yield $\mathscr{C}^+ = \frac{1}{2}(Id - \mathscr{A})$.

Remark 2.9 $\mathscr{C}^2 = \mathscr{C} \iff \mathscr{A}^2 = Id$

Using the Calderon projector boundary integral equations of first and second kind for mixed boundary and transmission problems can be derived.

Following e.g. [127, 405], consider the mixed boundary value problem

$$\Delta u = 0 \text{ in } \Omega, \quad u = g_1 \text{ on } \Gamma_1, \quad \frac{\partial u}{\partial \vec{n}} = g_2 \text{ on } \Gamma_2,$$

where

$$\Gamma = \bar{\Gamma}_1 \cup \bar{\Gamma}_2, \quad \Gamma_1 \cap \Gamma_2 = \emptyset$$

for given g_1, g_2.

Sought are $v := u$ on Γ_2, $\psi := \frac{\partial u}{\partial \vec{n}}$ on Γ_1. Obtain following system of boundary integral equations

$$\begin{pmatrix} W_{22} & K'_{12} \\ -K_{12} & V_{11} \end{pmatrix} \begin{pmatrix} v \\ \psi \end{pmatrix} = \begin{pmatrix} -W_{12} & I - K'_{22} \\ I + K_{11} & V_{21} \end{pmatrix} \begin{pmatrix} g_1 \\ g_2 \end{pmatrix}$$

where W_{ij} etc. means integration on Γ_j and evaluation on Γ_i.

2.6 Use of Complex Function Theory

In this section we provide another elementary approach in $\mathbb{R}^2$ which is based on complex function theory using in particular the Cauchy formula and the Cauchy-Riemann differential equations. Here again we derive the representation formula and give an applicable representation. of the hypersingular operator.

2.6.1 *Representation Formula Again*

Let the boundary Γ of a two dimensional domain Ω be a sufficiently smooth, simply connected curve given by a 1-periodic vector function

$$\mathbf{x} = \mathbf{x}(t) = \begin{cases} x = x(t) \\ y = y(t), \end{cases} \quad \text{for} \quad 0 \le t \le 1.$$

Let us consider the interior Dirichlet problem for the Laplacian: Given u_0 on Γ, find u satisfying

$$\begin{cases} \Delta u = 0 & \text{in} \quad \Omega \\ u = u_0 & \text{on} \quad \Gamma. \end{cases} \tag{2.34}$$

In order to reduce this boundary value problem to an integral equation on Γ, firstly we will deduce the representation formula for harmonic functions from the Cauchy formula of complex function theory.

Due to $\Delta u = 0$ the function u is the real part of a holomorphic function

$$u = \Re\, F(z), \; F(z) = u + iv, \; z = x + iy$$

For u and v there hold the Cauchy-Riemann differential equations

$$u_x = v_y, \; -u_y = v_x.$$

Hence

$$v(p) - v(p_0) = \int_{p_0}^{p} dv = \int_{p_0}^{p} (-u_y\, dx + u_x\, dy).$$

Here the integral on the right hand side is path independent, since

$$u_{xx} = -u_{yy}, \quad \text{i.e.} \quad \Delta u = 0$$

We apply the Cauchy integral formula to $F(z)$ and compute its real part in order to express u

$$u(z) = \Re\, F(z) = \Re \left(\frac{1}{2\pi i} \oint_{\Gamma} \frac{F(\zeta)}{z-\zeta} d\zeta \right), \quad \zeta = \xi + i\eta, \quad \zeta \in \Gamma,\; z \in \Omega \quad (2.35)$$

Due to the Cauchy Riemann equations also the boundary values of u and v are connected by

$$\dot{v} = \left.\frac{dv}{ds}\right|_{\Gamma} = v_x\dot{\xi} + v_y\dot{\eta} = u_x\dot{\eta} - u_y\dot{\xi} = \mathbf{n}\cdot\nabla u = \left.\frac{\partial u}{\partial n}\right|_{\Gamma}$$

Here and in the following s denotes the arc length on Γ and $\dot{\xi}$ the derivative of ξ w.r.t. s; $\mathbf{n}$ is the normal on Γ pointing outward of Ω.

We change (2.35) to

$$u(z) = \Re \frac{1}{2\pi i} \left(\oint_{\Gamma} \frac{u(\zeta)\frac{d\zeta}{ds} ds}{z-\zeta} + i \oint_{\Gamma} \frac{v(\zeta)\frac{d\zeta}{ds} ds}{z-\zeta} \right) \quad (2.36)$$

Now we have

$$\begin{aligned}
\Re \frac{1}{2\pi} \oint_{\Gamma} \frac{v(\zeta)\frac{d\zeta}{ds} ds}{z-\zeta} &= \Re \frac{1}{2\pi} \int_{\Gamma} v(\zeta) \frac{d}{ds} \ln(z-\zeta(s)) ds \qquad (2.37)\\
&= \Re \frac{1}{2\pi} (v(\zeta)\ln(z-\zeta))\big|_0^L - \Re \int_{s=0}^{L} \frac{dv(\zeta)}{ds} \ln(z-\zeta(s)) ds\\
&= \frac{1}{2\pi} v(\zeta) \ln|\zeta - z|\big|_0^L - \frac{1}{2\pi} \int_0^L \frac{\partial u}{\partial n}(\zeta) \Re \ln(z-\zeta) ds\\
&= 0 - \frac{1}{2\pi} \int_0^L \frac{\partial u}{\partial n}(\zeta) \ln|z-\zeta| ds_{\zeta}.
\end{aligned}$$

We rewrite the other integral in (2.36) as

$$\Re\,\frac{1}{2\pi i}\oint_\Gamma \frac{u(\zeta)\frac{d\zeta}{ds}ds}{z-\zeta} = \Re\,\frac{1}{2\pi i}\int_\Gamma u(\zeta)\frac{d}{ds}\ln(z-\zeta(s))ds$$
$$= \Re\,\frac{1}{2\pi i}\int_\Gamma u(\zeta)\frac{d}{ds}\{\ln|z-\zeta| + i\arg(z-\zeta)\}ds.$$

For the normal derivative

$$\frac{\partial}{\partial n}\ln|\zeta - z| = (\ln|z-\zeta|)_\xi\,\dot\eta - (\ln|z-\zeta|)_\eta\,\dot\xi$$

we use the Cauchy-Riemann equations for $\ln(z-\zeta)$:

$$\frac{\partial}{\partial n}\ln|\zeta - z| = \arg(z-\zeta)_\eta\,\dot\eta + \arg(z-\zeta)_\xi\,\dot\xi = \frac{d}{ds}\arg(z-\zeta)$$

Thus, we obtain

$$\Re\,\frac{1}{2\pi i}\oint_\Gamma \frac{u(\zeta)\frac{d\zeta}{ds}ds}{z-\zeta}) = \Re\,\frac{1}{2\pi i}\int_\Gamma u(\zeta)\{\frac{d}{ds}(\ln|z-\zeta|) + i\frac{\partial}{\partial n_\zeta}(\ln|z-\zeta|)\}ds$$
$$= \frac{1}{2\pi}\int_\Gamma u(\zeta)\frac{\partial}{\partial n_\zeta}(\ln|\zeta - z|)ds\ . \tag{2.38}$$

Next, we combine (2.37) and (2.38) and obtain the representation formula

$$u(x,y) = \frac{1}{2\pi}\int_\Gamma u(s)\frac{\partial}{\partial n_\zeta}(\ln|z-\zeta|) - \frac{1}{2\pi}\int_\Gamma \frac{\partial u}{\partial n_\zeta}\ln|z-\zeta|ds\ .$$

Hence the solution of (2.34) is a combination of the potential of a double layer (with density u) and the potential of a single layer (with density $\frac{\partial u}{\partial n}$).

We apply the formulas of Plemelj-Sochozki (see e.g. [43, 203]) to this representation of u which yield for $z \to \Gamma$

$$\frac{1}{2}u(z) = \frac{1}{2\pi}\int_\Gamma u(\zeta)\frac{\partial}{\partial n_\zeta}\ln|z-\zeta(s)|ds - \frac{1}{2\pi}\int_\Gamma \frac{\partial u(\zeta)}{\partial n_\zeta}\ln|z-\zeta(s)|ds, \quad z \in \Gamma$$

Since $u|_\Gamma = u_0$ must hold, this yields an equation with $\frac{\partial u}{\partial n} = q$ as unknown function on the boundary $\Gamma (z \in \Gamma)$:

$$-\int_\Gamma \ln|z-\zeta(s)|q(s)ds = f(z)\ = \pi u_0(z) - \int_\Gamma u_0(\zeta)\frac{\partial}{\partial n_\zeta}\ln|z-\zeta(s)|ds).$$

2.6.2 Applicable Representation of the Hypersingular Integral Operator

In this subsection we provide a representation of the hypersingular integral operator W introduced in Definition 2.6 as a Cauchy principal value and further a representation via the single layer potential on the boundary which is applicable to numerical simulations with the boundary element method. To this end we work in the complex plane

$$x = (x_1, x_2) \leftrightarrow \zeta = x_1 + ix_2;\; y = (y_1, y_2) \leftrightarrow z = y_1 + iy_2,$$

use the complex logarithm

$$\phi = \ln(\zeta - z) = \ln(|x - y|) + i\vartheta(\zeta - z)$$

involving the argument or angle $\vartheta(z)$ of z, and apply the Cauchy–Riemann differential equations

$$(\Re\phi)_{x_1} = (\Im\phi)_{x_2}\,, \qquad (\Im\phi)_{x_1} = -(\Re\phi)_{x_2}\,,$$

where $\Re\phi$ and $\Im\phi$ are respectively, the real part and imaginary part of ϕ.

Thus we can show the following

Proposition 2.1 *Let $f \in C^1(\Gamma)$. Then with $\mathbf{t}(x)$ denoting the tangent unit vector in x, there holds in the weak sense*

$$\int_\Gamma f(y)\left(\frac{\partial}{\partial n_x}\frac{\partial}{\partial n_y}\ln(|x-y|)\right)ds_y = \int_\Gamma \frac{df(y)}{ds}\frac{\langle x-y, \mathbf{t}(x)\rangle}{|x-y|^2}\,ds_y\,. \tag{2.39}$$

Here the first integral has to be understood in the weak (distributional) sense, the second integral exists as a Cauchy principal value, that is with Γ parametrized by $x = Z(\tau)$, respectively by $y = Z(\sigma)$ with $\tau, \sigma \in (0, 2\pi)$,

$$\begin{aligned}\int_\Gamma \frac{df(y)}{ds}\frac{\langle x-y, \mathbf{t}(x)\rangle}{|x-y|^2}\,ds_y &= \int_0^{2\pi} \frac{df(Z(\sigma))}{d\sigma}\frac{\langle Z(\tau)-Z(\sigma), \mathbf{t}(Z(\tau))\rangle}{|Z(\tau)-Z(\sigma)|^2}\,d\sigma \\ &= \lim_{\varepsilon\to 0}\left(\int_0^{\tau-\varepsilon}\ldots\,d\sigma + \int_{\tau+\varepsilon}^{2\pi}\ldots\,d\sigma\right).\end{aligned}$$

Moreover, the hypersingular operator W can be represented for any $\varphi, \psi \in H^{1/2}(\Gamma)$ via the single layer potential V on the boundary,

$$\langle W\psi, \varphi\rangle = \int_\Gamma \frac{d\varphi(x)}{ds_x} V(\frac{d\psi}{ds_x})(x)\, ds_x = \langle \dot{\varphi}, V\dot{\psi}\rangle\,. \tag{2.40}$$

Proof Let $g \in C_0^\infty(\mathbb{R}^2)$ arbitrarily. Then we have by partial integration on the curve Γ without (topological) boundary, hence without boundary terms,

$$\begin{aligned} I &:= \int_\Gamma\int_\Gamma g(x)\, f(y)\Big(\frac{\partial}{\partial n_x}\frac{\partial}{\partial n_y}\ln(|x-y|)\Big)\, ds_y\, ds_x \\ &= \int_\Gamma\int_\Gamma -\frac{\partial g(x)}{\partial n} f(y)\frac{\partial}{\partial n_y}\ln(|x-y|)\, ds_y\, ds_x\,. \end{aligned}$$

For fixed $x \in \Gamma$ there holds with the tangent unit vector $\mathbf{t} = (-n_2, n_1)$

$$\begin{aligned} \frac{\partial}{\partial n_y}\ln|x-y| &= \langle\big(\frac{\partial}{\partial y_1}\Re\ln(\zeta - z), \frac{\partial}{\partial y_2}\Re\ln(\zeta - z)\big), (n_1(y), n_2(y))\rangle \\ &= \langle\big(\frac{\partial}{\partial y_2}\Im\ln(\zeta - z), -\frac{\partial}{\partial y_1}\Im\ln(\zeta - z)\big), (n_1(y), n_2(y))\rangle \\ &= \langle\nabla_y\Im\ln(\zeta - z), t(y)\rangle = \frac{d}{ds_y}\Im\ln(\zeta - z). \end{aligned}$$

Again by partial integration, we obtain

$$I = \int_\Gamma\int_\Gamma \frac{\partial g(x)}{\partial n_x}\frac{df(y)}{ds_y}\Im\ln(\zeta - z)\, ds_y\, ds_x\,.$$

Now we rewrite for fixed $y \in \Gamma$, by partial integration and by Cauchy–Riemann equations,

$$\begin{aligned} \int_\Gamma \frac{\partial g(x)}{\partial n_x}\Im\ln(\zeta - z)ds_x &= \int_\Gamma \langle(\frac{\partial g}{\partial x_1}, \frac{\partial g}{\partial x_2}), (n_1, n_2)\rangle\cdot\Im\ln(\zeta - z)ds_x \\ &= -\int_\Gamma g(x)\langle\big(\frac{\partial}{\partial x_1}\Im\ln(\zeta - z), \frac{\partial}{\partial x_2}\Im\ln(\zeta - z)\big), (n_1, n_2)\rangle ds_x \\ &= -\int_\Gamma g(x)\langle\big(-\frac{\partial}{\partial x_2}\Re\ln(\zeta - z), \frac{\partial}{\partial x_1}\Re\ln(\zeta - z)\big), (n_1, n_2)\rangle ds_x \\ &= \int_\Gamma g(x)\langle\nabla_x\Re\ln(\zeta - z), (-n_2, n_1)\rangle)\, ds_x \end{aligned}$$

Hence we arrive at

$$
\begin{aligned}
I &= \int\limits_{\Gamma}\int\limits_{\Gamma} g(x)\frac{df(y)}{ds}\,\langle \nabla_x \ln|x-y|, \mathbf{t}(x)\rangle\, ds_y\, ds_x \\
&= \int\limits_{\Gamma}\int\limits_{\Gamma} g(x)\frac{df(y)}{ds}\,\frac{\langle x-y, \mathbf{t}(x)\rangle}{|x-y|^2}\, ds_y\, ds_x\,,
\end{aligned}
$$

what shows (2.39) in the weak sense. and setting $\varphi = g$, $\psi = f$ with $\dot{\varphi} = \frac{\partial\varphi}{\partial s}$,

$$
\begin{aligned}
\langle W\psi, \varphi\rangle &= \int_\Gamma (W\psi)(x)\varphi(x)ds_x = \int_\Gamma \frac{1}{2\pi}\int_\Gamma \frac{\partial}{\partial n_x}\frac{\partial}{\partial n_y}\ln|x-y|\psi(y)ds_y\varphi(x)ds_x \\
&= -\frac{1}{2\pi}\int_\Gamma\int_\Gamma \frac{d\varphi(x)}{ds_x}\frac{d\psi(y)}{ds_y}\ln|x-y|ds_y ds_x \\
&= \int_\Gamma \frac{d\varphi(x)}{ds_x} V(\frac{d\psi}{ds_x})(x)\,ds_x = \langle\dot{\varphi}, V\dot{\psi}\rangle\,.
\end{aligned}
$$

Of course, by completion, this formula holds for $\phi, \psi \in H^{1/2}(\Gamma)$. □

This latter representation (2.40) in 2D extends to an analogous representation of the hypersingular operator in 3D via the single layer potential, see [322].

For further reading concerning the topics of this chapter see [206, 225, 259, 276].

Chapter 3
A Fourier Series Approach

The aim of this chapter is to guide the reader from elementary Fourier series expansion to periodic Sobolev spaces on a simply connected smooth curve in $\mathbb{R}^2$. In this tour we detail on dual spaces and compact embedding. This leads to the compactness of the double-layer operator and its adjoint. Moreover in the scale of Sobolev spaces we prove the mapping property of the single-layer and hypersingular operators. Then we treat the exterior Dirichlet problem for the Laplacian and derive its explicit solution on the unit circle in terms of the Fourier coefficients. The Fourier tour concludes with the first Gårding inequality for a bilinear form which is basic in the BEM.

3.1 Fourier Expansion—The Sobolev Space $H^s[0, 2\pi]$

The Hilbert space $L^2[0, 2\pi]$ is the completion of the space $C[0, 2\pi]$ of the 2π-periodic complex-valued continuous function on $[0, 2\pi]$ with respect to the square mean norm that is defined by the $L^2[0, 2\pi]$ scalar product

$$(f, g) := \int_0^{2\pi} f(x)\, \bar{g}(x)\, dx\,.$$

Any function $\varphi \in L^2[0, 2\pi]$ can be expanded into the *Fourier series*

$$\sum_{k=-\infty}^{\infty} a_k e^{ikt}$$

© Springer International Publishing AG, part of Springer Nature 2018

J. Gwinner, E. P. Stephan, *Advanced Boundary Element Methods*,
Springer Series in Computational Mathematics 52,
https://doi.org/10.1007/978-3-319-92001-6_3

with the *Fourier coefficients*

$$a_k := \frac{1}{2\pi} \int_0^{2\pi} \varphi(t)\, e^{-ikt}\, dt\,.$$

This involves the trigonometric monoms

$$\chi_k(t) := e^{ikt} \quad (t \in \mathbb{R},\ k \in \mathbb{Z})$$

Note that $\{\chi_k\}_{k\in\mathbb{Z}}$ builds up an orthogonal system in $L^2[0, 2\pi]$:

$$(\chi_k, \chi_l) = \int_0^{2\pi} e^{i(k-l)t}\, dt = \begin{cases} 0 & (k \neq l) \\ 2\pi & (k = l) \end{cases}$$

In virtue of the Weierstrass approximation theorem the trigonometric polynomials are dense in $C[0, 2\pi]$ with respect to the maximum norm, hence with respect to the square mean norm. Hence by construction of $L^2[0, 2\pi]$, the orthogonal system $\{\chi_k\}_{k\in\mathbb{Z}}$ is complete and the above Fourier series converges to φ with respect to the square mean norm.

Since $\|\chi_k\|_2^2 = 2\pi$, *Parseval's equation* for any function $\varphi \in L^2[0, 2\pi]$, $\varphi = \sum_k a_k\, \chi_k$ reads as follows

$$\frac{1}{2\pi}\|\varphi\|_2^2 = \frac{1}{2\pi}\int_0^{2\pi} |\varphi(t)|^2\, dt = \frac{1}{2\pi}\sum_{k,l}(a_k\, \chi_k, a_l\, \chi_l) = \sum_{k=-\infty}^{\infty} |a_k|^2\,. \tag{3.1}$$

Moreover, classical analysis tells that the Fourier series of a continuously differentiable 2π-periodic function even converges absolutely and uniformly.

Integration by parts for such a function φ gives for $k \neq 0$,

$$2\pi a_k = \int_0^{2\pi} \varphi(t)\, e^{-ikt}\, dt = \frac{1}{ik}\int_0^{2\pi} \varphi'(t)\, e^{-ikt}\, dt \tag{3.2}$$

whereas for $k = 0$, $\frac{1}{2\pi}\int_0^{2\pi} \varphi'(t)\, dt = 0$ by periodicity of φ. Hence (3.1) applied to φ' leads to

$$\sum_{k=-\infty}^{\infty} k^2 |a_k|^2 = \frac{1}{2\pi}\int_0^{2\pi} |\varphi'(t)|^2\, dt < \infty\,.$$

This suggests to define function spaces of 2π-periodic functions by prescribing a certain decay of the Fourier coefficients a_k for $|k| \to \infty$.

Definition 3.1 Let $0 \leq s < \infty$. Then the Sobolev space $H^s[0, 2\pi]$ is the space of all functions $\varphi \in L^2[0, 2\pi]$ with the property

$$\sum_{k=-\infty}^{\infty} (1 + k^2)^s |a_k|^2 < \infty$$

for the Fourier coefficients a_k of φ.

Example 3.1 For $s = 0$, clearly $H^0[0, 2\pi] = L^2[0, 2\pi]$.

By the very definition of the function space $H^s[0, 2\pi]$ it makes sense to define the norm

$$\|\varphi\|_s = \left\{ \sum_k (1 + k^2)^s |a_k|^2 \right\}^{1/2}$$

and the associated scalar product

$$(\varphi, \psi)_s := \sum_{k \in \mathbb{Z}} (1 + k^2)^s a_k \overline{b}_k,$$

where a_k, respectively b_k are the Fourier coefficients of φ, respectively of ψ. Note that in view of the simple inequalities $k^2 \leq 1 + k^2 \leq 2k^2$, this norm is equivalent to the norm of the space H^s on $[0, 1]$ that was already given in the introduction.

Proposition 3.1 *$H^s[0, 2\pi]$ endowed with the scalar product $(., .)_s$ is a Hilbert space. The trigonometric polynomials are dense in $H^s[0, 2\pi]$.*

Sketch of proof: Clearly, $H^s[0, 2\pi]$ is a linear space. $(\varphi, \psi)_s$ is finite by the Cauchy-Schwarz inequality for sums and thus defines a scalar product. To verify the completeness of $H^s[0, 2\pi]$ use the completeness of $\mathbb{C}$.

For an arbitrary $\varphi \in H^s$ with Fourier coefficients a_k, consider the partial sum $\varphi_n = \sum_{|k| \leq n} a_k \chi_k$ of the Fourier series. Then

$$\|\varphi - \varphi_n\|_s^2 = \sum_{|k|=n+1}^{\infty} (1 + k^2)^s |a_k|^2 \to 0$$

for $n \to \infty$. Thus the trigonometric polynomials are dense in H^s. □

Theorem 3.1 *For $s_2 > s_1$ (≥ 0) there holds*

$$H^{s_2}[0, 2\pi] \subset H^{s_1}[0, 2\pi]$$

with dense and compact embedding.

Proof From

$$(1 + k^2)^{s_1} \leq (1 + k^2)^{s_2}$$

(since $s_1 \log(1+k^2) \leq s_2 \log(1+k^2)$) it follows $H^{s_2} \subset H^{s_1}$ with $\|\varphi\|_{s_1} \leq \|\varphi\|_{s_2}$ for $\varphi \in H^{s_2}$; thus the identity $I : H^{s_2} \to H^{s_1}$ is continuous. Density of the embedding is a consequence of the density of the trigonometric polynomials $\subset H^{s_2}$ in H^{s_1}.

To verify the compactness of I, introduce the operator $I_n : H^{s_2} \to H^{s_1}$ by

$$I_n\varphi := \sum_{k=-n}^{n} a_k \chi_k \quad \text{for any} \quad \varphi \in H^{s_2} \text{ with Fourier coefficients } a_k.$$

I_n has finite dimensional range and thus I_n is compact. Now

$$\begin{aligned}
\|(I_n - I)\varphi\|_{s_1}^2 &= \sum_{|k|>n} (1 + k^2)^{s_1} |a_k|^2 \\
&\leq \frac{1}{(1 + n^2)^{s_2 - s_1}} \sum_{|k|>n} (1 + k^2)^{s_2} |a_k|^2 \\
&\leq \frac{1}{(1 + n^2)^{s_2 - s_1}} \|\varphi\|_{s_2}^2 .
\end{aligned}$$

Hence $I_n - I \to 0$ in $\mathscr{L}(H^{s_2}, H^{s_1})$ for $n \to \infty$ and I is compact. □

In the following we provide another norm equivalent to $\|\cdot\|_s$. Let at first $s = l \in \mathbb{N}$ and let $C^l[0, 2\pi]$ denote the space of l times continuously differentiable 2π-periodic functions.

Theorem 3.2 *For $\varphi \in C^l[0, 2\pi] \subset H^l[0, 2\pi]$ with $l \in \mathbb{N}$,*

$$\|\varphi\|_{l,0} := \left\{ \int_0^{2\pi} [|\varphi(t)|^2 + |\varphi^{(l)}(t)|^2] \, dt \right\}^{1/2}$$

defines an equivalent norm to $\|\cdot\|_l$.

Proof Analogously to the calculation (3.2) , we apply l-times integration by parts

$$\int_0^{2\pi} \varphi^{(l)}(t) \, e^{-ikt} \, dt = (+ik)^l \int_0^{2\pi} \varphi(t) \, e^{-ikt} \, dt$$

and use (3.1) for φ and $\varphi^{(l)}$ to obtain

$$\|\varphi\|_{l,0}^2 = 2\pi \sum_{k=-\infty}^{\infty} (1 + k^{2l})\, |a_k|^2 .$$

Since for $k \neq 0$,

$$(1 + k^{2l}) \leq (1 + k^2)^l \leq (2k^2)^l \leq 2^l (1 + k^{2l}),$$

the claimed norm equivalence follows. □

Proposition 3.2 *For $\varphi \in C^1[0, 2\pi]$, $0 < p < 1$*

$$\|\varphi\|_{0,p} = \left\{ \int_0^{2\pi} |\varphi(t)|^2\, dt + \int_0^{2\pi}\int_0^{2\pi} \frac{|\varphi(t) - \varphi(\tau)|^2}{|\sin\frac{t-\tau}{2}|^{2p+1}}\, d\tau\, dt \right\}^{1/2} .$$

defines an equivalent norm to $\|\cdot\|_p$.

Exercise: Prove Proposition 3.2, see [Theorem 8.5][276].

Remark 3.1 The right hand integral exists, since $C^1[0, 2\pi] \subset C^{0,1}[0, 2\pi]$, that is,

$$\forall \varphi \in C^1[0, 2\pi] \quad \exists C_\varphi : |\varphi(t) - \varphi(\tau)| \leq C_\varphi |t - \tau|$$

and because of

$$\frac{2x}{\pi} \leq \sin x \quad \text{for } 0 \leq x \leq \frac{\pi}{2}$$

there holds

$$\frac{|\varphi(t) - \varphi(\tau)|^2}{|\sin\frac{t-\tau}{2}|^{2p+1}} \leq \tilde{C}_\varphi \frac{|t-\tau|^2}{|t-\tau|^{2p+1}} = \tilde{C}_\varphi |t - \tau|^{-1+\varepsilon}$$

for some $\varepsilon > 0$.

Corollary 3.1 *Let $s = k + p,\ k \in \mathbb{N},\ 0 < p < 1$. Then for $\varphi \in C^{k+1}[0, 2\pi]$*

$$\|\varphi\|_{k,p} := \left\{ \|\varphi\|_0^2 + \|\varphi^{(k)}\|_{0,p}^2 \right\}^{1/2} .$$

defines an equivalent norm to $\|\cdot\|_s$.

3.2 The Sobolev Space $H^s(\Gamma)$

In what follows, let Γ be the boundary of a simply connected bounded domain $\Omega \subset \mathbb{R}^2$ of class C^k for some $k \in \mathbb{N}$. Thus there exists a parameter representation

$$(x_1, x_2) = Z(t), \qquad t \in [0, 2\pi],$$

that is k-times continuously differentiable with $\left|\frac{dZ}{dt}\right| \geq \gamma_0 > 0$ for all $t \in [0, 2\pi]$. Given the parameter representation Z, we define for any $s \in [0, k]$ the Sobolev space

$$H^s(\Gamma) = \{\varphi \in L^2(\Gamma) : \varphi \circ Z \in H^s[0, 2\pi]\}.$$

where the scalar product on $H^s(\Gamma)$ stems from the scalar product via

$$(\varphi, \phi)_{H^s(\Gamma)} := (\varphi \circ Z, \phi \circ Z)_{H^s[0,2\pi]}.$$

Of course we want to admit different regular parameter representations for Γ. Therefore we have to show the invariance of our definition with respect of a change of the parameter representation.

Theorem 3.3 *Let $x = Z(t)$, $x = \tilde{Z}(t)$ ($t \in [0, 2\pi]$) (the same parameter interval without reduction of generality) be two different parameter representations for Γ. Then for any $s \in [0, k]$*

$$\tilde{H}^s(\Gamma) := \{\varphi \in L^2(\Gamma) : \varphi \circ \tilde{Z} \in H^s[0, 2\pi]\}$$

is homeomorphic to $H^s(\Gamma)$ defined above.

To show this theorem it is enough to apply the following lemma in the case $f = Z^{-1} \circ \tilde{Z}$.

Lemma 3.1 *Let f be a diffeomorphism of the interval $[0, 2\pi]$ onto itself of class $k \in \mathbb{N}$, that is, f is a bijection, f and f^{-1} belong to $C^k[0, 2\pi]$. Let $0 \leq s \leq k$. Then for any $\varphi \in H^s[0, 2\pi]$, we have $\varphi \circ f \in H^s[0, 2\pi]$ with*

$$\|\varphi \circ f\|_s \leq C\|\varphi\|_s,$$

where the constant C only depends on f, k and s.

For a proof of this lemma using appropriate equivalent norms we refer to [276, Lemma 8.14].

3.3 Interior Dirichlet Problem

Similar to [409], where the exterior Dirichlet problem is considered, we apply Fourier series techniques to solve the interior Dirichlet problem for the Laplacian in the unit sphere.

$$-\Delta u = 0 \text{ in } \Omega = \{z \in \mathbb{C} : |z| < 1\}$$
$$u = u_0 \text{ on } \Gamma = \partial\Omega$$

We start from Symm's boundary integral equation of the first kind

$$Vq = (I + K)u_0$$

for the Cauchy data $q = \frac{\partial u}{\partial n}|_\Gamma$, $u_0 = u|_\Gamma$, that is more detailed

$$-\frac{1}{\pi}\int_\Gamma q(\zeta) \ln|z - \zeta| \, d\gamma(\zeta) = f(z) \quad (z \in \Gamma),$$

where

$$f(z) := u_0(z) - \frac{1}{\pi}\int_\Gamma u_0(\zeta)\frac{\partial}{\partial n(\zeta)} \ln|z - \zeta| \, d\gamma(\zeta).$$

Since $(K'1)(z) = -1 \, (z \in \Gamma)$, (Exercise, use Fourier expansion and Calderon projector) we have

$$\int_\Gamma f(z) \, d\gamma(z) = \int_\Gamma u_0(z)\Big[1 - \frac{1}{\pi}\int_\Gamma \frac{\partial}{\partial n(z)} \ln|\zeta - z| d\gamma(\zeta)\Big] d\gamma(z) = 0. \qquad (3.3)$$

Now we use complex coordinates

$$z = x + iy, \; \zeta = \xi + i\eta,$$

the standard parametrization of the unit sphere $S(0, 1)$

$$\begin{matrix} x = \cos\tau, \; \xi = \cos t \\ y = \sin\tau, \; \eta = \sin t \end{matrix} \quad 0 \le t, \tau \le 2\pi,$$

and Fourier expansion of the sought 2π-periodic function $q : [0, 2\pi] \to \mathbb{C}$,

$$q(t) = \sum_{k\in\mathbb{Z}} \hat{q}_k \, e^{ikt}, \; \hat{q}_k = \int_0^{2\pi} q(\tau) \, e^{-ik\tau} \, d\tau.$$

Thus we can rewrite the left hand side $\frac{1}{\pi}L$ of the boundary integral equation as follows:

$$
\begin{aligned}
L &:= -\int_0^{2\pi} q(t)\, \ln|z-\zeta(t)|\, dt \\
&= -\int_0^{2\pi} \sum_{k\in\mathbb{Z}} \hat{q}_k\, e^{ikt}\, \ln|z-\zeta(t)|\, dt \\
&= -\int_0^{2\pi} \sum_{k\in\mathbb{Z}} \hat{q}_k\, e^{ikt}\, \ln\left|2\sin\frac{t-\tau}{2}\right| dt\,;
\end{aligned}
$$

since

$$
\begin{aligned}
|z-\zeta|^2 &= (x-\xi)^2 + (y-\eta)^2 \\
&= (\cos\tau - \cos t)^2 + (\sin\tau - \sin t)^2 \\
&= 4\sin^2\frac{t+\tau}{2}\sin^2\frac{\tau-t}{2} + 4\cos^2\frac{\tau+t}{2}\sin^2\frac{\tau-t}{2} \\
&= 4\sin^2\frac{t-\tau}{2}
\end{aligned}
$$

we have indeed

$$
|z-\zeta| = 2\left|\sin\frac{t-\tau}{2}\right|. \tag{3.4}
$$

The substitution $t' = t - \tau$ leads to

$$
L = -\sum_{k\in\mathbb{Z}} e^{ik\tau}\, \hat{q}_k \int_0^{2\pi} e^{ikt'}\, \ln\left|2\sin\frac{t'}{2}\right| dt'\,.
$$

Since by Fourier expansion

$$
\sum_{\nu=1}^{\infty} \frac{\cos\nu x}{\nu} = -\ln\left(2\sin\frac{x}{2}\right) \quad \text{für} \quad 0 < x \leq \pi
$$

we have for $k \in \mathbb{Z}$

$$-\int_0^{2\pi} e^{ikt'} \ln(2\sin\frac{t'}{2})\,dt' = \int_0^{\pi} \left(e^{ikt} + e^{-ikt}\right) \sum_{\nu=1}^{\infty} \frac{\cos \nu t}{\nu}\,dt$$

$$= \frac{1}{2}\sum_{\nu=1}^{\infty} \frac{1}{\nu} \int_0^{2\pi} \left(\cos\frac{k-\nu}{2}t' + \cos\frac{k+\nu}{2}t'\right) dt'$$

$$= \begin{cases} \frac{\pi}{|k|} & \text{if } |k| \in \mathbb{N}, \\ 0 & \text{if } k = 0 \end{cases} \tag{3.5}$$

Consequently

$$L = \sum_{\substack{k\in\mathbb{Z}\\ k\neq 0}} e^{ik\tau}\,\hat{q}_k \frac{\pi}{|k|}\,.$$

With

$$f(\tau) = \sum_{k\in\mathbb{Z}} \hat{f}_k e^{ik\tau}$$

we obtain (by (3.5) we have $\hat{f}_0 = 0$)

$$\frac{1}{\pi}L = \sum_{\substack{k\in\mathbb{Z}\\ k\neq 0}} e^{ik\tau}\,\hat{q}_k \frac{1}{|k|} = \sum_{\substack{k\in\mathbb{Z}\\ k\neq 0}} e^{ik\tau}\,\hat{f}_k$$

what results in

$$\hat{q}_k = |k|\,\hat{f}_k \quad \text{for } k \neq 0\,.$$

Finally we note that by the first Green formula we have

$$0 = \int_{\Omega} \Delta u\,dx = \int_{\Gamma} q(z)\,d\gamma(z) = \hat{q}_0\,.$$

Thus the unique solution of the boundary integral equation is completely determined by its Fourier sum as above.

Exercise Let $\hat{u}_k$ be the Fourier coefficients of u. Then for $s \in \mathbb{R}$

$$\|u\|^2_{H^s[0,2\pi]} = \sum_{\substack{k\in\mathbb{Z}\\ k\neq 0}} |k|^{2s}\, |\hat{u}_k|^2 + |\hat{u}_0|^2 \tag{3.6}$$

is an equivalent norm on $H^s[0, 2\pi]$. For $l \in \mathbb{N}_0$,

$$\|u\|_{H^l(\Gamma)} \cong \Big\{ \sum_{|\alpha|\le l} \int_\Gamma |D^\alpha u|^2 \, d\gamma \Big\}^{1/2}$$

and for $s \in \mathbb{R},\ s = l + p,\ 0 < p < 1,\ l \in \mathbb{N}$ there holds

$$\|u\|_{H^s(\Gamma)} \cong \Big\{ \|u\|^2_{H^l(\Gamma)} + \sum_{|\alpha|\le l} \int_\Gamma\int_\Gamma \frac{|D^\alpha u(x) - D^\alpha u(y)|^2}{|x-y|^{1+2p}}\, d\gamma(x)\, d\gamma(y) \Big\}^{1/2},$$

which is known as the (Aronszajn -) Slobodeckij norm, see e.g. [171, 259].

3.4 The Boundary Integral Operators in a Scale of Sobolev Spaces

3.4.1 The Operators V and W

Theorem 3.4 *Let $\Gamma \in C^\infty$. Then for any $\sigma \in \mathbb{R}$, the boundary integral operators*

$$V : H^\sigma(\Gamma) \to H^{\sigma+1}(\Gamma),$$
$$W : H^\sigma(\Gamma) \to H^{\sigma-1}(\Gamma)$$

are continuous.

Proof Let Γ be parametrized by $x = Z(t)$, respectively by $y = Z(\tau)$ with $t, \tau \in [0, 2\pi]$. For any $\varphi \in H^\sigma(\Gamma)$, that is $\tau \mapsto \varphi(Z(\tau)) \in H^\sigma[0, 2\pi]$, we have to show $V[\varphi] \circ Z \in H^{\sigma+1}[0, 2\pi]$. We start from

$$\begin{aligned}(V[\varphi] \circ Z)(t) &= -\frac{1}{\pi} \int_\Gamma \varphi(y)\, \ln|Z(t) - y|\, d\gamma(y) \\ &= -\frac{1}{\pi} \int_0^{2\pi} \varphi(Z(\tau))\, |\dot{Z}(\tau)| \ln|Z(t) - Z(\tau)|\, d\tau ,\end{aligned}$$

and decompose

$$(V[\varphi] \circ Z)(t) = -\frac{1}{\pi} \int_0^{2\pi} f(\tau) \ln|2 \sin \frac{\tau - t}{2}| \, d\tau - \frac{1}{\pi} \int_0^{2\pi} f(\tau) \ln \left| \frac{Z(t) - Z(\tau)}{2 \sin \frac{\tau - t}{2}} \right| d\tau$$

$$=: V_0[f](t) + V_1[f](t) .$$

Here

$$f(\tau) := |\dot{Z}(\tau)| \, \varphi(Z(\tau))$$

is a product of a C^∞-function and a H^σ-function, hence lies in $H^\sigma[0, 2\pi]$ what can be seen by using an appropriate equivalent norm of H^σ.

In virtue of the above calculation (see in particular the above calculation of L) we have

$$V_0[f](t) = V_0\Big[\sum_{k \in \mathbb{Z}} \hat{f}_k e^{ik\cdot}\Big](t)$$

$$= \sum_{k \neq 0} \frac{1}{|k|} \hat{f}_k \, e^{ikt} ,$$

with

$$\left[\widehat{V_0[f]}\right]_0 = 0 .$$

To analyse V_0 we use appropriate equivalent norms (see the exercise above!) and obtain

$$\|V_0[f]\|_{\sigma+1}^2 \cong \sum_{\substack{k \in \mathbb{Z} \\ k \neq 0}} |k|^{2(\sigma+1)} \left| \frac{\hat{f}_k}{|k|} \right|^2$$

$$\leq \sum_{k \neq 0} |\hat{f}_k|^2 |k|^{2\sigma} + |\hat{f}_0|^2 \cong \|f\|_\sigma^2 .$$

To treat V_1 we use

$$Z(t) - Z(\tau) = \int_\tau^t \dot{Z}(\eta) \, d\eta = (t - \tau) \int_0^1 \dot{Z}(\tau + \xi(t - \tau)) \, d\xi$$

and the identity (see the above calculation of L)

$$|e^{it}-e^{i\tau}|^2=(\cos t-\cos\tau)^2+(\sin t-\sin\tau)^2=4\sin^2\frac{t-\tau}{2}.$$

Thus we can see that the function

$$\zeta(t,\tau):=\frac{Z(t)-Z(\tau)}{2\sin\frac{\tau-t}{2}}=\frac{t-\tau}{2\sin\frac{\tau-t}{2}}\int\limits_0^1\dot Z(\tau+\xi(t-\tau))\,d\xi,$$

which is 2π-periodic in t and τ, belongs to C^∞. Furthermore by the regularity of Γ, we have

$$|\zeta(t,\tau)|=\left|\frac{Z(t)-Z(\tau)}{e^{it}-e^{-i\tau}}\right|\neq 0,$$

hence also $\in C^\infty$, what results in $V_1[f]\in C^\infty$. This proves the claim concerning V.

By Proposition 2.1

$$W[\varphi](x)=-\frac{1}{\pi}\int\limits_\Gamma\frac{d\varphi(y)}{ds}\frac{\langle x-y,t(x)\rangle}{|x-y|^2}\,ds_y$$

we obtain with $y=Z(\tau)$, $ds_y=d\gamma(y)=|\dot Z(\tau)|\,d\tau$

$$\begin{aligned}(W[\varphi]\circ Z)(t)&=-\frac{1}{\pi}\int\limits_0^{2\pi}\frac{d(\varphi\circ Z)(\tau)}{d\tau}\frac{\langle Z(t)-Z(\tau),t(Z(t))\rangle}{|Z(t)-Z(\tau)|^2}\,d\tau\\&=\frac{1}{|\dot Z(t)|}\frac{d}{dt}\Big\{-\frac{1}{\pi}\int\limits_0^{2\pi}\frac{d(\varphi\circ Z)(\tau)}{d\tau}\ln|Z(t)-Z(\tau)|\,d\tau\Big\}\\&=\frac{1}{|\dot Z(t)|}\frac{d}{dt}\Big\{V_0\Big[\frac{d(\varphi\circ Z)}{d\tau}\Big](t)+V_1\Big[\frac{d(\varphi\circ Z)}{d\tau}\Big](t)\Big\}.\end{aligned}$$

Now we can apply the mapping properties of V_0, and V_1 above and conclude

$$\begin{aligned}\varphi\in H^\sigma(\Gamma)&\mapsto\varphi\circ Z\in H^\sigma[0,2\pi]\mapsto\frac{d}{d\tau}(\varphi\circ Z)\in H^{\sigma-1}[0,2\pi]\\&\mapsto V_0\Big[\frac{d(\varphi\circ Z)}{d\tau}\Big]\in H^\sigma[0,2\pi]\mapsto\frac{d}{dt}V_0\Big[\frac{d(\varphi\circ Z)}{d\tau}\Big]\in H^{\sigma-1}\\&\mapsto\frac{1}{|\dot Z(t)|}\frac{d}{dt}V_0\Big[\frac{d(\varphi\circ Z)}{d\tau}\Big]\in H^{\sigma-1}[0,2\pi]\end{aligned}$$

respectively

$$\frac{1}{|\dot{Z}(t)|}\frac{d}{dt}V_1\Big[\frac{d(\varphi\circ Z)}{d\tau}\Big]\in C^\infty .$$

This shows the claim concerning D. □

The mapping properties

$$\begin{aligned} V &: H^{-1/2}(\Gamma) \to H^{1/2}(\Gamma), \\ W &: H^{1/2}(\Gamma) \quad \to H^{-1/2}(\Gamma) \end{aligned}$$

hold true also under weaker assumptions, e.g. $\Gamma \in C^2$ (see [276], Theorem 8.21), even on Lipschitz curves.

3.4.2 *The Operators K and K′*

We already know that K and K' are adjoint operators in $L^2(\Gamma)$. By a density argument this extends to arbitrary dual pairs $(H^s(\Gamma), H^{-s}(\Gamma))_{L^2(\Gamma)}$ for $s \in \mathbb{R}$, in particular for $s = \frac{1}{2}$

$$(K\varphi, \psi)_{L^2(\Gamma)} = (\varphi, K'\psi)_{L^2(\Gamma)} \quad (\forall \varphi \in H^{1/2}(\Gamma),\ \psi \in H^{-1/2}(\Gamma)) .$$

Let us note that the parameter $s = \frac{1}{2}$ plays a particular role with the boundary integral approach to elliptic boundary value problems of second order.

3.4.2.1 A Geometric Interpretation of the Kernel of the Double Layer Potential

Proposition 3.3 *For any $x, y \in \Gamma$ there holds*

$$\frac{\partial}{\partial n_y}(\ln|y-x|)\, ds_y = d\vartheta_x(y)\,,$$

where the polar coordinates $y - x = r(\cos\vartheta, \sin\vartheta)$ are used.

Proof Any parametrization $y = Z(\tau)$ leads to functions $r = r(\tau)$, $\vartheta = \vartheta(\tau)$ with respect to the above polar coordinates. With the components

$$\begin{aligned} y &= \binom{y_1}{y_2}, \\ \tfrac{dy}{d\tau} &= \binom{\dot{y}_1}{\dot{y}_2} \end{aligned}$$

we have

$$\vec{t} = \begin{pmatrix} \dot{y}_1 \\ \dot{y}_2 \end{pmatrix} / \sqrt{\dot{y}_1^2 + \dot{y}_2^2} = \begin{pmatrix} -n_2(y) \\ n_1(y) \end{pmatrix},$$

$$\vec{n} = \begin{pmatrix} \dot{y}_2 \\ -\dot{y}_1 \end{pmatrix} / \sqrt{\dot{y}_1^2 + \dot{y}_2^2}$$

and

$$ds = \sqrt{\dot{y}_1^2 + \dot{y}_2^2}\, d\tau \,.$$

Hence

$$\begin{aligned}
\frac{\partial}{\partial n_y} (\ln |y - x|)\, ds_y &= \frac{1}{r^2}\Big(n_1(y)(y_1 - x_1) + n_2(y)(y_2 - x_2)\Big)\, ds_y \\
&= \frac{1}{r^2}\Big((y_1 - x_1)\, dy_2 - (y_2 - x_2)\, dy_1\Big) \\
&= \frac{1}{r^2}\Big(r \cos\vartheta\, d(r \sin\vartheta) - r \sin\vartheta\, d(r \cos\vartheta)\Big) \\
&= \frac{1}{r^2}\Big(r \cos\vartheta\, \sin\vartheta\, dr + r^2 \cos^2\vartheta\, d\vartheta \\
&\quad - r \sin\vartheta \cos\vartheta\, dr + r^2 \sin^2\vartheta\, d\vartheta\Big) \\
&= d\vartheta_x(y)
\end{aligned}$$

□

One can show for $\Gamma \in C^2$ that

$$\frac{d\vartheta_x}{ds} =: k(s, \sigma)\,,$$

(where $x = z(\sigma),\ y = z(s)$ are related to the arc lenght parameter s) is continuous on $\mathbb{R}^2$ and $\lim\limits_{\sigma \to s} k(s, \sigma) = \frac{1}{2}\cdot$ (curvature of Γ in s). Consequently, for $\Gamma \in C^\infty$, the integral kernel k belongs to C^∞ and the integral operators $K, K' : H^s(\Gamma) \to H^t(\Gamma)$ are continuous for any $s, t \in \mathbb{R}$) and because of compact embedding, $K, K' : H^s(\Gamma) \to H^s(\Gamma)$ ($\forall s \in \mathbb{R}$) moreover compact.

Here we like to mention the fundamental article by Radon [347] where he introduces the boundaries of bounded rotation. In $\mathbb{R}^n$ with $n \geq 2$ this idea is further developed by J. Kral in [272] and Maz'ya in [301].

3.5 Solution of Exterior Dirichlet Problem by BIE

In this section we treat the following exterior Dirichlet problem for the Laplacian by boundary integral methods (BIE) and provide its explicit solution on the unit circle via Fourier series (see [409]).

Example 3.2 For sufficiently smooth curve $\Gamma = \partial\Omega$ find $u \in C^2(\Omega^c) \cap C^0(\overline{\Omega}^c)$ with

$$\Delta u = 0 \text{ in } \Omega^c := \mathbb{R}^2 \setminus \overline{\Omega}, \quad u|_\Gamma = g \tag{3.7}$$

where we demand the *decaying* condition at infinity: For given $B \in \mathbb{R}$ there exist some constants c, k such that for all $z \in \mathbb{R}^2$ with $|z| \geq k$:

$$|u(z) - B \ln |z|\,| \leq c < \infty. \tag{3.8}$$

This means that for some $a \in \mathbb{R}$

$$u(z) = a + B \ln |z| + o(1) \quad \text{as } |z| \to \infty .$$

Note that this decaying condition is weaker than that used in the representation theorem. The exterior boundary value problem (3.7) arises in many applications: potential flow, solid mechanis, conformal mappings.

Due to the considerations above, we can try to find the solution u in the form

$$u(z) = -\frac{1}{\pi}\int_\Gamma \psi(\zeta) \ln |z - \zeta| ds_\zeta - \omega. \tag{3.9}$$

with an unknown constant ω and the unknown density ψ.

Therefore with

$$u(z) = -\frac{1}{\pi} \ln |z| \int_\Gamma \psi(\zeta) ds - \frac{1}{\pi}\int_\Gamma \psi(\zeta) \ln |1 - \frac{\zeta}{z}| ds_\zeta - \omega$$

and $|z| \to \infty$ we have due to (3.8)

$$-\frac{1}{\pi}\int_\Gamma \psi(\zeta) ds_\zeta = B. \tag{3.10}$$

The BVP (3.7) with the decaying condition is thus reduced to the system of boundary integral equations

$$\begin{aligned} g &= -\tfrac{1}{\pi}\textstyle\int_\Gamma \psi(\zeta) \ln |z - \zeta| ds_\zeta - \omega \text{ on } \Gamma \\ \textstyle\int_\Gamma \psi(\zeta) ds_\zeta &= -\pi B \end{aligned} \tag{3.11}$$

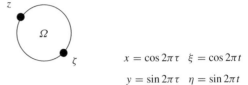

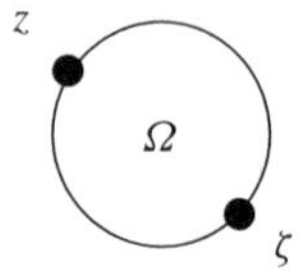

Fig. 3.1 Setting of unit circle

$x = \cos 2\pi\tau \quad \xi = \cos 2\pi t$

$y = \sin 2\pi\tau \quad \eta = \sin 2\pi t$

for the unknown function ψ on Γ and the *unknown constant* $\omega \in \mathbb{C}$ with a given function g on Γ and a *given constant* B. If ψ and ω in (3.11) are found then (3.9) yields for all $z \in \Omega^c$ the desired potential u.

For Γ, the unit circle, (3.11) can be solved explicitly via Fourier series. This will be the basis of the later on given analysis of (3.16) for general boundaries Γ and the finite element approximation. With the coordinates for $z = x + iy$, $\zeta = \xi + i\eta$ on Γ (Fig. 3.1) we obtain as before by the addition theorem of the sin function, $|z - \zeta| = |2\sin\pi(t - \tau)|$.

Next we expand $\psi^* = 2\pi\psi$ into a Fourier series:

$$\psi^*(t) = \sum_{k=-\infty}^{\infty} e^{ik2\pi t}\hat{\psi}_k, \quad \hat{\psi}_k = \int_0^1 \psi^*(t)e^{-ik2\pi t}dt. \tag{3.12}$$

Substitution into (3.11) yields with $ds = 2\pi dt, t' = t - \tau$,

$$\begin{aligned} -\int_\Gamma \psi(\zeta)\ln|z - \zeta|ds_\zeta - \pi\omega &= -\int_0^1 \sum_{k=-\infty}^{\infty} \hat{\psi}_k e^{2\pi ikt}\ln|2\sin\pi(t-\tau)|dt - \pi\omega \\ &= -\sum_{k=-\infty}^{\infty} \hat{\psi}_k \int_0^1 e^{2\pi ikt'}\ln|2\sin\pi t'|e^{2\pi ik\tau}dt' - \pi\omega. \end{aligned}$$

For the integral we compute explicitly as above using periodicity

$$\int_0^1 \ln|2\sin\pi t'|e^{ik2\pi t'}dt' = \begin{cases} -\frac{1}{2|k|}, & k \neq 0 \\ 0, & k = 0 \end{cases} \tag{3.13}$$

Hence (3.11) becomes

$$\sum_{k\neq 0} \frac{1}{2|k|}\hat{\psi}_k e^{ik2\pi\tau} - \pi\omega = \pi g, \quad \hat{\psi}_0 = -\pi B. \tag{3.14}$$

Inserting the Fourier expansion

$$g = \sum_{k=-\infty}^{\infty} e^{2\pi ik\tau}\hat{g}_k, \quad \hat{g}_k = \int_0^1 g(t)e^{-2\pi ikt}dt$$

of the given function g into (3.14) and equating coefficients in the expansion yields

$$-\omega = \hat{g}_0, \quad \hat{\psi}_k = 2|k|\pi \hat{g}_k \quad \text{for } k \neq 0, \ \hat{\psi}_0 = -\pi B. \tag{3.15}$$

Thus for the unit circle the solution ψ of (3.16) is completely determined by its Fourier series.

On a smooth curve Γ the integral equations look like

$$(V + C)\psi(z) = g(z) + \omega, \quad z \in \Gamma; \ \frac{-1}{\pi}\int_\Gamma \psi ds = B \tag{3.16}$$

$$V\psi(z) := -\frac{1}{\pi}\int_\Gamma \ln|z - \zeta|\psi(\zeta)ds_\zeta,$$

$$C\psi(z) := \int_\Gamma L(z, \zeta)\psi(\zeta)ds_\zeta, \quad z \in \Gamma.$$

with some smoother kernel $L(z, \zeta)$.

In order to analyze the solvability of (3.16) we introduce the Sobolev spaces $H^m(\Gamma)$ even for non-integers m. Therefore, we define a norm via the Fourier coefficients of the corresponding functions:

$$\|f\|_{H^m(\Gamma)} := \left\{ \sum_{j=-\infty}^{\infty} |j|^{2m}|\hat{f}_j|^2 + |\hat{f}_0|^2 \right\}^{1/2}. \tag{3.17}$$

For the integral equations in (3.16) we have the following well posedness result in Sobolev spaces:

Theorem 3.5 ([409]) *Let $s \in \mathbb{R}$ be fixed and let the solution of*(3.16) *be unique. Then for any $s \in \mathbb{R}$*

$$\left.\begin{array}{r} (V + C)\psi - \omega = g \\ -\frac{1}{\pi}\int_\Gamma \psi g ds = B \end{array}\right\} \tag{3.18}$$

is a bijective, continuous mapping from $H^s(\Gamma) \times \mathbb{R}$ onto $H^{s+1}(\Gamma) \times \mathbb{R}$.

Proof The Fourier approach allows to give the following short argument for the continuity of V for $s = -1/2$, $L = 0$ and Γ being the unit circle. With (3.17) and (3.14) and a generic constant $c > 0$ we have

$$\|V\psi\|^2_{H^{1/2}(\Gamma)} = \sum_{\substack{j=-\infty \\ j\neq 0}}^{\infty} |j|\,|\hat{g}_j|^2 \leq c \sum_{\substack{j=-\infty \\ j\neq 0}}^{\infty} |j|^{-1}|\hat{\psi}_j|^2 \leq c\|\psi\|^2_{H^{-1/2}(\Gamma)}.$$

The general case can be derived from the Fredholm alternative see [257]. □

3.6 A First Gårding Inequality

Due to the above mapping properties the bilinear form

$$a(\psi, \phi) := ((V + C)\psi, \phi)_{L^2(\Gamma)} \tag{3.19}$$

is continuous on $H^{-1/2}(\Gamma) \times H^{-1/2}(\Gamma)$. It is also coercive in the sense of a Gårding's inequality as seen from the following.

Theorem 3.6 ([409]) *For sufficiently smooth Γ there exists constants $\gamma > 0, c > 0$ such that for all $\psi \in H^{-1/2}(\Gamma)$*

$$a(\psi, \psi) \geq \gamma \|\psi\|^2_{H^{-1/2}(\Gamma)} - c\|\psi\|^2_{H^{-1}(\Gamma)}. \tag{3.20}$$

Proof First we show (3.20) for $L = 0$ and $\Gamma = \{z \in \mathbb{C} : |z| = 1\}$. Hence

$$a(\psi, \psi) = (V\psi, \psi)_{L^2(\Gamma)} = \int_0^1 (V\psi)(\tau)\overline{\psi}(\tau)d\tau.$$

As in (3.12) – (3.15) there holds

$$\begin{aligned}(V\psi, \psi)_{L^2(\Gamma)} &= \int_0^1 \sum_{k\in\mathbb{Z},\ k\neq 0} \frac{1}{2|k|\pi}\hat{\psi}_k e^{ik2\pi\tau} \cdot \sum_{\ell=-\infty}^{\infty} \overline{e^{-i\ell 2\pi\tau}\hat{\psi}_\ell} d\tau \\ &= \sum_{k\neq 0}\sum_{\ell=-\infty}^{\infty} \frac{1}{2\pi|k|}\hat{\psi}_k\overline{\hat{\psi}_\ell}\int_0^1 e^{i2\pi\tau(k-\ell)}d\tau\,.\end{aligned}$$

Thus

$$(V\psi, \psi)_{L^2(\Gamma)} = \frac{1}{2\pi}\|\psi\|^2_{H^{-1/2}(\Gamma)} - \frac{1}{2\pi}|\int_0^1 \psi dt|^2\,.$$

But since

$$|\int_0^1 \psi(t)dt|^2 \leq |\hat{\psi}_0|^2 + \sum_{k\neq 0}\frac{1}{k^2}|\hat{\psi}_k|^2 = \|\psi\|^2_{H^{-1}(\Gamma)}$$

we obtain from (3.16) the inequality

$$(V\psi, \psi)_{L^2(\Gamma)} \geq \frac{1}{2\pi}\|\psi\|^2_{H^{-1/2}(\Gamma)} - \frac{1}{2\pi}\|\psi\|^2_{H^{-1}(\Gamma)}. \tag{3.21}$$

In order to prove (3.20) in the general case we use (3.18) and proceed as follows with a smooth kernel $L(z,\zeta)$:

$$((V+C)\psi,\psi)_{L^2(\Gamma)} = -\frac{1}{\pi}\int_0^1\left(\int_0^1 \psi(t)\log|2\sin\pi(t-\tau)|dt\right)\psi(\tau)d\tau$$
$$+\int_0^1\left(-\frac{1}{\pi}\int_0^1 \psi(t)\log\left|\frac{z(t)-\zeta(t)}{2\sin\pi(t-\tau)}\right|dt\right.$$
$$\left.+\int_0^1 L(z,\zeta)\psi(t)dt\right)\psi(\tau)dt.$$

For the first integral, we have already derived (3.21). In order to obtain an estimate for the last two integrals we write them as

$$\int_0^1\int_0^1 \psi(t)\eta(t,\tau)\psi(\tau)dtd\tau$$

and estimate:

$$\left|\left(\psi,\int_0^1 \eta(t,\tau)\psi(t)dt\right)_{L^2(\Gamma)}\right| \le c\|\psi\|_{H^{-1}(\Gamma)}\left\|\int_0^1 \eta(t,\tau)\psi(t)dt\right\|_{H^1(\Gamma)}$$

$$\left\|\int_0^1 \eta(t,\tau)\psi(t)dt\right\|^2_{H^1(\Gamma)} := \int_0^1\left\{\int_0^1 \eta(t,\tau)\psi(t)dt\right\}^2 d\tau$$
$$+\int_0^1\left\{\int_0^1 \frac{\partial\eta}{\partial\tau}(t,\tau)\psi(t)dt\right\}^2 d\tau$$
$$\le c\left(\int_0^1\left\{\|\psi\|^2_{H^{-1}(\Gamma)}\left[\int_0^1|\eta|^2dt+\int_0^1\left|\frac{\partial\eta}{\partial t}\right|^2dt\right]\right\}d\tau\right.$$
$$\left.+\int_0^1\left\{\|\psi\|^2_{H^{-1}(\Gamma)}\left[\int_0^1\left|\frac{\partial\eta}{\partial\tau}\right|^2dt+\int_0^1\left|\frac{\partial^2\eta}{\partial\tau\partial t}\right|^2dt\right]\right\}d\tau\right)$$
$$= c\|\psi\|^2_{H^{-1}(\Gamma)}\left\{\int_0^1\int_0^1\left[\eta^2+\left|\frac{\partial\eta}{\partial t}\right|^2+\left|\frac{\partial\eta}{\partial\tau}\right|^2+\left|\frac{\partial^2\eta}{\partial\tau\partial t}\right|^2\right]dtd\tau\right\}.$$

Hence, since for sufficiently smooth Γ the terms in the brackets are bounded, we have

$$\left\|\int_0^1 \eta(t,\tau)\psi(t)dt\right\|_{H^1(\Gamma)} \le c\|\psi\|_{H^{-1}(\Gamma)}$$

and therefore

$$-\left|\int_0^1\int_0^1 \psi(t)\eta(t,\tau)\psi(t)dtd\tau\right| \geq -c^2\|\psi\|^2_{H^{-1}(\Gamma)}.$$

This yields together with (3.21) the desired estimate (3.20). □

For further reading see [276, 324, 343, 356, 409].

Chapter 4
Mixed BVPs, Transmission Problems and Pseudodifferential Operators

This chapter uses Fourier transform and the modern theory of pseudodifferential operators, see Appendix B. It brings a deeper insight in mixed boundary value problems in the interior and exterior of a connected surface in 3D. In particular, the Helmholtz interface problem from acoustics is studied in the presence of an obstacle in 3D. In Sect. 4.1 we consider a direct boundary integral equation method for the mixed boundary value problems (bvp). Then in Sect. 4.2 we look at the transmission problem and first treat it by the indrect method based on a single layer potential ansatz yielding a Riesz-Schauder system of second kind integral equations. Then we treat the transmission problem by the direct method giving a strongly elliptic system of boundary integral operators on the transmission manifold. In Sect. 4.3 we consider screen problems. In Sect. 4.4 the smoothness assumption of an analytic interface is relaxed to only Lipschitz continuity. in Sect. 4.5 we present a strongly elliptic system of pseudodifferential operators for the exterior Maxwell's equations.

4.1 Mixed Boundary Value Problems

In this section we report on the paper [397]. Let Ω_1 denote a bounded simply connected domain in $\mathbb{R}^3$ and $\Omega_2 = \mathbb{R}^3 \backslash \overline{\Omega}_1$, where $\Gamma = \partial\Omega_1 = \partial\Omega_2$ is assumed to be a sufficiently smooth, connected surface, for brevity C^∞. Γ is divided into two disjoint pieces Γ_1 and Γ_2 such that $\overline{\Gamma}_1 \cap \overline{\Gamma}_2 = \partial\Gamma_1 = \partial\Gamma_2 = \gamma$ defines a simple closed, smooth curve on Γ (see Fig. 4.1).

The interior $j = 1$ (exterior: $j = 2$) *mixed boundary value problem* reads as: *To given* g_1 *on* Γ_1 *and* g_2 *on* Γ_2 *find complex-valued* u_j *in* Ω_j *such that*

$$(\Delta + k_j^2)u_j = 0 \text{ in } \Omega_j, \qquad u_j = g_1 \text{ on } \Gamma_1 \text{ and } \frac{\partial u_j}{\partial n} = g_2 \text{ on } \Gamma_2 \tag{4.1}$$

© Springer International Publishing AG, part of Springer Nature 2018

J. Gwinner, E. P. Stephan, *Advanced Boundary Element Methods*,
Springer Series in Computational Mathematics 52,
https://doi.org/10.1007/978-3-319-92001-6_4

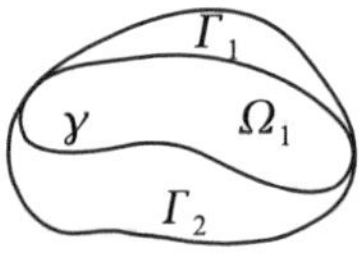

Fig. 4.1 Geometrical setting [397]

holds, where the solution u_2 *of the exterior problem has to satisfy at infinity:*

$$\text{If } k_2 \neq 0: \frac{\partial u_2(x)}{\partial |x|} - ik_2 u_2(x) = o\left(\frac{1}{|x|}\right) \text{ as } |x| \to \infty. \tag{4.2}$$

(SOMMERFELD'S radiation condition)

$$\text{If } k_2 = 0: u_2(x) = 0\left(\frac{1}{|x|}\right) \text{ as } |x| \to \infty. \tag{4.3}$$

Here $k_1, k_2 \in \mathbb{C}$ with $\Im k_j \geq 0$. $\frac{\partial u}{\partial n}$ means the normal derivative with respect to the outward unit normal n to Ω_1. We make the general assumption

$$k_1^2, k_2^2 \text{ are different from the eigenvalues of the interior and the exterior mixed boundary value problem, respectively.} \tag{4.4}$$

Before we give the variational formulation of the mixed boundary value problem (4.1)–(4.3) let us introduce the Sobolev spaces $H^s(\Omega_j)$, $H^s(\Gamma)$, $H^s(\Gamma_j)$ for smooth Γ, Γ_j as defined in the usual way [284],

$$\begin{aligned}
H^s(\Omega_j) &= \left\{u|_{\Omega_j} : u \in H^s(\mathbb{R}^3)\right\} \qquad (s \in \mathbb{R}) \\
H^s(\Gamma) &= \begin{cases} \{u|_\Gamma : u \in H^{s+1/2}(\mathbb{R}^3)\} & (s > 0) \\ L_2(\Gamma) & (s = 0) \\ \left(H^{-s}(\Gamma)\right)' \text{ (dual space)} & (s > 0) \end{cases} \\
H^s(\Gamma_j) &= \left\{u|_{\Gamma_j} : u \in H^s(\Gamma)\right\} \qquad (s \geq 0) \\
\tilde{H}^s(\Gamma_j) &= \left\{u \in H^s(\Gamma) : \operatorname{supp} u \subset \overline{\Gamma}_j\right\}, \quad H^s(\Gamma_1) = H^s(\Gamma)/\tilde{H}^s(\Gamma_2) \\
H^s(\Gamma_j) &= \left(\tilde{H}^{-s}(\Gamma_j)\right)' (s < 0), \; \tilde{H}^s(\Gamma_j) = \left(H^{-s}(\Gamma_j)\right)' (s < 0).
\end{aligned} \tag{4.5}$$

The spaces are endowed with their natural norms [284] which we shall recall later.

The most general case where (4.1)–(4.3) can be converted into a variational problem is the following:

$$g_1 \in H^{1/2}(\Gamma_1),\, g_2 \in H^{-1/2}(\Gamma_2) \text{ are given, and we look for } u \in H^1_{loc}(\Omega_j).$$

In this case $\frac{\partial u}{\partial n} \in \tilde{H}^{-1/2}(\Gamma_2) \subset H^{-1/2}(\Gamma)$ is defined by GREEN'S formula:

Lemma 4.1 *Let $u \in H^1_{loc}(\Omega_j)$ with $\Delta u \in L^2_{loc}(\Omega_j)$ and $v \in H^1(\Omega_j)$ with bounded support. Then $\frac{\partial u}{\partial n}\big|_\Gamma \in H^{-1/2}(\Gamma)$ is defined by*

$$\int_{\Omega_j} v \cdot \Delta u \, dx + \int_{\Omega_j} \nabla v \cdot \nabla u \, dx = (-1)^{j+1} \left\langle \frac{\partial u}{\partial n}\bigg|_\Gamma, v|_\Gamma \right\rangle_\Gamma \quad (j = 1, 2) \tag{4.6}$$

Here $\langle \cdot, \cdot \rangle_\Gamma$ is the duality between $H^{-1/2}(\Gamma) = (H^{1/2}(\Gamma))'$ and $H^{1/2}(\Gamma)$, given by $\langle f, g \rangle_\Gamma = \int_\Gamma f(z) g(z) ds_z$ for smooth functions f and g.

The mapping $u \mapsto \frac{\partial u}{\partial n}\big|_\Gamma$ is an extension by continuity of the corresponding trace mapping for smooth functions.

Now, let $u \in \mathscr{L}_j = \left\{ u_j \in H^1(\Omega_j) : (\Delta + k_j^2) = u_j \text{ in } \Omega_j \right\}$ be the variational solution of (4.1)–(4.3) with $u|_{\Gamma_1} = g_1$ and $\frac{\partial u}{\partial n}\big|_{\Gamma_2} = g_2$. Then with arbitrary extensions $lg_1 \in H^{1/2}(\Gamma)$ and $lg_2 \in H^{-1/2}(\Gamma)$ the Cauchy data $\binom{v}{\psi} = \binom{u|_\Gamma}{\frac{\partial u}{\partial n}|_\Gamma}$ admit the form

$$v = v^0 + lg_1, \quad \psi = \psi^0 + lg_2 \text{ with } v^0 \in \tilde{H}^{1/2}(\Gamma_2) \text{ and } \psi^0 \in \tilde{H}^{-1/2}(\Gamma_1),$$

because $v^0|_{\Gamma_1} = 0$ and $\psi^0|_{\Gamma_2} = 0$.

Definition 4.1 $\binom{v^0}{\psi^0} \in \tilde{H}^{1/2}(\Gamma_2) \times \tilde{H}^{-1/2}(\Gamma_1)$ are called the unknown CAUCHY data of the variational solution of (4.1)–(4.3).

They are the unknown layers of our system of boundary integral equations in the following.

In order to formulate the jump relations for the single and double layer potentials we define the following boundary integral operators.

Definition 4.2 Let $\psi \in C^\infty(\Gamma)$, Γ a bounded closed C^∞-surface. Then for $z \in \Omega_j$, $j = 1, 2$ we define with

$$\varphi_j(z, \zeta) = \begin{cases} \frac{-1}{4\pi|z-\zeta|} & \text{for } k_j = 0 \\ \frac{e^{ik_j|z-\zeta|}}{-4\pi|z-\zeta|} & \text{for } k_j \neq 0 \end{cases}$$

$$V_{\Omega_j}\psi(z) := -2 \int_\Gamma \psi(\zeta) \varphi_j(z, \zeta) ds_\zeta$$

$$K_{\Omega_j}\psi(z) := -2 \int_\Gamma \psi(\zeta) \frac{\partial}{\partial n_\zeta} \varphi_j(z, \zeta) ds_\zeta$$

and for $z \in \Gamma$ the operators

$$V_j\psi(z) := -2\int_\Gamma \psi(\zeta)\varphi_j(z,\zeta)ds_\zeta \tag{4.7}$$

$$K_j\psi(z) := -2\int_\Gamma \psi(\zeta)\frac{\partial}{\partial n_\zeta}\varphi_j(z,\zeta)ds_\zeta \tag{4.8}$$

$$K_j'\psi(z) := -2\int_\Gamma \psi(\zeta)\frac{\partial}{\partial n_z}\varphi_j(z,\zeta)ds_\zeta \tag{4.9}$$

$$W_j\psi(z) := -\frac{\partial}{\partial n_z}K_{\Omega_j}\psi(z). \tag{4.10}$$

For a distribution ψ we define $V_j\psi$ and $K_j\psi$ by approximating ψ by smooth functions and $K_j'\psi$ by duality using the relation

$$\left\langle K_j'\psi, \omega\right\rangle_\Gamma = \langle \psi, K_j\omega\rangle_\Gamma, \qquad \text{for all } \omega \in C^\infty(\Gamma)$$

which for smooth ψ is obviously valid.

The extension to distributions makes sense since all arising operators are pseudodifferential operators (see Sect. 4.2).

Since Γ is assumed to be C^∞ it follows that the function X is C^∞ and that an asymptotic expansion holds

$$|X(U) - X(u)| \sim \sum_{\nu=1}^{\infty} M_\nu(U, u-U) \text{ for } |u-U| \to 0$$

where M_ν is positive homogeneous of degree ν in $u - U$. We have

$$\varphi_j(r) = \frac{-1}{4\pi r}\sum_{l=0}^{\infty}\frac{\delta_l}{l!}r^l, \quad \delta_l(k_j) \in \mathbb{C}, \quad r = |X(U) - X(u)| \tag{4.11}$$

and therefore

$$\varphi_j\left(\left|X(U) - X(u)\right|\right) \sim \frac{-1}{4\pi|u-U|} + \sum_{l=0}^{\infty} L_j^l(U, u-U) \text{ for } |u-U| \to 0 \tag{4.12}$$

with L_j^l positive homogeneous of degree l in $u - U$. Substituting (4.12) into (4.7) and applying Fourier transform shows that for smooth Γ, V_j is a pseudodifferential operator of order -1 with principal symbol

$$\sigma(V_j)(\xi) = |\xi|^{-1}. \tag{4.13}$$

Using expansion (4.11) together with surface polar coordinates one can show that the operator of the double layer potential K_j is a pseudodifferential operator of order -1, too (see [259, 287]).

Now we define the matrix of operators

$$A_j := \begin{pmatrix} -K_j & V_j \\ W_j & K_j' \end{pmatrix}. \tag{4.14}$$

Due to the mapping properties of the operators we have

$$A_j : \begin{matrix} H^{1/2}(\Gamma) \\ \times \\ H^{-1/2}(\Gamma) \end{matrix} \rightarrow \begin{matrix} H^{1/2}(\Gamma) \\ \times \\ H^{-1/2}(\Gamma) \end{matrix} \quad \text{is continuous,}$$

and by the Calderon projector there holds the following result [129]:

Theorem 4.1

a) The statements (i) and (ii) on $\binom{v}{\psi} \in H^{1/2}(\Gamma) \times H^{-1/2}(\Gamma)$ *are equivalent:*

$$\begin{aligned} &(i)\ \tbinom{v}{\psi} \text{ are CAUCHY data of some } u \in \mathcal{L}_j \\ &(ii)\ \big(I + (-1)^j A_j\big)\tbinom{v}{\psi} = 0 \end{aligned} \tag{4.15}$$

b) The operators $\frac{1}{2}\big(I - (-1)^j A_j\big)$ *are projection operators, the so-called "CALDERON-projectors". They project* $H^{1/2}(\Gamma) \bigoplus H^{-1/2}(\Gamma)$ *onto the CAUCHY data of the weak solutions in* $\mathcal{L}_j$*. This means in particular* $A_j^2 = I$*, yielding the relations*

$$\begin{aligned} K_j^2 + V_j W_j = I = W_j V_j + K_j'^2 \\ -K_j V_j + V_j K_j' = 0 = -W_j K_j + K_j' W_j. \end{aligned} \tag{4.16}$$

Whereas the operators V_j, K_j and K_j' are weakly singular integral operators on Γ, the operator W_j of the normal derivative of the double layer potential is a hypersingular integral operator, its kernel is $0(|z - \zeta|^{-3})$ as $z \rightarrow \zeta$. W_j is a pseudodifferential operator of order $+1$ [129, 395]. The relation (4.16) shows that V_j is a regularizer to W_j since K_j^2, $K_j'^2$ are lower order pseudodifferential operators and therefore compact perturbations. Thus (4.16) together with (4.13) gives the principal symbol of W_j as

$$\sigma(W_j)(\xi) = |\xi| \tag{4.17}$$

with $|\xi| = \sqrt{\xi_1^2 + \xi_2^2}$, $\xi \in \mathbb{R}^2\backslash\{0\}$. Obviously $\sigma(W_j)(\xi)$ can also be computed by using local coordinate systems on the smooth manifold Γ and transformation to the case $\Gamma = \mathbb{R}^2$.

Now we give a solution procedure for the interior mixed boundary value problem via the direct method by inserting the boundary data into the system

$$u = \frac{1}{2}\left\{(I - K)u + V\frac{\partial u}{\partial n}\right\},$$

$$\frac{\partial u}{\partial n} = -\frac{1}{2}\frac{\partial}{\partial n}Ku + \frac{1}{2}\left(I + K'\right)\frac{\partial u}{\partial n}.$$

This gives

$$\begin{pmatrix} W_{22} & K'_{12} \\ -K_{21} & V_{11} \end{pmatrix}\begin{pmatrix} v \\ \psi \end{pmatrix} = \begin{pmatrix} -W_{12} & I - K'_{22} \\ I + K_{11} & -V_{21} \end{pmatrix}\begin{pmatrix} g_1 \\ g_2 \end{pmatrix}, \tag{4.18}$$

where the subscripts at W_{jk} etc. mean: integration over Γ_j and evaluation on Γ_k.

In order to describe the solvability of the above system we first give results on the mapping properties of the involved operators W_{jk}, V_{jk}, K_{jk}, K'_{jk}, $(j, k = 1, 2)$. We want to use the fact that W, V, K, K' are pseudodifferential operators (and hence bounded mappings in Sobolev spaces). Now the operators W_{jk} etc. act only on pieces of the manifold Γ, therefore their layers v, ψ have to be extended by zero on the remaining part of Γ.

Lemma 4.2 *For $s \in \mathbb{R}$ and $i, k = 1, 2$ the mappings are continuous:*

$$\begin{array}{ll} V_{ik} : \tilde{H}(\Gamma_i) \to H^{s+1}(\Gamma_k) & K_{ik} : \tilde{H}^s(\Gamma_i) \to H^{s+1}(\Gamma_k) \\ W_{ik} : \tilde{H}^{s+1}(\Gamma_i) \to H^s(\Gamma_k) & K'_{ik} : \tilde{H}^s(\Gamma_i) \to H^{s+1}(\Gamma_k). \end{array} \tag{4.19}$$

Proof By definition of $\tilde{H}^s(\Gamma_1)$ in (4.5) the extension of $\psi \in \tilde{H}^s(\Gamma_1)$ by zero

$$\psi^* := \begin{cases} \psi & \text{on } \Gamma_1 \\ 0 & \text{on } \Gamma_2 \end{cases} \quad \text{belongs to } H^s(\Gamma).$$

Therefore the continuity of the mappings (4.19) is seen by estimating the symbols of the pseudodifferential operators V, K, K' and W from above: Neglecting the local charts we have that the simple layer potential V is a continuous mapping from $H^s(\Gamma)$ into $H^{s+1}(\Gamma)$ (see definition of Sobolev spaces via Fourier transform in

Appendix B) since $|\sigma(V)(\xi)| \leq c(1+|\xi|)^{-1}$:

$$\begin{aligned}\|V\psi^*\|^2_{H^{s+1}(\Gamma)} &= \int (1+|\xi|^2)^{s+1}|\widetilde{V\psi}^*(\xi)|^2 d\xi \\ &\leq c \int (1+|\xi|^2)^{s+1}(1+|\xi|)^{-2}|\psi^*(\xi)|^2 d\xi \\ &\leq \tilde{c} \int (1+|\xi|^2)^s |\tilde{\psi}^*(\xi)|^2 d\xi = \tilde{c}\|\psi\|^2_{\tilde{H}^s(\Gamma_1)}.\end{aligned}$$

Matching up the local results and restriction to Γ_k yields (4.19) since

$$\|\psi\|_{\tilde{H}^s(\Gamma_1)} = \|\psi^*\|_{H^s(\Gamma)}. \tag{4.20}$$

The other assertions in (4.19) are shown analogously. This is standard in the theory of pseudodifferential operators [376] and [415]. □

In order to use Lemma 4.2 to obtain information on the solvability of the system (4.18) we rewrite it in a form appropriate to apply the result (4.19).

Substituting $v = v^0 + lg_1$, $\psi = \psi^0 + lg_2$ into the system we obtain

$$\begin{aligned} A_1 U^0 &:= \begin{pmatrix} W_{22} & K'_{12} \\ -K_{21} & V_{11} \end{pmatrix} \begin{pmatrix} v^0 \\ \psi^0 \end{pmatrix} = \begin{pmatrix} -W_{\Gamma 2} & (I-K'_\Gamma)_2 \\ (I+K_\Gamma)_1 & -V_{\Gamma 1} \end{pmatrix} \begin{pmatrix} lg_1 \\ lg_2 \end{pmatrix} \\ &:= B_1 lG \end{aligned} \tag{4.21}$$

Here $W_{\Gamma 2}$ etc. denotes integration on Γ and evaluation on Γ_2.

Theorem 4.2 *The mappings*

$$\begin{aligned} A_1 &: \tilde{H}^s(\Gamma_2) \times \tilde{H}^{s-1}(\Gamma_1) \to H^{s-1}(\Gamma_2) \times H^s(\Gamma_1); \\ B_1 &: H^s(\Gamma) \times H^{s-1}(\Gamma) \to H^{s-1}(\Gamma_2) \times H^s(\Gamma_1) \end{aligned} \tag{4.22}$$

are continuous for any real s.

Proof The mapping property of A_1 is a direct consequence of Lemma 4.2 whereas that of B_1 follows directly from the continuity of the extension lg_i in $H^s(\Gamma)$ for $g_i \in H^s(\Gamma_i)$ together with the mapping properties of the simple and the double layer potential and their respective normal derivatives. □

The system satisfies a Gårding inequality because it is a strongly elliptic system of pseudodifferential equations in appropriate Sobolev spaces.

Theorem 4.3 *There exists a constant $\gamma_1 > 0$ such that for all* $U = \begin{pmatrix} v^0 \\ \psi^0 \end{pmatrix}$

$$\langle (A_1 + C_1)U, U \rangle_0 \geq \gamma_1 \left\{ \|v^0\|^2_{\tilde{H}^{1/2}(\Gamma_2)} + \|\psi^0\|^2_{\tilde{H}^{(-1/2)}(\Gamma_1)} \right\}, \tag{4.23}$$

here

$$C_1 : \tilde{H}^{1/2}(\Gamma_2) \times \tilde{H}^{-1/2}(\Gamma_1) \to H^{-1/2}(\Gamma_2) \times H^{1/2}(\Gamma_1)$$

is compact and

$$\langle A_1 U, U \rangle_0 := \left\langle W_{22} v^0 + K'_{12} \psi^0, v^0 \right\rangle_{L^2(\Gamma_2)} + \left\langle -K_{21} v^0 + V_{11} \psi^0, \psi^0 \right\rangle_{L^2(\Gamma_1)}$$

Proof We use a partition of unity to reduce the global inequalities to local ones, i.e. to the inequality (4.23) for the individual terms $\chi_k v^0, \chi_k \psi^0 (k = 1, \ldots, N)$ (with $\chi_k \in C_0^\infty(S_k)$ and patches S_k covering Γ) instead of v^0, ψ^0 (see [397] for details). Since $K'_{12} : \tilde{H}^{-1/2}(\Gamma_1) \to H^{1/2}(\Gamma_2)$ and $K_{21} : \tilde{H}^{1/2}(\Gamma_2) \to H^{3/2}(\Gamma_1)$ are continuous mappings, they are compact mappings $\tilde{H}^{-1/2}(\Gamma_1) \to H^{-1/2}(\Gamma_2)$ and $\tilde{H}^{1/2}(\Gamma_2) \to H^{1/2}(\Gamma_1)$, respectively, by Rellich's embedding theorem. Therefore K'_{12} and K_{21} are compact perturbations and the principal symbol of A_1 has the form

$$\sigma(A_1)(\xi) = \begin{pmatrix} |\xi| & 0 \\ 0 & \frac{1}{|\xi|} \end{pmatrix}, \qquad \xi \in \mathbb{R}^2 \setminus \{(0,0)\}.$$

Now standard arguments yields the assertion (see [397, Theorem 3.3]). □

4.2 The Helmholtz Interface Problems

If a sound wave meets an obstacle, it is partially reflected from it and partially transmitted through it. Let us consider a steady-state sound wave that is set up in a homogeneous medium Ω characterized by a density ρ, a damping coefficient α and sound velocity c in which there is a homogeneous body Ω' of density ρ_i, damping coefficient β and sound velocity c_i. We shall characterize the sound wave by the pressure v and the angular frequency ω of the acoustic vibrations. Let the medium occupy all space $\mathbb{R}^3$ with the exception of the bounded domain Ω' occupied by the obstacle. We denote by v_o, v_i, v_e the complex-valued pressure of the incident, refracted and scattered wave, respectively, satisfying the homogeneous Helmholtz equations

$$\begin{aligned} \Delta v_i + k_i^2 v_i = 0,\ k_i^2 &= \frac{\omega(\omega + i\beta)}{c_i^2} \quad \text{in } \Omega', \\ \Delta v_e + k^2 v_e = 0,\ k^2 &= \frac{\omega(\omega + i\alpha)}{c^2} \quad \text{in } \Omega = \mathbb{R}^3 \setminus \overline{\Omega'}. \end{aligned} \tag{4.24}$$

Both the total acoustic field $v = v_e + v_o$ and the incident field v_o satisfy the homogeneous Helmholtz equation in the exterior domain Ω. At infinity the scattered wave v_e fulfills the Sommerfeld radiation condition

$$\lim_{r\to\infty} r\left(\frac{\partial v_e}{\partial r} - ikv_e\right) = 0, \quad \lim_{r\to\infty} v_e = 0. \tag{4.25}$$

Finally, on the boundary S of the obstacle, the pressure and the velocity of vibrations in the body and the medium must coincide, yielding the transmission conditions

$$v_i = v_e + v_o, \quad \frac{1}{\rho_i}\frac{\partial v_i}{\partial n} = \frac{1}{\rho}\left(\frac{\partial v_e}{\partial n} + \frac{\partial v_o}{\partial n}\right), \quad \text{on } S \tag{4.26}$$

where $\frac{\partial}{\partial n}$ denotes differentiation with respect to the outer normal n to S. Thus the scattering of sound is described by the interface problem (4.24)–(4.26).

For higher damping the constant β is usually large leading to the total reflection of a plane wave at an absolutely rigid immovable obstacle. Formally this means solving only the Helmholtz equation $(4.24)_2$ in Ω for the scattered field and requiring that the normal derivative of the total acoustic field vanishes on S, that is

$$\begin{aligned} \Delta v_e + k^2 v_e &= 0 && \text{in } \Omega = \mathbb{R}^3 \setminus \overline{\Omega'} \\ \frac{\partial v_e}{\partial n} &= -\frac{\partial v_o}{\partial n} && \text{on } S \end{aligned} \tag{4.27}$$

where v_e satisfies (4.25) at infinity.

In the following we assume for simplicity that S is a closed analytic surface which divides $\mathbb{R}^3$ into simply connected domains, an interior Ω' (bounded) and an exterior Ω (unbounded).

In order to avoid additional difficulties we assume:

$$k^2 \neq 0 \text{ is \underline{not} an eigenvalue of the interior Dirichlet problem.} \tag{4.28}$$

The uniqueness of the solution of the interface problem (4.24)–(4.26) and of the exterior Neumann problem (4.27) is wellknown. For brevity we give here only the uniqueness result for the interface problem (see [129, 395]).

Theorem 4.4 *Let $k, k_i \in \mathbb{C} \setminus \{0\}$ with $0 \leq \arg k, \arg k_i \leq \pi$ and let $\mu = \frac{1}{\rho}$, $\mu_i = \frac{1}{\rho_i} \in \mathbb{C} \setminus \{0\}$ be such that*

$$\kappa = \frac{\mu_i \overline{k_i}^2}{\mu \overline{k}^2} = \frac{\rho \overline{k_i}^2}{\rho_i \overline{k}^2} \in \mathbb{R}$$

where $\kappa \geq 0$ (< 0) if $\Re k \cdot \Re k_i \geq 0$ (< 0). Then the only solution of the homogeneous transmission problem (4.24)–(4.26) is $v_e = v_i = 0$.

In the following we first give a boundary integral equation method based on simple layers for solving both the interface problem (4.24)–(4.26) and the exterior Neumann problem (4.27). Then we give the corresponding double layer procedure. To this end we introduce the simple layer V_γ with the continuous density ψ on the surface S by

$$V_\gamma(\psi)(x) = \int_S \psi(y)\phi_\gamma(|x-y|)dS_y,\ x \in \mathbb{R}^3 \tag{4.29}$$

Here

$$\phi_\gamma(|x-y|) = \frac{e^{i\gamma|x-y|}}{4\pi|x-y|} \tag{4.30}$$

is the fundamental solution of the Helmholtz equation $\Delta w = -\gamma^2 w$ satisfying the Sommerfeld radiation condition for $\Re\gamma \neq 0$. There hold the following well-known properties of the simple layer potential [287].

Lemma 4.3 *For any complex γ, $0 \le \arg\gamma \le \frac{\pi}{2}$ and any continuous ψ on S:*

(i) $V_\gamma(\psi)$ *is continuous in* $\mathbb{R}^3$
(ii) $\Delta V_\gamma(\psi) = -\gamma^2 V_\gamma(\psi)$ *in* $\Omega \cup \Omega'$
(iii) $V_\gamma(\psi)(x) = O\left(|x|^{-1}e^{i\gamma|x|}\right)$ *as* $|x| \to \infty$
(iv) $\left(\frac{\partial}{\partial n}V_\gamma(\psi)\right)^{\pm}(x) = \mp\frac{1}{2}\psi(x) + \int_S K_\gamma(x,y)\psi(y)dS_y$ *on* S

where the kernel K_γ is $O\left(|x-y|^{-1}\right)$ as $y \to x$ and $\pm$ denotes the limit to S from Ω and Ω', respectively.

In order to describe the mapping properties of V_γ and K_γ as pseudodifferential operators acting in Sobolev spaces we first discuss some geometric ideas (see [287]). We introduce coordinate systems for S. These consist of a finite number of coordinate patches $S_1, \ldots, S_N$ covering S. For each patch there is a region $\Gamma_k \subset \mathbb{R}^2$ and a map X_k such that $x = X_k(u)$, $u = (u_1, u_2) \in \mathbb{R}^2$, covers S_k. The mappings are compatible on overlapping regions. To say that S is a regular analytic surface means that the individual maps from Γ_k to Γ_e on overlaps are analytic and that X_{k,u_1} and X_{k,u_2} are linearly independent.

We use the X_k to generate local coordinate systems in $\mathbb{R}^3$ and set

$$\tilde{e}_1(u) = X_{u_1},\ \tilde{e}_2(u) = X_{u_2},\ \tilde{e}_3(u) = \tilde{e}_1(u) \times \tilde{e}_2(u) \tag{4.31}$$

Then the equations

$$x = X(u) + u_3\tilde{e}_3(u),\ u \in \Gamma,\ |u_3| < \delta$$

will define a coordinate system for a region $U_k \subset \mathbb{R}^3$ with $u_3 = 0$ corresponding to S_k. We will assume that $u_3 > 0$ corresponds to Ω.

For simplification we further assume that the coordinate systems are orthonormal, that is, $\tilde{e}_i(u) \cdot \tilde{e}_j(u) = \delta_{ij}$.

Following the ideas of [376] we introduce a partition of unity $\sum_k \xi_k \equiv 1$ subordinate to the S_k and define $V_\gamma(\psi)$ by

$$V_\gamma(\psi)(x) = \sum_k \int_{\Gamma_k} \psi\,(X_k(u))\,\xi_k(u)\phi_\gamma\,(|x - X_k(u)|)\,du \tag{4.32}$$

Here the orthonormality of the coordinate system implies that the surface element is unity. For $x \in S$, (4.32) gives

$$V_\gamma(\psi)(x) = \sum_j \sum_k \xi_j \int_{\Gamma_k} \psi\,(X_k(u))\,\xi_k(X_k(u))\phi_\gamma\,(|x - X_k(u)|)\,du \tag{4.33}$$

Formula (4.33) is the basis for the idea of pseudodifferential operators on S. If $\psi \in C_0^\infty(S_k)$ for some patch S_k then $V_\gamma(\psi)$ will be in $C^\infty(S_k)$. The idea is to extend that definition to ψ's which need not to be C^∞ but lie in some Sobolev space on S. It is clear from (4.33) that one needs concentrate only on the quantities $\chi V_\gamma(\psi)$ where χ and ψ have support in the same patch S_k.

Let $\chi, \psi \in C_0^\infty(S_k)$. Then we have

$$\begin{aligned} \chi V_\gamma(\psi) &= \chi(X(U)) \int_{\Gamma_k} \psi(X(u))\phi_\gamma(|X(U) - X(u)|)du \\ &= \int_{\mathbb{R}^2} \widetilde{\psi}(u) K_\gamma(U, u - U) du \end{aligned} \tag{4.34}$$

with the kernel

$$K_\gamma(U, u - U) = \chi(X(U))\phi_\gamma(|X(U) - X(u)|).$$

Introducing the Fourier transform $\widehat{\psi}$ of $\widetilde{\psi}$ by

$$\widehat{\psi}(\xi) = \int_{\mathbb{R}^2} \widetilde{\psi}(u)e^{-i\xi\cdot u}du \tag{4.35}$$

we can write

$$\chi V_\gamma(\psi) = (2\pi)^{-2} \int_{\mathbb{R}^2} e^{i\xi\cdot x}\widehat{\psi}(\xi)a_\gamma(U, \xi)d\xi \tag{4.36}$$

with

$$a_\gamma(U,\xi) = \chi(X(U)) \int_{\mathbb{R}^2} e^{-i\xi\cdot\eta} K_\gamma(U,\eta)d\eta$$

Now, $a_\gamma(U,\xi)$ is called the symbol of V_γ.

Suppose that $K_\gamma(U,\eta)$ has an asymptotic expansion of the form

$$K_\gamma(U,\eta) \sim \sum_{n=r}^{\infty} K_\gamma^n(U,\eta) \tag{4.37}$$

where K_γ^n is homogenous of degree n in η. Then a_γ, the (distributional) Fourier transform of K_γ, has the form

$$a_\gamma(U,\xi) \sim \sum_{n=r}^{\infty} a_\gamma^n(U,\xi)$$

where a_γ^n is homogeneous of degree $-n-2$ in ξ. If (4.37) holds then V_γ obtained by (4.36) is called a pseudodifferential operator of order r and $a_\gamma^r(U,\xi)$ is its principal symbol. V_γ is called elliptic if $a_\gamma^r(U,\xi) \neq 0$ for $\xi \neq 0$.

Before we cite some results from [376] on pseudodifferential operators on S we recall the definition of Sobolev spaces on compact manifolds S. Via diffeomorphism χ mapping any domain $U \subset S$ onto open sets U_χ in $\mathbb{R}^2$ the Sobolev space $H^r(S)$ is the completion of $C^\infty(S)$, the space of infinitely differentiable functions on S, in the norm

$$\|\chi\psi\|_{H^r}^2 = \int_{\mathbb{R}^2} (1+|\xi|^2)^r |\widehat{\chi\psi}(\xi)|^2 d\xi, \ \psi \in C_0^\infty(S). \tag{4.38}$$

defined by a partition of unity subordinate to a covering of S by domains of charts [253].

Lemma 4.4 ([376]) *Let A be a pseudodifferential operator of order r on S. Then*

(i) A is a continuous map from $H^t(S)$ into $H^{t-r}(S)$ for any t
(ii) If A is elliptic the map $A : H^t(S) \to H^{t-r}(S)$ is Fredholm
(iii) If A is elliptic then $\psi \in H^t(S)$ and $A\psi \in H^s(S)$ implies $\psi \in H^{s+r}(S)$ and there is a constant $C_{t,s}$ such that

$$\|\psi\|_{s+r} \leq C_{t,s} (\|A\psi\|_s + \|\psi\|_t)$$

Now we apply the above ideas to V_γ and show first that the expansion (4.37) holds. Since S is assumed to be analytic it follows that the functions X are analytic and that

$$|X(U) - X(u)| = \sum_{\nu=1}^{\infty} M_\nu(U, u - U)$$

where M_ν is homogeneous of degree ν in $u - U$. Moreover the orthonormality of the coordinate system yields

$$M_1(U, u - U) = |u - U|$$

Then (4.30) gives

$$\phi_\gamma(r) = r^{-1} \sum_{j=0}^{\infty} \frac{\delta^j}{j!} r^j, \ \delta \in \mathbb{C}, \ r = |x - y| \tag{4.39}$$

Thus we obtain

$$\phi_\gamma\left(|X(U) - X(u)|\right) = |u - U|^{-1} + \sum_{\nu=0}^{\infty} k_\gamma^\nu(U, u - U) \tag{4.40}$$

with k_γ^ν homogeneous of degree ν in $u - U$. Substituting (4.40) into (4.34) yields (4.37) with $r = -1$ and

$$K_\gamma^{-1}(U, \eta) = \chi(X(U))|\eta|^{-1} \tag{4.41}$$

Hence application of Fourier transform (4.35) ($\eta \to \xi$) gives the principal symbol

$$a_\gamma^{-1}(U, \xi) = \chi(X(U))\frac{1}{2}|\xi|^{-1} \tag{4.42}$$

of the pseudo-differential operator V_γ. From (4.39) follows

$$\phi_\gamma(r) = \phi_i(r) + (i\gamma + 1) + \Phi_\gamma(r), \ \Phi_\gamma(r) = \sum_{k=1}^{\infty} \frac{\tilde{\delta}^k}{k!} r^k \tag{4.43}$$

yielding the following result (cf. [287, 395]), where V_i has kernel $\varphi_i(r)$.

Lemma 4.5 *There holds $V_\gamma = V_i + \widetilde{W}_\gamma$ where $\widetilde{W}_\gamma$ is a continuous map from $H^t(S)$ into $H^{t+3}(S)$. V_i maps bijectively $H^r(S)$ onto $H^{r+1}(S)$ for any $r \in \mathbb{R}$.*

Proof Due to (4.43) the first assertion follows from the decomposition

$$V_\gamma(\psi) = V_i(\psi) + \Gamma_\gamma(\psi) + W_\gamma(\psi) \tag{4.44}$$

with

$$\Gamma_\gamma = (i\gamma + 1)\frac{1}{4\pi}\int_S \psi dS_\psi, \; W_\gamma(\psi)(x) = \frac{1}{4\pi}\int_S \psi(y)\Phi_\gamma(|x-y|)dS_\gamma,$$

since W_γ is a pseudodifferential operator of order -3 and Γ_γ takes $H^r(S)$ into $H^t(S)$ for any t.

From (4.41) and (4.42) we see that V_i is an elliptic pseudodifferential operator of order -1. Thus by Lemma 4.4 V_i is a Fredholm operator from $H^r(S)$ into $H^{r+1}(S)$ for any r. Moreover V_i is self-adjoint from $H^{-\frac{1}{2}}(S)$ to $H^{\frac{1}{2}}(S)$ $\left(= \text{dual space of } H^{-\frac{1}{2}}(S)\right)$ since for any $\psi,\ \chi \in C_0^\infty(S)$ there holds

$$\int_S \psi(x)V_i(\chi)(x)dS_x = \int_S \chi(x)V_i(\psi)(x)dS_x$$

because ϕ_i depends only on $|x-y|$. Therefore V_i is bijective from $H^{-\frac{1}{2}}(S)$ onto $H^{\frac{1}{2}}(S)$ if $V_i(\psi) = 0$ implies $\psi = 0$. Then by Lemma 4.4 (iii) the assertion holds for any r. The injectivity of V_i follows by standard arguments: Suppose $V_i(\psi) = 0$ for $\psi \in H^{-\frac{1}{2}}(S)$. Then by Lemma 4.4 (iii) we have $\psi \in H^r(S)$ for any r and hence ψ is continuous. Thus due to Lemma 4.3 the potential $v(x) = \int_S \psi(y)\phi_i(|x-y|)dS_y$ is continuous in $\mathbb{R}^3$ satisfying $\Delta v - v = 0$ in $\Omega \cup \Omega'$, moreover $v = O\left(|x|^{-1}e^{-|x|}\right)$ as $|x| \to \infty$ and $v \equiv 0$ on S. Application of Green's theorem over $\Omega_R = \Omega' \cup \{x,\ |x| < R\}$ gives

$$0 = \int_{\Omega_R} (\Delta v - v)\, v dx = -\int_{\Omega_R}\left(|\text{grad }v|^2 + |v|^2\right)dx + \int_{\Gamma_R} v\frac{\partial v}{\partial n}$$

Thus

$$\int_{\Omega_R}\left(|\text{grad}\, v|^2 + |v|^2\right)dx = \int_{\Gamma_R} v\frac{\partial v}{\partial n}R^2 d\omega$$

and the integral on the right side vanishes as $R \to \infty$, because v and $\frac{\partial v}{\partial n}$ are both $O\left(\frac{e^{-R}}{R}\right)$ as $R \to \infty$. Hence $||v||_{H^1(\mathbb{R}^3)} \equiv 0$ implies $v \equiv 0$ in $\mathbb{R}^3$ and $\frac{\partial v}{\partial n} = 0$ on S. Now the jump relations (Lemma 4.3 (iv)) give $\psi = \left(\frac{\partial v}{\partial n}\right)^- - \left(\frac{\partial v}{\partial n}\right)^+ = 0$.

□

Via Lemma 4.3 (iv) there is defined an operator K_γ by

$$K_\gamma(\psi)(x) = -\frac{1}{4\pi}\int_S \frac{\partial}{\partial n_x}\frac{e^{i\gamma|x-y|}}{|x-y|}\psi(y)dS_y \tag{4.45}$$

which is the adjoint to the operator of the double layer potential

$$N_\gamma(\psi)(x) = -\frac{1}{4\pi}\int_S \frac{\partial}{\partial n_y}\frac{e^{i\gamma|x-y|}}{|x-y|}\psi(y)dS_y \tag{4.46}$$

Lemma 4.6 *$I + 2K_\gamma$ is bijective from $H^r(S)$ onto $H^r(S)$. Moreover,*

$$(I + 2K_\gamma)^{-1} = I + R_\gamma \tag{4.47}$$

where R_γ is continuous from $H^r(S)$ into $H^{r+1}(S)$.

Proof Lemma 4.3 (iv) shows that K_γ is a pseudodifferential operator of order -1 hence takes $H^r(S)$ into $H^{r+1}(S)$. Thus $I + 2K_\gamma$ is a Riesz-Schauder operator. To show that it is bijective it suffices to show that $(I + 2K_\gamma)\psi = 0$ implies $\psi = 0$. If it is bijective the formula (4.47) follows from the theory in [376].

Suppose, then, that $(I + 2K_\gamma)\psi = 0$. As before, we can use Lemma 4.4 to conclude that ψ is smooth. Now define v by $v = V_\gamma(\psi)$. We will have $(\Delta + \gamma^2)\, v = 0$ in Ω and $(I + 2K_\gamma)\psi = 0$ on S implies $\frac{\partial v}{\partial n} = 0$ on S. Now, uniqueness of this exterior Neumann problem gives $v \equiv 0$ in Ω. But we can also set $v = V_\gamma(\psi)$ in Ω'. Assuming that $\gamma^2 \neq 0$ is not an eigenvalue of the interior Dirichlet problem we deduce $v \equiv 0$ in Ω'. Then by the jump relations in Lemma 4.3 (iv) we have

$$\psi(x) = \left(\frac{\partial}{\partial n}V_\gamma\psi\right)^-(x) - \left(\frac{\partial}{\partial n}V_\gamma\psi\right)^+(x) = 0,\ x \in S$$

□

Now we are in the position to solve (4.27) by a simple layer method. Namely, setting $v_e = V_\gamma(\psi)$ the exterior Neumann problem (4.27) is transformed into a Fredholm integral equation of the second kind on S for the unknown layer ψ,

$$\psi(x) + 2\int_S K_\gamma(x, y)\psi(y)dS_y = 2\frac{\partial v_o}{\partial n}(x),\ x \in S, \tag{4.48}$$

which we abbreviate with the notation (4.45) by

$$(I + 2K_\gamma)\psi = 2\frac{\partial v_o}{\partial n}. \tag{4.49}$$

As a consequence of Lemma 4.3 and Lemma 4.6 there holds the following result.

Theorem 4.5 *If $\psi \in C^0(S)$ is a solution of (4.48) then $v_e = V_\gamma(\psi)$ yields a (classical) solution of (4.27). For any real r there exists exactly one solution of (4.48) for given data $\frac{\partial v_o}{\partial n} \in H^r(S)$.*

With formula $v_e = V_\gamma(\psi)$ for the exterior pressure we set for the total accoustic field

$$v = V_\gamma(\psi) + v_0 \text{ in } \Omega, \quad v = V_{\gamma_i}(\chi) \text{ in } \Omega'. \tag{4.50}$$

We obtain from the boundary conditions (4.26) a coupled system of pseudodifferential equations for the unknown layers (ψ, χ) on S:

$$\begin{aligned} &V_{\gamma_i}(\chi) = V_\gamma(\psi) + v_0, \\ &(I - 2K_{\gamma_i})\chi + \nu(I + 2K_\gamma)\psi = 2\nu\frac{\partial v_o}{\partial n}, \quad \nu = \frac{\rho_i}{\rho} \in \mathbb{R}. \end{aligned} \tag{4.51}$$

But by evaluating the kernel function $r^{-1}e^{i\gamma r}$ for small r one verifies as above that both V_{γ_i} and V_γ are pseudodifferential operators of order -1. Hence there holds (cf. Lemma 4.5)

$$V_\gamma(\psi) = V_{\gamma_i}(\psi) + W(\psi)$$

with a pseudodifferential operator W of order -3. Therefore multiplication of $(4.51)_1$ with the bijective operator $V_{\gamma_i}^{-1}$ yields

$$\chi - \psi = V_{\gamma_i}^{-1}W(\psi) + V_{\gamma_i}^{-1}(v_0). \tag{4.52}$$

Since furthermore

$$K_\gamma(\psi) = K_{\gamma_i}(\psi) + L(\psi)$$

with a pseudodifferential operator L of order -2 the second equation in (4.51) gives

$$\chi + \nu\psi = 2K_{\gamma_i}(\chi - \nu\psi) - 2L(\psi) + 2\frac{\partial v_o}{\partial n}\nu. \tag{4.53}$$

The equation (4.52) and (4.53) form a Riesz-Schauder system on $H^r(S) \times H^r(S)$, $r \in \mathbb{R}$. Each of the operators occuring on the right sides is of order at most -1 and the forcing terms $V_{\gamma_i}^{-1}(v_0)$ and $\frac{\partial v_o}{\partial n}$ belong to $H^r(S)$ for given $v_0 \in H^{r+1}(S)$. A reversal of the steps shows that if (ψ, χ) satisfy (4.52), (4.53), then they also satisfy (4.51). But the uniqueness result for (4.24)–(4.26) (Theorem 4.4) shows that the only solution of the homogeneous equations (4.52), (4.53) vanishes identically. Hence we have the following existence result for the interface problem (4.24)–(4.26) governing the scattering of sound (for a corresponding approach to Maxwell's interface problem see [287]):

Theorem 4.6 *Let $v_0 \in H^{r+1}(S)$ for arbitrary $r \in \mathbb{R}$. Then the equations (4.52) and (4.53) have a unique solution with $\chi, \psi \in H^r(S)$.*

Now we relax the regularity assumption on the interface Γ, whereas above for the treatment of the interface problem with pseudodifferential operators we assumed Γ to be analytic. This allowed to apply Riesz-Schauder theory for the existence proof of the solution of second kind integral equations (Theorem 4.6); we now only require Γ to be Lipschitz.

Next following [129], we convert the interface problem (4.24)–(4.26) via the direct method to an equivalent strongly elliptic system of pseudodifferential equations on the interface Γ. For simplicity of notation we write (4.24)–(4.26) as

$$(\Delta + k_j^2)u_j = 0 \qquad \text{in } \Omega_j \ (j = 1, 2)$$

$$u_1 = u_2 + v_0, \quad \mu\frac{\partial u_1}{\partial n} = \frac{\partial u_2}{\partial n} + \psi_0 \qquad \text{on } \Gamma$$

$$u_2(x) = O\left(\frac{1}{|x|}\right), \quad \frac{\partial u_2}{\partial |x|} - ik_2u_2(x) = O\left(\frac{1}{|x|}\right), \qquad |x| \to \infty,$$

where Ω_1 a bounded simply connected domain ($= \Omega'$) in $\mathbb{R}^3$, $\Omega_2 = \mathbb{R}^3 \setminus \overline{\Omega_1}$ and given $v_0 = u_0|_\Gamma$, $\psi_0 = \frac{\partial u_0}{\partial n}|_\Gamma$ with $(\Delta + k_2^2)u_0 = 0$ in Ω_1. By Theorem 4.1 this transmission problem is equivalent to the following relations for the Cauchy data $\begin{pmatrix} v_j \\ \psi_j \end{pmatrix}$ of u_j:

$$(1 - A_1)\begin{pmatrix} v_1 \\ \psi_1 \end{pmatrix} = 0, \tag{4.54}$$

$$(1 + A_2)\begin{pmatrix} v_2 \\ \psi_2 \end{pmatrix} = 0, \tag{4.55}$$

$$\begin{pmatrix} v_2 \\ \psi_2 \end{pmatrix} = M\begin{pmatrix} v_1 \\ \psi_1 \end{pmatrix} - \begin{pmatrix} v_0 \\ \psi_0 \end{pmatrix}, \quad \text{with } M = \begin{pmatrix} 1 & 0 \\ 0 & \mu \end{pmatrix} \tag{4.56}$$

$$\text{and } (1 - A_2)\begin{pmatrix} v_0 \\ \psi_0 \end{pmatrix} = 0, \tag{4.57}$$

with A_j as in (4.14). Now from the above system of six equations for four unknows we derive a system of two equations for two unknows: Writing $\begin{pmatrix} v \\ \psi \end{pmatrix} := \begin{pmatrix} v_1 \\ \psi_1 \end{pmatrix}$ and inserting (4.56) into (4.55) gives

$$(1 + A_2)M\begin{pmatrix} v \\ \psi \end{pmatrix} = (1 + A_2)\begin{pmatrix} v_0 \\ \psi_0 \end{pmatrix}.$$

Then multiplying by M^{-1} from the left and subtracting (4.54) gives the boundary integral equation

$$H\begin{pmatrix} v \\ \psi \end{pmatrix} := \frac{1}{2}(A_1 + M^{-1}A_2M)\begin{pmatrix} v \\ \psi \end{pmatrix} = \frac{1}{2}M^{-1}(1 + A_2)\begin{pmatrix} v_0 \\ \psi_0 \end{pmatrix}. \tag{4.58}$$

If $\begin{pmatrix} v_0 \\ \psi_0 \end{pmatrix}$ satisfy (4.57), this simplifies to

$$H\begin{pmatrix} v \\ \psi \end{pmatrix} = M^{-1}\begin{pmatrix} v_0 \\ \psi_0 \end{pmatrix}.$$

Now any solution $\begin{pmatrix} v \\ \psi \end{pmatrix}$ of (4.58) generates a solution of the original transmission problem (see [129] for details).

For the system (4.58) there holds the following Gårding inequality: There exists a compact operator $C : H^{\frac{1}{2}}(\Gamma) \bigoplus H^{-\frac{1}{2}}(\Gamma) \to H^{-\frac{1}{2}}(\Gamma) \bigoplus H^{\frac{1}{2}}(\Gamma)$ and a constant $\gamma > 0$ such that for $\mu \neq -1$ and smooth Γ there holds

$$\left|\left\langle (H + C)\begin{pmatrix} v \\ \phi \end{pmatrix}, \begin{pmatrix} \overline{v} \\ \overline{\phi} \end{pmatrix}\right\rangle_\Gamma\right| \geq \gamma\left(\|v\|^2_{\frac{1}{2}} + \|\phi\|^2_{-\frac{1}{2}}\right), \tag{4.59}$$

for all $v \in H^{\frac{1}{2}}(\Gamma)$, $\phi \in H^{-\frac{1}{2}}(\Gamma)$. Furthermore for $\Re(1 + \frac{1}{\mu}) > 0$ and $\Re(1 + \mu) > 0$ there holds also for a polygon Γ in $\mathbb{R}^2$

$$\Re\left\langle (H + C)\begin{pmatrix} v \\ \phi \end{pmatrix}, \begin{pmatrix} \overline{v} \\ \overline{\phi} \end{pmatrix}\right\rangle_\Gamma \geq \gamma\left(\|v\|^2_{\frac{1}{2}} + \|\phi\|^2_{-\frac{1}{2}}\right).$$

Here the operator

$$H = \begin{pmatrix} -\frac{1}{2}(K_1 + K_2) & \frac{1}{2}(V_1 + \mu V_2) \\ \frac{1}{2}\left(W_1 + \frac{1}{\mu}W_2\right) & \frac{1}{2}(K_1' + K_2') \end{pmatrix}$$

is elliptic in the Agmon-Douglas-Nirenberg sense with order $\begin{pmatrix} 0 & -1 \\ 1 & 0 \end{pmatrix}$ and principal symbol

$$\sigma(H)(\xi) = \begin{pmatrix} 0 & \frac{1}{2}(1 + \mu)\frac{1}{|\xi|} \\ \frac{1}{2}\left(1 + \frac{1}{\mu}\right)|\xi| & 0 \end{pmatrix}.$$

For $\mu \neq -1$ H is strongly elliptic yielding the Gårding inequality (4.59). Now for smooth Γ system (4.58) of boundary integral equations is an elliptic system of pseudodifferential equations.

The standard regularity theory for pseudodifferential operators shows that for given $\begin{pmatrix} v_0 \\ \psi_0 \end{pmatrix} \in H^s(\Gamma) \bigoplus H^{s-1}(\Gamma)$, any solution $\begin{pmatrix} v \\ \psi \end{pmatrix}$ of (4.58) is in $H^s(\Gamma) \bigoplus H^{s-1}(\Gamma)$ for any $s \in \mathbb{R}$. Under the assumption of Theorem 4.4 the solution of the transmission problem is unique. This implies that the operator H is injective. Now, by Gårding's inequality the operator H is Fredholm of index zero, and hence bijective yielding the existence of the solution of (4.58) (see [129]). Therefore insertion of that solution in the representation formula gives the solution of the original interface problem for the Helmholtz equation.

4.3 Screen Problems

For open boundary curves or surfaces S the correct setting of integral equations needs a refined analysis where the solutions of the integral equations must (in a weak sense) be extendable by zero from the open surface S to a closed surface $\tilde{S}$ (including S), i.e. for real s [253]

$$\tilde{H}^s(S) = \{\varphi : \varphi^* = \varphi \text{ on } S, \varphi^* = 0 \text{ on } \tilde{S} \backslash S, \varphi^* \in H^s(\tilde{S})\}.$$

Note $\tilde{H}^{1/2}(S) = H_{00}^{1/2}(S)$ in [284].

For given $g(h)$ we consider the Dirichlet (Neumann) screen problem ($k \in \mathbb{C} \setminus \{0\}$):

$$\begin{aligned} (\Delta + k^2)u &= 0 \quad \text{in } \Omega_S := \mathbb{R}^3 \backslash \overline{S} \\ u &= g \quad \text{on } S \\ (\frac{\partial u}{\partial n} &= h \quad \text{on } S) \\ \frac{\partial u}{\partial r} - iku &= o(\frac{1}{r}) \quad \text{as } r = |x| \to \infty \end{aligned}$$

where S is a bounded, simply connected, orientable,open surface in $\mathbb{R}^3$ with a smooth boundary curve γ which does not intersect itself. Extend S to an arbitrary smooth, simply connected, closed, orientable manifold ∂G_1 enclosing a bounded domain G_1 (see Fig. 4.2).

Let $\frac{\partial}{\partial n}$ denote the exterior normal derivative to ∂G_1. Let $[v]$ denote the jump $v_- - v_+$ where the subscript $+(-)$ means the limit from $\mathbb{R}^3 \backslash G_1$ (from G_1) to ∂G_1. Furthermore, let B denote a sufficiently large ball with radius R including $\overline{G_1}$ and let $G_2 := B \cap (\mathbb{R}^3 \backslash \overline{G_1})$ and ∂B denote the boundary of B.

Fig. 4.2 Geometrical setting [398]

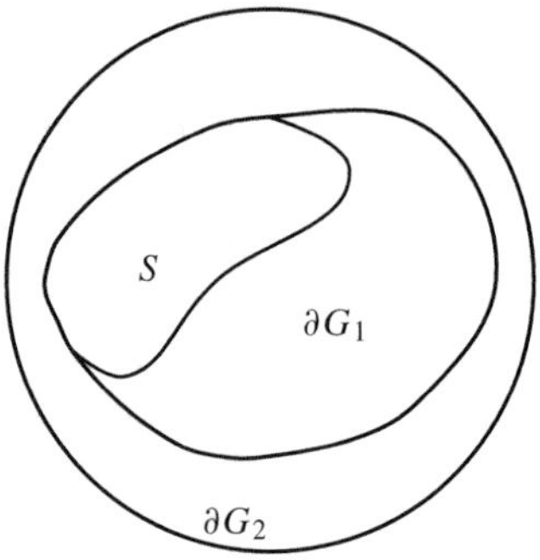

Let $\varphi(x,y) = \frac{1}{4\pi}\frac{e^{ik|x-y|}}{|x-y|}$ and

$$V_{G_j}u(x) := -2\int_\Gamma \varphi(x,y)u(y)ds_y \quad (x \in G_j),$$

$$K_{G_j}u(x) := -2\int_\Gamma \frac{\partial}{\partial n_y}\varphi(x,y)u(y)ds_y \quad (x \in G_j, \Gamma = \partial G_j)$$

Application of the representation formula

$$u(x) = (-1)^j 1/2(K_{G_j}u(x) - V_{G_j}\frac{\partial u}{\partial n}(x))$$

gives for $x \in G_1$

$$u(x) = -\frac{1}{2}\left(K_{G_1}u(x) - V_{G_1}\frac{\partial u}{\partial n}(x)\right) \tag{4.60}$$

$$0 = -\frac{1}{2}\left(K_{G_2}u(x) - V_{G_2}\frac{\partial u}{\partial n}(x)\right)$$

Since $[\frac{\partial u}{\partial n}]|_{\partial G_j\setminus\overline{S}} = 0$, addition yields with the outer boundary $\partial B = \{y \in \mathbb{R}^3, |y| = R\}$

$$u(x) = \int_{|y|=R} u(y)\frac{\partial}{\partial n_y}\varphi(x,y)ds_y - \int_{|y|=R}\frac{\partial u}{\partial n}(y)\varphi(x,y)ds_y - \int_S [\frac{\partial u}{\partial n}](y)\varphi(x,y)ds_y$$

For $x \to S$ the trace theorem yields with $u|_S = g$

$$g(x) = \int_{|y|=R}\{u(y)\frac{\partial}{\partial n_y}\varphi(x,y) - \frac{\partial u}{\partial n}(y)\varphi(x,y)\}ds_y - \int_S [\frac{\partial u}{\partial n}](y)\varphi(x,y)ds_y$$

Since the radiation condition holds for u and φ, the integral over $|y| = R$ vanishes as $R \to \infty$ and therefore the foregoing expression becomes

$$2g(x) = -2 \int_S [\frac{\partial u}{\partial n}]\varphi(x, y)ds_y =: V_S[\frac{\partial u}{\partial n}](x) \, , \; x \in S \tag{4.61}$$

Taking in (4.60) the normal derivative gives for $x \in G_1$ (note $[u]|_{\partial G_j \setminus \overline{S}} = 0$)

$$\frac{\partial u}{\partial n}(x) = \int_{|y|=R} \{u(y)\frac{\partial}{\partial n_x}\frac{\partial}{\partial n_y}\varphi(x, y) - \frac{\partial u}{\partial n}(y)\frac{\partial}{\partial n_x}\varphi(x, y)\}ds_y$$
$$- \int_S [u](y)\frac{\partial^2}{\partial n_x \partial n_y}\varphi(x, y)ds_y$$

Since φ and its derivatives satisfy the decay condition, letting $x \to S$, $R \to \infty$ and taking $\frac{\partial u}{\partial n}|_S = h$ gives

$$-2h(x) = 2 \int_S [u](y)\frac{\partial^2}{\partial n_x \partial n_y}\varphi(x, y)ds_y =: W_S[u](x) \, , \; x \in S \tag{4.62}$$

In [398] it is shown that (4.61) is equivalent to the Dirichlet screen problem and (4.62) to the Neumann screen problem and that for $\Im k \geq 0$ these integral equations are uniquely solvable with $[\frac{\partial u}{\partial n}]|_S \in \widetilde{H}^{-1/2}(S)$ for given $g \in H^{1/2}(S)$ and $[u]|_S \in \widetilde{H}^{1/2}(S)$ for given $h \in H^{-1/2}(S)$, respectively (for Lipschitz screens see [124]).

Now we come to the singularity of the densities of the integral equations (4.61) and (4.62) near the edge γ of the screen S. The analysis in [398] follows the procedure in [167] by (i) mapping locally S onto $\mathbb{R}^2_+$,(ii) applying the Wiener-Hopf technique in the halfspace $\mathbb{R}^2_+$ and (iii) patching together the local results.

Theorem 4.7 (Theorem 2.9 in [398])

(i) Let $g \in H^{3/2+\sigma}(S)$ be given. Then the solution of the integral equation (4.61) *has the form*

$$[\frac{\partial u}{\partial n}] = \beta(s)\rho^{-1/2}\chi(\rho) + \psi_r \quad on\ S \tag{4.63}$$

with $\beta \in H^{1/2+\sigma}(\gamma)$, $\psi_r \in \tilde{H}^{1/2+\sigma'}(S)$, $0 < \sigma' < \sigma < 1/2$

(ii) Let $h \in H^{1/2+\sigma}(S)$ be given. Then the solution of the integral equation (4.62) *has the form*

$$[u] = \alpha(s)\rho^{1/2}\chi(\rho) + v_r \quad on\ S \tag{4.64}$$

with $\alpha \in H^{1/2+\sigma}(\gamma)$, $v_r \in L^2(I; H^{1/2+\sigma}(\gamma)) \cap \tilde{H}^{3/2+\sigma'}(I; L^2(\gamma))$, $0 < \sigma' < \sigma < 1/2$, where S is identified with $I \times \gamma$, $I = [0, 1]$.

(Here s denotes the parameter of arclength of γ, ρ corresponds to the Euclidean distance to γ, χ is a C^∞ cut-off function with $\chi \equiv 1$ for $|\rho| < 1/2$ and $\chi \equiv 0$ for $|\rho| > 1$).

This result on the singularity of the screen problem provides the basis for the augmented BEM described in section 7.5. Crack problems can be dealt with like the screen problems above yielding an efficient solution procedure with boundary integral equations and boundary elements (see [130, 432]). Recently there has been intensive research on multiple screens (see the work of Claeys and Hiptmair [107, 108]).

4.4 Interface Problem in Linear Elasticity

Next, we want to relax the smoothness assumptions on the interface Γ and only require $\Gamma \in Lip$. We will show that the above approach still works and derive a Gårding inequality for the boundary integral operators related to linear elasticity problems. The reported results are taken from the paper [137] by Costabel and Stephan. Here we like to mention the celebrated fundamental book on three-dimensional potential theory of linearized elasticity [278].

The transmission problem in 3D in steady state elastodynamics reads (TMP):

For given vector fields u_0 and t_0 on the boundary Γ find vector fields u_j in Ω_j, $j = 1, 2$, satisfying the equations of linear elasticity

$$P_j u_j - \rho_j \omega^2 u_j = 0 \text{ in } \Omega_j,\, j = 1, 2$$

and the transmission conditions

$$u_1 = u_2 + u_0 \ , \quad t_1 = t_2 + t_0 \text{ on } \Gamma$$

Here the differential operators P_j are given by

$$P_j u = -(\mu_j \Delta u + (\lambda_j + \mu_j) \operatorname{grad} \operatorname{div} u)$$

$\rho_j > 0$ is the density of the medium Ω_j, and $\omega > 0$ is the frequency of the incident wave. We are interested in solutions $u_j \in H^1_{loc}(\Omega_j)$ and define

$$\mathscr{L}_1 = \{u_1 \in H^1(\Omega_1) : P_1 u_1 = \rho_1 \omega^2 u_1 \text{ in } \Omega_1\}$$

$$\mathscr{L}_2 = \{u_2 \in H^1_{loc}(\Omega_2) : P_2 u_2 = \rho_2 \omega^2 u_2 \text{ in } \Omega_2, u_2 \text{ satisfies a decay condition [137]}\}$$

Here and in the following all function spaces, including all Sobolev spaces are considered vectorial containing 3D vector fields.

Lemma 4.7 *Let $u \in H^1_{loc}(\overline{\Omega}_j)$ with compact support satisfy $P_j u \in L^2_{loc}(\Omega_j)$ and let $v \in H^1(\Omega_j)$ with bounded support. Then $T_j u\big|_\Gamma \in H^{-1/2}(\Gamma)$ is defined with $\langle f, g\rangle := \int_\Gamma f \cdot g ds$ by*

$$\int_{\Omega_j} P_j u \cdot v dx = (-1)^j \langle T_j u, v\rangle + \Phi_j(u, v) \tag{4.65}$$

with

$$\Phi_j(u, v) = \int_{\Omega_j} \sum_{i,h,k,l=1}^{3} a^j_{ihkl} \epsilon_{kl}(u)\epsilon_{ih}(v) dx\ , \quad a^j_{ihkl} = \lambda_j \delta_{ih}\delta_{kl} + \mu_j(\delta_{ik}\delta_{hl} + \delta_{il}\delta_{hk}),$$

where λ_j and μ_j denote the Lame constants in Ω_j $(j = 1, 2)$.

From (4.65) one obtains, with the symmetry of Φ_j, the second Green formula

$$\int_{\Omega_j} (P_j u \cdot v - u \cdot P_j v) dx = (-1)^j \int_\Gamma (v \cdot T_j(u) - u \cdot T_j(v)) ds \tag{4.66}$$

This gives in Ω_j with the fundamental solution $G_j(x, y, \omega)$ of $(P_j - \rho\omega^2)u_j = 0$, the Somigliana representation formula for $x \in \Omega_j$:

$$u_j(x) = (-1)^j \int_\Gamma \big\{T_j(x, y, \omega)v_j(y) - G_j(x, y, \omega)\phi_j(y)\big\} ds(y) \tag{4.67}$$

where $v_j = u_j, \phi_j = T_j(u_j) = t_j$ on Γ. Here G_j is the 3×3 matrix function

$$(G_j)_{ik} = \frac{-1}{4\pi\mu_j}\left\{\frac{1}{r} e^{ik^T_j r}\delta_{ik} + (k^T_j)^{-2}\partial_i\partial_k[(e^{ik^T_j r} - e^{ik^L_j r})r^{-1}]\right\}$$

with $r = |x - y|$ and $T_j(x, y, \omega) = T_{j,y}(G_j(x, y, \omega))^T$, k^L_j longitudinal (dilational) wave number, k^T_j transverse (shear) wave number.

Lemma 4.8 *Let $u_j \in \mathscr{L}_j$. Then (4.67) holds for u_j in Ω_j. For any $v_j \in H^{1/2}(\Gamma)$ and any $\phi \in H^{-1/2}(\Gamma)$ the formula (4.67) defines a vector field $u_j \in \mathscr{L}_j$.*

Taking Cauchy data in (4.67) yields on Γ

$$\begin{pmatrix} v_j \\ \phi_j \end{pmatrix} = \mathscr{C}_j \begin{pmatrix} v_j \\ \phi_j \end{pmatrix}$$

where the Calderón projector

$$\mathscr{C}_j = \begin{pmatrix} 1/2 + (-1)^j \Lambda_j & -(-1)^j V_j \\ -(-1)^j W_j & 1/2 - (-1)^j \Lambda'_j \end{pmatrix}$$

is defined via the boundary integral operators

$$V_j v(x) = \int_\Gamma G_j(x,y,\omega)v(y)ds(y) \qquad \Lambda_j v(x) = \int_\Gamma T_j(x,y,\omega)v(y)ds(y)$$

$$W_j v(x) = -T_{j,x}\int_\Gamma T_j(x,y,\omega)v(y)ds(y) \quad \Lambda'_j v(x) = \int_\Gamma T_j(y,x,\omega)^T v(y)ds(y)$$

Lemma 4.9

(a) The statements (i) and (ii) on $(v,\psi) \in \mathcal{H} := H^{1/2}(\Gamma) \times H^{-1/2}(\Gamma)$ are equivalent:

(i) (v,ψ) are Cauchy data of some $u_j \in \mathcal{L}_j$
(ii) $(I - \mathcal{C}_j)\binom{v}{\psi} = 0$

(b) The operators $\mathcal{C}_j$ are projection operators mapping $\mathcal{H}$ on its subspace of Cauchy data of weak solutions in $\mathcal{L}_j$

Thus we can write the transmission problem (TMP) in the equivalent form

$$(I - \mathcal{C}_1)\begin{pmatrix} v_1 \\ \phi_1 \end{pmatrix} = 0 \tag{4.68}$$

$$(I - \mathcal{C}_2)\begin{pmatrix} v_2 \\ \phi_2 \end{pmatrix} = 0 \tag{4.69}$$

$$\begin{pmatrix} v_2 \\ \phi_2 \end{pmatrix} = \begin{pmatrix} v_1 \\ \phi_1 \end{pmatrix} - \begin{pmatrix} v_0 \\ \phi_0 \end{pmatrix} \tag{4.70}$$

This is a system of 6 vector equations for 4 vector unknowns. From it we can extract a square subsystem by inserting $\binom{v_2}{\phi_2}$ from (4.70) in (4.69) and subtracting (4.68) from the resulting equation. We obtain

$$A\begin{pmatrix} v_1 \\ \phi_1 \end{pmatrix} = (I - \mathcal{C}_2)\begin{pmatrix} v_0 \\ \phi_0 \end{pmatrix} \quad \text{with } A := \mathcal{C}_1 - \mathcal{C}_2 \tag{4.71}$$

We have the following theorem.

Theorem 4.8 *Let $\binom{v_0}{\phi_0} \in \mathcal{H} = H^{1/2}(\Gamma) \times H^{-1/2}(\Gamma)$ be given. Then there holds*

(i) If $u_j \in \mathcal{L}_j$ solve the (TMP), then $\binom{v}{\phi} := \binom{v_1}{\phi_1} = \binom{u_1|_\Gamma}{T(u_1)|_\Gamma} \in \mathcal{H}$ solves (4.71)
(ii) If $\binom{v}{\phi} \in \mathcal{H}$ solves (4.71), then, with

$$\begin{pmatrix} v_1 \\ \phi_1 \end{pmatrix} := \mathcal{C}_1\begin{pmatrix} v \\ \phi \end{pmatrix} \text{ and } \begin{pmatrix} v_2 \\ \phi_2 \end{pmatrix} := \mathcal{C}_2\left(\begin{pmatrix} v \\ \phi \end{pmatrix} - \begin{pmatrix} v_0 \\ \phi_0 \end{pmatrix}\right)$$

and u_j defined by (4.67), $u_j \in \mathcal{L}_j$ $(j = 1, 2)$ solves the (TMP).

Proof

(i) follows from the derivation of (4.71).
(ii) From the definition of $\binom{v_j}{\phi_j}$ and the projection property of $\mathscr{C}_j$ follows

$$(I - \mathscr{C}_j)\begin{pmatrix} v_j \\ \phi_j \end{pmatrix} = 0,$$

hence $\binom{v_j}{\phi_j}$ are Cauchy data of certain $u_j \in \mathscr{L}_j$ which are then given by (4.67). It remains to show that the transmission condition is satisfied:

$$\begin{pmatrix} v_2 \\ \phi_2 \end{pmatrix} - \begin{pmatrix} v_1 \\ \phi_1 \end{pmatrix} = (\mathscr{C}_2 - \mathscr{C}_1)\begin{pmatrix} v \\ \phi \end{pmatrix} - \mathscr{C}_2\begin{pmatrix} v_0 \\ \phi_0 \end{pmatrix} = -A\begin{pmatrix} v \\ \phi \end{pmatrix} - \mathscr{C}_2\begin{pmatrix} v_0 \\ \phi_0 \end{pmatrix}$$
$$= (I - \mathscr{C}_2)\begin{pmatrix} v_0 \\ \phi_0 \end{pmatrix} - \mathscr{C}_2\begin{pmatrix} v_0 \\ \phi_0 \end{pmatrix} = -\begin{pmatrix} v_0 \\ \phi_0 \end{pmatrix}$$

□

Theorem 4.9 *The operator A satisfies a Gårding inequality: There exist $\gamma > 0$ and a compact operator $T : \mathscr{H} \to \mathscr{H}$ with*

$$\Re\left\langle (A+T)\begin{pmatrix} v \\ \phi \end{pmatrix}, \begin{pmatrix} v \\ \phi \end{pmatrix}\right\rangle \geq \gamma(\|v\|^2_{H^{1/2}(\Gamma)} + \|\phi\|^2_{H^{-1/2}(\Gamma)}) \quad \forall \begin{pmatrix} v \\ \phi \end{pmatrix} \in \mathscr{H} \tag{4.72}$$

Here the brackets denote the natural (anti)-duality of $\mathscr{H}$ with itself:

$$\left\langle \begin{pmatrix} v \\ \phi \end{pmatrix}, \begin{pmatrix} w \\ \psi \end{pmatrix}\right\rangle := \int_\Gamma (\overline{v}\psi + w\overline{\phi})ds \ \text{for} \ \begin{pmatrix} v \\ \phi \end{pmatrix}, \begin{pmatrix} w \\ \psi \end{pmatrix} \in \mathscr{H}.$$

Proof We write

$$A = A_1 + A_2 \ \text{with} \ A_j = (-1)^j\left(1/2\, I - \mathscr{C}_j\right) = \begin{pmatrix} -\Lambda_j & V_j \\ W_j & \Lambda'_j \end{pmatrix}$$

Since the sum of two strongly elliptic operators is strongly elliptic, it suffices to show the strong ellipticity of the operators A_1 and A_2. Proof for A_1: Due to density arguments, one needs to show the Gåding inequality (4.72) only for smooth (v, ϕ). Let then u_j, $j = 1, 2$, be defined by

$$u_j(x) = \chi(x)\int_\Gamma \{T_1(x,y)v(y) - G_1(x,y)\phi(y)\}ds(y), \quad x \in \Omega_j.$$

Here we choose $\chi \in C_0^\infty(\mathbb{R}^3)$ satisfying $\chi \equiv 1$ is a neighborhhood of $\overline{\Omega}_1$. Then, by definition of the Calderón projectors (i.e. the classical jump relations for the elastic potentials), the Cauchy data $v_j := u_j|_\Gamma$ and $\phi_j := T(u_j)$ satisfy

$$\begin{pmatrix} v_j \\ \phi_j \end{pmatrix} = (-1)^j \left(1/2\, I - (-1)^j A_1\right) \begin{pmatrix} v \\ \phi \end{pmatrix}.$$

By adding and subtracting these two equations, we find

$$A_1 \begin{pmatrix} v \\ \phi \end{pmatrix} = -\begin{pmatrix} v_1 \\ \phi_1 \end{pmatrix} - \begin{pmatrix} v_2 \\ \phi_2 \end{pmatrix}$$

$$\begin{pmatrix} v \\ \phi \end{pmatrix} = -\begin{pmatrix} v_1 \\ \phi_1 \end{pmatrix} + \begin{pmatrix} v_2 \\ \phi_2 \end{pmatrix}$$

Thus the bilinear form defined by A_1 is given by

$$\begin{aligned} 2\left\langle A_1 \begin{pmatrix} v \\ \phi \end{pmatrix}, \begin{pmatrix} v \\ \phi \end{pmatrix} \right\rangle &= \left\langle \begin{pmatrix} v_1 \\ \phi_1 \end{pmatrix} + \begin{pmatrix} v_2 \\ \phi_2 \end{pmatrix}, \begin{pmatrix} v_1 \\ \phi_1 \end{pmatrix} - \begin{pmatrix} v_2 \\ \phi_2 \end{pmatrix} \right\rangle \\ &= \left\langle \begin{pmatrix} v_1 \\ \phi_1 \end{pmatrix}, \begin{pmatrix} v_1 \\ \phi_1 \end{pmatrix} \right\rangle - \left\langle \begin{pmatrix} v_2 \\ \phi_2 \end{pmatrix}, \begin{pmatrix} v_2 \\ \phi_2 \end{pmatrix} \right\rangle \\ &\quad + \left\langle \begin{pmatrix} v_2 \\ \phi_2 \end{pmatrix}, \begin{pmatrix} v_1 \\ \phi_1 \end{pmatrix} \right\rangle - \left\langle \begin{pmatrix} v_1 \\ \phi_1 \end{pmatrix}, \begin{pmatrix} v_2 \\ \phi_2 \end{pmatrix} \right\rangle \\ &= 2\Re \int_\Gamma (\overline{v}_1 \phi_1 - \overline{v}_2 \phi_2) ds + 2i \Im \int_\Gamma (\overline{v}_2 \phi_1 - \overline{v}_1 \phi_2) ds \end{aligned}$$

Hence

$$\Re \left\langle A_1 \begin{pmatrix} v \\ \phi \end{pmatrix}, \begin{pmatrix} v \\ \phi \end{pmatrix} \right\rangle = \Re \int_\Gamma (\overline{v}_1 \phi_1 - \overline{v}_2 \phi_2) ds \tag{4.73}$$

Now we need the first Green formulas for P_1 in Ω_1 and Ω_2. This leads to

$$\tilde{\Phi}_j(u_j, u_j) - \int_{\Omega_j} \overline{u}_j \cdot (P_1 - \rho_1 \omega^2) u_j dx = -(-1)^j \int_\Gamma \overline{v}_j \phi_j ds \tag{4.74}$$

where

$$\tilde{\Phi}_j(u_j, u_j) := \int_{\Omega_j} \left(\sum a^1_{ihkl} \overline{\epsilon_{kl}(u_j)} \epsilon_{ih}(u_j) - \rho_1^2 \omega |u_j|^2 \right) dx.$$

Now $P_1 u_1 - \rho_1^2 \omega u_1 = 0$ and $P_1 u_2 - \rho_1^2 \omega u_2 = f_2$, where $f_2 \in C_0^\infty(\Omega_2)$. ($f_2 \equiv 0$ whenever $\chi \equiv 1$ or $\chi \equiv 0$ holds.) From (4.73) and (4.74) together we find

$$\Re \left\langle A_1 \begin{pmatrix} v \\ \phi \end{pmatrix}, \begin{pmatrix} v \\ \phi \end{pmatrix} \right\rangle = \Re \{ \tilde{\Phi}_1(u_1, u_1) + \tilde{\Phi}_2(u_2, u_2) - \int_{\Omega_2} \overline{u_2} \cdot f_2 \, dx \} \tag{4.75}$$

As the support of f_2 is disjoint from Γ, there is a compact operator T_1 on $H^{1/2}(\Gamma)\times H^{-1/2}(\Gamma)$ such that

$$\left|\int_{\Omega_2} \overline{u_2}\cdot f_2\; dx\right| \le \left\langle T_1\begin{pmatrix} v\\ \phi\end{pmatrix}, \begin{pmatrix} v\\ \phi\end{pmatrix}\right\rangle$$

From Korn's inequality and the trace lemma we find that there exist compact quadratic forms k_j on $H^1(\Omega_j)$ and hence a compact operator T_2 on $\mathscr{H} = H^{1/2}(\Gamma)\times H^{-1/2}(\Gamma)$ such that

$$\begin{aligned}\tilde{\Phi}_1(u_1,u_1)+\tilde{\Phi}_2(u_2,u_2) &\ge \gamma_1\left(\|u_1\|^2_{H^1(\Omega_1)}+\|u_2\|^2_{H^1(\Omega_1)}\right)-k_1(u_1)-k_2(u_2)\\ &\ge \gamma_2\left(\|v\|^2_{H^{1/2}(\Gamma)}+\|\phi\|^2_{H^{-1/2}(\Gamma)}\right)-\left\langle T_2\begin{pmatrix} v\\ \phi\end{pmatrix}, \begin{pmatrix} v\\ \phi\end{pmatrix}\right\rangle\end{aligned}$$

Finally we get

$$\Re\left\langle (A_1+T_1+T_2)\begin{pmatrix} v\\ \phi\end{pmatrix}, \begin{pmatrix} v\\ \phi\end{pmatrix}\right\rangle \ge \gamma_2\left(\|v\|^2_{H^{1/2}(\Gamma)}+\|\phi\|^2_{H^{-1/2}(\Gamma)}\right)$$

□

4.5 A Strongly Elliptic System for Exterior Maxwell's Equations

This section reports of an approach by [286] and [287]. In [285] a simple layer potential method for the three-dimensional eddy current problem is introduced. In [286] solution procedures for the perfect conductor problem are given. E.g. different sets of Maxwell equations are solved in the obstacle and outside while the tangential components of both electric and magnetic fields are continuous across the obstacle surface. In [287] it is shown, that the integral equation system resulting from the three-dimensional Maxwell's equations in air in the exterior of a perfect conductor is coercive and thus asymptotic convergence of Galerkin's method is established.

The purpose of this section is to show the coercivity of the system of equations belonging to the three-dimensional conductivity problem in an exterior unbounded domain using pseudodifferential operators.

4.5.1 A Simple Layer Procedure

We consider the eddy current problem: Let Ω' be a bounded interior and $\Omega = \mathbb{R}^3 \setminus \Omega'$. Ω' is to represent a perfect conductor characterized by constants ε, μ and

$\sigma = \infty$ denoting permitivity, permeabilty and conductivity. Ω is to represent air characterized by ε_0, μ_0 and $\sigma_0 = 0$. $S = \partial\Omega = \partial\Omega'$ is a closed analytic surface dividing $\mathbb{R}^3$ into the disjoint domains Ω and Ω'. The total electromagnetic field $(\mathbf{E}, \mathbf{H})$ consists of the sum of the incident $(\mathbf{E}^0, \mathbf{H}^0)$ and the scattered $(\mathbf{E}^S, \mathbf{H}^S)$ field. Thereby, $(\mathbf{E}^0, \mathbf{H}^0)$ is assumed to originate in Ω.

The time harmonic Maxwell's equations are given by

$$\text{curl}\,\mathbf{E} = i\omega\mu_0\mathbf{H}\ ,\quad \text{curl}\,\mathbf{H} = -i\omega\varepsilon_0 \qquad \mathbf{E} \text{ in } \Omega \tag{4.76}$$

$$\text{curl}\,\mathbf{E} = i\omega\mu\ \mathbf{H}\ ,\quad \text{curl}\,\mathbf{H} = (-i\omega\varepsilon + \sigma)\mathbf{E} \text{ in } \Omega'. \tag{4.77}$$

By appropriate rescaling of (4.76) and (4.77) one obtains

$$\text{curl}\,\mathbf{E} = \mathbf{H}\ ,\quad \text{curl}\,\mathbf{H} = \alpha^2\mathbf{E} \text{ in } \Omega \tag{4.78}$$

$$\text{curl}\,\mathbf{E} = \mathbf{H}\ ,\quad \text{curl}\,\mathbf{H} = i\beta\,\mathbf{E} \text{ in } \Omega', \tag{4.79}$$

with $\alpha^2 = \omega^2\varepsilon_0\mu_0$ and $\beta = (\omega\mu\sigma - i\omega^2\mu\varepsilon)$ and $\beta = \omega\mu\sigma > 0$, if $\varepsilon = 0$. Across S the tangential components of the fields

$$(\mathbf{n} \times \mathbf{E})^+ = (\mathbf{n} \times \mathbf{E})^-\ ,\ (\mathbf{n} \times \mathbf{H})^+ = (\mathbf{n} \times \mathbf{H})^- \tag{4.80}$$

must be continuous. At higher conductivity β is large, which leads to the perfect conductor approximations. This means solving only (4.78) and requiring that the tangential component of the total electric field $\mathbf{n} \times \mathbf{E} = 0$ vanishes on S, leading to

$$\left.\begin{array}{lr} \text{curl}\,\mathbf{E}^S = \mathbf{H}^S\ ,\quad \text{curl}\,\mathbf{H}^S = \alpha^2\mathbf{E}^S & \text{in } \Omega \\ (\mathbf{n} \times \mathbf{E}^S) = -(\mathbf{n} \times \mathbf{E}^0) & \text{on } S. \end{array}\right. \tag{4.81}$$

In [285] it is shown that for (4.78)–(4.80) at most one solution exists for any $\alpha > 0$ and $0 < \beta \le \infty$.

By introducing the simple layer potential

$$V_\alpha(M)(x) = \frac{1}{4\pi}\int_S M(y)\frac{e^{i\alpha|x-y|}}{|x-y|}\,dS_y \tag{4.82}$$

one can display the electric and magnetic fields in the Stratton-Chu representation formulas [413]

$$\left.\begin{array}{l} \mathbf{E} = V_\alpha(\mathbf{n} \times \mathbf{H}) - \text{curl}\,V_\alpha(\mathbf{n} \times \mathbf{E}) + \text{grad}\,V_\alpha(\mathbf{n} \cdot \mathbf{E}) \\ \mathbf{H} = \text{curl}\,V_\alpha(\mathbf{n} \times \mathbf{H}) - \text{curl}\,\text{curl}\,V_\alpha(\mathbf{n} \times \mathbf{E}) \end{array}\right\} \text{ in } \Omega. \tag{4.83}$$

Now setting $\mathbf{n} \times \mathbf{E} = 0$ in (4.83) and replacing $\mathbf{n} \times \mathbf{H}$ and $\mathbf{n} \cdot \mathbf{E}$ by unknowns $\mathbf{J}$ and M, yields

$$\mathbf{E} = V_\alpha(\mathbf{J}) + \mathrm{grad}_T\, V_\alpha(M) \ , \quad \mathbf{H} = \mathrm{curl}\, V_\alpha(\mathbf{J}). \tag{4.84}$$

Now div $\mathbf{H}$ is automatically zero, whereas we must guarantee that div $\mathbf{E} = 0$ in Ω. It suffices to require div $\mathbf{E} = 0$ on S. It follows from (4.84) and the kernel $\frac{e^{i\alpha|x-y|}}{|x-y|}$ that $\Delta\mathbf{E} = -\alpha^2\mathbf{E}$. Hence Δdiv $\mathbf{E} = -\alpha^2$div $\mathbf{E}$ in Ω. Moreover div $\mathbf{E}$ satisfies the radiation condition. Hence, by uniqueness for the scalar exterior Dirichlet problem div $\mathbf{E} = 0$ on S implies div $\mathbf{E} = 0$ in Ω. Hence we require div $\mathbf{E} = 0$ on S. Therefore applying the boundary condition of (4.81) and div $\mathbf{E} = 0$ on S in (4.84) one obtains a coupled system of pseudodifferential equations on the boundary surface S with the unknowns $\mathbf{J}$ and M:

$$\begin{aligned} V_\alpha(\mathbf{J})_T + \ \mathrm{grad}\, V_\alpha(M) &= -(\mathbf{n} \times \mathbf{E}^0) = -\mathbf{E}^0_T \\ V_\alpha(\mathrm{div}_T\ \mathbf{J}) - \alpha^2 V_\alpha(M) &= 0, \end{aligned} \tag{4.85}$$

where $V_\alpha(\mathbf{J})_T$ denotes the tangential component of the vector function $V_\alpha(\mathbf{J})$ and div $V_\alpha(\mathbf{J}) = V_\alpha(\mathrm{div}_T\ \mathbf{J})$.

4.5.2 Modified Boundary Integral Equations

Furthermore in [285] it is shown that there exists a continous map $J_\alpha(\mathbf{J})_T$ from $\mathbf{H}^r(S)$ into $H^{r+1}(S)$, $r \in \mathbb{R}$, such that

$$\mathrm{div}_T\ V_\alpha(\mathbf{J}) = V_\alpha(\mathrm{div}_T\ \mathbf{J}) + J_\alpha(\mathbf{J})_T. \tag{4.86}$$

Therefore by applying div_T onto $(4.85)_1$ and subtracting the result from $(4.85)_2$ one gets a new equivalent system:

$$\begin{aligned} V_\alpha(\mathbf{J})_T + \ \mathrm{grad}_T\, V_\alpha(M) &= -\mathbf{E}^0_T \\ -J_\alpha(\mathbf{J})_T - (\Delta_T + \alpha^2) V_\alpha(M) &= \mathrm{div}_T\ \mathbf{E}^0_T. \end{aligned} \tag{4.87}$$

In [287] it is mentioned, that (4.85) is not satisfying the Gårding's inequality but (4.87) does, so convergence for Galerkin's procedure is guaranteed.

In order to show the claimed Gårding inequality for the system (4.87), we consider the half-space case as in [287]. The equation system (4.87) becomes in the half-space case $\Omega = \{x \in \mathbb{R}^3 | x_3 > 0\}$:

$$\begin{aligned} \phi_\alpha * \mathbf{J} + \frac{\partial}{\partial x_1}\phi_\alpha * M\underline{e}_1 + \frac{\partial}{\partial x_2}\phi_\alpha * M\underline{e}_2 &= -4\pi(\underline{e}_3 \times \mathbf{E}^0) \\ -\left(\frac{\partial^2}{\partial x_1^2} + \frac{\partial^2}{\partial x_2^2} + \alpha^2\right)\phi_\alpha * M &= 4\pi \operatorname{div} \mathbf{E}_T^0. \end{aligned} \tag{4.88}$$

Here

$$\phi_\alpha(|x-y|) = \frac{e^{i\alpha|x-y|}}{|x-y|} \tag{4.89}$$

is the fundamental solution of the Helmholtz equation. In [285] it is shown that the series expansion

$$\phi_\alpha(r) = \frac{1}{r} + i\alpha + \sum_{j=1}^{\infty} \frac{\delta^j}{j!} r^j, \; \delta \in \mathbb{C} \tag{4.90}$$

with $r = |x-y|$ leads to the existence of a smoothing pseudodifferential operator W_α of the order -3 such that

$$V_\alpha = V_0(M) + W_\alpha(M). \tag{4.91}$$

The system (4.88) can be written as a 3×3-matrix of operators

$$A_\alpha U := \begin{pmatrix} V_\alpha|_1 & 0 & \operatorname{grad}_1 V_\alpha \\ 0 & V_\alpha|_2 & \operatorname{grad}_2 V_\alpha \\ 0 & 0 & -(\Delta + \alpha^2)V_\alpha \end{pmatrix} \begin{pmatrix} J^1 \\ J^2 \\ M \end{pmatrix} = \begin{pmatrix} -\mathbf{E}_\mathbf{T}^\mathbf{0}|_1 \\ -\mathbf{E}_\mathbf{T}^\mathbf{0}|_2 \\ \operatorname{div} \mathbf{E}_T^0 \end{pmatrix} := F \tag{4.92}$$

with $\mathbf{J} = J^1\underline{e}_1 + J^2\underline{e}_2$. One can show that the difference between A_α and

$$A_0 = \begin{pmatrix} V_0|_1 & 0 & \operatorname{grad}_1 V_0 \\ 0 & V_0|_2 & \operatorname{grad}_2 V_0 \\ 0 & 0 & -\Delta V_0 \end{pmatrix} \tag{4.93}$$

is compact. The principle symbol $\sigma(A_\alpha)(\xi)$ is obtained by the two-dimensional Fourier transformation $\tilde{F} : (x_1, x_2) \to (\xi_1, \xi_2)$ of A_0. In [287] it is shown that

$$\hat{\phi}_\alpha(\xi) = (\tilde{F}\phi_\alpha)(\xi) = (|\xi|^2 - \alpha^2)^{-\frac{1}{2}}. \tag{4.94}$$

Therefore the principle symbol can be displayed as

$$\sigma(A_\alpha)(\xi) = \begin{pmatrix} \frac{1}{|\xi|} & 0 & i\xi_1\frac{1}{|\xi|} \\ 0 & \frac{1}{|\xi|} & i\xi_2\frac{1}{|\xi|} \\ 0 & 0 & |\xi| \end{pmatrix} \tag{4.95}$$

with $|\xi|^2 = \xi_1^2 + \xi_2^2$. Finally it can be proven that there exist constants $\gamma' > 0$ and $\kappa > 4$ such that

$$\Re(\zeta_1, \zeta_2, \zeta_3) \begin{pmatrix} 1 & 0 & 0 \\ 0 & 1 & 0 \\ 0 & 0 & \kappa \end{pmatrix} \sigma(A_\alpha)(\xi) \begin{pmatrix} \bar\zeta_1 \\ \bar\zeta_2 \\ \bar\zeta_3 \end{pmatrix} \geq \gamma'(\zeta_1\bar\zeta_1 + \zeta_2\bar\zeta_2 + \zeta_3\bar\zeta_3) \tag{4.96}$$

for all $\zeta \in \mathbb{C}^3$ and all $\xi \in \mathbb{R}^3$ with $|\xi| = 1$. From that it follows that A_α is strongly elliptic (see Definition B.7 and [259]) and hence satisfies a Gårding inequality. The Galerkin procedure for the modified system (4.87) is analyzed in [286]. For a different approach see [41, 42].

In [135] a boundary integral equation method for transmission problems for strongly elliptic differential operators is analysed, which yields a strongly elliptic system of pseudodifferential operators and which therefore can be used for numerical computations with Galerkin's procedure. The method is shown to work for the vector Helmholtz equation with electromagnetic transmission conditions. The system of boundary values is slightly modified so that the corresponding bilinear form becomes coercive over H^1. The concept of the principal symbol of a system of pseudodifferential operators is used to derive existence and regularity results for the solution.

Chapter 5
The Signorini Problem and More Nonsmooth BVPs and Their Boundary Integral Formulation

In this chapter we deal with unilateral and nonsmooth boundary value problems, in particular Signorini problems without and with Tresca friction and nonmontone contact problems from adhesion/delamination in the range of linear elasticity. We show how the boundary integral techniques developed in the previous chapters can be used to transform those problems to boundary variational inequalities. This opens the way to the numerical treatment of these nonlinear problems by the BEM as detailed in Chap. 11.

5.1 The Signorini Problem in Its Simplest Form

In this section we follow [214] and introduce the Signorini boundary value problem in its simplest form taking the Laplace equation as elliptic equation. The Signorini problem is a unilateral boundary value problem, where the unilateral constraint lives on the boundary. Since the domain is governed by a linear pde with constant coefficients, a fundamental solution is available and integral equation methods apply. Here modifying the approach of H. Han [227] we derive an equivalent boundary variational inequality in the Cauchy data as unknows, where the associated bilinear form is shown to satisfy a Gårding inequality in appropriate Sobolev spaces on the boundary. Finally we turn to the convex cone of feasible solutions and provide a density result that is useful for the convergence analysis of the boundary element method to follow in Sect. 11.1.

Let $\Omega \subset \mathbb{R}^2$ be a bounded plane domain with the Lipschitz boundary Γ [327]. Then n, the outward normal to Γ, exists almost everywhere and $n \in [L^\infty(\Gamma)]^2$ (see [327, Lemma 2.4.2]). Here we consider the simple elliptic equation

$$-\Delta u = 0 \quad \text{in } \Omega . \tag{5.1}$$

© Springer International Publishing AG, part of Springer Nature 2018

J. Gwinner, E. P. Stephan, *Advanced Boundary Element Methods*,
Springer Series in Computational Mathematics 52,
https://doi.org/10.1007/978-3-319-92001-6_5

Thus we have the Cauchy data u and $\varphi := \frac{\partial u}{\partial n}$ on Γ.

To formulate the boundary conditions, let $\Gamma = \overline{\Gamma}_D \cup \overline{\Gamma}_N \cup \overline{\Gamma}_S$, where the open parts Γ_D, Γ_N, and Γ_S are mutually disjoint. We prescribe

$$u = 0 \quad \text{on } \Gamma_D\,, \tag{5.2}$$

$$\varphi = g \quad \text{on } \Gamma_N\,. \tag{5.3}$$

On the remaining part Γ_S, Signorini boundary conditions are imposed, i.e.

$$u \leq 0\,, \quad \varphi \leq h\,, \quad u(\varphi - h) = 0\,, \tag{5.4}$$

where $g \in H^{-1/2}(\Gamma_N)$ and $h \in H^{-1/2}(\Gamma_S)$ are given. We point out that a priori it is not known where $u = 0$ changes to $\varphi = h$ and the boundary part Γ_S is only taken large enough to contain this free boundary. Thus to make this free bounday problem meaningful we assume meas $(\Gamma_S) > 0$, but we do not require meas$(\Gamma_D) > 0$. Note there is no loss of generality to assume homogeneous conditions above. Indeed, more general conditions can be reduced to the form given above by a superposition argument that uses the solution of the linear boundary value problem

$$-\Delta u = f \quad \text{in } \Omega$$

$$u = u_D^0 \quad \text{on } \Gamma_D\,, \quad \varphi = 0 \quad \text{on } \Gamma_N\,, \quad u = u_S^0 \quad \text{on } \Gamma_S$$

and an appropriately redefined right hand side h in (5.4). To give the variational formulation of the boundary value problem (5.1)–(5.4) we introduce the bilinear form

$$\beta(v, w) := \int_\Omega \text{grad}\, v \cdot \text{grad}\, w\, dx = \sum_{k=1}^{2} \int_\Omega \frac{\partial v}{\partial x_k} \frac{\partial w}{\partial x_k}\, dx$$

and the linear form

$$\ell(v) := \int_{\Gamma_N} g\, v\, ds + \int_{\Gamma_S} h\, v\, ds$$

on the function space

$$H^1_{\Gamma_D,0}(\Omega) := \{v \in H^1(\Omega) : v = 0 \quad \text{on } \Gamma_D\} \tag{5.5}$$

and the convex cone

$$\mathscr{K} := \{v \in H^1_{\Gamma_D,0}(\Omega) : v \le 0 \quad \text{on } \Gamma_S\} \,. \tag{5.6}$$

Then the variational formulation of (5.1)–(5.4) in the domain Ω is easily obtained by Green's formula (see e.g. [266, 267] for more details) as the following variational inequality:

(P) Find $u \in \mathscr{K}$ such that

$$\beta(u, v - u) \ge \ell(v - u) \quad \forall v \in \mathscr{K} \,.$$

To derive a boundary integral formulation let a fundamental solution of (5.1) by given by

$$\mathscr{F}(x, y) := \frac{1}{2\pi} \ln |x - y| \,.$$

Now let $u \in \mathscr{K}$ be a solution of (P), hence $-\Delta u = 0$. According to the representation formula (Sect. 2.1, Theorem 2.1) we have

$$u(x) = \int_\Gamma \frac{\partial \mathscr{F}(x, y)}{\partial n_y} u(y)\, ds_y - \int_\Gamma \mathscr{F}(x, y)\, \varphi(y)\, ds_y \quad \forall x \in \Omega \,. \tag{5.7}$$

By the jump relations, respectively continuity properties of the simple layer potential, respectively of the double layer potential (see Sects. 2.2.2, 2.4.1, 4.1), (5.7) implies

$$\frac{1}{2} u(x) = \int_\Gamma \frac{\partial \mathscr{F}(x, y)}{\partial n_y} u(y)\, ds_y - \int_\Gamma \mathscr{F}(x, y)\, \varphi(y)\, ds_y \quad \forall x \in \Gamma \,, \tag{5.8}$$

$$\frac{1}{2} \varphi(x) = \int_\Gamma \frac{\partial^2 \mathscr{F}(x, y)}{\partial n_x \partial n_y} u(y)\, ds_y - \int_\Gamma \frac{\partial \mathscr{F}(x, y)}{\partial n_x} \varphi(y)\, ds_y \quad \forall x \in \Gamma \,. \tag{5.9}$$

Here, the first integral in (5.9) is a hypersingular integral (partie finie following Hadamard); by partial integration twice using the Cauchy–Riemann equations (see Sect. 2.6 , Proposition 2.1) one obtains

$$\int_\Gamma \frac{\partial^2 \mathscr{F}(x, y)}{\partial n_x\, \partial n_y} u(y)\, ds_y = \frac{d}{ds_x} \int_\Gamma \mathscr{F}(x, y) \frac{du(y)}{ds_y}\, ds_y \,.$$

Testing (5.8) by $\psi \in H^{-1/2}(\Gamma)$ leads to

$$-\frac{1}{2}\int_{\Gamma} u(x)\psi(x)\,ds_x + \int_{\Gamma}\int_{\Gamma} \frac{\partial \mathscr{F}(x,y)}{\partial n_y}\,u(y)\,\psi(x)\,ds_y\,ds_x$$

$$-\int_{\Gamma}\int_{\Gamma} \mathscr{F}(x,y)\,\varphi(y)\,\psi(x)\,ds_y\,ds_x = 0\,,$$

or shortly,

$$-\,b(\psi,u) + a(\psi,\varphi) = 0\,,\quad \forall \psi \in H^{(-1/2)}(\Gamma) \tag{5.10}$$

where

$$a(\psi,\varphi) := -\int_{\Gamma}\int_{\Gamma} \mathscr{F}(x,y)\,\varphi(y)\,\psi(x)\,ds_y\,ds_x$$

$$b(\psi,u) := \frac{1}{2}\int_{\Gamma} u(x)\,\psi(x)\,ds_x - \int_{\Gamma}\int_{\Gamma} \frac{\partial \mathscr{F}(x,y)}{\partial n_y}\,u(y)\,\psi(x)\,ds_y\,ds_x.$$

On the other hand, for any $v \in H^1_{\Gamma_D,0}(\Omega)$, by Green's formula,

$$\beta(u,v) = \int_{\Gamma} \nabla u \cdot \nabla u\,dv = \int_{\Gamma} \varphi v\,ds\,. \tag{5.11}$$

Testing (5.9) by v and plugging in (5.11) yields

$$\begin{aligned}\beta(u,v) \;&=\; -\int_{\Gamma}\int_{\Gamma} \mathscr{F}(x,y)\,\frac{du(y)}{ds_y}\,\frac{dv(x)}{ds_x}\,ds_y ds_x \\ &\quad - \int_{\Gamma}\int_{\Gamma} \frac{\partial \mathscr{F}(x,y)}{\partial n_x}\,\varphi(y)\,v(x)\,ds_y ds_x + \frac{1}{2}\int_{\Gamma} \varphi \cdot v\,ds \\ &=: a\left(\frac{du}{ds}\,,\frac{dv}{ds}\right) + b(\varphi,v)\,. \end{aligned} \tag{5.12}$$

By (5.10) and (5.12), introducing the convex cone

$$K := \{v \in H^{1/2}(\Gamma):\;\; v = 0 \text{ on } \Gamma_D\,, v \le 0 \text{ on } \Gamma_S\}$$

we arrive at the following variational problem: Find $[u, \varphi] \in K \times H^{-1/2}(\Gamma)$ such that

$$(\pi) \begin{cases} a\left(\frac{du}{ds}, \frac{dv}{ds} - \frac{du}{ds}\right) + b(\varphi, v-u) \geq l(v-u) & \forall v \in K\,, \\ -b(\psi, u) \qquad\qquad + a(\psi, \varphi) & = 0 \quad \forall \psi \in H^{-1/2}(\Gamma)\,. \end{cases}$$

This problem (π) is equivalent to the former variational problem (P), since conversely, for any solution $[u, \varphi]$ to (π), we can define u in Ω by means of (5.7), and for any $v \in H^1_{\Gamma_D,0}(\Omega)$ we can consider its trace $v|\Gamma$ to obtain $\beta(u, v-u) \geq \ell(v-u)$. Note that (π) is equivalent to the single boundary variational inequality: Find $[u, \varphi] \in K \times H^{-1/2}(\Gamma)$ such that for all $[v, \psi] \in K \times H^{-1/2}(\Gamma)$,

$$A([u, \varphi]\,, [v, \psi] - [u, \varphi]) \geq \ell(v-u)\,, \tag{5.13}$$

where the bilinear form A is given by

$$A([u, \varphi], [v, \psi]) := a\Big(\frac{du}{ds}\,, \frac{dv}{ds}\Big) + a(\psi, \varphi) + b(\varphi, v) - b(\psi, u)\,.$$

Indeed, since the variational equality in (π) is equivalent to the variational inequality

$$a(\psi - \varphi, \varphi) - b(\psi - \varphi, u) \geq 0$$

on the space $H^{-1/2}(\Gamma)$, the implication $(\pi) \Rightarrow$ (5.13) is immediate. On the other hand, (π) follows from (5.13) by the choices $\psi = 0$, $v = u$.

Remark A is not symmetric (although a, β are symmetric), hence the problem (π) is not equivalent to a minimization problem on K. A is positive semidefinite; indeed

$$A([u, \varphi], [u, \varphi]) = a\left(\frac{du}{ds}, \frac{du}{ds}\right) + a(\varphi, \varphi) \geq 0\,.$$

Now our aim is to establish a Gårding inequality for the bilinear form $A(\cdot, \cdot)$ in the space $H^{1/2}(\Gamma) \times H^{-1/2}(\Gamma)$, i.e. positive definiteness up to a compact perturbation term. The boundary integral operators that give rise to the bilinear form $A(\cdot, \cdot)$ can be understood as pseudodifferential operators. Since coordinate transformations do not affect their principal symbol, thus contribute only to compact perturbation terms (see e.g. [256] for more detailed arguments of this kind) we need only consider the case of a smooth domain in the subsequent reasoning.

Lemma 5.1 *There exist a constant $c_0 > 0$ and a compact operator $C_0 : H^{1/2}(\Gamma) \to H^{-1/2}(\Gamma)$ such that*

$$\left\|\frac{dv}{ds}\right\|^2_{-1/2,\Gamma} \geq c_0 \|v\|^2_{1/2,\Gamma} - \langle C_0 v, v\rangle_{H^{-1/2}\times H^{1/2}}\,, \quad \forall v \in H^{1/2}(\Gamma)\,. \tag{5.14}$$

Proof Let $\theta = 2\pi s/L$, where L is the boundary length, and we can assume without loss of generality that Γ is the unit circle. Then we can argue similar to [227] with the only difference that due to the nontrivial kernel of β an extra term enters. More detailed using the Fourier expansion for a smooth function v – what by density suffices to consider –

$$v = \frac{a_0}{2} + \sum_{n=1}^{\infty}(a_n \cos n\theta + b_n \sin n\theta)\,,$$

$$\frac{dv}{d\theta} = \sum_{n=1}^{\infty}(nb_n \cos n\theta - na_n \sin n\theta)\,,$$

one finds

$$\|v\|^2_{1/2,\Gamma} = \frac{a_0^2}{2} + \sum_{n=1}^{\infty}(1+n^2)^{1/2}\,(a_n^2 + b_n^2)\,,$$

$$\left\|\frac{dv}{d\theta}\right\|^2_{-1/2,\Gamma} = \sum_{n=1}^{\infty}(1+n^2)^{-1/2}\,n^2(a_n^2 + b_n^2)$$

$$\geq \frac{1}{2}\sum_{n=1}^{\infty}(1+n^2)^{1/2}\,(a_n^2 + b_n^2)\,,$$

$$a_0^2 = \left[\frac{1}{2\pi}\int_0^{2\pi} v(\theta)d\theta\right]^2 \leq c\,\|v\|^2_{0,\Gamma} \quad (c > 0)\,.$$

Hence

$$\left\|\frac{dv}{d\theta}\right\|^2_{-1/2,\Gamma} \geq \frac{1}{2}\,\|v\|^2_{1/2,\Gamma} - \frac{c}{4}\,\|v\|^2_{0,\Gamma}\,. \tag{5.15}$$

Since

$$H^{1/2}(\Gamma) \subset H^0(\Gamma) \equiv L^2(\Gamma) \subset H^{-1/2}(\Gamma)$$

forms a Gelfand triple with *compact* and dense embeddings, the last term in (5.15) can be replaced by $\langle C_0 v, v\rangle$ with $C_0 : H^{1/2}(\Gamma) \to H^{-1/2}(\Gamma)$ compact concluding the proof. □

Lemma 5.2 *The bilinear form $A(\cdot,\cdot)$ is bounded in $[H^{1/2}(\Gamma)\times H^{-1/2}(\Gamma)]^2$; moreover satisfies a Gårding inequality, i.e. there exist a positive constant c and a compact operator $C : H^{1/2}(\Gamma)\times H^{-1/2}(\Gamma)\to H^{-1/2}(\Gamma)\times H^{1/2}(\Gamma)$ such that*

$$\begin{aligned}&A([v,\psi],[v,\psi]) + \langle C[v,\psi],[v,\psi]\rangle_{[H^{-1/2}(\Gamma)\times H^{1/2}(\Gamma)]\times[H^{1/2}(\Gamma)\times H^{-1/2}(\Gamma)]}\\ &\geq c\|[v,\psi]\|^2_{H^{1/2}(\Gamma)\times H^{-1/2}(\Gamma)} := c\{\|v\|^2_{H^{1/2}(\Gamma)} + \|\psi\|^2_{H^{-1/2}(\Gamma)}\}\\ &\forall [v,\psi]\in H^{1/2}(\Gamma)\times H^{-1/2}(\Gamma)\,. \end{aligned}\tag{5.16}$$

Proof We have

$$A([v,\psi],[v,\psi]) = a\Big(\frac{dv}{ds},\frac{dv}{ds}\Big) + a(\psi,\psi)\,.$$

By Theorem 2.5, [114, Theorem 1]

$$|a(\psi,\psi)| \leq \text{const}\,\|\psi\|^2_{H^{-1/2}(\Gamma)}\,.$$

Since for any $v\in H^{1/2}(\Gamma)$, $\frac{dv}{ds} = \sum_i \frac{\partial v}{\partial x_i}\,\dot{x}_i \in H^{-1/2}(\Gamma)$, it follows

$$\Big\|a\Big(\frac{dv}{ds},\frac{dv}{ds}\Big)\Big\| \leq \text{const}\,\|v\|^2_{H^{1/2}(\Gamma)}\,.$$

Therefore it remains to prove (5.16). By Theorem 3.6, [114, Theorem 2] the bilinear form $a(\cdot,\cdot)$ satisfies a Gårding inequality on $[H^{-1/2}(\Gamma)]^2$ in the general case of a Lipschitz domain, i.e.

$$a(\psi,\psi) \geq c_a\|\psi\|^2_{H^{-1/2}(\Gamma)} - \langle C_A\psi,\psi\rangle_{H^{1/2}(\Gamma)\times H^{-1/2}(\Gamma)} \quad \forall\psi\in H^{-1/2}(\Gamma)\,,\tag{5.17}$$

where $c_a > 0$, $C_A : H^{-1/2}(\Gamma)\to H^{1/2}(\Gamma)$ is compact. Hence

$$a\Big(\frac{dv}{ds},\frac{dv}{ds}\Big) \geq c_a\Big\|\frac{dv}{ds}\Big\|^2_{H^{-1/2}(\Gamma)} - \langle C_A\frac{dv}{ds},\frac{dv}{ds}\rangle \quad \forall v\in H^{1/2}(\Gamma)\,.\tag{5.18}$$

Combining (5.17) and (5.18) with Lemma 5.1 yields (5.16). □

Finally we are concerned with the density relation

$$\overline{K\cap C^\infty(\Gamma)} = K\,,\tag{5.19}$$

which is essential for our convergence analysis to come. Since the embedding $H^{1/2}(\Gamma)\subset L^1(\Gamma)$ is continuous and L^1-convergence implies pointwise convergence almost everywhere for a subsequence, K is closed. Therefore it remains to show

$$K \subset \overline{K \cap C^\infty(\Gamma)}\,.$$

To this end one uses the continuity and surjectivity of the trace operator γ : $H^1(\Omega) \to H^{1/2}(\Gamma)$ and applies the analogous inclusion

$$\mathscr{K} \subset \overline{\mathscr{K} \cap C^\infty(\Gamma)}\,,$$

which in [215, section 4] is proved using Friedrich's regularization and the fact that with Ω a Lipschitz domain, $H = H^1(\Omega)$ is a Dirichlet space and hence in particular the map $w \in H \mapsto w^+ = \max(0, w)$ is a continuous map into H.

To conclude this section we refer the interested reader to [222] to see how the boundary integral approach described above extends to unilateral contact of a linear elastic body against a rigid foundation in the range of linear elasticity.

5.2 A Variational Inequality of the Second Kind Modelling Unilateral Frictional Contact

Let $\Omega \subset \mathbb{R}^d$ $(d \geq 2)$ be a bounded Lipschitz domain with its boundary $\partial\Omega$ and mutually disjoint parts Γ_D, Γ_N, and Γ_C such that $\partial\Omega = \overline{\Gamma}_D \cup \overline{\Gamma}_N \cup \overline{\Gamma}_C$ and meas $(\Gamma_C) > 0$. Let the data $f \in H^{-1/2}(\Gamma_N \cup \Gamma_C)$, $\gamma \in H^{1/2}(\Gamma_D) \cap C^0[\overline{\Gamma_D}]$, $\chi \in H^{1/2}(\Gamma_C) \cap C^0[\overline{\Gamma}_C]$, $g \in L^\infty(\Gamma_C)$ be given, where $g \geq 0$ and $\gamma|\overline{\Gamma}_D \cap \overline{\Gamma}_C \leq \chi|\overline{\Gamma}_D \cap \overline{\Gamma}_C$. Introduce

$$a(\hat{u}, \hat{v}) := \int_\Omega \nabla\hat{u} \cdot \nabla\hat{v}\, dx\,,$$

the bilinear form associated to the Laplacian, the convex closed set

$$\hat{K} := \left\{ \hat{v} \in H^1(\Omega) : \hat{v}|\Gamma_D = \gamma \text{ a.e. and } \hat{v}|\Gamma_C \leq \chi \text{ a.e.} \right\},$$

the linear form

$$l(\hat{v}) := \int_{\Gamma_N \cup \Gamma_C} f\hat{v}\, ds\,,$$

and the continuous, positively homogeneous and sublinear, hence convex functional

$$j(\hat{v}) := \int_{\Gamma_C} g|\hat{v}|\, ds$$

that describes Tresca friction. Then consider the variational inequality problem (π) of the second kind: Find $\hat{u} \in \hat{K}$ such that for all $\hat{v} \in \hat{K}$,

$$a(\hat{u}, \hat{v} - \hat{u}) + j(\hat{v}) - j(\hat{u}) \geq l(\hat{v} - \hat{u}). \tag{5.20}$$

There exists a unique solution $\hat{u}$ (see e.g [145, 146, 249]), if Γ_D has positive measure and hence the bilinear form is coercive by the Poincaré inequality. In the semicoercive case, when $\Gamma_D = \emptyset$, a necessary condition for existence of a solution is the recession condition

$$j(\rho) \geq l(\rho),\ \forall \rho = \text{const.} \leq 0\,,$$

what is equivalent to

$$\int_{\Gamma_C} g\, ds + \int_{\Gamma_N \cup \Gamma_C} f\, ds \geq 0\,.$$

If this condition is strengthened to

$$\int_{\Gamma_C} g\, ds + \int_{\Gamma_N \cup \Gamma_C} f\, ds > 0\,,$$

then existence of a solution is guaranteed (see Appendix C.3.2, [146, 201]).

With $g \equiv 0$ the variational problem (5.20) specializes to the domain variational inequality of the first kind that is studied in [297]. As is well-known, this latter variational inequality is the variational formulation of the mixed unilateral Dirichlet-Neumann-Signorini boundary value problem for the Laplacian. On the other hand, with $\chi \equiv +\infty$ formally, the unilateral constraint disappears and we arrive at the variational inequality of the second kind in [221] and in a similar form (with the Laplacian replaced by the Helmholtz operator) in [205]. To exhibit the relation of (5.20) to unilateral contact with friction in linear elasticity we insert the following remark.

Remark 5.1 In linear elasticity, instead of the unknown scalar field $\hat{u}$, there is the displacement field u which decomposes in its normal component $u_n = u \cdot n$ and its tangential component $u_t = u - u_n n$. Similarly as dual variable, the flux $\frac{\partial u}{\partial n}$ is to be replaced by the stress field T with its normal component T_n and its tangential component T_t. Then unilateral contact with a rigid foundation together with friction according to Coulomb's friction law requires the following conditions (see [249, 266]) on the contact surface Γ_C:

$$u_n \leq \chi, T_n \leq 0, (u_n - \chi)T_n = 0$$

and

$$|T_t| \leq \mathscr{F}|T_n|, \Big(\mathscr{F}|T_n| - |T_t|\Big)u_t = 0, u_t \cdot T_t \leq 0,$$

where $\mathscr{F} \geq 0$ is the friction coefficient. The latter condition expresses the obvious law that the modulus of the tangential component is limited by a multiple of the modulus of the normal component; if it is attained, then the body can slip off in the direction opposite to T_t; otherwise, the body sticks.

The fixed point approach to unilateral frictional contact as employed in the existence proofs [261, 319] leads to a approximating sequence of unilateral problems with given friction. In these approximations the unknown normal component is replaced by a given slip stress $g_n \geq 0$, such that the latter condition above reduces to

$$|T_t| \leq \mathscr{F} g_n, \Big(\mathscr{F} g_n - |T_t|\Big)u_t = 0, u_t \cdot T_t \leq 0\,.$$

The weak formulation of the unilateral contact problem with given friction (also known as unilateral Tresca friction problem) is the following variational inequality (see [249, section 7] for the proof of the formal equivalence of the classical and weak formulation): Find $u \in K$ such that for all $v \in K$

$$a(u, v - u) + \int\limits_{\Gamma_C} \mathscr{F} g_n\Big(|v_t| - |u_t|\Big)\, ds \geq \int\limits_{\Gamma_N} f \cdot (v - u)\, ds\,,$$

where f is the surface force, $a(\cdot, \cdot)$ is the bilinear form of strain energy in linear elasticity, and K is the appropriately defined convex set. In this sense, (π), (5.20) gives a simplified (scalar) model of the unilateral contact problem with given friction.

Here we use potential theory and reduce our variational problem (5.20) on the domain to the boundary $\Gamma = \partial\Omega$. We shall obtain two different, but equivalent boundary variational inequalities of the second kind: a mixed variational inequality in the Cauchy data $\left(u|_\Gamma, \dfrac{\partial u}{\partial n}\right)$ as unknowns and a primal variational inequality in the unknown $u|_\Gamma$ involving the Poincaré–Steklov operator (the Dirichlet-to-Neumann map).

To this end we list the relevant boundary integral operators and recall their mapping properties. With the fundamental solution for the Laplacian,

$$G(x,y) = -\frac{1}{2\pi} \ln|x-y| \ \text{ if } d=2,$$

$$G(x,y) = \frac{1}{4\pi}\frac{1}{|x-y|} \quad \text{ if } d=3,$$

the operators of the single layer potential V, the double layer potential K, its formal adjoint K', and the hypersingular integral operator W can be defined for $z \in \Gamma, \phi \in C^\infty(\Gamma)$ as follows:

$$V\phi(z) := 2\int_\Gamma G(z,x)\phi(x)\,ds_x,\ K\phi(z) := 2\int_\Gamma \frac{\partial}{\partial n_x} G(z,x)\phi(x)\,ds_x,$$

$$K'\phi(z) := 2\int_\Gamma \frac{\partial}{\partial n_z} G(z,x)\phi(x)\,ds_x,\ W\phi(z) := -\frac{\partial}{\partial n_z} K\phi(z).$$

From Sects. 2.3, 2.4 we know that the linear operators

$$V : H^{-1/2+\sigma}(\Gamma) \to H^{1/2+\sigma}(\Gamma),\ K : H^{1/2+\sigma}(\Gamma) \to H^{1/2+\sigma}(\Gamma)$$

$$K' : H^{-1/2+\sigma}(\Gamma) \to H^{-1/2+\sigma}(\Gamma),\ W : H^{1/2+\sigma}(\Gamma) \to H^{-1/2+\sigma}(\Gamma)$$

are well-defined and continuous for $|\sigma| < \frac{1}{2}$.

Similarly as with Han [227] for the Signorini problem with the Helmholtz operator in 3D and with Gwinner and Stephan [222] for the unilateral contact problem in 2D elasticity, we obtain as an equivalent reformulation of (π,(5.20)) the following boundary variational equality: Find $(u,\varphi) \in K^\Gamma \times H^{-1/2}(\Gamma)$ such that for all $(v,\psi) \in K^\Gamma \times H^{-1/2}(\Gamma)$

$$\frac{1}{2} B(u,\varphi; v-u, \psi) + j(v) - j(u) \geq l(v-u), \tag{5.21}$$

where

$$K^\Gamma := \left\{ v \in H^{1/2}(\Gamma) : v|\Gamma_D = \gamma|\Gamma_D, v|\Gamma_C \leq \chi \right\}$$

and the bilinear form B is given by

$$B(u,\varphi; v,\psi) := \langle Wu, v\rangle - \langle (I+K)\,u, \psi\rangle + \langle V\varphi, \psi\rangle + \langle (I+K)'\,\varphi, v\rangle\,.$$

Note that B is positive semidefinite on $H^{1/2}(\Gamma)/\mathbb{R}\times H^{-1/2}(\Gamma)$, but non-symmetric. Indeed, this mixed variational inequality characterizes a saddle point in $K^{\Gamma} \times H^{-1/2}(\Gamma)$ with a Lagrangian function appropriately defined.

Note that using the duality $\big(L^1(\Gamma), L^\infty(\Gamma)\big)$ there holds (see [151, chapter 4.3])

$$j(v) = \int_{\Gamma_C} g|v|\,ds = \max\left\{ \int_{\Gamma_C} gvw\,ds \mid w \in L^\infty(\Gamma),\ |w| \leq 1 \text{ a.e.} \right\}.$$

This leads to another saddle point characterization with a Lagrange multiplier $w \in L^\infty(\Gamma)$, $|w| \leq 1$ and a suitable Lagrangian. This latter duality relation will be a key argument in the convergence analysis to come.

Here we eliminate φ in (5.21) (as in [85] for unilateral problems, see [136, 257] for earlier application of the Schur complement and its boundary integral operator representation) and obtain as another equivalent reformulation of (π,(5.20)) and (π,(5.21)) the primal boundary variational inequality: Find $u \in K^{\Gamma}$ such that for all $v \in K^{\Gamma}$

$$\langle Su, v - u\rangle + j(v) - j(u) \geq l(v - u) \tag{5.22}$$

with the symmetric Poincaré–Steklov operator S for the interior problem,

$$S := \frac{1}{2}\Big[W + \big(K' + I\big)\, V^{-1}\,(K + I)\Big] : H^{1/2}(\Gamma) \to H^{-1/2}(\Gamma)$$

which is positive definite on $H^{1/2}(\Gamma)/\mathbb{R}$.

For further related boundary variational inequalities that arise from unilateral contact without friction and with Tresca friction for hemitropic solids in micropolar elasticity see [182–184].

5.3 A Nonmonotone Contact Problem from Delamination

In this section we describe a nonmonotone contact problem that models the delamination behaviour in bonded lightweight structures. We treat such nonlinear boundary value problems by a combination of boundary integral methods and regularization techniques from nondifferentiable optimization based on the investigations in [332, 335].

Let $\Omega \subset \mathbb{R}^d$ ($d = 2, 3$) be a bounded domain with Lipschitz boundary $\partial\Omega$. We assume that the boundary is decomposed into three disjoint parts Γ_D, Γ_N, and Γ_C such that $\partial\Omega = \overline{\Gamma}_D \cup \overline{\Gamma}_N \cup \overline{\Gamma}_C$ and, moreover, the measures of Γ_C and Γ_D are positive. Zero displacements are prescribed on Γ_D, surface tractions $\mathbf{t} \in (L^2(\Gamma_N))^d$

act on Γ_N, and on the part Γ_C a nonmonotone, generally multivalued boundary condition holds. The elastic body $\overline{\Omega}$ is subject to a volume force $\mathbf{f} \in [L^2(\Omega)]^d$ and $g \in H^{1/2}(\Gamma_C)$, $g \geq 0$, is a gap function associating every point $x \in \Gamma_C$ with its distance to the rigid obstacle measured in the direction of the unit outer normal vector $\mathbf{n}(x)$.

Further, $\varepsilon(\mathbf{u}) = \frac{1}{2}(\nabla\mathbf{u} + \nabla\mathbf{u}^T)$ denotes the linearized strain tensor and $\sigma(\mathbf{u}) = \mathscr{C} : \varepsilon(\mathbf{u})$ stands for the stress tensor, where $\mathscr{C}$ is the Hooke tensor, assumed to be uniformly positive definite with L^∞ coefficients. The stress vector on the surface can be decomposed further into the normal, respectively, the tangential stress:

$$\sigma_n = \sigma(\mathbf{u})\mathbf{n} \cdot \mathbf{n}, \qquad \sigma_t = \sigma(\mathbf{u})\mathbf{n} - \sigma_n\mathbf{n}.$$

Our benchmark problem is a two- or three-dimensional symmetric laminated structure with an interlayer adhesive under loading (see Fig. 5.1 below for the 2D benchmark problem). Because of the symmetry of the structure and the same forces applied to the upper and lower part, it suffices to consider only the upper half of the specimen, represented by $\Omega \subset \mathbb{R}^d$, $d = 2, 3$.

Problem (P) Find a displacement $\mathbf{u} \in \mathbf{H}^1(\Omega) := [H^1(\Omega)]^d$ such that

$$-\text{div}\,\sigma(\mathbf{u}) = \mathbf{f} \text{ in } \Omega \tag{5.23}$$

$$\mathbf{u} = 0 \text{ on } \Gamma_D;\ \sigma(\mathbf{u})\mathbf{n} = \mathbf{t} \text{ on } \Gamma_N$$

$$u_n \leq g, \sigma_t(\mathbf{u}) = 0 \text{ on } \Gamma_c,$$

$$-\sigma_n(\mathbf{u}) \in \partial f(u_n) \text{ on } \Gamma_c \tag{5.24}$$

The contact law (5.24), written as a differential inclusion by means of the Clarke subdifferential ∂f (see Appendix C.2) of a locally Lipschitz function f, describes the nonmonotone, multivalued behaviour of the adhesive. More precisely, ∂f is the physical law between the normal component σ_n of the stress boundary vector and the normal component $u_n = \mathbf{u} \cdot \mathbf{n}$ of the displacement $\mathbf{u}$ on Γ_C. A typical zig-zagged nonmonotone adhesion law is shown in Fig. 5.2 below.

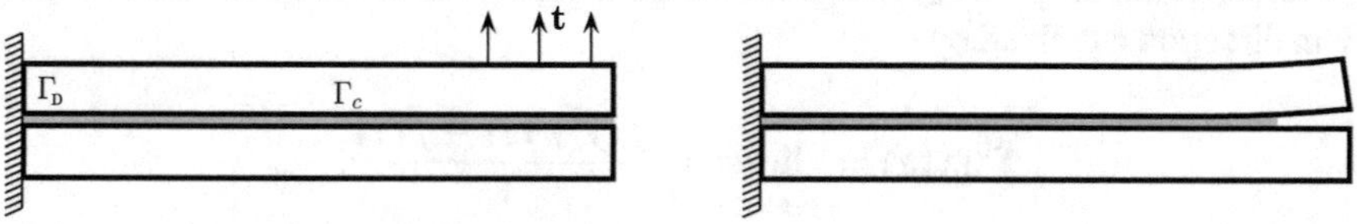

Fig. 5.1 Reference configuration for the 2D benchmark under loading [333]

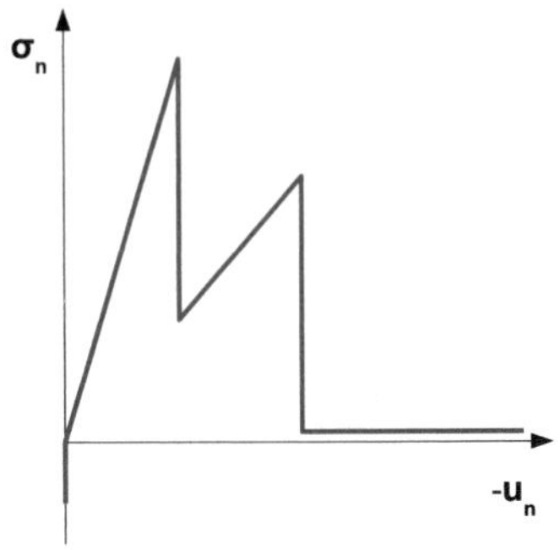

Fig. 5.2 A nonmonotone adhesion law [333]

To give a variational formulation of the above boundary value problem we define

$$\mathbf{H}^1_{\Gamma_D,0} = \{\mathbf{v} \in \mathbf{H}^1(\Omega) \ : \ \mathbf{v}|_{\Gamma_D} = 0\},$$

$$\mathscr{K} = \{\mathbf{v} \in \mathbf{H}^1_{\Gamma_D,0} \ : v_n \le g \text{ a. e. on } \Gamma_C\}$$

and introduce the $\mathbf{H}^1_{\Gamma_D,0}(\Omega)$-coercive and continuous bilinear form of linear elasticity

$$a(\mathbf{u}, \mathbf{v}) = \int_\Omega \sigma(\mathbf{u}) : \varepsilon(\mathbf{v})\, dx.$$

Multiplying the equilibrium equation (5.23) in Problem (P) by $\mathbf{v} - \mathbf{u}$, integrating over Ω and applying the divergence theorem yields

$$\int_\Omega \sigma(\mathbf{u}) : \varepsilon(\mathbf{v} - \mathbf{u})\, dx = \int_\Omega \mathbf{f} \cdot (\mathbf{v} - \mathbf{u})\, dx + \int_\Gamma \sigma(\mathbf{u})\mathbf{n} \cdot (\mathbf{v} - \mathbf{u})\, ds.$$

From the definition of the Clarke subdifferential (see Appendix C.2), the nonmonotone boundary condition (5.24) is equivalent to

$$-\sigma_n(u_n)(v_n - u_n) \le f^0(u_n; v_n - u_n).$$

Here, the notation $f^0(x; z)$ stands for the generalized directional derivative of f at x in direction z defined by

$$f^0(x; z) = \limsup_{y \to x, t \to 0^+} \frac{f(y + tz) - f(y)}{t}.$$

Substituting $\sigma(\mathbf{u})\mathbf{n}$ by $\mathbf{t}$ on Γ_N, using on Γ_C the decomposition

$$\sigma(\mathbf{u})\mathbf{n} \cdot (\mathbf{v} - \mathbf{u}) = \sigma_t(\mathbf{u}) \cdot (\mathbf{v_t} - \mathbf{u_t}) + \sigma_n(\mathbf{u})(v_n - u_n)$$

and taking into account that on Γ_C no tangential stresses are assumed, we obtain the hemivariational inequality: Find $\mathbf{u} \in \mathscr{K}$ such that

$$a(\mathbf{u}, \mathbf{v}-\mathbf{u}) + \int_{\Gamma_C} f^0(u_n(s); v_n(s) - u_n(s))\, ds \geq \int_{\Omega} \mathbf{f} \cdot (\mathbf{v}-\mathbf{u})\, dx$$
$$+ \int_{\Gamma_N} \mathbf{t} \cdot (\mathbf{v}-\mathbf{u})\, ds \quad \forall \mathbf{v} \in \mathscr{K}. \tag{5.25}$$

Using the Poincaré-Steklov operator S we can give a boundary integral formulation and rewrite (5.25) as a hemivariational inequality defined only on the boundary. To this end, we introduce the free boundary part $\Gamma_0 = \Gamma \backslash \overline{\Gamma}_D = \Gamma_N \cup \Gamma_C$ and recall the Sobolev spaces [259]:

$$\begin{aligned} H^{1/2}(\Gamma) &= \{v \in L^2(\Gamma) : \exists\, v' \in H^1(\Omega),\ \operatorname{tr} v' = v\}, \\ H^{1/2}(\Gamma_0) &= \{v = v'|_{\Gamma_0} : \exists\, v' \in H^{1/2}(\Gamma)\}, \\ \tilde{H}^{1/2}(\Gamma_0) &= \{v = v'|_{\Gamma_0} : \exists\, v' \in H^{1/2}(\Gamma),\ \operatorname{supp} v' \subset \Gamma_0\} \end{aligned}$$

with the standard norms

$$\|u\|_{H^{1/2}(\Gamma_0)} = \inf_{v \in H^{1/2}(\Gamma), v|_{\Gamma_0} = u} \|v\|_{H^{1/2}(\Gamma)} \quad \text{and} \quad \|u\|_{\tilde{H}^{1/2}(\Gamma_0)} = \|u_0\|_{H^{1/2}(\Gamma)},$$

where u_0 is the extension of u onto Γ by zero. The Sobolev space of negative order on Γ_0 are defined by duality as

$$H^{-1/2}(\Gamma_0) = (\tilde{H}^{1/2}(\Gamma_0))^* \quad \text{and} \quad \tilde{H}^{-1/2}(\Gamma_0) = (H^{1/2}(\Gamma_0))^*.$$

Moreover, from [259, Lemma 4.3.1] we have the inclusions

$$\tilde{H}^{1/2}(\Gamma_0) \subset H^{1/2}(\Gamma_0) \subset L^2(\Gamma_0) \subset \tilde{H}^{-1/2}(\Gamma_0) \subset H^{-1/2}(\Gamma_0).$$

For the solution $\mathbf{u}(\mathbf{x})$ of (5.23) with $\mathbf{x} \in \Omega \backslash \Gamma$ we have the Somigliana representation formula, see e.g. [229]:

$$\mathbf{u}(\mathbf{x}) = \int_{\Gamma} G(\mathbf{x}, \mathbf{y}) \left(\mathbf{T}_y \mathbf{u}(\mathbf{y})\right) ds_y - \int_{\Gamma} \mathbf{T}_y G(\mathbf{x}, \mathbf{y}) \mathbf{u}(\mathbf{y})\, ds_y + \int_{\Omega} G(\mathbf{x}, \mathbf{y}) \mathbf{f}(\mathbf{y})\, d\mathbf{y}, \tag{5.26}$$

where $G(\mathbf{x}, \mathbf{y})$ is the fundamental solution of the Navier-Lamé equation defined by

$$G(\mathbf{x}, \mathbf{y}) = \begin{cases} \frac{\lambda+3\mu}{4\pi\mu(\lambda+2\mu)} \left(\log|\mathbf{x}-\mathbf{y}|\mathbf{I} + \frac{\lambda+\mu}{\lambda+3\mu} \frac{(\mathbf{x}-\mathbf{y})(\mathbf{x}-\mathbf{y})^\top}{|\mathbf{x}-\mathbf{y}|^2}\right), & \text{if d=2} \\ \frac{\lambda+3\mu}{8\pi\mu(\lambda+2\mu)} \left(|\mathbf{x}-\mathbf{y}|^{-1}\mathbf{I} + \frac{\lambda+\mu}{\lambda+3\mu} \frac{(\mathbf{x}-\mathbf{y})(\mathbf{x}-\mathbf{y})^\top}{|\mathbf{x}-\mathbf{y}|^3}\right), & \text{if d=3} \end{cases}$$

with the Lamé constants $\lambda, \mu > 0$ depending on the material parameters, i.e. the modulus of elasticity E and the Poisson's ratio ν:

$$\lambda = \frac{E\nu}{1-\nu^2}, \qquad \mu = \frac{E}{1+\nu}.$$

Here, $\mathbf{T}_y$ stands for the traction operator with respect to $\mathbf{y}$ defined by $\mathbf{T}_y(\mathbf{u}) := \sigma(\mathbf{u}(\mathbf{y})) \cdot \mathbf{n}_y$, where $\mathbf{n}_y$ is the unit outer normal vector at $\mathbf{y} \in \Gamma$.

Thus we have the symmetric Poincaré–Steklov operator $S : \mathbf{H}^{\frac{1}{2}}(\Gamma) \to \mathbf{H}^{-\frac{1}{2}}(\Gamma)$ represented by

$$S = \frac{1}{2}\{W + (K' + I)V^{-1}(K + I)\}$$

Here, for $x \in \Gamma$

$$V\mathbf{v}(x) = 2\int_\Gamma G(x,y)\mathbf{v}(y)ds_y, \qquad K\mathbf{w}(x) = 2\int_\Gamma \left(\mathbf{T}_y G(x,y)\right)^\top \mathbf{w}(y)ds_y$$

$$K^\top\mathbf{v}(x) = 2\mathbf{T}_x\int_\Gamma G(x,y)\mathbf{v}(y)ds_y, \; W\mathbf{w}(x) = -2\mathbf{T}_x\int_\Gamma \left(\mathbf{T}_y G(x,y)\right)^\top \mathbf{w}(y)ds_y$$

denote the single layer potential, the double layer potential, its adjoint operator, and the hypersingular operator, respectively.

The Newton potential N is given by

$$N\mathbf{f} = \left(K' + \frac{1}{2}I\right)V^{-1}N_0\mathbf{f} - N_1\mathbf{f},$$

where N_0, N_1 are given for $x \in \Gamma$ by

$$N_0\mathbf{f} = \int_\Gamma G(\mathbf{x},\mathbf{y})\mathbf{f}(\mathbf{y})\,ds_y, \quad N_1\mathbf{f} = \mathbf{T}_x\int_\Gamma G(\mathbf{x},\mathbf{y})\mathbf{f}(\mathbf{y})\,ds_y.$$

Using the boundary function space, respectively the subset

$$\mathscr{V} = \widetilde{\mathbf{H}}^{1/2}(\Gamma_0) \quad \text{and} \quad \mathscr{K}^\Gamma = \{\mathbf{v} \in \mathscr{V} : v_n \le g \text{ a.e. on } \Gamma_C\},$$

multiplying $S\mathbf{u}$ by $\mathbf{v}-\mathbf{u}$, integrate on Γ_0, and using thereby again the decomposition of $\sigma\mathbf{n}$ on Γ_C into a tangential and normal part,we obtain as in the domain based case the boundary hemivariational inequality, Problem ($\mathscr{P}$): Find $\mathbf{u} \in \mathscr{K}^\Gamma$ such that

$$\langle S\mathbf{u}, \mathbf{v}-\mathbf{u}\rangle_{\Gamma_0} + \int_{\Gamma_C} f^0(u_n(s); v_n(s) - u_n(s))ds \ge \langle \mathbf{t}, \mathbf{v}-\mathbf{u}\rangle_{\Gamma_N}$$
$$+\langle N\mathbf{f}, \mathbf{v}-\mathbf{u}\rangle_{\Gamma_0} \quad \forall \mathbf{v} \in \mathscr{K}^\Gamma. \tag{5.27}$$

To shorten the right hand side we introduce the linear functional

$$\langle \mathbf{F}, \mathbf{v} \rangle_{\Gamma_0} = \int_{\Gamma_N} \mathbf{t} \cdot \mathbf{v} \, ds + \langle N\mathbf{f}, \mathbf{v} \rangle_{\Gamma_0}.$$

To settle the existence of solutions to such hemivariational inequalities we impose the following growth condition on ∂f: there exist positive constants c_1 and c_2 such that for all $\xi \in \mathbb{R}$ and $\eta \in \partial j(\xi)$ the following inequalities hold

(a) $|\eta| \leq c_1(1 + |\xi|)$;
(b) $\eta^T \xi \geq -c_2|\xi|$

Under these growth condition the functional $\varphi : \mathbf{H}^1(\Omega) \times \mathbf{H}^1(\Omega) \to \mathbb{R}$ defined by

$$\varphi(\mathbf{u}, \mathbf{v}) = \int_{\Gamma_C} f^0(u_n(s); v_n(s) - u_n(s)) \, ds, \quad \forall \mathbf{u}, \mathbf{v} \in \mathbf{H}^1(\Omega). \tag{5.28}$$

is well-defined, and as proved in the Appendix C.3.3, $\varphi(\cdot, \cdot)$ is pseudomonotone and weakly upper semicontinuous. We recall that the functional $\varphi : X \times X \to \mathbb{R}$, where X is a real reflexive Banach space, is pseudomonotone, if $u_n \rightharpoonup u$ (weakly) in X and $\liminf_{n\to\infty} \varphi(u_n, u) \geq 0$ imply that, for all $v \in X$, we have $\limsup_{n\to\infty} \varphi(u_n, v) \leq \varphi(u, v)$. Thus the existence of solutions follows from the theory of pseudomonotone variational inequalities (see C.8, also[163, Theorem 3.1], [212, Theorem 3])

To treat the nonsmoothness in the variational problem we sketch the regularization techniques of nondifferentiable optimization that are described in more detail in Appendic C.2 In order to smooth the functional φ we first approximate a locally Lipschitz function $f : \mathbb{R} \to \mathbb{R}$ via convolution by the function $\tilde{f} : \mathbb{R}_{++} \times \mathbb{R}$ given by

$$\tilde{f}(\varepsilon, x) = \int_{\mathbb{R}} f(x - \varepsilon y)\rho(y) \, dy,$$

where $\varepsilon > 0$ is a small regularization parameter and $\rho : \mathbb{R} \to \mathbb{R}_+$ is a probability density function such that

$$\kappa = \int_{\mathbb{R}} |t| \, \rho(t) \, dt < \infty$$

and

$$\mathbb{R}_+ = \{\varepsilon \in \mathbb{R} : \varepsilon \geq 0\}, \quad \mathbb{R}_{++} = \{\varepsilon \in \mathbb{R} : \varepsilon > 0\}.$$

In general, the smoothing function $\tilde{f}$ is not easily applicable in practice, since multivariate numerical quadrature is in generally involved, but for a special class of functions like a maximum, a minimum or a nested max-min function - what is sufficient for our applications -, it can be explicitly computed.

For example, if $f(x) = \max\{g_1(x), g_2(x)\}$, then $f(x) = g_1(x) + \mathrm{p}[g_2(x) - g_1(x)]$, where $\mathrm{p} : \mathbb{R} \to \mathbb{R}_+$ is the plus function defined by $\mathrm{p} = x^+ = \max\{x, 0\}$. Using, for example, the Zang probability density function

$$\rho(t) = \begin{cases} 1 \text{ if } -\frac{1}{2} \le t \le \frac{1}{2} \\ 0 \text{ otherwise} \end{cases}$$

for the smoothing approximation $\tilde{p}(\varepsilon, t)$ of $\mathrm{p}(t)$ defined via convolution, we obtain

$$\tilde{p}(\varepsilon, t) = \begin{cases} 0 & \text{if } \ t < -\frac{\varepsilon}{2} \\ \frac{1}{2\varepsilon}(t + \frac{\varepsilon}{2})^2 & \text{if } -\frac{\varepsilon}{2} \le t \le \frac{\varepsilon}{2} \\ t & \text{if } \ t > \frac{\varepsilon}{2}. \end{cases}$$

Hence, the smoothing function $\tilde{S} : \mathbb{R}_{++} \times \mathbb{R} \to \mathbb{R}$ of f defined by

$$\tilde{S}(\varepsilon, x) = g_1(x) + P(\varepsilon, g_2(x) - g_1(x))$$

takes the explicit form

$$\tilde{S}(\varepsilon, x) := \begin{cases} g_1(x) & \text{if } (i) \text{ holds} \\ \frac{1}{2\varepsilon}[g_2(x) - g_1(x)]^2 + \frac{1}{2}(g_2(x) + g_1(x)) + \frac{\varepsilon}{8} & \text{if } (ii) \text{ holds} \\ g_2(x) & \text{if } (iii) \text{ holds,} \end{cases}$$

where the cases (i), (ii), (iii) are defined below, respectively, by

(i) $g_2(x) - g_1(x) \le -\frac{\varepsilon}{2}$
(ii) $-\frac{\varepsilon}{2} \le g_2(x) - g_1(x) \le \frac{\varepsilon}{2}$
(iii) $g_2(x) - g_1(x) \ge \frac{\varepsilon}{2}$.

Also in the more general case of a maximum function $f(x) = \max\{g_1(x), \ldots, g_m(x)\}$ of smooth functions g_j, the smoothing approximation $\tilde{S}$ can be explicitly constructed (see Appendix C.2).

Thus we introduce $J_\varepsilon : \mathbf{H}^1(\Omega) \to \mathbb{R}$ by

$$J_\varepsilon(\mathbf{u}) = \int_{\Gamma_C} \tilde{S}(u_n(s), \varepsilon)\, ds.$$

and arrive at the regularized problem $(\mathscr{P}_\varepsilon)$ of (5.27): Find $\mathbf{u}_\varepsilon \in \mathscr{K}^\Gamma$ such that

$$\langle S\mathbf{u}_\varepsilon, \mathbf{v} - \mathbf{u}_\varepsilon \rangle_{\Gamma_0} + \langle DJ_\varepsilon(\mathbf{u}_\varepsilon), \mathbf{v} - \mathbf{u}_\varepsilon \rangle_{\Gamma_C} \ge \langle \mathbf{F}, \mathbf{v} - \mathbf{u}_\varepsilon \rangle_{\Gamma_0} \quad \forall \mathbf{v} \in \mathscr{K}^\Gamma, \tag{5.29}$$

where $DJ_\varepsilon : \mathbf{H}^{1/2}(\Gamma) \to \mathbf{H}^{1/2}(\Gamma)$ is the Gâteaux derivative of the functional J_ε and is given by

$$\langle DJ_\varepsilon(\mathbf{u}), \mathbf{v}\rangle_{\Gamma_C} = \int_{\Gamma_C} \frac{\partial}{\partial x}\tilde{S}(u_n(s), \varepsilon) v_n(s)\, ds.$$

We conclude this section with the following uniqueness result. Let c_S be the coerciveness constant of S. Assume now that there exists an $\alpha_0 \in [0, c_S)$ such that for any $\mathbf{u}, \mathbf{v} \in \mathscr{V}$ it holds

$$\varphi(\mathbf{u}, \mathbf{v}) + \varphi(\mathbf{v}, \mathbf{u}) \le \alpha_0 \|\mathbf{u} - \mathbf{v}\|^2_{\mathscr{V}}. \tag{5.30}$$

Theorem 5.1 *Under the assumption (5.30), there exists a unique solution of problem ($\mathscr{P}$), which depends Lipschitz continuously on the linear form given by the right hand side.*

Proof Assume that $\mathbf{u}, \tilde{\mathbf{u}}$ are two solutions of ($\mathscr{P}$). Then the inequalities below hold:

$$\langle S\mathbf{u} - \mathbf{F}, \mathbf{v} - \mathbf{u}\rangle_{\Gamma_0} + \varphi(\mathbf{u}, \mathbf{v}) \ge 0 \quad \forall \mathbf{v} \in \mathscr{K}^\Gamma$$
$$\langle S\tilde{\mathbf{u}} - \mathbf{F}, \mathbf{v} - \tilde{\mathbf{u}}\rangle_{\Gamma_0} + \varphi(\tilde{\mathbf{u}}, \mathbf{v}) \ge 0 \quad \forall \mathbf{v} \in \mathscr{K}^\Gamma.$$

Setting $\mathbf{v} = \tilde{\mathbf{u}}$ in the first inequality and $\mathbf{v} = \mathbf{u}$ in the second one, and summing up the resulting inequalities, we get

$$\langle S\mathbf{u} - S\tilde{\mathbf{u}}, \tilde{\mathbf{u}} - \mathbf{u}\rangle_{\Gamma_0} + \varphi(\mathbf{u}, \tilde{\mathbf{u}}) + \varphi(\tilde{\mathbf{u}}, \mathbf{u}) \ge 0. \tag{5.31}$$

From the coercivity of the operator S and the assumption (5.30) we obtain

$$c_S \|\mathbf{u} - \tilde{\mathbf{u}}\|^2_{\mathscr{V}} \le \varphi(\mathbf{u}, \tilde{\mathbf{u}}) + \varphi(\tilde{\mathbf{u}}, \mathbf{u}) \le \alpha_0 \|\mathbf{u} - \tilde{\mathbf{u}}\|^2_{\mathscr{V}}.$$

Hence, since $\alpha_0 \in [0, c_S)$, if $\mathbf{u} \neq \tilde{\mathbf{u}}$ we receive a contradiction.

Now let $\mathbf{F}_i \in \mathscr{V}^*$ and denote $\mathbf{u}_i = \mathbf{u}_{\mathbf{F}_i}$, $i = 1, 2$. Analogously to (5.31), we find that

$$\langle S\mathbf{u}_1 - \mathbf{F}_1 - S\mathbf{u}_2 + \mathbf{F}_2, \mathbf{u}_2 - \mathbf{u}_1\rangle_{\Gamma_0} + \varphi(\mathbf{u}_1, \mathbf{u}_2) + \varphi(\mathbf{u}_2, \mathbf{u}_1) \ge 0.$$

Hence,

$$c_S \|\mathbf{u}_1 - \mathbf{u}_2\|^2_{\mathscr{V}} \le \varphi(\mathbf{u}_1, \mathbf{u}_2) + \varphi(\mathbf{u}_2, \mathbf{u}_1) + \langle \mathbf{F}_2 - \mathbf{F}_1, \mathbf{u}_2 - \mathbf{u}_1\rangle_{\Gamma_0}$$

and by (5.30),

$$(c_S - \alpha_0)\|\mathbf{u}_1 - \mathbf{u}_2\|^2_{\mathscr{V}} \le \langle \mathbf{F}_2 - \mathbf{F}_1, \mathbf{u}_2 - \mathbf{u}_1\rangle_{\Gamma_0} \le \|\mathbf{F}_1 - \mathbf{F}_2\|_{\mathscr{V}^*} \|\mathbf{u}_1 - \mathbf{u}_2\|_{\mathscr{V}}.$$

Also, since $\alpha_0 < c_S$ we deduce that

$$\|\mathbf{u}_1 - \mathbf{u}_2\|_{\mathscr{V}} \leq \frac{1}{c_S - \alpha_0} \|\mathbf{F}_1 - \mathbf{F}_2\|_{\mathscr{V}^*},$$

which concludes the proof of the theorem. □

Next we asssume that the assumption that there exists a constant $\alpha_0 \geq 0$ (in general depending on $\varepsilon > 0$) such that

$$\left(\frac{\partial}{\partial x}\tilde{S}(x_1, \varepsilon) - \frac{\partial}{\partial x}\tilde{S}(x_2, \varepsilon)\right)(x_1 - x_2) \geq -\alpha_0 |x_1 - x_2|^2 \quad \forall x_1, x_2 \in \mathbb{R}. \tag{5.32}$$

Hence, for any $\mathbf{u}, \mathbf{v} \in \mathscr{V}$, we have

$$\begin{aligned}
&\langle DJ_\varepsilon(\mathbf{u}) - DJ_\varepsilon(\mathbf{v}), \mathbf{v} - \mathbf{u}\rangle_{\Gamma_C} \\
&= \int_{\Gamma_C} \left(\frac{\partial}{\partial x}\tilde{S}(u_n(s), \varepsilon) - \frac{\partial}{\partial x}\tilde{S}(v_n(s), \varepsilon)\right)(v_n(s) - u_n(s))\, ds \\
&\leq \alpha_0 \|u_n - v_n\|^2_{L^2(\Gamma_C)} \leq \alpha_0 \|\mathbf{u} - \mathbf{v}\|^2_{\mathscr{V}}.
\end{aligned}$$

Due to Theorem 5.1, we have the following uniqueness result for the regularized problem.

Theorem 5.2 *Under the assumption* (5.32) *with* $\alpha_0 < c_S$, *there exists a unique solution to the regularized problem* ($\mathscr{P}_\varepsilon$), *which depends Lipschitz continuously on the right hand side* $\mathbf{F} \in \mathscr{V}^*$.

The solution of unilateral nonsmooth boundary value problems with monotone/nonmonotone boundary conditions via multivalued boundary integral equations, boundary variational inequalites, respectively boundary hemivariational inequalities can be traced back to the work of Haslinger and Panagiotopoulos. While Haslinger et al. [232] study the unilateral Poisson problem of steady-state flow through a semipermeable membrane of infinite thickness, the vectorial linear elastic contact problem is treated by a reciprocal (dual) approach in [336]. In all their work, without using potential theory, the Poincaré–Steklov operator (or rather its inverse) has to be constructed by the solution of appropriate linear boundary value problems in the domain.

Chapter 6
A Primer to Boundary Element Methods

This chapter introduces the BEM in its $h-$version. First we make Fourier expansion of Chap. 3 more precise by asymptotic error estimates. Then we prove direct and inverse approximation estimates for periodic spline approximation on curves. Hence we develop the analysis of Galerkin methods and collocation methods for Symm's integral equation towards optimal a priori error estimates. Moreover, we subsume Galerkin and collocation methods as general projection methods. To this end we extend the above treatment of positive definite bilinear forms to the analysis of a sequence of linear operators that satisfy a uniform Gårding inequality and establish stability and optimal a priori error estimates in this more general setting. Interpreting several variants of collocation methods that combine collocation and quadrature as extended Galerkin methods we include their numerical analysis as well. Then augmenting the boundary element ansatz spaces by known singularity functions the Galerkin method is shown to converge with higher convergence rates. Finally to obtain higher convergence rates in weaker norms than the energy norm the Aubin–Nitsche duality estimates of FEM are extended to BEM so that it allows the incorporation of the singular solution expansion for nonsmooth domains. Sections 6.1–6.4 are based on the classroom notes by M. Costabel [116] whereas Sects. 6.5.1–6.5.6 are based on the classroom notes by W.L. Wendland [430]. Improved estimates of local type, pointwise estimates and postprocessing with the K-operator are considered in Sects. 6.5.7–6.5.9. Discrete collocation with trigonometric polynomials, where the concept of finite section operators is used, is a subject of Sect. 6.6. In Sect. 6.7 the standard BEM is enriched by special singularity functions modelling the behaviour of the solution near corners, thus yielding improved convergence. In Sect. 6.8 Galerkin-Petrov methods are considered. Section 6.9 presents the Arnold-Wendland approach to reformulate a collocation method as a Galerkin method whereas qualocation is investigated in Sect. 6.10. In Sect. 6.11 the use of radial basis functions (a meshless method) and of spherical splines in the Galerkin scheme is demonstrated for problems on the unit sphere. Integral equations of the first kind with the single layer and double layer

© Springer International Publishing AG, part of Springer Nature 2018

J. Gwinner, E. P. Stephan, *Advanced Boundary Element Methods*,
Springer Series in Computational Mathematics 52,
https://doi.org/10.1007/978-3-319-92001-6_6

potentials are our main subject. Integral equations of the second kind are studied only briefly, e.g. at the end of Sect. 6.4.

There has been a tremendous amount of research on spline collocation and Galerkin methods. We want to mention the works by J. Schmidt [365, 366], S. Prössdorf and B. Silbermann [345] and J. Saranen and A. Vainikko [356]. The hp-version of the BEM is one of the main subjects of this book and therefore considered separately in Chaps. 7 and 8.

For further reading we refer to the seminal papers by Hsiao and Wendland [257] and Nedelec and Planchard [324] and to the survey articles by W.L. Wendland (Part III of [362, 429] and [428]) and the lecture notes of J.C. Nedelec [320].

6.1 Galerkin Scheme for Strongly Elliptic Operators

As before let X, Y be Hilbert spaces and $A : X \longrightarrow Y$ a continuous, linear and bijective operator. In Chap. 2 we saw that $\operatorname{cap}(\Gamma) < 1$ is a sufficient condition for the operator V to be positive definite. In order to cover also examples like $A = I + C$, C compact, or $A = V$ with $\operatorname{cap}(\Gamma) > 1$ we need a more general sufficient criterion and therefore the notion of a compact operator, see Definition A.6 in Appendix A. We shall now consider again the situation of the general Galerkin method.

Theorem 6.1 *Under the assumptions:*

1. $\exists P_N : Y' \longrightarrow T_N$ *, bounded and linear, converging on Y' strongly to the identity operator, i.e.*

$$\|P_N\eta - \eta\|_{Y'} \overset{N\to\infty}{\longrightarrow} 0 \qquad \forall \eta \in Y',$$

2. $\exists Q_N : X_N \longrightarrow T_N,\ \exists M$ *(independent of N)* :
$|\langle Q_N v, Aw\rangle| \leq M \ \|v\|_X \ \|w\|_X,\ \forall v \in X_N,\ w \in X,\ N \in \mathbb{N},$
3. $\exists C : X \longrightarrow X'$ *compact* , $\exists \alpha > 0$ *(independent of N)*:
$|\langle Q_N v, Av\rangle + \langle Cv, v\rangle| \geq \alpha \|v\|_X^2$

the following holds:

(i) ***Existence of a unique solution***
$\exists N_0 \in \mathbb{N}\ \forall N \geq N_0\ \exists!\, u_N : \langle t, Au_N\rangle = \langle t, Au\rangle = \langle t, f\rangle \quad \forall t \in T_N,$
(ii) ***Stability of the method***
$\exists c > 0,\ N_0 \in \mathbb{N}\ \forall N \geq N_0 \ :\ \|u_N\| \ \leq \ c\|u\| \quad \forall u,$
(iii) ***Quasioptimal error estimate***
$\exists c > 0\, ,\, N_0 \in \mathbb{N}\ \forall N \geq N_0 \ :\ \|u - u_N\| \ \leq \ c \cdot \inf_{\chi \in X_N} \|u - \chi\|,$

(iv) ***Convergence of the method***
$d(u, X_N) \overset{N\to\infty}{\longrightarrow} 0 \implies \|u - u_N\|_X \overset{N\to\infty}{\longrightarrow} 0$.

Proof In Sect. 1.2 we have already proved the theorem for the case $C \equiv 0$. We now want to reduce the general case to this, i.e. we want to have

$$|\langle \tilde{Q}_N v, Av\rangle| \geq \tilde{\alpha}\ \|v\|^2, \quad \forall\, v \in X_N\,. \tag{6.1}$$

This would clearly be satisfied due to assumption *3.*, if we had

$$\langle \tilde{Q}_N v, Av\rangle = \langle Q_N v, Av\rangle \ + \ \langle Cv, v\rangle\,. \tag{6.2}$$

However, this would imply

$$\begin{aligned} A'\tilde{Q}_N &= A'Q_N \ + \ C \\ \Leftrightarrow \quad \tilde{Q}_N &= Q_N \ + \ (A')^{-1}C \end{aligned}$$

but raises the problem that we do not have a mapping from X_N into T_N . Here we define the operator

$$\tilde{Q}_N := Q_N + P_N(A')^{-1}C.$$

Then,

$$\tilde{Q}_N = Q_N + (A')^{-1}C + (P_N - 1)(A')^{-1}C\,,$$

where

$$(P_N - 1)(A')^{-1}C \Rightarrow 0 \text{ (in operator norm)}$$

by Lemma A.2 in Appendix A, since $(A')^{-1}C$ is compact and $P_N - 1 \to 0$ by assumption. Thus we have

(a)

$$\begin{aligned} |\langle \tilde{Q}_N v, Aw\rangle| &\leq |\langle Q_N v, Aw\rangle| \ + \ |\langle P_N(A')^{-1}Cv, Aw\rangle| \\ &\leq \tilde{M}\|v\|\ \|w\|,\ \forall\, N\,. \end{aligned}$$

(b)

$$\begin{aligned} |\langle \tilde{Q}_N v, Av\rangle| &= |\langle Q_N v, Av\rangle \ + \ \langle Cv, v\rangle \ + \ \langle (P_N - 1)(A')^{-1}Cv, Av\rangle| \\ &\geq |\langle Q_N v, Av\rangle \ + \ \langle Cv, v\rangle| \ - \ |\langle (P_N - 1)(A')^{-1}Cv, Av\rangle| \\ &\geq \alpha\, \|v\|_X^2 \ - \ \|(P_N - 1)(A')^{-1}C\| \cdot \|A\| \cdot \|v\|_X^2 \\ &\geq \tfrac{\alpha}{2}\|v\|_X^2\,, \quad N \geq N_0. \end{aligned}$$

□

Corollary 6.1 *Let $A,\ B:\ X \longrightarrow Y$ be two bijective, continuous operators with the operator $A - B$ being compact. If the general Galerkin method of Sect. 1.2 converges for the operator A, it converges for B, too.*

Definition 6.1 An operator $A : X \longrightarrow X'$ is said to be strongly elliptic, if there exists a decomposition $A = D + C$ with D positive definite and C compact.

Remark 6.1 The operator A is strongly elliptic if and only if there exists a constant $\alpha > 0$ and a compact operator $C : X \longrightarrow X'$ such that

$$\langle Av, v\rangle \geq \alpha \|v\|_X^2 + \langle Cv, v\rangle \quad \forall\, v \in X. \tag{6.3}$$

In the case of complex functions we have:

$$\Re\langle Av, v\rangle \geq \alpha \|v\|_X^2 + \Re\langle Cv, v\rangle \quad \forall\, v \in X. \tag{6.4}$$

The above inequalities are due to *Gårding* and thus usually referred to as *Gårding inequalities*.

As a consequence of Theorem 6.1 we have

Theorem 6.2 *Let $A : X \longrightarrow X'$ be a strongly elliptic and bijective operator. Then every Galerkin scheme for A is convergent.*

Example 6.1 Consider the single layer potential V for $\Gamma \in$ Lip. Then we may choose

$$X = H^{-\frac{1}{2}}(\Gamma), \quad \text{and } X' = H^{\frac{1}{2}}(\Gamma)\,.$$

Theorem 6.3 *Let Γ be a Lipschitz curve with cap$(\Gamma) \neq 1$. Then every Galerkin scheme for $A = V$ is convergent.*

Proof Consider first the case that cap$(\Gamma) < 1$. Then the operator $V : H_V := H^{-\frac{1}{2}}((\Gamma) \longrightarrow H_V'$ is positive definite.

For cap$(\Gamma) > 1$, use the integral mean $m(\varphi)$, see Definition 2.7. Then setting $\varphi_0 = \varphi - m(\varphi)$ yields

$$\langle V\varphi, \varphi\rangle = \langle \varphi_0, V\varphi_0\rangle + \langle C\varphi, \varphi\rangle$$

with $C\varphi = \frac{m(\varphi)}{L}$. Hence, the assertion of the theorem follows by application of Theorem 6.2. □

6.2 Galerkin Methods for the Single-Layer Potential

6.2.1 Approximation with Trigonometric Polynomials

We now want consider an approximation on $\Gamma = \partial B_1(0)$ with trigonometric polynomials. Note, that for $u(t) = \sum_{k=-\infty}^{\infty} u_k e^{2\pi ikt}$ with coefficients $u_k := \int_0^1 u(t)e^{-2\pi ikt}\,dt$ we have

$$\|u\|_s^2 := \sum_{k=-\infty}^{\infty} (1+k^2)^s |u_k|^2 .$$

We then define

- $T_N := \operatorname{span}\{e^{2\pi ikx} : |k| \le N\}$ with $\dim(T_N) = 2N+1$
- $\Pi_N : \begin{cases} H^s \longrightarrow T_N \\ u \mapsto (\Pi_N u)(x) := \sum_{|k|\le N} u_k e^{2\pi ikx} \in T_N . \end{cases}$

Theorem 6.4 *With the above definitions we have*

i) $\|u - \Pi_N u\|_s \overset{N\to\infty}{\longrightarrow} 0$ *for* $u \in H^s(\Gamma)$.
ii) **Approximation Property**
$\forall r \le s\ \exists c_{r,s}$ *(independent of N)* :
$\|u - \Pi_N u\|_r \le c_{r,s} \cdot N^{r-s} \|u\|_s, \quad \forall u \in H^s\ \forall N \in \mathbb{N}.$
iii) **Inverse Estimate**
$\forall r \le s\ \exists \tilde{c}_{r,s}$ *(independent of N)* :
$\|v\|_s \le \tilde{c}_{r,s} \cdot N^{s-r} \|v\|_r \quad \forall v \in T_N,\ \forall N \in \mathbb{N}.$

Proof

i) clear.
ii) $$\|u - \Pi_N u\|_r^2 = \sum_{|k|>N} (1+k^2)^r |u_k|^2 \le (1+N^2)^{r-s} \sum_{|k|>N} (1+k^2)^s |u_k|^2 \le 2^{r-s} N^{2(r-s)} \|u\|_s^2 .$$
iii) $$\|v\|_s^2 = \sum_{|k|\le N} (1+k^2)^s |v_k|^2 \le (1+N^2)^{s-r} \|v\|_r^2 .$$ □

As we have seen in Chap. 3 there holds

Theorem 6.5 *For* $\Gamma = \partial B_1(0)$ *the mapping* $V : H^s(\Gamma) \longrightarrow H^{s+1}(\Gamma)$ *is continuous.*

For the Galerkin scheme with trigonometric polynomials for $Vu = f$ on $\partial B_R(0)$, $R \neq 1$, we now have by application of Theorem 6.1 (or Theorem 6.2) the following result.

Theorem 6.6 *Let u_N be a solution to $Vu = f$ using the Galerkin method with trigonometric polynomials T_N as trial and test functions. For $\Gamma \in C^\infty$, $cap(\Gamma) \neq 1$ and $u \in H^s(\Gamma)$, $s \geq -\frac{1}{2}$ we then have*

$$\|u - u_N\|_{-\frac{1}{2}} \leq N^{-(s+\frac{1}{2})} \|u\|_s \, .$$

Furthermore let $r, s \in \mathbb{R}$ with $r \leq s$ be arbitrary.Then there exists a constant c independent of N such that $\|u - u_N\|_r \leq c \cdot N^{r-s} \|u\|_s$

Proof Using the approximation property in Theorem 6.4, we have

$$\|u - u_N\|_{-\frac{1}{2}} \leq c \cdot \inf_{w \in T_N} \|u - w\|_{-\frac{1}{2}} \leq c \cdot N^{-\left(s+\frac{1}{2}\right)} \|u\|_s \, .$$

□

We may also think about estimates in other norms, for instance

a) Norms above the energy-norm $(r \geq -\frac{1}{2},\ s \geq r)$:

$$\begin{aligned} \|u - u_N\|_r &\leq \|u - \Pi_N u\|_r + \|\Pi_N u - u_N\|_r \\ &\leq c \cdot N^{r-s} \|u\|_s + c \cdot N^{r+\frac{1}{2}} \cdot \underbrace{\|\Pi_N u - u_N\|_{-\frac{1}{2}}}_{c \cdot \|u\|_s N^{-\left(s+\frac{1}{2}\right)}} \\ &\leq c \cdot N^{r-s} \|u\|_s \, . \end{aligned}$$

b) Norms below the energy-norm $(r \leq -\frac{1}{2})$:

$$\|u - u_N\|_r = \sup_{v \in H^{|r|}} \frac{\langle u - u_N, v\rangle}{\|v\|_{|r|}}$$

$$\begin{aligned} |\langle u - u_N, v\rangle| &= |\langle V(u - u_N), V^{-1} v\rangle| \qquad \left(\|V^{-1}v\|_{|r|-1} \cong \|v\|_{|r|}\right) \\ &= |\langle V(u - u_N), V^{-1} v - t\rangle| \, , \quad \forall t \in \Pi_N \\ &\leq \|V(u - u_N)\|_{r'+1} \cdot \|V^{-1} v - t\|_{-r'-1} \\ &\quad \left(\text{since: } -r' \leq |r| \Leftrightarrow r' \geq r\right) \\ &\leq c \cdot \|u - u_N\|_{r'} \cdot c\, \underbrace{N^{-|r|-r'}}_{N^{r-r'}} \underbrace{\|V^{-1} v\|_{|r|-1}}_{\|v\|_{-r}} \\ &\quad \left(\text{by the approximation property of the above theorem}\right) \end{aligned}$$

$$\Rightarrow \|u - u_N\|_r \leq c \cdot N^{r-r'} \|u - u_N\|_{r'} \, , \quad \forall r' \geq r \, .$$

For $r' = -\frac{1}{2}$ it follows $\forall s \geq -\frac{1}{2}$:

$$\|u - u_N\|_r \leq c \cdot N^{r-s} \|u\|_s \, .$$

Corollary 6.2 *For any real number s there exists a constant c such that for $u \in H^s(\Gamma)$,*

$$|m(u) - m(u_N)| \le c \cdot N^{-s} \quad \text{and} \quad |cap(\Gamma) - cap_N(\Gamma)| \le c \cdot N^{-s} .$$

Note, for $Ve = 1$ we have: $cap(\Gamma) = e^{-\frac{2\pi}{Lm(e)}}$, $cap_N(\Gamma) = e^{-\frac{2\pi}{Lm(e_N)}}$.

Proof With the above theorem we have

$$\begin{aligned} |m(u) - m(u_N)| &= |\langle 1, u - u_N\rangle| \\ &\le \|1\|_0 \|u - u_N\|_0 \\ &\le cN^{-s}\|u\|_s . \end{aligned}$$

□

6.2.2 Approximation with Splines

Let a mesh Δ_N on Γ be defined analogously to section 6.3, i.e.

$$x_j = e^{2\pi i \frac{j}{N}} = x(jh) =: x(s_j), \quad h = \frac{1}{N}$$

and let $S^d_{\Delta_N} \equiv S^d_h$, as defined there with $d \ge -1$. We may assume that N is odd, e.g. $N = 2M + 1$. Each $\phi \in S^d_h$ is a polynomial of degree d on each interval, so $d + 1$ coefficients have to be determined. As ϕ is $(d-1)$-times differentiable in the nodes of Δ_N, we get that $\dim S^d_h = N$.

Theorem 6.7 *With the above definitions there holds:*

i) $v \in S^d_h \;\Leftrightarrow\; v_k \cdot k^{d+1} = v_{k+N} \cdot (k+N)^{d+1}, \quad \forall k \in \mathbb{Z}$.

ii) $v \in S^d_{\Delta_N}$ *is uniquely determined by* $\{v_k : |k| \le M\}$ *, i.e.*

$\Pi_M : S^d_{\Delta_N} \to T_N$ *is a bijection with inverse* $Q_M := \Pi_M^{-1} : T_N \to S^d_{\Delta_N}$.

iii) For $s < d + \frac{1}{2}$ *there exists constants* c_1, c_2 *such that*

$$c_1\|\Pi_M v\|_s \le \|v\|_s \le c_2\|\Pi_M v\|_s , \quad \forall v \in S^d_{\Delta_N} .$$

Proof We first note that $w \in S^{-1}_{\Delta_N} \Leftrightarrow w_{k+N} = w_k \;\; \forall k \in \mathbb{N}$.

To prove i), we have : $v \in S^d_{\Delta_N}$ implies $\left(\frac{d}{ds}\right)^{d+1} v \in S^{-1}_{\Delta_N}$
$\Leftrightarrow (2\pi i k)^{d+1} v_k$ is N-periodic $\Leftrightarrow (2\pi i(k+N))^{d+1} v_{k+N} = (2\pi i k)^{d+1} v_k$.

To prove ii) write $k \in \mathbb{Z}$ as $k = r + l \cdot N$, $|r| \le M$. Hence

$$v_k = v_r \cdot \frac{r^{d+1}}{(r + lN)^{d+1}} = \left(\frac{r}{k}\right)^{d+1} v_r .$$

This gives the desired mapping

$$Q_N : \sum_{|r|\leq M} v_r e^{2\pi ikx} \longrightarrow \sum_{k\in\mathbb{Z}} v_k e^{2\pi ikx} .$$

To prove iii) first note $\|\Pi_M v\|_s \leq \|v\|_s$. Moreover,

$$\begin{aligned}
\|v\|_s^2 &= \sum_{k\in\mathbb{Z}} (1+k^2)^s |v_k|^2 \\
&= \sum_{r=-M}^{M} \sum_{l\in\mathbb{Z}} \left(1+(r+lN)^2\right)^s |v_{r+lN}|^2 \\
&= \sum_{r=-M}^{M} \sum_{l\in\mathbb{Z}} \left(1+(r+lN)^2\right)^s \cdot \tfrac{r^{2(d+1)}}{(r+lN)^{2(d+1)}} |v_r|^2 \\
&= \sum_{r=-M}^{M} |v_r|^2 \sum_{l\in\mathbb{Z}} \left(1+(r+lN)^2\right)^s \cdot \tfrac{r^{2(d+1)}}{(r+lN)^{2(d+1)}}
\end{aligned}$$

Now for $r \neq 0$,

$$\sum_l \frac{(1+(r+lN)^2)^s}{(r+lN)^{2(d+1)}} r^{2(d+1)} \leq 2\sum_l (r+lN)^{2(s-d-1)} r^{2d+2}$$

is finite , if and only if, $s-d-1 < -1/2$ □

We define

$$P_N := Q_N \Pi_M , \tag{6.5}$$

i.e., $v = P_N u \in S^d_{\Delta_N}$ is uniquely determined by $v_r = u_r , \quad \forall\, |r| \leq M$.

Theorem 6.8 *With the above definition (6.5) the following holds:*

i) **Approximation Property**
$\forall r \leq s , \quad r < d + \frac{1}{2} , \quad s \leq d+1 \ \exists\, c_{r,s}$ *(independent of N)* :

$$\|u - P_N u\|_r \leq c_{r,s} N^{r-s} \|u\|_s \quad \forall u \in H^s .$$

ii) **Inverse Property**
$\forall r \leq s < d + \frac{1}{2} \ \exists\, c_{r,s}$ *(independent of N)* :

$$\|v\|_s \leq c_{r,s} N^{s-r} \|v\|_r \quad \forall v \in S^d_{\Delta_N} .$$

Proof ii) follows directly from assertion (i) of Theorem 6.7 combined with the inverse property for trigonometric polynomials.

To prove i) let $k = r + lN$. Then

$$\|u - P_N u\|_\tau^2 = \sum_{r\neq 0, |r|\leq M} \sum_{l\neq 0, l\in\mathbb{Z}} \left(1+k^2\right)^\tau \left| u_k - u_r \left(\frac{r}{k}\right)^{d+1} \right|^2$$

and $\|u - P_N u\|_\tau \leq \underbrace{\|u - \Pi_M u\|_\tau}_{\leq c M^{\tau-s}\|u\|_s} + \|\Pi_M u - P_N u\|_\tau^2$

$$\Longrightarrow \quad \|\Pi_M u - P_N u\|_\tau^2 = \sum_{r\neq 0, |r|\leq M} \sum_{l\neq 0, l\in\mathbb{Z}} |u_r|^2 \left(1+k^2\right)^\tau \left(\frac{r}{k}\right)^{2(d+1)} .$$

With
$$\begin{aligned}
\left(1+k^2\right)^\tau \left(\tfrac{r}{k}\right)^{2(d+1)} &\leq c \cdot k^{2\tau-2d-2} \cdot r^{2d+2} \\
&= c \cdot r^{2\tau} \cdot \left(\tfrac{k}{r}\right)^{2\tau-2d-2} \\
&= c \cdot r^{2\tau} \left(\tfrac{N}{r}\right)^{2\tau-2d-2} \cdot \left(l + \tfrac{r}{N}\right)^{2\tau-2d-2} \\
&= c \cdot r^{2s} N^{2(\tau-s)} \underbrace{\left(\frac{N}{r}\right)^{2s-2d-2}}_{\leq c \ \text{ for } s\leq d+1} \cdot \left(l + \tfrac{r}{N}\right)^{2\tau-2d-2}
\end{aligned}$$

it follows

$$\|\Pi_M u - P_N u\|_\tau^2 \leq c \cdot \underbrace{\sum_{r\neq 0} |u_r|^2 r^{2s}}_{\leq \|\Pi_M u\|_s^2} N^{2(\tau-s)} \cdot \underbrace{\sum_{l\neq 0} \left(l + \frac{r}{N}\right)^{2\tau-2d-2}}_{\leq c \ \text{ for } \tau < d+\frac{1}{2}} .$$

Thus, we finally obtain

$$\|\Pi_M u - P_N u\|_\tau^2 \leq c \|u\|_s^2 \cdot N^{2(\tau-s)} \quad \Longrightarrow \quad (i) .$$

□

For the Galerkin method using spline functions to approximate the solution of the integral equation of first kind $Vu = f$ on $\Gamma = \partial B_R(0)$, $R \neq 1$ we obtain

Theorem 6.9

i) Let $f \in H^{s+1}(\Gamma)$, $s \geq -\frac{1}{2}$. Then, for N sufficiently large there exists a unique $u_N \in S_h^d$, $d \geq 0$ such that:

$$\langle v, V u_N \rangle = \langle v, f \rangle \quad \forall\, v \in S_h^d .$$

ii) For $r \leq s$, $-d-2 \leq r < d + \frac{1}{2}$ and $-d - \frac{3}{2} < s \leq d+1$ there holds for $u \in H^s$

$$\|u - u_N\|_r \leq c \cdot N^{r-s} \|u\|_s .$$

iii) For $f \in H^{d+2}$ *we have*

$$\|u - u_N\|_{-\frac{1}{2}} \le c \cdot N^{-\frac{1}{2}-d} \|u\|_{d+1}$$
$$\text{and} \quad \|u - u_N\|_{-d-2} \le c \cdot N^{-2d+3} \cdot \|u\|_{d+1} \, .$$

iv) The above also holds for $\Gamma \in C^\infty$ *with* $cap(\Gamma) \neq 1$.

Proof We only want to prove (iv) and leave the other items as an exercise to the reader.

We have

$$\log|x(s) - x(t)| = \log|e^{2\pi is} - e^{2\pi it}| + \log \underbrace{\left| \frac{x(s) - x(t)}{e^{2\pi is} - e^{2\pi it}} \right|}_{\neq 0, \, \in C^\infty}$$
$$\text{implies } V_\Gamma = V_{\partial B_0} + T\,, \quad \text{with } T : H^s \longrightarrow C^\infty \quad \forall s$$

Thus, $T : H^s(\Gamma) \longrightarrow H^t(\Gamma) \quad \forall s,\ t \in \mathbb{R}$ is compact. □

6.3 Collocation Method for the Single-Layer Potential

For further reading see [305, 356] and also for general strongly elliptic equations [345]. As an example of the general Galerkin scheme in Sect. 6.1 we want to consider in this section a collocation method for the single layer potential V.

Let $\Gamma = \partial B_r(0)$ be a circle of positive radius $r < 1$ with the centre in the origin. Then we know that V is positive definite in $H^{-\frac{1}{2}}(\Gamma)$. We define a mesh Δ_N on Γ as follows:

$$x_j = r \cdot e^{2\pi i \frac{j}{N}} = x(jh)\,, \quad h = \frac{1}{N}$$
$$x(s) = r \cdot e^{2\pi is}\,, \quad |\dot{x}| = r \cdot 2\pi \neq 1$$

Example 6.2 For $N = 8$ we would have Fig. 6.1.

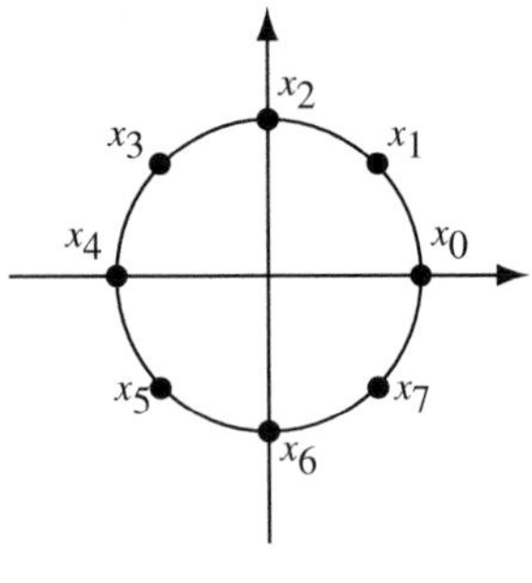

Fig. 6.1 Break points

We define the spaces

$$\begin{aligned} S^d_{\Delta_N} \equiv S^d_h &:= \{\varphi \in C^{d-1}(\Gamma) : \varphi|[s_i, s_{i+1}] \text{ is a polynomial of degree } d\} \\ S^{-1}_{\Delta_N} &:= \text{span}\{\delta_{x_j} : j = 0, \cdots, N-1\} \end{aligned}$$

and consider the nodal collocation for linear continuous functions for the equation

$$Vu = f$$

i.e. we are looking for a solution $u_N \in S^1_{\Delta_N}$ of

$$(Vu_N)(x_j) = f(x_j)\,, \quad j = 0, \cdots, N-1\,. \tag{6.6}$$

Theorem 6.10 *Let $f \in H^{\frac{3}{2}}(\Gamma)$ be given. Then the collocation method (6.6) converges in $H^{\frac{1}{2}}(\Gamma)$, i.e.*

$$\begin{aligned} &\exists\, N_0 \in \mathbb{N}\ \forall\, N \geq N_0\ \exists!\, u_N \in S^1_{\Delta_N} : && \|u_N\|_{H^{\frac{1}{2}}(\Gamma)} \leq c\|u\|_{H^{\frac{1}{2}}(\Gamma)} \\ &\textit{and} && \|u - u_N\|_{H^{\frac{1}{2}}(\Gamma)} \leq \tilde{c} \inf_{v \in S^1_{\Delta_N}} \|u - v\|_{H^{\frac{1}{2}}(\Gamma)} \end{aligned}$$

for some constants c and $\tilde{c}$.

Proof Consider V as a mapping $X := H^{\frac{1}{2}}(\Gamma) \longrightarrow H^{\frac{3}{2}}(\Gamma) =: Y$. Since the mapping V: $H^s(\Gamma) \longrightarrow H^{s+1}(\Gamma)$ is continuous and bijective for smooth Γ, we know that V is an isomorphism. We therefore first have to show that

(i) $X_N := S^1_{\Delta_N} \subseteq H^{\frac{1}{2}}(\Gamma)$
(ii) $T_N := S^{-1}_{\Delta_N} \subseteq H^{-\frac{3}{2}}(\Gamma)$.

For (i) note $S^1_{\Delta_N} \subseteq H^1(\Gamma) \subseteq H^{\frac{1}{2}}(\Gamma)$. For (ii) by *Sobolev's* embedding theorem, $H^s(\Gamma) \subseteq C(\Gamma)$ for $s > \frac{1}{2}$. Thus,

$$\|u\|^2_s = c \cdot \sum_{k \in \mathbb{Z}} |u_k|^2 (1 + k^2)^s\,, \quad \text{with } u_k = \int_0^1 u(s) e^{-2\pi i k s}\, ds\,.$$

In a second step we now have to construct the operators $Q_N : X_N \to T_N$. In the distributional sense, let $Q_N = -(\frac{d}{ds})^2$. Then $Q := Q_N$ gives $Q_N S^1_{\Delta_N} \subseteq S^{-1}_{\Delta_N}$.

To verify the assumptions of Theorem 6.1 we have to show that

1. $|\langle Q_N v, Aw\rangle| \leq M\|v\|_X \|w\|_X$, for $v \in X_N$, $w \in X$
2. $|\langle Q_N v, Av\rangle + \langle Cv, v\rangle| \geq \alpha \|v\|^2_X$, for $v \in X$

with $A = V$. The assertion 1. follows, since

$$\begin{aligned}|\langle Qv, Vw\rangle| &= |-\int_0^1 v''(s)(Vw)(s)\,ds| \\ &= |\int_0^1 v'(s)(Vw)'(s)\,ds| \\ &= |\int_0^1 v'(s)(Vw')(s)\,ds| \le M\|v'\|_{H^{-\frac{1}{2}}} \cdot \|w'\|_{H^{-\frac{1}{2}}}\end{aligned}$$

Setting $w := v$ above further yields 2., since

$$\langle v', Vv'\rangle \ge \gamma\|v'\|^2_{H^{-\frac{1}{2}}(\Gamma)} \ge \tilde{\gamma}\|v\|^2_{H^{\frac{1}{2}}(\Gamma)} ,$$

which completes the proof of the theorem. □

6.4 Collocation Methods—Revisited

After having briefly discussed the collocation method for the single layer potential in the previous section, we now want to investigate this method in some more detail and a more general setting. In the previous chapters we have already seen that the single layer potential V, given by

$$Vu(x) = -\frac{1}{\pi}\int_\Gamma \ln|x-y|u(y)\,ds_y ,$$

yields a solution of the Laplace equation. To solve the corresponding Dirichlet problem, we have to find a function u such that the boundary condition $Vu(x) = f(x)$ is satisfied for all $x \in \Gamma$. Since an analytic solution can rarely be given, one is looking for a approximate solution in a finite dimensional trial space of dimension N. Of course, we will not achieve that $Vu_N = f$ holds for all boundary points. If we choose N boundary points $x_1, x_2, \ldots, x_N$ and find a function u_N such that at these so-called *collocation points* the equation

$$Vu_N(x_j) = f(x_j) , \quad j = 1, \ldots, N$$

holds, we obtain an approximate solution for the problem, which may approach the exact solution u with N growing. This method is called *collocation.*

Although this method is not very sophisticated and good results have been attained by its practical application, the proof of convergence is delicate and for some simple cases it still remains open. For convergence estimates for regions with corners see Sect. 6.10.

In this chapter, we now want to prove the convergence of the collocation method for the single layer potential with smooth boundary, i.e. $\Gamma \in C^\infty$.

Let $\{\mu_1, \ldots, \mu_N\}$ be a basis of the trial space V_N with $x_1, \ldots, x_N$ being the collocation points. Then, the approximate solution u_N has a representation of the form

$$u_N = \sum_{i=1}^{N} \alpha_k \mu_k .$$

The problem to be solved is given as follows:

$$\boxed{\begin{array}{c}\text{Find } u_N \in V_N \text{ such that} \\ V u_N(x_j) \;=\; f(x_j), \qquad \text{for } j = 1, \ldots, N\end{array}}$$

which can be rewritten by

$$\begin{aligned}
& -\frac{1}{\pi} \int_\Gamma \ln|x_j - y| u_N(y)\, ds_y = f(x_j) && j = 1, \ldots, N \\
\Longleftrightarrow \quad & -\frac{1}{\pi} \int_\Gamma \ln|x_j - y| \sum_{k=1}^{N} \alpha_k \mu_k(y)\, ds_y = f(x_j) && j = 1, \ldots, N \\
\Longleftrightarrow \quad & \sum_{k=1}^{N} \alpha_k \left[-\frac{1}{\pi} \int_\Gamma \ln|x_j - y| \mu_k(y)\, ds_y \right] = f(x_j) && j = 1, \ldots, N
\end{aligned}$$

This yields a linear system of equations for the unknowns $\alpha_1, \ldots, \alpha_N$. Thus, for the calculation of the approximate solution, the collocation method leads to well-known numerical tasks.

For the further studies we want to recall again the δ-distribution, as already introduced in Chap. 2. The δ-distribution was defined by

$$\delta_{x_j}(f) \;=\; f(x_j) \qquad \forall f \in C_0^\infty(\Omega) .$$

However, one may conceive the δ-distribution δ_{x_j} as the derivative of a piecewise constant function with jump at x_j. This enables to link with the Galerkin-method as follows:

Galerkin:	Find u_N such that	$\langle t, V u_N \rangle \;=\; \langle t, f \rangle$	$\forall t \in V_N$
Collocation:	Find u_N such that	$V u_N(x_j) \;=\; f(x_j)$	$j = 1, \ldots, N$

Thus, both methods are projection methods of the form

$$\boxed{\begin{array}{c}\text{Find } u_N \in V_N \text{ such that} \\ \langle t, V u_N \rangle \;=\; \langle t, f \rangle \qquad \forall t \in T_N\end{array}} , \tag{6.7}$$

where an approximate solution u_N in a trial space V_N is to be found, i.e. the exact solution u is projected onto the trial space V_N and tested against a testfunction $t \in T_N$. While for the Galerkin-method the test and trial space are identical, for the collocation-method the test space T_N is spanned be the $\delta-$distributions of the collocation points.

Remark 6.2 Choosing in (6.7) for the testfunction t a δ-distribution δ_{x_j}, we obtain

$$\langle t, Vu_N \rangle \;=\; \delta_{x_j}(Vu_N) \;=\; Vu_N(x_j)$$

for the left hand side; and for the right hand side

$$\langle t, f \rangle \;=\; \delta_{x_j}(f) \;=\; f(x_j)\ .$$

In the subsections to follow we now want to discuss some important properties of the test and trial space of the collocation-method, before proving a convergence theorem for projection methods. Finally, we will show that the collocation method satisfies the assumptions of Theorem 6.11 and is therefore convergent.

6.4.1 Periodic Splines as Test and Trial Functions

We want to consider a region Ω with smooth boundary $\Gamma = \partial\Omega \in C^\infty$. It is well known that such a region can be transformed to the unit circle, apart from a compact perturbation of the solution. Hence, we will consider the problem only on the unit circle. To simplify the calculations further on, the boundary of the unit circle is mapped onto the interval $[0, 1]$. Let the boundary of the circle and the unit interval be partitioned uniformly according to

$$\begin{aligned} \underline{\Delta}_N &= \{\underline{x}_0, \dots, \underline{x}_{N-1}\} \qquad & \underline{x}_j &= e^{2\pi i j/N} \\ \Delta_N &= \Delta = \{x_0, \dots, x_{N-1}\} \qquad & x_j &= j/N\ . \end{aligned}$$

Let the number of grid points be odd, i.e. $N = 2M + 1$ and let the trial functions be splines of degree d. We may therefore define

$$S_\Delta^d := \left\{ \varphi : \begin{array}{l} \varphi \in C^{d-1} \text{ (globally) is a spline function of degree } d \\ \text{with respect to } \Delta_N = \Delta,\ \text{continued periodically on } \mathbb{R} \end{array} \right\}$$

Since inside every interval the spline function is a polynomial of degree d, i.e. $d + 1$ coefficients have to be determined, and it is further $d - 1$-times differentiable in the grid points, the dimension of the trial spaces will be

$$\dim S_\Delta^d \;=\; N \cdot (d+1) - N \cdot d \;=\; N\ .$$

Let $S_\Delta^{-1} := \operatorname{span}\{\delta(x_j) |\ j = 1, \ldots, N\}$ be the space being spanned by the δ-distributions corresponding to the grid points x_j. Here, $\delta(x_j)$ is to be understood as the derivative of a piecewise constant function with a jump at x_j. Since the δ-distributions are tempered, we have the following properties of the Fourier coefficients of spline functions as defined above:

(i)

$$(\widehat{\delta_{x_j}})_k = \int_0^1 e^{-2\pi ikx} \cdot \delta_{x_j}\, dx = e^{-2\pi ikx_j} = e^{-2\pi ikj/N}$$
$$(\widehat{\delta_{x_j}})_{k+N} = e^{-2\pi i(k+N)j/N} = e^{-2\pi ikj/N} \cdot \underbrace{e^{-2\pi ij}}_{=1} = e^{-2\pi ikj/N}$$

Thus, for $j = 1, \ldots, N$ there holds

$$(\widehat{\delta_{x_j}})_k = (\widehat{\delta_{x_j}})_{k+N} ,$$

and therefore also

$$w \in S_\Delta^{-1} \implies \widehat{w}_k = \widehat{w}_{k+N} ,$$

i.e. we have periodic Fourier coefficients.

(ii)

$$v \in S_\Delta^d \Rightarrow \left(\frac{d}{ds}\right)^{d+1} v \in S_\Delta^{-1} \tag{6.8}$$
$$\Rightarrow \widehat{[(\frac{d}{ds})^{d+1} v]}_k = \widehat{[(\frac{d}{ds})^{d+1} v]}_{k+N} .$$

For any arbitrary function f there holds

$$\widehat{\left[\left(\frac{d}{ds}\right)^m f\right]}_k = (2\pi ik)^m \widehat{f}_k$$

and thus

$$(2\pi ik)^{d+1} \widehat{v}_k = (2\pi i(k+N))^{d+1} \widehat{v}_{k+N} , \quad \text{hence } \widehat{v}_{k+N} = \frac{k^{d+1}}{(k+N)^{d+1}} \widehat{v}_k .$$

(iii) Any integer $k \in \mathbb{Z}$ has a representation of the form

$$k = r + lN \quad \text{with } |r| \le M ,\ l \in \mathbb{Z} .$$

Thus, there holds $\forall v \in S_\Delta^d$:

$$\widehat{v}_k = \widehat{v}_{r+lN} = \frac{(r+(l-1)N)^{d+1}}{(r+lN)^{d+1}}\widehat{v}_{r+(l-1)N} = \frac{r^{d+1}}{(r+lN)^{d+1}}\widehat{v}_r \,. \tag{6.9}$$

Here, one can easily see the reason for N being chosen to be odd and the fact that the spline functions are already uniquely determined by their first r Fourier coefficients $\widehat{v}_r$, $|r| \le M$.

(iv) For the Sobolev norm we recall

$$\|v\|_{H^s}^2 = \sum_{k\neq 0} |k|^{2s}|\widehat{v}_k|^2 + |\widehat{v}_0|^2 \,.$$

Since the spline functions are determined by only a few coefficients, it may be sufficient to consider only these in the definition of the norm. This abbreviated Sobolev-norm can be estimated trivially by the original one. Both norms would be equivalent, if the original norm could also be estimated by the abbreviated one. For the sake of simplicity we will only write:

$$\|v\|_{H^s}^2 = \sum_{k} |k|^{2s}|\widehat{v}_k|^2 \,.$$

Here, we note that one obtains the same results by treating the coefficient $\widehat{v}_0$ seperately. Now (6.9) gives for any $v \in S_\Delta^d$,

$$\begin{aligned}
\|v\|_{H^s}^2 = \sum_k |k|^{2s}|\widehat{v}_k|^2 &= \sum_{r=-M}^{M}\sum_{l\in\mathbb{Z}} |r+lN|^{2s}|\widehat{v}_{r+lN}|^2 \\
&= \sum_{r=-M}^{M}\sum_{l\in\mathbb{Z}} |r+lN|^{2s}\frac{r^{2(d+1)}}{(r+lN)^{2(d+1)}}|\widehat{v}_r|^2 \\
&= \sum_{r=-M}^{M} |\widehat{v}_r|^2 r^{2s} \sum_{l\in\mathbb{Z}} \frac{(r+lN)^{2s}}{r^{2s}} \cdot \frac{r^{2d+2}}{(r+lN)^{2d+2}} \\
&= \sum_{r=-M}^{M} |\widehat{v}_r|^2 r^{2s} \sum_{l\in\mathbb{Z}} \left(\frac{r+lN}{r}\right)^{2(s-d-1)} .
\end{aligned}$$

The second factor of the last term, i.e. $\sum_{l\in\mathbb{Z}} \left(\frac{r+lN}{r}\right)^{2(s-d-1)}$, is only bounded in the case

$$2(s-d-1) < -1 \,,$$

since $\frac{r+lN}{r} = 1 + l\frac{N}{r} \ge 1 + 2l$ is not bounded. In this case, both norms are equivalent for $v \in S_\Delta^d$ and $s < d + \frac{1}{2}$. In the following we will always use the abbreviated norm without changing the notation.

As already mentioned above, the collocation points are distributed uniformly over the circle and the interval, respectively, i.e. let

$$\begin{aligned} \underset{\sim}{\Delta}_{N,\varepsilon} &= \{\underset{\sim}{x}_0, \ldots, \underset{\sim}{x}_{N-1}\}, & \varepsilon &\in [0,1) \\ \widetilde{\Delta}_{N,\varepsilon} &= \widetilde{\Delta} = \{\tilde{x}_0, \ldots, \tilde{x}_{N-1}\}, & \tilde{x}_j &= x_j + \varepsilon/N = \tfrac{j+\varepsilon}{N} \\ S^{-1}_{\widetilde{\Delta}} &:= \operatorname{span}\{\delta(x-\tilde{x}_j) \mid j = 1, \ldots, N\} \end{aligned}$$

We therefore have

$$\widehat{\delta(x-\tilde{x}_j)}_k = \int_0^1 e^{-2\pi i k x}\delta(x-\tilde{x}_j)\,dx = e^{-2\pi i k \tilde{x}_j}$$

and

$$\begin{aligned} \widehat{\delta(x-\tilde{x}_j)}_{k+N} &= e^{-2\pi i (k+N)\tilde{x}_j} = e^{-2\pi i k \tilde{x}_j} \cdot e^{-2\pi i N \frac{j+\varepsilon}{N}} \\ &= e^{-2\pi i k \tilde{x}_j} \cdot e^{-2\pi i j} \cdot e^{-2\pi i \varepsilon} = \widehat{\delta(x-\tilde{x}_j)}_k \cdot e^{-2\pi i \varepsilon} \end{aligned}$$

and thus altogether

$$\widehat{v}_{r+lN} = \widehat{v}_r \cdot e^{-2\pi i l \varepsilon} \qquad \forall\, v \in S^{-1}_{\widetilde{\Delta}}\,.$$

6.4.2 Convergence Theorem for Projection Methods

Here we need some results on the convergence of projection methods, including compact perturbations and spaces with two norms. Here the concept of collectively compact operators (see Appendix A, Definition A.7) introduced by Anselone [2] is an important tool. Such results are well-known [202, 273, 345] but we present a formulation from [131] that is particularly adapted to the present case. This version includes Theorems 1.1, 1.2, 1.3, and 6.1 as special cases.

Let X and Y be Banach-spaces and $A : X \to Y$ a continuous and bijective operator. Let $(T_N)_N$ and $(V_N)_N$ be sequences of test and trial spaces with $V_N \subset X$, $T_N \subset Y'$ and $\dim V_N = \dim T_N < \infty \quad \forall N \in \mathbb{N}$. We consider the problem:

$$\boxed{\begin{array}{l} \text{Find } u_N \in V_N \text{ such that} \\ \langle t, Au_N\rangle = \langle t, f\rangle, \qquad \forall t \in T_N \end{array}} \tag{6.10}$$

Theorem 6.11 (Lemma 1.1 in [131]) *Let the following assumptions be satisfied:*

1. *There exist bounded linear operators $P_N : Y' \to T_N$, converging on Y' strongly to the identity operator, i.e.*

$$\|P_N v - v\|_{Y'} \overset{N\to\infty}{\longrightarrow} 0 \qquad \forall v \in Y'\,.$$

2. *Let X_0 be a Banach space, continuously embedded in X with the norm $\|\cdot\|_{X_0}$, i.e.*

$$\|x\|_X \leq C\|x\|_{X_0} \qquad \forall\, x \in X_0\,.$$

3. *Let $V_N \subset X_0 \qquad \forall N \in \mathbb{N}$.*
4. *For all $N \in \mathbb{N}$ we are given a mapping $Q_N : V_N \to T_N$ and a constant M independent of N such that*

$$| < Q_N v, Aw > | \leq M\|v\|_X\|w\|_{X_0} \qquad \forall\, v \in V_N,\ \forall\, w \in X_0\,.$$

5. *There exist a collectively compact sequence of operators $C_N : X \to Y'$ and a constant γ such that*

$$| < Q_N v, Av > + < C_N v, v > | \geq \gamma\|v\|_X^2 \qquad \forall v \in V_N, \forall N \in \mathbb{N}.$$

Then we have:

i) **Existence of a unique solution**
There exists an $N_0 \in \mathbb{N}$ such that $\forall\, N \geq N_0$ the system (6.10) has a unique solution $u_N \in V_N$ for any $f \in Y'$.
ii) **Stability of the method**
$\exists\, C$ *(independent of N)* $: \ \|u_N\|_X \leq C\|u\|_{X_0} \quad \forall\, N \geq N_0,\ \forall\, u \in X_0\,.$
iii) **Quasi-optimal error estimate**
$\exists\, C$ *(independent of N)* $: \ \|u - u_N\|_X \leq C \cdot \inf_{\chi \in V_N} \|u - \chi\|_{X_0} \quad \forall\, N \geq N_0\,.$

Proof We first want to prove the unique solvability of (6.10) and the stability property for the case $C_N = 0$ for all $N \in \mathbb{N}$, then for arbitrary C_N. Eventually, the quasi-optimality is to be derived from the first two statements.

Uniqueness: Let $\langle t, Av\rangle = 0$ for all $t \in T_N$. Then there holds with $Q_N v \in T_N$:

$$\gamma\|v\|_X^2 \overset{5.}{\leq} |\langle Q_N v, Av\rangle| = 0 \ \Rightarrow\ v \equiv 0\,,$$

thus, the homogeneous problem has only the trivial solution. Hence the solution is unique.

Existence: Testing $\langle t, Au_N\rangle = \langle t, f\rangle$ only for a basis of T_N and representing u_N in terms of a basis of V_N, we obtain a $N \times N$-system of linear equations, which has to be regular by the uniqueness of the solution. Hence, it is solvable.

For $u_N \in V_N$ there holds assumption 5.:

$$|\langle Q_N u_N, Au_N\rangle| \geq \gamma\|u_N\|_X^2\,.$$

We therefore have:

$$\begin{aligned}\|u_N\|_X^2 \le \tfrac{1}{\gamma}|\langle Q_N u_N, Au_N\rangle| \quad &= \tfrac{1}{\gamma}|\langle Q_N u_N, f\rangle| \\ = \tfrac{1}{\gamma}| < Q_N u_N, Au > | &\overset{4.}{\le} \tfrac{M}{\gamma}\|u_N\|_X \cdot \|u\|_{X_0},\end{aligned}$$

Hence

$$\|u_N\|_X \le \frac{M}{\gamma}\|u\|_{X_0} \,. \tag{6.11}$$

Now we consider the general case with nonvanishing perturbations $C_N \not\equiv 0$. By assumption there exists operators C_N and Q_N satisfying 4. and 5. With these, we will now define operators $\widetilde{Q}_N$ and $\widetilde{C}_N$ that also satisfy the two estimates such that now $\widetilde{C}_N \equiv 0$, i.e. we reduce the general case to the special case considered above.

Let A'^{-1} be the inverse of the adjoint $A' : \; Y' \to X'$ to A which is also continuous and bijective. We define

$$\begin{aligned}\widetilde{Q}_N &:= Q_N + P_N A'^{-1} C_N \\ &= Q_N + A'^{-1} C_N - (1 - P_N) A'^{-1} C_N\end{aligned}$$

Then $\widetilde{Q}_N$ satisfies assumption 4.:

$$\begin{aligned}|\langle \widetilde{Q}_N v, Aw\rangle| &\le |\langle Q_N v, Aw\rangle| + |\langle P_N A'^{-1} C_N v, Aw\rangle| \qquad \text{(by 4. for } Q_N) \\ &\le M\|v\|_X \cdot \|w\|_{X_0} + \|P_N A'^{-1} C_N\| \cdot \|A\| \cdot \|v\|_X \cdot \|w\|_X \,.\end{aligned}$$

Furthermore, $\|P_N A'^{-1} C_N\| \cdot \|A\|$ is bounded, since P_N is bounded, A and A'^{-1} are both bounded and C_N is compact, hence bounded, too. With assumption 2. we therefore obtain:

$$\begin{aligned}|\langle \widetilde{Q}_N v, Aw\rangle| &\le M\|v\|_X\|w\|_{X_0} + \underbrace{\|P_N A'^{-1} C_N\| \cdot \|A\|}_{=:M_1} \cdot \|v\|_X \cdot C \cdot \|w\|_{X_0} \\ &\le (M + M_1 C)\|v\|_X \cdot \|w\|_{X_0}\end{aligned}$$

We now have to show that $\widetilde{Q}_N$ satisfies assumption 5.:

$$\begin{aligned}|\langle \widetilde{Q}_N v, Av\rangle| &= |\langle Q_N v, Av\rangle + \langle A'^{-1} C_N v, Av\rangle - \langle (1 - P_N) A'^{-1} C_N v, Av\rangle| \\ &= |\underbrace{\langle Q_N v, Av\rangle + \langle A' A'^{-1} C_N v, v\rangle}_{\ge \gamma\|v\|_X^2 \text{ by 5. for } Q_N} - \langle A'(1 - P_N) A'^{-1} C_N v, v\rangle|\end{aligned}$$

As $1 - P_N \to 0$ strongly on Y' and the operators C_N and thus $A'^{-1}C_N$ are collectively compact, by Lemma A.2 the sequence $\left(\|A'(1 - P_N)A'^{-1}C_N\|\right)_N$ tends to zero. Defining $\delta_N := \|A'(1-P_N)A'^{-1}C_N\|$, we obtain:

$$|\langle \widetilde{Q}_N v, Av\rangle| \geq (\gamma - \delta_N)\cdot \|v\|_X^2 .$$

If we now choose N_0 such that $\gamma - \delta_N > 0$ for all $N \geq N_0$, e.g. $\gamma - \delta_N \geq \tilde{\gamma} > 0$, there holds:

$$|\langle \widetilde{Q}_N v, Av\rangle| \geq \tilde{\gamma}\|v\|_X^2 .$$

Let the solution operator $u \longrightarrow u_N$ be denoted by G_N. By (6.11), $G_N : X_0 \to (V_N, \|\cdot\|_X)$ is a projection operator with bounded norm:

$$\|G_N\| = \sup_u \frac{\|G_N u\|_X}{\|u\|_{X_0}} = \sup_u \frac{\|u_N\|_X}{\|u\|_{X_0}} \leq \sup_u \frac{C\|u\|_{X_0}}{\|u\|_{X_0}} = C .$$

Furthermore, there holds for all $v \in V_N$ that $G_N(v) = v$.

Now, in order to prove quasi-optimality, let $\tilde{u} \in V_N$ be arbitrary. We then have:

$$\|u - u_N\|_X = \|u - \tilde{u} - (u_N - \tilde{u})\|_X \leq \|u - \tilde{u}\|_X + \|u_N - \tilde{u}\|_X .$$

There further holds:

$$\|u_N - \tilde{u}\|_X = \|G_N(u - \tilde{u})\|_X \leq \|G_N\|\cdot\|u - \tilde{u}\|_{X_0} = C\|u - \tilde{u}\|_{X_0}$$

and thus by assumption 2. :

$$\begin{aligned}
\|u - u_N\|_X &\leq \|u - \tilde{u}\|_X + \|u_N - \tilde{u}\|_X \leq \|u - \tilde{u}\|_X + C\|u - \tilde{u}\|_{X_0} \\
&\leq \tilde{C}\|u - \tilde{u}\|_{X_0} + C\|u - \tilde{u}\|_{X_0} = \hat{C}\|u - \tilde{u}\|_{X_0}, \\
\|u - u_N\|_X &\leq \hat{C} \inf_{\tilde{u}\in V_N} \|u - \tilde{u}\|_{X_0} .
\end{aligned}$$

□

For the rest of this section, we want to show that the assumptions of the theorem are satisfied for the collocation method as described above.

For a smooth boundary ($\Gamma \in C^\infty$), the operator V maps H^s continuously and bijectively onto H^{s+1}. Furthermore, we have already seen the relationship for the Fourier coefficients:

$$\widehat{(Vu)}_m = \frac{\widehat{u}_m}{|m|} . \tag{6.12}$$

We choose S_Δ^d as trial space V_N and $S_{\tilde{\Delta}}^{-1}$ as the testspace. For the equivalence of the Sobolev-norm and the abbreviated Sobolev-norm there are two constraints to the spaces $X = H^s$ and $Y = H^{s+1}$, resp. $Y' = H^{-s-1}$, namely:

$$S^d_\Delta \subset H^s \text{ for } s < d + 1/2$$
$$S^{-1}_{\tilde{\Delta}} \subset H^{-s-1} \text{ for } -s-1 < -1 + 1/2 \Leftrightarrow s > -1/2 .$$

The dimension of the spaces S^d_Δ and $S^{-1}_{\tilde{\Delta}}$ is in both cases N. We will now show that the five assumptions of the Theorem 6.11 hold:

ad 1.: The operator $P_N : H^{-s-1} \to S^{-1}_{\tilde{\Delta}}$ is the projection onto the space of periodical splines, by density satisfying:

$$\|P_N v - v\| \to 0 \ \forall v \in H^{-s-1} .$$

ad 2.: We have $X_0 := X$. However, it should be noted that for problems with corners X and X_0 will be different (see Sect. 6.10).

ad 3.: There trivially holds $V_N \subseteq X_0 \subset X$, since $s < d + \frac{1}{2}$.

We now have to find operators Q_N such that assumptions 4. and 5. will hold.

ad 4.: We have

$$|\langle Q_N v, V w\rangle| \leq \|Q_N v\|_{H^{-s-1}} \|V w\|_{H^{s+1}} \quad \forall v \in V_N, \ \forall w \in H^s .$$

Making use of $\|V w\|_{H^{s+1}} \leq C \cdot \|w\|_{H^s}$, we only have to show:

$$\|Q_N v\|_{H^{-s-1}} \leq C\|v\|_{H^s} \quad \forall v \in X_N .$$

Now, this especially holds if we have for the Fourier coefficients:

$$|m|^{-2s-2} |\widehat{(Q_N v)}_m|^2 \leq C^2 \cdot |m|^{2s} |\hat{v}_m|^2$$
$$\text{resp.} \qquad |\widehat{(Q_N v)}_m| \leq C|m|^{2s+1} |\hat{v}_m| .$$

Hence, if we set $\widehat{(Q_N v)}_m := |m|^{2s+1} \hat{v}_m \ \forall |m| \leq M$, and in particular $\widehat{(Q_N v)}_0 := 0$, the operator Q_N is uniquely defined and satisfies the required property 4.

ad 5.: To prove the last assumption we will show:

$$\Re(\langle Q_N v, \overline{V v}\rangle) \geq \gamma \|v\|^2_{H^s} - \textit{compact perturbation}$$
$$\Longrightarrow \ |\langle Q_N v, V v\rangle| \geq \gamma \|v\|^2_{H^s} - \textit{compact perturbation} :$$

$$\langle Q_N v, \overline{V} v\rangle = \sum_k \widehat{(Q_N v)}_k \widehat{(\overline{V v})}_k (\text{by change of counting and } \widehat{(Q_N v)}_0 = 0)$$

$$\overset{k=m+lN}{=} \sum_{\substack{m=-M \\ m \neq 0}}^{M} \sum_{l \in \mathbb{Z}} \widehat{(Q_N v)}_{m+lN} \widehat{(\overline{V v})}_{m+lN} (\text{since } Q_N v \in S^{-1}_{\tilde{\Delta}} \text{ cf. 6.8})$$

$$= \sum_{\substack{m=-M \\ m\neq 0}}^{M} \sum_{l\in\mathbb{Z}} \widehat{(Q_N v)}_m e^{2\pi i l \varepsilon} \frac{\widehat{(\overline{v})}_{m+lN}}{|m+lN|} \text{(by definition of } Q_N \text{ and } v \in S_\Delta^d)$$

$$= \sum_{\substack{m=-M \\ m\neq 0}}^{M} \sum_{l\in\mathbb{Z}} |m|^{2s+1} \hat{v}_m e^{2\pi i l \varepsilon} \frac{1}{|m+lN|} \left(\frac{m}{m+lN}\right)^{d+1} \widehat{\overline{v}}_m$$

$$= \sum_{\substack{m=-M \\ m\neq 0}}^{M} |m|^{2s} |\hat{v}_m|^2 \sum_{l\in\mathbb{Z}} e^{2\pi i l \varepsilon} \frac{|m|}{|m+lN|} \left(\frac{m}{m+lN}\right)^{d+1}$$

$$\Longrightarrow \quad |\langle Q_N v, \overline{V v}\rangle| \geq \left(\|v\|_{H^s}^2 - |\hat{v}_0|^2\right)$$

$$\times \min_{-M\leq m\leq M} \Bigg| 1 + \underbrace{\sum_{\substack{l\in\mathbb{Z} \\ l\neq 0}} e^{2\pi i l \varepsilon} \frac{|\frac{m}{N}|}{|\frac{m}{N}+l|} \left(\frac{\frac{m}{N}}{\frac{m}{N}+l}\right)^{d+1}}_{\mathscr{Z}_\varepsilon^d(\frac{m}{N})} \Bigg|$$

To prove 5. for the compact perturbation $|\hat{v}_0|^2$ which is not depending on N, i.e. with collectively compact sequence $(C_N)_N$, it only remains to show:

$$\left|1 + \mathscr{Z}_\varepsilon^d\left(\frac{m}{N}\right)\right| \geq \gamma > 0 \quad \text{for arbitrary } \frac{m}{N} \tag{6.13}$$

With $N = 2M + 1$, the term $\frac{m}{N}$ will take on values in the interval $[-1/2, 1/2]$. For $x \in [-1/2, 1/2]$ we have:

$$\mathscr{Z}_\varepsilon^d(x) = \sum_{\substack{l\in\mathbb{Z} \\ l\neq 0}} e^{2\pi i l \varepsilon} \frac{|x|}{|l+x|} \left(\frac{x}{l+x}\right)^{d+1}$$

$$= |x| \cdot x^{d+1} \left(\sum_{l=1}^{\infty} e^{2\pi i l \varepsilon} \underbrace{\frac{1}{|l+x|}}_{=(l+x)} \frac{1}{(l+x)^{d+1}} + \sum_{l=-1}^{-\infty} e^{2\pi i l \varepsilon} \underbrace{\frac{1}{|l+x|}}_{=-(l+x)} \frac{1}{(l+x)^{d+1}} \right)$$

$$= \frac{x \cdot x}{|x|} x^{d+1} \left(\sum_{l=1}^{\infty} e^{2\pi i l \varepsilon} \frac{1}{(l+x)^{d+2}} - \sum_{l=-1}^{-\infty} e^{2\pi i l \varepsilon} \frac{1}{(l+x)^{d+2}} \right)$$

$$= \frac{x}{|x|} x^{d+2} \left(\sum_{l=1}^{\infty} e^{2\pi i l \varepsilon} \frac{1}{(l+x)^{d+2}} \pm \sum_{l=1}^{\infty} e^{-2\pi i l \varepsilon} \frac{1}{(l-x)^{d+2}} \right)$$

$$= \frac{x}{|x|} x^{d+2} \sum_{l=1}^{\infty} \left(e^{2\pi i l \varepsilon} \frac{1}{(l+x)^{d+2}} \pm e^{-2\pi i l \varepsilon} \frac{1}{(l-x)^{d+2}} \right),$$

with the '+'–sign for d being odd and the '−'–sign for d being even. Now, for both, odd and even d, there holds

$$\mathscr{Z}_\varepsilon^d(-x) = \overline{\mathscr{Z}_\varepsilon^d(x)}\,.$$

Therefore, with $\mathscr{Z}_\varepsilon^d(0) = 0$, we only have to examine $\mathscr{Z}_\varepsilon^d(x)$ for $0 < x \le 1/2$:

$$\mathscr{Z}_\varepsilon^d(x) = x^{d+2}\sum_{l=1}^{\infty}\left(e^{2\pi i l\varepsilon}(l+x)^{-d-2} \pm e^{-2\pi i l\varepsilon}(l-x)^{-d-2}\right) .$$

Here, we only want to consider the cases $\varepsilon = 0$ and $\varepsilon = 1/2$, i.e. the collocation points are chosen to be either the grid points or the midpoints of the intervals.

i) d even, $\varepsilon = 0$:

$$\begin{aligned}
\mathscr{Z}_0^d(x) &= x^{d+2}\sum_{l=1}^{\infty}\left((l+x)^{-d-2} - (l-x)^{-d-2}\right)\\
&= x^{d+2}\left(\sum_{l=1}^{\infty}(l+x)^{-d-2} - \sum_{l=2}^{\infty}(l-x)^{-d-2}\right) - x^{d+2}(1-x)^{-d-2}\\
&= x^{d+2}\left(\sum_{l=1}^{\infty}(l+x)^{-d-2} - \sum_{l=1}^{\infty}(l+1-x)^{-d-2}\right) - x^{d+2}(1-x)^{-d-2}
\end{aligned}$$

$$\mathscr{Z}_0^d(1/2) = \left(\tfrac{1}{2}\right)^{d+2}\cdot 0 - \left(\tfrac{1}{2}\right)^{d+2}\left(\tfrac{1}{2}\right)^{-d-2} = -1 .$$

Thus, for d even and $\varepsilon = 0$ (6.13) does not hold.

ii) d even, $\varepsilon = 1/2$:

$$\begin{aligned}
\mathscr{Z}_{1/2}^d(x) &= x^{d+2}\left(\sum_{l=1}^{\infty}e^{\pi i l}(l+x)^{-d-2} - \sum_{l=1}^{\infty}e^{-\pi i l}(l-x)^{-d-2}\right)\\
&= x^{d+2}\left(\sum_{l=1}^{\infty}e^{\pi i l}(l+x)^{-d-2} - \sum_{l=2}^{\infty}e^{-\pi i l}(l-x)^{-d-2}\right) - x^{d+2}e^{-\pi i}(1-x)^{-d-2}\\
&= x^{d+2}\left(\sum_{l=1}^{\infty}e^{\pi i l}(l+x)^{-d-2} - \sum_{l=1}^{\infty}e^{-\pi i(l+1)}(l+1-x)^{-d-2}\right) + x^{d+2}(1-x)^{-d-2}\\
&= x^{d+2}\sum_{l=1}^{\infty}\left((-1)^l(l+x)^{-d-2} + (-1)^l(l+1-x)^{-d-2}\right) + \left(\frac{x}{1-x}\right)^{d+2}\\
&= x^{d+2}\sum_{l=1}^{\infty}(-1)^l\Big(\underbrace{(l+x)^{-d-2} + (l+1-x)^{-d-2}}_{\ge 0}\Big) + \left(\frac{x}{1-x}\right)^{d+2},
\end{aligned}$$

making use of $e^{\pi i} = -1$.

Thus, we have an alternating, monotonously decreasing series which is therefore convergent. Since the first term of the sum is of negative sign, there holds:

$$\mathscr{Z}^d_{1/2}(x) \le \left(\frac{x}{1-x}\right)^{d+2} ,$$

and further, taking into account this first term:

$$\mathscr{Z}^d_{1/2}(x) \ge -x^{d+2}(1+x)^{-d-2} - x^{d+2}(2-x)^{-d-2} + \left(\frac{x}{1-x}\right)^{d+2}$$

$$= \underbrace{-x^{d+2}(1+x)^{-d-2} + x^{d+2}(1-x)^{-d-2}}_{\ge 0 \text{ for } 1/2 \ge x > 0} - x^{d+2}(2-x)^{-d-2}$$

$$\ge -x^{d+2}(2-x)^{-d-2} .$$

$$\Longrightarrow \quad \mathscr{Z}^d_\varepsilon(x) \in \left[-\left(\frac{x}{2-x}\right)^{d+2}, \left(\frac{x}{1-x}\right)^{d+2} \right] \quad \forall x \in [-1/2, 1/2],$$

i.e. in particular we have

$$\mathscr{Z}^d_\varepsilon(x) \le 1 \quad \text{and} \quad \mathscr{Z}^d_\varepsilon(x) \ge -\left(\frac{1/2}{2-1/2}\right)^{d+2} = -\left(\frac{1}{3}\right)^{d+2} > -1 .$$

Thus, to sum up, for even d the mid-point-collocation ($\varepsilon = 1/2$) is convergent whereas the break- (or grid-) point-collocation ($\varepsilon = 0$) is not.

iii) d odd, $\varepsilon = 1/2$:

$$\mathscr{Z}^d_{1/2}(x) = x^{d+2} \left(\textstyle\sum_{l=1}^\infty e^{\pi i l} (l+x)^{-d-2} + \displaystyle\sum_{l=1}^\infty e^{-\pi i l} (l-x)^{-d-2} \right)$$

$$= x^{d+2} \left(\sum_{l=1}^\infty e^{\pi i l} (l+x)^{-d-2} + \sum_{l=1}^\infty e^{-\pi i (l+1)} (l+1-x)^{-d-2} \right)$$

$$+ x^{d+2} e^{-\pi i} (1-x)^{-d-2}$$

$$\Longrightarrow \quad \mathscr{Z}^d_{1/2}(1/2) = \left(\tfrac{1}{2}\right)^{d+2} \underbrace{\left(\sum_{l=1}^\infty (-1)^l (l+1/2)^{-d-2} - \sum_{l=1}^\infty (-1)^l (l+1/2)^{-d-2} \right)}_{=0}$$

$$+ \left(\tfrac{1}{2}\right)^{d+2} (-1) \left(\tfrac{1}{2}\right)^{-d-2} = -1 .$$

Thus, for $\varepsilon = 1/2$ we do not attain convergence.

iv) d odd, $\varepsilon = 0$:

$$\mathscr{Z}^d_0(x) = x^{d+2} \left(\sum_{l=1}^\infty (l+x)^{-d-2} + \sum_{l=1}^\infty (l-x)^{-d-2} \right) \ge \gamma > 0 \quad \forall x > 0 .$$

Thus, for odd d the mid-point-collocation ($\varepsilon = 1/2$) is not convergent whereas the break-point-collocation ($\varepsilon = 0$) is convergent.

Now, all four cases can be summarized as follows:

The ε-collocation with spline functions of degree d for $Vv = f$ converges in the space $H^s(\Gamma)$ with $s \in (-1/2, d + 1/2)$, if for

1. d even the condition $\varepsilon = 1/2$
2. d odd the condition $\varepsilon = 0$

is satisfied. This result can be generalized to the following (proof omitted):

The ε-collocation with spline functions of degree d for $Vv = f$ converges in the space $H^s(\Gamma)$ with $s \in [-1, d + 1/2]$ if and only if there holds for

1. d even the condition $\varepsilon \neq 0$,
2. d odd the condition $\varepsilon \neq 1/2$.

Making use of the convergence estimates deduced from the quasi-optimality property

$$\|u - v_N\| \le C \cdot \inf_{v \in X_N} \{\|u - v\|\} ,$$

i.e. for d being the polynomial degree of the splines and h the grid-size we have

$$\|u - u_N\|_0 \le C \cdot h^{d+1} \|u\|_{d+1} \text{ resp. } \|u - u_N\|_r \le C \cdot h^{d-r+1} \|u\|_{d+1} \,, -1 \le r \le d+1$$

with $\|u\|_{d+1} \le C \cdot \|f\|_{d+2}$ Here we used the continuity of the inverse

$$\|u\|_s = \|V^{-1} f\|_s \le C \cdot \|f\|_{s+1}$$

and $\|u - u_N\|_r \le C \cdot h^{d-r+1} \|f\|_{d+2}$.

The highest order of convergence in the H^{-1} norm is given by:

$$\|u - u_N\|_{-1} \le C \cdot h^{d+2} \|f\|_{d+2} .$$

At the end of this section we briefly look at second kind equations. Banach algebra techniques play a dominant role in the convergence analysis of numerical methods for second kind integral equations on curves with corners and singular integral equations with discontinuous coefficients see S. Prössdorf and A. Rathsfeld [344]. Chapter 7 of the book [345] by S. Prössdorf and B. Silbermann gives a good introduction to and demonstration of the power of Banach algebra techniques in numerical analysis. The paper [122] is concerned with approximation methods for Neumann's integral equation

$$(I - K)u = f \quad \text{on } \Gamma$$

with the double layer potential

$$Ku(x) = -\frac{1}{\pi}\int_\Gamma u(y)\frac{\partial}{\partial n_y}\ln|x-y|ds_y$$

on curves Γ with corners. In [122] necessary and sufficient conditions for the stability of the piecewise constant $\epsilon-$ collocation and for the quadrature method, using the rectangular rule, are given using Banach algebra techniques together with Mellin-techniques as introduced in Chap. 9.

6.5 BEM on Quasiuniform Meshes

In Sects. 6.5.1–6.5.5 we follow [430].

6.5.1 Periodic Polynomial Splines

Let $\Delta = \{t_k\}_{k\in\mathbb{Z}}$ be a partition of $\mathbb{R}$ with grid points t_k, with $t_0 = 0$ and $t_{k+N} = t_k+1$ for a fixed $N \in \mathbb{N}$ and for all $k \in \mathbb{Z}$, i.e. $\{t_k\}_{k=0}^N$ is a partition on $[0, 1]$, which is extended 1-periodically.

$$h := \max\{t_{k+1} - t_k\}$$

is called the mesh size.

For simplicity we consider 1-periodic smoothest splines of degree d : $\mathcal{S}^d(\Delta)$ $(d \in \mathbb{N}_0)$, i.e. $\Phi \in \mathcal{S}^d(\Delta) \iff \varphi$ with all derivatives up to order $\leq d-1$ is 1-periodic and continuous on $\mathbb{R}$ and $\varphi|(t_k, t_{k+1})$ is a polynomial at most of degree d $(\forall k \in \mathbb{Z})$.

For any $d \in \mathbb{N}_0$ $\mathcal{S}^d(\Delta)$ is a N-dimensional space and has as basis $\{B_{j,d}\}_{j=0}^{N-1}$, the *B-Splines* due to de Boor, which are defined recursively as follows:

Let Q_j be the characteristic function of $[t_j, t_{j+1})$. Definier $B_{j,0}$ $(j \in \mathbb{Z})$ by

$$B_{j,0}(t) = Q_j(t) \quad \text{für} \quad t \in [t_j, t_{j+N})$$

and extend $B_{j,0}$ 1-periodically on $\mathbb{R}$.

Then for $d \geq 1,\ j \in \mathbb{Z}$

$$B_{j,d}(t) = \frac{t-t_j}{t_{j+1}-t_j}B_{j,d-1}(t) + \frac{t_{j+d+1}-t}{t_{j+d+1}-t_{j+1}}B_{j+1,d-1}(t)$$

for $t \in [t_j, t_{j+N})$ (hence for support $\text{supp}\,(B_{j,d}) \subset [t_j, t_{j+d+1}]$) and extend $B_{j,d}$ 1-periodically on $\mathbb{R}$.

There holds $\mathcal{S}^d(\Delta) \subset H^s(\mathbb{R}) \iff s < d + \frac{1}{2}$.

6.5.2 *The Approximation Theorem*

Theorem 6.12 *Let*

$$-\infty < s \le r \le d+1, \quad s < d + \tfrac{1}{2}. \tag{6.14}$$

Then there esists a constant $C = C(r, s, d)$ and to any $u \in H^r$ and to any partition Δ there exists a $\varphi \in \mathcal{S}^d(\Delta)$ such that

$$\|u - \varphi\|_s \le C h^{r-s} \|u\|_r \tag{6.15}$$

For the proof we need the **interpolation theorem**

A: Let the linear operator $L : H^\sigma \to H^j$ be continuous with the operator norm $\|L\|_{\sigma,j}$, as well as $\|L\|_{\sigma,j-1}$ for $L : H^\sigma \to H^{j-1}$ (also continuous due to embedding).
Then there holds for $\varrho \in [j-1, j]$ and $L : H^\sigma \to H^\varrho$

$$\|L\|_{\sigma,\varrho} \le (\|L\|_{\sigma,j-1})^{j-\varrho} (\|L\|_{\sigma,j})^{\varrho-j+1} \tag{6.16}$$

(log-convexity, $j - \varrho \in [0, 1]$ und $\varrho = (j-\varrho)(j-1) + (\varrho - j + 1)j$)

B: $L : H^{j-1} \to H^\varrho$ be bounded (and linear) with operator norm $\|L\|_{j-1,\varrho}$, as well as $\|L\|_{j,\varrho}$ for $L : H^j \to H^\varrho$. Then there holds for $j - 1 \le \sigma \le j$ and $L : H^\sigma \to H^\varrho$

$$\|L\|_{\sigma,\varrho} \le (\|L\|_{j-1,\varrho})^{j-\sigma} (\|L\|_{j,\varrho})^{\sigma-j+1} \tag{6.17}$$

Proof (Sketch, More Detailed in [276]) **A.** follows from the definition of the Sobolev spaces via Fourier series and from the Hölder inequality. **B.** follows from A. by use of adjoint operators:

$$L^* : H^{-\varrho} \to H^{1-j} \quad \text{resp.} \quad L^* : H^{-\varrho} \to H^{-j}$$

with

$$\|L^*\|_{-\varrho,1-j} = \|L\|_{j-1,\varrho}, \quad \|L^*\|_{-\varrho,-j} = \|L\|_{j,\varrho} \quad .$$

□

Proof (Approximation Theorem)

1. $d = 0 = s,\ r = 1$
   We apply the equivalent norms

$$\|u\|_\sigma^2 = \|u\|_0^2 + \|u^{(\sigma)}\|_0^2 \quad \text{if } \sigma \in \mathbb{N}$$

$$\|u\|_\sigma^2 = \|u\|_0^2 + \int_0^1 \int_0^1 \frac{|u^{(m)}(t) - u^{(m)}(\tau)|}{|t-\tau|^{1+2\mu}}\, dt\, d\tau$$

for $\sigma = m + \mu,\ m \in \mathbb{N}_0,\ 0 < \mu < 1.$
Let $u \in H^1$. Set for $k = 0, 1, \ldots, N-1$

$$\varphi(t) := \frac{1}{t_{k+1} - t_k} \int_{t_k}^{t_{k+1}} u(\tau)\, d\tau \quad \text{für} \quad t \in [t_k, t_{k+1})$$

Now with Cauchy Schwarz inequality for $t \in (0, 1)$

$$\begin{aligned} |u(t) - \varphi(t)|^2 &= \sum_{k=0}^{N-1} Q_k(t) \left| \frac{1}{t_{k+1} - t_k} \int_{t_k}^{t_{k+1}} 1[u(t) - u(\tau)]\, d\tau \right|^2 \\ &\le \sum_{k=0}^{N-1} Q_k(t)\, h_k^{-1} \int_{t_k}^{t_{k+1}} |u(t) - u(\tau)|^2\, d\tau \end{aligned} \tag{6.18}$$

Now we estimate further for $t, \tau \in (t_k, t_{k+1})$

$$\begin{aligned} |u(t) - u(\tau)|^2 &= \left| \int_\tau^t u'(\sigma)\, d\sigma \right|^2 \\ &\le |t - \tau| \int_{t_k}^{t_{k+1}} |u'|^2\, d\sigma \\ &\le h_k \int_{t_k}^{t_{k+1}} |u'(\sigma)|^2\, d\sigma \end{aligned} \tag{6.19}$$

Now integration yields

$$\int_{t_k}^{t_{k+1}} |u(t) - u(\tau)|^2\, d\tau \le h_k^2 \int_{t_k}^{t_{k+1}} |u'|^2\, d\tau$$

and inserting in (6.18) and integration gives

$$\begin{aligned} \|u - \varphi\|_0^2 &= \int_0^1 |u(t) - \varphi(t)|^2\, dt \le \int_0^1 \sum_{k=0}^{N-1} Q_k(t) h_k \int_{t_k}^{t_{k+1}} |u'|^2\, d\sigma\, dt \\ &\le \sum_{k=0}^{N-1} h_k^2 \int_{t_k}^{t_{k+1}} |u'|^2\, d\tau \le h^2 \int_0^1 |u'|^2 d\tau \end{aligned} \tag{6.20}$$

2. $d = 0 < s < \frac{1}{2}$, $r = 1$
In this case (6.15) follows from the estimate that we show next

$$J := \int_0^1 \int_0^1 |u(t) - \varphi(t) - (u(\tau) - \varphi(\tau))|^2 |t - \tau|^{-1-2s} \, dt \, d\tau$$
$$\leq c\, h^{2-2s} \|u'\|_0^2$$

Here and in the following c denotes different positive constants constants which are independent of u and Δ.
It is

$$J = \sum_{0 \leq k,l < N} J_{k,l}$$

where $J_{k,l}$ is the corresponding integral on $(t_k, t_{k+1}) \times (t_l, t_{l+1})$.
With (6.19) there follows (due to $d = 0 \Rightarrow \varphi(t) = \varphi(\tau)$ for $t, \tau \in (t_k, t_{k+1})$)

$$J_{k,k} = \int_{I_{kk}} |u(t) - u(\tau)|^2 |t - \tau|^{-1-2s} \, dt \, d\tau$$
$$\leq \int_{t_k}^{t_{k+1}} |u'|^2 \, d\tau \int_{I_{kk}} |t - \tau|^{-2s} \, dt \, d\tau$$
$$= (t_{k+1} - t_k)^{2-2s} \int_{(0,1)^2} |\tilde{t} - \tilde{\tau}|^{-2s} \, d\tilde{t} \, d\tilde{\tau} \int_{t_k}^{t_{k+1}} |u'|^2 \, d\tau$$
$$\leq c\, h^{2-2s} \int_{t_k}^{t_{k+1}} |u'|^2 \, d\tau \tag{6.21}$$

With(6.18) and (6.19) there follows for $k \neq l$ due to $(|A| + |B|)^2 \leq 2(|A|^2 + |B|^2)$

$$J_{kl} \leq 2\Big\{ \int_{t_k}^{t_{k+1}} \int_{t_l}^{t_{l+1}} \frac{|u(t) - \varphi(t)|^2}{|t - \tau|^{1+2s}} \, dt \, d\tau + \int_{t_k}^{t_{k+1}} \int_{t_l}^{t_{l+1}} \frac{|u(\tau) - \varphi(\tau)|^2}{|t - \tau|^{1+2s}} \, dt \, d\tau \Big\}$$
$$\leq 2h \Big\{ \int_{t_k}^{t_{k+1}} |u'|^2 \, d\tau + \int_{t_l}^{t_{l+1}} |u'|^2 \, d\tau \Big\} \int_{t_k}^{t_{k+1}} \int_{t_l}^{t_{l+1}} \frac{dt \, d\tau}{|t - \tau|^{1+2s}}$$

Now summation gives (due to symmetry)

$$\sum_{k\neq l} J_{kl} \leq 8h \sum_{k=0}^{N-1} \Big(\sum_{l=k+1}^{N-1} c_{kl} \Big) \int_{t_k}^{t_{k+1}} |u'|^2 \, d\tau \tag{6.22}$$

where

$$\begin{aligned} c_{kl} &= \int_{t_k}^{t_{k+1}} \int_{t_l}^{t_{l+1}} \frac{dt\, d\tau}{(\tau - t)^{1+2s}} \\ &= -\frac{1}{2s} \int_{t_k}^{t_{k+1}} (\tau - t)^{-2s} \Big|_{\tau=t_l}^{t_{l+1}} dt \\ &= \frac{1}{(1-2s)2s} \Big\{ (t_{l+1} - t_{k+1})^{1-2s} - (t_{l+1} - t_k)^{1-2s} \\ &\quad -(t_l - t_{k+1})^{1-2s} + (t_l - t_k)^{1-2s} \Big\} \end{aligned}$$

Hence with $c_s = \frac{1}{(1-2s)2s}$

$$\begin{aligned} \sum_{l=k+1}^{N-1} c_{kl} &= c_s \Big\{ (t_N - t_{k+1})^{1-2s} - (t_N - t_k)^{1-2s} + (t_{k+1} - t_k)^{1-2s} \Big\} \\ &\leq c_s (t_{k+1} - t_k)^{1-2s} \leq c_s h^{1-2s} \end{aligned}$$

which leads with (6.21) and (6.22) to the asserted estimate.

3. $r = d + 1, \quad s \in [d, d + \frac{1}{2})$
   Introduce the following spaces

$$\overset{\circ}{H}{}^s = \{u \in H^s : \int_0^1 u \, dt = 0\} \quad (s \geq 0),$$

$$\overset{\circ}{S}{}^d (\Delta) := \mathcal{S}^d (\Delta) \cap \overset{\circ}{H}{}^s$$

Then

$$\begin{aligned} H^s &= \overset{\circ}{H}{}^s \oplus \mathbb{C}, \quad \mathcal{S}^d (\Delta) = \overset{\circ}{\mathcal{S}}{}^d \oplus \mathbb{C}, \\ D^d &:= \left(\tfrac{d}{dt}\right)^m : \begin{cases} \overset{\circ}{H}{}^{s+d} \to \overset{\circ}{H}{}^s \\ \overset{\circ}{\mathcal{S}}{}^d (\Delta) \to \overset{\circ}{\mathcal{S}}{}^0 (\Delta) \end{cases} \end{aligned}$$

Now it suffices to show that $\forall u \in \overset{\circ}{H}^{d+1}$ $\exists \psi \in \overset{\circ}{\mathscr{S}}^0(\Delta)$ such that

$$\|D^d u - \psi\|_{s-d} \le c\, h^{d+1-s} \|D^d u\|_1$$

With $s - d \in [0, 1/2)$ this follows from 1. and 2. .

4. The case of an integer: $s \le r \le d+1,\ s \le d,\ (s, r \in \mathbb{Z})$
Since the case $s = r$ is trivial, it remains to show the assertion for $s < r$ what we do inductively for $s = d - k$ ($k \in \mathbb{N}_0$).
$k = 0$, i.e. $s = d,\ r = d + 1$ is contained in 3.
induction step from $k \to k + 1$:
We have to show the assertion for $s = d - k - 1 = j - 1$ with $j := d - k$. For this we use the orthogonal projection

$$P_{\varrho,\Delta} : H^{\varrho} \to \mathscr{S}^d(\Delta)$$

with

$$(\forall u \in H^{\varrho},\ \varphi \in \mathscr{S}^d(\Delta)) \quad (u - P_{\varrho,\Delta} u, \varphi)_{\varrho} = 0$$

$$\iff \quad \|u - P_{\varrho,\Delta} u\|_s = \min\{\|u - \varphi\|_s : \varphi \in \mathscr{S}^d(\Delta)\}$$

and obtain

$$\begin{aligned}
\|u - P_{j-1,\Delta} u\|_{j-1} &\le \|u - P_{j,\Delta} u\|_{j-1} \\
&= \sup\left\{|(\psi, u - P_{j,\Delta} u)_j| \ :\ \|\psi\|_{j+1} \le 1\right\} \\
&= \sup\left\{|(\psi - P_{j,\Delta}\psi, u - P_{j,\Delta} u)_j| \ :\ \|\psi\|_{j+1} \le 1\right\} \\
&= \sup\left\{\|\psi - P_{j,\Delta}\psi\|_j\ \|u - P_{j,\Delta} u\|_j \ :\ \|\psi\|_{j+1} \le 1\right\} \\
&\le c\, h\, \|\psi\|_{j+1} \cdot c\, h^{r-j} \|u\|_r \\
&\le c\, h^{r-(j-1)} \|u\|_r
\end{aligned} \tag{6.23}$$

hence the assertion of the induction.

5. Completion of the proof by interpolation

i) With 4. and (6.23) we obtain for $j = r$ (trivial), $j = r - 1, r - 2, \dots,$ $r = d + 1, d, \dots,$ that

$$I - P_{j,\Delta} : H^r \to H^{j-1}$$

is continuous with norm $\leq c\, h^{r+1-j}$ and

$$I - P_{j,\Delta} : \; H^r \to H^j$$

is continuous with norm $\leq c\, h^{r-j}$
Now Interpolation Theorem Part A gives: For $s \in (j-1, j)$, $r \leq d$ and $r \in \mathbb{Z}$ with $s \leq r$ there holds

$$\begin{aligned}\|I - P_{j,\Delta}\|_{r,s} &\leq (c\, h^{r+1-j})^{j-s}\, (c\, h^{r-j})^{s-j+1} \\ &\leq c\, h^{j-s+r-j} = c\, h^r\end{aligned}$$

where

$$I - P_{j,\Delta} : \; H^r \to H^s,$$

also

$$I : \; H^r \to H^s,$$

hence also $P_{j,\Delta}$, i.e. for $r \in \mathbb{Z}$ and $s \in \mathbb{R}$ with $s \leq r$ and $s \leq d$ there holds

$$\forall u \in H^r \; \exists \varphi := P_{j,\Delta} u \in \mathscr{S}^d(\Delta) \;\text{ such that }\; \|u - \varphi\|_s \leq c\, h^{r-s} \|u\|_r \tag{6.24}$$

ii) Case $r \notin \mathbb{Z}$, due to assumption $r \in [s, d+1]$
Consider

$$P_{s,\Delta} : \; H^s \to \mathscr{S}^d(\Delta) \subset H^s$$

due to $s \leq d < d + \frac{1}{2}$. Now

$$I - P_{s,\Delta} : \; H^s \to H^s$$

is continuous with norm ≤ 1 and

$$I - P_{s,\Delta} : \; H^{d+1} \to H^s$$

is continuous with norm $\leq c\, h^{d+1-s}$ by (6.24)
Interpolation due to B. gives for $r = \frac{r-d-1}{s-d-1} s + \frac{s-r}{s-d-1}(d+1)$

$$I - P_{s,\Delta} : \; H^r \to H^s$$

is continuous with norm $\leq 1^{(r-d-1/s-d-1)}\, (ch^{d+1-s})^{\frac{s-r}{s-d-1}} = c' h^{r-s}$

Finally

iii) Case $d < s < d + \frac{1}{2}$, $r \in [s, d+1]$
Consider again

$$I - P_{s,\Delta} \colon H^s \to H^s$$

and

$$I - P_{s,\Delta} : H^{d+1} \to H^s$$

is continuous with norm $\leq c\, h^{d+1-s}$ by 3.

Finally interpolation by B. analogously to (ii) yields the assertion. □

6.5.3 Stability and Inverse Estimates

In this section we consider in more detail 1-periodic B-splines as basis for $S^d(\Delta)$ for special families of grids Δ n. We find a norm which is equivalent to the L_2-norm on $S^d(\Delta)$ where the stability constants are independent of the mesh width. Further we show for the trial space $S^d(\Delta)$ the so-called inverse property, which is important for the convergence analysis.

Let $\Delta = \left\{\frac{j}{N}\right\}_{j\in\mathbb{Z}}$ be the family of equidistant meshes with mesh width $h = \frac{1}{N} (N \in \mathbb{N})$.

For fixed degree $d \in \mathbb{N}_0$ of trial functions we introduce

$$\mu(\tau) = \left\{ \begin{array}{cccl} d=0 & d=1 & d=2 & \text{for} \\ \hline 1 & \tau & \frac{1}{2}\tau^2 & 0 \leq \tau < 1, \\ 0 & 2-\tau & -\tau^2+3\tau-\frac{3}{2} & 1 \leq \tau < 2, \\ 0 & 0 & \frac{1}{2}\tau^2-3\tau+\frac{9}{2} & 2 \leq \tau < 3, \\ 0 & 0 & 0 & \text{else} \end{array} \right. \tag{6.25}$$

[compare for example: $d = 2 : \mu = B_{2,0}$ with $h = 1$, $\tau_i = ih (i = 0, \ldots, 3)$] where there holds

$$\int_{\mathbb{R}} \mu(\tau) d\tau = 1. \tag{6.26}$$

Further we set for $l \in \mathbb{Z}$, $N \geq 1 + d$

$$\mu_l(t) := \mu\left(\frac{t}{h} - l + 1\right) \quad \text{for } h(l-1) \leq t \leq h(l+d), \tag{6.27}$$

$$\mu_l(t+k) = \mu_l(t) \quad \text{for } k \in \mathbb{Z}; \tag{6.28}$$

i.e. the functions μ_l are built from the function μ by squeezing by the factor $\frac{1}{h}$ (dilatation), together with shifting of $l-1$ (translation) on the t-axis and 1-periodic extension.

Theorem 6.13 *The splines μ_l ($l = 1, \dots, N$) defined in* (6.25)–(6.28) *build a basis of $S^d(\Delta)$. Furthermore for all*

$$\varphi = \sum_{j=1}^{N} \gamma_j \mu_j (\gamma_j \in \mathbb{R})$$

there holds

$$c_1 h \sum_{j=1}^{N} \gamma_j^2 \le \| \sum_{j=1}^{N} \gamma_j \mu_j \|^2_{L^2(0,1)} \le c_2 h \sum_{j=1}^{N} \gamma_j^2 \tag{6.29}$$

with constants $0 < c_1 < c_2$ *independent of* $h = \frac{1}{N}$ *and* γ_j.

Proof For the first assertion and the proof of (6.29) we consider the symmetric Gram matrix

$$\begin{aligned} M_{jk} = M_{kj} &:= \tfrac{1}{h}(\mu_j, \mu_k)_{L^2(0,1)} \\ &= \tfrac{1}{h}\int_0^1 \mu\left(\tfrac{t}{h} - j + 1\right)\mu\left(\tfrac{t}{h} - k + 1\right)\, dt \\ &= \int_{-j+1}^{N-j+1} \mu(\sigma)\mu(\sigma - k + 1)\, d\sigma. \end{aligned}$$

We find:

$$\begin{aligned} &d = 0 : M_{jk} = \delta_{jk}; \\ &d = 1 : M_{jj} = \tfrac{2}{3}, M_{j,j+1} = \tfrac{1}{6}, \text{ zero else;} \\ &d = 2 : M_{jj} = \tfrac{17}{20}, M_{j,j+1} = \tfrac{13}{60}, M_{j,j+2} = \tfrac{1}{30}, \text{ zero else;} \end{aligned}$$

For these values there holds

$$\sum_{k \ne j} \frac{M_{kj}}{M_{jj}} = \sum_{k \ne j} \frac{M_{jk}}{M_{jj}} = q < 1, \quad (j = 1, \dots, N) \tag{6.30}$$

i.e. the Gram matrix M is strongly diagonally dominant. Consequently M is regular, since for the row-sum-norm $\left\| \frac{1}{M_{11}} M - I \right\|_\infty = q < 1$.

With the help of (6.30) we estmate

$$\left\| \sum_{j=1}^{N} \gamma_j \mu_j \right\|^2_{L^2(0,1)} = h \sum_{j,k=1}^{N} \gamma_j \gamma_k M_{jk}$$

$$= h \left\{ \sum_{j=1}^{N} \gamma_j^2 M_{jj} + \sum_{j=1}^{N} \sum_{\substack{k=1 \\ k \neq 1}}^{N} \gamma_j \gamma_k M_{jk} \right\}$$

$$= hM_{11} \left\{ \sum_{j=1}^{N} \gamma_j^2 + \sum_{j \neq k} \gamma_j \gamma_k \frac{M_{jk}}{M_{jj}} \right\}$$

from above by

$$\leq hM_{11} \left\{ \sum_{j=1}^{N} \gamma_j^2 + \frac{1}{2} \sum_{j=1}^{N} \sum_{\substack{k=1 \\ k \neq j}}^{N} \gamma_j^2 \frac{M_{jk}}{M_{jj}} + \frac{1}{2} \sum_{k=1}^{N} \sum_{\substack{j=1 \\ j \neq k}}^{N} \gamma_k^2 \frac{M_{jk}}{M_{kk}} \right\}$$

$$= hM_{11}(1+q) \sum_{j=1}^{N} \gamma_j^2$$

and from below by

$$\geq hM_{11} \left\{ \sum_{j=1}^{N} \gamma_j^2 - \sum_{j \neq k} \frac{1}{2} (\gamma_j^2 + \gamma_k^2) \frac{M_{jk}}{M_{jj}} \right\}$$

$$= hM_{11}(1-q) \sum_{j=1}^{N} \gamma_j^2,$$

which shows (6.29). □

Theorem 6.14 (Inverse Property) *For arbitrary $r, s \in \mathbb{R}$ with*

$$r \leq s < d + \frac{1}{2}$$

there exists a constant $c = c(r, s)$, such that for all $h = \frac{1}{N}$ and for all $\varphi \in S^d(\Delta)$ there holds

$$\|\varphi\|_s \leq c h^{r-s} \|\varphi\|_r. \tag{6.31}$$

Remark Again $S^d(\Delta) \subset H^s$ implies the assumption $s < d + \frac{1}{2}$.

Proof

1. Step: We show (6.31) for $d < r \leq s < d+\frac{1}{2}$. Since $D^d\varphi \in S^0(\Delta)$ for $\varphi \in S^d(\Delta)$ it suffices to consider the case $d = 0$. Using the equivalent Sobolev-Slobodecki-Norm gives then with $r = d + \rho, s = d + \sigma, 0 < \rho \leq \sigma < \frac{1}{2}; h \leq 1$

$$\begin{aligned}\|\varphi\|_s^2 &= \|D^d\varphi\|_\sigma^2 + \|D^d\varphi\|_0^2 \\ &\leq ch^{2(\rho-\sigma)}\|D^d\varphi\|_\rho^2 + \|D^d\varphi\|_0^2 \\ &\leq ch^{2(r-s)}\|\varphi\|_r^2.\end{aligned}$$

For $0 < \rho \leq \sigma < 1/2$ we have to show for arbitrary $h, \varphi \in S^0(\Delta)$

$$J_\sigma(\varphi) \leq ch^{2(\rho-\sigma)}J_\rho(\varphi),$$

where

$$J_\sigma(\varphi) := \int_0^1\int_0^1 |\varphi(x) - \varphi(y)|^2|x-y|^{-1-2\sigma}\, dx\, dy.$$

With the representation $\varphi = \sum_{l=1}^{N} \gamma_l\mu_l^{(0)}$ we find

$$J_\sigma(\varphi) = \sum_{1\leq l,k\leq N} J_{lk;\sigma},\ J_{lk;\sigma} := \int_{(l-1)h}^{lh}\int_{(k-.1)h}^{kh} \frac{|\gamma_l-\gamma_k|^2}{|x-y|^{1+2\sigma}}\, dx\, dy.$$

Especially there holds $J_{ll;\sigma} = 0$, while $k = l + 1$

$$J_{l,l+1;\sigma} = \frac{|\gamma_{l+1}-\gamma_l|^2}{(1-2\sigma)2\sigma}\left\{2 - 2^{1-2\sigma}\right\}h^{1-2\sigma},$$

consequently

$$\frac{J_{l,l+1;\sigma}}{J_{l,l+1;\rho}} = c(\sigma,\rho)h^{-2(\sigma-\rho)}.$$

For the remaining $|l - k| \geq 2$ we estimate as

$$\begin{aligned}J_{lk;\sigma} &= \int_{(l-1)h}^{lk}\int_{(k-1)h}^{kh} \frac{|\gamma_l-\gamma_k|^2}{|x-y|^{1+2\rho}}\frac{dx\,dy}{|x-y|^{2(\sigma-\rho)}} \\ &\leq h^{2(\rho-\sigma)}J_{lk;\rho}.\end{aligned}$$

Hence (6.31) is shown in this case.

2. Step: We extend the validility of (6.31) to arbitrary $r \leq s$ with the help of interpolation.
First let $d < r < s < d + \frac{1}{2}$. With $r = \frac{1}{2}s + \frac{1}{2}(2r - s)$, $2r - s < s$ follows (by direct application of the Hölder inequality or with the interpolation theorem applied to the embedding $H^s \subset H^{2r-s}$)

$$\|\varphi\|_r \leq \|\varphi\|_s^{1/2} \ \|\varphi\|_{2r-s}^{1/2},$$

consequently with the inequality (6.31) shown in the 1. step

$$\|\varphi\|_r^2 \leq ch^{r-s}\|\varphi\|_r \ \|\varphi\|_{2r-s}.$$

The inequality $\|\varphi\|_s \leq ch^{r-s}\|\varphi\|_r$ once more applied yields $\|\varphi\|_s \leq ch^{2(r-s)}\|\varphi\|_{2r-s}$. By induction we obtain

$$\|\varphi\|_s \leq ch^{(k+1)(r-s)}\|\varphi\|_{r+k(r-s)} \quad (k \in \mathbb{N}).$$

Hence there holds (6.31) for $s \in \left(d, d + \frac{1}{2}\right)$ and arbitrary $r \leq s$. Is $s \leq d$, then we choose a $t \in \left(d, d + \frac{1}{2}\right)$, interpolate

$$s = \frac{t-s}{t-r}r + \frac{s-r}{t-r}t$$

and obtain

$$\begin{aligned}\|\varphi\|_s &\leq \|\varphi\|_r^{(t-s)/(t-r)}\|\varphi\|_t^{(s-r)/(t-r)} \\ &\leq c\|\varphi\|_s^{(t-s)/(t-r)}h^{(r-t)(s-r)/(t-r)}\|\varphi\|_r^{(s-r)/(t-r)} \\ &= ch^{r-s}\|\varphi\|_r.\end{aligned}$$

□

Remark 6.3 Both previous theorems can be extended to quasiuniform meshes, where there exists a constant $\gamma \geq 1$ such that $\gamma^{-1}h \leq t_{k+1} - t_k \leq h$ for all grid points t_h (see [345]).

6.5.4 Aubin-Nitsche Duality Estimate and Superapproximation

Now we improve the error estimates with the help of the orthogonality of the Galerkin scheme

$$\forall\, \chi \in S^m(\Delta) : (Au_h, \chi)_{L^2} = (f, \chi)_{L^2} \tag{6.32}$$

Let $A^* : H^\alpha \to H^{-\alpha}$ be the adjoint operator to A.

Lemma 6.1 ([258]) *Let u solve $Au = f$ and u_h be the corresponding Galerkin solution and*

$$\rho < m + \frac{1}{2}, -m - 1 + 2\alpha \leq \rho \leq \alpha \tag{6.33}$$

and $A^* : H^\alpha \to H^{-\alpha}$ continuous und $A^* : H^{2\alpha-\rho} \to H^{-\rho}$ continuous and bijective. Then there holds the error estimate

$$\|u - u_h\|_\rho \leq c \cdot h^{\alpha-\rho} \|u - u_h\|_\alpha. \tag{6.34}$$

Proof

$$\begin{aligned} \|u - u_n\|_\rho &= \sup_{\|\varphi\|_{-\rho} \leq 1} |(u - u_h, \varphi)_{L^2}| \\ &= \sup_{\|\varphi\|_{-\rho} \leq 1} |(u - u_h, A^* w)_{L^2}| \end{aligned}$$

where

$$\begin{aligned} w &= {A^*}^{-1} \varphi \in H^{2\alpha-\rho}, \|w\|_{2\alpha-\rho} \leq \|\varphi\|_{-\rho} \\ \|u - u_h\|_\rho &= \sup_{\|\varphi\|_{-\rho} \leq 1} \left\{ \left| \Big(u - u_h, A^*(w - \chi)\Big)_{L^2} + (u - u_h, A^* \chi)_{L^2} \right| \right\} \\ &= \sup_{\|\varphi\|_{-\rho} \leq 1} \inf_{\chi \in H_h} \left| \Big(u - u_h, A^*(w - \chi)\Big)_{L^2} + \Big(A(u - u_h), \chi\Big)_{L^2} \right| \\ &\leq \sup_{\|\varphi\|_{-\rho} \leq 1} \inf_{\chi \in H_h} c \|u - u_h\|_\alpha \ \|A^*(w - \chi)\|_{-\alpha} \\ &\leq \sup_{\|\varphi\|_{-\rho} \leq 1} \inf_{\chi \in H_h} c \|u - u_h\|_\alpha \ \|w - \chi\|_\alpha \\ &\leq c'' h^{\alpha-\rho} \sup_{\|\varphi\|_{-\rho} \leq 1} \|w\|_{2\alpha-\rho} \ \|u - u_h\|_\alpha \\ &\leq c''' h^{\alpha-\rho} \sup_{\|\varphi\|_{-\rho} \leq 1} \|\varphi\|_{-\rho} \ \|u - u_h\|_\alpha \end{aligned}$$

□

Theorem 6.15 *Let $A : H^\alpha \to H^{-\alpha}$ be continuous, bijective and H^α-coercive without compact perturbation. $A^* : H^{2\alpha-\rho} \to H^{-\rho}$ be continuous and bijective with $-m - 1 + 2\alpha \leq \rho \leq \alpha \leq \sigma \leq m + 1, \alpha < m + \frac{1}{2}$. Then there exists a $h_0 > 0$, such that for all h mit $0 < h \leq h_0$ the Galerkin equations (6.32) are uniquely solvable and there holds*

$$\|u - u_h\|_\rho \leq c h^{\sigma-\rho} \|u\|_\sigma \tag{6.35}$$

with a constant c, which does not depend on h, u, u_h.

Proof The assumption of Lemma 6.1 are all satisfied, since with A also $A^* : H^\alpha \to H^{-\alpha}$ is continuous. Then (6.34) gives the assertion. □

Example 6.3 Application of the Galerkin scheme to Symm's integral equation:

$$Au := -\frac{1}{\pi} \int_\Gamma u(y) \log|x-y| \, ds_y = f(x) \text{ on } \Gamma$$

under the assumption diam $(\Gamma) < 1$ gives due to $\alpha = -\frac{1}{2}$ the best order of convergence for $\rho = -m-1-1, \sigma = m+1$:

$$\|u - u_h\|_{-m-2} \le ch^{2m+3} \|u\|_{m+1}$$

The Galerkin scheme for the Fredholm integral equation of the second kind

$$Au := u(x) - \frac{1}{\pi} \int_{\Gamma\{x\}} u(y) \frac{\partial}{\partial \gamma_x} (\log(x-y)) \, ds_y = f(x) \text{ auf } \Gamma$$

gives due to $\alpha = 0$ the highest convergence order

$$\|u - u_h\|_{-m-1} \le ch^{2m+2} \|u\|_{m+1}.$$

If $S^m(\Delta)$ has the inverse property, then (6.35) can be extended to the indices $\alpha < \rho \le \sigma$.

Theorem 6.16 *Let $S^m(\Delta)$ have the inverse property* (6.31)*. Then for the Galerkin solutions there holds additionally under the assumptions of Theorem 6.15*

$$\|u - u_h\|_\rho \le ch^{\sigma-\rho} \|u\|_\sigma \tag{6.36}$$

for $\alpha < \rho \le \sigma \le m+1$ and $\rho < m + \frac{1}{2}$.

Proof Let $\tilde{u}_\Delta$ the best approximation of u due to Theorem 6.12 with $\rho = -m-2+2\alpha$. Then there holds (6.34) and (6.35)

$$\begin{aligned}
\|u - u_h\|_\rho &\le \|u - \tilde{u}_\Delta\|_\rho + \|\tilde{u}_\Delta - u_h\|_\rho \\
&\le ch^{\sigma-\rho}\|u\|_\sigma + Mh^{\alpha-\rho}\|\tilde{u}_\Delta - u_h\|_\alpha \\
&\le ch^{\sigma-\rho}\|u\|_\sigma + Mh^{\alpha-\rho}\|\tilde{u}_\Delta - u\|_\alpha + Mh^{\alpha-\rho}\|u - u_h\|_\alpha \\
&\le ch^{\sigma-\rho} + cMh^{\alpha-\rho+\sigma-\alpha}\|u\|_\sigma + c'Mh^{\alpha-\rho+\sigma-\alpha}\|u\|_\sigma
\end{aligned}$$

□

In the proof of the following lemma we use the Aubin-Nitsche duality estimate.

Lemma 6.2 (Superapproximation Property) *Let $\omega \in C_0^\infty(I_0')$ for any interval I_0' and $t_0 < d + \frac{1}{2}$ with $-q \le t \le t_0$, $s \le d$,. Then there exists for any spline*

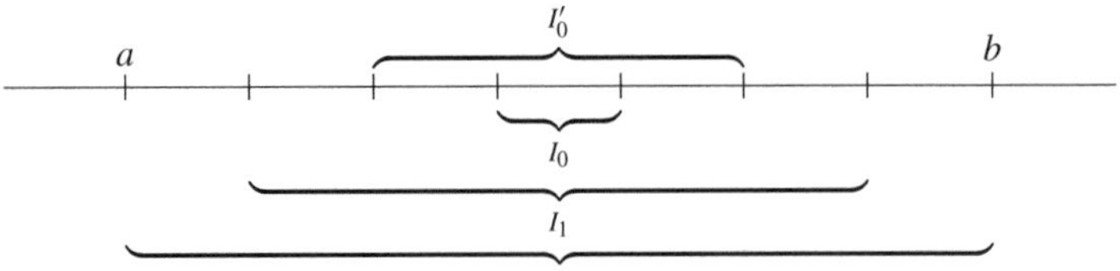

Fig. 6.2 Geometrical setting

$\psi \in S^d$ a spline $\xi \in S^d$ such that

$$\|\omega\psi - \xi\|_{H^t(\Gamma)} \leq c \cdot h^{s+1-t} \|\psi\|_{H^s(I_0')} .$$

Proof We only want to consider the case $d = 0$. For $d \neq 0$ the proof is left to the reader as an exercise.

First, the intervalls I_0, I_0', I_1 and I shall be given according to Fig. 6.2.

Then, for $\omega \in C_0^\infty(I_0')$ there exists a $\eta \in S^0$ such that

$$v := \omega\psi + \eta \ \in H^1(I)$$

with $\eta|_{I\backslash I_0'} \in \mathbb{P}_0$ being a polynomial of degree 0. Defining the integral operator

$$D^{-1} f(x) := \int_a^x f(\tau)\, d\tau \qquad \text{for } a < x < b , \quad I = (a, b)$$

we have

$$w := D^{-q} v \in H^{1+q} \quad \text{with} \quad w|_{I\backslash I_0'} \in \mathbb{P}_q .$$

Fixing now I_2 such that $I_1 \subset\subset I_2 \subset\subset I$ we have

$$\exists \xi \in S^q(I) : \quad \|D^k(w - \xi)\|_{L^2(I)} \leq c \cdot h^{q+1-k} \|D^{q+1} w\|_{L^2(I)} ,$$

with $\xi = w$ in $I_2\backslash I_1$. If we define now

$$\varphi := D^q \xi \text{ and } \rho \in S^0 , \quad \text{given by } \rho(x) = \begin{cases} -(\eta(x) - \varphi(x)) , & x \in I_2 \\ 0 , & x \in I\backslash I_2 \end{cases} ,$$

we have in the interval I_2:

$$\omega\psi - \rho \ = \ \underbrace{D^q w}_{=v} - \underbrace{D^q \xi}_{=\varphi} \ = \omega\psi + \eta - \varphi$$

Thus,

$$
\begin{aligned}
\|\omega\psi - \rho\|_{-q(I_2)} :=& \sup_{0\neq f\in H_0^q(I_2)} \frac{\langle D^q(w-\xi), f\rangle_{L^2(I_2)}}{\|f\|_{q(I_2)}} \\
=& \sup_{0\neq f\in H_0^q(I_2)} \frac{\langle (w-\xi), D^q f\rangle_{L^2(I_2)}}{\|f\|_{q(I_2)}} \\
\leq& \ \|w-\xi\|_{L^2(I_2)} \ \leq \ c\cdot h^{q+1}\|D^{q+1}w\|_{L^2(I_2)} \\
\leq& \ c\cdot h^{q+1}\|\psi\|_{L^2(I_0')},
\end{aligned}
$$

since $D^{q+1}w \ = \ v' \ = \ (\omega\psi+\eta)' \ = \ \omega'\psi$. □

6.5.5 Numerical Quadrature

Projection and variational methods like Galerkin and collokation methods do not only lead to the error by the method as analyzed above but to further errors given by the numerical quadratures for the boundary and domain integrals involved, as well as due to the approximation of a curved boundary by a polygon. The effect of such variational crimes on the error asymptotic can be analysed similarly as in the finite element method.

In this section we restrict ourselves to error estimates of the numerical quadrature for boundary element Galerkin schemes on families Δ of quasiuniform meshes.

Starting point for the following error analysis is the following theorem on the condition of the Galerkin method. Here let A be a pseudodifferential operator of order 2α.

Theorem 6.17 *Let Δ be quasiuniform and let all assumptions of Theorem 6.15 be satisfied and additionally assume $A : H^{2\alpha}(\Gamma) \to L^2(\Gamma)$ is continuous. Then for the Galerkin solutions $u_h \in S^d(\Delta)$ with right hand side $f \in S^d(\Delta)$ there holds*

$$
\|u_h\|_0 \leq ch^{2\alpha'} \ \|f\|_0 \ \text{ and } \ \|f\|_0 \leq ch^{2\alpha'-2\alpha}\|u_h\|_0, \tag{6.37}
$$

where $\alpha' = \min\{0, \alpha\}$. Consequently the condition of the Galerkin equations is of order $O(h^{-2|\alpha|})$.

Proof i) Case $\alpha \leq 0$. Then (6.35) gives with $d \geq 0$

$$
\begin{aligned}
&\|u - u_h\|_\alpha \leq c\|u\|_\alpha, \\
&\|u_h\|_\alpha \leq (1+c)\|u\|_\alpha \leq \tilde{c}\|f\|_{-\alpha}.
\end{aligned}
$$

Due to Thorem 6.15 and Remark 6.3 follows

$$
\|u_h\|_0 \leq ch^\alpha \ \|u_h\|_\alpha \leq c'h^\alpha \ \|f\|_{-\alpha} \leq c''h^{2\alpha} \ \|f\|_0.
$$

On the other hand for $f = P_h A u_h$ there holds with L_2-orthogonal projection P_h

$$\|f\|_0 \le \|Au_h\|_0 \le c\|u_h\|_{2\alpha} \le c\|u_h\|_0.$$

ii) Case $\alpha \ge 0$. Now we estimate

$$\begin{aligned}\|u_h\|_0 &\le \|u_h\|_\alpha \le (1+c)\|u\|_\alpha \le \tilde{c}\ \|f\|_{-\alpha} \le \tilde{c}\ \|f\|_0,\\ \|f\|_0 &\le \|Au_h\|_0 \le c\ \|u_h\|_{2\alpha} \le ch^{-2\alpha}\|u_h\|_0.\end{aligned}$$

Hence we obtain in both cases (6.37). □

Next we start from the numerically computed values $\tilde{a}_{j_k}$ for the entries $(A\mu_j, \mu_k)$ of the Galerkin matrix and from the approximate values $\tilde{f}_k$ for the right hand sides (f, μ_k) of the Galerkin equations. We assume that

$$\begin{aligned}|(A\mu_j, \mu_k)_{L_2} - \tilde{a}_{jk}| &\le ch^R, \\ |(f, \mu_k)_{L_2} - \tilde{f}_k| &\le ch^{R'}.\end{aligned} \tag{6.38}$$

Lemma 6.3 *Under the assumptions (6.38) there hold on $S^d(\Delta)$ the estimates*

$$\begin{aligned}\|A - \tilde{A}\|_{L_2,L_2} &\le ch^{R-2}, \\ \|f - \tilde{f}\|_{L_2} &\le ch^{R'-1},\end{aligned} \tag{6.39}$$

where $\tilde{A} : S^d(\Delta) \to S^d(\Delta)$ and $\tilde{f} \in S^d(\Delta)$ are defined by

$$(\tilde{A}w, v)_{L_2} = \sum_{j,k=1}^N \lambda_j \tilde{a}_{jk} \rho_k, (\tilde{f}, w)_{L_2} = \sum_{j=1}^N \lambda_j \tilde{f}_j$$

for all $w = \sum_{j=1}^N \lambda_j \mu_j$ und $v = \sum_{k=1}^N \rho_k \mu_k$.

Proof i) Estimates for the $L_2 - L_2$ operator norm:

$$|(A - \tilde{A})w, v)_{L_2}| = |\sum_{j,k} \lambda_j \rho_k \left\{(A\mu_j, \mu_k)_{L_2} - \tilde{a}_{jk}\right\}|$$

$$\le \left\{\sum_j |\lambda_j|^2 \cdot \sum_k |\rho_k|^2 \cdot \sum_{j,k} |(A\mu_j, \mu_k)_{L_2} - \tilde{a}_{jk}|^2\right\}^{1/2}$$

with (6.29) and (6.38)

$$\le c\left\{\frac{1}{h} \cdot \|w\|_{L_2}\ \|v\|_{L_2}\right\} \cdot h^R \cdot \{N^2\}^{1/2} = ch^{R-2}\ \|w\|_{L_2}\ \|v\|_{L_2}.$$

ii) The estimate for the L_2 vector norm follows analoguously:

$$
\begin{aligned}
|(f - \tilde{f}, w)_{L_2}| &= |\sum_{j=1}^{N} \lambda_j \{(f, \mu_j)_{L_2} - \tilde{f}_j\}| \\
&\leq \left\{ \sum_{j=1}^{N} |\lambda_j|^2 \cdot \sum_{j=1}^{N} |(f, \mu_j)_{L_2} - \tilde{f}_j|^2 \right\}^{1/2} \\
&\leq c \frac{1}{h^{1/2}} \|w\|_{L_2} h^{R'} \cdot N^{1/2} = c h^{R'-1} \, \|w\|_{L_2}.
\end{aligned}
$$

□

Lemma 6.4 (Second Strang Lemma) *Let (6.38) be satisfied and $R-2+2\alpha' > 0$. Then there exists a $h_0 > 0$, such that the equations*

$$\sum_{j=1}^{N} \tilde{\gamma}_j \tilde{a}_{jk} = \tilde{f}_j$$

are uniquely solvable for any $h \in (0, h_0)$, and for $\tilde{u}_k = \sum_{j=1}^{N} \tilde{\gamma}_j \mu_j$ there hold the asymptotic error estimates

$$\|\tilde{u}_h - u_h\|_{L_2} \leq (c_1 h^{R'-1} + c_2 h^{R-2} \|u_h\|_{L_2}) h^{2\alpha'}. \tag{6.40}$$

Proof 1. Step: There holds

$$
\begin{aligned}
\tilde{A} &= P_h A P_h + (\tilde{A} - P_h A P_h) \\
&= P_h A P_h \{I + (P_h A P_h)^{-1} (\tilde{A} - P_h A P_h)\}.
\end{aligned}
$$

From (6.37) and (6.39) follows

$$\|(P_h A P_h)^{-1} (\tilde{A} - P_h A P_h)\|_{L_2, L_2} \leq c h^{2\alpha'} \cdot h^{R-2} \to 0$$

for $h \to 0$. Consequently $\tilde{A}^{-1}$ can be represented as a convergent Neumann series for sufficiently small $h > 0$. This gives

$$\|\tilde{A}^{-1}\|_{L_2, L_2} \leq c' \|(P_h A P_h)^{-1}\|_{L_2, L_2} \leq c'' h^{2\alpha'} \tag{6.41}$$

2. Step: From

$$P_h A P_h u_h = P_h f \;\; , \;\; \tilde{A} \tilde{u}_h = \tilde{f}$$

we conclude

$$\tilde{A}(\tilde{u}_h - u_h) = \tilde{f} - P_h f + (P_h A P_h - \tilde{A})u_h,$$

and with (6.41) and with (6.40)

$$\|\tilde{u}_h - u_h\|_{L_2} \leq c'' h^{2\alpha'} \{c_1 h^{R'-1} + c_2 h^{R-2} \|u_h\|_{L^2}\}.$$

□

Now we can collect the above error estimates in the following result.

Theorem 6.18 *Let the assumptions of Theorem 6.15 hold. Furthermore let $S^d(\Delta)$ have the inverse property and let $R - 2 + 2\alpha' > 0$. Then there holds for*

$$-d - 1 + 2\alpha \leq \rho \leq \sigma \leq d + 1, \rho < d + \frac{1}{2}, \alpha < d + \frac{1}{2}$$

the asymptotic error estimate including quadrature error

$$\|u - \tilde{u}_h\|_\rho \leq c_0 h^{\sigma-\rho} \|u\|_\sigma + c_1 h^{R'-1+2\alpha'+(-\rho)'} + c_2 h^{R-2+2\alpha'-\rho'} \|u\|_{\alpha^+}, \tag{6.42}$$

where $\alpha^+ = \max\{\alpha, 0\}$, *where R' as in Lemma 6.3*

Proof First we obtain with Theorem 6.15 and 6.16 and the inverse property, according to the sign of ρ

$$\begin{aligned}\|u - \tilde{u}_h\|_\rho &\leq \|u - u_h\|_\rho + \|u_h - \tilde{u}_h\|_\rho \\ &\leq c_0 h^{\sigma-\rho} \|u\|_\sigma + \tilde{c} h^{(-\rho)'} \|u_h - \tilde{u}_h\|_{L_2},\end{aligned}$$

where analogously to above $(-\rho)' = \min\{0, -\rho\}$. This gives with Lemma 6.4

$$\begin{aligned}\|u - \tilde{u}_h\|_\rho \leq\ & c_0 h^{\sigma-\rho} \|u\|_\sigma + c_1 h^{R'-1+2\alpha'+(-\rho)'} \\ &+ c_2 h^{R-2+2\alpha'+(-\rho)'} \|u_h\|_{L_2}.\end{aligned}$$

It only remains to estimate $\|u_h\|_0$. In case $\alpha \geq 0$ we apply Theorem 6.15 with $\sigma = \rho = \alpha$

$$\|u_h\|_0 \leq \|u_h\|_\alpha \leq \|u - u_h\|_\alpha + \|u\|_\alpha \leq (c + 1)\|u\|_\alpha.$$

On the other hand in case $\alpha < 0$ application of Theorem 6.16 gives due to $0 < d + \frac{1}{2}$ with $\sigma = \rho = 0$

$$\|u_h\|_0 \leq \|u - u_h\|_0 + \|u\|_0 \leq (c + 1)\|u\|_0.$$

□

6.5.6 Local $H^{-1/2}$-Error Estimates

We consider the problem

$$\langle L\,(u-u_h)\,,\,\chi\rangle_{L^2(\Gamma)}=0 \quad \text{for } \chi\in S_h\,,$$

where $L = V$ denotes the single-layer potential on a smooth boundary Γ and $u_h \in S_h$ the Galerkin approximation of u. We first want to show that for

$$e:=u-u_h\,,\quad I_0\subset\subset I_0'\subseteq\Gamma \quad \text{and} \quad s=k+\frac{1}{2},\ \ k\in\mathbb{Z}$$

there holds the estimate in the energy-norm:

$$\|e\|_{-\frac{1}{2},I_0}\leq c\cdot\left\{\min_{\chi\in S_h}\|u-\chi\|_{-\frac{1}{2},I_0'}+\|e\|_{H^{-s}(\Gamma)}\right\} \tag{6.43}$$

The derivation of (6.43) is motivated by the approach for the FEM Galerkin solution of the Laplace in [363]

First we note

$$\|Lv\|_{\frac{1}{2}}\leq c\cdot\|v\|_{-\frac{1}{2}} \tag{6.44}$$

$$\langle Lv,v\rangle\cong\|v\|^2_{-\frac{1}{2}},\qquad (\text{with}\quad \operatorname{cap}(\Gamma)<1) \tag{6.45}$$

and

$$L^*\psi=\varphi:\ \ \|\psi\|_{-\frac{1}{2}+l}\leq c\cdot\|\varphi\|_{\frac{1}{2}+l}\,,\qquad l=1,2,\ldots,s-\frac{3}{2} \tag{6.46}$$

We consider a localisisation $I_0\subset\subset I_1\subset\subset I_2\subset\subset I_3\subset\subset I_4\subset\subset I_0'\subset\Gamma$ with

$$\begin{aligned}\omega&\equiv 1 \quad \text{auf } I_0,\quad supp(\omega)\subseteq I_1\\ \tilde\omega&\equiv 1 \quad \text{auf } I_3,\quad supp(\tilde\omega)\subseteq I_4\,,\end{aligned}$$

for $\omega,\tilde\omega\in C^\infty$, such that

$$\|(\omega L-L\omega)(\tilde\omega v)\|_{\frac{1}{2}-l}\leq c\cdot\|\tilde\omega v\|_{-\frac{3}{2}-l} \tag{6.47}$$

$$\|L((1-\tilde\omega)v)\|_{\frac{1}{2},I_2}\leq c\cdot\|v\|_{-s}\,, \tag{6.48}$$

for $\operatorname{supp}(1-\tilde\omega)\cap I_2=\emptyset$. The approximation is given by:
$\forall\, v\in H^{\frac{1}{2}}(\Gamma)\ \exists\,\chi_h\in S_h$ with $\operatorname{supp}(\chi_h)\subseteq I_2$, such that

$$\|\omega v-\chi_h\|_{-\frac{1}{2}}\leq c\cdot h\|v\|_{\frac{1}{2},I_1}\,. \tag{6.49}$$

For $v_h \in S_h$ with supp $(v_h) \subset I_2$, we have (superapproximation):

$$\|\omega v_h - \chi_h\|_{-\frac{1}{2}} \leq c \cdot h \|v_h\|_{-\frac{1}{2}, I_2} \,. \tag{6.50}$$

It should be noted that we may take for χ_h the local L_2-Projection. Eventually, if we assume that I_0' is a part of the mesh, there holds the inverse property

$$\|\chi\|_{0,I_0'} \leq c \cdot h^{-k} \|\chi\|_{-s,I_0'}, \quad \forall \chi \in S_h. \tag{6.51}$$

Remark 6.4 All the above approximation assumptions are local. Since in (6.51) $k \neq s$ is possible, we even need not to have a quasiuniform grid; we only need $h_{max,loc} \leq h^{\beta}_{min,loc}$ for a fixed $\beta > 0$.

We now want to prove the assertion (6.43):
Let u_h be the Galerkin solution, i.e.

$$u \to u_h \in S_h \quad \text{with} \quad \langle L(u - u_h), \chi \rangle = 0 \quad \forall \chi \in S_h$$

Then we have

$$\begin{aligned} \|e\|_{-\frac{1}{2}, I_0} = \|u - u_h\|_{-\frac{1}{2}, I_0} &\leq \|\omega e\|_{-\frac{1}{2}} \\ &\leq \underbrace{\|\omega u - (\omega u)_h\|_{-\frac{1}{2}}}_{=:J_1} + \underbrace{\|(\omega u)_h - \omega u_h\|_{-\frac{1}{2}}}_{=:J_2} . \end{aligned}$$

We have

$$J_1 \leq c \cdot \|\omega u\|_{-\frac{1}{2}} \leq c \cdot \|u\|_{-\frac{1}{2}, I_1} \,. \tag{6.52}$$

For J_2 we may further expand

$$\omega u_h - (\omega u)_h = (\omega u_h - (\omega u_h)_h) + ((\omega u_h)_h - (\omega u)_h) = \Theta_1 + \Theta_2 \,.$$

Thus, with (6.44), (6.45) and the superapproximation (6.50) it follows

$$\begin{aligned} \|\Theta_1\|_{-\frac{1}{2}} \leq c \cdot \min_{\chi \in S_h} \|\omega u_h - \chi\|_{-\frac{1}{2}} &\leq c \cdot h \|u_h\|_{-\frac{1}{2}, I_1} \\ \leq c \cdot \|u\|_{-\frac{1}{2}, I_1} &+ c \cdot h \|e\|_{-\frac{1}{2}, I_1} \,. \end{aligned} \tag{6.53}$$

Now, for $\chi \in S_h$ we have

$$\begin{aligned} \langle L\Theta_2, \chi \rangle = -\langle L(\omega e)_h, \chi \rangle &= -\langle L(\omega \tilde{\omega} e), \chi \rangle \\ = \underbrace{\langle (\omega L - L\omega) \tilde{\omega} e, \chi \rangle}_{=:T_1} &- \underbrace{\langle \omega L(\tilde{\omega} e), \chi \rangle}_{=:T_2} \end{aligned} \tag{6.54}$$

Now, assuption (6.47) with $l = 0$ yields

$$|T_1| \leq c \cdot \|\tilde{\omega} e\|_{-\frac{3}{2}} \cdot \|\chi\|_{-\frac{1}{2}} \leq c \cdot \|e\|_{-\frac{3}{2},I_4} \cdot \|\chi\|_{-\frac{1}{2}} \,,$$

while we have for T_2:

$$T_2 = \underbrace{\langle \omega L\left((1-\tilde{\omega})e\right), \chi\rangle}_{=:T_2'} - \underbrace{\langle Le, \omega\chi\rangle}_{=:T_2''} .$$

For the first term we have by (6.48):

$$|T_2'| \leq c \cdot \|e\|_{-s} \cdot \|\chi\|_{-\frac{1}{2}} .$$

For the second term we use $e = u - u_h$ and thus

$$\begin{aligned} |T_2''| &= |\langle Le, \omega\chi - \psi\rangle| \\ &\leq |\langle L(\tilde{\omega}e), \omega\chi - \psi\rangle| + |\langle L\left((1-\tilde{\omega})e\right), \omega\chi - \psi\rangle| \\ &\leq c \cdot \|e\|_{-\frac{1}{2},I_4} \cdot h\|\chi\|_{-\frac{1}{2}} + c \cdot \|e\|_{-s} \cdot h \cdot \|\chi\|_{-\frac{1}{2}} \,, \end{aligned}$$

making use of (6.48) and (6.50) and considering that $\omega\chi - \psi$ has support in I_2.

If we now take $\chi = \Theta_2$ in (6.54) and use (6.45), we obtain

$$\|\Theta_2\|_{-\frac{1}{2}} \leq c \cdot \left(h \cdot \|e\|_{-\frac{1}{2},I_4} + \|e\|_{-\frac{3}{2},I_4} + \|e\|_{-s}\right) . \tag{6.55}$$

Thus, by (6.52) and (6.53)

$$\|e\|_{-\frac{1}{2},I_0} \leq c \cdot \left(\|u\|_{-\frac{1}{2},I_1} + h \cdot \|e\|_{-\frac{1}{2},I_4} + \|e\|_{-\frac{3}{2},I_4} + \|e\|_{-s}\right) \tag{6.56}$$

For the rest of the proof we need the following

Lemma 6.5 *For $l = 1, 2, \ldots, s - \frac{3}{2}$ there holds*

$$\|e\|_{-l-\frac{1}{2},I_0} \leq c \cdot \left(h \cdot \|e\|_{-\frac{1}{2},I_0'} + \|e\|_{-l-\frac{3}{2},I_0'} + \|e\|_{-s}\right) .$$

Proof First of all we have

$$\|e\|_{-l-\frac{3}{2},I_0} \leq \|\omega e\|_{H^{-l-\frac{1}{2}}(\Gamma)} = \sup_{\substack{\varphi\in H^{l+\frac{1}{2}}(\Gamma) \\ \|\varphi\|_{l+\frac{1}{2}}=1}} \langle \omega e, \varphi\rangle$$

By (6.46), for all those φ with $L^*\psi = \varphi$ there holds: $\psi \in H^{l-\frac{1}{2}}(\Gamma)$. Thus,

$$\begin{aligned}
\langle \omega e, \varphi \rangle &= \langle L(\omega e), \psi \rangle = (\langle L(\omega\tilde{\omega}e), \psi \rangle \\
&= \langle (L\omega - \omega L)(\tilde{\omega}e), \psi \rangle + \langle \omega L(\tilde{\omega}e), \psi \rangle \\
&= \langle (L\omega - \omega L)(\tilde{\omega}e), \psi \rangle + \langle \omega L((\tilde{\omega} - 1)e), \psi \rangle + \langle Le, \omega\psi \rangle \\
&=: S_1 + S_2 + S_3 .
\end{aligned}$$

Here, we have

$$|S_1| \le \|(L\omega - \omega L)\tilde{\omega}e\|_{-l+\frac{1}{2}} \cdot \|\psi\|_{l-\frac{1}{2}} \le c \cdot \|e\|_{-l-\frac{3}{2},I_0'}$$

and

$$|S_2| \le c \cdot \|e\|_{-s}$$

Finally, by direct application of the approximation property (6.49), for $l \ge 1$ we now have

$$\begin{aligned}
|S_3| &\le |\langle Le, \omega\psi - \chi \rangle| \\
&\le |\langle L(\tilde{\omega}e), \omega\psi - \chi \rangle| + |\langle L\left((1 - \tilde{\omega})e\right), \omega\psi - \chi \rangle| \\
&\le c \cdot \left(\|e\|_{-\frac{1}{2},I_0} + \|e\|_{-s} \right) \cdot \|\omega\psi - \chi\|_{-\frac{1}{2}} \\
&\le c \cdot \left(\|e\|_{-\frac{1}{2},I_0} + \|e\|_{-s} \right) \cdot h \cdot \|\psi\|_{\frac{1}{2}}
\end{aligned}$$

For the rest of the proof of (6.43) we may now apply the above lemma inductively in (6.56):

$$\|e\|_{-\frac{1}{2},I_0} \le c \left(\|u\|_{-\frac{1}{2},I_0'} + h\|e\|_{-\frac{1}{2},I_0'} + \|e\|_{-s} \right) .$$

Iterating the estimation $k-$times then yields

$$\|e\|_{-\frac{1}{2},I_0} \le c \left(\|u\|_{-\frac{1}{2},I_0'} + h^k\|e\|_{-\frac{1}{2},I_0'} + \|e\|_{-s} \right) .$$

Now, using (6.51) we obtain

$$\begin{aligned}
h^k\|e\|_{-\frac{1}{2},I_0'} &\le c\|u\|_{-\frac{1}{2},I_0'} + c \cdot h^k \|u_h\|_{0,I_0'} \\
&\le c\|u\|_{-\frac{1}{2},I_0'} + c\|u_h\|_{-s,I_0'} \\
&\le c \left(\|u\|_{-\frac{1}{2},I_0'} + \|e\|_{-s} \right) .
\end{aligned}$$

With $u - \chi = u - u_h + (u_h - \chi)$ the assertion (6.43) finally follows. For further reading see [355, 417]

6.5.7 Local L^2-Error Estimates

Again we assume Γ to be smooth and $L = V$ Analoguous to the proof in the last section we may show for a global quasiuniform mesh that

$$\|e\|_{0,I_0} \leq c \left(\min \|u - \chi\|_{0,I_0'} + \|e\|_{H^{-s}(\Gamma)} \right) \tag{6.57}$$

Proof The proof is similar to the above proof, but now for $L^2(\Gamma)$ instead of $H^{-1/2}(\Gamma)$. First, we use the stability in $L^2(\Gamma)$ and $H^{-1}(\Gamma)$:

$$\|v - v_h\|_i \leq c \cdot \min_{\chi \in S_h} \|v - \chi\|_i \,, \qquad i = 0, -1 \tag{6.58}$$

We consider the problem

$$Lv = f \quad \text{on } \Gamma \text{ with } L := V$$

Thus, by the orthogonality property of the Galerkin method we have

$$\langle L(v - v_h), \chi \rangle \;=\; 0 \qquad \forall \chi \in S_h \,.$$

Then we have

$$\|\omega u - (\omega u)_h\|_0 \;\leq\; c \cdot \|\omega u\|_0 \;\leq\; c \cdot \|u\|_{0,I_1} \,.$$

For $\Theta_1 := \omega u_h - (\omega u)_h$ we obtain

$$\begin{aligned} \|\Theta_1\|_0 \;&\leq\; \inf_{\chi \in S_h} \|\omega u_h - \chi\|_0 \\ &\overset{(6.50)}{\leq}\; c \cdot h \|u_h\|_{0,I_1} \;\leq\; c \|u_h\|_{-1,I_1} \\ &\leq\; c \cdot \|u\|_{0,I_1} \;+\; c \cdot \|e\|_{-\frac{1}{2},I_1} \,. \end{aligned} \tag{6.59}$$

Here, we use that the inverse property holds in this case.

For $\Theta_2 := (\omega e)_h = (\omega u)_h - (\omega u_h)_h$ let $L^*\psi = (\omega e)_h$. Then, there holds:

$$\begin{aligned} \|((\omega e)_h\|_0^2 &= \langle (\omega e)_h, L^*\psi \rangle \;=\; \langle L(\omega e)_h, \psi \rangle \;=\; \langle L(\omega e)_h, \psi_h \rangle \\ &= \langle L(\omega e), \psi_h \rangle \;=\; \langle L\,(\omega \tilde{\omega} e)\,, \psi_h \rangle \\ &= \underbrace{\langle (L\omega - \omega L)(\tilde{\omega} e), \psi_h \rangle}_{=:T_1} \;+\; \underbrace{\langle L(\tilde{\omega} e, \omega \psi_h \rangle}_{=:T_2} \end{aligned}$$

Here, with (6.58) for $i = -1$ we have for T_1:

$$\begin{aligned} T_1 &\leq \|(L\omega - \omega L)(\tilde{\omega} e)\|_1 \cdot \|\psi_h\|_{-1} \\ &\leq c \cdot \|\tilde{\omega} e\|_{-1} \cdot \|\psi\|_{-1} \\ &\leq c \cdot \|e\|_{-1,I_0'} \cdot \|(\omega e)_h\|_0 \,. \end{aligned}$$

Correspondingly, we have for T_2:

$$\begin{aligned} T_2 &= \langle L(\tilde{\omega}e), \omega\psi_h\rangle \\ &= \underbrace{\langle \omega L(\tilde{\omega}-1)e, \psi_h\rangle}_{=:T_2'} + \underbrace{\langle Le, \omega\psi_h\rangle}_{=:T_2''} . \end{aligned}$$

Since $\omega(\tilde{\omega}-1)=0$, we have that $\omega L(\tilde{\omega}-1)$ is of order k for arbitrary k. Thus,

$$\begin{aligned} |T_2'| &\le \|\omega L(\tilde{\omega}-1)e\|_1 \cdot \|\psi_h\|_{-1} \\ &\le c \cdot \|e\|_{-s} \cdot \|(\omega e)_h\|_0 \end{aligned}$$

and for T_2'' we have

$$T_2'' \le \underbrace{|\langle L(\tilde{\omega}e), \omega\psi_h - \chi|}_{=:\tau_1} + \underbrace{|\langle L(1-\tilde{\omega})e, \omega\psi_h - \chi\rangle|}_{=:\tau_2} .$$

For τ_1 we have

$$\begin{aligned} \tau_1 &\le \|(L(\tilde{\omega}e)\|_1 \cdot \|\omega\psi_h - \chi)\|_{-1} \\ &\overset{(6.50)}{\le} c\|\tilde{\omega}e\|_0 \cdot h\|\psi_h\|_{-1} \\ &\overset{\text{Stability}}{\le} c \cdot h\|\tilde{\omega}e\|_0 \cdot \|(\omega e)_h\|_0 \\ &\le c \cdot h\|e\|_{0,I_0} \cdot \|(\omega e)_h\|_0 \end{aligned}$$

and for τ_2

$$\begin{aligned} \tau_2 &\le \|\omega L((1-\tilde{\omega})e)\|_{1,I_0'} \cdot h\|\psi_h\|_{-1} \\ &\le c \cdot \|e\|_{-s} \cdot \|(\omega e_h)\|_0 \cdot h \end{aligned}$$

Combining the above one obtains

$$\|(\omega e_h)\|_0 \le c\left(\|e\|_{-1,I_0'} + \|e\|_{-s} + h \cdot \|e\|_{0,I_0'}\right)$$

Finally, together with the results for the $H^{-\frac{1}{2}}$–case, we have

$$\begin{aligned} \|e\|_{0,I_0} &\le c\left(\|u\|_{0,I_1} + \|e\|_{-\frac{1}{2},I_0'} + \|e\|_{-s} + h \cdot \|e\|_{0,I_0'}\right) \\ &\le c\left(\|u\|_{0,I_1} + \|e\|_{-s} + h\|e\|_{0,I_0'}\right) . \end{aligned}$$

This yields the desired estimate. □

6.5.8 The K-Operator-Method

The method to be presented in this section is due to a work of J.H Bramble and A.H. Schatz [61]. They considered the finite element method for the Dirichlet problem for the Laplacian in a plane domain Ω. Here we consider again this problem with given g on a smooth boundary $\partial\Omega = \Gamma$ with the single-layer potential ansatz

$$U(t) = -\frac{1}{\pi} \int\limits_{\Gamma} \ln|t-s| z(s)\, ds\,, \quad t \in \Omega \tag{6.60}$$

for some unknown density $\varphi = z(s)$. Thus,

$$g(t') = -\frac{1}{\pi} \int\limits_{\Gamma} \ln|t'-s| \cdot z(s)\, ds \tag{6.61}$$

A parametrisation of the curve Γ shall be given by

$$\gamma : \begin{cases} [0,1] \mapsto \Gamma & \\ \quad x \quad \ \ \gamma(x) = t' & \text{with} \quad |\gamma'| > 0 \\ \qquad \to & \\ \quad y \quad \ \ \gamma(y) = s & \end{cases}$$

We then have to solve the integral equation

$$Lu(x) = -2 \int_0^1 \ln|\gamma(x) - \gamma(y)| \cdot u(y)\, dy \stackrel{!}{=} f(x) \tag{6.62}$$

with $u(x) := \frac{1}{2\pi} z[\gamma(x)] \cdot |\gamma'(x)|$ and $f(x) := g[\gamma(x)]$ for $0 \le x < 1$, which can both be extended periodically on $\mathbb{R}^2$.
Now, for the partition $\Delta:\ 0 = x_0 < x_1 < \ldots < x_{N-1} < x_N = 1$ of the interval $[0,1]$ and the space of test– and trialfunctions $S_h := \{\varphi : \ \varphi$ is 1-periodical and piecewise constant on $\Delta\}$ the *Galerkin*–method gives:

$$\boxed{\begin{array}{c} \text{Find } u_h \in S_h \text{ such that} \\ \langle Lu_h, \varphi\rangle = \langle f, \varphi\rangle\,, \quad \forall \varphi \in S_h \end{array}}\,, \tag{6.63}$$

where the scalar product and corresponding norm are given by $\langle v, w\rangle = \int\limits_0^1 v(x) \cdot w(x)\, dx$ and $\|w\|_0^2 = \langle w, w\rangle$, respectively.

Here, we have

$$u(t) = -2 \int\limits_0^1 \ln|t - \gamma(y)| u(y)\, ds_y\,, \qquad t \in \Omega \cup \Gamma$$

$$\text{and } u_h(t) = -2 \int\limits_0^1 \ln|t - \gamma(y)| u_h(y)\, ds_y, \quad t \in \Omega \cup \Gamma$$

and for $t \in \Omega$:

$$
\begin{aligned}
|u(t) - u_h(t)| &= 2|\int_0^1 \underbrace{\ln|t-\gamma(y)|}_{G(t-\gamma(y))}(u(y) - u_h(y))\, ds_y| \\
&\leq c \cdot \|G(t-\gamma(\cdot))\|_2 \cdot \underbrace{\|u - u_h\|_{-2}}_{=O(h^3)} \\
&\leq c \cdot h^3 \|G(t-\gamma(\cdot))\|_2
\end{aligned}
$$

For $t \in \Gamma$ we have

$$|u(t) - u_h(t)| \leq c \cdot h \|G(t-\gamma(\cdot))\|_0$$

Now, the K-operator-method is a post-processing such that

$$|u(t) - \tilde{u}_h(t)| = O\left(h^3\right) \qquad \text{for } t \in \Omega \cup \Gamma .$$

We then approximate u by $\tilde{u_h} = K_h * u_h$ with K_h being a combination of B-Splines such that polynomials up to a certain degree are produced by convolution:

$$K_h(x) := \frac{1}{h} K\left(\frac{x}{h}\right) \qquad \text{with}$$

$$
\begin{aligned}
K(x) &:= -\tfrac{1}{12}[\psi(x+1) + \psi(x-1)] + \tfrac{7}{6}\psi(x) \\
\psi(x) &= \begin{cases} x+1 \,, & -1 \leq x \leq 0 \\ 1-x \,, & 0 \leq x \leq 1 \\ 0 \,, & \text{else} \end{cases}
\end{aligned}
$$

We now want to prove some important properties of the functions K_h:

Lemma 6.6 ([61]) *For the functions K_h as defined above there holds:*

i) $\|K_h * u - u\|_0 \leq c \cdot h^s \|u\|_s$, $\quad 0 \leq s \leq 4$
ii) $D^\alpha(K_h * u) = V_h^\alpha * \partial_h^\alpha u,$ $\quad \alpha = 1, 2$,
where V_h^α denotes a combination of B-splines of lower degree and ∂_h^α is the central difference operator.

Using this lemma, we shall now study the application of the K_h and will therefore distinguish two cases:

Case 1: **Uniform grid**
Let the grid be given by:

$$\Delta : \quad 0 = x_0 < x_1 < \ldots < x_{N-1} < x_N = 1, \quad \text{with } x_i = \frac{i}{N}, \quad h = \frac{1}{N}$$

If then $u_h \in S_h$ satisfies $\langle Lu_h, \varphi\rangle = \langle f, \varphi\rangle \ \forall \varphi \in S_h$, and if furthermore $u \in H^3 := \{u : \ D^\alpha u \in L^2, \quad |\alpha| \le 3\}$, then there holds

$$\|K_h * u_h - u\|_0 \le c \cdot h^3 \|u\|_3$$

Here, we have $\|K_h * u_h - u\| \le \|K_h * u_h - K_h * u\|_0 + \|K_h * u - u\|$,

where $\qquad u_h^*(t) := -2 \int\limits_0^1 \ln|t - \gamma(y)| K_h * u_h(y)\, dy,$

and thus $\qquad |u_h(t) - u_h^*(t)| = 2 |\int\limits_0^1 \ln|t - \gamma(y)| \, (K_h * u_h(y) - u(y))\, dy|$

$$\le c \cdot h^3 \|G(t - \gamma(\cdot))\|_0, \quad t \in \Omega \cup \Gamma .$$

Note that if we only have $u \in H^3(I_1) \cap H^1(\Gamma)$, we shall apply local estimates.

Case 2: **Quasiuniform grid**

Let a quasiuniform grid be given according to Fig. 6.3.

Then, there holds:

$\exists h_0 > 0 \ \forall h \in (0, h_0) \ \forall \varphi \in S_h$ with $\operatorname{supp}(\varphi) \subset\subset I_0$:

$$\varphi(\cdot - 2h) \in S_h \text{ with } \operatorname{supp}(\varphi(\cdot - 2h)) \subset\subset I_1$$

Now, if $u \in H^3(I_1) \cap H^1$ we have

$$\|K_h * u_h - u\|_{L^2(I_0)} \le c \cdot h^3 \left(\|u\|_{H^3(I_1)} + \|u\|_{H^1}\right),$$

where in this case

$$\tilde{u}_h^* := -2 \int\limits_{I_0} \ln|t - \gamma(y)| K_h * u_h(y)\, dy \ - \ 2 \int\limits_{I - I_0} \ln|t - \gamma(y)| u_h(y)\, dy$$

and thus

$$|\tilde{u}_h^* - u(t)| \le c \cdot h^3 \left(\|G(t - \gamma(\cdot))\|_{L^2(I_0)} \ + \ \|G(t - \gamma(\cdot))\|_{H^2(I \setminus I_0)}\right).$$

For further reading we refer to [407, 416].

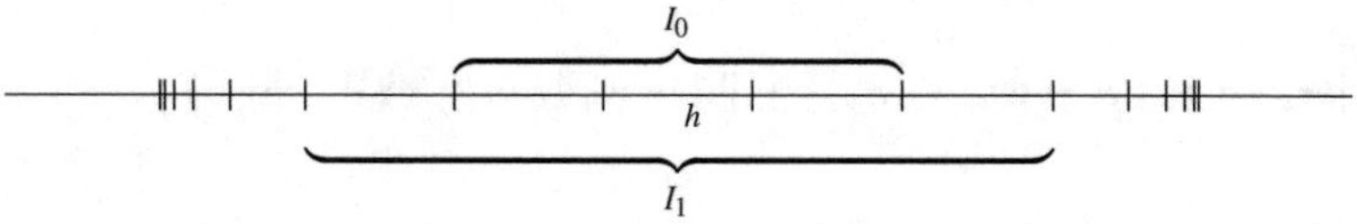

Fig. 6.3 Geometrical setting

6.5.9 L^∞-Error Estimates for the Galerkin Approximation

Here we present from Rannacher and Wendland [349] the estimates for the single layer potential. In this section, let Γ be a closed smooth curve or surface in $\mathbb{R}^2$ or $\mathbb{R}^3$, respectively. We want to consider L^∞-estimates for the Galerkin-error of the single-layer potential operator, i.e. $Vu = f$ on Γ. For $\phi_h \in S_h^{k,m}(\Gamma) \Longleftrightarrow \phi_h \in H^m(\Gamma)$ and $\phi_h|_\Delta \in \mathbb{P}_{k-1}$, the Galerkin method yields

$$\langle Vu_h, \phi_h\rangle_{L^2(\Gamma)} = \langle Vu, \phi_h\rangle_{L^2(\Gamma)} \quad \forall \phi_h \in S_h^{k,m}(\Gamma)$$

Now, for the Galerkin-error $e_h := u - u_h$ we want to prove the estimate

Theorem 6.19 *For a function $u \in L^\infty(\Gamma)$ and a space of testfunctions $S_h^{k,m}(\Gamma)$ with $-k \le -\frac{1}{2}$ there holds:*

$$\|u - u_h\|_{L^\infty(\Gamma)} \le c\left(\log\frac{1}{h}\right)^{\frac{n}{2}-1} \cdot \inf_{\phi_h \in S_h} \|u - \phi_h\|_{L^\infty(\Gamma)} .$$

Let $z \in \Gamma$ be fixed and define a weight function

$$\sigma(x) := \left(|x - z|^2 + \kappa^2 h^2\right)^{\frac{1}{2}}, \quad \kappa \ge 1$$

and weighted Sobolev norms by

$$\|v\|_{r;\beta}^2 := \sum_{|j|\le r} \sum_{K\in\Pi_h} \int_K \sigma^\beta(x)|D^j v(x)|^2\, dx ,$$

with Π_h being a regular triangulation. Define the Galerkin-error by $e := u - u_h$, then for a smooth approximation δ of the Dirac-distribution on Γ, we will consider the inner product $\langle e, \delta\rangle$, i.e. we will solve the equation $V^*g = \delta$ on Γ:

$$\begin{aligned}
\langle e, \delta\rangle &= \langle e, V^*g\rangle = \langle e, V^*(g - g_h + g_h)\rangle && (g_h \text{ being the Galerkin-solution}) \\
&= \langle e, V^*\eta\rangle + \langle e, V^*g_h\rangle && \text{for } \eta := g - g_h \\
&= \langle u - \phi_h, V^*\eta\rangle + \underbrace{\langle e, V^*g_h\rangle}_{=0} && \text{with } e = u - u_h + \phi_h - \phi_h \\
\Longrightarrow\ |\langle e, \delta\rangle| &\le |\langle u - \phi_h, V^*\eta,\rangle| \le \|u - \phi_h\|_{L^\infty(\Gamma)} \cdot \|V^*\eta\|_{L^1(\Gamma)} .
\end{aligned}$$

For the last term we have the estimate

$$\|V^*\eta\|_{L^1(\Gamma)} \le c\left(\log\frac{1}{h}\right)^{\frac{n}{2}-1} h^{\frac{n}{2}-\frac{3}{2}}\|V\eta\|_{0;2} ,$$

where
$$V\eta(x) := \begin{cases} -\frac{1}{2\pi} \int\limits_\Gamma \ln|x-y|\eta(y)\,ds_y\,, \; n=2 \\ -\frac{1}{4\pi} \int\limits_\Gamma \frac{1}{|x-y|}\eta(y)\,ds_y\,, \; n=3 \end{cases} .$$

Now, for the remainder of the proof we need the following auxiliary lemmata:

Lemma 6.7 *Defining* $\xi_i := x_i - z_i$ *for* $1 \le i \le n$, *there holds for* $V\xi_i - \xi_i V$ *:*

$$\|(V\xi_i - \xi_i V)\phi\|_{r+1} \le c \cdot \|\phi\|_{r-1}\,, \quad 1 \le i \le n\,, \quad -\frac{1}{2} \le r \le k\,.$$

Lemma 6.8 *For* $\eta := g - g_h$ *there holds:*

$$\|\xi_i\eta\|_{k;0} + \|\xi_i^2\eta\|_{k;-2} \le c \cdot \|g\|_{k;2}\,, \quad 1 \le i \le n\,.$$

Lemma 6.9 *For* $\alpha \le \beta \le m$ *there exists a constant* $c > 0$ *such that*

$$\|\xi_i\eta\|_\beta \le c \cdot h^{\alpha-\beta}\|\xi_i\eta\|_\alpha + h^{k-\beta}\|g\|_{k;2}\,, \quad 1 \le i \le n\,.$$

Note, in our case we always have $\alpha = -\frac{1}{2}$.

Lemma 6.10 *With* $V\eta(x)$ *as defined in the proof of the theorem above there holds*

$$\|V\eta\|_{0,2} \le c \cdot h^{\frac{1}{2}} \sum_{i=1}^n \|\xi_i\eta\|_{-\frac{1}{2}} + c \cdot h^{k+1}\|g\|_{k;2}\,.$$

Lemma 6.11 *There exist constants* $\epsilon > 0$ *and* $c > 0$ *(independent of* ϵ*) such that*

$$\|\xi_i\eta\|_{-\frac{1}{2}} \le \epsilon \cdot h^{-\frac{1}{2}}\|V\eta\|_{0;2} + c\left(1 + \frac{1}{\epsilon}\right)h^{k+\frac{1}{2}}\|g\|_{k;2}\,.$$

Note that for ϵ sufficiently small the last two lemmata yield $\|V\eta\|_{0;2} \le c \cdot h^{k+1}\|g\|_{k;2}$, giving

$$\begin{aligned} |\langle e, \delta\rangle| \le & c\left(\log\frac{1}{h}\right)^{\frac{n}{2}-1} h^{\frac{n}{2}-\frac{3}{2}}h^{k+1}\|g\|_{k;2} \inf_{\phi_h\in S_h^{k,m}} \|u-\phi_h\|_{L^\infty(\Gamma)} \\ & + c\{h^k\|g\|_k + \|g\|\}\|e\|_{-2}\,. \end{aligned} \tag{6.64}$$

Lemma 6.12 *For* δ *as defined above there holds*

i) $\|g\|_r \le c\|\delta\|_{r+1}\,, \quad 0 \le r \le k$

ii) $\|g\|_{k;2} \le c\left\{\|\delta\|_k + h\|\delta\|_{k+1} + \sum_{i=1}^n \|\xi_i\delta\|_{k+1}\right\}$

Lemma 6.13 ([349]) *With $z \in K$ there exists a function $\delta \in C_0^\infty(K)$ such that $\exists\, c > 0$ (independent of z) : $\phi_h(z) = \langle \phi_h, \delta \rangle \quad \forall \phi_h \in S_h^{k,m}$ satisfying*

1. $\|\delta\|_{L^1} \le c$
2. $h^r \|\delta\|_{r+1} \le c \cdot h^{-\frac{n}{2}-\frac{1}{2}}, \quad 0 \le r \le k$
3. $h^k \|\xi_i \delta\|_{k+1} \le c \cdot h^{\frac{1}{2}-\frac{n}{2}}, \quad 1 \le i \le n$.

Before proving the above lemmata we first want to complete the proof of Theorem 6.19:

For $z \in K$ we have

$$\begin{aligned} |e(z)| &\le |(u-\phi_h)(z)| \;+\; |(\phi_h - u_h)(z)| \\ &\le |(u-\phi_h)(z)| \;+\; |\langle \phi_h - u_h, \delta\rangle| \\ &\le |(u-\phi_h)(z)| \;+\; |\langle e, \delta\rangle| \;+\; \|u-\phi_h\|_{L^\infty(K)} \cdot \|\delta\|_{L^1(K)} \end{aligned}$$

Thus, by Lemma 6.13 we have

$$|e(z)| \le c \cdot \inf_{\phi_h \in S_h^{k,m}} \|u-\phi_h\|_{L^\infty} \;+\; |\langle e, \delta\rangle| \tag{6.65}$$

With Lemma 6.12,(6.64) it follows

$$|\langle e, \delta\rangle| \;\le\; c\left(\log\frac{1}{h}\right)^{\frac{n}{2}-1} \inf_{\phi_h \in S_h^{k,m}} \|u-\phi_h\|_{L^\infty} + c h^{-\frac{1}{2}-\frac{n}{2}} \|e\|_{-2}$$

completing the proof of the theorem.

In the remainder of this section we now prove the above lemmata:

Proof of Lemma 6.7: Standard property of pseudodifferential operators.

Proof of Lemma 6.8: First, we have $D_i(\xi_i;\eta) = \delta_{i,j}\eta \;+\; \xi_j D_i \eta$ and thus

$$\begin{aligned} |D_i(\xi_j\eta)|^2 &\le c\cdot \xi_j^2 |D_i\eta|^2 \;+\; |\eta|^2 \\ &\le c\cdot\sigma^2 |D_i\eta|^2 \;+\; |\eta|^2 \\ \Longrightarrow \quad |D_k D_i \xi_j^2 \eta|^2 &\le c\cdot\sigma^4 |D_k D_j \eta|^2 + c\cdot|\eta|^2 + c\cdot\sigma^2|D_i\eta| + c\cdot\sigma^2|D_k\eta| \,. \end{aligned}$$

Furthermore, there holds

$$\begin{aligned} \|\xi_i\eta\|_{k;0}^2 &\le c\{\|\eta\|_{k;2}^2 + \|\eta\|_{k-1;0}^2\} \\ \|\xi_i^2\eta\|_{k;-2}^2 &\le c\|\eta\|_{k;2}^2 \;+\; c'\|\eta\|_{k-2;-2}^2 \;+\; c''\|\eta\|_{k-1;0}^2 \end{aligned}$$

If now $P_h : H^r \longrightarrow S_h$ is the interpolation operator, there holds

$$\|v - P_h v\|_{j;\beta} \le c\cdot h^{r-j}\|v\|_{r;\beta} \,. \tag{6.66}$$

With ϕ_h being a spline we further have

$$\begin{aligned} \|\phi_h\|_{k,\beta} &\leq c\|\phi_h\|_{k-1,\beta} \\ \|\phi_h\|_{k,\beta} &\leq c\|\phi_h\|_{k,\alpha} \end{aligned} \tag{6.67}$$

and thus:

$$\begin{aligned} \|\eta\|_{k;2} &\leq \|\eta - P_h\eta\|_{k;2} + \|P_h\eta\|_{k;2} \\ &\leq c\|g\|_{k;2} + \|P_h\eta\|_{k;2} \\ &\leq c\|g\|_{k;2} + c\|\eta\|_{k-1;0} \quad \text{using (6.66) and (6.67)}. \\ \|\eta\|_{k-j;0} &\leq \|g - P_h g\|_{k-j;0} + \|P_h(g - g_h)\|_{k-j;0} \leq c\|g\|_{k;0} \cdot h^j \end{aligned}$$

$$\Longrightarrow \quad \|\eta\|_{k-2;-2} \leq \|\eta\|_{k-2;0} c \cdot h^{-1} \leq c \cdot h^{-1} h^2 \|g\|_k$$

The assertion of the lemma finally follows, since

$$\|g\|_k^2 = \sum_{|j|\leq r} \sum_{K\in\Pi_h} \int_{\Pi_h} |D_j g|^2 \sigma^2 \sigma^{-2}\, dx \leq c \cdot h^{-2} \|g\|_{k;2}^2 \qquad \text{with } \sigma^{-2} \leq c \cdot h^{-2}.$$

Proof of Lemma 6.9: The assertion follows by the three arguments (for $\beta = 0$):

1. $\|\xi_j\eta\|_0 \leq \|\xi_j\eta - P_h(\xi_j\eta)\|_0 + \|P_h(\xi_j\eta)\|_0$
2. $\|\xi_j\eta - P_h(\xi_j\eta)\|_0 \leq c \cdot \|\xi_j\eta\|_k h^k \leq c \cdot h^k \|g\|_{k,2}$. (using Lemma 6.8)
3. $\|P_h(\xi_j\eta)\|_0 \leq c \cdot h^{-\frac{1}{2}} \|P_h(\xi_j\eta)\|_{-\frac{1}{2}}$ by the inverse property
   $\leq c \cdot h^{-\frac{1}{2}} \|\xi_j\eta\|_{-\frac{1}{2}}$

Proof of Lemma 6.10: For the proof of this lemma we further need the following estimates:

a) $\|g\|_k^2 \leq c \cdot h^{-1} \|g\|_{k;2}^2$
b) $\|f\|_{1;-2}^2 \leq c \sum_i \|\xi_i f\|_1^2 + c\|f\|_0^2 + c \cdot h^2 \|f\|_1^2$
c) $\|f\|_{k;2} \leq c \cdot \sum_i \|\xi_i f\|_k + \|f\|_{k-1} + \|f\|_k$
d) $\|\sigma^2 f\|_{1;-2} \leq c\|f\| + \|f\|_{1,2}$
   with $|D_i \sigma^2 f|^2 \leq c \cdot \sigma^4 |D_i f|^2 + c \cdot \sigma^2 |f|^2$.

Now, we have for lemma 6.10:

$$\begin{aligned} \|V^*\eta\|_{0;2}^2 &= \langle \sigma^2 V^*\eta, V^*\eta \rangle \\ &= \langle \sigma^2 V^*\eta - P_h\left(\sigma^2 V^*\eta\right), V^*\eta \rangle \\ &\leq \|\sigma^2 V^*\eta - P_h\left(\sigma^2 V^*\eta\right)\|_{0;-2} \|V^*\eta\|_{0;2} \end{aligned}$$

$$
\begin{aligned}
\Longrightarrow \quad \|V^*\eta\|_{0;2} &\le \|\sigma^2 V^*\eta - P_h\left(\sigma^2 V^*\eta\right)\|_{0;-2} \\
&\le c \cdot h\|\sigma^2 V^*\eta\|_{1;-2} \\
&\le c \cdot h\|V^*\eta\|_0 \;+\; c \cdot h\|\xi_j V^*\eta\|_1 \;+\; c \cdot h^2\|V^*\eta\|_1 \\
&\le 2c \cdot h\|\eta\|_{-1} \;+\; c \cdot h^{\frac{1}{2}}\|\xi_j\eta\|_{-\frac{1}{2}} \;+\; c \cdot h^{k+1}\|g\|_{k;2} \\
&\le c \cdot h \cdot h^{k+1}\|g\|_k \;+\; c \cdot h^{\frac{1}{2}}\|\xi_j\eta\|_{-\frac{1}{2}} \;+\; c \cdot h^{k+1}\|g\|_{k;2} \\
&\le c \cdot h^{k+1}\|g\|_{k,2} \;+\; c \cdot h^{\frac{1}{2}}\|\xi_j\eta\|_{-\frac{1}{2}} \;+\; c \cdot h^{k+1}\|g\|_{k;2}
\end{aligned}
$$

Proof of Lemma 6.11:The assertion of the lemma can be shown by

$$
\begin{aligned}
\|\xi_i\eta\|^2_{-\frac{1}{2}} &\le \frac{1}{\gamma}\langle\xi_i\eta, V^*\xi_i\eta\rangle \\
&\le \frac{1}{\gamma}\langle\xi_i\eta, V^*\xi_i\eta - \xi_i V^*\eta\rangle \;+\; \frac{1}{\gamma}\langle\xi_i^2\eta - P_h\left(\xi_i^2\eta\right), V^*\eta\rangle \\
&\le c\|\xi_i\eta\|_{-\frac{1}{2}} \cdot \|\eta\|_{-\frac{3}{2}} \;+\; c \cdot h^k\|g\|_{k,2}\|V^*\eta\|_{0;2} \\
&\le c\|\xi_i\eta\|_{-\frac{1}{2}} h^{\frac{1}{2}+k}\|g\|_{k,2} \;+\; c \cdot h^k\|g\|_{k,2}\|V^*\eta\|_{0;2}
\end{aligned}
$$

Proof of Lemma 6.12: The proof is straight forward. At the end of this section we now want to give a further estimate for the potential.

Theorem 6.20 ([349]) *Under the same assumptions as made above there holds*

$$
\|Vu - Vu_h\|_{L^\infty(\Gamma)} \le c \cdot h^{k+1} \cdot \left(\log\frac{1}{h}\right)^{\frac{n}{2}} \|u\|_{W^{k,\infty}(\Gamma)}
$$

For pointwise estimates of pseudodifferential equations of positive order see [350].

6.6 A Discrete Collocation Method for Symm's Integral Equation on Curves with Corners

Corner singularities of the solution (here of Symm's integral equation) yield only slow convergence for a numerical scheme like the Collocation method. This can be overcome by an appropriate mesh grading transformation. This procedure is described below and goes back to the initiating work [95] by Chandler and Graham. In this section we present from [159] a collocation method with trigonometric polynomials and its discrete counterpart for Symm's integral equation

$$
-\frac{1}{\pi}\int_\Gamma \ln|x-\xi|\,u(\xi)\,\mathrm{d}\Gamma(\xi) = f(x), \quad x \in \Gamma, \tag{6.68}
$$

on the boundary Γ of a simply connected bounded domain Ω in $\mathbb{R}^2$. Γ is assumed to be (infinitely) smooth, with the extension of a corner near at point x_0. Near x_0 Γ should consist of two straight lines intersecting with an interior angle $(1-\chi)\pi$, $0 < |\chi| < 1$. We assume cap $(\Gamma) \neq 1$.

For smooth Γ collocation and quadrature methods based on splines or trigonometric polynomials are analyzed in [379]. Here we show for a curve with a corner that collocation and discrete collocation with trigonometric polynomials converge with a rate as high as justified by the order of the mesh grading and the regularity of the data .

We rewrite (6.68) using an appropriate nonlinear parametrization $\gamma : [0,1] \to \Gamma$ which varies more slowly than arc-length parametrization in the vincinity of x_0. Consider a parametrization $\gamma_0 : [0,1] \to \Gamma$ such that $\gamma_0(0) = \gamma_1(0) = x_0$ and $|\gamma_0'(s)| > 0$ for all $0 \le s \le 1$. Choosing a grading exponent $q \in \mathbb{N}$ and selecting a function υ such that

$$\upsilon \in C^\infty[0,1], \quad \upsilon(0) = 0, \quad \upsilon(1) = 1, \quad \upsilon'(s) > 0, \quad 0 \le s \le 1, \tag{6.69}$$

we define the mesh grading transformation near the corner (see also [95])

$$\gamma(s) = \gamma_0(\omega(s)), \quad \text{where} \quad \omega(s) = \frac{\upsilon^q(s)}{\upsilon^q(s) + \upsilon^q(1-s)}. \tag{6.70}$$

The parametrization γ is graded with exponent q near the corner. With $x = \gamma(s)$, $\xi = \gamma(\sigma)$, equation (6.68) becomes

$$Kw(s) := -2\int_0^1 \ln|\gamma(s) - \gamma(\sigma)|\, w(\sigma)\, d\sigma = g(s), \quad s \in [0,1], \tag{6.71}$$

where

$$w(\sigma) = \frac{1}{2\pi}|\gamma'(\sigma)|\, u(\gamma(\sigma)), \quad g(s) = f(\gamma(s)). \tag{6.72}$$

The solution w of the transformed equation (6.71) may be made as smooth as desired on $[0,1]$ provided f is smooth and the grading exponent is sufficiently large. Therefore w can be optimally approximated using trigonometric polynomials as basis functions.

We rewrite (6.71) as

$$Aw + Bw = g \tag{6.73}$$

with

$$Aw(s) = -2\int_0^1 \ln|2e^{-\frac{1}{2}} \sin(\pi(s-\sigma))|\, w(\sigma)\, d\sigma, \tag{6.74}$$

$$Bw(s) = \int_0^1 b(s,\sigma)\,\mathrm{d}\sigma, \tag{6.75}$$

$$b(s,\sigma) := -2\ln\left|\frac{\gamma(s)-\gamma(\sigma)}{2\mathrm{e}^{-\frac{1}{2}}\sin(\pi(s-\sigma))}\right|, \quad 0 < s,\sigma < 1, \quad s \neq \sigma. \tag{6.76}$$

The kernel (6.76) is 1-periodic in both variables and C^∞ for $0 < s,\sigma < 1$, but Γ has fixed singularities at the four corners of the square $[0,1]\times[0,1]$.

Next we consider **trigonometric collocation**: Let H^t, $t \in \mathbb{R}$, be the usual Sobolev spaces of 1-periodic functions on the real line, with norm given by

$$\| v \|_t^2 = |\hat{v}(0)|^2 + \sum_{m\neq 0} |m|^{2t} |\hat{v}(m)|^2,$$

where the Fourier coefficients of v are defined by

$$\hat{v}(m) = \left(v, \mathrm{e}^{i2\pi ms}\right) = \int_0^1 v(s)\,\mathrm{e}^{-i2\pi ms}\,\mathrm{d}s. \tag{6.77}$$

Introduce the collocation points

$$s_j = jh + \frac{h}{2}, \quad j \in \mathbb{Z}, \quad h := \frac{1}{n}, \tag{6.78}$$

and let $\mathscr{T}_h$ denote the n-dimensional space of trigonometric polynomials with the standard basis

$$\varphi_k(s) = \mathrm{e}^{i2\pi ks}, \quad k \in \Lambda_n := \left\{ j \in \mathbb{Z} : -\frac{n}{2} < j \le \frac{n}{2} \right\}, \tag{6.79}$$

i.e.

$$\mathscr{T}_h = \operatorname{span}\left\{\mathrm{e}^{i2\pi ks},\ k \in \Lambda_n,\ s \in [0,1]\right\}.$$

Then, for any continuous 1-periodic function v, the interpolatory projection $\mathscr{Q}_h v$ onto $\mathscr{T}_h$ is well defined by

$$(\mathscr{Q}_h v)(s_j) = v(s_j), \quad j = 0,\ldots,n-1. \tag{6.80}$$

The following lemma shows that $\mathscr{Q}_h$ has optimal convergence properties:

Lemma 6.14 ([5]) *For $r \ge t \ge 0$ and $r > \frac{1}{2}$ there exists $c > 0$ such that*

$$\| v - \mathscr{Q}_h v \|_t \le c\,h^{r-t} \| v \|_r \quad \text{if } v \in H^r. \tag{6.81}$$

With (6.79) we have

$$\mathcal{Q}_h v(s) = \sum_{k \in \Lambda_n} \alpha_k \varphi_k(s), \quad \alpha_k := h \sum_{j \in \Lambda_n} v(s_j) \overline{\varphi_k(s_j)}.$$

Then the collocation method for (6.73) seeks $w_h \in \mathcal{T}_h$ such that

$$\mathcal{Q}_h (A + B) w_h = \mathcal{Q}_h g.$$

Since $\mathcal{Q}_h$ commutes with A on $\mathcal{T}_h$, there holds

$$(A + \mathcal{Q}_h B) w_h = \mathcal{Q}_h g, \quad w_h \in \mathcal{T}_h. \tag{6.82}$$

Following [4, 155], we rewrite (6.73) as the second kind equation

$$(I + M) w = e, \quad \text{with } M = A^{-1} B, \quad e = A^{-1} g. \tag{6.83}$$

Due to [4] the operator A in (6.74) can be written as

$$Av(s) = \sum_{m \in \mathbb{Z}} \frac{\hat{v}(m)}{\max(1, | m |)} \varphi_m(s), \qquad v \in H^t.$$

and

$$A^{-1} v(s) = \sum_{m \in \mathbb{Z}} \max(1, | m |) \, \hat{v}(m) \varphi_m(s), \qquad v \in H^{t+1}.$$

Therefore the integral operator A is an isomorphism of H^t onto H^{t+1} for any real t. In addition

$$A^{-1} = -\mathcal{H} D + \mathcal{J} = -D\mathcal{H} + \mathcal{J} \tag{6.84}$$

where $\mathcal{J} v(s) = \hat{v}(0)$, $Dv(s) = v'(s)$ and $\mathcal{H}$ the (suitably normalized)Hilbert transform

$$\mathcal{H} v(s) = -\frac{1}{2\pi} \, p.v. \int_0^1 \cot(\pi(s - \sigma)) v(\sigma) \, d\sigma,$$

which is bounded in L^2. Therefore the operator M of (6.83) becomes

$$M = -\mathcal{H} DB + \mathcal{J} B.$$

We now recall some analytical results on equations (6.71) and (6.83) which are needed in the convergence analysis of the trigonometric collocation method. The first theorem was proved in [155], using a decomposition of M into a Mellin convolution operator local to each corner and a compact operator on H^0.

Theorem 6.21 ([155]) *The operators $I + M : H^0 \to H^0$ and $K : H^0 \to H^1$ are continuously invertible, and there holds the strong ellipticity estimate*

$$Re\,((I + M + T)\,v, v) \geq c \parallel v \parallel_0^2, \qquad v \in H^0,$$

with some compact operator T on H^0.

The next theorem shows that the unique solution of (6.71) is smooth provided the given data f in (6.68) is smooth and the grading exponent q is sufficiently large. Let $H^l\,(\Gamma)\,, l > 0$, denote the restriction of the usual Sobolev space $H^{l+\frac{1}{2}}\left(\mathbb{R}^2\right)$ to Γ.

Theorem 6.22 ([155]) *Let $l \in \mathbb{N}$, $q > \left(l + \frac{1}{2}\right)(1+ \mid \chi \mid)$, and suppose that $f \in H^{l+\frac{5}{2}}\,(\Gamma)$.Then the unique solution of (6.71) satisfies $w \in H^l$. Moreover, there exists $\delta < \frac{1}{2}$ such that*

$$D^m w\,(s) = O\left(\mid s \mid^{l-m-\delta}\right), \quad as\ s \to 0, \quad m = 0, \ldots, l. \tag{6.85}$$

The following theorem describes the properties of the kernel function $b\,(s, \sigma)$ defined in (6.76).

Theorem 6.23 ([156]) *On each compact subset of $\mathbb{R} \times \mathbb{R} \setminus (\mathbb{Z} \times \mathbb{Z})$, the derivates $D_s^i\, D_\sigma^m\, b\,(s, \sigma)$ of order $i + m \leq q$ are bounded and 1-periodic. Moreover, for $s, \sigma \in \left[-\frac{1}{2}, \frac{1}{2}\right] \setminus \{0\}$, we have the estimates*

$$\mid b\,(s, \sigma) \mid \leq c \mid \ln\,(\mid s \mid + \mid \sigma \mid) \mid,$$
$$\mid D_s^i\, D_\sigma^m\, b\,(s, \sigma) \mid \leq c\,(\mid s \mid + \mid \sigma \mid)^{-i-m}, \qquad 1 \leq i + m \leq q.$$

Next we rewrite the collocation method (6.82) as a projection method for (6.83). For $v \in H^0$, let $R_h v \in \mathcal{T}_h$ be the solution of the collocation equations $A R_h v = \mathcal{Q}_h A v$. Then $R_h = A^{-1} \mathcal{Q}_h A$ is a well-defined projection operator of H^0 onto $\mathcal{T}_h$.

Note that (6.82) is equivalent to

$$(I + R_h M)\, w_h = R_h e.$$

It is well-known that the use of Mellin convolution operators implies that stability can only be shown for a slightly modified Collocation method (see [95, 157]). We introduce, for $\tau > 0$ sufficiently small, the truncation operator

$$T_\tau v = \begin{cases} v\,(s)\,, & \text{if } s \in (\tau, 1 - \tau) \\ 0, & \text{if } s \in (0, \tau) \cup (1 - \tau, 1) \end{cases}$$

and consider the modified collocation method

$$(A + \mathcal{Q}_h B T_{i^\star h})\, w_h = \mathcal{Q}_h g, \quad w_h \in \mathcal{T}_h, \tag{6.86}$$

where $i^\star$ is a fixed natural number independent of h. If $i^\star = 0$ then (6.86) coincides with (6.82). Otherwise, (6.86) can be obtained from (6.82) by a slight change to the coefficient matrix of the corresponding linear system. Now (6.86) is equivalent to

$$(I + R_h M T_{i^\star h})\, w_h = R_h e, \quad w_h \in \mathcal{T}_h. \tag{6.87}$$

The following theorem provides the convergence of the (modified) collocation method with optimal order in the L^2 norm.

Theorem 6.24 ([159]) *Let $q \geq 2$, and suppose that $i^\star$ is sufficiently large.*

(i) The method (6.87) *is stable, that is the estimate*

$$\| \,(I + R_h M T_{i^\star h})\, v \,\|_0 \geq c \,\| \,v\, \|_0, \quad v \in \mathcal{T}_h \tag{6.88}$$

holds for all h sufficiently small, where c is independent of h and v.

(ii) If, in addition, the hypothesis of Theorem 6.22. holds, then (6.86) *has a unique solution for all h sufficiently small and*

$$\| \,w - w_h\, \|_0 \leq c\, h^l, \tag{6.89}$$

where c is a constant which depends on w and $i^\star$ but is independent of h.

Proof As in [155] we first verify the stability estimate (6.88). Since, by Theorem 6.21, $I + M$ is strongly elliptic and invertible on H^0, we have stability of the finite section operators $T_\tau\,(I + M)\,T_\tau$ as $\tau \to 0$, i.e.

$$\| \,(I + M T_\tau)\, v\, \|_0 \geq c \,\| \,v\, \|_0, \quad v \in H^0, \quad \tau \leq \tau_0. \tag{6.90}$$

To prove (6.88), we need (6.90) and the following perturbation result:

For fixed $q \geq 2$ and each $\varepsilon > 0$, there exists $i^\star \geq 1$ such that for all h sufficiently small

$$\| \,(I - R_h)\, M T_{i^\star h} v\, \|_0 \leq \varepsilon \,\| \,v\, \|_0, \quad v \in H^0. \tag{6.91}$$

Observe that the operator M takes the form $M = -HDB + JB$, where JB is a compact operator on H^0. Since $\| \,T_{i^\star h}\, \|_0 = 1$ and $R_h \to I$ pointwise on H^0, it is sufficient to prove (6.91) with M replaced by $-HDB$. From (6.93) and the fact that $I - R_h$ annihilates the constants, we obtain the estimate

$$\begin{aligned} \| \,(I - R_h)\, M T_{i^\star h} v\, \|_0 &\leq c h \,\| \,M T_{i^\star h} v\, \|_1 = c h \,\| \,D M T_{i^\star h} v\, \|_0 \\ &\leq c h \,\| \,D^2 B T_{i^\star h} v\, \|_0 \,. \end{aligned}$$

To prove (6.91), it is sufficient to verify that

$$\| D^2 BT_{i^\star h} v \|_0 \le \frac{c}{i^\star h} \| v \|_0, \quad v \in H^0, \tag{6.92}$$

where c is indepent of $i^\star$, h and v. Using Theorem 6.23. , we now obtain

$$\begin{aligned} | D^2 BT_{i^\star h} v(s) | &\le \int_{J_{i^\star h}} | D_s^2 b(s,\sigma) | \, | v(\sigma) | \, d\sigma \\ &\le c \int_{J_{i^\star h}} (| s | + | \sigma |)^{-2} \, | v(\sigma) | \, d\sigma \\ &\le \frac{c}{i^\star h} \int_{J_{i^\star h}} \frac{| \sigma |}{(| s | + | \sigma |)^2} \, | v(\sigma) | \, d\sigma, \quad s \in \left(-\frac{1}{2}, \frac{1}{2}\right), \end{aligned}$$

where $J_{i^\star h} = \left(-\frac{1}{2}, -i^\star h\right) \cup \left(i^\star h, \frac{1}{2}\right)$. Then (6.92) follows by taking L^2 norms and using the fact that the integral operator with Mellin convolution kernel $\sigma (s+\sigma)^{-2}$ is bounded on $L^2(0,\infty)$ (see [155] Appendix).

The proof of (6.88) is complete since with (6.90) and (6.91) there holds

$$\| (I + R_h M T_{i^\star h}) v \|_0 \ge \| v + M T_{i^\star h} v \|_0 - \| M T_{i^\star h} v - R_h M T_{i^\star h} v \|_0 \ge c \| v \|_0 .$$

To prove the error estimate (6.89), we note that

$$\| w - w_h \|_0 \le \| (I - R_h) w \|_0 + \| w_h - R_h w \|_0,$$

where the first term is of order h^l by Theorem 6.22. and

$$\| v - R_h v \|_0 \le c h^l \| v \|_l, \quad \text{if } v \in H^l. \tag{6.93}$$

Using (6.88), (6.87) with (6.83), and the uniform boundedness of R_h on H^0 gives

$$\begin{aligned} \| w_h - R_h w \|_0 &\le c \| (I + R_h M T_{i^\star h})(w_h - R_h w) \|_0 \\ &= c \| R_h \{(I + M) w - (I + M T_{i^\star h}) R_h w\} \|_0 \\ &\le c \| (I - R_h) w \|_0 + c \| (I - T_{i^\star h}) w \|_0 . \end{aligned}$$

The proof is complete since by (6.85) (with $m = 0$) and $\| (I - T_{i^\star h}) w \|_0 \le c h^l$ the last term is of order h^l again. □

The following corollary shows that the collocation solutions to the transformed equation yield superconvergent approximations to interior potentials.

Corollary 6.3 ([159]) *Under the hypothesis of Theorem 6.24.(ii), we have*

$$\| w - w_h \|_{-1} \leq c\, h^{l+\beta},$$

where $\beta = 1$ *if* $i^\star = 0$ *and* $\beta = \frac{1}{2}$ *if* $i^\star \geq 1$.

In the following we consider **discrete collocation**. To define a fully discrete version of the collocation method (6.82), introduce the nodes

$$\sigma_r = r\,h, \quad r \in \mathbb{Z}, \quad \text{where } h := \frac{1}{n}. \tag{6.94}$$

To evaluate the integral

$$I\,(v) = \int_0^1 v\,(\sigma)\,\mathrm{d}\sigma$$

for a 1-periodic continuous function v, approximate it by the trapezoidal rule

$$I_h\,(v) = h \sum_{r=0}^{n-1} v\,(\sigma_r)\,. \tag{6.95}$$

The integral operator B of (6.75) is now approximated by

$$B_h v\,(s) := I_h\,(b\,(s,\cdot)\,v\,(\cdot)) = h \sum_{r=0}^{n-1} b\,(s,\sigma_r)\,v\,(\sigma_r)\,, \tag{6.96}$$

and replacing B with B_h in (6.82), the discrete collocation method can be written in the form

$$(A + \mathcal{Q}_h B_h)\,w_h = \mathcal{Q}_h g, \quad w_h \in \mathcal{T}_h. \tag{6.97}$$

To obtain a linear system for finding w_h, let

$$w_h\,(s) = \sum_{k\in\Lambda_n} \alpha_k \varphi_k\,(s)$$

and calculate the coefficients α_k from (6.97) and the definitions of A $\left(A\varphi_k = \frac{1}{\max(1,|k|)}\varphi_k\right)$, $\mathcal{Q}_h$ and B_h :

$$\sum_{k\in\Lambda_n} \left[\frac{\varphi_k\,(s_j)}{\max\,(1, \mid k \mid)} + (B_h\varphi_k)\,(s_j)\right] \alpha_k = g\,(s_j)\,, \quad j = 0, \ldots, n-1. \tag{6.98}$$

Using nodal values of w_h as unknows, the following system is obtained which is computationally less expensive:

$$\sum_{k=0}^{n-1} \left[\beta_{jk} + h\, b\left(s_j, \sigma_k\right)\right] w_h\left(\sigma_k\right) = g\left(s_j\right), \quad j = 0, \ldots, n-1, \tag{6.99}$$

where

$$\beta_{jk} = \sum_{r \in \Lambda_n} \frac{h\, \varphi_r\left(s_j\right) \overline{\varphi_k\left(\sigma_r\right)}}{\max\left(1, \mid r \mid\right)}.$$

For the computation of the coefficients β_{jk} one can use the fast Fourier transform.

Our convergence analysis follows the same lines as above. That is, instead of (6.97) we consider the modified method

$$\left(A + \mathcal{Q}_h B_h T_{i^\star h}\right) w_h = \mathcal{Q}_h g, \quad w_h \in \mathcal{T}_h. \tag{6.100}$$

Setting $M_h = A^{-1} B_h$ and using (6.83) and the projection R_h, (6.100) can be written as

$$\left(I + R_h M_h T_{i^\star h}\right) w_h = R_h e, \quad w_h \in \mathcal{T}_h. \tag{6.101}$$

For our analysis, the following standard estimate for the trapezoidal rule (6.95) is needed.

Lemma 6.15 *Let $l \in \mathbb{N}$, and suppose that v has 1-periodic continuous derivates of order $< l$ on $\mathbb{R}$ and that $D^l v$ is integrabel on $(0, 1)$. Then*

$$\mid I\left(v\right) - I_h\left(v\right) \mid \leq c\, h^l \int_0^1 \mid D^l v\left(\sigma\right) \mid d\sigma,$$

where c does not depend on v and h.

The following lemma is the key to the stability of (6.101); it is used in the proof of Theorem 6.25

Lemma 6.16 ([159]) *For fixed $q \geq 2$ and for each $\varepsilon > 0$, there exists $i^\star \geq 1$ independent of h such that, for all $v \in \mathcal{T}_h$ and all sufficiently small h,*

$$\| \left(M - M_h\right) T_{i^\star h} M_h T_{i^\star h} v \|_0 \leq \varepsilon \| v \|_0, \tag{6.102}$$

$$\| \left(M - M_h\right) T_{i^\star h} \left(I - R_h\right) M_h T_{i^\star h} v \|_0 \leq \varepsilon \| v \|_0 . \tag{6.103}$$

Theorem 6.25 *Assume $q \geq 2$, and suppose that $i^\star$ is sufficiently large. Then the estimate*

$$\| \left(I + R_h M_h T_{i^\star h}\right) v \|_0 \geq c \| v \|_0, \quad v \in \mathcal{T}_h \tag{6.104}$$

holds for all h sufficiently small, where c is independent of v and h.

Proof Due to Theorem 6.24. (i) , the operators

$$(I + R_h M T_{i^\star h})^{-1} : \mathcal{T}_h \to \mathcal{T}_h, \quad h \leq h_0$$

exists and are uniformly bounded with respect to the H^0 operator norm if $i^\star$ is large enough. Setting

$$C_h := I - (I + R_h M T_{i^\star h})^{-1} R_h M_h T_{i^\star h}$$

and

$$D_h := (I + R_h M T_{i^\star h})^{-1} R_h (M_h - M) T_{i^\star h} R_h M_h T_{i^\star h}$$

gives

$$C_h (I + R_h M_h T_{i^\star h}) = I - D_h. \tag{6.105}$$

Now, using

$$\| (M - M_h) T_{i^\star h} v \|_0 \leq c \| v \|_0, \quad v \in \mathcal{T}_h$$

and the uniform boundedness of R_h on H^0, we see that $R_h M_h T_{i^\star h}$ and hence C_h are also uniformly bounded. Furthermore, Lemma 6.16 yields for some $\varepsilon \in (0, 1)$ provided that $i^\star$ is sufficiently large

$$\begin{aligned} &\| D_h v \|_0 \leq c \| R_h (M_h - M) T_{i^\star h} R_h M_h T_{i^\star h} v \|_0 \\ &\leq c \{ \| (M - M_h) T_{i^\star h} M_h T_{i^\star h} v \|_0 + \| (M - M_h) T_{i^\star h} (I - R_h) M_h T_{i^\star h} v \|_0 \} \\ &\leq \varepsilon \| v \|_0, \qquad v \in \mathcal{T}_h, \qquad h \leq h_0 \end{aligned}$$

Hence $(I - D_h)^{-1}$ exists and is uniformly bounded, and (6.105) gives

$$\| (I + R_h M_h T_{i^\star h})^{-1} \|_0 = \| (I - D_h)^{-1} C_h \|_0 \leq c\,, \qquad h \leq h_0\,,$$

which yields (6.104). □

Finally, we show in [159] that (6.100) converges with the same optimal order as the collocation method.

Theorem 6.26 ([159]) *Let* $l \in \mathbb{N}$, $q > \left(l + \frac{1}{2}\right)(1 + | \chi |)$, *and suppose that* $f \in H^{l+\frac{5}{2}}(\Gamma)$. *Suppose further that* $i^\star$ *is sufficiently large. Then* (6.100) *has a unique solution for all* h *sufficiently small and*

$$\| w - w_h \|_0 \leq c\, h^l, \tag{6.106}$$

where c *is independent of* h.

Table 6.1 L^2 error of the density [159]

	$q = 2$		$q = 3$		$q = 4$		$q = 5$	
n	$\|w_h - w^*\|_0$	EOC	$\|w_h - w^*\|_0$	EOC	$\|w_h - w^*\|_0$	EOC	$\|w_h - w^*\|_0$	EOC
16	8.25-2	0.69	5.08-2	1.35	4.80-2	2.25	6.37-2	3.22
32	5.11-2	0.68	1.98-2	1.22	1.00-2	1.81	6.82-3	2.39
64	3.18-2	0.72	8.47-3	1.23	2.85-3	1.80	1.29-3	2.39
128	1.91-2	0.85	3.60-3	1.28	8.16-4	1.82	2.45-4	2.40
256	1.06-2	1.24	1.47-3	1.58	2.30-4	2.05	4.67-5	2.54
512	4.46-3		4.91-4		5.53-5		8.00-6	

As a numerical example we choose

$$\gamma_0(s) = \sin \pi s \left(\cos (1-\chi)\pi s, \sin (1-\chi)\pi s\right), \qquad s \in [0,1], \qquad 0 < |\chi| < 1.$$

Here Γ is the boundary of a "teardrop-shaped" region with a single corner at $s = 0$ (or $s = 1$) and smooth elsewhere. We solve (6.68) with the quadrature-collocation shema (6.82). The right hand side f in (6.68) is

$$f(x) = \exp(x_1)\cos(x_2) + \operatorname{Re}\left\{(x_1 + \mathrm{i}x_2)^{\frac{1}{(1-\chi)}}\right\},$$

with $\chi = -0.76$, which corresponds to a re-entrant corner. Since the exact solution of (6.68) is unknown, we compute an approximation w^* with (6.82) for $n = 1024$ and use it as the exact solution (Table 6.1).

Remark 6.5 For discrete collocation of the hypersingular integral equation governing the Neumann problem see [231]. For collocation with Chebyshev/Jacobi polynomials for first kind integral equations on an interval see [166, 230, 381]. Collocation and discrete collocation with trigonometric polynomials for the mixed Dirichlet-Neumann problem for the Laplacian are investigated in [158, 262] and in [406].

6.7 Improved Galerkin Method with Augmented Boundary Elements

If the boundary curve has corners, the boundary charges and solutions develop singularities. In this section the augmentation of boundary elements by special singular functiones near the corners is used to derive higher convergence rate for the Galerkin method for integral equations on polygons. The use of singularity functions as additional test/trial functions was originally introduced in the finite element method by G. Fix [180]. For including the singular behaviour, parts of the pioneering work by Costabel and Stephan in [128] are presented.

Let by A denote the set of all exceptional exponents, i.e., $A=\Big\{\alpha_{jk} | j=\{1,\dots,J\}, k\in\mathbb{N}\Big\}\cap(0,2)$. The space Z^s is defined for all s with $s-1/2\in(0,2)\setminus A$:

$$u\in Z^s \Leftrightarrow u=u_0+\Big(\sum_{j=1}^{J}\sum_{\alpha_{jk}<s-1/2} c_{jk}u_{jk}\Big)\chi_j, \tag{6.107}$$

where

(i) $u_0\in H^{s-1}(\Gamma)$ for $s\in[1/2,3/2)$
$u_0|_{\Gamma^j}\in H^{s-1}(\Gamma^j)$ for $s\in[3/2,5/2]$

(ii) $c_{jk}=0$ if u_{jk} is not defined.

(iii) $u_{jk}=x^{\alpha_{jk}}$

and

$$\|u\|^2_{Z^s}=\|u_0\|^2_{\tilde H^{s-1}(\Gamma)}+\sum_{j=1}^{J}\sum_{\alpha_{jk}<s-1/2}|c_{jk}|^2 \quad \text{if } s\in[1/2,3/2)\setminus A.$$

Note $Z^{1/2}=\tilde H^{-1/2}(\Gamma)$.

Next, we construct augmented finite element spaces such that

$$S_h^{p,t,k}\subset Z^k\subset Z^s \quad (s\leqslant k).$$

Definition 6.2

$$\tilde U\in S_h^{p,t,k}\Leftrightarrow \tilde U=\tilde u_0+\sum_{j=1}^{J}\sum_{\alpha_{jk}<p-1/2} c_{jk}u_{jk}\chi_j \tag{6.108}$$

where $c_{jk}\in\mathbb{R}$ is arbitrary, and u_{jk}, χ_j are as in the definition of Z^p above and $\tilde u_0\in S_h^{t,k}$.

Lemma 6.17 *The finite element spaces $S_h^{p,t,k}$ have the following approximation property* (6.109) *and inverse property* (6.110)*:*
For any $U\in Z^r$ there exists a $\tilde U\in S_h^{p,t,k}$ with $t\geqslant r$ and $p\geqslant r$ and a constant $c>0$, independent of h and U, such that for $q\leqslant\min\{k,r\}$

$$\|U-\tilde U\|_{Z^q}\leqslant c\,h^{r-q}\|U\|_{Z^r} \tag{6.109}$$

For $q\leqslant r$, $\varepsilon>0$, $k\geqslant r$ there exists a constant $M>0$, independent of h, such that for all $\tilde U\in S_h^{p,t,k}$

$$\|\tilde U\|_{Z^r}\leqslant M\,h^{k-r-\varepsilon}\|\tilde U\|_{Z^q} \tag{6.110}$$

with $\varepsilon=0$ if $A\cap[q-1/2,r'-1/2]=\varnothing$ where $r'=\max\{p,r\}$.

Proof For showing (6.109) we choose $\tilde{c}_{jk} = c_{jk}$ for $\alpha_{jk} < r - 1/2$ and $\tilde{c}_{jk} = 0$ for $\alpha_{jk} \in (r - 1/2, p - 1/2)$, where $\tilde{c}$ are the coefficients in (6.108) and c_{jk} those in (6.107). Thus it remains to estimate the smooth parts, i.e. $\|u_0 - \tilde{u}_0\|_{Z^q}$ which reduces to an ordinary Sobolev norm where we can apply the convergence property of a $S_h^{t,k}$-system. Hence

$$\|U - \tilde{U}\|_{Z^q} = \|u_0 - \tilde{u}\|_{Z^q} \leqslant c\, h^{r-q} \|u_0\|_{Z^r} \leqslant c\, h^{r-q} \|U\|_{Z^r}.$$

The inverse property (6.110) for $q \leq r \leq k$ and $A \cup [q - 1/2, r' - 1/2] = \emptyset$ $\big(r' = \max(r, p)\big)$ follows immediately from the definition of norms and the inverse property of $S_h^{t,k}$-systems [14]

$$\|\tilde{U}\|^2_{Z^r} = \|\tilde{u}_0\|^2_{Z^r} + \sum_{j=1}^{J} \sum_{\alpha_{jk} < r-1/2} |\tilde{c}_{jk}|^2 \leq M^2\, h^{2(q-r)} \|\tilde{u}_0\|_{Z^q} + \sum_{j=1}^{J} \sum_{\alpha_{jk} < q-1/2} |\tilde{c}_{jk}|^2$$
$$\leqslant M^2\, h^{2(q-r)} \|\tilde{U}\|^2_{Z^q}.$$

For the proof in the remaining cases see [128].

Now, we can perform the augmented Galerkin method for the following integral equation with the hypersingular operator W:
For given $f \in H^{-1/2}(\Gamma)$ find $u \in H^{1/2}(\Gamma)$ such that

$$Wu = (I + K')f. \tag{6.111}$$

For given $f \in H^{-1/2}(\Gamma)$ find $U_h^* \in S_h^{p,t,k}$ such that

$$\langle W\, U_h^*, V\rangle = \langle (1 + K')f, V\rangle \quad \forall V \in S_h^{p,t,k}.$$

As shown in [128] we have

Theorem 6.27 *There holds the improved error estimate*

$$\|U - U_h^*\|_{Z^r} \leqslant c\, h^{s-r-\varepsilon} \|U\|_{Z^s} \leqslant c'\, h^{s-r-\varepsilon} \|f\|_{H^{s-1}(\Gamma)}$$

where $U \in Z^s$ solves integral equation (6.111).

Remark 6.6 For sufficiently large s the above theorem gives explicit estimates for the error of the stress intensity factors $c_{jk} - \tilde{c}_{jk}$ (see [128]).

Remark 6.7 For corresponding results on augmented BEM applied to 2D mixed bvp's see [128, 433], (for numerical results see [279]) and to 2D crack problems see [256, 432].

6.8 Duality Estimates for Projection Methods

For higher rates of convergence in negative norms (or norms involving less derivates than the energy norm), Aubin-Nitsche-type duality estimates are a standard tool in the analysis of finite element Galerkin methods [7, 106, 328, 412]. Such duality arguments are also available for Galerkin methods for certain integral equations and pseudodifferential equations on smooth manifolds [258] and also for some collocation methods that can be reduced to such Galerkin methods [3, 357].

Standard formulations [258] require unique solvability of the adjoint equation in the negative norm, and this is in general not satisfied in the presence of corners and edges of discontinuous boundary conditions, due to singular solutions of the homogeneous equation.

It is useful to write projection methods in the form of Galerkin-Petrov methods [202], which is always possible. Also collocation methods can easily be written in this form [131].

Let X and Y be Banach spaces and $A : X \to Y$ an isomorphism of norm $\| A \|$. We assume that we have the subspaces

$$V_N \subset X, \qquad T_N \subset Y',$$

and we replace the equation for $u \in X$:

$$Au = f \tag{6.112}$$

by the equation for $u_N \in V_N$:

$$< t, Au_N > = < t, f > \quad \forall t \in T_N, \tag{6.113}$$

the brackets denoting the duality between the space Y and its dual Y'. Here $f \in Y$ is given and u_N in (6.113) is considered as an approximate solution to (6.112). One usually has in addition

$$\dim V_N = \dim T_N < \infty$$

and uses a whole sequence of V_N, T_N, u_N to approximate u, but for the moment we do not need this.

We assume that we somehow know an estimate for the error $u - u_N$ in the norm of X and want to estimate it in a smaller norm.

Thus we have a second norm $\| \cdot \|_{X_1}$ on X that satisfies with some $M > 0$

$$\| x \|_{X_1} \leqq M \| x \|_X \quad \forall x \in X. \tag{6.114}$$

By completion of X one thus has a Banach space $X_1 \supset X$ with X densely embedded, and hence the dual space X_1' is contained in X' in a natural way:

$$X_1' = \left\{ \xi \in X' \mid \xi \text{ is continuous with respect to } \| \cdot \|_{X_1} \right\}.$$

Note that the adjoint $A' : Y' \to X'$ is bijective and continuous, so that A'^{-1} is well-defined on X_1'.

The following lemma from [134] contains the abstract version of the Aubin-Nitsche duality estimate.

Lemma 6.18 *Assume that there is a constant* $c_N > 0$ *such that*

$$inf\left\{\| A'^{-1}\xi - \tau \|_{Y'} | \, \tau \in T_N\right\} \leqq \varepsilon_N \| \xi \|_{X_1'} \quad \forall \xi \in X_1'. \tag{6.115}$$

Then for all $f \in Y$, *and* u *and* u_N *satisfying* (6.112) *and* (6.113), *respectively, there holds*

$$\| u - u_N \|_{X_1} \leqq \varepsilon_N \| A \| \| u - u_N \|_X . \tag{6.116}$$

Proof We have

$$\| u - u_N \|_{X_1} = \sup\left\{|< \xi, u - u_N >| \mid \xi \in X_1', \| \xi \|_{X_1'} = 1\right\}.$$

Choose $\xi \in X_1'$ with $\| \xi \|_{X_1'} = 1$ and $\tau \in T_N$. Then

$$\begin{aligned} |< \xi, u - u_N >| &= |< A'^{-1}\xi, A(u - u_N) >| = |< A'^{-1}\xi - \tau, A(u - u_N) >| \\ &\leqq \| A'^{-1}\xi - \tau \|_{Y'} \| A(u - u_N) \|_Y \leqq \| A'^{-1}\xi - \tau \|_{Y'} \| A \| \| u - u_N \|_X . \end{aligned}$$

Minimizing over $\tau \in T_N$ we find with (6.115)

$$|< \xi, u - u_N >| \leqq \varepsilon_N \| A \| \| u - u_N \|_X .$$

□

Remark 6.8 This is only a small modification of the statement and the proof of the Aubin-Nitsche lemma for Galerkin methods as stated by Ciarlet [106]. Even for Galerkin methods and for a Hilbert space X_1, however, it turns out to be useful to distinguish between X_1 and X_1', as we shall see now.

6.8.1 Application to Galerkin Methods

The well-known quasioptimality result (Céa lemma)

$$\| u - u_N \|_X \leqq C_0 \inf\{\| u - v \|_X | \, v \in V_N\} \tag{6.117}$$

is usually applied in the following way.

One uses an approximation result for the spaces V_N in X that holds for u in some subspace of X. An example is the order of convergence of best spline (or trigonometric) approximation in the norm of a Sobolev space (i.e. one of higher regularity).

Additionally, one uses regularity results about the operator A that for f in Eq. (6.112) given in a subspace Y_1 of Y,the solution u is contained in a subspace of X where the above approximation property holds, i.e

$$Y_1 \subset Y \quad \text{with } \| \cdot \|_Y \leqq M \| \cdot \|_{Y_1} . \tag{6.118}$$

and an estimate

$$\inf\{\| u - v \|_X | \; v \in V_N\} \leqq \delta_N \| f \|_{Y_1}, \quad \text{for all } u \in X \quad \text{with } f = Au \in Y_1. \tag{6.119}$$

Of course, (6.117) and (6.119) together give a convergence rate $\mathcal{O}(\delta_N)$ for the error in the norm of X.

Now suppose we have the situation of a Galerkin method:

$$X' = Y, \quad Y' = X, \quad V_N = T_N. \tag{6.120}$$

We assume further that A is selfadjoint:

$$A = A'.$$

Then we see immediately that the two conditions (6.115) and (6.119) are identical, if we have $\varepsilon_N = \delta_N$ and $X_1' = Y_1$. We can take the latter as a definition for X_1' and for X_1 :

$$\| x \|_{X_1} := \sup_{y \in Y_1 \setminus \{0\}} \frac{|< y, x >|}{\| y \|_{Y_1}} \quad \text{for all } x \in X. \tag{6.121}$$

We see from (6.118) that (6.114) holds, and therefore we can apply Lemma 6.18.

Theorem 6.28 *Let* (6.117)–(6.121) *be satisfied. Then there holds*

$$\| u - u_N \|_{X_1} \leqq \delta_N \| A \| \; \| u - u_N \|_X \leqq \delta_N^2 \; C_0 \| A \| \; \| f \|_{Y_1} \tag{6.122}$$

for all $f \in Y_1$ *and* u *and* u_N *satisfying* (6.112) *and* (6.113).

Remark 6.9 If A is not selfadjoint, one has to require the same regularity and approximation result for A' as for A :

$$\inf\{\| u - v \|_X | \; v \in V_N\} \leqq \delta_N \| A'u \|_{Y_1} \quad \text{for all } u \in X \quad \text{with } f = A'u \in Y_1.$$

We now present an example of a Galerkin scheme for a boundary integral equations where singularities of the solution are incorporated into the space of trial functions.

We consider the integral equation of the first kind with the single layer potential on a polygon Γ

$$Vu(z) = -\frac{1}{\pi}\int\limits_{\Gamma} \ln \mid z-\zeta \mid u(\zeta)\, ds_{\zeta} = f(z), \quad z \in \Gamma.$$

If Γ is sealed such that its capacity is different from 1, then $V : H^{-\frac{1}{2}+\sigma}(\Gamma) \to H^{\frac{1}{2}+\sigma}(\Gamma)$ is bijective for $\mid \sigma \mid \leq \frac{1}{2}$, where $H^s(\Gamma)$ denotes the Sobolev space of order s on Γ.

There holds the following regularity result: There exist real numbers α_{jk}, natural numbers r_{jk} and explicitly known singular functions u_{jk} which behave like $\mid z - z_j \mid^{\alpha_{jk}} \ln \mid z - z_j \mid^{r_{jk}}$ near the corner z_j and are C^{∞} elsewhere. If $f \in H^{s+1}(\Gamma)$ $(s \geq -1,\ s \notin A(\Gamma))$, where $A(\Gamma) \subset \mathbb{R}$ is a certain discrete set. Then there exist numbers K_{js} depending on s and c_{jk} depending on f, such that

$$u = \sum_{j=1}^{J}\sum_{k=1}^{K_{js}} c_{jk}u_{jk} + u^0$$

with $u^0 \in H^s(\Gamma)$ .There is an a-priori estimate

$$\sum_{j=1}^{J}\sum_{k=1}^{K_{js}} \mid c_{jk} \mid + \| u^0 \|_{H^s(\Gamma)} \leq c\|f\|_{H^{s+1}(\Gamma)}.$$

The trial spaces $S_h^{d,s}$ contain a regular finite element space S_h^d on a grid with meshwidth h, namely the smoothest splines of degree d, plus the singular functions u_{jk}, their number depending on s. Then one has the following approximation property:

$$\begin{aligned} \inf\left\{\| u - v \|_{H^t(\Gamma)} \middle|\, v \in S_h^{d,s}\right\} &= \inf\left\{\| u^0 - v^0 \|_{H^t(\Gamma)} \middle|\, v^0 \in S_h^{d}\right\} \\ &\leqq Ch^{s-t} \| u^0 \|_{H^s(\Gamma)} \\ &\leqq Ch^{s-t} \| f \|_{H^{s+1}(\Gamma)} \end{aligned} \tag{6.123}$$

for all $t \leqq s$ with $t < -\frac{1}{2} + \sigma_0$, $s \leqq d+1$, and $s \notin A(\Gamma)$, and C not depending on f and h. The Galerkin scheme reads as:

Find $u_h \in S_h^{d,s}$ such that

$$< \tau, Vu_h > = < \tau, f > \quad \text{for all } \tau \in S_h^{d,s}. \tag{6.124}$$

Then (6.123) implies the error estimate

$$\| u - u_h \|_{H^{-\frac{1}{2}}(\Gamma)} \leqq Ch^{s+\frac{1}{2}} \| f \|_{H^{s+1}(\Gamma)}. \tag{6.125}$$

Now we make the following identifications for $-\frac{1}{2} \leqq s \leqq d+1,\ s \notin A(\Gamma)$:

$$X = H^{-\frac{1}{2}}(\Gamma), \quad Y = H^{\frac{1}{2}}(\Gamma), \quad Y_1 = H^{s+1}(\Gamma).$$

Then (6.123) implies (6.119) with $\delta_N = Ch^{s+\frac{1}{2}}$.

The norm in X_1 is given by $\| \cdot \|_{H^{-s-1}(\Gamma)}$, and Theorem 6.28 can be applied.

Theorem 6.29 *For the Galerkin scheme* (6.124) *there holds the error estimate*

$$\| u - u_h \|_{H^{-s-1}(\Gamma)} \leqq Ch^{s+\frac{1}{2}} \| u - u_h \|_{H^{-\frac{1}{2}}(\Gamma)} \leqq Ch^{2s+1} \| f \|_{H^{s+1}(\Gamma)} \cdot$$

For fixed degree d of the piecewise polynomials, the highest possible order is $\mathcal{O}\left(h^{2d+3}\right)$ which is obtained for $s = d+1$. The number $\sum_{j=1}^{J} K_{js}$ of singular functions has to be chosen correspondingly.

6.8.2 Application to Collocation Methods

In collocation methods, the space T_N of test functions is generated by Dirac delta functions supported by the collocation points. Let us consider a one-dimensional problem, e.g. a boundary integral equation belonging to a two-dimensional boundary value problem. Thus let $\Gamma \subset \mathbb{R}^2$ be a Lipschitz curve.

Let N collocation points

$$\Delta_N = \{x_0, x_1, \ldots, x_N\} \subset \Gamma \quad \text{with } x_N = x_0$$

be chosen and

$$h := \sup\left\{ | x_j - x_{j+1} | \mid j = 0, \ldots, N-1 \right\}.$$

Let

$$S^{-1}(\Delta_N) := \operatorname{span}\{\delta(x - x_N) \mid n = 1, \ldots, N\}.$$

Then $S^{-1}(\Delta_N) \subset H^s(\Gamma)$ for all $s < -\frac{1}{2}$. Here the Sobolev spaces $H^s(\Gamma)$ are defined by transfer from the parameter interval through a fixed periodic parameter representation. It is well known that the definition of $H^s(\Gamma)$ is independent of the specific parameter representation for $| s | \leqq 1$ in the case of a Lipschitz curve Γ, for $| s | < \frac{3}{2}$ if Γ is piecewise smooth, and for all s if Γ is smooth.

In order to satisfy condition (6.115) for $T_N = S^{-1}(\Delta_N)$ we need the following approximation result [134]

Lemma 6.19 *For all $p \leq q \leq 0$ with $p < -\frac{1}{2}$ there exists a constant M, independent of Δ_N, such that*

$$\inf\left\{\|\varphi-\tau\|_{H^p(\Gamma)} \mid \tau \in S^{-1}(\Delta_N)\right\} \leq M h^{q-p}\|\varphi\|_{H^q(\Gamma)} \tag{6.126}$$

for all $h > 0$ and all $\varphi \in H^q(\Gamma)$.

Suppose now that

$$A : H^s(\Gamma) \to H^t(\Gamma) \tag{6.127}$$

is a continuous bijective linear operator of norm $\| A \|$, for some $s, t \in \mathbb{R}$. Let $f \in H^t(\Gamma)$ be given and let $u \in H^s(\Gamma)$ solve

$$Au = f, \tag{6.128}$$

whereas $u_h \in V_N$ solves the collocation equations

$$Au_h(x_n) = f(x_n) \quad (n = 1, \ldots, N). \tag{6.129}$$

The trial function space V_N is supposed to be a N-dimensional subspace of $H^s(\Gamma)$. The Eq. (6.129) can equivalently be written in the form

$$< t, Au_h > = < t, f > \quad \text{for all } t \in S^{-1}(\Delta_N), \tag{6.130}$$

if t is such that $S^{-1}(\Delta_N) \subset H^{-t}(\Gamma)$, i.e. $t > \frac{1}{2}$.

We want to apply Lemma 6.18 with

$$X = H^s(\Gamma), \quad Y = H^t(\Gamma), \quad T_N = S^{-1}(\Delta_N).$$

In order to make (6.126) and (6.115) equivalent, we thus have to put $p = -t$. If we assume $-t < q \leqq 0$, then $H^t(\Gamma) \subset H^{-q}(\Gamma)$, and hence the definition

$$\| v \|_{X_1} := \| Av \|_{H^{-q}(\Gamma)} \quad \text{for } v \in H^s(\Gamma) \tag{6.131}$$

makes sense and defines a norm, because A is injective. The dual norm is given by

$$\| \xi \|_{X_1'} = \| A'^{-1}\xi \|_{H^q(\Gamma)}, \tag{6.132}$$

as can be seen from

$$\| \xi \|_{X_1'} = \sup\left\{\frac{|< \xi, v >|}{\| v \|_{X_1}} \mid v \in X\right\} = \sup\left\{\frac{|< \xi, v >|}{\| Av \|_{H^{-q}(\Gamma)}} \mid v \in H^s(\Gamma)\right\}$$

$$= \sup \left\{ \frac{|< \xi, A^{-1}w >|}{\| w \|_{H^{-q}(\Gamma)}} \mid w \in H^t(\Gamma) \right\} = \sup \left\{ \frac{|< A'^{-1}\xi, w >|}{\| w \|_{H^{-q}(\Gamma)}} \mid w \in H^t(\Gamma) \right\}$$

$$= \| A'^{-1}\xi \|_{H^q(\Gamma)}.$$

Here we used the fact that $H^t(\Gamma)$ is dense in $H^{-q}(\Gamma)$.

Now all hypotheses of Lemma 6.18 are satisfied, and we obtain the following theorem.

Theorem 6.30 *Let A be as in* (6.127) *and let u and u_h satisfy* (6.128) *and* (6.129), *respectively. Let t and q satisfy*

$$t > \frac{1}{2} \quad and \ -t \leqq q \leqq 0,$$

and $\| \cdot \|_{X_1}$ be defined by (6.131). *Then*

$$\| u - u_h \|_{X_1} \leqq Ch^{t+q} \| u - u_h \|_{H^s(\Gamma)}, \tag{6.133}$$

where $C = M \| A \|$ with the constant M from *Lemma* 6.19 *for $p = -t$.*

Remark 6.10 If A satisfies an a-priori estimate of the form

$$\| v \|_{H^r(\Gamma)} \leqq \gamma \| Av \|_{H^{-q}(\Gamma)} \tag{6.134}$$

for some r and γ, then the error estimate (6.133) can also be written in the form

$$\| u - u_h \|_{H^r(\Gamma)} \leqq Ch^{t+q} \| u - u_h \|_{H^s(\Gamma)} . \tag{6.135}$$

The a-priori estimate (6.135)holds in particular if A is a bijective elliptic pseudodifferential operator of oder $s - t$. Then $r = s - t - q$, and one obtains

$$\| u - u_h \|_{H^r(\Gamma)} \leqq Ch^{s-r} \| u - u_h \|_{H^s(\Gamma)} \tag{6.136}$$

for $s - t \leqq r \leqq s$.

The highest order $\mathcal{O}(h^t)$ in Theorem 6.30 is obtained for $q = 0$. This corresponds to $r = s - t$ in (6.136).

This version of the duality argument based on the a-priori estimate (6.134) is equivalent to the arguments in [258]. Note, however, that it could not be applied in the previous section for the augmented Galerkin procedures. There, due to the presence of singular solutions, the estimate corresponding to (6.134) does not hold.

As an example we present from [131] error estimates for the collocation method applied to the integral equation of the second kind with the double layer pontential on a piecewise smooth curve:

$$Au := (I + K)u = f,$$

with

$$Ku(z) = -\frac{1}{\pi} \int_{\Gamma} u(\zeta) \frac{\partial}{\partial n_\zeta} \ln | z - \zeta | \, ds_\zeta \quad (z \in \Gamma).$$

It is known that $I+K : H^s(\Gamma) \to H^s(\Gamma)$ is continuous and bijective for $0 \le s \le 1$ if Γ is Lipschitz, and for $| s - \frac{1}{2} | < \sigma_0$ with

$$\sigma_0 := \min \left\{ \frac{\pi}{\pi + | \pi - \omega_j |} \mid j = 1, \ldots, J \right\} \in \left(\frac{1}{2}, 1 \right)$$

if Γ is piecewise smooth .

In [131] for the nodal collocation method with piecewise linear trial functions it is shown that

$$\| u - u_h \|_{H^1(\Gamma)} \le c h^\sigma \| f \|_{H^{1+\sigma}(\Gamma)},$$

if $f \in H^{1+\sigma}(\Gamma)$ and $\sigma < \sigma_0 - \frac{1}{2}$. Now Theorem 6.30 yields $\| u - u_h \|_{L^2(\Gamma)} = \mathcal{O}\left(h^{\frac{1}{2}+\sigma_0-\varepsilon} \right) \quad \forall \varepsilon > 0$.

6.9 A Collocation Method Interpreted as (GM)

Following Arnold and Wendland [3] we consider a planar Jordan curve Γ with a regular parameter transformation,

$$\Gamma : z = \left(z_1(t), z_2(t) \right) \cong z_1(t) + i \, z_2(t)$$

where z is 1-periodic on $\mathbb{R}$ and $\left| \frac{dz}{dt} \right| \ne 0$. Thus there is a 1–1 correspondence between functions defined on Γ and 1-periodic functions on $\mathbb{R}$. Therefore the analysis to follow is based on the periodic Sobolev spaces $H^s (s \in \mathbb{R})$ that are defined as the closure of all smooth real-valued 1-periodic functions with respect to

$$\|f\|_s := \|f\|_{H^s} := \{ |\hat{f}_0|^2 + \sum_{0 \ne k \in \mathbb{Z}} |\hat{f}_k|^2 \, |2\pi k|^{2s} \}^{1/2}$$

where $\hat{f}_k = \int_0^1 e^{-2\pi i k t} f(t) dt \ \ (k \in \mathbb{Z})$ (as in Chap. 3).

The associated scalar product is

$$< f, g >_s := \hat{f}_0 \cdot \bar{\hat{g}}_0 + \sum_{0 \ne k \in \mathbb{Z}} \hat{f}_k \bar{\hat{g}}_K \, |2\pi k|^{2s}$$

what can be extended to a duality pairing on $H^{s+\alpha} \times H^{s-\alpha}$ for arbitrary $\alpha \in \mathbb{R}$. By this duality, $v \in H^{s+\alpha}$

$$\|v\|_{s+\alpha} = \sup_{w \in H^{s-\alpha}} \frac{< v, w >_s}{\|w\|_{s-\alpha}}$$

holds for any $v \in H^{5+\alpha}$.

Remark 6.11 Note that for $j \in \mathbb{N},\ f^{(j)} = (\frac{d}{dt})^j$ we have by partial integration

$$\begin{aligned} \hat{f^{(j)}}_k &= \int_0^1 e^{-2\pi ikt} f^{(j)}(t)dt \\ &= (2\pi ik)^j \int_0^1 e^{-2\pi ikt} f(t)dt = (2\pi ik)^j \hat{f}_k\,, \end{aligned}$$

by periodicity, in particular $\hat{f^{(j)}}_0 = 0$. Hence by the Parseval identity

$$\begin{aligned} \|f\|_j^2 &= |\hat{f}_0|^2 + \sum_{0 \neq k \in \mathbb{Z}} |\hat{f}_k|^2\, |2\pi k|^{2j} \\ &= |\hat{f}_0|^2 + \sum_{k \in \mathbb{Z}} |\hat{f^{(j)}}_k|^2 = \|f\|_{L^2}^2 + \|f^{(j)}\|_{L^2}^2\,. \end{aligned}$$

Again, we want to approximate the equation $Au = f$. Here we take $A : H^{j+\alpha} \to H^{j-\alpha}$;

the number 2α is called the *order* of A.

(e.g. in the case $A = V$, we have $\alpha = -\frac{1}{2}$, and the oder is -1),

whereas $j \in \mathbb{R}$ is specified later.

To begin with the approximation, we fix a mesh $\Delta = \{t_1 = 0 < t_2 \ldots < t_N < 1\}$ (N fixed $\in \mathbb{N}$), which is periodically extended by $t_{l+N} = t_l + 1$ $\quad (\forall l \in \mathbb{Z})$.

Then $S_d(\Delta)$ denotes the space of 1-periodic, $(d-1)$times continuous differentiable spline functions of degree d subordinated to the mesh Δ.

In the following we assume that the degree d is odd (> 0), $j := (d+1)/2 \in \mathbb{N}$. Then the collocation method (CM) reads: Find $u_\Delta\ in S_d(\Delta)$ such that

$$(Au_\Delta)(t_l) = f(t_l) \quad l = 1, \ldots, N$$

which is a linear equation system in the unknown ξ_l in the ansatz $u_\Delta = \sum_{l=1}^N \xi_l \mu_l$.

Now let us discuss the setting of (CM) in the scale of Sobolev spaces. We have

$$Au_\Delta \in AS_d(\Delta) \subset AH^s \subset H^{s-2\alpha}$$

provided $s < d + \frac{1}{2}$.

(CM) makes only sense, if $Au_\triangle$ is continuous in the mesh points t_l, that is, for some $s < d + \frac{1}{2}$ the space $H^{s-2\alpha}$ should be embedded in the space of continuous functions. Sobolev's embedding theorem forces $s - 2\alpha > 0 + \frac{1}{2}$ $(n = 1)$, that is, the following assumption

$$(A1) \quad d = 2j - 1 > 2\alpha \Leftrightarrow j - \alpha > \frac{1}{2}$$

(that is in the case $A = V$ the simplest trial space is $S_1(\triangle)$ consisting of piecewise linear spline functions)

Further we impose the following assumptions:

$$\text{(A2)} \quad A : \; H^{j+\alpha} \to H^{j-\alpha} \quad \text{is bijective}$$

$$\text{(A3)} \quad < Au, u >_j \geq \gamma \; \|u\|^2_{j+\alpha} - < Ku, u >_j, \quad \forall u \in H^{j+\alpha},$$

where $\gamma > 0$, K: $H^{j+\alpha} \to H^{j-\alpha}$ is compact (linear), what is called a Gårding inequality.

The key of the convergence analysis of Arnold and Wendland [3] is to reformulate (CM) as a Galerkin method (GM) using partial integration. To this end, introduce

$$Ju := \int_0^1 u(t)\,dt, \quad J_{\triangle u} := \sum_{l=1}^{N} \frac{1}{2}(t_{l+1} - t_{l-1})\,u(t_l).$$

Note that the latter functional is the numerical approximation of the first integral by the trapezoid rule, since

$$\sum_{l=1}^{N} \frac{1}{2}\big(u(t_{l+1})+u(t_l)\big)\big(t_{l+1}-t_l\big) = \frac{1}{2}\sum_{l=1}^{N} u(t_l)(t_{l+1}-t_l) + \frac{1}{2}\sum_{k=2}^{N+1} u(t_k)(t_k - t_{k-1}) = J_{\triangle u}$$

Theorem 6.31 *[3] Let $w \in H^{j-\alpha}$. Then*

$$(i) \quad w(t_l) = 0 \; for \; l = 1, \ldots, N$$

if and only if

$$(ii) \quad < w - Jw + J_{\triangle w}, \; \chi >_j = 0 \; for \; all \; \chi \in S_d(\triangle).$$

Let us point out that in virtue of (A1), the values w in (i) and (ii) are defined; by $S_d(\triangle) \subset H^{j+\alpha}$, the scalar product in (ii) is defined.

Proof For any real-valued (no restriction of generality) $f, g \in H^j$, (see the first remark above)

$$< f, g >_j = \int_0^1 f\, dt \int_0^1 g\, dt + \int_0^1 \left[\left(\frac{d}{dt}\right)^j f(t)\right]\left[\left(\frac{d}{dt}\right)^j g(t)\right] dt\,.$$

Hence by partial integration in (ii),

$$< w - Jw + J_{\Delta w}, \chi >_j = J_{\Delta w} J\chi + (-1)^{j-1} \int_0^1 w' \chi^{(2j-1)}(t)\, dt.$$

On the other hand,

$$\left(\frac{d}{dt}\right)^{2j-1} : \{v \in S_d(\Delta) \mid Jv = 0\} \to \{\tilde{v} \in S_0(\Delta) \mid J\tilde{v} = 0\}$$

is an isomorphism onto the space of piecewise constants subordinated to the mesh Δ with integral mean zero. In the range space we specify

$$\tilde{\chi}(t) := \begin{cases} -h_i^{-1} & \text{for } t \in [t_{i-1}, t_i) \\ h_{i+1}^{-1} & t \in [t_i, t_{i+1}) \\ 0 & \text{else on } [0, 1], \text{ periodically extended to } \mathbb{R} \end{cases}$$

where $h_i = t_i - t_{i-1}$. Let $\chi_i \in S_d(\Delta)$ such that $\tilde{\chi} = \chi_i^{(2j-1)}$ and $J\chi_i = 0$. and plug these special functions in the formula above to obtain

$$< w - Jw + J_{\Delta w}, \chi_i >_j = (-1)^{j-1}\left[h_{i+1}^{-1}\big(w(t_{i+1}) - w(t_i)\big) - h_i^{-1}\big(w(t_i) - w(t_{i-1})\big)\right]$$

Hence and from (ii) it follows for some constant κ

$$h_i^{-1}\big(w(t_i) - w(t_{i-1})\big) = \kappa \qquad \forall i \in \mathbb{Z}$$

and $i = N + 1, N, \ldots, 1$ gives

$$\begin{aligned} w(t_{N+1}) &= \kappa h_{N+1} + w(t_N) \\ &= \ldots \\ &= \kappa \underbrace{(h_{N+1} + h_N + \ldots + h_1)}_{t_{N+1} - t_0 = 1} + w(t_0) \; = \kappa + w(t_0) \end{aligned}$$

Since w is 1-periodic, in particular $w(t_0) = w(0) = w(1) = w(t_{N+1})$, κ vanishes.

Therefore (ii) implies

$$w(t_i) = w(t_0) \qquad \forall i \in \mathbb{Z}\,. \tag{6.137}$$

Now (ii) does not hold only for these special χ_i, but also for $\chi = 1 \in S_d(\triangle)$, hence

$$< w - Jw + J_{\triangle w}, 1 >_j = J_{\triangle w} \cdot 1 = 0\,. \tag{6.138}$$

Evidently (i) follows from (6.137) and (6.137).

Conversely (i) implies (6.137) and (6.137), hence (ii), since $\{\chi_i, 1\}$ is a basis of $S_d(\triangle)$. □

We can understand J and $j_\triangle$ in (ii) as operators, since Ju, $J_{\triangle u}$ can be considered as constant functions with the values Ju, $J_{\triangle u}$ respectively.

Thus formula (ii) gives rise to the operator $A_\triangle := (1 - J + J_\triangle)A : H^{j+\alpha} \to H^{j-\alpha}$ and by the theorem above, we have $u_\triangle \in S_d(\triangle)$ solves the (CM) equations

$$\begin{array}{llll} \Leftrightarrow & \big(A(u_\triangle - u)\big)(t_l) & = 0 & (\forall l) \\ \Leftrightarrow & < A_\triangle(u_\triangle - u), v >_j & = 0 & \forall v \in S_d(\triangle) \\ \Leftrightarrow & < A_\triangle u_\triangle, v >_j & = < A_\triangle u, v >_j & \forall v \in S_d(\triangle) \end{array}$$

what is a Galerkin method!

Lemma 6.20 *The operator $A_\triangle = (1 - J + J_\triangle)A$ is invertible with the inverse*

$$A_{\triangle^{-1}} = A^{-1}(1 + J - J_\triangle)\,.$$

Furthermore, there exists a positive constant C such that

$$\|A_\triangle\|_{j+\alpha, j-\alpha} + \|A_{\triangle^{-1}}\|_{j-\alpha, j+\alpha} \le C$$

for every mesh $\triangle$.

Proof For any $s \in \mathbb{R}$, $\|J_\triangle\|_s^2 = (\int_0^1 J_{\triangle u}\, dt)^2 = (J_{\triangle u})^2 \le \|u\|_\infty^2$, where the latter estimate is a consequence of the above formula of $J_\triangle$ (trapezoid rule).

Moreover by $j - \alpha > \frac{1}{2}$, $H^{j-\alpha} \subset C^0$ and $J_\triangle : H^{j-\alpha} \to H^{j-\alpha}$, are uniformly bounded. By continuity of A, the operators $A_\triangle$ are uniformly bounded, too.

Since

$$JJ = J_{\triangle J} = J \quad \text{and} \quad J_\triangle J_\triangle = JJ_\triangle = J_\triangle$$

we verify

$$\begin{aligned} A_\triangle \quad A_{\triangle^{-1}} &= (1 - J + J_\triangle)(1 + J - J_\triangle) \\ &= 1 - J + J_\triangle + J - J + J - J_\triangle + J_\triangle - J_\triangle = 1 \end{aligned}$$

Thus again the uniform boundedness of the operators $A_{\triangle^{-1}}$ follows from the continuity of A^{-1}. □

Theorem 6.32 *There exist positive constants C and h_0 such that, for any mesh $\triangle$ with $h_\triangle := \max(t_l - l_{l-1}) \le h_0$ there holds the stability estimate*

$$\inf_{\substack{v \in S_d(\triangle) \\ \|v\|_{j+\alpha} = 1}} \sup_{\substack{z \in S_d(\triangle) \\ \|z\|_{j+\alpha} = 1}} < A_{\triangle v}, z >_j \ \ge C\,.$$

Proof Since according to (A3), A satisfies a Gårding inequality, the fundamental Theorem 6.1 on the geneneral Galerkin method applies and yields the estimate

$$\|w_\triangle\|_{j+\alpha} \le C\ \|w\|_{j+\alpha} \tag{6.139}$$

for the solution $w_\triangle \ inS_d(\triangle)$ of the Galerkin equations

$$< Aw_\triangle, v_\triangle >_j = < Aw, v_\triangle >_j \qquad \forall v_\triangle \in S_d(\triangle)\,,$$

where the mesh $\triangle$ is arbitrary, but $h_\triangle \in (0, h_0)$ with appropriate positive constans C and h_0. Thus we obtain the continuity of the Galerkin operator. Now we consider the solution $u_\triangle \ inS_d(\triangle)$ of the equations

$$< A_\triangle u_\triangle, v_\triangle >_j = < A_{\triangle u}, v_\triangle >_j \qquad \forall v_\triangle \in S_d(\triangle)$$

and rewrite these equations as follows:

$$\begin{aligned} < Au_\triangle, v_\triangle >_j &= < A_{\triangle u} - \{(J_\triangle - J)Au_\triangle\}, v_\triangle >_j \\ &= < Au - \{(J_\triangle - J)Au_\triangle - (J_\triangle - J)Au\}, v_\triangle >_j \\ &= < A(u - A^{-1}\{(J_\triangle - J)Au_\triangle - (J_\triangle - J)Au\}), v_\triangle >_j\ . \end{aligned}$$

Thus by (6.138) we arrive at

$$\begin{aligned} \|u_\triangle\ |_{j+\alpha} &\le c\,(\|u\|_{j+\alpha} + \|A^{-1}(J_\triangle - J)Au_\triangle\|_{j+\alpha} + \|A^{-1}(J_\triangle - J)Au\|_{j+\alpha})) \\ &\le c\,(\|u\|_{j+\alpha} + c_1\big(\|(J_\triangle - J)Au_\triangle\|_{j-\alpha} + \|(J_\alpha - J)Au\|_{j-\alpha})\big) \\ &\le c\ \ \|u\|_{j+\alpha} + c_3 h_\triangle^\mu \|u_\triangle\|_{j-\alpha} + c_3 h_\triangle^\mu \|u\|_{j+\alpha}\,, \end{aligned}$$

where the approximation by the trapezoid rule provides the estimate

$$\|(J_\triangle - J)v\|_{j-\alpha} \le c_2 h_\triangle^\mu \|v\|_{j-\alpha}$$

for $v = Au_\triangle, Au$ respectively with $\mu > \min(2, j - \alpha) > 0$. Therefore for any $h \in (0, h_0]$, where $c_3 h_\triangle^\mu \le \min(c, \frac{1}{2})$,, there holds uniformly with respect to $h_\triangle$

$$\|u_\triangle\|_{j+\alpha} \le 3c \, \|u\|_{j+\alpha} \, .$$

Finally the fundamental Theorem 6.1 entails the claimed stability estimate. □

Thus in virtue of Theorem 6.1, Part iii) we obtain the following convergence result.

Theorem 6.33 *There exist positive constants C and h_0 such that, for any mesh $\triangle$ with $h_\triangle \le h_0$, there exists a unigue solution $u_\triangle \in S_d(\triangle)$ of the (CM) equations and there holds the quasioptimal error estimate*

$$\|u - u_\triangle\|_{j+\alpha} \le C \ \mathit{inf}\{\|u - v\|_{j+\alpha} \, : \, v \in S_d(\triangle)\} \, .$$

6.10 Modified Collocation and Qualocation

Following Costabel and Stephan [131] we again consider odd degree spline functions in the collocation method, but we dispense with the smoothness of the boundary Γ. Instead more generally, Γ is assumed to be a connected closed planar curve patched together from smooth arcs $\Gamma^j (j = 1, \ldots, J)$, that intersect each other in the corners z_j at the inner angles $\omega_j \in (0, 2\pi)$.

In what follows, we use the subsequent definition of the Sobolov spaces $H^s(\Gamma)$:

for any $s > 0$, the set of the restrictions of functions in $H^{s+\frac{1}{2}}(\mathbb{R}^2)$ to Γ
(this makes sense by the embedding theorem that ensures $u \in H^s(\Gamma) \Leftrightarrow$
$\exists$ extension $\tilde{u} \in H^{s+\frac{1}{2}}(\mathbb{R}^2)$)
for any $s < 0$, by duality $H^s(\Gamma) := H^{-s}(\Gamma)'$ and $H^0(\Gamma) := L^2(\Gamma)$.

Here as a simple instance of the convergence analysis we study "Symm's" integral equation $Vu = f$ where V ist the simple layer potential

$$Vu(z) := -\frac{1}{\pi} \int_\Gamma u(\zeta) \ln |z - \zeta| \, ds(\zeta)$$

and $V : H^s(\Gamma) \to H^{s+1}(\Gamma)$ is known to be continuous and bijective.

To describe the collocation method, let $\triangle_N = \{x_1, \ldots, x_N\} \subset \Gamma$. be a mesh, which contains the corner points and where the points x_j are nodal points of the ansatz functions and collocation points as well.

Let $S^1(\Delta_N)$ denote the N dimensional space of spline functions of order 1 subordinate to the mesh Δ_N; that is, $u \in S^1(\Delta_N)$, if and only if u is continuous on Γ and is a linear function of the arc length on each segment $\overline{x_n\, x_{n+1}}$, $n = 0, \ldots, N-1$, where $x_0 := x_N$. Let the mesh parameter h :=max$\{|x_{n+1} - x_n| : n = 0, \ldots, N - 1\} \to 0 \quad (N \to \infty)$.

Then the collocation method (CM) reads:

Find $u_N \in S^1(\Delta_N)$ such that

$$Vu_N(x_n) = f(x_n) \quad n = 1, \ldots, N$$

that is

$$< Vu_N, t_N > = < f, t_N > \quad \forall t_N \in S^{-1}(\Delta_N)\,,$$

where we use $S^{-1}(\Delta_N) := \text{span}\{\delta(x - x_n) \bigm| n = 1, \ldots, N\}$, the linear space spanned up by the Dirac functionals in the mesh points.

To obtain convergence results, we intend to apply the principal Theorem 6.11.

If we define $Q_N = Q := D^2$, that is, the second derivative (in the sense of distributions) with respect to the arc length, then we have

$$QS^1(\Delta_N) \subset S^{-1}(\Delta_N).$$

Thus in this setting of Q we are led to define

$$X := H^{1/2}(\Gamma),\, L_N := S^1(\Delta_N),\, T_N := S^{-1}(\Delta_N),\, A = V$$

however, generally $< Qv, Av >= \infty$, since Q does *not* map into $Y' = (AX)' \subset H^{-3/2}(\Gamma)$ (compare, in contrast, assumption 2)!

Therefore we have to modify the setting and introduce the space

$$\overset{0\,1/2}{H}(\Gamma) := \{u \in H^{1/2}(\Gamma) \bigm| \forall\, j = 1, \ldots, J\; \exists \tilde{u}_j \in H^{1/2}(\Gamma) : \\ \tilde{u}_j \bigm| \Gamma^j = u \bigm| \Gamma^j;\; \tilde{u}_j \bigm| \Gamma \backslash \Gamma^j = 0\}\,;$$

with the norm

$$\|u\|^2_{\overset{0\,1/2}{H}(\Gamma)} := \sum_{j=1}^{J} \|\tilde{u}_j\|^2_{H^{1/2}(\Gamma)}\,.$$

Then $\overset{0\,1/2}{H}(\Gamma)$ is the completion of $C_0^\infty(\Gamma \backslash \{z_1, \ldots, z_J\})$ with respect to this norm and $\overset{0\,1/2}{H}(\Gamma) \overset{dense}{\subset} H^{1/2}(\Gamma)$. The associated ansatz space is now

$$\overset{0}{S}(\Delta_N) := \{v \in S^1(\Delta_N) \bigm| v(z_j) = 0 \quad (j = 1, \ldots, J)\}$$

where we need that $\{z_1, \ldots, z_J\} \subset \triangle_N$. Thus $\dim \overset{0}{S}(\triangle_N) = N - J$. Let us fix J functions $\eta_1, \ldots, \eta_J \in H^{3/2}(\Gamma)$ such that

$$\eta_j(z_k) = \delta_{jk} \qquad (j, k = 1, \ldots, J)$$

and introduce the projection $R : H^{3/2} \to H^{3/2}(\Gamma) \cap \overset{0}{H}{}^{1/2}(\Gamma)$ by

$$Rg(z) := g(z) - \sum_{j=1}^{J} g(z_j)\eta_j(z) .$$

Hence the adjoint operator R' acts on $S^{-1}(\triangle_N)$ as follows

$$\begin{aligned} < R'\delta(z - x_k), g > &= < \delta(z - x_k), Rg >= (Rg)(x_k) \\ &= g(x_k) - \sum_{j=1}^{J} g(z_j)\eta_j(x_k) \\ &= < \delta(z - x_k) - \sum_{j=1}^{J} \eta_j(x_k)\delta(z - z_j), g > \ . \end{aligned}$$

With $L_N := \overset{0}{S}(\triangle_N)$, $T_N := R'S^{-1}(\triangle_N)$, this fits in our abstract setting of Sect. 6.4, and the *modified* collocation method reads: Find $u_N \in \overset{0}{S}(\triangle_N)$ such that

$$Vu_N(x_n) - \sum\nolimits_{j=1}^{J} Vu_N(z_j)\eta(x_n) = f(x_n) - \sum\nolimits_{j=1}^{J} f(z_j)\eta_j(x_n) \quad (n = 1, \ldots, N) .$$

The reader is cautioned that these modified collocation equations follow from the above collocation equations, but not vice versa!

With $X := \overset{0}{H}{}^{1/2}(\Gamma)$, $A := V$, $Y := AX$, $Q_N \equiv Q = R'D^2$ the principal theorem 6.11 (Proof of the Gårding inequality by localization to a reference angle Γ_ω and by application of the Mellin transformation) yields:

Theorem 6.34 $\forall N \geq N_0 \quad \exists^1 u_N \in \overset{0}{S}(\triangle_N)$ *that solves the modified collocation equations and satisfies* $\|u_N\|_{\overset{0}{H}{}^{1/2}(\Gamma)} \leq C\|u\|_{\overset{0}{H}{}^{1/2}(\Gamma)}$;

$\|u - u_N\|_{\overset{0}{H}{}^{1/2}(\Gamma)} \leq C \inf\{\|u - v\|_{\overset{0}{H}{}^{1/2}(\Gamma)} : v \in \overset{0}{S}(\triangle_N)\}.$

A drawback of this convergence analysis is the required smoothness assumption $u \in \overset{0}{H}{}^{1/2}(\Gamma) \subset H^{1/2}(\Gamma)$ for the solution u of the considered integral equation. In view of the corners, this is not always a realistic assumption, even with a smooth right hand side f. Instead introducing

$$\alpha_j := \min\{\frac{\pi}{\omega_j}, \frac{\pi}{2\pi - \omega_j}\} \in (\frac{1}{2}, 1) \text{ for } j = 1, \dots, J$$

$$\alpha_0 := \min\{\alpha_j | j = 1, \dots, J\} \in (\frac{1}{2}, 1)$$

the solution u behaves like $O(|z - z_j|^{\alpha_j - 1})$ near the corners.

Therefore in general, $u \notin H^s(\Gamma)$ for $s \geq \alpha_0 - \frac{1}{2} \in (0, \frac{1}{2})$.

To cope with this local loss of smoothness, one chooses a weight function $\rho \in C^\infty(\mathbb{R}^2 \backslash \{z_1, \dots, z_J\})$ with $\rho(z) = |z - z_j|$ in a neighborhood of z_j $(j = 1, \dots, J)$ and introduces the weighted Sobolev space

$$H_\rho^{1/2}(\Gamma) := \frac{1}{\rho} H^{\frac{1}{2}}(\Gamma) = \{u \mid \rho u \in H^{1/2}(\Gamma)\}$$

and accordingly the ansatz space

$$S_\rho(\Delta_N) := \frac{1}{\rho} \overset{0}{S}(\Delta_N)$$

and the test space

$$\overset{0\ -1}{S}(\Delta_N) = \{\varphi \in S^{-1}(\Delta_N) | \operatorname{supp} \varphi \cap \{z_1, \dots, z_J\} = \emptyset\}.$$

By doing so, one obtains convergence and the asymptotic convergence estimate with respect to the norm $\|.\|_{H_\rho^{1/2}}$.

In 1988, I. H. Sloan [378] presented a quadrature-modified collocation method and coined the term "qualocation method" as a short name for this new method. If for collocation methods the number of collocation points surmounts the degrees of freedom of the trial functions then appropriate projection composed with the overdetermined system of linear equations leads to qualocation equations [431] where high rates of convergence can be elaborated [383–385].

In the following, we briefly sketch this method for the solution of integral equations $Au = f$. In addition to the ansatz space L_h, and the test space T_h a quadrature formula Q_h, comes into play, and the qualocation method reads:

Find $u_h \in L_h$ such that

$$< Au_h, t >_h = < f, t >_h \qquad \forall t \in T_h\,,$$

where $< v, w >_h := Q_h(v\overline{w})$ (and as usual, $\overline{w}$ denotes the conjugate complex function to w).

More explicitly with $L_h = \text{span}\{v_1, \ldots, v_N\}$, $T_h = \text{span}\{t_1, \ldots, t_N\}$, one has to solve the subsequent linear system:

$$\sum_{j=1}^{N} < Av_j, t_k >_h \xi_j = < f, t_k >_h \qquad (k = 1, \ldots, N) .$$

This extends the collocation method, for using the quadrature formula

$$Q_h g = \sum_{l=1}^{M} w_l g(x_l) \qquad w_l > 0, \sum w_l = 1, \ x_l \in \Gamma$$

the qualocation equations can be rewritten in the case $M = N$ as

$$\sum_{l=1}^{N} w_l [Au_N(x_l) - f(x_l)] \bar{t}_k(x_l) = 0 \quad (k = 1, \ldots, N) ,$$

what is equivalent to the equations

$$Au_N(x_l) = f(x_l) \quad (l = 1, \ldots, N) ,$$

only provided the matrix $\{\bar{t}_k(x_l)\}_{k,l=1,\ldots,N}$ is nonsingular.
In [385] strongly elliptic boundary integral equations

$$Lu := (b_+ L_+ + b_- L_- + K)u = f$$

are considered on a smooth curve Γ in a space of $1-$periodic functions, where $b_\pm \in \mathbb{C}$ and $L_\pm$ are operators given in Fourier series form by

$$L_+ u(x) = \hat{u}(0) + \sum_{n \neq 0} |n|^\beta \hat{u}(n) e^{2\pi i n x}$$

$$L_+ u(x) = \hat{u}(0) + \sum_{n \neq 0} (sign)|n|^\beta \hat{u}(n) e^{2\pi i n x}$$

with $\beta \in \mathbb{R}$ (the order of the pseudodifferential operator) and a smoothing operator K. Note, for $\beta = -1$ L_+ is the logarithmic-kernel operator. The qualocation method in [385] is defined with a uniform grid $\{x_k = kh, k \in \mathbb{Z}\}$ and the trial space S_N of smoothest periodic splines of order $r \geq 1$ (i.e. degree $\leq r - 1$ and $r - 2$ continuous derivatives) and the test space S'_N of order $r' \geq 1$. Together with a quadratic rule

$$Q_N g = h \sum_{k=0}^{N-1} \sum_{j=1}^{J} \omega_j g(x_k + h\xi_j)$$

where $0 \leq \xi_1 < \xi_2 < \ldots < \xi_J < 1$, $\sum_{j=1}^{J} \omega_j = 1, \omega_j > 0$ there is associated a discrete inner product $\langle u, v \rangle = Q_N(u\overline{v})$ which approximates the exact inner product

$$(u, v) = \int_0^1 u(x)\overline{v(x)}dx.$$

With this notation the qualocation method reads: Find $u_N \in S_N$ such that

$$\langle Lu_h, v' \rangle = \langle f, v' \rangle \quad \text{for all } v' \in S'_N \tag{6.140}$$

Note if $J = 1$ this is equivalent to $\epsilon-$collocation, analyzed by Schmidt [365].

Note, the Galerkin method (obtained by setting $S'_N = S_N$ and replacing $\langle u, v \rangle$ by (u, v)) is stable and convergent for a subset of the above equations namely for strongly elliptic operator equations. The following definition, adapted from [408] is taken appropriate in [385].

Definition 6.3 $L_0 := b_+L_+ + b_-L_-$ is strongly elliptic if there exists $\theta \in \mathbb{C}$ such that

$$\Re[\theta(b_+ + b_-)] > 0 \text{ and } \Re[\theta(b_+ - b_-)] > 0$$

As a basis of S_N take $\{\psi_\mu : \mu \in \Lambda_N\}$, where

$$\psi_\mu = a_\mu \sum_{k=0}^{N-1} e^{2\pi i \mu x_k} b_k \ , \quad \mu \in \Lambda_N$$

and

$$\Lambda_N = \{\mu \in \mathbb{Z} : -\frac{N}{2} < \mu \leq \frac{N}{2}\},$$

similarly $\{\psi'_N : \mu \in \Lambda_N\}$ for S'_N. Then the qualocation method becomes: Find $u_h = \sum_{\mu \in \Lambda_N} \hat{u}_h(\mu)\psi_N$ such that

$$\langle Lu_h, \psi'_\mu \rangle = \langle f, \psi'_\mu \rangle \quad \text{for all } \mu \in \Lambda_N.$$

To derive an analysis of (6.140) first the simplified equation $L_0 u = f$ is considered. As worked out in [96] there holds

$$\langle L_\pm \psi_\mu, \psi'_\mu \rangle = \begin{cases} 1 & \text{if } \mu = 0 \\ [\mu]_\beta^\pm D_\pm(\frac{\mu}{N}) & \text{if } \mu \in \Lambda_N \backslash \{0\} \end{cases}$$

where

$$[n]_\beta^+ = \begin{cases} 1, & n = 0 \\ |n|^\beta, & n \neq 0 \end{cases} \quad , \quad [n]_\beta^- = \begin{cases} 1, & n = 0 \\ (sign)|n|^\beta, & n \neq 0 \end{cases}$$

and

$$D_{\pm}(y) = \sum_{j=1}^{J} \omega_j [1 + \Omega_{\pm}(\xi_j, y)][1 + \overline{\Delta'}(\xi_j, y)], \quad y \in [-1/2, 1/2]$$

with

$$\Delta'(\xi, y) = \begin{cases} y^{r^0} F^{+}_{r^0}(\xi, y), & \text{if } r' \text{ even} \\ y^{r^0} F^{-}_{r^0}(\xi, y) & \text{if } r' \text{ odd} \end{cases}$$

$$F^{+}_{\alpha}(x, y) = \sum_{l \neq 0} \frac{e^{2\pi i l x}}{|l + y|^{\alpha}} \qquad F^{-}_{\alpha}(x, y) = \sum_{l \neq 0} \frac{sign\ l}{|l + y|^{\alpha}} e^{2\pi i l x}$$

and

$$\Omega_{\substack{+\\(-)}}(\xi, y) = \begin{cases} |y|^{r-\beta} F^{+}_{r-\beta}(\xi, y) & \text{if } r \text{ even } +, r \text{ odd } (-) \\ (sign\ y)|y|^{r-\beta} F^{-}_{r-\beta}(\xi, y) & \text{if } r \text{ odd } +, r \text{ even } (-) \end{cases}$$

With

$$D(y) := b_{+} D_{+}(y) + b_{-}(sign\ y) D_{-}(y), \quad y \in [-1/2, 1/2]$$

the qualocation method (6.140) is called stable if

$$\inf\{|D(y)| : y \in [-1/2, 1/2]\} > 0$$

The qualocation method is of order $r - \beta + b$ if

$$E(y) := b_{+} E_{+}(y) + b_{-}(sign\ y) E_{-}(y) = O(|y|^{r-\beta+b}) \quad \text{for } y \in [-1/2, 1/2]$$

where $E_{\pm}(y) = \Omega_{\pm}(\xi_j, y)[1 + \overline{\Delta'}(\xi_j, y)]$.

As a generalization of [96] Sloan and Wendland derive in [385] the following convergence result by first analysing the Fourier coefficients of L_0 and then applying a standard perturbation argument to include K.

Theorem 6.35 *Let the qualocation method* (6.140) *be stable and of order* $r - \beta + b, b \geq 0$, *then for all* N *sufficiently large* u_h *is uniquely defined. Moreover, for all* s, t *satisfying* $s < r - 1/2, \beta + 1/2 < t, \beta - b \leq s \leq t \leq r$ *there holds*

$$\|u_h - u\|_s \leq c h^{t-s} |u|_{t+\max\{\beta-s,0\}}$$

The following theorem characterises qualocation methods that are stable for strongly elliptic operators.

Theorem 6.36 ([385]) *For the qualocation method* (6.140) *with a symmetric quadrature rule with positive weights there holds:*
Assume r, r' are of the same parity and if $J = 1$ that $\xi_1 \neq 1/2$ if r, r' even and $\xi_1 \neq 0$ if r, r' odd. The method (6.140) *is stable for all strongly elliptic operators L_0 if and only if*

$$D_+(y) \geq |D_-(y)| \quad \textit{for all } y \in [0, 1/2].$$

In [385][Section 5] a list of qualocation methods is given that are stable for all strongly elliptic operators: For example for the logarithmic-kernel operator (single layer potential) one can take for $r = r' = b = 1$ the 2 point rule $G_{2,1,2}$ of order 3 with $\xi_1 = 0.2113248654051872$, $\xi_2 = 0.7886751345948128$ and $\omega_1 = \omega_2 = 1/2$ (This rule integrates exactly all polynomials of degree ≤ 2). For qualocation under reduced regularity see [382, 419].

Finally, note that as long as the order α of the operator is not zero, the condition numbers of the discrete equations are unbounded independent of the sign of α. Hence in order to use iteration schemes for scaling the discrete conditions, suitable preconditioners must be applied (see e.g [255, 420]). For related work of Langer and Steinbach on boundary element tearing and interconnecting methods we refer to [280].

6.11 Radial Basis Functions and Spherical Splines

Radial basis functions are used in [418] to define approximate solutions to boundary integral equations on the unit sphere. These equations arise from the integral reformulation of the Laplace equation in the exterior of the sphere, with given Dirichlet or Neumann data, and a vanishing condition at infinity. Radial basis functions yield a meshless method which is especially suitable to handle sattelite data.

In the following we consider boundary integral equations on the unit sphere. Let $\mathbb{S}$ denote the unit sphere in $\mathbb{R}^3$, i.e., $\mathbb{S} := \{\mathbf{x} \in \mathbb{R}^3 : \|\mathbf{x}\| = 1\}$, and $\mathbb{B}_e$ the exterior of the sphere, i.e., $\mathbb{B}_e := \{\mathbf{x} \in \mathbb{R}^3 : \|\mathbf{x}\| > 1\}$, where $\|\mathbf{x}\|$ denotes the Euclidean norm in $\mathbb{R}^3$. We now follow [418] and consider the Laplace equation

$$\Delta U = 0 \quad \text{in } \mathbb{B}_e, \tag{6.141}$$

with either a Dirichlet boundary condition

$$U = U_D \quad \text{on } \mathbb{S}, \tag{6.142}$$

or else a Neumann boundary condition

$$\partial_\nu U = Z_N \quad \text{on } \mathbb{S}, \tag{6.143}$$

where $\partial_\nu = \partial/\partial\nu$ denotes differentiation in the direction of the outward unit normal ν, and the vanishing condition at infinity for both the Dirichlet and Neumann cases is

$$U(\mathbf{x}) = O(1/\|\mathbf{x}\|) \quad \text{as } \|\mathbf{x}\| \to \infty. \tag{6.144}$$

The solutions of these problems can be represented in terms of spherical harmonics $Y_{l,m}$, $m = -l, \ldots, l$ and $l = 0, 1, \ldots$. They form an orthonormal basis for $L^2(\mathbb{S})$. For any function $v \in L^2(\mathbb{S})$, its associated Fourier series,

$$v = \sum_{l=0}^{\infty} \sum_{m=-l}^{l} \widehat{v}_{l,m} Y_{l,m}(\theta, \varphi), \quad \text{where} \quad \widehat{v}_{l,m} = \int_{\mathbb{S}} v(\theta, \varphi) \overline{Y_{l,m}}(\theta, \varphi) d\sigma, \tag{6.145}$$

converges in $L^2(\mathbb{S})$. Here $d\sigma$ is the element of surface area.

It is well-known that if the Dirichlet data U_D has an expansion as a sum of spherical harmonics

$$U_D(\theta, \varphi) = \sum_{l=0}^{\infty} \sum_{m=-l}^{l} \widehat{(U_D)}_{l,m} Y_{l,m}(\theta, \varphi),$$

then (see [323, Theorem 2.5.1]) the Dirichlet problem (6.141), (6.142) and (6.144) has the unique solution

$$U(r, \theta, \varphi) = \sum_{l=0}^{\infty} \sum_{m=-l}^{l} \frac{1}{r^{l+1}} \widehat{(U_D)}_{l,m} Y_{l,m}(\theta, \varphi). \tag{6.146}$$

Similarly, if

$$Z_N(\theta, \varphi) = \sum_{l=0}^{\infty} \sum_{m=-l}^{l} \widehat{(Z_N)}_{l,m} Y_{l,m}(\theta, \varphi),$$

then (see [323, Theorem 2.5.2]) the Neumann problem (6.141), (6.143) and (6.144) has the unique solution

$$U(r, \theta, \varphi) = -\sum_{l=0}^{\infty} \sum_{m=-l}^{l} \frac{1}{(l+1) r^{l+1}} \widehat{(Z_N)}_{l,m} Y_{l,m}(\theta, \varphi). \tag{6.147}$$

Note that the spherical harmonic basis functions in (6.146), (6.147) are global. In contrast, in [418] we use spherical basis functions obtained from compactly supported radial basis functions, which are better able to capture local properties of the solutions. We shall propose a solution process in which the boundary value problems are reformulated in terms of boundary integral equations on $\mathbb{S}$, the solutions of which are then approximated by spherical basis functions.

Next we reformulate the boundary value problems (6.141)–(6.144) as boundary integral equations. For $s \in \mathbb{R}$, the Sobolev space $H^s(\mathbb{S})$ is defined as usual (see e.g. [323]) with norm and Hermitian product given by

$$\|v\|_s := \left(\sum_{l=0}^{\infty} \sum_{m=-l}^{l} (l+1)^{2s} |\widehat{v}_{l,m}|^2 \right)^{1/2} \tag{6.148}$$

and

$$\langle v, w \rangle_s := \sum_{l=0}^{\infty} \sum_{m=-l}^{l} (l+1)^{2s} \widehat{v}_{l,m} \overline{\widehat{w}_{l,m}}.$$

Note that

$$|\langle v, w \rangle_s| \le \|v\|_s \|w\|_s \quad \forall v, w \in H^s(\mathbb{S}), \forall s \in \mathbb{R}, \tag{6.149}$$

and

$$\|v\|_{s_1} = \sup_{\substack{w \in H^{s_2}(\mathbb{S}) \\ w \neq 0}} \frac{\langle v, w \rangle_{\frac{s_1+s_2}{2}}}{\|w\|_{s_2}} \quad \forall v \in H^{s_1}(\mathbb{S}), \forall s_1, s_2 \in \mathbb{R}. \tag{6.150}$$

The single-layer potential S and the double-layer potential D are defined by

$$Sv(\mathbf{x}) = \frac{1}{2\pi} \int_{\mathbb{S}} v(\mathbf{y}) \frac{1}{\|\mathbf{x} - \mathbf{y}\|} \, d\sigma_{\mathbf{y}} \quad , \quad Dv(\mathbf{x}) = \frac{1}{2\pi} \int_{\mathbb{S}} v(\mathbf{y}) \frac{\partial}{\partial \nu_{\mathbf{y}}} \frac{1}{\|\mathbf{x} - \mathbf{y}\|} \, d\sigma_{\mathbf{y}},$$

for $\mathbf{x} \in \mathbb{B}_e$. Associated with these potentials, we define the following boundary integral operators

$$\begin{aligned}
Vv(\mathbf{x}) &= \frac{1}{2\pi} \int_{\mathbb{S}} v(\mathbf{y}) \frac{1}{\|\mathbf{x} - \mathbf{y}\|} \, d\sigma_{\mathbf{y}} \\
Kv(\mathbf{x}) &= \frac{1}{2\pi} \int_{\mathbb{S}} v(\mathbf{y}) \frac{\partial}{\partial \nu_{\mathbf{y}}} \frac{1}{\|\mathbf{x} - \mathbf{y}\|} \, d\sigma_{\mathbf{y}} \\
K^*v(\mathbf{x}) &= \frac{1}{2\pi} \frac{\partial}{\partial \nu_{\mathbf{x}}} \int_{\mathbb{S}} v(\mathbf{y}) \frac{1}{\|\mathbf{x} - \mathbf{y}\|} \, d\sigma_{\mathbf{y}} \\
Wv(\mathbf{x}) &= -\frac{1}{2\pi} \frac{\partial}{\partial \nu_{\mathbf{x}}} \int_{\mathbb{S}} v(\mathbf{y}) \frac{\partial}{\partial \nu_{\mathbf{y}}} \frac{1}{\|\mathbf{x} - \mathbf{y}\|} \, d\sigma_{\mathbf{y}},
\end{aligned}$$

for $\mathbf{x} \in \mathbb{S}$. The traces and normal derivatives on $\mathbb{S}$ of S and D are given by (see Lemma 2.2 noting that the limits are taken from the exterior of $\mathbb{S}$)

$$(Sv)|_{\mathbb{S}} = Vv \quad \text{and} \quad \partial_\nu(Sv) = -v + K^*v, \quad \text{if } v \in H^{-1/2}(\mathbb{S}),$$

and

$$(Dv)|_{\mathbb{S}} = v + Kv \quad \text{and} \quad \partial_\nu(Dv) = -Wv \quad \text{if } v \in H^{1/2}(\mathbb{S}).$$

If $U \in H^1_{\text{loc}}(\mathbb{B}_e)$ satisfies (6.141) and (6.144), then using the single-layer and double-layer potentials, and Green's theorem we can represent U as (see Chap. 2)

$$U = \frac{1}{2}D(U|_{\mathbb{S}}) - \frac{1}{2}S(\partial_\nu U) \quad \text{in } \mathbb{B}_e, \tag{6.151}$$

allowing us to compute U from a knowledge of both $U|_{\mathbb{S}}$ and $\partial_\nu U$. In fact, by taking the trace on both sides of (6.151) we obtain, after rearranging the equation,

$$V(\partial_\nu U) = -U|_{\mathbb{S}} + K(U|_{\mathbb{S}}) \quad \text{on } \mathbb{S}.$$

Similarly, by taking the normal derivative of both sides of (6.151) we find

$$W(U|_{\mathbb{S}}) = -\partial_\nu U - K^*(\partial_\nu U) \quad \text{on } \mathbb{S}.$$

Therefore, the Dirichlet problem (6.141), (6.142) and (6.144) is equivalent to

$$Vz = f \quad \text{on } \mathbb{S}, \quad \text{where } f = -U_D + KU_D, \tag{6.152}$$

and the Neumann problem (6.141), (6.143) and (6.144) is equivalent to

$$Wu = g \quad \text{on } \mathbb{S}, \quad \text{where } g = -Z_N - K^*Z_N. \tag{6.153}$$

Due to (6.151), the solution U of the Dirichlet problem can be computed from the solution z of (6.152) by

$$U = \frac{1}{2}DU_D - \frac{1}{2}Sz,$$

and the solution of the Neumann problem can be computed from the solution u of (6.153) by

$$U = \frac{1}{2}Du - \frac{1}{2}SZ_N.$$

Equation (6.152) is a weakly singular integral equation and equation (6.153) is a hypersingular integral equation. In the following we present efficient algorithms to solve these equations . Note that V and W are pseudo-differential operators of order -1 and 1, respectively. They have the following representations in terms of spherical harmonics (see [323, page 122]):

$$Vv = 2\sum_{l=0}^{\infty}\sum_{m=-l}^{l}\frac{1}{2l+1}\widehat{v}_{l,m}Y_{l,m}, \tag{6.154}$$

$$Wv = -2\sum_{l=0}^{\infty}\sum_{m=-l}^{l}\frac{l(l+1)}{2l+1}\widehat{v}_{l,m}Y_{l,m}. \tag{6.155}$$

Now we define weak solutions to (6.152) and (6.153). It is well-known [304, 323] that $V : H^{-1/2}(\mathbb{S}) \to H^{1/2}(\mathbb{S})$ and $W : H^{1/2}(\mathbb{S})/\mathbb{R} \to H^{-1/2}(\mathbb{S})$ are bijective, implying that (6.152) has a unique solution for all $f \in H^{1/2}(\mathbb{S})$, and (6.153) has a unique solution up to a constant for all $g \in H^{-1/2}(\mathbb{S})$. Defining the bilinear forms

$$a_V(v, w) := \langle Vv, w\rangle_0 \quad \forall v, w \in H^{-1/2}(\mathbb{S})$$

and

$$a_W(v, w) := -\langle Wv, w\rangle_0 \quad \forall v, w \in H^{1/2}(\mathbb{S}),$$

we seek weak solutions to equations (6.152) and (6.153) respectively as follows:

$$z \in H^{-1/2}(\mathbb{S}) : a_V(z, v) = \langle f, v\rangle_0 \quad \forall v \in H^{-1/2}(\mathbb{S}), \tag{6.156}$$

and

$$u \in H^{1/2}(\mathbb{S}) : \int_{\mathbb{S}} u(\mathbf{x})\, d\sigma_{\mathbf{x}} = 0 \quad \text{and} \quad a_W(u, v) = -\langle g, v\rangle_0 \quad \forall v \in H^{1/2}(\mathbb{S}). \tag{6.157}$$

We note from (6.154) and (6.155) that

$$a_V(v, v) \simeq \|v\|_{-1/2}{}^2 \quad \forall v \in H^{-1/2}(\mathbb{S}) \quad , \quad a_W(v, v) \simeq \|v\|_{1/2}{}^2 \quad \forall v \in H^{1/2}(\mathbb{S})/\mathbb{R}. \tag{6.158}$$

Next we shall approximate the solutions of the above equations with spherical basis functions. These functions are defined via positive definite kernels.

A continuous function $\Phi : \mathbb{S} \times \mathbb{S} \to \mathbb{C}$ is called a *positive definite kernel* on $\mathbb{S}$ if it satisfies

(i) $\Phi(\mathbf{x}, \mathbf{y}) = \overline{\Phi(\mathbf{y}, \mathbf{x})}$ for all $\mathbf{x}, \mathbf{y} \in \mathbb{S}$;
(ii) for every set of distinct points $\{\mathbf{x}_1, \ldots, \mathbf{x}_M\}$ on $\mathbb{S}$, the $M \times M$ matrix A with entries $A_{i,j} = \Phi(\mathbf{x}_i, \mathbf{x}_j)$ is positive semi-definite.

If the matrix A is positive definite then Φ is called a *strictly positive definite* kernel.

We shall define the kernel Φ in terms of a univariate function $\phi : [-1, 1] \to \mathbb{R}$,

$$\Phi(\mathbf{x}, \mathbf{y}) = \phi(\mathbf{x} \cdot \mathbf{y}) \quad \forall \mathbf{x}, \mathbf{y} \in \mathbb{S}.$$

If ϕ has a series expansion in terms of Legendre polynomials P_l,

$$\phi(t) = \frac{1}{4\pi} \sum_{l=0}^{\infty} (2l + 1)\widehat{\phi}(l) P_l(t), \tag{6.159}$$

where

$$\widehat{\phi}(l) = 2\pi \int_{-1}^{1} \phi(t) P_l(t) dt, \tag{6.160}$$

then due to the addition formula [323]

$$\sum_{m=-l}^{l} Y_{l,m}(\mathbf{x})\overline{Y_{l,m}(\mathbf{y})} = \frac{2l + 1}{4\pi} P_l(\mathbf{x} \cdot \mathbf{y}) \quad \forall \mathbf{x}, \mathbf{y} \in \mathbb{S}, \tag{6.161}$$

the kernel Φ can be represented as

$$\Phi(\mathbf{x}, \mathbf{y}) = \sum_{l=0}^{\infty} \widehat{\phi}(l) \sum_{m=-l}^{l} Y_{l,m}(\mathbf{x})\overline{Y_{l,m}(\mathbf{y})}. \tag{6.162}$$

This kernel is called a *zonal* kernel. The kernel Φ is strictly positive definite if and only if $\widehat{\phi}(l) \geq 0$ for all $l \geq 0$, and $\widehat{\phi}(l) > 0$ for infinitely many even values of l and infinitely many odd values of l; see [97]. In the following we shall assume that $\widehat{\phi}(l) > 0$ for all $l \geq 0$.

The native space associated with ϕ is defined by

$$\mathcal{N}_\phi := \{v \in \mathcal{D}'(\mathbb{S}) : \|v\|_\phi{}^2 = \sum_{l=0}^{\infty} \sum_{m=-l}^{l} \frac{|\widehat{v}_{l,m}|^2}{\widehat{\phi}(l)} < \infty\},$$

where $\mathcal{D}'(\mathbb{S})$ is the space of distributions defined on $\mathbb{S}$. This space is equipped with an inner product and a norm defined by

$$\langle v, w \rangle_\phi = \sum_{l=0}^{\infty} \sum_{m=-l}^{l} \frac{\widehat{v}_{l,m}\overline{\widehat{w}_{l,m}}}{\widehat{\phi}(l)} \quad \text{and} \quad \|v\|_\phi = \left(\sum_{l=0}^{\infty} \sum_{m=-l}^{l} \frac{|\widehat{v}_{l,m}|^2}{\widehat{\phi}(l)} \right)^{1/2}.$$

If the coefficients $\widehat{\phi}(l)$ for $l = 0, 1, \ldots$ satisfy

$$c_1(l + 1)^{-2\tau} \leq \widehat{\phi}(l) \leq c_2(l + 1)^{-2\tau} \tag{6.163}$$

for some positive constants c_1 and c_2, and some $\tau \in \mathbb{R}$, then the native space $\mathcal{N}_\phi$ can be identified with the Sobolev space $H^\tau(\mathbb{S})$, and the corresponding norms are equivalent.

Let $X = \{\mathbf{x}_1, \ldots, \mathbf{x}_M\}$ be a set of data points on the sphere. Two important parameters characterising the set X are the *mesh norm* h_X and *separation radius* q_X, defined by

$$h_X := \sup_{\mathbf{y}\in\mathbb{S}} \min_{1\le i\le M} \theta(\mathbf{x}_i, \mathbf{y}) \quad \text{and} \quad q_X := \frac{1}{2}\min_{i\ne j}\theta(\mathbf{x}_i, \mathbf{x}_j),$$

where $\theta(\mathbf{x}, \mathbf{y}) := \cos^{-1}(\mathbf{x}\cdot\mathbf{y})$. The *spherical basis functions* Φ_i, $i = 1, \ldots, M$, associated with X and the kernel Φ are defined by

$$\Phi_i(\mathbf{x}) := \Phi(\mathbf{x}, \mathbf{x}_i) = \sum_{l=0}^{\infty}\sum_{m=-l}^{l} \widehat{\phi}(l)\overline{Y_{l,m}(\mathbf{x}_i)}Y_{l,m}(\mathbf{x}). \tag{6.164}$$

Note that if (6.163) holds then $\Phi_i \in H^s(\mathbb{S})$ for all s satisfying $s < 2\tau - 1$.

Let

$$V_X^\phi := \operatorname{span}\{\Phi_1, \ldots, \Phi_M\}. \tag{6.165}$$

We assume that (6.163) holds for some $\tau > 1$ so that $V_X^\phi \subset \mathcal{N}_\phi = H^\tau(\mathbb{S}) \subset C(\mathbb{S})$ and study the approximation property of V_X^ϕ as a subspace of Sobolev spaces. The following lemma, proven in [418] shows the boundedness of the interpolation operator in the native space.

Lemma 6.21 ([418]) *The interpolation operator $I_X : C(\mathbb{S}) \to V_X^\phi$ defined by*

$$I_X v(\mathbf{x}_j) = v(\mathbf{x}_j), \quad j = 1, \ldots, M, \quad v \in C(\mathbb{S}), \tag{6.166}$$

is well-defined, and is a bounded operator in $\mathcal{N}_\phi$. In fact, this operator is the $\mathcal{N}_\phi$-orthogonal projection from $\mathcal{N}_\phi$ onto V_X^ϕ.

Proposition 6.1 ([418]) *Assume that* (6.163) *holds for some $\tau > 1$. For any $s, t \in \mathbb{R}$ satisfying $0 \le t \le \tau \le s \le 2\tau$, if $v \in H^s(\mathbb{S})$ then the following estimate holds*

$$\|I_X v - v\|_t \le C h_X^{\min\{s-t,\, 2(\tau-t)\}}\|v\|_s.$$

The convergence analysis for the approximate solutions to (6.156) and (6.157) requires the following approximation property of V_X^ϕ.

Theorem 6.37 ([418]) *Assume that* (6.163) *holds for some $\tau > 1$. For any $s, t \in \mathbb{R}$ satisfying $t \le \tau$ and $t \le s \le 2\tau$, if $v \in H^s(\mathbb{S})$ then there exists $\eta \in V_X^\phi$ such that*

$$\|v - \eta\|_t \le C h_X^\mu \|v\|_s, \tag{6.167}$$

where $\mu = \min\{s-t, 2(\tau-t), 2\tau+|s|\}$*, and where the constant* C *is independent of* v *and* h_X*.*

Proof We prove the result by considering different cases of values of s and t.

Case 1: $0 \le t \le \tau \le s \le 2\tau$ Note that in this case $\mu = \min\{s-t, 2(\tau-t)\}$. We can choose $\eta = I_X v$ yielding (6.167) with $s = 2\tau$ (and with $s < 2\tau$ by interpolation).
In the following cases, it is easy to see that $s - t \le 2(\tau - t)$ and thus $\mu = \min\{s-t, 2\tau+|s|\}$.

Case 2: $0 \le t \le s < \tau$
Let $L = \lfloor \frac{1}{h_X} \rfloor$. We define for each $v \in H^s(\mathbb{S})$ a polynomial of degree L by

$$P_L v = \sum_{l=0}^{L} \sum_{m=-l}^{l} \widehat{v}_{l,m} Y_{l,m}.$$

With $\eta = I_X P_L v$ we have

$$\begin{aligned}
\|v-\eta\|_t^2 &\le 2\|v - P_L v\|_t^2 + 2\|P_L v - I_X P_L v\|_t^2 \\
&\le 2 \sum_{l=L+1}^{\infty} \sum_{m=-l}^{l} (l+1)^{2t} |\widehat{v}_{l,m}|^2 + c h_X^{2(\tau-t)} \|P_L v\|_\tau^2 \\
&= 2 \sum_{l=L+1}^{\infty} \sum_{m=-l}^{l} (l+1)^{2(t-s)} (l+1)^{2s} |\widehat{v}_{l,m}|^2 \\
&\quad + c h_X^{2(\tau-t)} \sum_{l=1}^{L} \sum_{m=-l}^{l} (l+1)^{2(\tau-s)} (l+1)^{2s} |\widehat{v}_{l,m}|^2 \\
&\le c(L+1)^{2(t-s)} \|v\|_s^2 + c L^{2(\tau-s)} h_X^{2(\tau-t)} \|v\|_s^2,
\end{aligned}$$

where in the second step we have used the result given in Case 1. Here c is a generic constant which may take different values at different occurrences. Since $L \le h_X^{-1}$ and $(L+1)^{-1} \le h_X$, we deduce (6.167) with $\mu = s - t$.

Case 3: $t < 0 \le s \le 2\tau$ and **Case 4:** $t \le s < 0$ see [418]. □

Optimal estimates are afterwards obtained in[338].

For the approximation of the hypersingular equation (6.157) we use radial basis functions suggested by [427, page 128]. First we define a smoothing operator I on the space $C_K[0, \infty)$ of continuous functions in $[0, \infty)$ with compact supports by

$$I : C_K[0,\infty) \to C_K[0,\infty), \quad Iv(r) = \int_r^\infty s v(s) ds, \quad r \ge 0.$$

For any non-negative integer m, let

$$\tilde{\rho}_m(r) = \begin{cases} (1-r)^{m+2}, & 0 < r \le 1, \\ 0, & r > 1, \end{cases}$$

and

$$\rho_m(r) = I^m \tilde{\rho}_m(r), \quad r \ge 0.$$

We define

$$\phi^{(W)}(t) = \rho_m(\sqrt{2-2t}), \quad t \in [-1, 1], \tag{6.168}$$

and denote by $\Phi_i^{(W)}$, $i = 1, \ldots, M$, the corresponding spherical basis functions; see (6.164). Here the superscript N indicates that the functions are specifically chosen for equation (6.157) arising from the Neumann problem. We suppress the dependence on m in the notation of $\phi^{(W)}$ and $\Phi_i^{(W)}$ because m will be chosen once and for all during the whole solution process. The functions $\Phi_i^{(W)}$, $i = 1, \ldots, M$, are locally supported radial basis functions. It is proved in [315, Proposition 4.6] that $\widehat{\phi}^{(W)}(l)$ satisfies (6.163) with

$$\tau^{(W)} = m + 3/2. \tag{6.169}$$

In Fig. 6.4 we plotted $l^{2m+3}\widehat{\phi}^{(W)}(l)$ to observe the asymptotic behaviour of $\widehat{\phi}^{(W)}(l)$ for $m = 0, 1, 2, 3$, with $\widehat{\phi}^{(W)}(l)$ computed by the MATLAB function `quadl` which uses an adaptive Lobatto quadrature.

For given $X = \{\mathbf{x}_1, \ldots, \mathbf{x}_M\} \subset \mathbb{S}$, let $V_N := V_X^{\phi^{(W)}}$. We will solve (6.157) approximately by solving the Galerkin scheme

$$u_X \in V_N: \quad \int_{\mathbb{S}} u_X(\mathbf{x})\, d\sigma_{\mathbf{x}} = 0 \quad \text{and} \quad a_W(u_X, v_X) = -\langle g, v_X \rangle_0 \quad \forall v_X \in V_N. \tag{6.170}$$

Using (6.155) and (6.161), one obtains the following formula to compute the entries of the stiffness matrix from (6.170):

$$\begin{aligned} a_W(\Phi_i^{(W)}, \Phi_j^{(W)}) &= 2 \sum_{l=0}^{\infty} \frac{l(l+1)}{2l+1} |\widehat{\phi^{(W)}}(l)|^2 \sum_{m=-l}^{l} \overline{Y_{l,m}(\mathbf{x}_i)} Y_{l,m}(\mathbf{x}_j) \\ &= \frac{1}{2\pi} \sum_{l=0}^{\infty} l(l+1) |\widehat{\phi^{(W)}}(l)|^2 P_l(\mathbf{x}_i \cdot \mathbf{x}_j). \end{aligned} \tag{6.171}$$

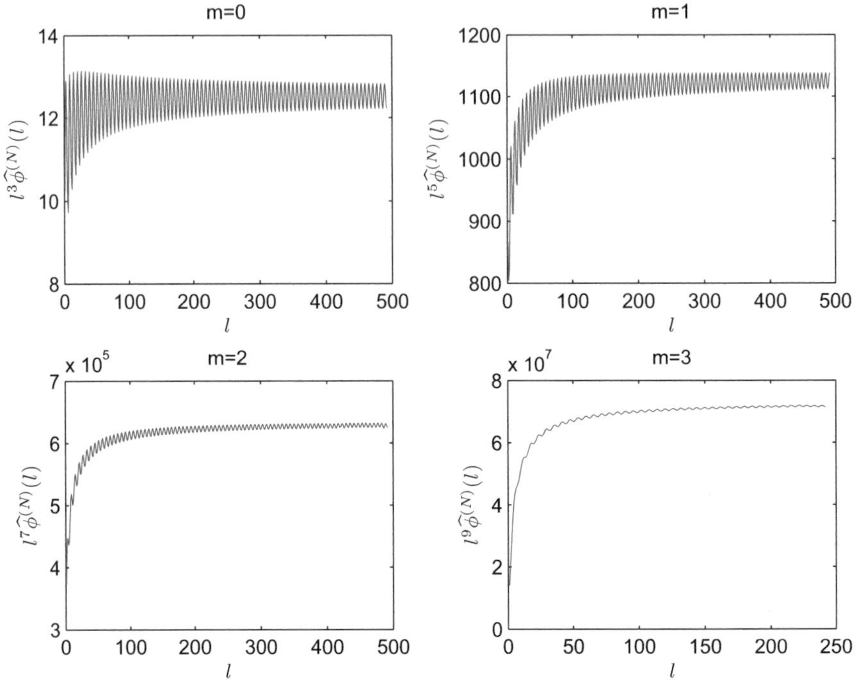

Fig. 6.4 Asymptotic behaviour of $\widehat{\phi}^{(W)}(l)$ [418]

The right-hand side in (6.170) is computed by using (6.153), noting $K^* = -S/2$ (see [323, page 122]),

$$\langle g, \Phi_i^{(W)} \rangle_0 = \sum_{l=0}^{\infty} \sum_{m=-l}^{l} \frac{l}{2l+1} \widehat{(Z_N)}_{l,m} \widehat{\phi}^{(W)}(l) \overline{Y_{l,m}(\mathbf{x}_i)}. \tag{6.172}$$

Theorem 6.37 yields the following *a priori* error estimate.

Theorem 6.38 ([418]) *Let* $\phi^{(W)}$ *be defined by* (6.168) *for some non-negative integer* m, *and let* $\tau^{(W)} = m + 3/2$. *If* u *is the solution to* (6.157) *satisfying* $u \in H^s(\mathbb{S})$, $1/2 \le s \le 2\tau^{(W)}$, *and* u_X *the solution to* (6.170) *then*

$$\|u - u_X\|_{1/2} \le C h_X^{\min\{s-1/2,\ 2\tau^{(W)}-1\}} \|u\|_s.$$

Proof The condition

$$\int_{\mathbb{S}} u(\mathbf{x})\, d\sigma_{\mathbf{x}} = \int_{\mathbb{S}} u_X(\mathbf{x})\, d\sigma_{\mathbf{x}} = 0$$

yields $\widehat{u}_{0,0} = \widehat{(u_X)}_{0,0} = 0$, implying (together with (6.158))

$$\|u - u_X\|_{1/2}^2 \simeq a_W(u - u_X, u - u_X) = a_W(u - u_X, u - w_X)$$
$$\leq c\|u - u_X\|_{1/2}\|u - w_X\|_{1/2}$$

for all $w_X \in V_N$. Now the required estimate follows from Theorem 6.37. □

The corresponding optimal results can be found in Theorem 5.4 in [338].

In [418] the weakly singular integral equation is treated as follows. The following univariate function is used:

$$\phi^{(V)}(t) = \frac{1}{4\pi}\sum_{l=0}^{\infty}(2l+1)(l+1)\widehat{\phi}^{(W)}(l)P_l(t), \tag{6.173}$$

Let $\Phi_i^{(V)}$, $i = 1, \ldots, M$, denote the corresponding spherical basis functions. It is clear that $\widehat{\phi}^{(V)}(l)$ satisfies (6.163) with

$$\tau^{(V)} = \tau^{(W)} - 1/2 = m + 1;$$

see (6.169).

Letting

$$V_D := V_X^{\phi^{(V)}} = \operatorname{span}\{\Phi_1^{(V)}, \ldots, \Phi_M^{(V)}\},$$

we approximate the solution z of (6.156) by

$$z_X \in V_D : \quad a_V(z_X, v_X) = \langle f, v_X\rangle \quad \forall v_X \in V_D. \tag{6.174}$$

The resulting stiffness matrix has entries given as (cf. (6.171))

$$a_V(\Phi_i^{(V)}, \Phi_j^{(V)}) = 2\sum_{l=0}^{\infty}\frac{(l+1)^2}{2l+1}|\widehat{\phi^{(W)}}(l)|^2\sum_{m=-l}^{l}\overline{Y_{l,m}(\mathbf{x}_i)}Y_{l,m}(\mathbf{x}_j)$$
$$= \frac{1}{2\pi}\sum_{l=0}^{\infty}(l+1)^2|\widehat{\phi^{(W)}}(l)|^2 P_l(\mathbf{x}_i\cdot\mathbf{x}_j).$$

The right-hand side of (6.174) is computed by using (6.152) (see [323, page 122]),

$$\langle f, \Phi_i^{(V)}\rangle = -\sum_{l=0}^{\infty}\sum_{m=-l}^{l}\frac{(l+1)^2}{2l+1}\widehat{(U_D)}_{l,m}\widehat{\phi}^{(W)}(l)\overline{Y_{l,m}(\mathbf{x}_i)}. \tag{6.175}$$

A priori error estimates similar to those in Theorem 6.38 can be proved.

Theorem 6.39 ([418]) *Let $\phi^{(W)}$ be defined by* (6.168) *for some positive integer m, $\phi^{(V)}$ be defined by* (6.173), *and $\tau^{(V)} = m + 1$. If z is the solution to* (6.156) *satisfying $z \in H^s(\mathbb{S})$, $-1/2 \le s \le 2\tau^{(V)}$, and z_X the solution to* (6.174), *then*

$$\|z - z_X\|_{-1/2} \le C h_X^{s+1/2} \|z\|_s.$$

Proof We note that approximation property requires $m > 0$, and that

$$\min\{s + \frac{1}{2}, 2(\tau^{(V)} + \frac{1}{2}), 2\tau^{(V)} + |s|\} = s + \frac{1}{2}.$$

Since

$$\|z - z_X\|_{-1/2}^2 \simeq a_V(z - z_X, z - z_X),$$

the remainder of the proof is similar to that of Theorem 6.38, and is therefore omitted. □

Next we present numerical results obtained from experiments with the set of scattered points X generated by a simple algorithm [354] which partitions the sphere into equal areas; see Fig. 6.5; for detais see . The sets of points we used have number of points $M = 20, 30, 40, 50, 100, 500$, and 1000.

The spherical basis functions $\Phi_i^{(W)}$, $i = 1, \ldots, M$, are defined by (6.164) using the univariate function $\phi^{(W)}$ given by (6.168) with $m = 0, 1, 2$. The coefficients $\widehat{\phi}_N(l)$ with $l = 1, \ldots, 500$ are computed by the MATLAB function `quadl` which uses an adaptive Lobatto quadrature. The spherical basis functions $\Phi_i^{(V)}$, $i = 1, \ldots, M$, are defined by (6.164) with $\phi^{(V)}$ given by (6.173).

In [418] the exterior Neumann problem (6.141), (6.143) and (6.144) is considered with a boundary data given by

$$Z_N(\mathbf{x}) = \frac{0.5x_3 - 1}{(1.25 - x_3)^{3/2}},$$

so that the exact solution is

$$U(\mathbf{x}) = \frac{1}{\|\mathbf{x} - \mathbf{p}\|} \quad \text{with } \mathbf{p} = (0, 0, 0.5).$$

Here $\mathbf{x} = (x_1, x_2, x_3)$. Due to (6.151) and (6.153), the exact solution to (6.157) is given by $u = U|_{\mathbb{S}}$. Let $\mathbf{n} = (0, 0, 1)$. By using the identity (see [323, page 20])

$$(1 - 2t\cos\theta + t^2)^{-1/2} = \sum_{l=0}^{\infty} t^l P_l(\cos\theta), \quad t < 1,$$

and the addition formula (6.161), one obtains,

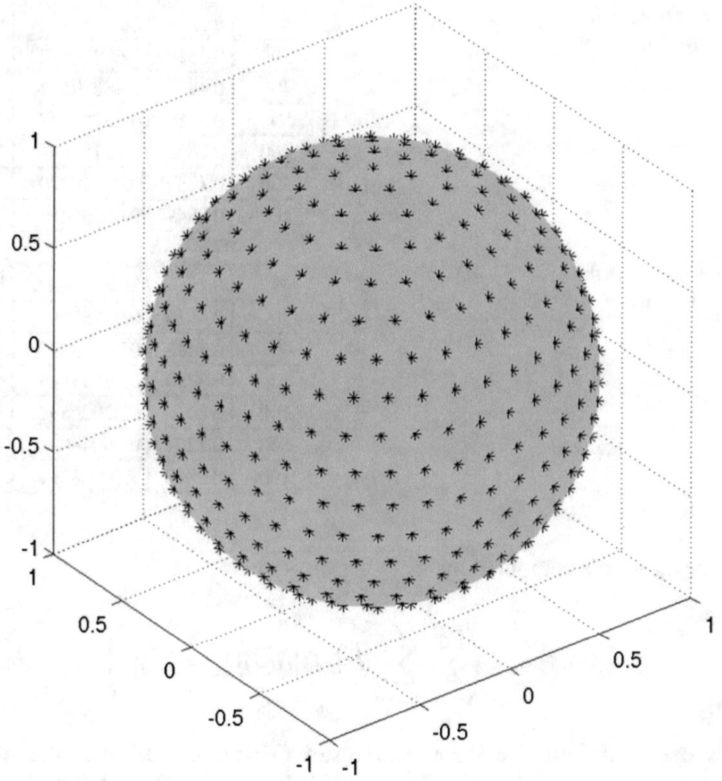

Fig. 6.5 Scattered points in equally partitioned areas

for $\mathbf{x} = (\sin\theta\cos\varphi, \sin\theta\sin\varphi, \cos\theta) \in \mathbb{S}$,

$$u(\mathbf{x}) = \frac{1}{\|\mathbf{x} - \mathbf{p}\|} = \frac{1}{\sqrt{1 - \cos\theta + 1/4}} = \sum_{l=0}^{\infty} \frac{1}{2^l} P_l(\mathbf{x} \cdot \mathbf{n})$$

$$= \sum_{l=0}^{\infty} \sum_{m=-l}^{l} \frac{4\pi}{2^l(2l+1)} Y_{l,m}(\mathbf{n}) Y_{l,m}(\mathbf{x}),$$

so that

$$\widehat{u}_{l,m} = \frac{4\pi}{2^l(2l+1)} Y_{l,m}(\mathbf{n}). \tag{6.176}$$

Now (6.170) is solved and the approximate solution u_X is compared with the exact solution u. Note that

$$\widehat{(u_X)}_{l,m} = \widehat{\phi^{(W)}}(l) \sum_{i=1}^{M} c_i \overline{Y_{l,m}(\mathbf{x}_i)}.$$

Table 6.2 Errors in the $H^{1/2}$-norm with $m = 0$

M	h_X	$\|u_X - u\|_{1/2}$	EOC
20	0.6514	0.60872542	
40	0.4418	0.18859512	3.0179
50	0.3750	0.13264247	2.1469
100	0.2672	0.05752634	2.4649
500	0.1237	0.00738320	2.6658
1000	0.0849	0.00303414	2.3627

Table 6.3 Errors in the $H^{1/2}$-norm with $m = 1$

M	h_X	$\|u_X - u\|_{1/2}$	EOC
20	0.6514	0.93582688	
40	0.4418	0.17405797	4.3322
50	0.3750	0.07695943	4.9784
100	0.2672	0.02026597	3.9369
500	0.1237	0.00044098	4.9701
1000	0.0849	0.00008591	4.3459

The error $u_X - u$ is computed by

$$\|u_X - u\|_{1/2} \approx \left(\sum_{l=1}^{500} \sum_{m=-l}^{l} (l+1) |\widehat{(u_X)}_{l,m} - \widehat{u}_{l,m}|^2 \right)^{1/2}.$$

It is expected from the theoretical result (Theorem 6.38) that the order of convergence for the $H^{1/2}$-norm of the error is $2(m + 1)$. The estimated orders of convergence (EOC) shown in Tables 6.2, 6.3 appear to agree with the theoretical results.

In [418] also the exterior Dirichlet problem (6.141), (6.142) and (6.144) is solved with boundary data

$$U_D(\mathbf{x}) = \frac{1}{(1.25 - x_3)^{1/2}}.$$

The exact solution is given by

$$U(\mathbf{x}) = \frac{1}{\|\mathbf{x} - \mathbf{p}\|} \quad \text{with } \mathbf{p} = (0, 0, 0.5),$$

and hence, due to (6.151) and (6.152), the exact solution to (6.156) is

$$z(\mathbf{x}) = \partial_\nu U(\mathbf{x}) = \frac{-1 + \mathbf{x} \cdot \mathbf{p}}{\|\mathbf{x} - \mathbf{p}\|^3} = \frac{(0.5x_3 - 1)}{(1.25 - x_3)^{3/2}}.$$

It follows from (6.146), (6.147), and (6.176) that

$$\widehat{z}_{l,m} = -\frac{4\pi(l+1)}{2^l(2l+1)} Y_{l,m}(\mathbf{n}).$$

Table 6.4 Errors in the $H^{-1/2}$-norm with $m = 0$

M	h_X	$\|u_X - u\|_{-1/2}$	EOC
20	0.6514	0.63633932	
40	0.4418	0.18846925	3.1339
50	0.3750	0.13291921	2.1301
100	0.2672	0.05752374	2.4712
500	0.1237	0.00738510	2.6654
1000	0.0849	0.00303428	2.3632

Table 6.5 Errors in the $H^{-1/2}$-norm with $m = 1$

M	h_X	$\|u_X - u\|_{-1/2}$	EOC
20	0.6514	0.93557363	
40	0.4418	0.18160356	4.2222
50	0.3750	0.07737696	5.2042
100	0.2672	0.02040513	3.9327
500	0.1237	0.00044099	4.9790
1000	0.0849	0.00008591	4.3460

Now (6.174) is solved and the approximate solution z_X is compared with the exact solution z. Note that

$$\widehat{(z_X)}_{l,m} = (l+1)\widehat{\phi^{(W)}}(l) \sum_{i=1}^{M} c_i \overline{Y_{l,m}(\mathbf{x}_i)}.$$

The error $z_X - z$ is approximated by

$$\|z_X - z\|_{-1/2} \approx \left(\sum_{l=0}^{500} \sum_{m=-l}^{l} \frac{|\widehat{(z_X)}_{l,m} - \widehat{z}_{l,m}|^2}{l+1} \right)^{1/2}.$$

The theoretical result (Theorem 6.39) requires $m > 0$ and an order of convergence of $2m + 5/2$ is shown in the $H^{-1/2}$-norm. The results for $m = 0, 1$ are listed in Tables 6.4, 6.5

In [243] the authors analyze the approximation by radial basis functions of a hyper singular integral equation on an open surface. In order to accommodate the homogenous essential boundary condition along the surface boundary, scaled radial basis functions on an extended surface and Lagrangian multipliers on the extension are used.

Spherical splines are used in [339] to define approximate solutions to strongly elliptic pseudodifferential equations on the unit sphere. These equations arise from geodesy. The approximate solutions are found by using the Galerkin method. The authors prove optimal convergence (in Sobolev norms) of the approximate solution by spherical splines to the exact solution. A pseudodifferential operator L is a linear operator that assigns to any $v \in \mathcal{D}'(\mathbb{S})$ a distribution

$$Lv := \sum_{\ell=0}^{\infty} \sum_{m=-\ell}^{\ell} \widehat{L}(\ell)\widehat{v}_{\ell,m} Y_{\ell,m}. \tag{6.177}$$

The sequence $\{\widehat{L}(\ell)\}_{\ell\geq 0}$ is referred to as the *spherical symbol* of L. Let $\mathscr{K}(L) := \{\ell : \widehat{L}(\ell) = 0\}$. Then

$$\text{kernel of L} = \text{span}\{Y_{\ell,m} : \ell \in \mathscr{K}(L),\ m = -\ell, \ldots, \ell\}.$$

For ease of presentation we consider here only the case kernel of $L = \emptyset$ (see [339] for the general case). We look for the problem:

Find $u \in H^\alpha$ such that

$$\langle Lu, v\rangle = \langle g, v\rangle \quad \forall v \in H^\alpha \tag{6.178}$$

where L is a strongly elliptic pseudodifferential operator of order 2α whose symbol is given by

$$\widehat{L}(\ell) = \begin{cases} \widehat{L}(\ell) & \text{if } \ell \notin \mathscr{K}(L) \\ (1+\ell)^{2\alpha} & \text{if } \ell \in \mathscr{K}(L). \end{cases}$$

Let $\{\mathbf{v}_1, \mathbf{v}_2, \mathbf{v}_3\}$ be linearly independent vectors in $\mathbb{R}^3$. The *triheron* T generated by $\{\mathbf{v}_1, \mathbf{v}_2, \mathbf{v}_3\}$ is defined by

$$T := \{\mathbf{v} \in \mathbb{R}^3 : \mathbf{v} = b_1\mathbf{v}_1 + b_2\mathbf{v}_2 + b_3\mathbf{v}_3 \text{ with } b_i \geq 0,\ i = 1, 2, 3\}.$$

The intersection $\tau := T \cap \mathbb{S}$ is called a *spherical triangle*. For each $\mathbf{v} \in \tau$, where τ is a spherical triangle having vertices $\mathbf{v}_1$, $\mathbf{v}_2$, $\mathbf{v}_3$, there exist unique $b_1(\mathbf{v})$, $b_2(\mathbf{v})$, $b_3(\mathbf{v})$ satisfying

$$\mathbf{v} = b_1(\mathbf{v})\mathbf{v}_1 + b_2(\mathbf{v})\mathbf{v}_2 + b_3(\mathbf{v})\mathbf{v}_3, \tag{6.179}$$

which are called the *spherical barycentric coordinates* of $\mathbf{v}$ with respect to τ.

Let $\Delta = \{\tau_i : i = 1, \ldots, \mathscr{T}\}$ be a set of spherical triangles. If Δ satisfies

- $\bigcup_{i=1}^{\mathscr{T}} \tau_i = \mathbb{S}$,
- each pair of distinct triangles in Δ are either disjoint or share a common vertex or an edge,

then Δ is called a *spherical triangulation* of the sphere $\mathbb{S}$.

Given $X = \{\mathbf{x}_1, \ldots, \mathbf{x}_N\}$ a set of points on $\mathbb{S}$, we can form a spherical triangulation Δ which contains triangles whose vertices are elements of X (see [339]).

Given nonnegative integers r and d, the set of spherical splines of degree d and smoothness r associated with Δ is defined by

$$S_d^r(\Delta) := \{s \in C^r(\mathbb{S}) : s|_\tau \in \mathscr{P}_d, \tau \in \Delta\}.$$

Here, $\mathscr{P}_d$ is the space of restrictions to $\mathbb{S}$ of homogeneous polynomials of degree d in $\mathbb{R}^3$. If τ has vertices $\mathbf{v}_1, \mathbf{v}_2, \mathbf{v}_3$, then $s|_\tau$ can be written as

$$s|_\tau(\mathbf{v}) = \sum_{i+j+k=d} c^\tau_{ijk} B^{d,\tau}_{ijk}(\mathbf{v}), \quad \mathbf{v} \in \tau,$$

where the coefficients c^τ_{ijk} are real numbers and the functions

$$B^{d,\tau}_{ijk}(\mathbf{v}) := \frac{d!}{i!j!k!} b^i_1(\mathbf{v}) b^j_2(\mathbf{v}) b^k_3(\mathbf{v}), \quad i + j + k = d,$$

are called the *spherical Bernstein–Bézier basis polynomials* of degree d relative to τ. Here, $b_i(\mathbf{v})$, $i = 1, 2, 3$, are given by (6.179). For more details, see [339]

For any spherical triangle τ, we denote by $|\tau|$ the diameter of the smallest spherical cap containing τ, and by ρ_τ the diameter of the largest spherical cap inside τ. Here the diameter of a cap is, as usual, twice its radius.We define

$$|\Delta| := \max\{|\tau|, \tau \in \Delta\}, \quad \rho_\Delta := \min\{\rho_\tau, \tau \in \Delta\} \quad \text{and} \quad h_\Delta := \tan\frac{|\Delta|}{2}. \tag{6.180}$$

Definition 6.4 Let β be a positive real number. A triangulation Δ is said to be β-*quasiuniform* provided that

$$\frac{|\Delta|}{\rho_\Delta} \le \beta.$$

In the following we briefly introduce the construction of a quasi-interpolation operator $Q : L_2(\mathbb{S}) \to S^r_d(\Delta)$ which is introduced in [318]. First we introduce the set of *domain points* of Δ to be

$$\mathscr{D} := \bigcup_{\tau=\langle \mathbf{v}_1,\mathbf{v}_2,\mathbf{v}_3\rangle \in \Delta} \left\{ \xi^\tau_{ijk} = \frac{i\mathbf{v}_1 + j\mathbf{v}_2 + k\mathbf{v}_3}{d} \right\}_{i+j+k=d}.$$

Here, $\tau = \langle \mathbf{v}_1, \mathbf{v}_2, \mathbf{v}_3\rangle$ denotes the spherical triangle whose vertices are $\mathbf{v}_1, \mathbf{v}_2, \mathbf{v}_3$. We denote the domain points by $\xi_1, \ldots, \xi_D$, where $D = \dim S^0_d(\Delta)$. Let $\{B_l : l = 1, \ldots, D\}$ be a basis for $S^0_d(\Delta)$ such that the restriction of B_l on the triangle containing ξ_l is the Bernstein-Bézier polynomial of degree d associated with this point, and that B_l vanishes on other triangles.

A set $\mathscr{M} := \{\zeta_l\}_{l=1}^M \subset \mathscr{D}$ is called a *minimal determining set* for $S^r_d(\Delta)$ if, for every $s \in S^r_d(\Delta)$, all the coefficients $\nu_l(s)$ in the expression $s = \sum_{l=1}^D \nu_l(s) B_l$ are uniquely determined by the coefficients corresponding to the basis functions which are associated with points in $\mathscr{M}$. Given a minimal determining set, the authors of [339] construct a basis $\{B^*_l\}_{l=1}^M$ for $S^r_d(\Delta)$ by requiring

$$\nu_{l'}(B^*_l) = \delta_{l,l'}, \quad 1 \le l, l' \le M.$$

Using the Hahn–Banach Theorem the linear functionals ν_l, $l = 1, \ldots, M$, are extended to all of $L_2(\mathbb{S})$. Now, the quasi-interpolation operator $Q : L_2(\mathbb{S}) \to S_d^r(\Delta)$ is defined by

$$Qv := \sum_{l=1}^{M} \nu_l(v) B_l^*, \quad v \in L_2(\mathbb{S}).$$

The following theorem is shown in [339] with an analysis similar to [418] but with Q instead of the interpolation operator I_X.

Theorem 6.40 *Assume that Δ is a β-quasiuniform spherical triangulation with $|\Delta| \le 1$, and that there holds*

$$\begin{cases} d \ge 3r + 2, & \text{if } r > 1 \\ d \ge 1, & \text{if } r = 0. \end{cases} \tag{6.181}$$

Then for any $v \in H^s$, there exists $\eta \in S_d^r(\Delta)$ satisfying

$$\|v - \eta\|_t \le C h_\Delta^{s-t} \|v\|_s,$$

where $t \le r + 1$ and $t \le s \le d + 1$. Here C is a positive constant depending only on d and the smallest angle in Δ.

We consider the Galerkin equation: Find $\widetilde{u} \in S_d^r(\Delta)$ such that

$$\langle L\widetilde{u}, v\rangle = \langle f, v\rangle \quad \forall v \in S_d^r(\Delta). \tag{6.182}$$

Theorem 6.41 *Assume that Δ is a β-quasiuniform spherical triangulation with $|\Delta| \le 1$ and that* (6.181) *hold. If the order 2α of the pseudodifferential operator L satisfies $\alpha \le r + 1$, and if u and $\widetilde{u}$ satisfy, respectively,* (6.178) *and* (6.182), *then*

$$\|u - \widetilde{u}\|_t \le C h_\Delta^{s-t} \|u\|_s,$$

where $s \le d + 1$ and $2\alpha - d - 1 \le t \le \min\{s, \alpha\}$. Here C is a positive constant depending only on d and the smallest angle in Δ.

Chapter 7
Advanced BEM for BVPs in Polygonal/Polyhedral Domains: h- and p-Versions

This chapter presents, $h-$, $p-$BEM on graded meshes and $hp-$BEM on quasiuniform meshes for the numerical treatment of boundary value problems in polygonal and polyhedral domains. For ease of presentation we also introduce here the $hp-$version on geometrically graded meshes (for details and proofs see Chap. 8). For the solutions of Dirichlet and Neumann problems we present decompositions into a sum of special singularity terms (describing their edge and corner behaviors) and in regular parts (see Theorem 7.3, Theorem 7.12 for two-dimensions and Theorem 7.7, Theorem 7.16 for three dimensions). These regularity results by von Petersdorff, Stephan [425] are based on the seminal works of Dauge [141] and Kondratiev [270]. Chapter 7 is organized as follows: The results for the single layer integral equation covering the Dirichlet problem are presented in Sect. 7.1 ; those for the hypersingular integral equation covering the Neumann problem in Sect. 7.2. Then in Sect. 7.3 the proofs for the results for the integral equations on curves are given, whereas in Sect. 7.4 the results for the integral equations on surfaces . We present approximation results for solutions of the integral equations on graded meshes in 2D and 3D from the PhD thesis by von Petersdorff [423], see also [426]. Also in detail we investigate the $hp-$version of BEM on quasi uniform meshes on polygons based on the paper by Suri and Stephan [405]. For the p-version BEM with quasi uniform meshes on polyhedra we refer to [51, 52, 374].

There has been much work on the regularity of elliptic problems. The interested reader might also look into the key papers by Maz'ya, Nazarov and Plamenevsky [302, 303], and into their text books [271, 317]. Recently the concept of detached asymptotics has proved to be very fruitful, see [316].

© Springer International Publishing AG, part of Springer Nature 2018

J. Gwinner, E. P. Stephan, *Advanced Boundary Element Methods*,
Springer Series in Computational Mathematics 52,
https://doi.org/10.1007/978-3-319-92001-6_7

7.1 The Dirichlet Problem

In this section we consider integral equation methods for solving boundary value problems in non-smooth domains Ω. First we deal with the standard Dirichlet problem

$$\begin{aligned} \Delta u &= 0 \quad \text{in } \Omega, \\ u &= g \quad \text{on } \Gamma = \partial\Omega. \end{aligned} \tag{7.1}$$

Here Ω is either a plane, curvilinear polygon or the exterior of an open arc or a polyhedron with a piecewise C^∞–surface or an open surface piece. We use the convention that the normal vector always points away from Ω, with corresponding modifications when Ω is the exterior of an arc or an open surface.

As shown in Chaps. 2 and 4, (7.1) is converted into the integral equation of the first kind for $\phi = \frac{\partial u}{\partial n}|_\Gamma$:

$$V\phi = (I + K)g \quad \text{on } \Gamma \tag{7.2}$$

The connection between the boundary value problem (7.1) and the integral equation (7.2) is as follows.

Theorem 7.1 *Let $g \in H^{1/2}(\Gamma)$.*

(i) Then there exists exactly one solution $\phi \in H^{-1/2}(\Gamma)$ of the integral equation (7.2).
(ii) The problem (7.1) *and the equation* (7.2) *are equivalent, i.e. let $u \in H^1_{loc}(\Omega)$ solve* (7.1) *then $\phi = \frac{\partial u}{\partial n}\big|_\Gamma$ solves* (7.2)*, conversely, let $\phi \in H^{-1/2}(\Gamma)$ solves* (7.2) *then u defined by* (2.2) *with $u|\Gamma = g$ solves* (7.1)

Remark 7.1 The above method remains valid for general elliptic boundary value problems and even transmission problems with differential operators with constant coefficients: For problems in elasticity, acoustics, electromagnetics etc. see Chap. 4 (These boundary integral equations have similar properties as the original differential equations; they allow a variational formulation and are strongly elliptic.)

We observe that the integral equation (7.2) can be solved approximately by the Galerkin's method using conforming subspaces $\{X_N\}$ of $H^{-1/2}(\Gamma)$:

For given $g \in H^{1/2}(\Gamma)$ find $\phi_N \in X_N$ such that

$$\langle V\phi_N, \chi\rangle = \langle (1 + K)g, \chi\rangle \qquad \forall \chi \in X_N. \tag{7.3}$$

Here the bracket $\langle w, v\rangle$ denotes the duality between $H^{1/2}(\Gamma)$ and $H^{-1/2}(\Gamma)$, which can be identified with the L^2–inner product. V is strongly elliptic by Theorems 4.3 and 6.3.

Thus Theorem 6.1 gives the following result for the Galerkin procedure (7.3):

Theorem 7.2 *Let $N \geqslant N_0$. Then the equation* (7.3) *has a unique solution $\phi_N \in X_N$. Furthermore there holds*

$$\|\phi - \phi_N\|_{H^{-1/2}(\Gamma)} \leqslant C \inf_{\chi \in X_N} \|\phi - \chi\|_{H^{-1/2}(\Gamma)} \tag{7.4}$$

where the constant C is independent of N,ϕ_N and the solution ϕ of (7.2).

Remark 7.2 The proof of Theorem 7.2 is based on the strong ellipticity of the integral operator V. This property means that V is coercive in the sense of a Gårding inequality, i.e. there exists a constant $\eta > 0$ and a compact operator T from $H^{-1/2}(\Gamma)$ into $H^{1/2}(\Gamma)$ such that

$$\langle (V+T)\psi, \psi \rangle \geqslant \eta \|\psi\|^2_{H^{-1/2}(\Gamma)} \qquad \forall \psi \in H^{-1/2}(\Gamma). \tag{7.5}$$

This inequality corresponds directly by integration by parts to the Dirichlet bilinear form for the variational solution $u \in H^1_{\mathrm{loc}}(\Omega)$ of (7.1).

Theorem 7.2 guarantees the convergence of <u>any</u> Galerkin scheme for solving integral equation (7.2) by use of conforming subspaces $X_N \subset H^{-1/2}(\Gamma)$. Due to the quasioptimality estimate (7.4) the rate of convergence of the used scheme is determined by the choice of approximating subspaces X_N and the regularity of the exact solution ϕ of the integral equation.

For an appropriate choice of X_N it is crucial to know the behavior of the solution ϕ near crack trips, corners and edges. There ϕ becomes singular which corresponds to the behavior of the solution u of the original problem (7.1).

7.1.1 Regularity on a Polygon

We consider next the case of a plane polygon Γ with straight line segments Γ^i. By t_j $(j = 0, \ldots, J)$ we denote the corner points where Γ^j and Γ^{j+1} meet $(t_J = t_0)$. The interior angle at t_j is denoted by ω_j.

The following explicit regularity result for ϕ is obtained in [128] using localization and Mellin transformation.

Theorem 7.3 *Let $g \in H^s(\Gamma)$, $1/2 \leqslant s < 3/2$, $s \notin A = \left\{\alpha_{jk} = \dfrac{k\pi}{\omega_j}, 1 \leqslant j \leqslant J,\ k \in \mathbb{N}\right\}$. The solution ϕ of* (7.2) *has the form*

$$\phi = \sum_{j=1}^{J} \left(\sum_{\alpha_{jk} < s-1/2} c_{jk}\, \rho_j^{\alpha_{jk}-1} \right) \chi_j + \phi_0 \quad , \quad \phi_0 \in H^{s-1}(\Gamma),\ c_{jk} \in \mathbb{R}.$$

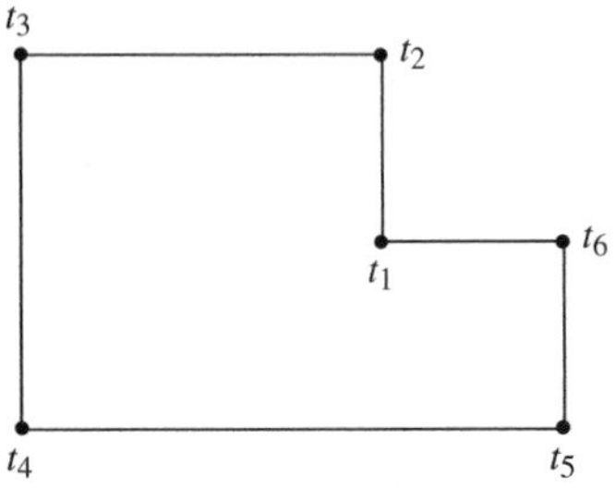

Fig. 7.1 $\Omega = L-$shaped domain with reentrant corner at vertex t_1 with $\omega_1 = \frac{3}{2}\pi$

Fig. 7.2 $\Omega = \mathbb{R}^2 \backslash \Gamma$, $\Gamma = [-1, 1]$

Here ρ_j denotes the Euclidean distance between $x \in \Gamma$ and the vertex t_j whereas χ_j is a C^∞-cut-off function concentrated near t_j. For $s > 3/2$ and $s - 1/2 \notin \mathbb{N}$, the corresponding theorem still holds.

Remark 7.3 The analogous result for a general curvilinear polygon Ω is given in [126]. There, additional singularity terms like $\rho^{\alpha_{jk}-1} \log \rho$ due to the curvature of the axis Γ^j may appear.

We illustrate the regularity result for some canonical domains.

Example 7.1 For Γ being the boundary of the L-shaped domain in Fig. 7.1 we obtain with $g \in H^{3/2-\varepsilon}(\Gamma)$, $\varepsilon > 0$ arbitrary,

$$\phi = c_{11}\, \rho_1^{-1/3} \chi_1 + \phi_0 \quad , \quad \phi_0 \in H^{1/2-\varepsilon}(\Gamma).$$

Here the singularity terms located at the vertices $t_2, \ldots, t_6$ are included in ϕ_0.

Example 7.2 For the slit Γ in Fig. 7.2 and for $g \in H^{3/2-\varepsilon}(\Gamma)$, $\varepsilon > 0$ arbitrary, we have $\phi \in \tilde{H}^{-1/2}(\Gamma)$ with

$$\phi = c_{11}\chi_1\rho_1^{-1/2} + c_{21}\chi_2\rho_2^{-1/2} + \phi_0, \qquad \phi_0 \in \tilde{H}^{1/2-\varepsilon}(\Gamma).$$

Next, we comment on various choices for the approximating subspaces X_N.

7.1.2 BEM on a Polygon

7.1.2.1 h-p Method with Quasiuniform Mesh on a Polygon

Construction: Let for each $\Gamma^j \subset \Gamma$ there be given a family of grids $\{T_h^j\}$ which partition each Γ^j into N_h^j pieces, $\Gamma^j = \bigcup_{i=1}^{N_h} \bar{\Gamma}_{h,i}^j$ such that $\Gamma_{h,i}^j$ is an open interval.

We assume that $\{T_h^j\}$ is quasi uniform, in the sense that with $h_i^j = \text{meas}\left(\Gamma_{h,i}^j\right)$ and $h = \max_{i,j} h_i^j$, there exists a constant τ independent of h such that

$$h/h_i^j \leqslant \tau \qquad \text{for all intervals } \Gamma_{h,i}^j \tag{7.6}$$

For $p \geqslant 0$, $S_{p,h}(\Gamma^j)$ will denote the set of all functions v defined on Γ^j such that the restriction $v|_{\Gamma_{h,i}^j}$ belongs to $P_p\left(\Gamma_{h,i}^j\right)$, the space of polynomials of degree $\leqslant p$ on $\Gamma_{h,i}^j$. For $p \geqslant 0$, $h > 0$, we define $S_{p,h}(\Gamma)$ to be the set of functions on Γ whose restrictions to $\Gamma^j \subset \Gamma$ belong to $S_{p,h}(\Gamma^j)$.

In the standard h-method one decreases the mesh size h and keeps $p = p_0$ fixed, i.e. one takes as approximating subspaces $X_N = S_{p_0,h}(\Gamma)$ where N is proportional to $1/h$.

On the other hand, in the p-version, one uses a fixed mesh $h = h_0$ and increases the degrees p of the approximating polynomials, i.e. one takes $X_N = S_{p,h_0}(\Gamma)$ where N is proportional to p.

In the h-p method one combines the two approaches, thus one takes as approximating subspaces $X_N = S_{p,h}(\Gamma)$ with two functions $p(N)$ and $h(N)$. For example, one may take $p(N) \sim N^{a_1}$ and $h(N) \sim N^{-a_2}$ where $a_1, a_2 > 0$. Note, that we do not impose continuity for the elements of $S_{p,h}(\Gamma)$ at the mesh points, since we need only $S_{p,h}(\Gamma) \subset H^{-1/2}(\Gamma)$ which is guaranteed already if $S_{p,h}(\Gamma) \subset L^2(\Gamma)$.

There holds the following convergence result for the Galerkin solution $\phi_{p,h} \in S_{p,h}(\Gamma)$ of (7.3). Its proof follows by combining Proposition 7.2, Theorem 7.20 and Theorem 7.2

Theorem 7.4 ([405]) *Let* (7.6) *hold and let p be sufficiently large and h be small enough. Then the Galerkin equations* (7.3) *are uniquely solvable in* $S_{p,h}(\Gamma)$. *Let* $\phi \in H^{-1/2}(\Gamma)$ *be the solution of the integral equation* (7.2) *with right hand side* $f = (1 + K)g \in H^s(\Gamma)$ *and* $\phi_{p,h} \in S_{p,h}(\Gamma)$ *be the Galerkin solution, then we have for* $s \geqslant 1/2$, $s \notin A$, *with* $\alpha = \min \alpha_{jk}$

$$\left\|\phi - \phi_{p,h}\right\|_{H^{-1/2}(\Gamma)} \leqslant C \max\{e_1, e_2\}$$

$$\text{with } \begin{cases} e_1 = \max\{h^\alpha p^{-2\alpha},\ h^{\min\{\alpha, p-\alpha+1/2\}} p^{-2\alpha}\} \log^{1/2} p \\ e_2 = h^{\min\{s-1/2,\ p+1/2\}} p^{-(s-1/2)} \log^{1/2} p \end{cases}$$

where the constant C depends on ϕ *but is independent of h and p.*

Again we illustrate the convergence rates when Ω is as in Example 7.1 and 7.2. For the h-version we obtain

$$\left\|\phi - \phi_{p,h}\right\|_{H^{-1/2}(\Gamma)} = \begin{cases} \mathcal{O}(h^{2/3}) & \text{if } \Omega \text{ is } L\text{-shaped} \\ \mathcal{O}(h^{1/2}) & \text{if } \Omega = \mathbb{R}^2 \setminus [-1, 1] \end{cases} .$$

For the $p-$version we have

$$\|\phi - \phi_{p,h}\|_{H^{-1/2}(\Gamma)} = \begin{cases} \mathcal{O}\left(p^{-4/3+\epsilon} \log^{1/2} p\right) & \text{if } \Omega \text{ is } L\text{-shaped} \\ \mathcal{O}\left(p^{-1+\epsilon} \log^{1/2} p\right) & \text{if } \Omega = \mathbb{R}^2 \setminus [-1, 1] \end{cases} .$$

In both cases (Example 7.1 and 7.2) for the h-p version the quantity e_1 dominates e_2 yielding

$$E = \mathcal{O}\left(h^\alpha p^{-2\alpha} \log^{1/2} p\right) \quad \text{where } \alpha = 2/3 \text{ in Example 7.1}$$
$$\alpha = 1/2 \text{ in Example 7.2.}$$

Note that increasing the dimension of the subspaces used by changing p is twice as efficient (in terms of the asymptotic rate of convergence) as changing h. In order to compare the h, p, and h-p-method we introduce the degree of freedom $N_f = \dim X_N$. For the h-method $N_f \sim h^{-1}$, hence the rate of convergence is $N_f^{-\alpha-\varepsilon}$. For the p-method we have $N_f \sim p$, thus the rate of convergence is $N_f^{-2\alpha-\varepsilon}$. In the case of the h-p-method the rate convergence depends on the choice of $h(N)$ and $p(N)$. For the above example $p(N) \sim N^{a_2}$, $h(N) \sim N^{-a_1}$ we have $N_f \sim p(N)h^{-1}(N) = N^{a_1} N^{a_2}$. Hence the convergence rate is

$$h(N)^\alpha p(N)^{-2\alpha} \sim N^{-\alpha(a_1+2a_2)} \sim N_f^{-\alpha\frac{a_1+2a_2}{a_1+a_2}-\varepsilon}$$

This convergence rate lies between the rate $N_f^{-\alpha-\varepsilon}$ of the $h-$version and the rate $N_f^{-2\alpha-\varepsilon}$ of the $p-$version. For example, for $a_1 = a_2 = 1$ we have $N_f^{-\frac{3\alpha}{2}-\varepsilon}$.

7.1.2.2 h-Method with Graded Mesh on a Polygon

The mesh and the boundary element space are constructed as follows: Bisect all sides of the polygon. For each corner $t_j (j = 1, \ldots, J)$ of the polygon identify each of the 2 adjacent parts with the interval $I = [0, 1]$ such that t_j corresponds to 0 and the endpoints correspond to 1. Then choose the mesh points corresponding to

$$x_k = \left(\frac{k}{N}\right)^{\beta_j}, \; k = 0, \ldots, N-1, \qquad x_k = (kh)^{\beta_j}, \; h = \frac{1}{N} \tag{7.7}$$

where $\beta_j \geq 1$ is called the grading exponent ($\beta_j = 1$ means a uniform mesh) (Fig. 7.3). Thus we obtain the graded meshes $\{Z^+_{j,k}; k = 0, \ldots, N-1\}$ and $\{Z^-_{j,k}; k = 0, \ldots, N-1\}$ on the adjacent sides of each corner t_j, $j = 1, \ldots, J$. For $p \geqslant 0$, $\beta = (\beta_1, \ldots, \beta_J)$, $S^\beta_{p,h}(\Gamma)$ denotes the set of piecewise polynomials of degree p on the graded mesh described above which might be refined differently

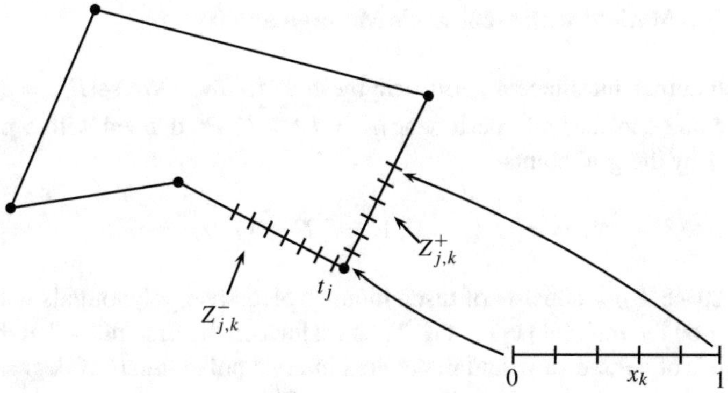

Fig. 7.3 Example of graded mesh

at various vertices. Thus the simplest choice for an approximating subspace in the Galerkin scheme (7.3) is $S^{\beta}_{0,h}(\Gamma)$, i.e. the set of piecewise constant functions defined on the graded mesh (for simplicity we consider only the case $\beta = \beta_j$). There holds as application of Proposition 7.3 to Theorem 7.2 the following convergence result.

Theorem 7.5 ([426]) *Let* (7.7) *hold and h be sufficiently small. Let* $\phi \in H^{-1/2}(\Gamma)$ *solve* (7.2) *with* $g \in H^s(\Gamma)$, $s \geqslant 1/2$, $s \notin A$, $s + 1/2 \notin \mathbb{N}$. *Then the Galerkin equations* (7.3) *are uniquely solvable in* $S^{\beta}_{p,h}(\Gamma)$. *Moreover we have for* $\alpha = \min \alpha_{jk}$

$$\|\phi - \phi_h\|_{H^{-1/2}(\Gamma)} \leqslant C \begin{cases} h^{\alpha\beta-\varepsilon} & \text{if } \beta \le \frac{2p+3}{2\alpha} \\ h^{\frac{2p+3}{2}} & \text{if } \beta > \frac{2p+3}{2\alpha} \end{cases}$$

where the constant $C = C(\beta)$ *is independent of* h, ϕ *and* ϕ_h.

We illustrate the convergence rates when Ω is as in Example 7.1 and 7.2. With $E = \|\phi - \phi_h\|_{H^{-1/2}(\Gamma)}$ we obtain with $p = 0$

$$E = \begin{cases} \mathcal{O}\left(h^{2\beta/3-\varepsilon}\right) & \text{if } \beta < 9/4 \\ \mathcal{O}\left(h^{3/2}\right) & \text{if } \beta > 9/4 \end{cases}, \qquad \text{if } \Omega \text{ is } L\text{-shaped}$$

and

$$E = \begin{cases} \mathcal{O}\left(h^{\beta/2-\varepsilon}\right) & \text{if } \beta < 3 \\ \mathcal{O}\left(h^{3/2}\right) & \text{if } \beta > 3 \end{cases}, \qquad \text{if } \Omega \text{ is } \mathbb{R}^2 \setminus [-1, 1]$$

7.1.2.3 h-p Method with Geometric Mesh on a Polygon

Near each corner introduce a geometric mesh as follows: We set $\Gamma^k = (0, 1)$ for simplicity and consider the mesh near $t_k = \overline{\Gamma^{k-1}} \cap \overline{\Gamma^k}$ defined with a parameter $0 < \sigma < 1$ by the grid points

$$x_j = \sigma^j \quad (j = 0, 1, \ldots, N-1), \; x_N = 0.$$

Now the space $X_{N,\sigma}$ consists of discontinuous piecewise polynomials with degree $N-1-j$ on the interval $[x_{j+1}, x_j]$. Thus we have in the first interval at the corner polynomials of degree zero, and in the next interval polynomials of degree one and so on, i.e. with the interval $I_j^k = [x_{j+1}, x_j]$ on Γ^k

$$X_{N,\sigma} = \left\{ v \in L^2(\Gamma) : v|_{I_j^k} \in P_{p_j} \;, \; p_j = N - 1 - j \right\}$$

Here the total degree of freedom $N_f = \dim X_N$ is proportional to N^2.

In many practical cases the given data g in (7.1) is analytic on each piece Γ^k of the polygon Γ Two different proofs of the following result are presented in Sects. 8.1 and 8.2.

Theorem 7.6 ([20, 240]) *Let N be sufficiently large. Then the Galerkin equations* (7.3) *are uniquely solvable in $X_{N,\sigma}$. Let $\phi_{N,\sigma}$ denote the Galerkin solution and $\phi \in H^{-1/2}(\Gamma)$ be the exact solution of the integral equation* (7.2) *with analytic g, then we achieve the exponential convergence*

$$\left\| \phi - \phi_{N,\sigma} \right\|_{H^{-1/2}(\Gamma)} \leqslant C\, e^{-\beta\sqrt{N}}$$

where the constants C and β are independent of N.

Note that the choice of the geometric mesh is independent of the order of singularity of the solution of (7.1). Therefore we use in our canonical examples (L-shaped domain, slit domain) the same mesh near the vertices and the same subspaces $X_{N,\sigma}$.

7.1.3 Regularity on a Polyhedron

Next we consider the case of a *polyhedron* Ω with the surface $\Gamma = \partial\Omega = \bigcup_{j=1}^{J} \overline{\Gamma^j}$ in $\mathbb{R}^3$ with plane faces Γ^j. We describe Dirichlet data $g \in H^s(\Gamma)$ where $H^s(\Gamma)$ is defined as follows:

$$H^s(\Gamma) := \left\{ u|_\Gamma \,|u \in H^{s+1/2}(\mathbb{R}^3) \right\}.$$

Then the Neumann data ϕ of the solution has regularity H^{s-1} away from the edges and corners. Near an edge with opening ω there are edge singularities of the form $c(y)\rho^{m\nu+2p-1}$. Here $\nu = \frac{\pi}{\omega}$, $m > 0$ and $p \geqslant 0$ integers and ρ denotes the distance to the edge, while the stress intensity factor $c(y)$ is a function defined on the edge. Near the corners we get additional corner singularities of the form

$$r^{\lambda_k} w_k(\xi), \ \xi \in \Gamma_0$$

where r denotes the distance to the vertex and w_k is a function on the spherical polygon $\Gamma_0 = \Gamma \cap S_2$. S_2 is a sphere centered in the vertex, θ and ϕ are polar coordinates on S_2. The exponent λ_k and the function w_k are obtained as follows: Consider the eigenvalue problem for the Laplace- Beltrami operator $\Delta_{\theta,\phi}$ on S_2, and let μ_k be the $k-$th eigenvalue with corrsponding eigenfunction v_k

$$\Delta_{\theta,\phi} v_k(\theta,\phi) = \mu_k \, v_k(\theta,\phi)$$
$$v_k(\theta,\phi)|_{\Gamma_0} = 0$$

then

$$\lambda_k := -1/2 + \sqrt{\mu_k^2 + \frac{1}{4}} \quad , \quad w_k := \frac{\partial}{\partial n} v_k(\theta,\phi)\Big|_{\Gamma_0} .$$

We state the decomposition theorem for the neighbourhood of a vertex t_0:

Theorem 7.7 ([425]) *Let ω_j denote the openings of the edges γ_j meeting in t_0, $\nu_j = \frac{\pi}{\omega_j}$ $(j = 1 \ldots J)$. Choose one of the edge exponents $s_1 = m\nu_j + 2p$ where m, j, p are some integers, let s_2 be the next larger value of the form $(m\nu_j + 2p)$ or $\lambda_k + 1/2$*

$$s_2 = \min\left\{(m\nu_j + 2p), \lambda_k + 1/2 \mid m\nu_j + 2p > s_1, \lambda_k + 1/2 > s_1, \ j, p, k \ \textit{integers}\right\}$$

This corresponds to the first singularity not occuring in the decomposition. Let $s > s_1$ with $s - 1/2 \notin \mathbb{N}$, $s \neq m\nu_j$, $s \neq \lambda_k + 1/2$. Then

$$\Delta u = 0 \qquad \textit{in } \Omega$$
$$u|_\Gamma = g \in H^{s+1/2}(\Gamma)$$

implies for $\phi := \frac{\partial u}{\partial n}\Big|_\Gamma$

$$\phi = \phi^0 + \chi(r) \sum_{0<\lambda_k<s_2-1/2} a_k r^{\lambda_k - 1} w_k + \sum_{j=1,\ldots,J} \chi_j(\theta) \sum_{m\nu_j+2p\leqslant s_1} C^j_{m,p} \theta_j^{m\nu_j+2p-1}$$

$$r \cdot \phi^0 \in \mathscr{H}^{\tilde{s}}(\Gamma), \quad C^j_{m,0} = \sum_{s_1 < \lambda_k + 1/2 < s} a^j_{m,k} r^{\lambda_k - 1} + \tilde{C}^j_{m,0}, \quad r \cdot \tilde{C}^j_{m,0} \in H^{s - m\nu_j}_{-m\nu_j}(\mathbb{R}^+)$$

$$r \cdot C^j_{m,p} \in H^{s - m\nu_j - 2p}_{-m\nu}(\mathbb{R}^+) \quad \textit{for } p > 0$$

$$\tilde{s} = \min\{s - s_1, -1/2 + s_2 - \varepsilon\} \ , \ \varepsilon > 0$$

Here θ_j denotes the angle to the edge γ_j, $\mathscr{H}^{\tilde{s}}(\Gamma)$ are the functions ϕ on Γ such that their restrictions $\phi|_{\Gamma^k}$ to each face Γ^k of the polygon have the regularity $\phi|_{\Gamma^k} \in H^{\tilde{s}}(\Gamma^k)$, i.e. $\mathscr{H}^{\tilde{s}}(\Gamma) = \prod_{k=1}^{J} H^s(\Gamma^k)$. $\mathscr{H}^{\tilde{s}}(\Gamma_0)$ is defined analogously.

Next we illustrate Theorem 7.7 with an example:

Example 7.3 Let Ω be the exterior of a square in $\mathbb{R}^3$. Then $\omega_j = 2\pi$ and the edge exponents are $\nu_j = 1/2$ and the vertex exponents λ_k are (see [235])

$$\lambda_1 = 0.297\,, \qquad \lambda_2 = 1.426\,, \qquad \lambda_3 = 2.06$$

Now assume for the given Dirichlet data $g \in H^{3-\varepsilon}(\mathbb{R}^3)$. Then we have near a corner of the square

$$\phi = \phi^0 + a_1 r^{.297-1} w_1(\theta) + C^1_1(r)\theta^{-1/2} + C^2_1(r)\left(\frac{\pi}{2} - \theta\right)^{-1/2}$$

$$\phi^0 \in H^{1-\varepsilon}(\Gamma) \quad C^j_1 = a^j_1 r^{1.426-1} + a^j_2 r^{2.06-1} + \tilde{C}^j_{1,0} \quad r\,\tilde{C}^j_{1,0} \in H^{3-\varepsilon}_{-1/2}(\mathbb{R}^+)$$

Note that the function $w_1(\theta)$ also becomes singular near the edges. We can state an alternative form of the decomposition with a smoother function $\tilde{w}_1(\theta)$ instead of $w_1(\theta)$ and where the edge singularities are expressed with the distance ρ_j to the edge rather than θ (Fig. 7.4).

$$\phi = \phi_0 + a_1 r^{.297-1} \tilde{w}_1(\theta) + e^1_1(y_1)\rho_1^{-1/2} + e^2_1(y_2)\rho_2^{-1/2}$$

$$\phi_0 \in H^{1-\varepsilon}(\Gamma), \ e^j_1(y_j) = a^j_1 y_j^{-.203} + a^j_2 y_j^{.926} + a^j_3 r^{1.56} + e^j_{1,0}, \ e^j_{1,0} \in H^{3-\varepsilon}_0(\mathbb{R}^+)$$

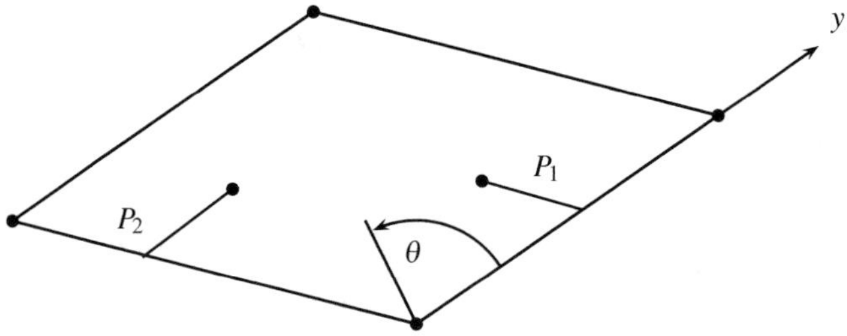

Fig. 7.4 Screen

Here $e_1^j(y_j)$ are the physically relevant stress intensity factors. They blow up towards the corners of the square.

Remark 7.4 Since $\rho^{\nu_j} \in H^{1/2+\nu_j-\varepsilon}(\Gamma)$ and $r^{\lambda_k} \in H^{1+\lambda_k-\varepsilon}(\Gamma)$ the regularity of ϕ for sufficiently smooth g is $\phi \in \mathcal{H}^{-1/2+\alpha}(\Gamma)$ with

$$\alpha = \min\left\{\nu_j, \lambda_k + 1/2\right\}. \tag{7.8}$$

Next, we consider various boundary element methods for the polyhedron.

7.1.3.1 h-p-Method with Quasi Uniform Mesh on a Polyhedron

Construct a quasi uniform mesh with width h analogously as above and define the space $S_{p,h}(\Gamma)$ of piecewise polynomials of degree p on this mesh. Then the rate of convergence is determined by the regularity of the solution, i.e. the parameter α in (7.8):

Theorem 7.8 ([51]) *Let p be sufficiently large and h be small enough. Then the Galerkin equations* (7.3) *are uniquely solvable in $S_{p,h}(\Gamma)$. Let $\phi \in H^{-1/2}(\Gamma)$ be the solution of the integral equation* (7.2) *with sufficiently smooth right hand side g, and let $\phi_{p,h}$ the Galerkin solution, then we have with α in* (7.8)

$$\left\|\phi - \phi_{p,h}\right\|_{H^{-1/2}(\Gamma)} \leqslant C \cdot h^{\alpha-\varepsilon} p^{\varepsilon-2\alpha}. \tag{7.9}$$

Remark 7.5

(i) If $N_f = \dim X_N$ denotes the degrees of freedom , then Theorem 7.8 gives for the h-method the rate $\mathcal{O}\left(N_f^{-\alpha}\right)$ and for the p-method the rate $\mathcal{O}\left(N_f^{-2\alpha}\right)$.

(ii) Method of the proof: By Theorem 7.2, we have to show, how fast the solution ϕ can approximated in the spaces X_N. Theorem 7.7 gives the decomposition of ϕ in singularities and a regular part. Then the approximation result for the h-method is contained in the proof of Theorem 7.9 for $\beta = 1$. The result for the p-method of FEM in 3D was shown in [16]. The estimate (7.9) can be proved by splitting the solution into regular/singular terms as in Theorem 7.7. Then one treats the various terms (edge, vertex and edge-vertex singularities) separately (modifiying the 2D analysis in section 7.4) by extending the 1D results for the quasi uniform hp-version from section 7.3 (see also [51, 405] and [50] for the p-version for a three-dimensional crack problem).

In Example 7.3 we have $\alpha = \min\{1/2, .297 + 1/2\} = 1/2$, hence Theorem 7.8 yields the convergence rate $\mathcal{O}\left(h^{1/2}\right)$ for the h-method and $\mathcal{O}\left(p^{-1}\right)$ for the p-method.

7.1.3.2 h-Method with Graded Mesh on a Polyhedron

For simplicity we only consider the case where all the faces meeting at the vertex are convex. Then each of these sectors can be mapped linearly on the quadrant $\mathbb{R}^+ \times \mathbb{R}^+ \subset \mathbb{R}^2$. On the quadrant we introduce a graded mesh given by the lines

$$x_1 = (i\,h)^\beta \quad , \quad x_2 = (j\,h)^\beta \quad , \quad h = 1/N\ ,\ i, j = 0 \dots N$$

Let $S^\beta_{0,h}$ be the space of piecewise constant functions on this mesh. As in Theorem 7.5 we can compensate the effect of the singularities by a appropriately graded mesh and get the convergence rate $h^{3/2}$. Theorem 7.9 follows by combining Theorem 7.2 with the 2D approximation results in Sect. 7.4.1.

Theorem 7.9 ([423, 426]) *Let $\psi \in H^{-1/2}(\Gamma^j)$ have on every corner of the edge Γ^j a decomposition of the form as in Theorem 7.7 with $\nu_{jm} > 0, \lambda_k > 0$. Let*

$$\alpha_0 := \min\{\lambda_k + \frac{1}{2}, \nu_{im}\}$$

Then we can approximate ψ for $\beta \geq 1$ by the spaces S^β_h on Γ^j in the following way:

Let $\phi_h \in S^\beta_h$ be the piecewise constant function which coincides on every subdomain with the mean value of ψ there. Then it holds for all $\epsilon > 0$

$$\|\psi - \phi_h\|_{H^{-1/2}(\Gamma^j)} \leq C h^{a-\epsilon}$$

with

$$a := \begin{cases} \min\{\alpha_0\beta, 3/2\} & \text{for } \alpha_0 \geq 1/2 \\ \min\{\alpha_0\beta\frac{2}{3}(1+\alpha_0), 1+\alpha_0\} & \text{for } \alpha_0 < 1/2 \end{cases}$$

Here C depends on β and ϵ but not on h.

For the construction of graded meshes on individual faces on Γ, we can assume that all faces of Γ are triangles. On general polygonal faces the construction is similar, or one can first subdivide the polygon into triangles. On a triangular face $F \subset \Gamma$, we first draw three lines through the centroid and parallel to the sides of F. This makes F divided into three parallelogams and three triangles (see Fig. 7.5). Each of the three parallelograms can be mapped onto the unit square $\hat{Q} = (0, 1)^2$ by a linear transformation such that the vertex $(0, 0)$ of $\hat{Q}$ is the image of a vertex of F. Analogously, each of the three sub-triangles can be mapped onto the unit triangle $\hat{T} = \{(x_1, x_2); 0 < x_1 < 1, 0 < x_2 < x_1\} \subset \hat{Q}$ such that the vertex $(1, 1)$ of $\hat{T}$ is the image of the centroid of F. Then, the graded mesh on $\hat{Q}$ (and therefore on $\hat{T}$) is

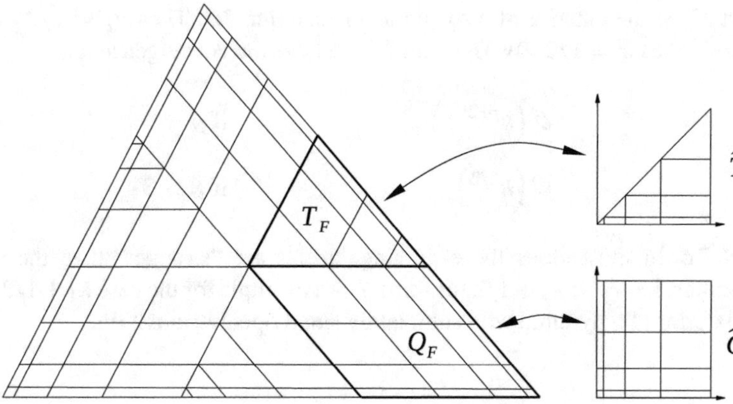

Fig. 7.5 Graded mesh on the triangular face $F \subset \Gamma$. The triangular (resp. parallelogram) block of elements T_F (resp. Q_F) is the image of the graded mesh on the unit triangle $\hat{T}$ (resp. the unit square $\hat{Q}$)

generated by the lines

$$x_1 = \left(\frac{i}{N}\right)^\beta \qquad x_2 = \left(\frac{j}{N}\right)^\beta \qquad i, j = 0, 1, \ldots, N.$$

Here $\beta \geq 1$ is the grading parameter, and $N \geq 1$ corresponds to the level of refinement. Mapping each cell of these meshes back onto the face F, we obtain a graded mesh of triangles and parallelograms on F (see Fig. 7.5). The diameter of the largest element of this mesh is proportional to βN^{-1}. Thus, $h = 1/N$ defines the mesh paramter.

We illustrate this result with two examples analogous to the examples after Theorem 7.5.

(i) Let Ω be the exterior of a cube. Then the angles of the edges are $\dfrac{3\pi}{2}$, i.e. $\nu_j = \dfrac{2}{3}$. The first vertex singularity is r^{λ_1} with $\lambda_1 + 1/2 \geq 2/3$, hence $\alpha = \min\{\nu_j, \lambda_1 + 1/2\} = \frac{2}{3}$ and Theorem 7.9 gives the convergence rate

$$\mathcal{O}\left(h^{2\beta/3-\varepsilon}\right) \qquad \text{if } \beta < 9/4$$

$$\mathcal{O}\left(h^{3/2}\right) \qquad \text{if } \beta > 9/4$$

(ii) Let Ω be the exterior of a square as in Example 7.3. Then $\nu_j = 1/2$, $\lambda_1 = 0.297$, thus $\alpha = 1/2$. By Theorem 7.9 we have the convergence rate

$$\begin{aligned} &\mathcal{O}\left(h^{\beta/2-\varepsilon}\right) && \text{if } \beta < 3 \\ &\mathcal{O}\left(h^{3/2}\right) && \text{if } \beta > 3 \end{aligned}$$

Remark 7.6 In most cases the edge singularities are "stronger" than the vertex singularities, i.e. $\nu_j < \lambda_1 + 1/2$ for some j. An example for the case $\lambda_1 + 1/2 < \nu_j$ is the exterior of a pyramid with sufficiently small opening at the tip.

7.2 The Neumann Problem

In this section we consider integral equation methods for the Neumann problem in non-smooth domains Ω above.

$$\begin{aligned} &\Delta u = 0 \quad \text{in } \Omega, \\ &\frac{\partial u}{\partial n} = f \quad \text{on } \Gamma = \partial\Omega. \quad \int_\Gamma f = 0 \end{aligned} \tag{7.10}$$

With the Cauchy data $v = u|_\Gamma$ and $\phi = \frac{\partial u}{\partial n}\big|_\Gamma$ using the jump relations (see Sect. 2.2.2) one obtains for $x \in \Gamma$ the equation

$$W\,v = (I - K')\phi \quad \text{on } \Gamma \tag{7.11}$$

with the integral operators

$$K'w(x) := 2\int_\Gamma \frac{\partial}{\partial n_x} G(x,y)w(y)ds(y), \; Ww(x) := -2\frac{\partial}{\partial n_x}\int_\Gamma \frac{\partial}{\partial n_y} G(x,y)w(y)ds(y).$$

Insertion of the given data $\frac{\partial u}{\partial n} = f$ in (7.10) into (7.11) gives the integral equation of the first kind for ϕ:

$$W\,v = (I - K')f \quad \text{on } \Gamma \tag{7.12}$$

The connection between the boundary value problem (7.10) and the integral equation (7.12) is as follows.

Theorem 7.10 *Let $f \in H^{-1/2}(\Gamma)$.*

(i) Then there exists exactly one solution $v \in H^{1/2}(\Gamma)$ of the integral equation (7.12).
(ii) The problem (7.10) *and the equation* (7.12) *are equivalent,*

We observe that the integral equation (7.12) can be solved approximately by Galerkin's method using conforming subspaces $\{Y_N\}$ of $H^{1/2}(\Gamma)$:

For given $f \in H^{-1/2}(\Gamma)$ find $v_N \in Y_N$ such that

$$\langle W\, v_N, \zeta\rangle = \langle (I - K')f, \zeta\rangle \qquad \forall \zeta \in Y_N. \tag{7.13}$$

Theorem 6.1 gives the following result for the Galerkin procedure (7.13):

Theorem 7.11 *Let $N \geqslant N_0$. Then the equation* (7.13) *has a unique solution $v_N \in Y_N$. Furthermore there holds*

$$\|v - v_N\|_{H^{1/2}(\Gamma)} \leqslant C \inf_{\zeta \in Y_N} \|v - \zeta\|_{H^{1/2}(\Gamma)} \tag{7.14}$$

where the constant C is independent of N, v_N and the solution v of (7.12).

Remark 7.7 The proof of Theorem 7.11 is based on the strong ellipticity of the integral operator W. This property means that W is coercive in the sense of a Gårding inequality, i.e. there exists a constant $\eta > 0$ and a compact operator T from $H^{1/2}(\Gamma)$ into $H^{-1/2}(\Gamma)$ such that (see Theorem 4.3)

$$\langle (W + T)v, v\rangle \geqslant \eta\, \|v\|^2_{H^{1/2}(\Gamma)} \qquad \forall v \in H^{1/2}(\Gamma). \tag{7.15}$$

This inequality corresponds directly by integration by parts to the Dirichlet bilinear form for the variational solution $u \in H^1_{\text{loc}}(\Omega)$ of (7.10).

Theorem 7.11 guarantees the convergence of any Galerkin scheme for solving integral equation (7.12) by use of conforming subspaces $Y_N \subset H^{1/2}(\Gamma)$. Due to the quasioptimality estimate (7.14) the rate of convergence of the used scheme is determined by the choice of approximating subspaces Y_N and the regularity of the exact solution v of the integral equation.

For an appropriate choice of Y_N it is crucial to know the behavior of the solution v near crack trips, corners and edges. There v becomes singular which corresponds to the behavior of the solution u of the original problem (7.10).

Now we consider the case of a *plane polygon* Γ with straight line segments Γ^i. By t_j $(j = 0, \ldots, J)$ we denote the corner points where Γ^j and Γ^{j+1} meet $(t_j = t_0)$. The interior angle at t_j is denoted by ω_j.

The following explicit regularity result for v is obtained in [128] using localization and Mellin transformation (see Lemma 9.5).

Theorem 7.12 *Let* $f \in H^{s-1}(\Gamma)$, $1/2 \leqslant s < 3/2$, $s \notin A = \left\{\alpha_{jk} = \frac{k\pi}{\omega_j}, 1 \leqslant j \leqslant J,\ k \in \mathbb{N}\right\}$. *The solution* v *of* (7.12) *has the form*

$$v = \sum_{j=1}^{J} \left(\sum_{\alpha_{jk} < s-1/2} c_{jk}\, \rho_j^{\alpha_{jk}} \right) \chi_j + v_0 \quad , \quad v_0 \in H^s(\Gamma),\ c_{jk} \in \mathbb{R}. \tag{7.16}$$

Next, we comment on various choices for the approximating subspaces Y_N.

h-p Method with Quasiuniform Mesh on a Polygon

On the mesh constructed as above we introduce the boundary element space as follows. For $p \geqslant 1$, $S_{p,h}(\Gamma^j)$ will denote the set of all functions v defined on Γ^j such that the restriction $v|_{\Gamma^j_{h,i}}$ belongs to $P_p\left(\Gamma^j_{h,i}\right)$, the space of polynomials of degree $\leqslant p$ on $\Gamma^j_{h,i}$. For $p \geqslant 1$, $h > 0$, we define $S_{p,h}(\Gamma)$ to be the set of functions on Γ whose restrictions to $\Gamma^j \subset \Gamma$ belong to $S_{p,h}(\Gamma^j)$.

In the standard h-method one decreases the mesh size h and keeps $p = p_0$ fixed, i.e. one takes as approximating subspaces $Y_N = S_{p_0,h}(\Gamma)$ where N is proportional to $1/h$.

On the other hand, in the p-version, one uses a fixed mesh $h = h_0$ and increases the degrees p of the approximating polynomials, i.e. one takes $Y_N = S_{p,h_0}(\Gamma)$ where N is proportional to p.

In the h-p method one combines the two approaches, thus one takes as approximating subspaces $Y_N = S_{p,h}(\Gamma)$ with two functions $p(N)$ and $h(N)$. Note, that we do impose continuity for the elements of $S_{p,h}(\Gamma)$ at the mesh points, since $S_{p,h}(\Gamma) \subset H^{1/2}(\Gamma)$. There holds the following convergence result for the Galerkin solution $v_{p,h} \in S_{p,h}(\Gamma)$ of (7.13). Its proof follows by combining Proposition 7.1 and Theorem 7.11.

Theorem 7.13 ([405]) *Let* (7.6) *hold and let p be sufficiently large and h be small enough. Then the Galerkin equations* (7.13) *are uniquely solvable in* $S_{p,h}(\Gamma)$. *Let* $v \in H^{1/2}(\Gamma)$ *be the solution of the integral equation* (7.12) *with right hand side* $g = (1 + K')f \in H^{s-1}(\Gamma)$ *and* $v_{p,h} \in S_{p,h}(\Gamma)$ *be the Galerkin solution, then we have for* $s \geqslant 1/2$, $s \notin A$, *with* $\alpha = \min \alpha_{jk}$

$$\left\| v - v_{p,h} \right\|_{H^{1/2}(\Gamma)} \leqslant C \max\{e_1, e_2\}$$

$$\text{with} \begin{cases} e_1 = \max\{h^{\alpha} p^{-2\alpha},\ h^{\min\{\alpha, p-\alpha+1/2\}} p^{-2\alpha}\} \log^{1/2} p \\ e_2 = h^{\min\{s-1/2,\ p+1/2\}} p^{-(s-1/2)} \log^{1/2} p \end{cases}$$

where the constant C depends on v but is independent of h and p.

The proof follows by combining Proposition 7.1 and Theorem 7.19 in Sect. 7.3.1. Optimal converges rates for the p-version without log-terms are derived in [208] and for the hp-version in [209].

h-Method with Graded Mesh on a Polygon

Mesh and boundary element space are constructed as above with $p \geq 1$. There holds as application of Theorem 7.11 the following convergence result.

Theorem 7.14 ([426]) *Let* (7.7) *hold and h be sufficiently small. Let* $v \in H^{1/2}(\Gamma)$ *solve* (7.12) *with* $f \in H^{s-1}(\Gamma)$, $s \geqslant 1/2$, $s \notin A$, $s+1/2 \notin N$. *Then the Galerkin equations* (7.13) *are uniquely solvable in* $S^{\beta}_{p,h}(\Gamma)$. *Then we have for* $\alpha = \min \alpha_{jk}$

$$\|v - v_h\|_{H^{1/2}(\Gamma)} \leqslant C \begin{cases} h^{\alpha\beta-\varepsilon} & \text{if } \beta < \frac{2p+3}{2\alpha} \\ h^{\frac{2p+3}{2}} & \text{if } \beta > \frac{2p+3}{2\alpha} \end{cases}$$

Proof *Left as exercise for the reader, c.f. Proposition 7.3.*

h-p Method with Geometric Mesh on a Polygon

Near each corner introduce a geometric mesh as in Sects. 8.1 or 8.3: We set $\Gamma^k = (0,1)$ for simplicity and consider the mesh near $t_k = \overline{\Gamma^{k-1}} \cap \overline{\Gamma^k}$ defined with $0 < \sigma < 1$ by the grid points

$$x_j = \sigma^j \quad (j = 0, 1, \dots, N-1), \; x_N = 0.$$

Now the space $Y_{N,\sigma}$ consists of continuous piecewise polynomials with degree $N - j$ on the interval $[x_{j+1}, x_j]$. Thus we have in the first interval at the corner polynomials of degree one, and in the next interval polynomials of degree 2 and so on, i.e. with the interval $I^k_j = [x_{j+1}, x_j]$ on Γ^k

$$Y_{N,\sigma} = \left\{ v \in C^0(\Gamma) : v|_{I^k_j} \in P_{p_j}, \; p_j = N - j \right\}$$

Here the total degree of freedom $N_f = \dim Y_N$ is proportional to N^2.

In many practical cases the given data f in (7.10) is analytic on each side Γ^k of the polygon Γ.

Theorem 7.15 ([21, 240]) *Let N be sufficiently large. Then the Galerkin equations* (7.13) *are uniquely solvable in* $Y_{N,\sigma}$. *Let* $v_{N,\sigma}$ *denote the Galerkin solution and* $v \in H^{1/2}(\Gamma)$ *be the exact solution of the integral equation* (7.12) *with piecewise* *analytic f, then we achieve the exponential convergence*

$$\left\| v - v_{N,\sigma} \right\|_{H^{1/2}(\Gamma)} \leqslant C\, e^{-\beta\sqrt{N}}$$

where the constants C and β are independent of N.

7.2.1 Regularity on a Polyhedron

Next we consider the Neumann problem on a *polyhedron* Ω. The following regularity result follows from [426] where the mixed Dirichlet-Neumann BVP is considered.

Theorem 7.16 *Choose $s_1 = m\nu_j + 2p$ where m, j, p are some integers, let s_2 be the next larger value of the form $(m\nu_j + 2p)$ or $\lambda_k + 1/2$. Let $s > s_1$ with $s - 1/2 \notin \mathbb{N}$, $s \neq m\nu_j$, $s \neq \lambda_k + 1/2$. Then*

$$\Delta u = 0 \qquad \text{in } \Omega, \qquad \left.\frac{\partial u}{\partial n}\right|_{\Gamma} = f \in H^{s-1/2}(\Gamma)$$

implies (if $s \geq s_1 + s_2 - 1/2 - \epsilon$)

$$v = v^0 + \chi(r) \sum_{0<\lambda_k<s_2-1/2} a_k\, r^{\lambda_k}\, w_k + \sum_{j=1,\dots,J} \chi_j(\theta) \sum_{m\nu_j+2p\leqslant s_1} h^j_{m,p} \theta_j^{m\nu_j+2p},$$

$$w_k \in \mathscr{H}^{\tilde{s}}(\Gamma), \quad h^j_{m,p}(r) = \sum_{\lambda_k+1/2<s} a^j_{m,k}\, r^{\lambda_k - m\nu_j - 2p} + k^j_{m,p}, \quad k^j_{m,p} \in H_0^{s-m\nu_j-2p}(\mathbb{R}^+)$$

$$v^0 \in H^{s_1+1/2-\epsilon}(\Gamma), \qquad \tilde{s} = \min\{s - s_1, -1/2 + s_2 - \varepsilon\}\ , \ \varepsilon > 0.$$

Next we illustrate Theorem 7.16 with an example:

Example 7.4 Let Ω be the exterior of a square in $\mathbb{R}^3$. Let $f \in H^1(\Gamma)$ given in (7.10). Then the unique solution of (7.12) has at the corner the decomposition

$$v = v^0 + a_1\, r^{.297} w_1(\theta) + h_1(r) r^{1/2} \theta^{1/2} + h_2(r) r^{1/2} \left(\frac{\pi}{2} - \theta\right)^{1/2}$$

$$v^0 \in \tilde{H}^{2-\varepsilon}(\Gamma) \quad h_i(r) = b_1^j\, r^{0.297-1/2} + b_2^j\, r^{1.426-1/2} + k_i(r)\,, b_i^j \in \mathbb{R}$$

$$k_i \in H^{3/2-\varepsilon} \text{ and } r^{3/2-\varepsilon} k_i \in L^2(\mathbb{R}^+)$$

Next, we consider various boundary element methods for the polyhedron.

7.2.1.1 h-p- Method with Quasiuniform Mesh on a Polyhedron

Construct a quasi uniform mesh with width h analogously as above and define the space $S_{p,h}(\Gamma)$ of piecewise polynomials of degree p on this mesh. Then the rate of convergence is determined by the regularity of the solution, i.e. the parameter α in (7.8):

Theorem 7.17 *Let p be sufficiently large and h be small enough. Then the Galerkin equations* (7.3) *are uniquely solvable in* $S_{p,h}(\Gamma)$*. Let* $v \in H^{1/2}(\Gamma)$ *be the solution of the integral equation* (7.2) *with sufficiently smooth right hand side* f*, and let* $v_{p,h}$ *the Galerkin solution, then we have with* α *in* (7.8)

$$\left\| v - v_{p,h} \right\|_{H^{1/2}(\Gamma)} \leqslant C \cdot h^{\alpha-\varepsilon} p^{\varepsilon-2\alpha}. \tag{7.17}$$

Remark 7.8 The proof for the $h-$version (p fixed) is given by the proof of Theorem 7.18 in Sect. 7.4.2 by setting $\beta = 1$. By a refined analysis based on [24] Schwab and Suri in [374] showed the $p-$version result (7.17) (h fixed) with $\epsilon = 0$. In [52] Heuer and Bespalov prove (7.17) with $\epsilon = 0$ and a possible $\log p/h$-term.

In Example 7.4 we have $\alpha = \min\{1/2, .297 + 1/2\} = 1/2$, hence Theorem 7.17 yields the convergence rate $\mathcal{O}\left(h^{1/2}\right)$ for the h-method and $\mathcal{O}\left(p^{-1}\right)$ for the p-method. Here there holds (7.17) with $\epsilon = 0$.

7.2.1.2 h-Method with Graded Mesh on a Polyhedron

For the construction of the graded mesh see above. Let S_h^β be the space of piecewise linear, continuous functions on this mesh. As in Theorem 7.14 we can compensate the effect of the singularities by an appropriately graded mesh and get the convergence rate $h^{3/2}$. The following theorem follows by combining Theorem 7.11 with 2D approximation results for the trace v in Sect. 7.4. There we present its proof from the Thesis [423] of T. von Petersdorff.

Theorem 7.18 ([426]) *Let h be sufficiently small and* α *as in* (7.8)*. Then we have*

$$\|v - v_h\|_{H^{1/2}(\Gamma)} \leqslant C \begin{cases} h^{\alpha\beta-\varepsilon} & \text{if } \beta < \frac{3}{2\alpha} \\ h^{3/2} & \text{if } \beta > \frac{3}{2\alpha} \end{cases}$$

where the constant $C = C(\beta)$ *is independent of h.*

7.3 1D-Approximation Results

7.3.1 hp-Method with Quasiuniform Mesh on Polygons

Here we report from [405]. Let $S^0_{h,p}(\Gamma^j)$ denote those functions in $S_{h,p}(\Gamma^j)$ which are continuous over Γ^j.

Proposition 7.1 *For $v \in H^r(\Gamma^j)$, $r > 1/2$ there exists $v^j_{h,p} \in S^0_{h,p}(\Gamma^j)$, $p \geqslant 1$, such that*

$$v^j_{h,p}(t_l) = v(t_l) \quad for\, l = j-1, j$$

$$\|v - v^j_{h,p}\|_{\tilde{H}^{1/2}(\Gamma^j)} \leqslant C\, h^{\mu-1/2} p^{-(r-1/2)} \log^{1/2} p\, \|v\|_{H^r(\Gamma^j)} \tag{7.18}$$

where $\mu = \min\{r, p+1\}$.

Proof Consider $\Gamma^j_{h,1}$, the first interval of Γ^j, assumed to be $I_h = [0, h]$. Take $\hat{v}(x) \equiv v(\frac{x}{h}) \in H^r(I)$, $I = [0, 1]$. By [25] there exists a projection $P^{1/2}_p$: $H^r(I) \to \mathbb{P}_p(I)$ s.t. $\forall \hat{w} \in H^r(I)$

$$P^{1/2}_p \hat{w} = \hat{w} \text{ at } x = 0,\ x = 1$$

$$P^{1/2}_p \hat{w} = \hat{w} \text{ for } \hat{w} \in P_p(I)$$

$$\|\hat{w} - P^{1/2}_p \hat{w}\|_{\tilde{H}^{1/2}(I)} \leqslant C\, p^{-(r-1/2)} \log^{1/2} p\, \|\hat{w}\|_{H^r(I)}$$

Hence for any $\hat{S} \in \mathbb{P}_p(I)$

$$\begin{aligned}
&\|\hat{v} - P^{1/2}_p \hat{v}\|_{\tilde{H}^{1/2}(I)} = \|(\hat{v} - \hat{S}) - P^{1/2}_p(\hat{v} - \hat{S})\|_{\tilde{H}^{1/2}(I)} \\
&\leqslant C\, p^{-(r-1/2)} \log^{1/2} p \inf_{\hat{S}\in\mathbb{P}_p(I)} \|\hat{v} - \hat{S}\|_{H^r(I)} \leqslant C\, p^{-(r-1/2)} h^{\mu-1/2} \log^{1/2} p\, \|v\|_{H^r(I_h)}
\end{aligned} \tag{7.19}$$

Repeating this over each subinterval gives the assertion. In (7.19) we have used that for $k \geq 0$

$$\inf_{\hat{p}\in\mathbb{P}_p(I)} \|\hat{v} - \hat{p}\|_{H^k(I)} \leq C h^{\mu-1/2} \|v\|_{H^k(I_h)}$$

where $\mu = \min(p+1, k)$ and C depends on k but is independent of p, h and μ (see [23]). □

Next, we consider the singular functions $v_{jk} = c_{jk} \rho_j^{\alpha_{jk}} \chi_j$. We look at the approximating polynomials, which vanish at the endpoints of Γ^j, of a function v defined on $\Gamma^j = (-1, 1)$ by $v(x) = (1+x)^\alpha \chi(x)$, where $\alpha > 0$ and χ is a C^∞

cut-off function satisfying $\chi = 1$ for $x \leqslant -1/2$, $\chi = 0$ for $x \geqslant 0$. Let us consider the weighted spaces $W^s(\mu, \nu)$ with $\mu, \nu \in \mathbb{R}$, integer $s > 0$ with norm

$$\|u\|^2_{W^s(\mu,\nu)} = \int_{-1}^{1} \left[(1-x^2)^{-\mu} \left[\frac{\partial^s u}{\partial x^s}\right]^2 + (1-x^2)^{-\nu} u^2 \right] dx$$

The use of $W^s(\mu, \nu)$ is essential to show that the p-version has twice the convergence rate of the h-version for singular functions.

Lemma 7.1 *Let $\hat{w}(x) = (1+x)^\alpha \chi(x)$ for $x \in I = (-1, 1)$ with $\alpha > 0$. Then there exists $\hat{w}_p \in \mathbb{P}_p(I)$ with*

$$\hat{w}_p(\pm 1) = \hat{w}(\pm 1) = 0,$$

$$\|\hat{w} - \hat{w}_p\|_{\tilde{H}^{1/2}(I)} \leqslant C\, p^{-2\alpha+\varepsilon}$$

with $p \geqslant 1$, $\min(\frac{3}{2}, 2\alpha) > \varepsilon > 0$and C independent of p.

Proof First, one sees that $\hat{w} \in \overset{\circ}{W}{}^s(\mu - s, \mu)$, $\mu = \frac{1}{2} + \frac{\varepsilon}{3}$, for any $s < 2\alpha + \frac{1}{2} - \frac{\varepsilon}{3}$ provided $\alpha > -\frac{1}{4} + \frac{\varepsilon}{6}$. Here the weighted Sobolev space $\overset{\circ}{W}{}^s(\mu, \nu)$ is the completion of the set $\{u \in C_0^\infty \,|\, \|u\|_{W^s(\mu,\nu)} < \infty\}$. Due to [22] $\forall u \in \overset{\circ}{W}{}^s(\mu - s, \mu)$, $s > \mu$, $\exists u_p \in \mathbb{P}_p$ s.t. $u_p = u = 0$ at ± 1 and

$$\|u - u_p\|_{\overset{\circ}{H}{}^{1/2+\tilde{\varepsilon}}(I)} \leqslant C\, p^{-(s-1/2)+\tilde{\varepsilon}} \|u\|_{\overset{\circ}{W}{}^s(\mu-s,\mu)}, \quad 0 < \tilde{\varepsilon} \leqslant \frac{1}{2}$$

Choosing $s = 2\alpha + \frac{1}{2} - 2\tilde{\varepsilon}$, $\tilde{\varepsilon} = \frac{\varepsilon}{3}$ yields with $u_p = \hat{w}_p$

$$\|\hat{w} - \hat{w}_p\|_{\overset{\circ}{H}{}^{1/2+\tilde{\varepsilon}}(I)} \leqslant C\, p^{-2\alpha+\varepsilon}$$

which gives the assertion. □

We note that in [22] u is expanded in a series of Jacobi polynomials and the weighted spaces $\overset{\circ}{W}{}^s(\mu - s, \mu)$ are the right setting.
Next let us consider

$$v(x) = x^\alpha \tilde{\chi}(x) \quad \text{on } I = (0, 1) \tag{7.20}$$

with $\tilde{\chi} \in C_0^\infty$ s.t. $\tilde{\chi} = 1$ for $x \leqslant 1/4$, $\tilde{\chi} = 0$ for $x \geqslant 1/2$.

Theorem 7.19 *For v in (7.20) on $\Gamma^j = (0, 1)$ and $\alpha > 0$ $\exists v_{h,p} \in \overline{S^0_{h,p}}(\Gamma^j)$ s.t.*

$$\|v - v_{h,p}\|_{\tilde{H}^{1/2}(\Gamma^j)} \leqslant C \max\{h^\alpha p^{-2\alpha+\varepsilon}, \min\{h^\alpha, h^{p+1/2} p^{-2\alpha} \log^{1/2} p\}\}$$

with $C > 0$ independent of h and p.

Proof Split v into w_1, w_2 with $w_1(x) = v(x)\chi(\frac{x}{h})$, $w_2(x) = v(x)(1-\chi(\frac{x}{h}))$. Then $w_1(x) = x^\alpha \chi(\frac{x}{h}) = h^\alpha \hat{w}(\frac{x}{h})$ with $\hat{w}$ as in lemma above. Then $\exists\, w_p(x) = \hat{w}_p(\frac{x}{h}) \in \mathbb{P}_p(I_h)$, $I_h = (0, h)$ with $w_p = 0$ at $x = 0$, $x = h$ and

$$\|w - w_p\|_{\tilde{H}^{1/2}(I_h)} \leqslant C\, p^{-2\alpha+\varepsilon}$$

Now taking $w^1_{p,h}(x) = h^\alpha w_p(x) \in \mathbb{P}_p(I_n)$ gives

$$\|w_1 - w^1_{p,h}\|_{\tilde{H}^{1/2}(I_h)} \leqslant C\, h^\alpha p^{-2\alpha+\varepsilon} \tag{7.21}$$

Extending $w^1_{p,h}$ by 0, we get a function in $\overline{S^0_{h,p}}(\Gamma^j)$ s.t. (7.21) holds in $\tilde{H}^{1/2}(\Gamma^j)$. Then approximating w_2 in $[h/4, 1]$ yields the assertion (see [405]). □

Next we observe that the antiderivative v of ψ_s can be approximated in the $\tilde{H}^{1/2}(\Gamma^i)$ norm by a polynomial $v_{p,h}$. Therefore the $\psi_{p,h}$, defined to be the derivative of $v_{p,h}$ (with respect to arc length) will approximate ψ_s in the $\tilde{H}^{-1/2}(\Gamma^i)$norm with the same accuracy.

Proposition 7.2 ([405]) *For $\psi \in H^r(\Gamma^j)$, $r > -1/2$ there exists $\psi^j_{h,p} \in S_{h,p}(\Gamma^j)$, such that*

$$\|\psi - \psi^j_{h,p}\|_{\tilde{H}^{-1/2}(\Gamma^j)} \leqslant C\, h^{\mu+1/2} p^{-(r+1/2)} \log^{1/2} p\, \|\psi\|_{H^r(\Gamma^j)} \tag{7.22}$$

where $\mu = \min\{r, p + 1\}$.

Proof Take $\Gamma^j_{h,1} = I_h$ and let $\psi \in H^r(I_h)$ with $r > 0$ and let $\bar{\psi} = \frac{1}{h}\int_0^h \psi(t)\, dt$ and define $v(x) = \int_0^x (\psi - \bar{\psi})(t)\, dt$.Then $v \in H^{r+1}(I_h) \cap \tilde{H}^{1/2}(I_h)$. By (7.19) there exists a polynomial $P^{1/2}_p v \in \mathbb{P}_{p+1}(I_h)$ such that

$$\|v - P^{1/2}_p v\|_{\tilde{H}^{1/2}(I_h)} \leqslant C\, p^{-(r+1/2)} h^{\mu+1/2} \log^{1/2} p\, \|v\|_{H^{r+1}(I_h)}$$

where $\mu = \min\{r, p + 1\}$. Now taking $\psi_p = (P^{1/2}_p v)' + \bar{\psi}$ we have

$$\|\psi - \psi_p\|_{\tilde{H}^{-1/2}(I_h)} \leq C\|v - P^{1/2}_p v\|_{\tilde{H}^{1/2}(I_h)} \leqslant C\, p^{-(r+1/2)} h^{\mu+1/2} \log^{1/2} p\, \|\psi\|_{H^r(I_h)}$$

Repeating this over each subinterval completes the proof. □

For the convenience of the reader we want to give a further detail where u is expanded in a series of Chebyshev polynomials. Let

$$u(x) = (x+1)^{1/2}\chi(x)\ ,\quad x \in I = [-1, 1] \tag{7.23}$$

with $\chi \in C^\infty$ satisfying $\chi(x) = 1$ for $-1 \le x \le -1/2$, $\chi(x) = 0$ for $1/2 < x < 1$. (Other singularity functions $(x+1)^\alpha$ can be treated similarly [22].) We consider the approximation of u in the $\tilde{H}^{1/2}(I)$-norm by functions in $\mathbb{P}_p(I)$. Let u be transformed to the periodic function $\hat{u}$ on $\hat{I} = [-\pi, \pi]$ by the mapping $x = \cos(\xi)$, i.e. $\hat{u}(\xi) = u(x)$. Then

$$\hat{u}(\xi) = (1+\cos(\xi))^{1/2}\ \chi(\cos(\xi)) = \sqrt{2}\ \chi(\cos(\xi))\cos(\xi/2)$$

Theorem 7.20 ([404]) *Let u be defined by* (7.23)*. Then for $p = 1, 2, \ldots$ there exists a polynomial $u_p^0 \in \mathbb{P}_p(I)$ s.t.*

$$u_p^0(\pm 1) = u(\pm 1) \tag{7.24}$$

$$\|u - u_p^0\|_{\tilde{H}^{1/2}(I)} \le Cp^{-1}\log^{1/2} p \tag{7.25}$$

Proof Write $\hat{u}(\xi) = \sum\limits_{k=0}^{\infty} a_k \cos(k\xi)$ and set $u_p^0 := u_p + \overline{u}$ where $\overline{u}$ is a linear function s.t u_p^0 satisfies (7.24) and $u_p \in \mathbb{P}_p(I)$ is defined in terms of Chebyshev polynomials $T_k(x) = \cos(k\cos^{-1}(x))$ of degrees $\le p$ by $\hat{u}_p = \sum\limits_{k=0}^{p} a_k T_k(\cos(\xi))$. Now

$$a_k = c\int\limits_0^\pi \hat{u}\cos(k\xi)d\xi$$

satisfies

$$|a_k| \le \frac{C}{k^2}.$$

Therefore

$$\|u - u_p\|^2_{H^{1/2}(I)} = \|\hat{u} - \hat{u}_p\|^2_{H^{1/2}(\hat{I})} = C\sum_{k=p+1}^{\infty} a_k^2(1+k^2)^{1/2} \le C\sum_{p+1}^{\infty}\frac{(1+k^2)^{1/2}}{k^4}$$

which behaves like

$$\int_{p+1}^{\infty}\frac{C}{x^3}dx = \frac{C}{p^2}.$$

Hence

$$\|u - u_p\|_{H^{1/2}(I)} \leq \frac{C}{p}.$$

Now for any x we have

$$|(u - u_p)(x)| \leq \sum_{k=p+1}^{\infty} |a_k| \leq \sum_{k=p+1}^{\infty} \frac{C}{k^2} \leq \frac{C}{p} \tag{7.26}$$

Furthermore

$$\|u - u_p^0\|_{\tilde{H}^{1/2}(I)} \leq \|u - u_p^0\|_{H^{1/2}(\hat{I})} + \|(1 - x^2)^{-1/2}(u - u_p^0)\|_{H^0(I)}$$

Hence we must bound the second term. We have

$$\int_{-1}^{1} (1 - x^2)^{-1}(u - u_p^0)^2 dx = (\int_0^{1/p} + \int_{1/p}^{\pi - 1/p} + \int_{\pi - 1/p}^{\pi})(\hat{u} - \hat{u}_p^0)^2 \sin(\xi))^{-1} d\xi$$

Now $1/\sin(\xi)$ is bounded on $[1/p, \pi - 1/p]$. Hence using (7.26)

$$\int_{1/p}^{\pi - 1/p} (\hat{u} - \hat{u}_p^0)^2 \sin(\xi))^{-1} d\xi \leq \frac{C}{p^2} \int_{1/p}^{\pi - 1/p} \sin(\xi))^{-1} d\xi \leq \frac{C}{p^2} \log p.$$

Furthermore (see [404], p. 38 for details) since

$$\int_0^{1/p} (\hat{u} - \hat{u}_p^0)^2 \sin(\xi))^{-1} d\xi \leq \frac{C}{p^2}$$

altogether we have

$$\|(1 - x^2)^{-1/2}(u - u_p^0)\|_{H^0(I)} \leq \frac{C \log^{1/2} p}{p}$$

completing the proof of the theorem.

□

In the framework of Jacobi-weighted Besov and Sobolev spaces in 1D Guo and Heuer analyze in [208] lower and upper bounds for approximation errors in the p-version BEM for hypersingular and weakly singular integral operators on polygons. They prove optimal convergence rates.

7.3.2 Approximation of the Normal Derivative on a One Dimensional Boundary—The h-Version on a Graded Mesh

Here we consider the approximation of singular functions $y^{\nu-1}$ on $I = [0,1]$. Similar results hold for $y^{\nu-1}\log(y)$.

Lemma 7.2 ([423]) *Let* $\nu > 0,\ \frac{1}{2} - \nu < \sigma \leqslant 1,\ \sigma \geqslant 0,\ p := h^{-1}\int_0^h y^{\nu-1}dy$. *Then* $\exists C_\nu \in \mathbb{R}$, *independent of* h, *such that*

$$\|y^{\nu-1} - p\|_{\tilde{H}^{-\sigma}([0,h])} \leqslant C_\nu h^{\nu+\sigma-1/2} \tag{7.27}$$

Proof Let $q := h^{-1}\int_0^h g\,dy$. Then

$$\|y^{\nu-1} - p\|_{\tilde{H}^{-\sigma}([0,h])} = \sup_{g\in H^\sigma([0,h])} \frac{\|\langle y^{\nu-1} - p, g\rangle\|}{\|g\|_{H^\sigma([0,h])}} = \sup_{g\in H^\sigma([0,h])} \frac{\|\langle y^{\nu-1}, g - q\rangle\|}{\|g\|_{H^\sigma([0,h])}}$$

For $\nu > 1/2$ there hold $y^{\nu-1} \in L^2([0,h])$ and hence

$$\|y^{\nu-1} - p\|_{\tilde{H}^{-\sigma}([0,h])} \leqslant C_\nu h^{\nu-1/2} C h^\sigma \tag{7.28}$$

For $0 < \nu \leqslant 1/2$ choose s with $\frac{1}{2} - \nu < s < \sigma$. Using $\tilde{y} := h^{-1}y$ gives

$$\|y^{\nu-1} - p\|_{\tilde{H}^{-\sigma}([0,h])} \leqslant C_\nu h^\nu \sup_{g\in H^\sigma([0,h])} \frac{\|g(h\,\tilde{y}) - q\|_{H^s([0,1])}}{\|g\|_{H^\sigma([0,h])}}$$

Furthermore interpolating between $L^2([0,1])$ and $H^1([0,1])$ one obtains

$$\|g(h\,\tilde{y}) - q\|_{H^s([0,1])} \leqslant c\,h^{\sigma-1/2}\|g\|_{H^\sigma([0,h])}.$$

□

Proposition 7.3 ([423]) *Let* $\psi \in \tilde{H}^{-1/2}([0,1])$ *have the form*

$$\psi(x) = \sum_{k=1}^{K} a_k x^{\nu_k - 1} + \psi^0(x) \ \textit{with}\ \nu_k > 0,\ \psi^0(x) \in H^1([0,1])$$

with $\nu_0 := \min\{\nu_k\}$.*Let* S_h^β *be space of piecewise constant functions on mesh* $x_k = \left(\frac{k}{N}\right)^\beta,\ k = 0,\ldots,N$ *with* $\beta \geqslant 1,\ h = 1/N$. *Then* $\forall\varepsilon > 0$ *and* $-1 \leqslant s \leqslant \nu_0 - 1/2$

there holds

$$\inf_{p_h\in S_h^\beta} \|\psi - p_h\|_{\tilde{H}^s([0,1])} \leqslant c \begin{cases} h^{(\nu_0-s-1/2)\beta-\varepsilon}, & 1 \leqslant \beta \leqslant \frac{1-s}{\nu_0-s-1/2} \\ h^{1-s}, & \beta > \frac{1-s}{\nu_0-s-1/2} \end{cases} \tag{7.29}$$

Proof Note for $1 - 1/\beta \leqslant \tilde{\gamma} \leqslant 1$ we have

$$h_k = x_k - x_{k-1} \leqslant \beta h^\beta k^{\beta-1} \leqslant h^{(1-\tilde{\gamma})\beta} \beta x_k^{\tilde{\gamma}}$$

We approximate seperately the regular and singular parts of ψ. Let p^0 denote on each $I_k = [x_{k-1}, x_k]$ the mean value of ψ^0. Then

$$\begin{aligned}\|\psi^0 - p^0\|^2_{\tilde{H}^s(I)} \leqslant \sum_{k=1}^N \|\psi^0 - p^0\|^2_{\tilde{H}^s(I_k)} &\leqslant C \sum_{k=1}^N h_k^{-2s+2} \|\frac{d\psi^0}{dx}\|_{L^2(I_k)} \\ &\leqslant C' h^{-2s+2} \|\psi^0\|^2_{H^1(I)}\end{aligned}$$

Next we consider $f(x) = x^{\nu_k - 1}$. On $I_k,\ k > 2$, we proceed as above and get:

$$\sum_{k=2}^N \|f - p\|^2_{\tilde{H}^s(I_k)} \leqslant C' \sum_{K=2}^N h^{(1-\tilde{\gamma})\beta(-2s+2)} x_k^{\tilde{\gamma}(-2s+2)} \|f'\|^2_{L^2(I_k)}$$

Since for $k \geqslant 2$, and $x \in I_k$, we have

$$x_k = (k\,h)^\beta = \left(\frac{k}{k-1}\right)^\beta x_{k-1} < 2^\beta x_{k-1} < 2^\beta x$$

this yields

$$\sum_{k=2}^N \|f - p\|^2_{\tilde{H}^s(I_k)} \leqslant c\,h^{(1-\tilde{\gamma})\beta(-2s+2)} \int_0^1 |f'(x)|^2 x^{\tilde{\gamma}(-2s+2)} dx$$

if the intergral exists, i.e. if

$$1 - \tilde{\gamma} < \frac{\nu_k - s - 1/2}{1 - s}. \tag{7.30}$$

Finally on the interval I_1 we take (7.27) on $[0, h_1]$ with $h_1 = h^\beta$ and obtain with $\sigma = -s$

$$\|x^{\nu_k - 1} - p\|_{\tilde{H}^s([0,h_1])} \leqslant C\,h^{\nu_k - s - 1/2)\beta}.$$

Now, choose $\tilde{\gamma}$ by $1-\tilde{\gamma} = \min\{\frac{\nu_0 - s - 1/2}{1-s} - \tilde{\varepsilon}, \frac{1}{\beta}\}$ with $\tilde{\varepsilon} > 0$. Hence $\tilde{\gamma} \geq 1 - \frac{1}{\beta}$ and (7.30) holds. Altogether we obtain (7.29). □

7.4 2D-Approximation Results

Lemma 7.3 ([423]) *Let* Q, Q_j $(j = 1, \ldots, N)$ *be Lipschitz domains with* $\overline{Q} = \bigcup_{j=1}^{N} \overline{Q}_j$, $\tilde{u} \in \tilde{H}^s(Q)$, $u \in H^s(Q)$, $s \in [-1, 1]$. *Then there holds*

$$\sum_{j=1}^{N} \|u\|^2_{H^s(Q_j)} \leqslant \|u\|^2_{H^s(Q)} \tag{7.31}$$

$$\|\tilde{u}\|^2_{\tilde{H}^s(Q)} \leqslant \sum_{j=1}^{N} \|\tilde{u}\|^2_{\tilde{H}^s(Q_j)} \tag{7.32}$$

Proof $0 \leqslant s \leqslant 1$ (larger s analogously). Consider the map $T : \prod_{j=1}^{N} \tilde{H}^s(Q_j) \to \tilde{H}^s(Q)$ which extends u_j on Q_j to u on Q. T is continuous for $s = 0,\ 1$ with norm $\leqslant 1$ when $\prod_{j=1}^{N} \tilde{H}^s(Q_j)$ $(s = 0,\ 1)$ has norm

$$\|(u_j)_{j=1,\ldots,N}\|^2 = \sum_{j=1}^{N} \|u_j\|^2_{\tilde{H}^s_{(Q_j)}}.$$

Now interpolation yields (7.32), since there holds

$$\|(u_j)_{j=1,\ldots,N}\|^2_{[s]} \leqslant \sum_{j=1}^{N} \|u_j\|^2_{[s]}$$

(Here $\|\cdot\|_{[s]}$ denotes the respective interpolation norm).
Inequality (7.31) is obvious for $s = 0,\ 1$. By duality this yields (7.32) for $s = 1$:

$$\begin{aligned}
\|u\|_{\tilde{H}^{-s}(Q)} &= \sup_{v \in H^s(Q)} \frac{\langle u, v\rangle_{\tilde{H}^{-s}(Q)\times H^s(Q)}}{\|v\|_{H^s(Q)}} \\
&= \sup_{v \in H^s Q} \frac{\sum_{j=1}^{N} \langle u|_{Q_j}, v|_{Q_j}\rangle_{\tilde{H}^{-s}(Q_j)\times H^s(Q_j)}}{\|v\|_{H^s(Q)}}
\end{aligned}$$

$$\leqslant \sup_{v\in H^s(Q)} \frac{\sum_{j=1}^{N} \|u\|_{\tilde{H}^{-s}(Q_j)} \|v\|_{H^s(Q_j)}}{\|v\|_{H^s(Q)}}$$

$$\leqslant \left(\sum_{j=1}^{N} \|u\|^2_{\tilde{H}^{-s}(Q_j)} \sup_{v\in H^s(Q)} \frac{\sum_{j=1}^{N} \|v\|^2_{H^s(Q_j)}}{\|v\|^2_{H^s(Q)}} \right)^{1/2}$$

$$\leqslant \left(\sum_{j=1}^{N} \|u\|^2_{\tilde{H}^{-s}(Q_j)} \right)^{1/2}$$

With interpolation one obtains now (7.32) for $-1 \leqslant -s \leqslant 0$. Finally duality yields (7.31) for $0 \leqslant s \leqslant 1$ (see also Remark 2.2 in [86]). □

The next lemma allows to estimate tensor product functions.

Lemma 7.4 ([423]) *Let* $I_j = [0, h_j],\ 0 \leqslant s_j \leqslant 1,\ f_j \in \tilde{H}^{-s_j}(I_j)$ *for* $j = 1, 2$. *Then there holds*

$$\|f_1(x) f_2(y)\|_{\tilde{H}^{-s_1-s_2}(I_1\times I_2)} \leqslant \|f_1\|_{\tilde{H}^{-s_1}(I_1)} \|f_2\|_{\tilde{H}^{-s_2}(I_2)}$$

Next we approximate $\tilde{H}^s$-functions on rectangles by constants.

Lemma 7.5 ([423]) *Let* $-1 \leqslant s \leqslant 0,\ R = [0, h_1] \times [0, h_2],\ u \in H^1(R)$ *and* $p = \frac{1}{h_1 h_2} \int_R u(x, y) dy\, dx$. *Then holds*

$$\|u - p\|_{\tilde{H}^s(R)} \leqslant C \max\{h_1, h_2\}^{-s} \left(h_1 \|u_x\|_{L^2(R)} + h_2 \|u_y\|_{L^2(R)}\right) \tag{7.33}$$

If $u(x, y) = u_1(x) u_2(y),\ u_j \in H^1([0, h_j])\ (j = 1, 2)$ *then*

$$\|u - p\|_{\tilde{H}(R)} \leqslant c \left(h_1^{1-s} \|u_x\|_{L^2}(R) + h_2^{1-s} \|u_y\|_{L^2}(R)\right)$$

7.4.1 Approximation of the Normal Derivative on a Two-dimensional Boundary—The h-Version on a Graded Mesh

Here we prove Theorem 7.9 (see the thesis by T. von Petersdorff [423]). The results are derived for the h-version on graded meshes and contain automatically the case of a quasiuniform mesh by setting the grading parameter $\beta = 1$.

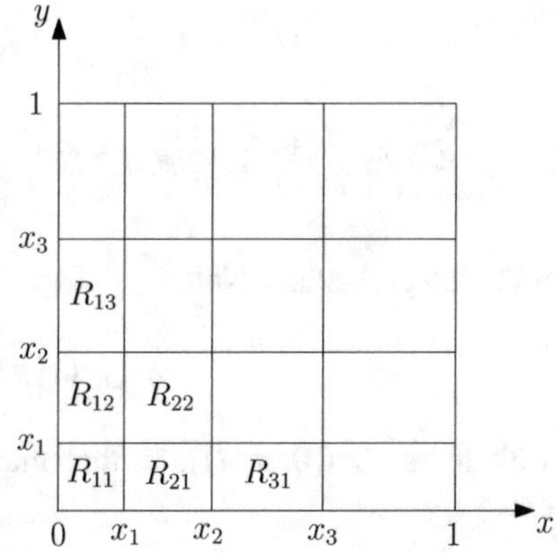

Fig. 7.6 Mesh on a square

Proof Approximation of regular part ψ^0: We approximate ψ^0 on each rectangle R_{kl}(see Fig. 7.6) by the L^2-projection $p|_{R_{kl}} = \frac{1}{h_k h_l} \int\limits_{R_{kl}} \psi^0$.

Then we use (7.32) for $s = -1/2$ and estimate the approximation error on each rectangle with (7.33) and $h_k \leq \beta h$:

$$\|\psi^0 - p\|^2_{\tilde{H}^{-1/2}(Q)} \leqslant c \sum_{k,l=1}^{N} \|\psi^0 - p\|^2_{\tilde{H}^{-1/2}(R_{kl})}$$

$$\leqslant c \sum_{k,l=1}^{N} \max\{h_k, h_l\} \left(h_k^2 \|\psi_x^0\|^2_{L^2(R_{kl})} + h_l^2 \|\psi_y^0\|^2_{L^2(R_{kl})} \right) \leqslant c' h^3 \|\psi^0\|^2_{H^1(Q)}$$

Approximation of corner singularity $f = r^{\lambda_k - 1} w_k(\theta)$, $w_k \in H^1([0, \pi/2])$:
With $p|_{R_{kl}} := \frac{1}{h_k k_l} \int\limits_{R_{kl}} f(x, y) dy\, dx$ and (7.32) and (7.33) one obtains

$$\|f - p\|^2_{\tilde{H}^{-1/2}(Q)} \leqslant c \sum_{\substack{k,l=1 \\ k+l \neq 2}}^{N} \max\{h_k, h_l\} \left(h_k^2 \|f_x\|^2_{L^2(R_{kl})} + h_l^2 \|f_y\|^2_{L^2(R_{kl})} \right)$$

$$+ c\|f - p\|_{\tilde{H}^{-1/2}(R_{11})} \tag{7.34}$$

Estimate for $k \geqslant 2,\ l \geqslant 2$: Note for $k \geqslant 2,\ \ x \in [x_{k-1}, x_k]$ there holds $|h_k| \leqslant \beta 2^{\beta\gamma} h\, x^\gamma$ with $\gamma = 1 - \frac{1}{\beta} > 0$. Therefore

$$\max\{h_k, h_l\} h_k^2 \|f_x\|^2_{L^2(R_{kl})} \leqslant c \|f_x \max\{x^\gamma, y^\gamma\}^{1/2} x^\gamma\|^2_{L^2(R_{kl})} \cdot h^3$$

and

$$\sum_{h,l=2}^{N} \|f-p\|^2_{\tilde{H}^{-1/2}(R_{kl})} \leqslant c\,h^3 \int\limits_0^1 \int\limits_0^1 |f_x(x,y)|^2 \max\{x^\gamma, y^\gamma\} x^{2\gamma}\, dy\, dx$$

if the integral exists. Note

$$|f_x(x,y)| \leqslant r^{\lambda_k - 2}\tilde{w}(\theta), \tag{7.35}$$

with $\tilde{w} \in L^2([0,\pi/2])$; further $\max\{x^\gamma, y^\gamma\} \leqslant r^\gamma$. Hence the above integral exists if

$$\beta > \frac{3}{2(\lambda_k + 1/2)}. \tag{7.36}$$

Proceeding analogously for f_y we get under this condition that

$$\sum_{k,l=2}^{N} \|f-p\|^2_{\tilde{H}^{-1/2}(R_{kl})} \leqslant c\,h^3$$

<u>Estimate for $k=1,\ l>1$</u>(analogously $k>1,\ l=1$):(7.34) gives

$$\sum_{l=2}^{N} \|f-p\|^2_{\tilde{H}^{-1/2}(R_{1l})} \leqslant c \sum_{l=2}^{N} \max\{h_1, h_l\} \left(h_1^2 \|f_x\|^2_{L^2(R_{1l})} + h_l^2 \|f_y\|^2_{L^2(R_{1l})} \right)$$

The term with f_x is bounded by

$$c \sum_{l=2}^{N} h_l^3 \|f_x\|^2_{L^2(R_{1l})} \leqslant c'\,h^3 \sum_{l=2}^{N} x_{l-1}^{3\gamma} \|f_x\|^2_{L^2(R_{1l})} \leqslant c\,h^3 \sum_{l=2}^{N} \|f_x(x,y) y^{3\gamma/2}\|^2_{L^2(R_{1l})}$$

$$\leqslant c\,h^3 \int\limits_{x=0}^{h_1} \int\limits_{y=0}^{1} |f_x(x,y)|^2 y^{3\gamma}\, dy\, dx$$

if the integral exists. With (7.35) this is bounded by

$$c\,h^3 \int\limits_{r=0}^{\sqrt{2}} \int\limits_{\phi=0}^{\pi/2} r^{2\lambda_k - 4} r^{3\gamma} r\, dr\, d\phi.$$

This integral exists for $\beta > \frac{3}{2(\lambda_k+1/2)}$. The f_y-term is handled analogously yielding

$$\sum_{l=2}^{N} \|f-p\|^2_{\tilde{H}^{-1/2}(R_{1l})} \leqslant c\,h^3.$$

Estimate for $k=1$, $l=1$: $f \in L^2(R_{11})$ because $\lambda_k > 0$. Now

$$\|f-p\|_{L^2(R_{11})} \leqslant \|f\|_{L^2(R_{11})} = c\,h_1^{\lambda_k}$$

For any constant q there holds

$$\|f-p\|_{\tilde{H}^{-1}(R_{11})} = \sup_{g\in H^1(R_{11})} \frac{\langle f-p, g\rangle}{\|g\|_{H^1(R_{11})}} = \sup_{g\in H^1(R_{11})} \frac{\langle f-p, g-q\rangle}{\|g\|_{H^1(R_{11})}}$$

Choose q as L^2 projection of g. Then

$$\|f-p\|_{\tilde{H}^{-1}(R_{11})} \leqslant \|f-p\|_{L^2(R_{11})} \sup_{g\in H^1(R_{11})} \frac{\|g-q\|_{L^2(R_{11})}}{\|g\|_{H^1(R_{11})}} \leqslant C\,h_1^{\lambda_k} h_1$$

Hence interpolation gives with (7.36)

$$\|f-p\|_{\tilde{H}^{-1/2}(R_{11})} \leqslant c\,h_1^{\lambda_k+1/2} = c\,h^{\beta(\lambda_k+1/2)} \leq Ch^{3/2}$$

Approximation of edge singularities: There are two types of edge singularities:

(1) $f(x,y) = \chi_i(\theta_i) b_{im} \rho_i^{\nu_{im}-1}$ with regular edge intensity factor $b_{im} \in H_0^1(\mathbb{R}^+)$
(2) $f(x,y) = \chi_i(\theta_i) y_i^{\lambda_k-\nu_{im}} \rho_i^{\nu_{im}-1}$ with corner singularities in the edge intensity factor.

We consider $Q = [0,1]^2$ with singularity at the x-axis, i.e. $y_i = x$, $\rho_i = y$. With (7.32) we have

$$\|f-p\|^2_{\tilde{H}^{-1/2}(Q)} \leqslant \sum_{k,l=1}^{N} \|f-p\|^2_{\tilde{H}^{-1/2}(R_{kl})} \tag{7.37}$$

First we consider case (2):

Estimate for $l \geqslant k, l \geqslant 2$: Define $\tilde{\chi} \in C^\infty([0,\pi/2])$ with $\tilde{\chi}(\theta) = 0$ for $\theta \in [0, \frac{\phi_0}{2}]$ and $= 1$ for $\theta \in [\phi_0, \pi/2]$ where ϕ_0 is sufficiently small, namely $\tan\phi_0 \leqslant 2^{-\beta}$. Then $g := \tilde{\chi} f = f$ on $R_{kl}, l \geqslant k,\ l \geqslant 2$. Now, $g = r^{\lambda_k-1} w(\theta)$, $w \in H^1([0,\pi/2])$, and the proof for the corner singularity yields an approximation q

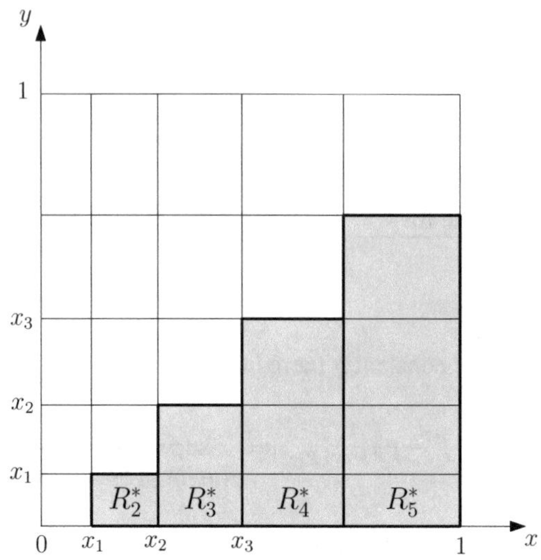

Fig. 7.7 The domains R_j^*

on R_{kl} with

$$\sum_{\substack{l\geqslant k\\ l\geqslant 2}} \|f-q\|^2_{\tilde H^{-1/2}(R_{kl})} \leqslant \sum_{k,l=1} \|g-q\|^2_{\tilde H^{-1/2}(R_{kl})} \leqslant c\,h^3.$$

<u>Estimate for $l<k$.</u> Here $\theta<\pi/4$ and $\chi_i(\theta)=1$. Hence $f(x,y)=f_1(x)f_2(y)$. Divide $R^*:=\bigcup_{1\leqslant l\leqslant k} R_{kl}$ into subdomains $R_j^*=\bigcup_{l=1}^{j-1} R_{jl},\ j=2,\dots,N$ (Fig. 7.7). Setting $f_1(x)=x^{\lambda-\nu},\ f_2(y)=y^{\nu-1},\ ,I_k=[x_{k-1},x_k],\ I_k^*=[0,x_k]$ we have

$$\|f-p\|^2_{\tilde H^{-1/2}(R^*)} \leqslant \sum_{j=2}^N \|f-p\|^2_{\tilde H^{-1/2}(R_j^*)}$$

For $\nu>1/2$ there holds (for $\nu<1/2$ see [423])

$$\|f-p\|_{\tilde H^{-1/2}(R_j^*)} \leqslant \|f_1-p_1\|_{\tilde H^{-1/2}(I_j)}\|f_2\|_{L^2(I^*_{j-1})}+\|f_2-p_2\|_{\tilde H^{-1/2}(I^*_{j-1})}\|p_1\|_{L^2(I_j)}$$

with $p_1,\ p_2$ mean value of $f_1,\ f_2$ on $I_j,\ I^*_{j-1}$ respectively.

Now

$$\sum_{j=2}^N \|f_2-p_2\|^2_{\tilde H^{-1/2}(I^*_{j-1})}\|p_1\|^2_{L^2(I_j)} \leqslant \sum_{j=2}^N \|f_1\|^2_{L^2(I_j)} \sum_{k=1}^{j-1} \|f_2-p_2\|^2_{\tilde H^{-1/2}(I_k)}$$

$$\leqslant c\,h^3 \int_{x=0}^{1} \int_{y=0}^{x} x^{2\lambda-2\nu} y^{2\nu-4+3\gamma}\,dy\,dx$$

$$= c'\,h^3 \int_{x=0}^{1} x^{(2\lambda-3+3\gamma)}dx \leq c''h^3$$

These integrals exist for $2\lambda_k - 3 + 3\gamma > -1$ and $2\nu_{im} - 4 + 3\gamma > -1$, i.e.

$$\beta = \frac{1}{1-\gamma} > \frac{3}{2\min\{\lambda_k + \frac{1}{2},\ \nu_{im}\}} \tag{7.38}$$

On the other hand for

$$\gamma > 1 - \frac{2\min\{(\lambda_k + 1/2, \nu_{im})\}}{3} \tag{7.39}$$

we have

$$\sum_{j=2}^{N} \|f_1 - p_1\|^2_{\tilde{H}^{-1/2}(I_j)} \|f_2\|^2_{L^2(I^*_{j-1})} \leqslant \sum_{j=2}^{N} C\,h_j^3 \|f_1'\|^2_{L^2(i_j)} |x_{j-1}^{2\nu-1}|$$

$$\leqslant c\,h^3 \sum_{j=2}^{N} \|x^{(\lambda-\nu-1)+3\gamma/2+(\nu-1/2)}\|^2_{L^2(I_j)}$$

$$\leqslant c\,h^3 \int_{x_1}^{1} x^{2\lambda+3\gamma-3}dx \ \leq c\,h^3$$

<u>Estimate for $k = l = 1$</u> Consider $\lambda - \nu > -1/2$ then

$$\|f-p\|_{\tilde{H}^{-1/2}(R_{11})} \leqslant \|x^{\lambda-\nu}y^{\nu-1} - q\|_{\tilde{H}^{-1/2}(R_{11})} + \|(1-\chi(\theta)x^{\lambda-\nu}y^{\nu-1} - \tilde{q}\|_{\tilde{H}^{-1/2}(R_{11})} \tag{7.40}$$

Here $q, \tilde{q}$ denote the respective mean values on R_{11}. Then for $0 < \varepsilon < \lambda$

$$\|x^{\lambda-\nu}y^{\nu-1} - q\|_{\tilde{H}^{-1/2}(R_{11})} \leqslant \|x^{\lambda-\nu}\|_{L^2(I_1)} \|y^{\nu-1} - q_2\|_{\tilde{H}^{-1/2}(R_{11})}$$
$$+ \|y^{\nu-1}\|_{\tilde{H}^{\nu-\frac{1}{2}-\varepsilon}(I_1)} \|x^{\lambda-\nu} - q_1\|_{\tilde{H}^{\varepsilon-\nu}(R_{11})} \leqslant C\,h_1^{\nu}$$

with the mean values q_1, q_2 of $x^{\lambda-\nu}$ and $y^{\nu-1}$ on I_1. Now $(1-\chi(\theta))x^{\lambda-\nu}y^{\nu-1} = r^{\lambda-1}\tilde{w}(\theta)$, $\tilde{w} \in L^2([0, \pi/2])$ hence, the second term in (7.40) is bounded by $C\,h_1^{\lambda+1/2}$. For $\lambda - \nu \leqslant 1/2$ there holds $\nu > 1/2$ and $f = \chi(\theta)x^{\lambda-\nu}y^{\nu} \in L^2(R_{11})$.

Hence

$$\|\chi(\theta)x^{\lambda-\nu}y^{\nu-1} - p\|_{\tilde{H}^{-1/2}(R_{11})} \leqslant h_1^{1/2}\|\chi(\theta)x^{\lambda-\nu}y^{\nu-1}\|_{L^2(R_{11})} \leqslant C\,h_1^{\lambda+1/2}$$

In both cases we thus have with $h_1 = h^\beta$ and (7.38)

$$\|f - p\|_{\tilde{H}^{-1/2}(R_{11})} \leqslant C\,h^{3/2}.$$

Next we consider case (1)(Edge singularity with regular edge function):

$$f(x,y) = b(x)y^{\nu-1} + (\chi(\theta) - 1)b(x)y^{\nu-1} =: f_1 + f_2$$

Note $f_2 \in H^1(Q)$, and thus is approximated with order $h^{3/2}$ like the regular part ψ^0. f_1 has tensor product form. Let q denote the mean value of f_1 on each R_{jk}, q_1 the mean value of $b(x)$ on $[x_{k-1}, x_k]$ and q_2 for $y^{\nu_{im}-1}$. Then for $\nu \leq 1/2$, any $\tilde{\varepsilon} > 0$ and (7.38)

$$\begin{aligned}\|b(x)y^{\nu-1} - q_1(x)q_2(y)\|_{\tilde{H}^{-1/2}(Q)} &\leqslant \|b\|_{L^2(I)}\|y^{\nu-1} - q_2\|_{\tilde{H}^{-1/2}(I)}\\ &+ \|y^{\nu-1}\|_{\tilde{H}^{\nu-1/2-\varepsilon}(I)}\|b - q_1\|_{\tilde{H}^{\varepsilon-\nu}(I)} \leqslant C\,h^{3/2-\tilde{\varepsilon}} + c\,h^{1+\nu-\varepsilon}\end{aligned}$$

For $\nu > 1/2$ there holds

$$\begin{aligned}\|b(x)y^{\nu-1} - q_1(x)q_2(y)\|_{\tilde{H}^{-1/2}(Q)} &\leqslant \|b\|_{L^2(I)}\|y^{\nu-1} - q_2\|_{\tilde{H}^{-1/2}(I)}\\ &+ \|y^{\nu-1}\|_{L^2(I)}\|b - q_1\|_{\tilde{H}^{-1/2}(I)} \leqslant c\,h^{3/2-\tilde{\varepsilon}} + C\,h^{3/2}.\end{aligned}$$

Collecting the estimates for the various parts of ψ on the subdomains gives with (7.37) the assertion of Theorem 7.9. □

Finally, we consider the case $\beta < 3/2\alpha_0$. Then

$$h_k \leqslant \beta\,h^{\beta(1-\tilde{\gamma})}x_k^{\tilde{\gamma}},\ 1 - \frac{1}{\beta} \leqslant \tilde{\gamma} \leqslant 1.$$

Now, we can take $\tilde{\gamma}$ instead of $\gamma = 1 - \frac{1}{\beta}$ and perform the proof analogously by choosing $\tilde{\gamma} = 1 - 2\alpha_0/3 + \varepsilon,\ \varepsilon > 0$. Before we obtained with $h_k \leqslant \beta x_k^\gamma$ the order $h^{3/2}$, now we get only

$$h^{\beta(1-\tilde{\gamma})3/2} = h^{\beta(2\alpha_0/3-\varepsilon)3/2} = h^{\alpha_0\beta-\tilde{\varepsilon}}.$$

7.4.2 Approximation of the Trace on a Two-Dimensional Boundary—The h-Version on a Graded Mesh

In this section we prove Theorem 7.18. Here it helps to use anisotropic Sobolev spaces.

Definition 7.1 Let $H^{(1,0)}(\Omega)$ denote the closure of $C^\infty(\Omega)$ in the norm

$$\|u\|^2_{H^{(1,0)}(\Omega)} := \|u\|^2_{L^2(\Omega)} + \|u_x\|^2_{L^2(\Omega)}.$$

We define the anisotropic space $H^{(s,0)}(\Omega)$ by interpolation:

$$H^{(s,0)}(\Omega) := \left(L^2(\Omega), H^{(1,0)}(\Omega)\right)_{[s]}, \quad H^{(0,s)}(\Omega) := \left(L^2(\Omega), H^{(0,1)}(\Omega)\right)_{[s]}$$

Note for the homogeneous Sobolev spaces we have for $s \geqslant 0$

$$H^s(\Omega) = H^{(s,0)}(\Omega) \cap H^{(0,s)}(\Omega)$$

and

$$\|u\|_{H^s(\Omega)} \leqslant C\|u\|_{H^{(s,0)}(\Omega)} + C\|u\|_{H^{(0,s)}(\Omega)}.$$

Further we need Sobolev spaces, where the functions satisfy only on a part of the boundary a condition like in the $\tilde{H}$-spaces.

Definition 7.2 Let Ω be Lipschitz, $\Gamma_0 \cap \overline{\Omega} \subset \partial\Omega$ part of the boundary,

$$\overset{\circ}{H}{}^1_{\Gamma_0}(\Omega) := \left\{u \in H^1(\Omega) | \; u|_{\Gamma_0} = 0\right\}$$

$$\tilde{H}^s_{\Gamma_0}(\Omega) := \left(L^2(\Omega), \overset{\circ}{H}{}^1_{\Gamma_0}(\Omega)\right)_{[s]}$$

Herewith for $s \neq 1/2$ we can estimate the norm of a function by the norms in the subdomains, but with constants depending on the domains.

Lemma 7.6 ([423]) *Let $\Omega, \Omega_1, \Omega_2$ be Lipschitz with $\overline{\Omega} = \overline{\Omega_1} \bigcup \overline{\Omega_2}, \overline{\Gamma_0} = \partial\Omega_1 \cap \partial\Omega_2, \partial\Omega_1 = \overline{\Gamma_0} \bigcup \overline{\Gamma_1}, \partial\Omega_2 = \overline{\Gamma_0} \bigcup \overline{\Gamma_2}, 0 \leqslant s \leqslant 1, s \neq \frac{1}{2}$. Then there exists a constant $c > 0$ such that $\forall u \in H^s(\Omega), \tilde{u} \in \tilde{H}^s(\Omega)$ there holds*

$$\|u\|_{H^s(\Omega)} \leqslant c\|u\|_{H^s(\Omega_1)} + c\|u\|_{H^s(\Omega_2)}$$

$$\|\tilde{u}\|_{\tilde{H}^s(\Omega)} \leqslant c\|u\|_{\tilde{H}^s_{\Gamma_1}(\Omega_1)} + c\|u\|_{\tilde{H}^s_{\Gamma_2}(\Omega_2)}$$

Proof Consider the mapping T which maps $u \in H^s(\Omega)$ into $(u|_{\Omega_1}, u|_{\Omega_2}) \in H^s(\Omega_1) \times H^s(\Omega_2)$. For $s \neq \frac{1}{2}$ the range of T is closed in $H^s(\Omega_1) \times H^s(\Omega_2)$ which is equivalent to $\overset{\circ}{H}{}^s(\Omega) = \tilde{H}^s(\Omega)$ for $s \neq \frac{1}{2}$ (cf. [415][(5.22)]). Hence the assertion of the lemma follows from the graph theorem. □

First, we consider approximation with bilinear functions on rectangles.

Lemma 7.7 ([423]) *Let $Q = [0, h_1] \times [0, h_2]$, $u \in H^3(Q)$, p the bilinear interpolant of u at the vertices of Q. Then there holds*

$$\|u - p\|_{L^2(Q)} \leqslant C\left(h_1^2\|u_{xx}\|_{L^2(Q)} + h_2^2\|u_{yy}\|_{L^2(Q)} + h_1^2 h_2\|u_{xxy}\|_{L^2(Q)}\right) \tag{7.41}$$

$$\|(u - p)_x\|_{L^2(Q)} \leqslant C\left(h_1\|u_{xx}\|_{L^2(Q)} + h_2^2\|u_{xyy}\|_{L^2(Q)}\right) \tag{7.42}$$

Proof First we note that for $u \in H^2(I)$, $I = [0, 1]$, with linear interpolant Πu at 0 and 1 there holds:

$$\|u - \Pi u\|_{L^2(I)} \leqslant \|u''\|_{L^2(I)} \tag{7.43}$$

$$\|\Pi u\|_{L^2(I)} \leqslant \|u\|_{L^2(I)} + C\|u'\|_{L^2(I)} \tag{7.44}$$

On $Q = [0, 1]^2$, Π_x denotes the partial interpolation operator in x, i.e. $(\Pi_x u)(x, y) := (\Pi u(\cdot, y))(x)$ for all $y \in I$. Now $p := \Pi_y \Pi_x u$ and

$$\|u - \Pi_y \Pi_x u\|_{L^2(Q)} \leqslant \|u - \Pi_y u\|_{L^2(Q)} + \|\Pi_y (u - \Pi_x u)\|_{L^2(Q)}$$

(7.43) yields

$$\|u - \Pi_y x\|_{L^2(Q)} \leqslant \int_0^1 \|u_{yy}(x, \cdot)\|_{L^2(I)}^2 \partial x \leqslant \|u_{yy}\|_{L^2(Q)}^2 \tag{7.45}$$

Further for fixed $x \in I$ we have with (7.44) and (7.45)

$$\begin{aligned}\|\Pi_y (u - \Pi_x u)(x, \cdot)\|_{L^2(I)} &\leq \|(u - \Pi_x u)(x, \cdot)\|_{L^2(I)} + C\|\frac{\partial}{\partial y}(u - \Pi_x u)(x, \cdot)\|_{L^2(I)} \\ &\leqslant c\|u_{xx}\|_{L^2(Q)}^2 + c\|u_{xxy}\|_{L^2(Q)}^2.\end{aligned}$$

Hence, (7.41) holds for $Q = [0, 1]^2$. The substitution $\tilde{x} = h_1 x$, $\tilde{y} = h_2 y$ gives (7.41) for $Q = [0, h_1] \times [0, h_2]$. (7.42) follows analogously. □

Now we prove Theorem 7.18.

Approximation of regular part v^0: Let $p \in S_h^\beta$ denote the interpolant of v^0, where S_h^β are the linear functions on the graded mesh (see Fig. 7.6) $x_k = (kh)^\beta$, $y_l = (lh)^\beta$. Hence

$$\|v^0 - p\|^2_{L^2(Q)} = \sum_{j,k=1}^{N} \|v^0 - p\|^2_{L^2(R_{jk})} \leqslant \sum_{j,k=1}^{N} \Big(h_j^4 \|v^0_{xx}\|^2_{L^2(R_{jk})} + h_k^4 \|v^0_{yy}\|^2_{L^2(R_{jk})} + h_j^4 h_k^2 \|v^0_{xxy}\|^2_{L^2(R_{jk})} \Big) \lesssim h^4 \|v^0\|^2_{H^3(Q)}$$

$$\|v^0 - p\|^2_{H^1(R_{jk})} = \sum_{j,k=1}^{N} \|v^0 - p\|^2_{H^1(R_{jk})} \leqslant \sum_{j,k=1}^{N} c\Big(h_j^2 \|v^0_{xx}\|^2_{L^2(R_{jk})} + h_k^2 \|v^0_{yy}\|^2_{L^2(R_{jk})} + h_j^4 \|v^0_{xxy}\|^2_{L^2(R_{jk})} + h_k^4 \|v^0_{xyy}\|^2_{L^2(R_{jk})} \Big) \lesssim h^2 \|v^0\|^2_{H^3(Q)}$$

Interpolation yields $\|v^0 - p\|_{H^{1/2}(Q)} \lesssim h^{3/2} \|v^0\|_{H^3(Q)}$.
Approximation of corner singularity $u = r^\lambda w(\theta)$ with $w \in H^3([0, \pi/2])$. We devide the square $Q = [0, 1]^2$ in $N - 1$ overlapping domains A_j, $j = 1, \ldots, N - 1$ (see Fig 7.8).

$$A_j = \bigcup \{R_{kl} | j \leqslant k \leqslant j + 1, l \leqslant k \text{ or } j \leqslant l \leqslant j + 1, k \leqslant l\}$$

Further we take a partition of unity $\{\chi_j\}$ on Q w.r.t. the sets A_j, e.g. let χ_{kl} the piecewise bilinear function on the mesh, which is 1 at (x_k, x_l) and 0 at all other nodes. We take $\chi_j = \sum \{\chi_{kl} | \text{supp}\chi_{kl} \subset A_j\}$. Then $|(\chi_j)_x| \leqslant h_j^{-1}$, $|(\chi_j)_y| \leqslant h_j^{-1}$.

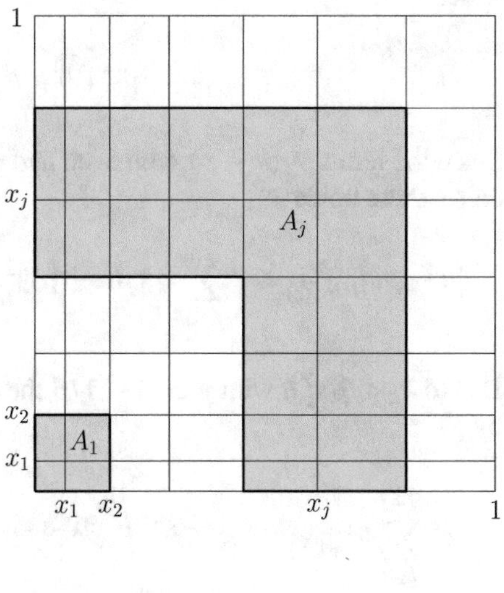

Fig. 7.8 The domains A_j

We estimate the approximation error in each A_j in L^2 and H^1; then interpolation gives an upper bound for $\chi_j(u-p)$ in $\tilde{H}^{1/2}$. This is then used to derive the estimate on Q.
For $j > 1$ with $h_k \le h_{j+1}$ when $R_{kl} \subset A_j$ we have

$$\|u-p\|^2_{L^2(A_j)} \leqslant \sum_{R_{kl}} c\left(h_k^4 \|u_{xx}\|^2_{L^2(R_{kl})} + h_l^4 \|u_{yy}\|^2_{L^2(R_{kl})} + h_k^4 h_l^2 \|u_{xxy}\|^2_{L^2(R_{kl})}\right)$$

$$\|u-p\|^2_{H^1(A_j)} \leqslant c\Big(h_{j+1}^2 \|u_{xx}\|_{L^2(A_j)} + h_{j+1}^2 \|u_{yy}\|_{L^2(A_j)} + h_{j+1}^4 \|u_{xxy}\|^2_{L^2(A_j)} + h_{j+1}^4 \|u_{xyy}\|^2_{L^2(A_j)}\Big)$$

This implies with $u = r^\lambda w(\theta)$

$$\|u-p\|^2_{L^2(A_j)} \leqslant \left(C\, h_{j+1} x_{j-1}\right)\left(h_{j+1}^4 x_{j-1}^{2\lambda-4} + h_{j+1}^6 x_{j-1}^{2\lambda-6}\right)$$

$$\|u-p\|^2_{H^1(A_j)} \leqslant \left(C\, h_{j+1} x_{j+1}\right)\left(h_{j+1}^2 x_{j-1}^{2\lambda-4} + h_{j+1}^4 x_{j-1}^{2\lambda-6}\right)$$

Note that χ_j vanishes on $\zeta_j := \partial A_j \setminus \partial Q$, hence interpolation yields

$$\|\chi_j(u-p)\|^2_{\tilde{H}^{1/2}_{\zeta_j}(A_j)} \leqslant \left(C\, h_{j+1} x_{j+1}\right)\left(h_{j+1}^3 x_{j-1}^{2\lambda-4}\right).$$

Since $r^\lambda \in H^1(Q)$ we have for $j = 1$

$$\|u-p\|^2_{L^2(A_1)} \leqslant C x_2^{2\lambda+2}, \qquad \|u-p\|^2_{H^1(A_1)} \leqslant C\, x_2^{2\lambda},$$

hence

$$\|\chi_1(u-p)\|^2_{\tilde{H}^{1/2}_{\zeta_1}(A_1)} \leqslant C\, x_2^{2\lambda+1}$$

Since the terms $\chi_j(u-p)$ with even and odd j have different supports and vanish on ζ_j, there holds

$$\|u-p\|^2_{H^{1/2}(Q)} \leqslant 2 \sum_{j \text{ odd}} \|\chi_j(u-p)\|^2_{\tilde{H}^{1/2}_{\zeta_j}(A_j)} + 2 \sum_{j \text{ even}} \|\chi_j(u-p)\|^2_{\tilde{H}^{1/2}_{\zeta_j}(A_j)}$$

Due to $h_k \leqslant \beta x_k^\gamma h$ with $\gamma := 1 - 1/\beta$ the terms for $j = 2, \ldots, N-1$ are bounded by

$$C \sum_{j=2}^{N-1} h_{j+1}^4 x_{j+1}^{2\lambda-3} \leqslant C\, h^3 \int_{x_2}^{1} x^{2\lambda-3+3\gamma}\, dx = \mathcal{O}(h^3) \quad \text{for } \beta > \frac{3}{2(\lambda+1/2)}.$$

Also the term with $j = 1$ is of same order. Hence in total for the corner singularity we have convergence rate $h^{3/2}$ in $H^{1/2}(Q)$.
Approximation of singular edge functions:

$$f(x, y) = x^{\lambda-\nu} y^{\nu} \chi(x) \tag{7.46}$$

with corner exponent λ, edge exponent ν and cut-off function χ with $\chi(\theta) = 1$ near $\theta = 0$. Divide $Q = (0, 1)^2$ into 2 triangles

$$A_{:} = \{(x, y) \in Q | y \leqslant x\}, \quad B := \{(x, y) \in Q | y > x\},$$

and estimate the interpolation error on each triangle in $H^{1/2+\varepsilon}$, $0 \le \epsilon \le 1/2$. Lemma 7.6 yields error bound in $H^{1/2+\varepsilon}(Q)$.

On triangle B there holds $f(x, y) = r^{\lambda} w(\theta)$ with $w(\theta)$ smooth on $[\frac{\pi}{4}, \frac{\pi}{2}]$. Extend $w(\theta)$ to a smooth function $\tilde{w}(\theta)$ on $[0, \frac{\pi}{2}]$ and define $\tilde{f}(x, y) := r^{\lambda} \tilde{w}(\theta)$. Now $\tilde{f}$ can be approximated on Q like a corner singularity and for its interpolant $\tilde{p}$ there holds

$$\|\tilde{f} - \tilde{p}\|_{H^{1/2+\varepsilon}(Q)} \leqslant c\, h^{3/2-\varepsilon}$$

yielding, by restriction to B,

$$\|f - p\|_{H^{1/2+\varepsilon}(B)} \leqslant c\, h^{3/2-\varepsilon}.$$

On triangle A we estimate the approximation error separately in the anisotropic spaces $H^{(s,0)}$ and $H^{(0,s)}$ with $s > 1/2$. Consider the larger domain $\tilde{A}$ with (see Fig. 7.9)

$$A \subset \tilde{A} := \bigcup \{R_{kl} | l \leqslant k + 1\}$$

Fig. 7.9 The rectangles A_j

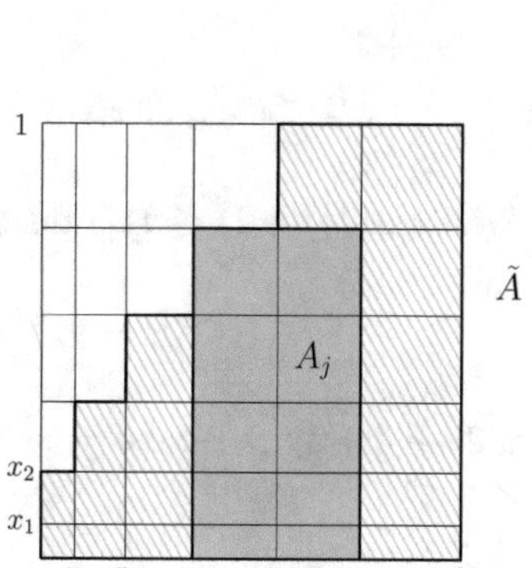

Let $\chi_j(x)$ be piecewise linear on $0 < x_1 < \cdots < x_{N-1} < 1$ with $\chi_j(x_j) = 1$ and vanishing in the other nodes, $j = 2, \ldots, N-2$. χ_1 pw. linear with $\chi_1(0) = \chi_1(x_1) = 1$ and $\chi_1(x_j) = 0$ for $j > 1$. χ_{N-1} pw. linear with $\chi_{N-1}(x_{N-1}) = \chi_{N-1}(1) = 1$ and $\chi_{N-1}(x_j) = 0$ for $j < N-1$. Divide $\tilde{A}$ into overlapping rectangles $A_j = [x_{j-1}, x_{j+1}] \times [0, x_{j+1}],\ j = 1, \ldots, N-1$. With

$$g_j := \begin{cases} (f(x,y) - p(x,y))\,\chi_j(x) \text{ on } A_j \\ 0 \text{ else} \end{cases} \tag{7.47}$$

there holds $f - p = \sum_{j=1}^{N-1} g_j$. With $\chi(\theta) \equiv 1$ on $A_2, \ldots, A_{N-1}$ in (7.46) we have $f(x,y) = f_1(x) f_2(y)$ with $f_1(x) := x^{\lambda-\nu}$, $f_2(y) := y^{\nu}$. Let ζ_j denote the "left and right" boundary of A_j for $j = 2, \ldots, N-2$; ζ_1 the "right boundary" of A_1, ζ_{N-1} the "left boundary" of A_{N-1}. Then

$$\|\sum_{j=1}^{N-1} g_j\|^2_{H^{(s,0)}(\tilde{A})} \leqslant 2 \sum_{j \text{ odd}} \|g_j\|_{\tilde{H}^{(s,0)}_{\zeta_j}(A_j)} + 2 \sum_{j \text{ even}} \|g_j\|_{\tilde{H}^{(s,0)}_{\zeta_j}(A_j)}$$

Now for $j \geqslant 2$ we have with the interpolant p_j of f_j with $s > 1/2$

$$\begin{aligned} \|g_j\|^2_{\tilde{H}^{(s,0)}_{\zeta_j}(A_j)} &\leqslant \|\chi_1 f_1\|^2_{\tilde{H}^s([x_{j-1},x_{j+1}])} \|f_2 - p_2\|^2_{L^2([0,x_{j+1}])} \\ &\quad + \|\chi_j(f_1 - p_1)\|^2_{\tilde{H}^s([x_{j-1},x_{j+1}])} \|p_2\|^2_{L^2([0,x_{j+1}])} =: B_1 + B_2 \end{aligned}$$

Let us first consider the term B_1: Interpolation gives

$$\|\chi_j f_1\|_{\tilde{H}^s([x_{j-1},x_{j+1}])} \leqslant c\, h_j^{1/2-s} x_j^{\lambda-\nu}$$

On the other hand we use the one dimensional approximation result

$$\|f_2 - p_2\|_{L^2([0,x_{j+1}])} \leqslant c \left(\frac{x_{j+1}}{j+1}\right)^2 x_{j+1}^{-2\gamma} \|f_2''(x) x^{2\gamma}\|_{L^2([0,x_{j+1}])}$$

With $x_{j+1}/(j+1) \leqslant h_{j+1}$ this implies for $f_2(x) = x^{\nu}$

$$\|f_2 - p_2\|^2_{L^2([0,x_{j+1}])} \leqslant c\, h_j^4 x_{j+1}^{-4\gamma} x_{j+1}^{2\nu-3+4\gamma}$$

if $2\nu - 4 + 3\gamma > -1$, i.e. $\gamma > 1 - 2\nu/3$, hence $\beta > 3/(2\nu)$. Altogether

$$\sum_{j-2}^{N-1} \|\chi_j f_1\|^2_{\tilde{H}^s([x_{j-1},x_{j+1}])} \|f_2 - p_2\|^2_{L^2([0,x_{j+1}])} \leqslant c' h^{4-2s} \int_0^1 x^{2\lambda-3+(4-2s)\gamma} dx < \infty \tag{7.48}$$

for $s < \frac{1}{2} + \varepsilon$ with ε sufficiently small and $\beta > \frac{3}{2(\lambda+1/2)}$.
Next, we consider the term B_2: First we have

$$\|\chi_j(f_1 - p_1)\|^2_{\tilde{H}^s([x_{j-1},x_{j+1}])} \leqslant c\, h_j^{4-2s} h_j x_j^{2(\lambda-\nu-2)}$$

Hence with $f_2(x)) = x^\nu$

$$\sum_{j=2}^{N-1} \|\chi_j(f_1 - p_1)\|^2_{\tilde{H}^s([x_{j-1},x_{j+1}])} \|p_2\|^2_{L^2([0,x_{j+1}])} \leqslant \sum_{j=2}^{N-1} C\, h_j^{5-2s} x_j^{2\lambda-2\nu-4} x_j^{2\nu+1},$$

which can be bounded like (7.48) above.
Finally, we must bound g_1 in (7.47) on A_1. This is done by a similarity argument: $A_1 = [0, x_2]^2$ is mapped with $\tilde{x} = x_2^{-1} x$, $\tilde{y} = x_2^{-1} y$ onto $[0, 1]^2$ and $|g_1(x, y)| = x_2^\lambda |\tilde{g}(x, y)|$ with $\tilde{g}(x, y) = \tilde{\chi}(x, y)\left(r^\lambda w(\theta) - \tilde{p}\right)$ with functions $\tilde{\chi}$, $\tilde{p}$ independent of h. Interpolation gives for $0 \leqslant s \leqslant 1$

$$\|g_1(x, y)\|_{H^{(s,0)}[0,x_2]^2} \leqslant C\, x_2^{1-s+\lambda} \leqslant C'\, h^{\beta(1-s+\lambda)}.$$

This yields for $s = \frac{1}{2} + \varepsilon$, $\beta > 3/(2\lambda + 1)$ and restriction of $\tilde{A}$ to A for $\tilde{\varepsilon} > 0$ (suff. small)

$$\|f - p\|_{H^{(1/2+\varepsilon,0)}(A)} \leqslant C\, h^{3/2-\tilde{\varepsilon}}$$

Next, we estimate $f - p$ in $H^{(0,s)}(\tilde{A})$. We subdivide $\tilde{A}$ into disjoint rectangles $\tilde{A}_j = [x_{j-1}, x_j] \times [0, x_{j_0}]$, $j_0 := \min\{j + 1, N\}$, $j = 1, \ldots, N$; see Fig. 7.10.

In x-direction only L^2-regularity is used (Fig. 7.10). Hence

$$\|f - p\|^2_{H^{(0,s)}(\tilde{A})} \leqslant \sum_{j=1}^{N} \|f - p\|^2_{H^{(0,s)}\left(\tilde{A}_j\right)}$$

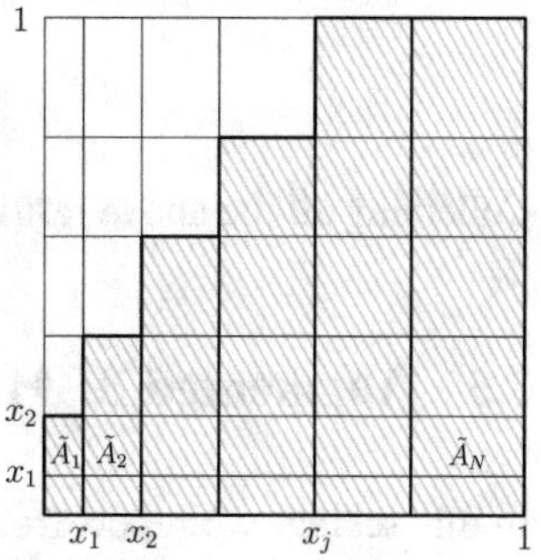

Fig. 7.10 The rectangles $\tilde{A}_j$

On the rectangles $\tilde{A}_j (j \geqslant 2)$ we have $f(x,y) = f_1(x)f_2(y), p(x,y) = p_1(x)p_2(y)$. Hence with

$$I_j = [x_{j-1}, x_j], I_j^* = [0, x_{j_0}],\ j_0 = \min\{j+1, N\},\ 0 \leqslant s \leqslant 1 :$$

$$\|f - p\|_{H^{(0,s)}\left(\tilde{A}_j\right)} \leqslant \|p_1\|_{L^2(I_j)} \|f_2 - p_2\|_{H^s(I_j^*)} + \|f_1 - p_1\|_{L^2(I_j)} \|f_2\|_{H^s(I_j^*)}$$

Hence

$$\sum_{j=2}^{N} \|f - p\|^2_{H^{(0,s)}\left(\tilde{A}_j\right)} \leqslant c\, h^{4-2s} \sum_{j=2}^{N} h_j x_j^{2\lambda - 3 + (4-2s)\gamma}$$

which we already estimated in (7.48). For details see [423], p. 83.
The remaining term on $\tilde{A}_1$ can be estimated by the above similarity arguments, leading to

$$\|f - p\|_{H^{(0,s)}(\tilde{A}_1)} \leqslant C\, h^{3/2 - \tilde{\varepsilon}}$$

Altogether we have $\|f - p\|_{H^{(0,1/2+\varepsilon)}(A)} \leqslant c\, h^{3/2-\tilde{\varepsilon}}$ and collecting the estimates we have with Lemma 7.6 on $Q = A \bigcup B$: $\|f - p\|_{H^{1/2+\varepsilon}(Q)} \leqslant c\, h^{3/2-\tilde{\varepsilon}}$.
Approximation of regular edge functions: $\tilde{f}(x,y) = \chi(\theta) b(x) y^\nu, b(x) \in H_0^3(I))$ with $\chi \equiv 1$ near $\theta = 0$. Due to $b(x) \in H_0^3(I)$ there holds $(1 - \chi(\theta)) b(x) y^\nu \in H^3(Q)$ and hence it can be approximated like the regular part. We set $f(x,y) = b(x) y^\nu =: f_1(x) f_2(y)$ and $p(x,y) = p_1(x) p_2(y)$ with pw. linear interpolants p_j of f_j. Hence for $0 \leqslant s < \frac{1}{2} + \nu$

$$\begin{aligned}\|f - p\|_{H^s(Q)} \leqslant & \|p_1\|_{L^2(I)} \|f_2 - p_2\|_{H^s(I)} + \|p_1\|_{H^s(I)} \|f_2 - p_2\|_{L^2(I)} \\ & + \|f_1 - p_1\|_{L^2(I)} \|f_2\|_{H^s(I)} + \|f_1 - p_1\|_{H^s(I)} \|f_2\|_{L^2(I)}\end{aligned}$$

yields (by application of 1D approximation results cf. Theorem 7.14, Proposition 7.3) for $\beta > 3/(2\nu)$

$$\|f - p\|_{H^{1/2}(Q)} \leqslant c\, h^{3/2}.$$

Collecting all the above results completes the proof of Theorem 7.18. □

7.5 Augmented BEM for Screen/Crack Problems

In this section we report from [398] and treat the screen problems of Sect. 4.4 by the augmented BEM. We solve the boundary integral equations (4.61) and (4.62) in finite dimensional subspaces S_h^1, S_h^2 of the Sobolev spaces $\tilde{H}^{-1/2}(S)$ and $\tilde{H}^{1/2}(S)$, respectively. For conformity we assume the families of finite element subspaces S_h^1, S_h^2 satisfy for integers t, k

$$\begin{aligned} S_h^1 &= S_h^{t-1,k-1}(S) \subset H^{k-1}(S) \subset \tilde{H}^{-1/2}(S) \\ S_h^2 &= S_h^{t,k}(S) \subset H^k(S) \cap \tilde{H}^1(S) \subset \tilde{H}^{1/2}(S), \quad t > k \geq 1 \end{aligned} \tag{7.49}$$

The surface S is given by local representations such that regular partitions in the parameter domains are mapped onto a corresponding partition of S. On the partitions in the parameter domains we use a regular (t, k)-system , called $S_h^{t,k}$ of finite elements. The parameters mean: h is the mesh size of the partition, e.g. longest side of a triangle of a triangulation; $t - 1$ is the degree of piecewise polynomials; k describes the conformity $S_h^{t,k} \subset H^k(S)$. Now the Galerkin procedures for the screen problems read:

For the Dirichlet problem find $\psi_h \in S_h^{t-1,k-1}(S)$ such that

$$\langle V_S \psi_h, \phi \rangle_{L^2(S)} = \langle 2g, \phi \rangle_{L^2(S)} \tag{7.50}$$

for all $\phi \in S_h^{t-1,k-1}(S)$ with t, k as in (7.49). For the Neumann problem find $v_h \in S_h^{t,k}(S)$ such that

$$\langle W_S v_h, w_h \rangle_{L^2(S)} = -\langle 2h, w_h \rangle_{L^2(S)} \tag{7.51}$$

for all $w_h \in S_h^{t,k}(S)$ with t, k as in (7.49). The solvability of the above Galerkin schemes and their convergence are based on the Gårding inequalities for V_S and W_S and the uniqueness of the integral equations. Application of the general results on the Galerkin procedure for strongly elliptic pseudodifferential operators yields the following Theorem.

Theorem 7.21 *There exists a $h_0 > 0$ such that* (7.50) *and* (7.51) *are uniquely solvable for any $h, 0 < h \leq h_0$ and*

$$\|\psi - \psi_h\|_{-1/2,S} \leq c \inf_{\chi \in S_h^{t-1,k-1}(S)} \|\psi - \chi\| \tag{7.52}$$

$$\|v - v_h\|_{1/2,S} \leq c \inf_{w \in S_h^{t,k}(S)} \|v - w\| \tag{7.53}$$

with c independent of h. Here $\|\cdot\|_{r,S}$ denotes the norm in $\tilde{H}^r(S)$

Due to Theorem 4.7 in Chap. 4 the exact solutions of $V_S \psi = 2g$ and of $W_S v = -2h$ behave like $\rho^{-1/2}$ and $\rho^{1/2}$,respectively, where ρ is the distance to the boundary γ of the screen S. Since $\rho^{1/2} \in H^{1-\epsilon}$ for some $\epsilon > 0$ the above estimates give at most convergence of order $1/2 - \epsilon$. As in the two-dimensional case [410, 411] we can improve the asymptotic rate of convergence by using so-called singular elements in the Galerkin procedures. This gives the augmented finite element spaces $Z_h^{1/2}(S)$, $Z_h^{3/2}(S)$ on S:

$$Z_h^{1/2}(S) := \{\tilde{\psi} = \tilde{\psi}_r + \tilde{\beta}\rho^{-1/2}\chi \,:\, \tilde{\beta} \in S_h^{t',l}(\gamma),\ \tilde{\psi}_r \in \overset{\circ}{S}{}_h^{t-1,k-1}(S)\} \tag{7.54}$$

$$Z_h^{3/2}(S) := \{\tilde{v} = \tilde{v}_r + \tilde{\alpha}\rho^{1/2}\chi \,:\, \tilde{\alpha} \in S_h^{t',l}(\gamma),\ \tilde{v}_r \in \overset{\circ}{S}{}_h^{t,k}(S)\} \tag{7.55}$$

where $\tilde{\alpha}, \tilde{\beta} \in S_h^{t',l}(\gamma) \subset H^1(\gamma)$; $\tilde{\psi}_r \in \overset{\circ}{S}{}_h^{t-1,k-1}(S) \subset H^{k-1}(S) \cap \tilde{H}^1(S)$;

$$\tilde{v}_r \in \overset{\circ}{S}{}_h^{t,k}(S) \subset H^k(S) \cap \tilde{H}^2(S)$$

with $t' > l \geq 1$; $t', l \in \mathbb{N}$ and t, k as in (7.49).
The improved Galerkin schemes now read as:
For the Dirichlet screen problem find $\psi_h = \beta_h\rho^{-1/2}\chi + \psi_h^r \in Z_h^{1/2}(S)$ such that

$$\langle V_S\psi_h, \tilde{\phi}\rangle_{L^2(S)} = \langle 2g, \tilde{\phi}\rangle_{L^2(S)} \tag{7.56}$$

for all $\tilde{\phi} = \tilde{\beta}\rho^{-1/2}\chi + \tilde{\phi}^r \in Z_h^{1/2}(S)$.
For the Neumann screen problem find $v_h = \alpha_h\rho^{1/2}\chi + v_h^r \in Z_h^{3/2}(S)$ such that

$$\langle W_S v_h, \tilde{w}\rangle_{L^2(S)} = -\langle 2h, \tilde{w}\rangle_{L^2(S)} \tag{7.57}$$

for all $\tilde{w} = \tilde{\alpha}\rho^{1/2}\chi + \tilde{v}^r \in Z_h^{3/2}(S)$.
The above Galerkin equations with test functions $\tilde{w} \in Z_h^{3/2}(S), \tilde{\phi} \in Z_h^{1/2}(S)$ define quadratic system of linear equations for the unknown coefficients of $\alpha_h, \beta_h \in S_h^{t',l}(\gamma)$, $v_h^r \in \overset{\circ}{S}{}_h^{t,k}(S)$ and $\psi_h^r \in \overset{\circ}{S}{}_h^{t-1,k-1}(S)$. In [398] the following result is proven:

Theorem 7.22 *The Galerkin equations* (7.56) *and* (7.57) *are uniquely solvable for sufficiently small h and there holds*

$$\|\psi - \psi_h\|_{-1/2,S} \leq c \inf_{\eta \in Z_h^{1/2}(S)} \|\psi - \eta\|_{-1/2,S} \leq ch^{1+\sigma}\|\psi\|_{Z^{1/2+\sigma}(S)} \tag{7.58}$$

$$\|v - v_h\|_{1/2,S} \leq c \inf_{w \in Z_h^{3/2}(S)} \|v - w\|_{1/2,S} \leq ch^{1+\sigma}\|v\|_{Z^{3/2+\sigma}(S)}. \tag{7.59}$$

with c independent of h . The arising norms are defined as follows ($\|\cdot\|_{q,\gamma}$ denotes the Sobolev norm in $H^q(\gamma)$):

$$\|v\|_{Z^q(S)} := \begin{cases} \|\alpha\sqrt{\rho}\chi(\rho) + v_r\|_{q,S} & , 1/2 \le q < 1-\epsilon\ ,\ \epsilon > 0 \text{ arbitrary} \\ \|\alpha\|_{q,\gamma} + \|v_r\|_{q,S} & , 1 \le q \le 3/2 + \sigma. \end{cases}$$

$$\|\psi\|_{Z^p(S)} := \begin{cases} \|-\frac{\alpha}{2}\rho^{-1/2} + \psi_r\|_{p,S} & , -1/2 \le p < -\epsilon\ , \\ \|\alpha\|_{p,\gamma} + \|\psi_r\|_{p,S} & , 0 \le p \le 1/2 + \sigma. \end{cases}$$

Similar results can be shown for crack problems in linear elasticity, see [396].

Chapter 8
Exponential Convergence of hp-BEM

The first section of this chapter collects results from [240] which gives a further contribution to the analysis of the *hp*-version of the boundary element method (BEM) by presenting a more general result for Dirichlet and Neumann problems than [21] allowing the use of a general geometric mesh refinement on the polygonal boundary Γ. Here as in [240] we prove the exponential convergence of the *hp*-version of the boundary element method by exploiting only features of the solutions of the boundary integral equations. The key result in this approach is an asymptotic expansion of the solution of the integral equations in singularity functions reflecting the singular behaviour of the solutions near corners of Γ. With such expansions we show that the solutions of the integral equations belong to countably normed spaces. Therefore these solutions can be approximated exponentially fast in the energy norm via the *hp*- Galerkin solutions of those integral equations. This result is not restricted to integral equations which stem from boundary value problems for the Laplacian but applies to Helmholtz problems as well. Further applications are 2D crack problems in linear elasticity. For numerical experiments with *hp*-version (BEM) see [165, 340].

In Sect. 8.2 we consider the *hp*-version of the boundary element method (BEM) for Dirichlet and Neumann screen problems of the Laplacian in $\mathbb{R}^3\setminus\overline{\Gamma}$, where Γ is a planar surface piece with polygonal boundary (for details see also the survey paper [400]).

For the pure Dirichlet and the pure Neumann problems of the Laplacian the exponential convergence of the corresponding hp-version of the Galerkin scheme was already shown in Babuška, Guo, Stephan [21]. Here in Sect. 8.3 we extend their analysis to the mixed bvp of the Laplace equation. A short version of this section is the conference paper [211]. By a further approach J. Elschner [154] has shown exponential convergence for the Galerkin solution for Mellin convolution equations (of second kind) on the interval (0, 1).

© Springer International Publishing AG, part of Springer Nature 2018

J. Gwinner, E. P. Stephan, *Advanced Boundary Element Methods*,
Springer Series in Computational Mathematics 52,
https://doi.org/10.1007/978-3-319-92001-6_8

8.1 The hp-Version of BEM on Polygons

We consider boundary integral equation methods for solving Dirichlet and Neumann boundary value problems for the Laplacian in a polygonal domain with boundary Γ. Let us assume that Γ has conformal radius less than one; this can always be achieved by an appropriate scaling. Then the problems under consideration are the following ones:

Dirichlet Problem For given $f \in H^{1/2}(\Gamma)$ find $u \in H^1(\Omega)$ such that

$$\Delta u = 0 \text{ in } \Omega, \ u = f \text{ on } \Gamma \tag{8.1}$$

Neumann Problem For given $g \in H^{-1/2}(\Gamma)$ with $\int_\Gamma g ds = 0$ find $u \in H^1(\Omega)$ such that

$$\Delta u = 0 \text{ in } \Omega, \ \frac{\partial u}{\partial n} = g \text{ on } \Gamma \tag{8.2}$$

Here $\frac{\partial u}{\partial n}$ denotes the normal derivative of u with respect to the outer normal n. It is well-known [114] that problems (8.1) and (8.2) can be converted into boundary integral equations of the first kind on Γ. With $v = u\big|_\Gamma, \psi = \frac{\partial u}{\partial n}\big|_\Gamma$ we have for (8.1) and (8.2), respectively,

$$V\psi = (1 + K)f \quad \text{on } \Gamma \tag{8.3}$$

$$Wv = (1 - K')g \quad \text{on } \Gamma \tag{8.4}$$

with the integral operators (for $w \in H^{-1/2}(\Gamma), z \in H^{1/2}(\Gamma)$)

$$Vw(x) := -\frac{1}{\pi}\int_\Gamma \ln|x-y| w(y) ds_y, \quad Kw(x) := -\frac{1}{\pi}\int_\Gamma \frac{\partial}{\partial n_y}(\ln|x-y|) w(y) ds_y$$

$$K'w(x) := -\frac{1}{\pi}\int_\Gamma \frac{\partial}{\partial n_x}(\ln|x-y|) w(y) ds_y, \ Wz(x) := \frac{1}{\pi}\frac{\partial}{\partial n_x}\int_\Gamma \frac{\partial}{\partial n_y}(\ln|x-y|) z(y) ds_y.$$

It is also well-known that there exist unique solutions $\psi \in H^{-1/2}(\Gamma)$ of (8.3) and $v \in H_0^{1/2}(\Gamma) = \left\{w \in H^{1/2}(\Gamma) : \int_\Gamma w ds = 0\right\}$ of (8.4). The boundary integral operators V and W are strongly elliptic pseudodifferential operators satisfying a Gårding inequality on $H^{-1/2}(\Gamma)$ and $H_0^{1/2}(\Gamma)$, respectively. Therefore due to [408] any conforming Galerkin scheme for (8.3) and (8.4) converges quasioptimally in the energy norm. Let X_N, Y_N denote subspaces of dimension N of $X := H^{-1/2}(\Gamma)$ and $Y := H_0^{1/2}(\Gamma)$ then the Galerkin schemes read:

Find $\psi_N \in X_N$ satisfying

$$\langle V\psi_N, \phi\rangle = \langle (1+K)f, \phi\rangle \quad \forall \phi \in X_N, \tag{8.5}$$

find $v_N \in Y_N$ satisfying

$$\langle W v_N, w\rangle = \langle (I-K')g, w\rangle \quad \forall w \in Y_N. \tag{8.6}$$

Then for the Galerkin solutions ψ_N, v_N and the true solutions ψ and v there holds [408], by Theorem 6.1

$$\|\psi - \psi_N\|_{H^{-1/2}(\Gamma)} \leq c_1 \|\psi - \phi\|_{H^{-1/2}(\Gamma)} \quad \forall \phi \in X_N \tag{8.7}$$

and

$$\|v - v_N\|_{H^{1/2}(\Gamma)} \leq c_2 \|v - w\|_{H^{1/2}(\Gamma)} \quad \forall w \in Y_N \tag{8.8}$$

where the constants $c_1, c_2 > 0$ are independent of N.

It is shown in [405], Chap. 7 that the h-version and the p-version of (8.5) and (8.6) on a quasiuniform mesh have only algebraic rate of convergence whereas it is shown in [21] that the h-p version on a geometric mesh converges exponentially fast. However in [21] the boundary element mesh is the trace on Γ of a geometric mesh in Ω and the boundary elements on Γ must be the traces or normal derivatives on Γ of finite element functions in Ω. This means a restriction on the choice of boundary elements and on the construction of the geometric mesh refinement on the boundary Γ. Here we give a new proof of the exponential convergence of the h- p version of the boundary element method which does not require these restrictions. The analysis given here can be extended, e.g. to curved polygons Γ and to the Helmholtz operator in (8.1), (8.2) (instead of the Laplacian) as shown in [240].

To describe the hp- version we introduce the geometric mesh Γ_σ^n on $\Gamma = \bigcup_{j=1}^J \Gamma^j$, Γ^j being open arcs, with endpoints t_{j-1}, t_j. First, we bisect each side Γ^j with length d_j into two pieces Γ_1^j (containing the vertex t_{j-1}) and Γ_2^j (containing the vertex t_j). Then for the distance of the subarc $\Gamma_1^{j,k}$ to the vertex t_{j-1} there holds $\operatorname{dist}\left(t_{j-1}, \Gamma_1^{j,k}\right) = \frac{d_j}{2}\sigma^{n-k+1}$ for $k \leq n+1$ where $\sigma \in (0,1)$ and n is an integer. On this geometric mesh Γ_σ^n the boundary element space $S^{P,\ell}(\Gamma_\sigma^n)(\ell = 1 \text{ or } 2)$ is given by

$$\begin{aligned} S^{P,\ell}(\Gamma_\sigma^n) := \big\{ &\psi \in H^{\ell-1}(\Gamma^j), 1 \leq j \leq J,\ \psi \in C^0(\Gamma) \\ &\text{if } \ \ell = 2;\ |\psi|_{\Gamma_k^{j,m}} \in P_p(\Gamma_k^{j,m}), k = 1, 2, m = 1, \dots, n+1 \big\} \end{aligned} \tag{8.9}$$

where $P_p(\Gamma_k^{j,m})$ denotes the space of polynomials of degree $\leq p$ on the subarc $\Gamma_k^{j,m}$.

With the choice $X_N := S^{P,1}(\Gamma_\sigma^n)$ in the Galerkin scheme (8.5) we have the following results on exponential convergence.

Theorem 8.1 *Provided the given data f in (8.1) is piecewise analytic, then there holds the estimate*

$$\|\psi - \psi_N\|_{H^{-1/2}(\Gamma)} \leq Ce^{-b\sqrt{N}} \tag{8.10}$$

for the error between the Galerkin $\psi_N \in S^{P,1}(\Gamma_\sigma^n)$ of (8.5) and the solution ψ of (8.3) if the degrees P are suitably chosen. Here the positive constants C and b depend on the mesh parameter σ but not on the dimension N of $S^{P,1}(\Gamma_\sigma^n)$.

With the choice $Y_N := S^{P,2}(\Gamma_\sigma^n)$ in the Galerkin scheme (8.6) we have

Theorem 8.2 *Provided the given data g in (8.2) is piecewise analytic, then there holds the estimate*

$$\|v - v_N\|_{H^{1/2}(\Gamma)} \leq Ce^{-b\sqrt{N}} \tag{8.11}$$

for the error between the Galerkin solution $v_N \in S^{P,2}(\Gamma_\sigma^n)$ of (8.6) and the solution v of (8.4) if the degrees P are suitably chosen here; the positive constants C, b depend on σ but are independent of $N = dim S^{P,2}(\Gamma_\sigma^n)$.

Remark 8.1 The functions in X_N need not to be continuous on Γ since $X_N \subset H^{-1/2}(\Gamma)$ whereas the constraint $Y_N \subset H^{1/2}(\Gamma)$ requires continuity for the functions in Y_N.

The proofs of Theorem 8.1 and 8.2 are based on regularity results for the solutions of the integral equations and on approximation results for splines on the geometric mesh.

From [128] we know that for $f \in H^t(\Gamma^j)$, $1 \leq j \leq J$ the solution ψ of (8.3) has the form (see Chap. 9)

$$\psi(x) = \sum_{j=1}^{J}\sum_{k=1}^{n} c_k^j x^{\alpha_{kj}-1}\chi_j(x) + \psi_0(x), \quad c_k^j \in \mathbb{R}, \alpha_{kj} = k\frac{\pi}{\omega_j}, \quad n \leq \frac{\omega_j}{\pi}(t - 3/2) \tag{8.12}$$

where $\psi_0\big|_{\Gamma^j} \in H^{t-1}(\Gamma^j)$ whereas the solution v of (8.4) for $g\big|_{\Gamma^j} \in H^{\tau-1}(\Gamma^j)$, $1 \leq j \leq J$, has the form

$$v(x) = \sum_{j=1}^{J}\sum_{k=1}^{n} d_k^j x^{\alpha_{kj}}\chi_j(x) + v_0(x) \tag{8.13}$$

with $v_0\big|_{\Gamma^j} \in H^\tau(\Gamma^j)$, $d_k^j \in \mathbf{R}$. Here χ_j is a C^∞– cut off-function concentrated at the jth corner, with opening angle ω_j. When $\frac{\pi}{\omega_j}$ is an integer then the singularity functions in (8.12), (8.13) have the forms $x^{\alpha_{kj}-1}\ln|x|$ and $x^{\alpha_{kj}}\ln|x|$, respectively.

Note if the boundary Γ is curvilinear there appear also terms of the form $x^{k\frac{\pi}{\omega_j}+m}$, m integer, in the above expansions (see [126]).

For the local singularity terms we have the following result using the countably normed spaces B^ℓ_β.

Lemma 8.1 *Let* $R > 0$ *and* $\varphi_\mu(x) := x^\mu,\ \varphi_{\mu,k}(x) := x^\mu \log^k x$ for $x \in (0, R)$, k *pos. integer. Then*

(i) $\varphi_\mu \in B^\ell_\beta(0, R)$ for $\mu > \ell - 1/2 - s$,

(ii) $\varphi_{\mu,k} \in B^\ell_\beta(0, R)$ for $\mu > \ell - 1/2 - \beta$, *where* $u \in B^\ell_\beta(0, R)$ *if and only if* $u \in H^{m,\ell}_\beta(0, R)$ $\forall m \ge \ell$ *and* $\exists C > 0, d > 1$ *such that*

$$\|x^{\beta+k-\ell} u^{(k)}\|_{L^2(0,R)} < C d^{k-\ell}(k-\ell)!, \quad k = \ell, \ell+1, \ldots$$

and $u \in H^{m,\ell}_\beta(0, R)$ *if and only if*

$$u \in H^{\ell-1}(0, R) \text{ and} |u|^{m,\ell}_{H_\beta(0,R)} := \|x^{\beta+k-\ell} u^{(k)}\|_{L^2(0,R)} < \infty, \ell \le k \le m.$$

The proof of the Lemma 8.1 follows immediately by inspection.

Lemma 8.2 *Let* $|c_n| < C < \infty$ *then*

$$\varphi(x) := \sum_{n=1}^{\infty} c_n x^{n\frac{\pi}{\omega}} \in B^\ell_\beta(0, R) \text{ for } \beta > \ell - \pi/\omega - 1/2$$

Proof We have to show $\|\varphi^{(k)} x^{\beta+k-\ell}\|_{L^2(0,R)} \le C d^{k-\ell}(k-\ell)!\quad (k \ge \ell)$.

With $\varphi_N := \sum_{n=1}^N c_n x^{\alpha_n}$, $\alpha_n = n\frac{\pi}{\omega}$ and $(\alpha)_k := \alpha(\alpha-1)\cdots(\alpha-k+1)$ we have

$$\|x^{\beta+k-\ell} \varphi_N^{(k)}\|^2_{L^2(0,R)} = \int_0^R |\varphi_N^{(k)}(x)|^2 x^{2(\beta+k-\ell)} dx$$

$$\le C \sum_{n=1}^{N} c_n^2 (\alpha_n)_k^2 \frac{R^{2(\alpha_n-\ell+\beta)+1}}{2(\alpha_n+\beta-\ell)+1}$$

$$\le C R^{2(\beta-\ell)+1} \left[\sum_{\substack{n\ge1\\ \alpha_n<k}} \Gamma(\alpha_n+1)^2 \Gamma(k-\alpha_n)^2 \frac{R^{2\alpha_n}}{2(\alpha_n+\beta-\ell)+1} \right.$$

$$\left. + \sum_{\substack{n\le N\\ \alpha_n\ge k}} \alpha_n^{2k} \frac{R^{2\alpha_n}}{2(\alpha_n+\beta-\ell)+1} \right]$$

$$\le C R^{2(\beta-\ell)+1} \left[\sum_{\substack{n\ge1\\ \alpha_n<k}} k!^2 \frac{R^{2\alpha_n}}{2(\alpha_n+\beta-\ell)+1} + \frac{C}{1-\alpha_\infty^2 R^{2\pi/\omega}} \right] \le C k!^2$$

Here we have used

(i) $|(\alpha_n)_k| \le (\alpha_n)^k$ for $\alpha_n \ge k$, $|(\alpha_n)_k| \le \Gamma(\alpha_n+1)\Gamma(k-\alpha_n)$ for $\alpha_n < k$,
(ii) $\Gamma(\alpha_n+1)\Gamma(k-\alpha_n) \le ck!$ for $\alpha_n < k$,
(iii) $\sum_{\substack{n\in\mathbb{N}\\ \alpha_n\ge k}} \alpha_n^{2k}\frac{R^{2\alpha_n}}{2(\alpha_n+\beta-\ell)+1} \le \frac{C}{1-\alpha_\infty^2 R^{2\pi/\omega}}$ with $\alpha_\infty := \lim_{n\to\infty}\left(\frac{\alpha_{n+1}}{\alpha_n}\right)^{\alpha_n} = e^{\pi/\omega}$.

For the proof of (iii) we observe that with $a_n := \alpha_n^{2k}\frac{R^{2\alpha_n}}{2(\alpha_n+\beta-\ell)+1}$ there holds $\lim_{n\to\infty}\sup\frac{a_{n+1}}{a_n} \le \alpha_\infty^2 R^{2\pi/\omega} < 1$ for suitably chosen R.

Hence with some $C > 0$

$$\sum_{\substack{u\in\mathbb{N}\\ \alpha_n\ge k}} \alpha_n^{2k}\frac{R^{2\alpha_n}}{2(\alpha_n+\beta-\ell)+1} \le C\sum_{n=0}^{\infty}(\alpha_\infty^2 R^{2\pi/\omega})^n = \frac{C}{1-\alpha_\infty^2 R^{2\pi/\omega}}. \qquad \square$$

Remark 8.2 For a more general version of Lemma 8.2 see [240]. Inspection of the proof of Lemma 8.2 shows $\sum_{n=1}^{\infty} c_n x^{n\pi/\omega}\ln|x| \in B_\beta^\ell(0,R)$ for $\beta > \ell-\pi/\omega-1/2$.

From [128] we know that the coefficients c_k^j and d_k^j in (8.12), (8.13) are continuous functionals on the given data f and g. Therefore these coefficients are bounded satisfying the assumption of Lemma 8.2. Hence if $f \in H^t(\Gamma^j)$, $1 \le j \le J$, for any t then the solution ψ of (8.3) has the form

$$\psi = \sum_{j=1}^{J}\sum_{k=1}^{\infty} c_k^j x^{\alpha_{kj}-1}\ln|x|\chi_j(x) + \psi_0,\ \psi_0\Big|_{\Gamma^j} \in H_{\Gamma^j}^{t-1}$$

with the notation in (8.12) and therefore with Lemma 8.2 we have $\psi \in B_\beta^1(\Gamma)$ for $\beta > \frac{1}{2} - \frac{\pi}{\omega}$. Analogously, if $g \in H^{\tau-1}(\Gamma^j)$, $1 \le j \le J$, for any τ then the solution v of (8.4) has the form

$$v = \sum_{j=1}^{J}\sum_{k=1}^{\infty} d_k^j x^{\alpha_{kj}}\chi_j(x) + v_0,\ v_0\Big|_{\Gamma^j} \in H^\tau(\Gamma^j)$$

with the notation in (8.13) and therefore with Lemma 8.2 we have $v \in B_\beta^2(\Gamma)$ with $\beta > \frac{3}{2} - \frac{\pi}{\omega}$. Next we need some properties of Legendre polynomials.

Lemma 8.3

(i) Let $I = (-1,1)$, $u(x) = \sum_{j=0}^{\infty} c_j\ell_j(x)$, ℓ_j *Legendre polynomial of degree* j. *Then*

$$\int_I |u^{(k)}(x)|^2(1-x^2)^k dx = \sum_{j\ge k} c_j^2\frac{2}{2j+1}\frac{(j+k)!}{(j-k)!}$$

(ii) *Let* $I = (-1, 1)$ *and* $u \in H^{k+1}(I)(k \in \mathcal{N}_0)$. *Then there exists a* $\varphi \in P_k(I)$ *and a constant* $c > 0$ *such that*

$$\left\| (u-\varphi)^{(m)} \right\|^2_{L^2(I)} \leq C \frac{(k-s)!}{(k+s+2-2m)!} \left\| u^{(s+1)} \right\|^2_{L^2(I)}$$

$(m = 0, 1, 0 \leq s \leq k, s \in \mathcal{N}_0, k > 0$ *or* $m = s = k = 0)$ *where* $\varphi(-1) = u(-1), \varphi(1) = u(1)$ *for* $k > 0$.

(iii) *Let* $J = (a, b), h = b - a$ *and* $u \in H^{k+1}(J)(k \in \mathcal{N}_0)$. *Then there exist a* $\varphi \in P_k(J)$ *and a constant* $C > 0$ *and that*

$$\left\| (u-\varphi)^{(m)} \right\|^2_{L^2(J)} \leq C h^{-2m} \left(\frac{h}{2}\right)^{2(s+1)} \frac{(k-s)!}{(k+s+2-2m)!} \left\| u^{(s+1)} \right\|^2_{L^2(J)}$$

$(m = 0, 1, 0 \leq s \leq k, k > 0$ or $m = s = k = 0)$ *and* $\varphi(a) = u(a), \varphi(b) = u(b)$ $(k > 0)$.

(iv) *Let* $I = (0, R)$ *for* $R > 0$, $J = (a, b)$, $J \subset I$ *and* $\lambda > 0$ *be a fixed number with* $h = b - a \leq \lambda a$. *Then for* $u \in H^{k+1,\ell}_\beta(I)$ *there exists a polynomial* $\varphi \in P_k(J)$ *and a constant* $c > 0$ *such that for* $n = 0(k = 0)$ *and* $n = 0, 1(k > 0)$, *respectively, there holds*

$$\left\| (u-\varphi)^{(n)} \right\|^2_{L^2(J)} \leq C a^{2(\ell-n-\beta)} \frac{\Gamma(k-s+1)}{\Gamma(k+s+3-2n)} \left(\frac{\lambda}{2}\right)^{2s} |u|^2_{H^{s+1,\ell}_\beta(I)} \tag{8.14}$$

$(n < s + 1, \quad 1 \leq \ell \leq s + 1 \leq k + 1, s \in \mathbf{R})$ *with* $\varphi(a) = u(a), \varphi(b) = u(b)(k > 0)$.

Proof Assertion (i) is wellknown (see e.g. [17]). (ii) follows from (i) by expanding u and u' in Legendre series (see [210]). (iii) follows from (ii) via affine transformation (see [210]). Assertion (iv) can be seen as follows.

By definition $|u|^2_{H^{s+1,\ell}_\beta(I)} \geq a^{2(\beta+s+1-\ell)} \left\| u^{(s+1)} \right\|^2_{L^2(J)}$.

By (iii) there exists $\varphi \in P_k(J)$ with

$$\left\| (u-\varphi)^{(n)} \right\|^2_{L^2(J)} \leq C h^{-2n} \frac{(k-s)!}{(k+s+2-2n)!} \left(\frac{h}{2}\right)^{2(s+1)} a^{-2(\beta+s+1-\ell)} |u|^2_{H^{s+1,\ell}_\beta(I)}$$

yielding (8.14). □

Next we consider a geometric mesh I^n_σ on $I = (0, 1)$ with n subintervals $I_j = [x_{j-1}, x_j]$,
$x_0 = 0, x_j = \sigma^{n-j}, h_j = x_j - x_{j-1}, 1 \leq j \leq n$.

For a degree vector $p = (p_1, \ldots, p_n)$ of nonnegative integers we set

$$S^{p,\ell}(I^n_\sigma) = \left\{ q \in H^\ell(I) : q|_{I_j} \in P_{p_j}(I_j) \right\} \tag{8.15}$$

Lemma 8.4 *Let* $I = (0,1), u \in B^{\ell}_{\beta}(I), \ell = 1, 2,$ *then there exists a* $\varphi \in S^{p,\ell-1}(I^n_\sigma)$ *with* $0 < \sigma < 1$ $p_1 = \ell - 1$, $p_i = \max\{\ell, [\mu i]\}(i = 2, \ldots, n)$ *such that*

$$\|u - \varphi\|_{H^{\ell-1}(I)} \le Ce^{-b\sqrt{N}} \tag{8.16}$$

where the positive constants C and b depend on σ *but are independent of* $N = \dim S^{p,\ell-1}(I^n_\sigma)$.

Proof First we use Lemma 8.3 (iv) on each subinterval $I_i (i > 1)$: Thus we have a $\varphi_i \in P_{p_i}(I_i)$ with

$$\left\|(u - \varphi_i)^{(n)}\right\|^2_{L^2(I_i)} \le Cx_{i-1}^{2(\ell-n-\beta)} \frac{\Gamma(p_i - s_i + 1)}{\Gamma(p_i + s_i + 3 - 2n)} \left(\frac{\lambda}{2}\right)^{2s_i} |u|^2_{H^{s_i+1,\ell}_\beta(I)}$$

$(n < s_i + 1, \quad 1 \le \ell \le s_i + 1 \le p_i + 1, s_i \in \mathbf{R}) \in H^{t-1}_{\Gamma j}$ since $u \in B^{\ell}_{\beta}(I)$ implies $u \in H^{s_i+1\ell}_{\beta}(I)$ $(s_i + 1 \ge \ell)$.

On the first interval $I_1 (i = 0)$ we have (see [210])

$$\|u - \varphi_1\|^2_{H^{\ell-1}(I_1)} \le Ch_1^{2(1-\beta)} |u|^2_{H^{\ell,\ell}_\beta(I_1)}.$$

Thus there exists $\varphi \in S^{p,\ell-1}(I^n_\sigma)$ with

$$\|u - \varphi\|^2_{H^{\ell-1}(I)} \le C\left[\sigma^{2(1-\beta)n} + \sum_{i=2}^{n} x_{i-1}^{2(1-\beta)} \frac{\Gamma(p_i - s_i + 1)}{\Gamma(p_i + s_i + 5 - 2\ell)} \left(\frac{\lambda}{2}\right)^{2s_i} |u|^2_{H^{s_i+1,\ell}_\beta(I)}\right]$$

With the estimate

$$|u|^{s+1,\ell}_{H_{\beta(I)}} \le C(\ell) d^s \Gamma(s+1) \; (s \in \mathbf{R}_+)$$

and $x_i - x_{i-1} \le \lambda x_{i-1} = \frac{1-\sigma}{\sigma}\sigma^{n-i+1} (2 \le i \le n)$ we obtain

$$\begin{aligned}
\|u - \varphi\|^2_{H^{\ell-1}(I)} &\le C\Big[\sigma^{2(1-\beta)n} \\
&\quad + \sum_{i=2}^{n} \sigma^{2(n-i+1)(1-\beta)} \frac{\Gamma(p_i - s_i + 1)}{\Gamma(p_i + s_i + 5 - 2\ell)} \Gamma(s_i + 1)^2 \left(\frac{\rho d}{2}\right)^{2s_i}\Big] \\
&\le C\left[\sigma^{2(1-\beta)n} + \sum_{i=2}^{n} \sigma^{2(n-i+1)(1-\beta)} p_i (F(\rho d, \alpha_i))^{p_i}\right]
\end{aligned}$$

where

$$F(d,\alpha) := \left(\frac{\alpha d}{2}\right)^{2\alpha} \frac{(1-\alpha)^{1-\alpha}}{(1+\alpha)^{1+\alpha}} \text{ and } \alpha_i := \max\left\{\frac{1}{p_i}, \alpha_{\min}\right\}, \alpha_{\min} := \frac{2}{\sqrt{4+\lambda^2 d^2}}.$$

There holds
$\inf_{\alpha\in(0,1)} F(d,\alpha) = F_{\min} = F\left(d, \frac{2}{\sqrt{4+d^2}}\right) < 1$ with $F_{\min} := F(\alpha_{\min})$.

Taking $p_i = \max\{\ell, [\mu i]\}$ ($[x]$ means the smallest integer greater or equal to x) $(i = 2, \dots, n)$ with

$$\mu > \max\left\{1, \frac{2(1-\beta)\log\sigma}{\log F_{\min}}\right\} \tag{8.17}$$

and defining i_0 by $p_{i_0} = \left[\frac{1}{\alpha_{\min}}\right] + 1$, then $p_{i_0} = [\mu i_0] \le \frac{1}{\alpha_{\min}} + 2$ and thus i_0 is bounded.

Hence

$$\|u-\varphi\|^2_{H^{\ell-1}(I)} \le C\Big[\sigma^{2(1-\beta)n} + \sum_{i=2}^{i_0} \sigma^{2(n-i+1)(1-\beta)} p_i F(\rho d, \alpha_i)^{p_i}$$

$$+ \sum_{i=i_0+1}^{n} \sigma^{2(n-i+1)(1-\beta)} p_i (F_{\min})^{p_i}\Big]$$

$$\le C\sigma^{2(1-\beta)n}\left[1 + \sum_{i=2}^{i_0} \sigma^{2(1-i)(1-\beta)} (F_{\min})^{p_i} p_i \max_{1\le i\le i_0} \left(\frac{F(\rho d, \frac{1}{p_i})}{F_{\min}}\right)^{p_i}\right.$$

$$\left. + \sum_{i=i_0+1}^{n} \sigma^{2(1-i)(1-\beta)} p_i (F_{\min})^{p_i}\right].$$

With $p_i = [\mu i]$ and $q := \frac{F^\mu}{\sigma^{2(1-\beta)}} < 1$ due to (8.17) we have $\sum_{i>i_0} i q^i < \infty$ since $(iq^i)^{1/i} \to q < 1$ as $i \to \infty$.

Hence the term in the bracket is bounded yielding with a positive constant c

$$\|u-\varphi\|^2_{H^{\ell-1}(I)} \le c\, e^{2(1-\beta)n}. \tag{8.18}$$

Next we observe for $\ell = 1$: $N = \dim S^{P,0}(K^n_\sigma) = 1 + \sum_{i=2}^n (p_i+1) = 1 + \sum_{i=2}^n ([\mu i]+1) \le c\mu n^2$ and for $\ell = 2$: $N = \dim S^{P,1}(I^n_\sigma) = 2 + \sum_{i=2}^n (p_i + 1) - n + 1 \le c\mu n^2$.

Hence we obtain from (8.18) ($\ell = 1, 2$)

$$\|u-\varphi\|_{H^{\ell-1}(I)} \le C e^{-b\sqrt{N}}$$

with

$$b = \frac{1-\beta}{\sqrt{\mu}} \log \frac{1}{\sigma}. \tag{8.19}$$

□

Corollary 8.1 *Let* $I = (0,1), u \in B_\beta^2(I)$ *for some* $\beta < 1$*, then there exists a* $\varphi \in S^{P,1}(I_\sigma^n)$ *with* $0 < \sigma < 1$, $p_1 = 1$, $p_i = [\mu i]$, $2 \le i \le n$*, such that*

$$\|u - \varphi\|_{H^{1/2}(I)} \le ce^{-b\sqrt{N}}$$

with constants $c, b > 0$ *independent of* $N = \dim S^{p,1}(I_\sigma^n)$.

Proof The assertion follows by interpolation directly from Lemma 8.4. □

The corollary can be generalised from the interval I to the polygon Γ in a straightforward manner.

Now the proofs of Theorem 8.1 and 8.2 are completed as follows:

Proof (of Theorem 8.2) First we observe with Lemma 8.2 that the analyticity of g on Γ^j implies $v \in B_\beta^2(\Gamma)$ for $1 > \beta > 3/2 - \pi/\omega$ where v satisfies (8.4). Hence by Lemma 8.4 there exists for each boundary piece Γ_k^j a $\varphi_k^j \in S^{P_{j,k},1}(\Gamma_k^j)$ with degree $P_{j,k,m}$ on $\Gamma_k^{j,m}$ such that ($\ell = 1$ or 2)

$$\left\| v - \varphi_k^j \right\|_{H^{\ell-1}(\Gamma_k^j)} \le Ce^{-b_{j,k}\sqrt{N_{j,k}}},$$
$$N_{j,k} = \dim S^{P_{j,k},1}(\Gamma_j^k),\ k = 1,2,\ j = 1,\dots,J$$

where φ_k^j coincides with v at the endpoints of Γ_k^j.

Let

$$\widetilde{\varphi}_k^j = \begin{cases} \varphi_k^j & \text{on } \Gamma_k^j \\ 0 & \text{elsewhere} \end{cases} \quad \text{and} \quad v_k^j = \begin{cases} v & \text{on } \Gamma_k^j \\ 0 & \text{elsewhere} \end{cases}.$$

Then for $\ell = 1$ and 2

$$\begin{aligned} \|v - \sum_{j=1}^{J}\sum_{k=1}^{2} \widetilde{\varphi}_k^j\|_{H^{\ell-1}(\Gamma)} &\le \sum_{j=1}^{J}\sum_{k=1}^{2} \left\| v_k^j - \widetilde{\varphi}_k^j \right\|_{H^{\ell-1}(\Gamma)} \\ &= \sum_{j=1}^{J}\sum_{k=1}^{2} \left\| v|_{\Gamma_k^j} - \varphi_k^j \right\|_{H^{l-1}(\Gamma_k^j)} \le Ce^{-b\sqrt{N}} \end{aligned} \tag{8.20}$$

with $b = \min_{\substack{1\le j\le J\\1\le k\le 2}} \{b_{j,k}\}$, $N = \min_{\substack{1\le j\le J\\1\le k\le 2}} \{N_{j,k}\}$. Note the estimate (8.20) holds since $v_k^j - \widetilde{\varphi}_k^j \in C^0(\Gamma)$ and $v_k^j - \widetilde{\varphi}_k^j \equiv 0$ on $\Gamma \setminus \Gamma_k^j$. Hence the assertion of Theorem 8.2 follows from (8.20) by interpolation. □

Proof (of Theorem 8.1) First we observe with Lemma 8.2 that the analyticity of f on Γ^j implies $\psi \in B_\beta^1(\Gamma)$ for $1 > \beta > 3/2 - \pi/\omega$ where ψ satisfies (8.3). Hence by Lemma 8.4 there exists for each boundary piece Γ_k^j a $\varphi_k^j \in S^{P_{j,k},0}(\Gamma_k^j)$ with degree $P_{j,k,m} - 1$ on $\Gamma_k^{j,m}$ such that

$$\left\|\psi - \varphi_k^j\right\|_{L^2(\Gamma_k^j)} \le Ce^{-b_{j,k}\sqrt{N_{j,k}}}, \quad N_{j,k} = \dim S^{P_{j,k},0}(\Gamma_j^k)$$

Hence the assertion of Theorem 8.1 follows. □

8.1.1 Application to Acoustic Scattering

We consider for $\mu, k_1, k_2 \in \mathbb{C}\backslash\{0\}$ and $\mu \neq -1$ the transmission problem

$$(\Delta + k_1^2)u_1 = 0 \text{ in } \Omega_1, \qquad (\Delta + k_1^2)u_2 = 0 \text{ in } \Omega_2 := \mathbf{R}^2\backslash\overline{\Omega}_1$$

$$u_1 = u_2 + v_0, \qquad \mu\frac{\partial u_1}{\partial u} = \frac{\partial u_2}{\partial u} + \psi_0 \text{ on } \Gamma \tag{8.21}$$

subject to the Sommerfeld radiation condition

$$\frac{\partial u_2}{\partial R} - ik_2u_2 = o(R^{-1/2}), \quad u_2 = O(R^{-1/2}) \text{ as } |x| = R \to \infty.$$

In the case of scattering problems, u_1 (u_2) denote the refracted (scattered) field and v_0 and ψ_0 are the boundary trace and the normal derivative of the incident field u_0. In [129] the above transmission problem is reduced on $\Gamma = \partial\Omega_1$ for the Cauchy data $v_1 = u_{1|\Gamma}$, $\psi_1 = \frac{\partial u_1}{\partial u}\big|_\Gamma$:

$$H\begin{pmatrix} v_1 \\ \psi_1 \end{pmatrix} := \frac{1}{2}\begin{pmatrix} -(K_1 + K_2) & V_1 + \mu V_1 \\ W_1 + \frac{1}{\mu}W_2 & K_1' + K_2' \end{pmatrix}\begin{pmatrix} v_1 \\ \psi_1 \end{pmatrix} = \begin{pmatrix} v_0 \\ \frac{1}{\mu}\psi_0 \end{pmatrix} \tag{8.22}$$

where ($j = 1$ or 2)

$$V_j\varphi(z) = -2\int_\Gamma \gamma_j(z_1\zeta)\varphi(\zeta)ds_\zeta, \quad K_j\varphi(z) = -2\int_\Gamma \varphi(\zeta)\frac{\partial}{\partial n_\zeta}\gamma_j(z,\zeta)ds_\zeta, z \in \Omega_j$$

$W_ju_j = -\frac{\partial}{\partial n}K_ju_j\big|_\Gamma$ and K_j' is the adjoint operator of K_j

and

$$\gamma_j(z,\zeta) = -\frac{i}{4}H_0^{(1)}(k_j|z-\zeta|) = \frac{1}{2\pi}\ln|z-\zeta| + O(|z-\zeta|^{-1}) \tag{8.23}$$

is the fundamental solution of the Helmholtz equation $\Delta w = -k_j^2 w$ in Ω_j where $H_0^{(1)}$ is the Hankel function of first order and degree zero.

It is shown in [129] that the operator H from $H^{1/2}(\Gamma) \times H^{-1/2}(\Gamma) \to H^{1/2}(\Gamma) \times H^{-1/2}(\Gamma)$ is bijective if and only if the homogeneous transmission problem (8.21) as well as the adjoint problem – obtained by interchanging Ω_1 and Ω_2 – have only the trivial solution. This is assumed in the following. From the regularity results in [129] follows that for piecewise analytic data v_0, ψ_0 the solution (v_1, ψ_1) of (8.22) has expansions of the form (8.12), (8.13) with $\alpha_{k_j} = k\alpha_j$ and α_j being a zero of the transcendental equation

$$\frac{\sin(\pi - \omega_j)}{\sin \pi\omega_j} = \pm\left(\frac{\mu+1}{\mu-1}\right). \tag{8.24}$$

The boundary element Galerkin scheme for (8.22) reads. Find $(v_N, \psi_N) \in Y_M \times X_N$ such that

$$\left\langle H\begin{pmatrix} v_N \\ \psi_N \end{pmatrix}, \begin{pmatrix} w \\ \phi \end{pmatrix}\right\rangle_\Gamma = \left\langle \begin{pmatrix} v_0 \\ \mu\psi_0 \end{pmatrix}, \begin{pmatrix} w \\ \phi \end{pmatrix}\right\rangle_\Gamma \qquad \forall (w,\phi) \in Y_M \times X_N \tag{8.25}$$

where Y_M, X_N are finite dimensional subspaces of $H^{1/2}(\Gamma)$ and $H^{-1/2}(\Gamma)$ with $\dim X_N = N$ and $\dim Y_M = M$.

Since the operator H satisfies a Gårding's inequality in $H^{1/2}(\Gamma)\times H^{-1/2}(\Gamma)$ this boundary element Galerkin scheme converges quasioptimally in the energy norm, i.e.

$$\begin{aligned} &\|v_N - v_1\|_{H^{1/2}(\Gamma)} + \|\psi_M - \psi_1\|_{H^{-1/2}(\Gamma)} \\ &\le C\left\{\inf_{w\in X_N}\|v_1 - w\|_{H^{1/2}(\Gamma)} + \inf_{\phi\in Y_M}\|\psi_1 - \phi\|_{H^{-1/2}(\Gamma)}\right\} \end{aligned} \tag{8.26}$$

Next we choose $X_N = S^{p-1,1}(\Gamma_\sigma^n)$ and $Y_M = S^{P,2}(\Gamma_\sigma^n)$ as in Section 2 and obtain the exponential convergence of the *hp*- version of the Galerkin scheme (8.25) for the transmission problem (8.21) (see [210, 241]).

Proposition 8.1 *Let v_0 and ψ_0 in* (8.21) *be piecewise analytic, then for the error between the Galerkin solution $v_N \in S^{P,2}(\Gamma_\sigma^n).\psi_N \in S^{P-1,1}(\Gamma_\sigma^n)$ and the exact solution of* (8.22) *there holds*

$$\|v_1 - v_N\|_{H^{1/2}(\Gamma)} + \|\psi_1 - \psi_M\|_{H^{-1/2}(\Gamma)} \le Ce^{-b\sqrt{N}}$$

if the degrees P are suitably chosen. Here N is the number of degrees of freedom of $S^{P,2}(\Gamma_\sigma^n)$*, C and b are constants depending on* σ *but not on N.*

Proof Firstly, we observe that for piecewise analytic data v_0, ψ_0 the solution (v_1, ψ_1) of (8.22) belong to $B^2_\beta(\Gamma) \times B^1_\beta(\Gamma)$ with $1 > \beta > 1/2 - \alpha_{\min}$ where $\alpha_{\min}$ is the smallest zero of (8.24). Therefore application of the above analysis yields the assertion of the proposition. □

Remark 8.3 For the transmission problem (8.21) with $k_1 = k_2 = 0$ the exponential convergence of the *hp*- version of the bem is shown in [210].

Two-dimensional crack problems in linear elasticity can be converted into first kind integral equations (see [256, 432]). for example, let us consider the Neumann crack problem for the domain Ω_Γ exterior to an arc Γ: find $u \in H^1_{\ell oc}(\Omega_p)$ such that $\Delta^* u \equiv \mu \Delta u + (\lambda + \mu)\text{grad div}\, u = 0$ in $\Omega_\Gamma = \mathbf{R}^2 \setminus \overline{\Gamma}$

$$T(u)\Big|_{\Gamma_1} = \psi_1, \quad T(u)\Big|_{\Gamma_2} = \psi_2$$

for given $\psi_i \in H^{\frac{-1}{2}}(\Gamma)$, $i = 1, 2$, where T denotes the traction operator on the sides Γ_1 and Γ_2 of Γ and λ,μ are the given Lamé constants. Under appropriate conditions, e.g. assuming a decaying condition for u at infinity, this problem can be converted into the integral equation

$$W\phi(x) = -T_x \int_\Gamma (T_y(E(x, y)))^T \phi(y) ds_y = f(x) \tag{8.27}$$

for the jump $\phi \equiv [u] = u|_{\Gamma_1} - u|_{\Gamma_2}$ with the fundamental solution of the Navier operator Δ^*

$$E(x, y) = \frac{\lambda + 3\mu}{4\pi\mu(\lambda + z\mu)} \left\{ \ln \frac{1}{|x-y|} \widetilde{I} + \frac{\lambda+\mu}{\lambda+3\mu} \frac{(x-y)(x-y)^T}{|x-y|^2} \right\}$$

Here T denotes the transposed tensor and $\widetilde{I}$ is the identity matrix and f is given via ψ_1 and ψ_2. It is shown in [432] that the solution ϕ of the hypersingular integral equation (8.27) behaves like $x^{\frac{1}{2}}(d_1 + d_2 x + d_3 x^2 + \cdots)$, $d_j \in \mathbf{R}$, near the crack tip, i.e. like v in (8.13) with $\alpha_k = \frac{1}{2} + k$, k, k integer > 0. Therefore $\phi \in B^2_\beta(\Gamma)$ for $\beta > 1$ since in the case of a crack $\omega = 2\pi$. The operator W in (8.27) satisfies a Gårding's inequality in $\widetilde{H}^{\frac{1}{2}}(\Gamma)$ (see [432]) and therefore the corresponding Galerkin scheme converges quasioptimally in $\widetilde{H}^{\frac{1}{2}}(\Gamma)$. Therefore the above analysis applies also to the integral equation (8.27) yielding exponentially fast convergence for the Galerkin solution of the *hp*- version for (8.27).

8.2 The hp-Version of BEM on Surfaces

In this section we report from [235] on the *hp*-version of the Galerkin boundary element method for Dirichlet and Neumann screen problems in $\mathbb{R}^3$ when the screen Γ is a smooth open surface piece with piecewise smooth boundary.

That is, given f or g on Γ find $u \in \mathbb{R}^3 \backslash \overline{\Gamma}$ satisfying

$$\Delta u = 0 \text{ in } \mathbb{R}^3 \backslash \bar{\Gamma}$$

$$u = f \in H^{1/2}(\Gamma) \text{ (Dirichlet)} \qquad \text{or} \qquad \frac{\partial u}{\partial n} = g \in H^{-1/2}(\Gamma) \text{ (Neumann)}$$

and

$$u = \mathcal{O}(|x|^{-1}) \quad \text{as } |x| \to \infty.$$

These exterior boundary value problems are called screen problems and can be formulated equivalently as first kind integral equations with weakly singular and hypersingular kernels, namely (see Sect. 4.3)

$$V\psi(x) := \frac{1}{2\pi} \int_\Gamma \frac{1}{|x-y|} \psi(y)\, ds_y = 2f(x), \ \ x \in \Gamma \text{ (Dirichlet)} \tag{8.28}$$

$$Wv(x) := -\frac{1}{2\pi} \frac{\partial}{\partial n_x} \int_\Gamma \frac{\partial}{\partial n_y} \frac{1}{|x-y|} v(y)\, ds_y = 2g(x), \ \ x \in \Gamma \text{ (Neumann).} \tag{8.29}$$

As we have shown in [398] (see also Section 5.3) these integral equations have unique solutions $\psi \in \tilde{H}^{-1/2}(\Gamma)$, $v \in \tilde{H}^{1/2}(\Gamma) = H_{00}^{1/2}(\Gamma)$.

The Galerkin boundary element schemes for (8.28) and (8.29) read with the L^2-duality on Γ $\langle \cdot, \cdot \rangle$:

Find $\psi_N \in S_{h,p}^0$

$$\langle V\psi_N, \phi_N \rangle = \langle 2f, \phi_N \rangle \quad \forall \phi_N \in S_{h,p}^0 \subset \tilde{H}^{-1/2}(\Gamma) \tag{8.30}$$

and find $v_N \in S_{h,p}^1$

$$\langle Wv_N, w_N \rangle = \langle 2g, w_N \rangle \quad \forall w_N \in S_{h,p}^1 \subset \tilde{H}^{1/2}(\Gamma) \tag{8.31}$$

Since the operators V and W define coercive, continuous bilinear forms we immediately have quasi-optimality of the Galerkin errors:

$$\|\psi - \psi_N\|_{\tilde{H}^{-1/2}(\Gamma)} \lesssim \operatorname{dist}\left(\psi, S^0_{h,p}(\Gamma)\right)$$

and

$$\|v - v_N\|_{\tilde{H}^{1/2}(\Gamma)} \lesssim \operatorname{dist}\left(v, S^1_{h,p}(\Gamma)\right).$$

In [235] we prove the Theorem 8.3 using the setting of countably normed spaces together with a detailed investigation of the special singular behaviour of the solutions of the screen problems for the Laplacian at corners and edges, see Examples 7.3 and 7.4. When these problems are converted via the direct method into boundary integral equations then the solutions of the latter possess these corner and corner-edge singularities. For the screen problems above these estimates yield only very low convergence rate of order $\mathcal{O}(h^{1/2-\varepsilon} p^{-1+2\varepsilon})$ with arbitrary $\varepsilon > 0$ (see [51, 374, 426] and Chap. 7).

The indices h and p in the notation for the trial spaces $S^0_{h,p}(\Gamma)$ and $S^1_{h,p}(\Gamma)$ refer to h- and p-versions, respectively; where in the h-version a more accurate Galerkin solution is obtained by mesh refinement (and the polynomial degree p is kept fixed) whereas in the p-version a higher accuracy is obtained by increasing the polynomial degree on the same mesh. The implementation of the h-version is standard. In the p-version BEM for the weakly singular integral equation we use tensor products of Legendre polynomials on rectangular meshes and for the hypersingular integral equation we take instead antiderivatives of Legendre polynomials.

If one uses a geometric mesh refinement together with a properly chosen polynomial degree distribution one obtains even exponentially fast convergence rates for the Galerkin errors of the above integral equations. Numerical experiments are presented at the end of this section which show exponential convergence. For application of our error analysis to Helmholtz screen problems see [250]. It is only for ease of presentation that we consider screen problems. In case of a closed surface $\Gamma = \partial\Omega$ a similar analysis can be performed leading also to exponential convergence; the interested reader might look at [251, 290, 295].

For the finite element method the exponential convergence of the hp-version was proposed for three dimensional problems in [207] making use of the setting of countably normed spaces. Whereas the analysis [207] requires the use of special meshes our analysis allows to use much simpler meshes due to the tensor product structure of our approximated subspaces of the boundary element hp-version. Our approach uses regularity results of the solutions of the underliying integral equations which follow from [425, 426] and are based on [141]. Those regularity results show that the solutions have decompositions (see Sects. 7.1.3, 7.2.1) into special edge and corner-edge singularities which on the other hand belong to countably normed spaces see [235]. For smooth given data the solutions of the screen problems admit improved decompositions into additional edge and corner-edge singularities plus arbitrarily smooth remainders. This is why we can show that the error of the

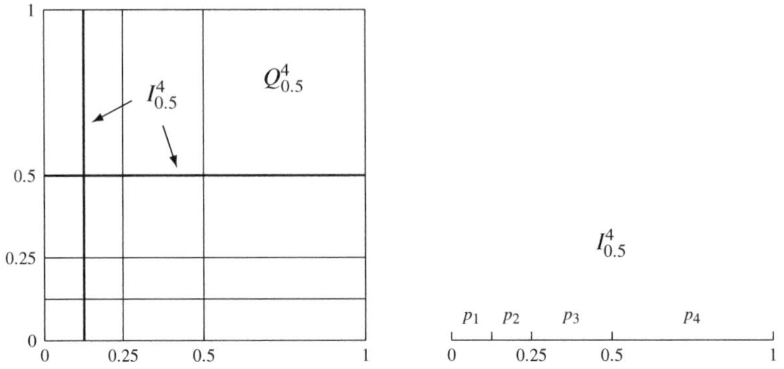

Fig. 8.1 Geometric mesh on the square plate ($\sigma = 0.5$, $n = 4$)

Galerkin solution in the hp-version of the boundary element method consists of one term which decays exponentially fast with a number of unknowns N and of a term $O(N^{-\alpha})$ with arbitrarily large positive α. The $O(N^{-\alpha})$–term results from the approximation of the smooth remainder of the solution of the integral equation. Our numerical experiments show no contribution of this $O(N^{-\alpha})$–term and clearly demonstrate the exponentially fast convergence of the hp–version of the BEM.

For simplicity we take $\Gamma = [0, 2]^2 \times \{0\}$ and introduce the geometric mesh Γ^n_σ (cf. Fig. 8.1) with the reference mesh Q^n_σ where I^n_σ consists of pieces $[x_{k-1}, x_k]$, $x_0 = 0$, $x_k = \sigma^{n-k}$, $k = 1, \ldots, n$. With Q^n_σ we associate a degree vector and define $S^0_{h,p-1}(\Gamma^n_\sigma)$ $(S^1_{h,p}(\Gamma^n_\sigma))$ as the vector space of all piecewise (continuous) polynomials on Γ^n_σ having degree p_k in x and p_l in y on $[x_{k-1}, x_k] \times [x_{l-1}, x_l]$, $1 \le k, l \le n$. We have $h_k = x_k - x_{k-1} \le (\frac{1}{\sigma} - 1)x$ $\forall x \in [x_{k-1}, x_k]$. The detailed proof of the following theorem is given in [235].

Theorem 8.3 *For given piecewise analytic functions f, g in* (8.28) *and* (8.29) *and corresponding Galerkin solutions $\psi_N \in S^0_{h,p-1}(\Gamma^n_\sigma)$, $v_N \in S^1_{h,p}(\Gamma^n_\sigma)$ of* (8.30) *and* (8.31) *on the geometric mesh Γ^n_σ there holds*

$$\left.\begin{array}{l} \|\psi - \psi_N\|_{\tilde{H}^{-1/2}(\Gamma)} \\ \|v - v_N\|_{\tilde{H}^{1/2}(\Gamma)} \end{array}\right\} \le C \exp(-b\sqrt[4]{N}) + \mathcal{O}(N^{-\alpha})$$

with constants C, $b > 0$ independent of the dimension N of the trial space and arbitrary $\alpha > 0$.

In order to give a flavour of the proof let us assume $B^2_\beta(Q)$ regularity; this is satisfied for the higher order terms in the expansions (see Examples 7.3 and 7.4), whereas the lower order terms must be treated separately (see [235] for details). As shown in [251] the solution of the Neumann problem (up to an additional term) has also this regularity.

The local mesh at a right angle corner of Γ is given in Fig. 8.1. The proof of the theorem is based on analysing the error in countably normed spaces and is based on the following lemma showing exponential convergence.

Lemma 8.5 *For $u \in B^2_\beta(Q)$, $0 < \beta < 1$, there exists a spline $u_N \in S^1_{h,p}(Q^n_\sigma)$ and constants C, $b > 0$ independent of N, but dependent on σ, μ, β such that*

$$\|u - u_N\|_{H^1(Q)} \leq C\, e^{-b\sqrt[4]{N}} \tag{8.32}$$

with $p_1 = 1$, $p_k = \max(2, [\mu(k-1)] + 1)$ $(k > 1)$ for $\mu > 0$.

In the above lemma we need the countably normed function space $B^2_\beta(Q)$ which we introduce now for the square $Q = [0, 1]^2$ with the help of weighted Sobolev spaces $H^{k,2}_\beta(Q)$ as

$$B^2_\beta(Q) = \Big\{u : u \in H^{k,2}_\beta(Q), \forall k \geq 2, \|\Phi_{\beta,\alpha,2} D^\alpha u\|_{L^2(Q)} \leq C\, d^{k-2}(k-2)! \\ \text{for } |\alpha| = k = 2, 3, \ldots, \text{ with } C \geq 1, d \geq 1 \text{ indpt. of } k\Big\}.$$

$$\Phi_{\beta,(\alpha_1,\alpha_2),2}(x, y) = \begin{cases} x^{\beta+\alpha_1-2}, & \alpha_1 \geq 2, \alpha_2 = 0 \\ x^\beta + y^\beta, & \alpha_1 = 1, \alpha_2 = 1 \\ x^{\beta+\alpha_1-2} y + x^{\beta+\alpha_1-1} + y^\beta, & \alpha_1 \geq 2, \alpha_2 = 1 \\ x^{\beta+\alpha_1-2} y_2^\alpha + (x^\beta + y^\beta) x^{\alpha_1-1} y^{\alpha_2-1} + x_1^\alpha y^{\beta+\alpha_2-2}, & \alpha_1 \geq 2, \alpha_2 \geq 2 \\ x^\beta + x y^{\beta+\alpha_2-2} + y^{\beta+\alpha_2-1}, & \alpha_1 = 1, \alpha_2 \geq 2 \\ y^{\beta+\alpha_2-2}, & \alpha_1 = 0, \alpha_2 \geq 2 \end{cases}$$

whereas the weighted Sobolev spaces $H^{k,2}_\beta(Q)$ are given by

$$|u|^2_{H^{k,2}_\beta(Q)} = \sum_{|\alpha|=2}^{k} \int_Q |\partial_x^{\alpha_1} \partial_y^{\alpha_2} u(x, y)|^2 \Phi^2_{\beta,\alpha,2}(x, y)\, dy\, dx.$$

$$\|u\|^2_{H^{k,2}_\beta(Q)} = \|u\|^2_{H^1(Q)} + |u|^2_{H^{k,2}_\beta(Q)}.$$

Proof

1.) In element R_{11} at the origin: Due to $u \in H^{2,2}_\beta(Q)$ there exists a bilinear interpolant $\phi_{11} \in \mathscr{P}_{11}(R_{11})$ with $u(0, 0) = \phi_{11}(0, 0)$, $u(0, h_1) = \phi_{11}(0, h_1)$, $u(h_1, 0) = \phi_{11}(h_1, 0)$, $u(h_1, h_1) = \phi_{11}(h_1, h_1)$ $(h_1 = x_1 = \sigma^{n-1})$

$$\|u - \phi_{11}\|^2_{H^1(R_{11})} \leq C\, h_1^{2(1-\beta)} \|u\|^2_{H^{2,2}_\beta(Q)}.$$

2.) On strips near edges $\{(x,y)\,|\,h_1 \le x \le 1, 0 \le y \le h_1\} \cup \{(x,y)\,|\,0 \le x \le h_1, h_1 \le y \le 1\}$ there exist polynomials $\phi_{k1} \in \mathscr{P}_{p_k 1}(R_{k1})$ and $\phi_{1l} \in \mathscr{P}_{1 p_l}(R_{1l})$, coinciding with u at vertices $(0 < \beta < 1)$:

$$\|u - \phi_{k1}\|^2_{H^1(R_{k1})} \le C\, h_1^{2(1-\beta)} |u|^2_{H^{2,2}_\beta(Q)}$$
$$+ C\, x_{k-1}^{2(1-\beta)} \frac{\Gamma(p_k - s_k + 1)}{\Gamma(p_k + s_k + 1)} \left(\frac{\lambda}{2}\right)^{2(s_k+1)} |u|^2_{H^{s_k+2,2}_\beta(Q)} \quad (k \ge 2)$$

$$\|u - \phi_{1l}\|^2_{H^1(R_{1l})} \le C\, h_1^{2(1-\beta)} |u|^2_{H^{2,2}_\beta(Q)}$$
$$+ C\, x_{l-1}^{2(1-\beta)} \frac{\Gamma(p_l - s_l + 1)}{\Gamma(p_l + s_l + 1)} \left(\frac{\lambda}{2}\right)^{2(s_l+1)} |u|^2_{H^{s_l+2,2}_\beta(Q)} \quad (l \ge 2). \tag{8.33}$$

Therefore (corresponding estimates hold away from the edges) on R_{kl} $(2 \le k, l \le n)$ with $1 \le s_k \le p_k$ for $0 \le \alpha_1, \alpha_2 \le 1$ there holds:

$$\|D^\alpha(u - \phi_{kl})\|^2_{L^2(R_{kl})} \le$$
$$\le C \Bigg\{ x_{k-1}^{2(2-\alpha_1-\beta)} \frac{\Gamma(p_k - s_k + 1)}{\Gamma(p_k + s_k + 3 - 2|\alpha|)} \left(\frac{\lambda}{2}\right)^{2s_k} |u|^2_{H^{s_k+3,2}_\beta(Q)}$$
$$+ x_{l-1}^{2(2-\alpha_2-\beta)} \frac{\Gamma(p_l - s_l + 1)}{\Gamma(p_l + s_l + 3 - 2|\alpha|)} \left(\frac{\lambda}{2}\right)^{2s_l} |u|^2_{H^{s_l+3,2}_\beta(Q)} \Bigg\}$$

3.) Combining 1.) and 2.) we obtain $(1 \le s_k \le p_k)$

$$\sum_{k,l=1}^n \|u - \phi_{kl}\|^2_{H^1(R_{kl})}$$
$$\le C\, h_1^{2(1-\beta)} \|u\|^2_{H^{2,2}_\beta(Q)} + (2n-2)\, C\, h_1^{2(1-\beta)} |u|^2_{H^{2,2}_\beta(Q)}$$
$$+ 2nC \sum_{k=2}^n x_{k-1}^{2(1-\beta)} \frac{\Gamma(p_k - s_k + 1)}{\Gamma(p_k + s_k + 1)} \left(\frac{\lambda}{2}\right)^{2(s_k+1)} |u|^2_{H^{s_k+3,2}_\beta(Q)}$$

Now with $h_1 = \sigma^{n-1}$ and

$$|u|_{H^{s_k+3,2}_\beta(Q)} \le C d^{s_k+1} \Gamma(s_k + 2) \tag{8.34}$$

we obtain (8.32). Note: $u \in B^2_\beta(Q)$ implies (8.34). □

Figures 8.2 and 8.3 show numerical experiments (cf. [300]) obtained with the integral equations for linear elasticity treating crack problems with the open surface piece Γ as crack surface. The operators are here given with the Green's function for the Navier–Lamé equation

$$G(x, y) = \frac{\lambda + 3\mu}{4\pi\mu(\lambda + 2\mu)} \left\{ \frac{1}{|x - y|} I + \frac{\lambda + \mu}{\lambda + 3\mu} \frac{(x - y)(x - y)^t}{|x - y|^3} \right\}$$

The legends for Figs. 8.2 and 8.3 have the following meanings: conf-uni-h-4 and conf-uni-p-4 mean conforming h-version of BEM and conforming uniform p-version of BEM on uniform rectangular meshes, respectively. conf-grad-h-4-beta=4.0 stands for conforming h-version of the BEM on graded meshes graded algebraically towards the edges of $\Gamma = [-1, 1]^2$ with grading parameter $\beta = 4$. geo-sigma=0.5-mu=0.5 and geo-sigma=0.17-mu=0.5 stand for two hp-versions of the BEM with geometric mesh parameter geo-sigma and parameter mu for the polynomial degree distribution. Figures 8.2 and 8.3 show clearly the exponentially fast convergence of the hp-version on the geometric mesh with optimal mesh grading parameter $\sigma = 0.17$. The parameter $\mu = 0.5$ describes the increase of the polynomial degree, namely (q, p), (q, p), $(q, p + 1)$, $(q, p + 1)$, $(q, p + 2)$, $(q, p + 2)$,... in the x_2-direction and correspondingly in the x_1-direction, for a geometric mesh consisting of rectangles only and refined towards the edges. Very

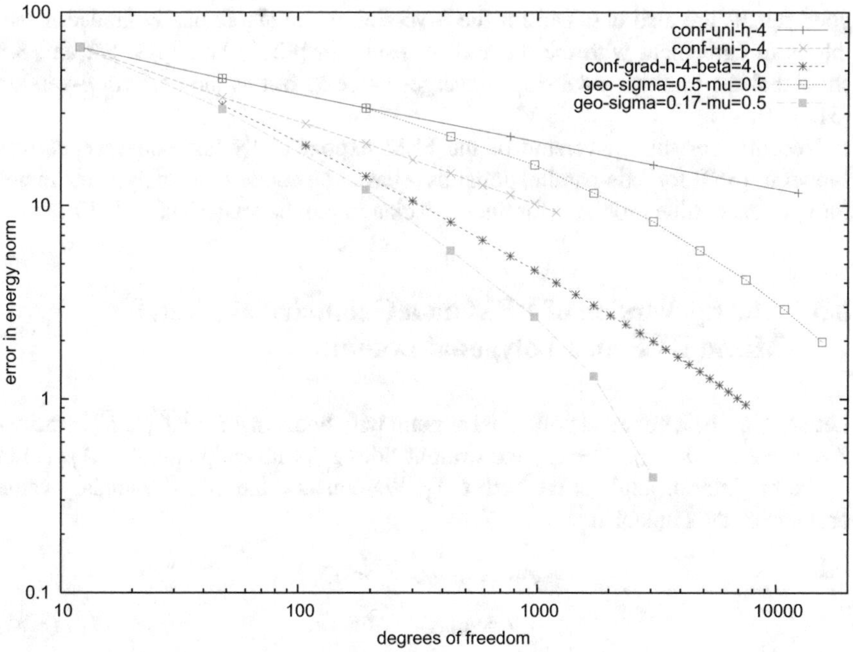

Fig. 8.2 Weakly singular integral equation (Lamé) [300].

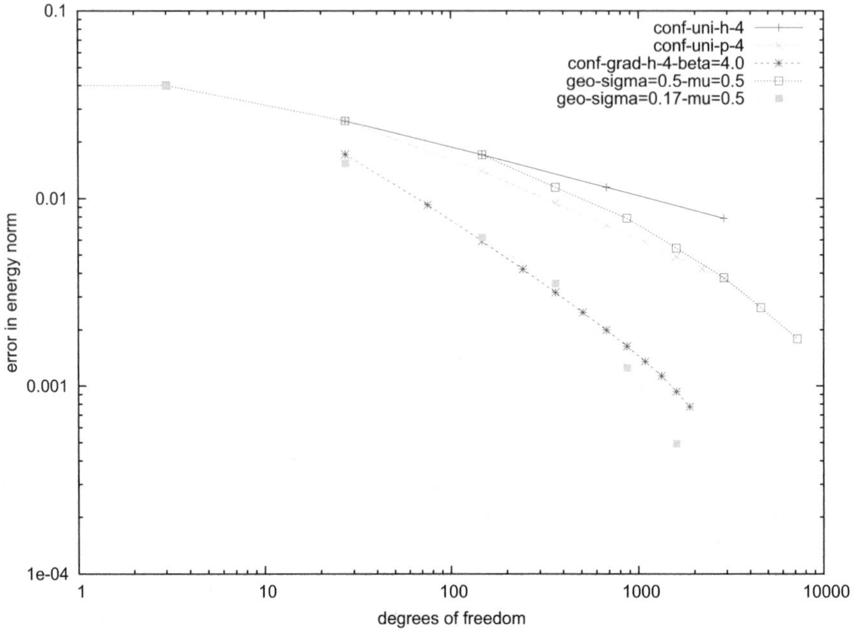

Fig. 8.3 Hypersingular integral equation (Lamé) [300]

good results are also obtained for the h-version on an algebraically graded mesh; this is in agreement with the theoretical results in [426]. Also Figs. 8.2 and 8.3 show that the uniform p-version converges twice as fast as the uniform h-version [51, 374].

Recently for the hp-version of the FEM exponentially fast convergence was shown in [368] for axis-parallel domains – based on anisotropic analytic estimates for boundary value problems for the Laplacian in polyhedra derived in [119].

8.3 The hp-Version of BEM on a Geometrical Mesh for Mixed BVP on a Polygonal Domain

Let $\Omega \subset \mathbb{R}^2$ be a bounded polygonal domain with boundary $\Gamma = \overline{\Gamma}_1 \cup \overline{\Gamma}_2$, vertices A_j, $j = 1 \dots M$, $A_{M+1} := A_1$ and straight sides Γ^j with endpoints A_j, A_{j+1}. Let ω_j be the interior angle at the vertex A_j. We consider the mixed boundary value problem of the Laplacian

$$\begin{aligned} \Delta u &= 0 && \text{in } \Omega, \\ u &= g_1 && \text{on } \Gamma_1, \\ \frac{\partial u}{\partial n} &= g_2 && \text{on } \Gamma_2. \end{aligned} \tag{8.35}$$

If $g_1 \in H^{1/2}(\Gamma)$, $g_2 \in H^{1/2}(\Gamma_2)$ then (8.35) has a unique solution $u \in H^1(\Omega)$. Here the spaces $H^s(\Gamma)$, $H^s(\Gamma_j)$, $\tilde{H}^s(\Gamma_j)$ for $j = 1, 2$ are defined as follows:

$$H^s(\Gamma) := \begin{cases} \{u|_\Gamma : u \in H^{s+1/2}(\mathbb{R}^2)\} & \text{for } s > 0 \\ L^2(\Gamma) & \text{for } s = 0 \\ \left(H^{-s}(\Gamma)\right)' & \text{for } s < 0 \end{cases}$$

$$H^s(\Gamma_j) := \begin{cases} u|_{\Gamma_j} : u \in H^s(\Gamma) & \text{for } s \geqslant 0 \\ (\tilde{H}^{-s}(\Gamma_j))' & \text{for } s < 0 \end{cases}$$

$$\tilde{H}^s(\Gamma_j) := \begin{cases} u|_{\Gamma_j} : u \in H^s(\Gamma), \ u|_{\Gamma/\overline{\Gamma}_j} = 0 & \text{for } s \geqslant 0 \\ \left(H^{-s}(\Gamma_j)\right)' & \text{for } s < 0 \end{cases}$$

and similarly for Γ^j instead of Γ_j. In order to formulate a regularity result for piecewise analytic data g_1, g_2 we introduce weighted Sobolev spaces on Ω and Γ.

Let $H^{m,l}_\beta(\Omega)$, $m \geq l \geq 0$ integers, be the completion of the set of all infinitely differentiable functions under the norm

$$\|u\|^2_{H^{m,l}_\beta(\Omega)} = \|u\|^2_{H^{l-1}(\Omega)} + \sum_{|\alpha|=k,k=l}^{k=m} \|\Phi_{\beta+k-l}|D^\alpha u|\|^2_{L^2(\Omega)}, \quad \text{for } l \geq 1 \tag{8.36}$$

$$\|u\|^2_{H^{m,0}_\beta(\Omega)} = \sum_{|\alpha|=k,k=0}^{k=m} \|\Phi_{\beta+k}|D^\alpha u|\|^2_{L^2(\Omega)}, \tag{8.37}$$

where

$\Phi_{\beta+k}(x) = \prod_{i=1}^M |r_i(x)|^{\beta_i+k}$, $x \in \Omega$ and $r_i(x) = dist(x, A_i) = |x - A_i|$, $x \in \Omega$, denotes the Euclidean distance between the point x and the vertex A_i.

Let

$$B^l_\beta(\Omega) = \{u \in H^{l,l}_\beta(\Omega), \|\Phi_{\beta+k-l}|D^\alpha u|\|_{L^2(\Omega)} \leq Cd^{k-1}(k-l)!, \tag{8.38}$$
$$k = l, l+1, \ldots, C \geq 1, d \geq 1, \text{independent of } k\}$$

For investigation of the singularities at corners we introduce weighted Sobolev spaces and countable normed spaces on the boundary Γ.

Let $I = (a, b)$ and for $x \in (a, b)$, $\hat{r}_1 = |x - a|$, $\hat{r}_2 = |x - b|$, $\hat{\Phi}_{\hat{\beta}+k}(x) = \prod_{i=1}^2 \hat{r}_i^{\hat{\beta}_i+k}(x)$, $\hat{\beta} = (\hat{\beta}_1, \hat{\beta}_2)$, $0 < \hat{\beta}_1, \hat{\beta}_2 < 1$, k integer. Now we define for $k \geq l \geq 0$ and integer $l \geq 0$ the spaces $H^{k,l}_{\hat{\beta}}(I)$, $B^l_{\hat{\beta}}(I)$ to (8.36),(8.38) with I instead of Ω (for details see [21]).

By $B^{l,l+1}_{\hat\beta}(\Gamma_j)$ we denote the space of all functions with restrictions on $\Gamma^i \subset \Gamma_j$, $1 \le i \le M$, $j = 1, 2$ belonging to $B^{l}_{\hat\beta_i}(\Gamma^i)$, $0 < \hat\beta_i < \frac{1}{2}$ or $B^{l+1}_{\hat\beta_i}(\Gamma^i)$, $\frac{1}{2} < \hat\beta_i < 1$.

Exploring the analysis by Babuška, Guo in [19],[18] one obtains the following regularity result for the mixed bvp (8.35), cf also [21].

Theorem 8.4 *Let $g_1 \in C(\Gamma_1) \cap B^{1,2}_{\hat\beta}(\Gamma_1)$, $g_2 \in B^{0,1}_{\hat\beta}(\Gamma_2)$ with $\hat\beta = (\hat\beta_1, \ldots, \hat\beta_M)$, $\hat\beta_i = (\hat\beta_{i,1}, \hat\beta_{i,2})$, $0 < \hat\beta_{i,j} \le 1$, $1 \le i \le M$, $1 \le j \le 2$. Then $u \in B^2_\beta(\Omega)$ with $\beta = (\beta_1, \ldots, \beta_M)$ satisfying* (8.39), *where*

$$\beta_i \begin{cases} = \beta_i^* & \text{if } \beta_i^* > 1 - \dfrac{\pi}{2\omega_i} \\ > 1 - \dfrac{\pi}{2\omega_i} & \text{if } \beta_i^* \leqslant 1 - \dfrac{\pi}{2\omega_i} \end{cases} \quad \text{for } A_i \in \overline{\Gamma}_1 \cap \overline{\Gamma}_2 \tag{8.39}$$

$$\beta_i \begin{cases} = \beta_i^* & \text{if } \beta_i^* > 1 - \dfrac{\pi}{\omega_i} \\ > 1 - \dfrac{\pi}{\omega_i} & \text{if } \beta_i^* \leqslant 1 - \dfrac{\pi}{\omega_i} \end{cases} \quad \text{for } A_i \notin \overline{\Gamma_1} \cap \overline{\Gamma_2}$$

with $\beta_i^* = \max(\overline{\beta}_{i-1,2}, \overline{\beta}_{i,1})$, $\overline{\beta}_{i,j} := \hat\beta - \frac{1}{2}\text{sign}(\hat\beta_{i,j} - 1/2)$

Next we give an equivalent boundary integral equation formulation of Problem (8.35). Let $G(z, \zeta)$ be the fundamental solution of the Laplacian

$$G(z, \zeta) = -\frac{1}{2\pi} \ln |z - \zeta|$$

and define the following boundary integral operators: Let $f_j \in C_0^\infty(\Gamma_j)$, $j, k = 1, 2$. Then for $z \in \Gamma_k$

$$V_{ji} f_j(z) := -2 \int_{\Gamma_j} f_j(\zeta) G(z, \zeta)\, ds_\zeta, \quad K_{ji} f_j(z) := -2 \int_{\Gamma_j} f_j(\zeta) \frac{\partial}{\partial n_\zeta} G(z, \zeta)\, ds_\zeta$$

$$K'_{ji} f_j(z) := -2 \int_{\Gamma_j} f_j(\zeta) \frac{\partial}{\partial n_z} G(z, \zeta) ds_\zeta, \; W_{ij} f_j(z) := 2 \frac{\partial}{\partial n_z} \int_{\Gamma_j} f_j(\zeta) \frac{\partial}{\partial n_\zeta} G(z, \zeta) ds_\zeta.$$

For the distribution f_j on Γ_j we define $V_{jk} f_j$ and $K_{jk} f_j$ by approximating f_j with smooth functions and $K'_{jk} f_j$ by duality using the relation

$$\langle K'_{jk} f_j, \tilde f_k \rangle_{L^2(\Gamma_k)} = \langle f_j, K_{kj} \tilde f_k \rangle_{L^2(\Gamma_j)}, \quad \forall \tilde f_k \in C^\infty(\Gamma_k).$$

Here the subscript of D_{jk} etc means integration over Γ_j and evaluation on Γ_k.

Define the extension operator $l : H^{1/2}(\Gamma_1) \to H^{1/2}(\Gamma)$in the following way: Assume that Γ is parametrized by a piecewise linear function $\phi(x) : [-1, 1]$ with $\Gamma_1 = \{\phi(x)|x \in [-1, 0]\}$, $\Gamma_2 = \{\phi(x)|x \in [0, 1]\}$ and let $v \in H^{1/2}(\Gamma_1)$ be expressed with respect to this parametrization. Then define

$$l\,v(x) := \begin{cases} v(x) & \text{if } x \in [-1, 0] \\ v(-x) & \text{if } x \in [0, 1]. \end{cases}$$

Next we introduce the extension operator $\tilde{l} : H^{-1/2}(\Gamma_2) \to H^{-\frac{1}{2}}(\Gamma)$ for $\psi \in H^{-\frac{1}{2}}(\Gamma_2)$ by

$$\tilde{l}\psi(x) := \begin{cases} -\psi(-x) & \text{if } x \in [-1, 0] \\ \psi(x) & \text{if } x \in [0, 1] \end{cases}$$

Then we have from [127, 128] the following result:Here we need for $s \in \mathbb{R}$ the space

$$\tilde{H}^s(\Gamma_k) = \{f \in H^s(\Gamma) : suppf \subset \Gamma_k\}$$

Theorem 8.5 ([127, 128].) *The boundary integral equation*

$$\mathscr{A}\begin{pmatrix} v^* \\ \psi^* \end{pmatrix} = \mathscr{B}\begin{pmatrix} g_1 \\ g_2 \end{pmatrix} - \mathscr{A}\begin{pmatrix} l\,g_1|_{\Gamma_2} \\ \tilde{l}\,g_2\big|_{\Gamma_1} \end{pmatrix} \tag{8.40}$$

with the boundary integral operators

$$\mathscr{A} := \begin{pmatrix} W_{22} & K'_{12} \\ -K_{21} & V_{11} \end{pmatrix}, \quad \mathscr{B} := \begin{pmatrix} -W_{12} & 1 - K'_{22} \\ 1 + K_{11} & -V_{21} \end{pmatrix}$$

has a unique solution $\begin{pmatrix} v^* \\ \psi^* \end{pmatrix}$ *in* $\tilde{H}^{1/2}(\Gamma_2) \times \tilde{H}^{-1/2}(\Gamma_1)$*. This solution yields the unknown boundary data* $u|_{\Gamma_2}$, $\left.\dfrac{\partial u}{\partial n}\right|_{\Gamma_1}$ *of the unique solution* $u \in H^1(\Omega)$ *of problem* (8.35) *by*

$$u|_{\Gamma_2} = v^* + l\,g_1|_{\Gamma_2}, \quad \left.\frac{\partial u}{\partial n}\right|_{\Gamma_1} = \psi^* + \tilde{l}\,g_2\big|_{\Gamma_1}. \tag{8.41}$$

Now we define the boundary element spaces for the h-p-method. Let d_j be the length of the side Γ^j. First bisect each side Γ^j into two parts Γ^j_- (containing A_j) and Γ^j_+ (containing A_{j+1}). Choose a mesh parameter $0 < \sigma < 1$, and an integer $n \geqslant 0$. Introduce the points $A_-^{j,k}$ on Γ_-^j with $\mathrm{dist}(A_-^{j,k}, A_j) = \sigma^{n-k}\frac{d_j}{2}$ for $k = 1 \dots n$ and $A_-^{j,0} := A_j$. This defines n subintervals $\Gamma_-^{j,k}$ with endpoints $A_-^{j,k-1}$ and $A_-^{j,k}$ for $k = 1 \dots n$. Analogously define the points $A_+^{j,k}$ on Γ_+^j with $\mathrm{dist}(A_+^{j,k}, A_{j+1}) = \sigma^{n-k}\frac{d_j}{2}$ yielding n subintervals $\Gamma_+^{j,k}$ with endpoints $A_+^{j,k-1}$ and $A_+^{j,k}$ for $k = 1 \dots n$.

Define the spaces $S_{n,2}$ and $S_{n,1}$ on Γ_2 and Γ_1 as follows

$$S_{n,2} := \left\{ v \in C^0(\Gamma_2) \,\middle|\, v|_{\Gamma_\pm^{j,k}} \in P_k(\Gamma_\pm^{j,k}) \text{ for } \Gamma_\pm^{j,k} \subset \Gamma^j \subset \Gamma_2,\ k = 1 \dots n \right\} \tag{8.42}$$

$$S_{n,1} := \left\{ \psi \mid \psi|_{\Gamma_\pm^{j,k}} \in P_k(\Gamma_\pm^{j,k}) \text{ for } \Gamma_\pm^{j,k} \subset \Gamma^j \subset \Gamma_1,\ k = 1 \dots n \right\} \tag{8.43}$$

where $P_k(\Gamma_\pm^{j,k})$ denotes the space of polynomials of degree $\leqslant k$ on $\Gamma_\pm^{j,k}$. Then let

$$S_n := S_{n,2} \times S_{n,1}. \tag{8.44}$$

There holds $\dim S_n \leqslant M(n+2)^2$. Then we obtain exponential convergence for the Galerkin method (8.45).

Next we describe the Galerkin method for the approximation of the solution of the integral equation (8.40). Choose a sequence of finite dimensional subspaces $S_n \subset \tilde{H}^{1/2}(\Gamma_2) \times \tilde{H}^{-1/2}(\Gamma_1)$ with

$$\overline{\bigcup_{n=1}^{\infty} S_n} = \tilde{H}^{1/2}(\Gamma_2) \times \tilde{H}^{-1/2}(\Gamma_1).$$

Then find a solution $(v_n^*, \psi_n^*) \in S_n$ satisfying

$$\forall \begin{pmatrix} w_n \\ \phi_n \end{pmatrix} \in S_n \quad \left\langle \mathscr{A} \begin{pmatrix} v_n^* \\ \psi_n^* \end{pmatrix}, \begin{pmatrix} w_n \\ \phi_n \end{pmatrix} \right\rangle = \left\langle \mathscr{B} \begin{pmatrix} g_1 \\ g_2 \end{pmatrix} - \mathscr{A} \begin{pmatrix} l\, g_1|_{\Gamma_2} \\ \tilde{l}\, g_2|_{\Gamma_1} \end{pmatrix}, \begin{pmatrix} w_n \\ \phi_n \end{pmatrix} \right\rangle. \tag{8.45}$$

Then the strong ellipticity of the operator $\mathscr{A}$ [127] gives the quasioptimality of the Galerkin solution:

Theorem 8.6 *For sufficiently large n the Galerkin equations* (8.45) *have a unique solution* $\begin{pmatrix} v_n^* \\ \psi_n^* \end{pmatrix} \in S_n$. *Furthermore there holds*

$$\left\| \begin{pmatrix} v_n^* \\ \psi_n^* \end{pmatrix} - \begin{pmatrix} v^* \\ \psi^* \end{pmatrix} \right\|_{\tilde{H}^{1/2}(\Gamma_2)\times\tilde{H}^{-1/2}(\Gamma_1)} \leqslant C \inf \left\| \begin{pmatrix} w_n \\ \phi_n \end{pmatrix} - \begin{pmatrix} v^* \\ \psi^* \end{pmatrix} \right\|_{\tilde{H}^{1/2}(\Gamma_2)\times\tilde{H}^{-1/2}(\Gamma_1)} \tag{8.46}$$

with C independent of n, where the infimum is taken for $\begin{pmatrix} w_n \\ \phi_n \end{pmatrix} \in \tilde{H}^{1/2}(\Gamma_2) \times \tilde{H}^{-1/2}(\Gamma_1)$.

Theorem 8.7 *Assume the boundary data in* (8.35) *satisfy* $g_1 \in B^{1,2}_{\hat{\beta}}(\Gamma_1)$, $g_2 \in B^{0,1}_{\hat{\beta}}(\Gamma_2)$. *Let* $\begin{pmatrix} v_n^* \\ \psi_n^* \end{pmatrix} \in S_n$ *be the solution of the Galerkin equation* (8.45) *for sufficiently large n, with S_n given by* (8.42)–(8.44) *and let u be the exact solution of the boundary value problem* (8.35). *Then* $v_n := l\, g_1|_{\Gamma_2} + v_n^*$, $\psi_n := \tilde{l}\, g_2\big|_{\Gamma_1} + \psi_n^*$ *satisfy*

$$\| v_n - u|_{\Gamma_2} \|_{\tilde{H}^{1/2}(\Gamma_2)} + \left\| \psi_n - \frac{\partial u}{\partial n}\bigg|_{\Gamma_1} \right\|_{\tilde{H}^{-1/2}(\Gamma_1)} \leqslant C\, e^{-b\sqrt{N}} \tag{8.47}$$

where $N = \dim S_n = \dim S_{n,z} + \dim S_{n,1}$ *and $C, b > 0$ are constants independent of n*

Proof By Theorem 8.4, $u \in B^2_{\beta}(\Omega)$. By [18, 19] there exists a function $\tilde{u}_n$ in an h-p finite element space $\tilde{S}_n$ on a geometric mesh in Ω satisfying

$$\|u - \tilde{u}_n\|_{H^1(\Omega)} \leqslant C\, e^{-\tilde{b}(\dim \tilde{S}_n)^{1/3}}$$

with $\tilde{b} > 0$, and C independent of n. The geometric mesh in Ω is here given such that its nodes on the boundary Γ create the above introduced geometric mesh. Now we will take the traces on Γ_2 and the normal derivatives on Γ_1. The mapping $T : H^1(\Omega) \to \tilde{H}^{1/2}(\Gamma_2) \times \tilde{H}^{-1/2}(\Gamma_1)$ given by

$$f \mapsto \left(f|_{\Gamma_2} - (l\ f|_{\Gamma_1})|_{\Gamma_2},\ \frac{\partial f}{\partial n}\bigg|_{\Gamma_1} - \left(\tilde{l}\ \frac{\partial f}{\partial n}\bigg|_{\Gamma_2} \right)\bigg|_{\Gamma_1} \right)$$

is continuous. This gives with $f := u - \tilde{u}_n$

$$\left\| \left(u|_{\Gamma_2} - (l\, u|_{\Gamma_1})|_{\Gamma_2} \right) - \left(\tilde{u}_n|_{\Gamma_2} - (l\, \tilde{u}_n|_{\Gamma_1})|_{\Gamma_2} \right) \right\|_{\tilde{H}^{1/2}(\Gamma_2)} \leqslant C\, e^{-\tilde{b}(\dim \tilde{S}_n)^{1/3}} \tag{8.48}$$

$$\left\| \left(\frac{\partial u}{\partial n}\bigg|_{\Gamma_1} - \left(\tilde{l}\, \frac{\partial u}{\partial n}\bigg|_{\Gamma_2} \right)\bigg|_{\Gamma_1} \right) - \left(\frac{\partial \tilde{u}_n}{\partial n}\bigg|_{\Gamma_1} - \left(\tilde{l}\, \frac{\partial \tilde{u}_n}{\partial n}\bigg|_{\Gamma_1} \right)\bigg|_{\Gamma_2} \right) \right\|_{\tilde{H}^{-1/2}(\Gamma_1)} \leq C e^{-\tilde{b}(\dim \tilde{S}_n)^{1/3}}. \tag{8.49}$$

By Theorem 8.5 we have for the exact solution (v^*, ψ^*) of (8.40)

$$v^* = u|_{\Gamma_2} - \left(l\, u|_{\Gamma_1}\right)\Big|_{\Gamma_2} \quad , \quad \psi^* = \left.\frac{\partial u}{\partial n}\right|_{\Gamma_1} - \left(\tilde{l}\left.\frac{\partial u}{\partial n}\right|_{\Gamma_2}\right)\Bigg|_{\Gamma_1} \tag{8.50}$$

using the boundary conditions $u|_{\Gamma_1} = g_1$, $\left.\frac{\partial u}{\partial n}\right|_{\Gamma_2} = g_2$ in (8.35). By the construction of the spaces $\tilde{S}_n$ and S_n there holds $\left(\tilde{u}_n|_{\Gamma_2}, \left.\frac{\partial \tilde{u}_n}{\partial n}\right|_{\Gamma_1}\right) \in S_n$ and we obtain with the definition of the operators $l, \tilde{l}$

$$v_n^* := \tilde{u}_n|_{\Gamma_2} - \left(l\, \tilde{u}_n|_{\Gamma_1}\right)\Big|_{\Gamma_2} \in S_{n,2} \quad , \quad \psi_n^* := \left.\frac{\partial \tilde{u}_n}{\partial n}\right|_{\Gamma_1} - \left(\tilde{l}\left.\frac{\partial \tilde{u}_n}{\partial n}\right|_{\Gamma_2}\right)\Bigg|_{\Gamma_1} \in S_{n,1} \tag{8.51}$$

Thus (8.48),(8.49),(8.50),(8.51) yield the existence of $(v_n^*, \psi_n^*) \in S_n$ with

$$\left\| v^* - v_n^* \right\|_{\tilde{H}^{1/2}(\Gamma_2)} + \left\| \psi^* - \psi_n^* \right\|_{\tilde{H}^{-1/2}(\Gamma_1)} \leqslant C\, e^{-\tilde{b}(\dim \tilde{S}_n)^{1/3}} \tag{8.52}$$

Using $\dim \tilde{S}_n \geqslant C_1 n^3$ and $\dim S_n \leqslant C_2 n^2$ with suitable constants C_1, C_2 we obtain

$$C\, e^{-\tilde{b}(\dim \tilde{S}_n)^{1/3}} \leqslant C\, e^{-b(\dim S_n)^{1/2}} \tag{8.53}$$

with $b > 0$ independent of n. Now the quasioptimality (8.46) gives the result (8.47) using the definitions of u_n, ψ_n and (8.41). □

Chapter 9
Mapping Properties of Integral Operators on Polygons

In this chapter we introduce the analysis of boundary integral operators on a polygon with the tool of the Mellin transformation from the original paper [128]. The interested reader may also look into [241] where the Mellin calculus is used to analyse the mapping properties of the integral operators in countably normed spaces. These results are crucial for deriving exponentially fast convergence of the $hp-$version of the boundary element method (see Chap. 8). The results of the subsection describing the regularity of the solution near the vertices were originally published in [138]. The Mellin calculus is used in Sect. 9.3 to analyze the regularity of the solution at the tip of an interface crack, in Sect. 9.4 to analyze the mixed boundary value problem for the Laplacian with the hypersingular operator and the singular behaviour of its solution at the point where Dirichlet and Neumann conditions meet and in Sect. 9.5 to analyze the mapping propeties of boundary integral operators with countably normed spaces. In the framework of these spaces the analysis of the exponential convergence of the hp Galerkin approximation is presented in Sect. 8.1.

9.1 Mellin Symbols

In this section (following [128]) we now have to cope with the problem that the integral operators are now defined on curves with corners so that we cannot directly apply the Fourier transformation. We will consider a polygon Γ as follows: We split it into sectors Γ^{ω} and pieces of straight lines (see Fig. 9.1) and apply now Mellin techniques on Γ^{ω} and pseudodifferential operators on the straight lines.

© Springer International Publishing AG, part of Springer Nature 2018

J. Gwinner, E. P. Stephan, *Advanced Boundary Element Methods*,
Springer Series in Computational Mathematics 52,
https://doi.org/10.1007/978-3-319-92001-6_9

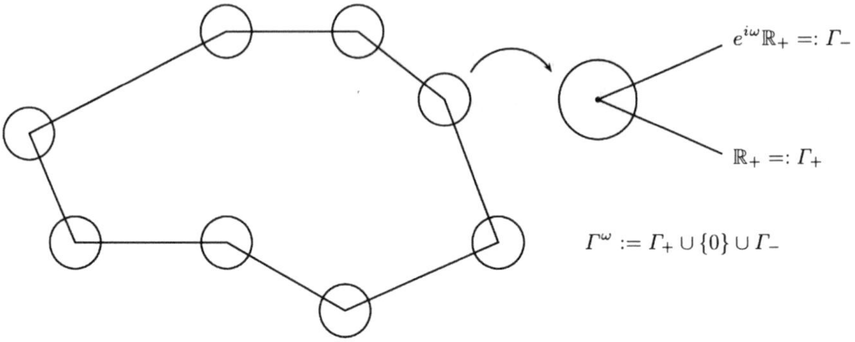

Fig. 9.1 Geometrical setting

For $\phi \in C_0^\infty(0, \infty)$ we define the *Mellin* transformed $\hat{\varphi}$ of ϕ by:

$$\hat{\phi}(\lambda) := \int_0^\infty x^{i\lambda-1}\phi(x)\,dx = \int_{-\infty}^\infty \phi(e^{-t})e^{-i\lambda t}\,dt\ ,$$

making use of the Euler transformation $\begin{cases} \mathbb{R} \to \mathbb{R}+ \\ t \mapsto e^{-t} \end{cases} := x\ .$

We have the inverse formula:

$$\phi_h(x) = \phi_h(e^{-t}) = \frac{1}{2\pi} \int_{\Im(\lambda)=h} e^{i\lambda t}\hat{\phi}(\lambda)\,d\lambda\ .$$

Defining now the operator (single-layer potential)

$$V \cong \begin{pmatrix} V_{--} & V_{+-} \\ V_{-+} & V_{++} \end{pmatrix}$$

with $u|_{\Gamma^\omega} \cong (u_-, u_+)$ on $\mathbb{R}_+$, we have for $\phi \in C_0^\infty[0, \infty)$

$$V_{++}\phi(x) := -\frac{1}{\pi}\int_0^\infty \ln|x-y|\phi(y)\,dy$$

$$= \underbrace{-\frac{1}{\pi}\int_0^\infty \ln(y)\phi(y)\,dy}_{=:l(\phi)} - \underbrace{\frac{1}{\pi}\int_0^\infty \ln\left|1-\frac{x}{y}\right|\phi(y)\,dy}_{V_0\phi(x)}\ , x \in \mathbb{R}_+ = \Gamma_+$$

$$V_{-+}\phi(x) := -\frac{1}{\pi}\int\limits_0^\infty \ln|x - e^{i\omega}y|\phi(y)\,dy \quad, x \in \mathbb{R}_+ = \Gamma_+$$

$$\begin{aligned} V_{+-}\phi(x) := &\quad -\frac{1}{\pi}\int\limits_0^\infty \ln|xe^{i\omega} - y|\phi(y)\,dy = \quad l(\phi) - \frac{1}{\pi}\int\limits_0^\infty \ln|1 - \frac{x}{y}e^{-i\omega}|\phi(y)\,dy \\ =: &\quad l(\phi) - V_\omega\phi(x) \quad, x \in \Gamma_- \end{aligned}$$

It is easily verified that there holds $V_{++} = V_{--}$, $V_{+-} = V_{-+}$ and on Γ^ω :

$$V = \begin{pmatrix} l & l \\ l & l \end{pmatrix} + \begin{pmatrix} V_0 & V_\omega \\ V_\omega & V_0 \end{pmatrix}.$$

Consider now the double-layer potential

$$K \cong \begin{pmatrix} K_{--} & K_{+-} \\ K_{-+} & K_{++} \end{pmatrix}.$$

Due to the geometric interpretation of K as the variation of the angle, i.e. $Kg(z) := -\frac{1}{\pi}\int_\Gamma g(\zeta)\frac{\partial}{\partial n_\zeta}\ln|z - \zeta|ds_\zeta = -\frac{1}{\pi}\int_\Gamma g(\zeta)d\theta_\zeta(z)$, where $\theta_\zeta(z)$ is the angle between $\zeta - z$ and some fixed direction, we have $K_{--} = K_{++} = 0$. Furthermore, there holds

$$\begin{aligned} K_{+-}\phi(x) &:= -\frac{1}{\pi}\int\limits_0^\infty \frac{\partial}{\partial_{n_y}}\ln|xe^{i\omega} - y|\phi(y)\,dy \\ &= \frac{1}{\pi}\int\limits_0^\infty \Im(xe^{i\omega} - y)^{-1}\phi(y)\,dy =: K_\omega\phi(x). \end{aligned}$$

With $K_{-+} = K_{+-} = K_\omega$ we therefore have on Γ_ω:

$$K = \begin{pmatrix} 0 & K_\omega \\ K_\omega & 0 \end{pmatrix}.$$

In the following lemma we provide the Mellin symbols $\widehat{V_0}(\lambda)$, $\widehat{V_\omega}(\lambda)$, $\widehat{K_\omega}(\lambda)$.

Lemma 9.1 *For the operators V and K as given above there holds:*

i) Let $\phi \in C_0^\infty[0,\infty)$. Then, for $\Im(\lambda) \in (-1,0)$ we have:

$$\widehat{V_0\phi}(\lambda) = \widehat{V_0}(\lambda)\cdot\widehat{\phi}(\lambda - i) := \frac{\cosh(\pi\lambda)}{\lambda\sinh(\pi\lambda)}\widehat{\phi}(\lambda - i),$$

$$\widehat{V_\omega\phi}(\lambda) = \widehat{V_\omega}(\lambda)\cdot\widehat{\phi}(\lambda - i) := \frac{\cosh(\pi-\omega)\lambda}{\lambda\sinh(\pi\lambda)}\widehat{\phi}(\lambda - i)\ .$$

ii) Let $\phi \in C_0^\infty(0,\infty)$. Then, for $\Im(\lambda) \in (-1,1)$ we have:

$$\widehat{K_\omega\phi}(\lambda) = \widehat{K_\omega}(\lambda)\cdot\widehat{\phi}(\lambda) := -\frac{\sinh(\pi-\omega)\lambda}{\sinh(\pi\lambda)}\widehat{\phi}(\lambda)\ .$$

Proof Before proving the statements of the lemma, we leave it as an exercise to the reader to show that for $\omega \in (0,2\pi)$ and $\Im(\lambda) \in (-1,0)$ the following identity holds (see [167]):

$$\int_{-\infty}^{\infty} \frac{e^{-i\lambda\tau}}{e^{-\tau\pm i\omega}-1}\,d\tau \;=\; i\pi\frac{e^{\pm\lambda(\omega-\pi)}}{\sinh(\pi\lambda)}\ . \tag{9.1}$$

We now want to start with the equation for the double-layer potential, i.e. ii):

$$\begin{aligned} K_\omega\phi(e^{-t}) &= \frac{1}{\pi}\int_{-\infty}^{\infty} \Im\left(\frac{1}{e^{-t+i\omega}-e^{-\tau}}\right)\phi(e^{-\tau})e^{-\tau}\,d\tau \\ &= \frac{1}{2\pi i}\int_{-\infty}^{\infty}\left(\frac{1}{e^{-(t-\tau)+i\omega}-1} - \frac{1}{e^{-(t-\tau)-i\omega}-1}\right)\phi(e^{-\tau})\,d\tau \\ &=: \int_{-\infty}^{\infty} f(t-\tau)\phi(e^{-\tau})\,d\tau\ . \end{aligned} \tag{9.2}$$

Defining $Ff(\lambda) := \int_{-\infty}^{\infty} e^{-i\lambda t} f(t)\,dt$, (9.1) yields:

$$\begin{aligned} Ff(\lambda) &= \tfrac{1}{2\pi i}\left(i\pi\tfrac{e^{\lambda(\omega-\pi)}}{\sinh(\pi\lambda)} - i\pi\tfrac{e^{-\lambda(\omega-\pi)}}{\sinh(\pi\lambda)}\right) \\ &= -\tfrac{\sinh(\pi-\omega)\,\lambda}{\sinh(\pi\lambda)} \\ &=: \widehat{K_\omega}(\lambda)\ . \end{aligned}$$

Thus, by the convolution theorem for the Fourier transformation and the above definition of the Mellin transformation, the assertion of ii) directly follows from (9.2).

i):

$$\begin{aligned} V_\omega\phi(e^{-t}) &= -\tfrac{1}{\pi}\int_{-\infty}^{\infty} \ln|1-e^{\tau-t-i\omega}|\phi(e^{-\tau})e^{-\tau}\,d\tau \\ &=: \int_{-\infty}^{\infty} f(t-\tau)g(\tau)\,d\tau\,. \end{aligned}$$

Here we have

$$Fg(\lambda) = \int_{-\infty}^{\infty} e^{-i\lambda\tau}e^{-\tau}\phi(e^{-\tau})d\tau = \widehat{\phi}(\lambda - i)\,.$$

Now, integration by parts yields

$$\begin{aligned} Ff(\lambda) &= -\tfrac{1}{\pi}\int_{-\infty}^{\infty} e^{-i\lambda t}\,\underbrace{\Re\Big(\ln(1-e^{-t-i\omega})\Big)}_{=f(t)}\,dt \\ &= -\int_{-\infty}^{\infty} \tfrac{e^{-i\lambda t}}{i\pi\lambda}\Re\Big(\tfrac{e^{-t-i\omega}}{1-e^{-t-i\omega}}\Big)\,dt \\ &= \frac{\cosh(\pi-\omega)\lambda}{\lambda\sinh(\pi\lambda)} =: \widehat{V_\omega}(\lambda)\,, \end{aligned}$$

for $\mathrm{Im}\lambda \in (-1, 0)$. For $\widehat{V_0\phi}$ one shall argue correspondingly and the assertion follows analogously to i). □

9.1.1 Mapping Properties in Weighted Sobolev Spaces

In this subsection we will consider in some more detail the weighted Sobolev spaces.

Definition 9.1 We define the *weighted Sobolev space* $\overset{\circ}{W}{}_0^s(\mathbb{R}_+)$ to be the completion of the space $C_0^\infty(0, \infty)$ with respect to the norm:

$$\|\phi\|^2_{\overset{\circ}{W}{}_0^s(\mathbb{R}_+)} := \frac{1}{2\pi}\int_{\Im(\lambda)=s-\frac{1}{2}} \left(1+|\lambda|^2\right)^s |\widehat{\phi}(\lambda)|^2\,d\lambda \qquad \text{for } s \in \mathbb{R}\,,$$

where $\widehat{\phi}(\lambda) := \int_0^\infty x^{i\lambda-1}\phi(x)\,dx\,.$

For $\phi \in \mathscr{S}$ we have the Fourier transformation:

$$\widetilde{\phi}(\xi) := \int_{-\infty}^{\infty} e^{-i\xi x}\phi(x)\,dx \quad \text{for } \xi \in \mathbb{R} \quad \Longrightarrow \widetilde{\phi} \in \mathscr{S}$$

Here, we recall the definition of the standard Sobolev space on $\mathbb{R}$ (see Appendix B)

Definition 9.2 The space $H^s(\mathbb{R})$ is defined as the completion of $C_0^\infty(\mathbb{R})$ with respect to the norm

$$\|\phi\|^2_{H^s(\mathbb{R})} := \int_{\mathbb{R}} \left(1 + |\xi|^2\right)^s |\widetilde{\phi}(\xi)|^2\,d\xi \quad \text{for } s \in \mathbb{R}\,.$$

In order to investigate the mapping properties of the integral operators (e.g. the single-layer potential V) on a smooth boundary Γ, it therefore suffices to consider its action on a function $\phi \in C_0^\infty(\mathbb{R})$. For the theorem to follow we will need the notion of the *principal symbol* of an operator which is defined as the leading term of the Fourier transformation of the kernel of the operator and, furthermore, the *convolution theorem*, stating that

$$\widetilde{\gamma * \phi}(\xi) = \widetilde{\gamma}(\xi) \cdot \widetilde{\phi}(\xi)$$

holds $\forall \phi \in C_0^\infty(\mathbb{R})$ (see Appendix B, Definition B.6).

Theorem 9.1 *For a smooth boundary Γ there holds:*

$$V : H^s(\Gamma) \overset{continuously}{\longrightarrow} H^{s+1}(\Gamma) \quad \forall s \in \mathbb{R} \tag{9.3}$$

(or equivalently:

$$V_{\mathbb{R}} : H^s(\mathbb{R}) \overset{continuously}{\longrightarrow} H^{s+1}(\mathbb{R}) \quad \forall s \in \mathbb{R}\,). \tag{9.4}$$

Proof We will show (9.4):

$$\|V_{\mathbb{R}}\phi\|^2_{H^{s+1}(\mathbb{R})} = \int_{\mathbb{R}} (1 + |\xi|^2)^{s+1} \underbrace{\left|(\widetilde{V_{\mathbb{R}}\phi})(\xi)\right|^2}_{=\frac{1}{|\xi|^2}|\widetilde{\phi}(\xi)|^2} d\xi$$

(by the convolution theorem)

Now, for $|\xi|$ large enough there holds: $\frac{1}{|\xi|^2} - \frac{1}{1+|\xi|^2} \sim \frac{1}{|\xi|^4}$. Let χ be C^∞- cut-off function with $\chi(\xi) = 0$ for $|\xi| < \epsilon$ and $\chi(\xi) = 1$ for $|\xi| > 2\epsilon$. Then, for the principal symbol of $\chi V_{\mathbb{R}}$ we have:

$$\sigma(\chi V_{\mathbb{R}})(\xi) = \chi \frac{1}{|\xi|} = \begin{cases} \frac{1}{|\xi|} \,, & |\xi| > 2\epsilon \\ 0 \,, & |\xi| < \epsilon \\ \text{smooth} \,, & \text{else} \end{cases} .$$

We note that $V_{\mathbb{R}}$ may be replaced by $\chi V_{\mathbb{R}}$, since the corresponding principal symbols are the same apart from $|\xi| < 2\epsilon$. We may therefore replace the symbol of $V_{\mathbb{R}}$ by $\frac{1}{(1+|\xi|^2)^{\frac{1}{2}}}$.

Thus, we have

$$\|V_{\mathbb{R}}\phi\|^2_{H^{s+1}(\mathbb{R})} \leq c^2 \|\phi\|^2_{H^s(\mathbb{R})} ,$$

completing the proof of the theorem. □

Corollary 9.1 *The following operators are continuous:*

(i)	$V_0 : \overset{\circ}{W_0}{}^{s}(\mathbb{R}^+)$	$\longrightarrow$	$\overset{\circ}{W_0}{}^{s+1}(\mathbb{R}^+)$	*for* $\|s\| < \frac{1}{2}$
(ii)	$V_\omega : \overset{\circ}{W_0}{}^{s}(\mathbb{R}^+)$	$\longrightarrow$	$\overset{\circ}{W_0}{}^{s+1}(\mathbb{R}^+)$	*for* $\|s\| < \frac{1}{2}$
(iii)	$K_\omega : \overset{\circ}{W_0}{}^{s}(\mathbb{R}^+)$	$\longrightarrow$	$\overset{\circ}{W_0}{}^{s}(\mathbb{R}^+)$	*for* $-\frac{1}{2} < s < \frac{3}{2}$

Proof We leave it as an exercise to the reader to show:

(i) $\widehat{V_\omega \phi}(\lambda) = \frac{\cosh(\pi-\omega)\lambda}{\lambda \sinh(\pi\lambda)} \widehat{\phi}(\lambda - i)$

$\Longrightarrow \left|\widehat{V_\omega}(\lambda)\right| \sim \frac{1}{1+|\lambda|}$ on every line $\Im(\lambda) = h \in (-1, 0)$.

(ii) $\exists C > 0$ (independent of λ) : $|\widehat{K_\omega}(\lambda)| < C$ on $\Im(\lambda) = h \in (-1, 1)$.

We then have for $\phi \in \overset{\circ}{W_0}{}^{s}(\mathbb{R}^+)$:

$$\begin{aligned}
\|V_\omega \phi\|^2_{\overset{\circ}{W_0}{}^{s+1}(\mathbb{R}^+)} &\sim \int_{\Im(\lambda)=s+\frac{1}{2}} \left(1+|\lambda|^2\right)^{s+1} |\widehat{V_\omega}(\lambda)|^2 \cdot |\widehat{\phi}(\lambda - i)|^2 \, d\lambda \\
&\lesssim \int_{\Im(\lambda)=s+\frac{1}{2}} \left(1+|\lambda|^2\right)^{s} |\widehat{\phi}(\lambda - i)|^2 \, d\lambda \\
&\lesssim \int_{\Im(\lambda)=s-\frac{1}{2}} \left(1+|\lambda|^2\right)^{s} |\widehat{\phi}(\lambda)|^2 \, d\lambda \simeq \|\phi\|^2_{\overset{\circ}{W_0}{}^{s}(\mathbb{R}^+)} ,
\end{aligned}$$

For the other operators one shows the assertions correspondingly. □

We want to show next the relation between the weighted Sobolev spaces $\overset{\circ}{W_0}{}^{s}(\mathbb{R}^+)$ and the Sobolev spaces $\widetilde{H}^s(\mathbb{R}_+)$ of $H^s(\mathbb{R}_+)$ functions that have a zero continuation on $\mathbb{R}_-$ in $H^s(\mathbb{R})$:

Lemma 9.2 ([284]) *Let* $\chi \in C_0^\infty[0,\infty)$ *. Then, the mapping*

$$\begin{cases} \overset{\circ}{W_0}{}^{s}(\mathbb{R}^+) \longrightarrow \widetilde{H}^s(\mathbb{R}_+) \\ \qquad u \mapsto \chi u \end{cases}$$

and its inverse $\left(\widetilde{H}^s(\mathbb{R}_+) \longrightarrow \overset{\circ}{W_0}{}^{s}(\mathbb{R}^+)\right)$ *are both continuous for* $s \geq 0$ *, i.e. the norms of* $\widetilde{H}^s(\mathbb{R}_+)$ *and* $\overset{\circ}{W_0}{}^{s}(\mathbb{R}^+)$ *for* $s \geq 0$ *are equivalent on compact intervals.*

Remark 9.1 It was shown in [204] that the mappings $u \mapsto \chi u$ from $H^s(\mathbb{R}_+)$ into $\overset{\circ}{W_0}{}^{s}(\mathbb{R}^+)$ and vice versa are continuous for $s \leq 0$, too.

Lemma 9.3 *For* $0 \leq s < \frac{3}{2}$ *there holds:*

i) $H^s(\Gamma^\omega) = \left\{u = (u_1, u_2) \in H^s(\mathbb{R}^+)^2 \mid u_- - u_+ \in \widetilde{H}^s(\mathbb{R}^+)\right\}$
ii) $H^{-s}(\Gamma^\omega) = \left\{u = (u_1, u_2) \in H^{-s}(\mathbb{R}^+)^2 \mid u_- + u_+ \in \widetilde{H}^{-s}(\mathbb{R}^+)\right\}$

Furthermore, the mappings

$$\begin{Bmatrix} R : H^s(\Gamma^\omega) \to H^s(\mathbb{R}^+) \times \widetilde{H}^s(\mathbb{R}^+) \\ (u_-, u_+) \mapsto (u_- + u_+, u_- - u_+) \end{Bmatrix} \text{ and } \begin{Bmatrix} R : H^{-s}(\Gamma^\omega) \to \widetilde{H}^{-s}(\mathbb{R}^+) \times H^{-s}(\mathbb{R}^+) \\ (u_-, u_+) \mapsto (u_- + u_+, u_- - u_+) \end{Bmatrix}$$

are isomorphisms.

Proof For the proof of the lemma we refer to [128, 204]. □

We now want to show that the mapping

$$\chi V \chi : H^{s-1}(\Gamma^\omega) \longrightarrow H^s(\Gamma^\omega) \tag{9.5}$$

is continuous for $s \in \left(-\frac{1}{2}, \frac{3}{2}\right)$, where χ shall denote the C^∞-cut-off-function, concentrated at $\{0\}$. By Lemma 9.3 we have that (9.5) is equivalent to the continuity of the mapping:

$$R\chi V \chi R^{-1} : \widetilde{H}^{s-1}(\mathbb{R}^+) \times H^{s-1}(\mathbb{R}^+) \longrightarrow H^s(\mathbb{R}^+) \times \widetilde{H}^s(\mathbb{R}^+) \text{ for } s \in \left(-\frac{1}{2}, \frac{3}{2}\right).$$

Note that $H^s = \widetilde{H}^s$ for $|s| < \frac{1}{2}$ and

$$R \widehat{=} \begin{pmatrix} 1 & 1 \\ 1 & -1 \end{pmatrix} \text{ and thus } R\chi V \chi R^{-1} \widehat{=} \chi \begin{pmatrix} 2I + V_0 + V_\omega & 0 \\ 0 & V_0 - V_\omega \end{pmatrix} \chi .$$

Hence, (9.5) will be shown by the following lemma:

Lemma 9.4 ([128]) *Let* $\chi \in C_0^\infty[0,\infty)$ *with* $supp(1-\chi) \subset\subset (0,\infty)$ *.*

Then the following mappings are continuous:

i) $\begin{cases} \widetilde{H}^s(\mathbb{R}^+) \longrightarrow H^{s+1}(\mathbb{R}^+) \\ \quad u \mapsto \chi(l + V_\omega)\chi u \end{cases}$ *for* $s \in \left(-\frac{3}{2}, \frac{1}{2}\right)$

ii) $\begin{cases} H^s(\mathbb{R}^+) \longrightarrow \widetilde{H}^{s+1}(\mathbb{R}^+) \\ \quad u \mapsto \chi(V_0 - V_\omega)\chi u \end{cases}$ *for* $s \in \left(-\frac{3}{2}, \frac{1}{2}\right)$

iii) $\begin{cases} \begin{cases} \widetilde{H}^s(\mathbb{R}^+) \\ H^s(\mathbb{R}^+) \end{cases} \longrightarrow \begin{cases} \widetilde{H}^s(\mathbb{R}^+) \\ H^s(\mathbb{R}^+) \end{cases} \\ \quad u \mapsto \chi K_\omega \chi u \end{cases}$ *for* $s \in \left(-\frac{1}{2}, \frac{3}{2}\right)$

In analogy to the above lemma and with the help of the following exercise one can show that the mapping

$$\chi(I + K)\chi : H^s(\Gamma^\omega) \longrightarrow H^s(\Gamma^\omega) \tag{9.6}$$

is continuous for $s \in \left(-\frac{1}{2}, \frac{3}{2}\right)$.

Exercise 9.1 *For the operators as defined above there holds*

$$\chi(I + K)\chi \hat{=} \chi \begin{pmatrix} 1 & K_\omega \\ K_\omega & 1 \end{pmatrix} \chi$$

and thus

$$R\chi(I + K)\chi R^{-1} \hat{=} \chi \begin{pmatrix} 1 + K_\omega & 0 \\ 0 & 1 - K_\omega \end{pmatrix} \chi .$$

Thus, (9.6) follows from the fact that the mapping

$$R\chi(I + K)\chi R^{-1} : H^s(\mathbb{R}^+) \times \widetilde{H}^s(\mathbb{R}^+) \longrightarrow H^s(\mathbb{R}^+) \times \widetilde{H}^s(\mathbb{R}^+) \tag{9.7}$$

is continuous for $s \in \left(-\frac{1}{2}, \frac{3}{2}\right)$, which itself is a consequence of Lemma 9.4.

Theorem 9.2 *For the single and double layer potential operators V and K there holds:*

i) $\exists \gamma = \gamma(I, \omega) > 0 \quad \forall v \in L^2(\Gamma^\omega)$ *with* $\operatorname{supp}(v) \subset\subset I \subset\subset \Gamma^\omega$:

$$\Re\left(\langle v, (I + K)v \rangle\right) \geq \gamma \|v\|^2_{L^2(\Gamma^\omega)} .$$

ii) $\exists \gamma = \gamma(I, \omega) > 0 \quad \forall \psi \in H^{-\frac{1}{2}}(\Gamma^\omega)$ *with* $\operatorname{supp}(\psi) \subset\subset I \subset\subset \Gamma^\omega$:

$$\Re\left(\langle \psi, V\psi \rangle\right) \geq \gamma \|\psi\|^2_{H^{-\frac{1}{2}}(\Gamma^\omega)} .$$

Proof We first note that on Γ^ω the operator K maps even (resp. odd) functions onto even (resp. odd) functions.

ad i):

We have:

$$\begin{aligned}
\langle v, (I+K)v\rangle_{L^2(\Gamma^\omega)} &= \left\langle v, R^{-1}\begin{pmatrix} I+K_\omega & 0 \\ 0 & I-K_\omega \end{pmatrix} Rv\right\rangle_{L^2(\Gamma^\omega)} \\
&= \frac{1}{2}\left\langle Rv, \begin{pmatrix} I+K_\omega & 0 \\ 0 & I-K_\omega \end{pmatrix} Rv\right\rangle_{L^2(\Gamma^\omega)} \\
&= \frac{1}{2}\Big\langle v_+ + v_-, (I+K_\omega)(v_+ + v_-)\Big\rangle_{L^2(\mathbb{R}^+)\times L^2(\mathbb{R}^+)} \\
&+ \frac{1}{2}\Big\langle v_- - v_+, (I-K_\omega)(v_- - v_+)\Big\rangle_{L^2(\mathbb{R}^+)\times L^2(\mathbb{R}^+)}\,.
\end{aligned}$$

Now, making use of *Parseval's* equation for the Mellin transformation we obtain for arbitrary $v \in C_0^\infty(0,\infty)$

$$\begin{aligned}
\Re\left(\langle v, (I\pm K_\omega)v\rangle_{L^2(\Gamma^\omega)}\right) &= \frac{1}{2\pi}\int\limits_{\Im(\lambda)=-\frac{1}{2}} \overline{\widehat{v(\lambda)}}\ \ \widehat{(I\pm K_\omega)v}(\lambda)\,d\lambda \\
&= \frac{1}{2\pi}\int\limits_{\Im(\lambda)=-\frac{1}{2}} \left(1 \mp \frac{\sinh(\pi-\omega)\lambda}{\sinh(\pi\lambda)}\right)|\widehat{v}(\lambda)|^2\,d\lambda\,.
\end{aligned}$$

There further holds

$$\left|\frac{\sinh(\pi-\omega)\lambda}{\sinh(\pi\lambda)}\right| \;\le\; \left|\sin\frac{\pi-\omega}{2}\right| \;=:\; q \overset{\omega\neq 0,2\pi}{<} 1 \quad \forall\lambda \text{ with } \Im(\lambda) = -\frac{1}{2}$$

and thus

$$\Re\left(\langle v, (I\pm K)v\rangle\right) \;\ge\; \frac{1-q}{2\pi}\int\limits_{\Im(\lambda)=-\frac{1}{2}} |\widehat{v}(\lambda)|^2\,d\lambda \;=\; \underbrace{c(1-q)}_{=:\gamma}\,\|v\|^2_{L^2(\Gamma^\omega)}\,.$$

Here, we note that on polygons the operator K is a contraction map, i.e.

$$\|Kv\|_{L^2(\Gamma^\omega)} \;\le\; \eta\|v\|_{L^2(\Gamma^\omega)} \text{ with } \eta < 1\,.$$

ad ii)

Analogously to i) we now have:

$$\begin{aligned}
\langle\psi, V\psi\rangle_{H^{-1/2}(\Gamma^\omega)\times H^{\frac{1}{2}}(\Gamma^\omega)} = 1/2 \Bigg(&\Big\langle(\psi_- + \psi_+), (V_0 + V_\omega)(\psi_- + \psi_+)\Big\rangle_{\widetilde{H}^{-\frac{1}{2}}(\mathbb{R}^+)\times H^{\frac{1}{2}}(\mathbb{R}^+)} \\
&+ \Big\langle(\psi_- - \psi_+), (V_0 - V_\omega)(\psi_- - \psi_+)\Big\rangle_{H^{-\frac{1}{2}}(\mathbb{R}^+)\times \widetilde{H}^{\frac{1}{2}}(\mathbb{R}^+)}\Bigg)
\end{aligned}$$

By *Parseval's* equation there holds for arbitrary $\psi \in C_0^\infty[0,\infty)$:

$$\begin{aligned}\langle \psi, (V_0 \pm V_\omega)\psi\rangle_{L^2(\Gamma^\omega)} &= \frac{1}{2\pi} \int\limits_{\Im(\lambda)=-\frac{1}{2}} \widehat{\psi}(\lambda)\left(\overline{\widehat{V_0}(\lambda)} \pm \overline{\widehat{V_\omega}(\lambda)}\right)\widehat{\psi}(\lambda - i)\, d\lambda \\ &= \frac{1}{2\pi} \int\limits_{\Im(\lambda)=0} \left(\overline{\hat{V}_0(\lambda)} \pm \overline{\hat{V}_\omega(\lambda)}\right)|\widehat{\psi}(\lambda - i)|^2\, d\lambda\end{aligned}$$

and thus

$$\langle \psi, (V_0 \pm V_\omega)\psi\rangle_{L^2(\Gamma^\omega)} = \frac{1}{2\pi} \int\limits_{\Im(\lambda)=0} m_\pm(\lambda)|\widehat{\psi}(\lambda - i)|^2\, d\lambda$$

$$\text{for} \quad m_\pm(\lambda) = \frac{\cosh(\pi\lambda) \pm \cosh(\pi - \omega)\lambda}{\lambda \sinh(\pi\lambda)} .$$

Note that for $\lambda \in \mathbb{R}$ we have

$$m_+(\lambda) \sim \frac{1 + |\lambda|}{|\lambda|^2} \qquad \text{and} \qquad m_-(\lambda) \sim \frac{1}{1 + |\lambda|} .$$

Hence,

$$\begin{aligned}\Re\left(\langle \psi, V\psi\rangle\right) &\geq \gamma \left(\int\limits_{\Im(\lambda)=0} \frac{1+|\lambda|}{|\lambda|^2} \left|\widehat{\psi}_-(\lambda - i) + \widehat{\psi}_+(\lambda - i)\right|^2 d\lambda \right. \\ &\qquad \left. + \int\limits_{\Im(\lambda)=-1} \frac{1}{1+|\lambda|} \left|\widehat{\psi}_-(\lambda) - \widehat{\psi}_+(\lambda)\right|^2 d\lambda \right) \\ &\geq \gamma \left(\|\psi_- + \psi_+\|^2_{\widetilde{H}^{-\frac{1}{2}}(\mathbb{R}^+)} + \|\psi_- - \psi_+\|^2_{\overset{\circ}{W}_0^{-\frac{1}{2}}(\mathbb{R}^+)} \right) \\ &\geq \gamma \cdot \|R\psi\|^2_{\widetilde{H}^{-\frac{1}{2}}(\mathbb{R}^+) \times H^{-\frac{1}{2}}(\mathbb{R}^+)} \\ &\geq \gamma \cdot \|\psi\|^2_{H^{-\frac{1}{2}}(\Gamma^\omega)} ,\end{aligned}$$

since there holds (see [128]):

$$\exists C > 0 \ \forall u \in C_0^\infty(0,\infty) : \ \|u\|^2_{\widetilde{H}^{-\frac{1}{2}}(\mathbb{R}^+)} \leq C \int\limits_{\Im(\lambda=0)} \frac{1 + |\lambda|}{|\lambda|^2} \cdot |\hat{u}(\lambda - i)|^2\, d\lambda ,$$

if the integral exists. □

9.2 Properties of the Mellin Transformation

Let $u \in C_0^\infty(0, \infty)$ be given. Then the Mellin transformed of u, defined by

$$\widehat{u}(\lambda) = \int_0^\infty x^{i\lambda-1} u(x)\, dx$$

is an entire, analytic function. Defining

$$u_h(x) := \frac{1}{2\pi} \int_{\Im(\lambda)=h} x^{-i\lambda} \hat{u}(\lambda)\, d\lambda ,$$

the residue theorem yields for u_{h_1} and u_{h_2} with $h_2 > h_1$ as in Fig. 9.2

$$u_{h_2} = u_{h_1} - i \sum_{\Im(\lambda)\in(h_1,h_2)} \text{Res}\, \hat{u}(\lambda) x^{-i\lambda} ,$$

for u_{h_2} being the meromorphic continuation of u_{h_1}.

Here, we assume that $\widehat{u}(\lambda)$ only has poles in the range $|\lambda| < M < \infty$ and that $\widehat{u}(\lambda)$ is rapidly decaying for $|\lambda| > \widetilde{M} > M$.

Lemma 9.5 ([128]) *Let $u \in \widetilde{H}^s_{comp}(\mathbb{R}_+)$ be given, i.e. for $s < t$, $\widehat{u}(\lambda)$ is meromorphic for $\Im(\lambda) < t - \frac{1}{2} =: k$ with poles of order $m_k + 1$ at the points $\lambda_l = i\alpha_l$ for $s - \frac{1}{2} < \alpha_l < k$, $1 \leq l \leq L$ and $\int_{\Im(\lambda)=s-1/2} \left(1 + |\lambda|^2\right)^s |\widehat{u}(\lambda)|^2\, d\lambda < \infty$.*

Then there holds:

i) For $u_k(x) \in \widetilde{H}^t(\mathbb{R}_+)$ and $\chi \in C_0^\infty[0, \infty)$ we have

$$u(x) = \sum_{l=1}^{L} \sum_{m=0}^{m_l} c_{lm} x^{\alpha_l} \log^m(x) \chi(x) + u_k(x) \tag{9.8}$$

with $\chi \equiv 1$ at $x = 0$ and $c_{lm} = \frac{i^{m+1}}{m!} Res_{\lambda=\lambda_l} \left\{(\lambda - \lambda_l)^m \hat{u}(\lambda)\right\}$.

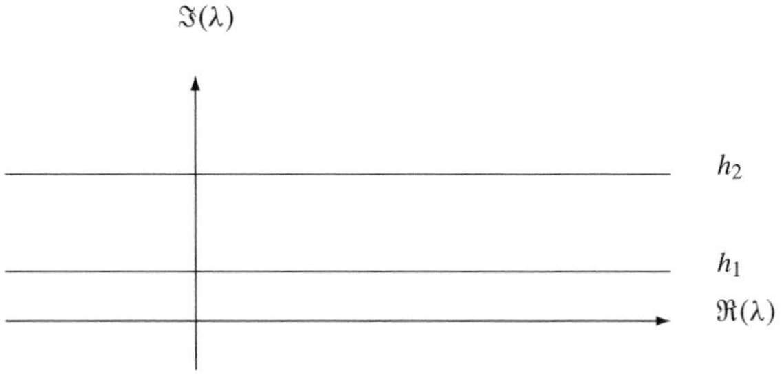

Fig. 9.2 Domain for Mellin transform

ii) $\sum_{l=1}^{L}\sum_{m=0}^{m_l} |c_{lm}|^2 \le c\left(\|u\|^2_{\widetilde{H}^s} + \|u_k\|^2_{\widetilde{H}^t}\right)$.

iii) On the contrary, if for all $u \in \widetilde{H}^s_{comp}(\mathbb{R}_+)$ *with a decomposition as in (9.8) there holds that* $u_k \in \widetilde{H}^t(\mathbb{R}_+)$, *then the Mellin-transformed function has the above properties.*

Proof Note that for $u(x) = x^\alpha \ln^l x\chi(x)$ we have

$$\widehat{u}(x) = \frac{\widehat{\phi}(x)}{(\lambda - i\alpha)^{l+1}}$$

where $\widehat{\phi}$ is an entire function of exponential type which is rapidly decreasing for $\Re\lambda \to \pm\infty$. Therefore the inverse Mellin transform

$$u_h(x) := \frac{1}{2\pi}\int\limits_{\Im\lambda=h} e^{i\lambda t}\hat{u}(\lambda)d\lambda \ \ (x = e^{-t} \in \mathbb{R}_+)$$

exists for $h \notin \{\alpha_1, \dots, \alpha_n\}$ and the path of integration may be shifted if we take into account the residues of $e^{i\lambda t}\hat{u}(\lambda)$. Thus we get $u_h = u$ for $h < \alpha_1$ and $u_{h_2}(x) - u_{h_1}(x) = -i\sum_{\Im\lambda\in(h_1,h_2)} \mathrm{Res}\{\hat{u}(\lambda)e^{i\lambda t}\}(h_1 < h_2)$.

Now

$$-i\mathrm{Res}_{\lambda=i\alpha_k}\{e^{i\lambda t}\hat{f}(\lambda)\} = \frac{-i}{(l_k)!}\left(\frac{d}{d\lambda}\right)^{l_k}[e^{i\lambda t}\hat{u}(\lambda)(\lambda - i\alpha_k)^{l_k+1}]\Big|_{x=i\alpha_k} = -\sum_{l=0}^{l_k} c_{kl}x^{\alpha_k}\log^l x$$

gives i). For further details see [128].

With this lemma we have:

$$u \in \widetilde{H}^s_{comp}(\mathbb{R}_+) \ \Rightarrow \ \widehat{u}(\lambda) \text{ is holomorphic for } \Im(\lambda) < s - \frac{1}{2},$$

where $u = u_h$ for $h < s - \frac{1}{2}$, i.e. if the solutions do lie in the energy-space, the parts below the energy-norm will cause smooth perturbations.

Lemma 9.6 *Let* $k,\ \phi \in C_0^\infty(0,\infty)$ *and* $\alpha,\ \beta \in \mathbb{C}$. *Then there holds:*

$$u(x) := \int\limits_0^\infty x^\alpha y^\beta k\left(\frac{x}{y}\right)\frac{\phi(y)}{y}\,dy \quad \Rightarrow \quad \widehat{u}(\lambda) = \widehat{k}(\lambda - i\alpha)\widehat{\phi}\Big(\lambda - i(\alpha+\beta)\Big).$$

Exercise 9.2 Prove Lemma 9.6 by using the result for the Fourier transform of the convolution of two functions in the Appendix together with the Euler transformation.

For the rest of this section we want to consider again the Dirichlet problem

$$\begin{aligned} -\Delta u &= 0 \qquad \text{in } \Omega, \\ u &= g \qquad \text{on } \Gamma := \partial\Omega, \end{aligned}$$

given as an integral equation of the form

$$V\frac{\partial u}{\partial n} = (I+K)g \quad \text{on} \quad \Gamma \tag{9.9}$$

with

$$\mathscr{A}\Psi = \mathscr{B}G \quad \text{on} \quad \Gamma^{\omega}$$

for

$$\mathscr{A} \widehat{=} \begin{pmatrix} V_0 & V_\omega \\ V_\omega & V_0 \end{pmatrix} + \begin{pmatrix} l & l \\ l & l \end{pmatrix}, \qquad \mathscr{B} \widehat{=} \begin{pmatrix} I & K_\omega \\ K_\omega & I \end{pmatrix},$$

$\Psi := (\psi_-, \psi_+)^t$, $G := (g_-, g_+)^t$ and Γ^ω corresponding to Fig. 9.3

We assume that $g_\pm \in C_0^\infty[0,\infty)$, $g_+(0) = g_-(0)$ and will then show that a solution Ψ of (9.9) has a representation of the form:

$$\psi_\pm(x) = \left(\sum_{k=1}^{n} \sum_{l=0}^{l_k} c_{kl}^{\pm} x^{\alpha_k - 1} \log^l x \right) \chi(x) + \underbrace{\psi_\pm^0(x)}_{\in \widetilde{H}^s(\mathbb{R}_+),\ s<3/2} .$$

Let $\mathscr{A}\Psi = \mathscr{B}G =: H$. We then have

$$\widehat{\mathscr{A}\Psi}(\lambda) = \widehat{\mathscr{A}}(\lambda)\widehat{\Psi}(\lambda - i) = \widehat{H}(\lambda)$$

with

$$\widehat{\mathscr{A}}(\lambda) = \begin{pmatrix} \widehat{V}_0(\lambda) & \widehat{V}_\omega(\lambda) \\ \widehat{V}_\omega(\lambda) & \widehat{V}_0(\lambda) \end{pmatrix} = \frac{1}{\lambda \sinh(\pi\lambda)} \begin{pmatrix} \cosh(\pi\lambda) & \cosh(\pi-\omega)\lambda \\ \cosh(\pi-\omega)\lambda & \cosh(\pi\lambda) \end{pmatrix}$$

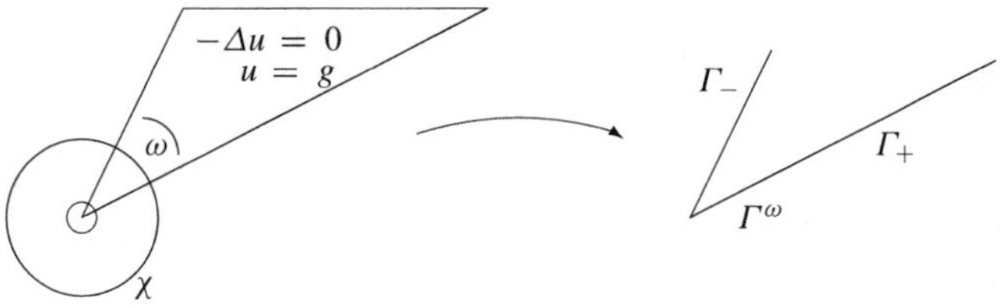

Fig. 9.3 Localization

for $\Im(\lambda) \in (-1,0)$. Here, $\widehat{\Psi}(\lambda-i)$ is the meromorphic extension of $\widehat{\chi\psi}(\lambda-i)$ for the region from $\Im(\lambda) \in (-1,0)$ to $\Im(\lambda) \in (-1, s-\frac{1}{2})$. For all $h \in \left(-1, s-\frac{1}{2}\right)$, $h \neq \Im(\lambda_p)$ with λ_p being a pole of $\widehat{\Psi}(\lambda-i) := \widehat{\mathscr{A}}(\lambda)^{-1} \cdot \widehat{H}(\lambda)$, the function

$$\Psi_h(x) = \frac{1}{2\pi} \int\limits_{\Im(\lambda)=h} \widehat{\Psi}(\lambda-i) x^{-i\lambda-1}\, d\lambda$$

exists. For $h \in (-1,0)$ there further holds that $\Psi_h(x) = \chi\psi(x)$. For all other h Cauchy's integral theorem yields:

$$\Psi_h(x) = \chi\psi(x) - i \sum_{\Im(\lambda)\in(-1,h)} \operatorname{Res}\left(\widehat{\Psi}(\lambda-i)x^{-i\lambda-1}\right).$$

The residuals at the poles of $\widehat{H}(\lambda)$ and $\widehat{\mathscr{A}}^{-1}(\lambda)$ will be given by:

(i) $\lambda = 0$ $\left(\text{pole of } \widehat{H}(\lambda)\right)$

(ii) $\lambda = ik\,,\ \ k \in \mathbb{N}$ $\left(\text{pole of } \widehat{H}(\lambda)\right)$

(iii) zeroes of $\det \widehat{\mathscr{A}}(\lambda)$ $\left(\text{poles of } \widehat{\mathscr{A}}^{-1}(\lambda)\right)$.

For the last item we have

$$\det \widehat{\mathscr{A}}(\lambda) = \frac{\sinh(2\pi-\omega)\lambda \cdot \sinh(\lambda\omega)}{\lambda^2 \sinh^2(\pi\lambda)}$$

and

$$\widehat{\mathscr{A}}^{-1}(\lambda) = \frac{\lambda \sinh(\pi\lambda)}{\sinh(2\pi-\omega)\lambda \cdot \sinh(\lambda\omega)} \begin{pmatrix} \cosh(\pi\lambda) & -\cosh(\pi-\omega)\lambda \\ -\cosh(\pi-\omega)\lambda & \cosh(\pi\lambda) \end{pmatrix}.$$

$$\Longrightarrow \qquad \sinh(2\pi-\omega)\lambda \cdot \sinh(\lambda\omega) \overset{!}{=} 0$$

$$\overset{\lambda=i\alpha,\ \alpha\in(0,2)}{\Longrightarrow} \begin{cases} \text{(i) } \sinh(2\pi-\omega)i\alpha = 0 \Leftrightarrow \alpha = \frac{l\pi}{2\pi-\omega} & , l = 1,2,3 \\ \text{(ii) } \qquad \sinh(i\alpha\omega) = 0 \Leftrightarrow \alpha = \frac{k\pi}{\omega} & , k = 1,2,3 \end{cases}$$

For more details the interested reader is referred to [128], where also the Neumann problem and the mixed Dirichlet-Neumann problem are considered.

9.2.1 Local Regularity at Vertices

In order to obtain a local representation of the solution of integral equations with singularity functions at the vertices of a polygon, we first need the regularity of the solution on the smooth parts of the boundary. This regularity is characterised by some standard a priori estimates making use of pseudodifferential operators. In this subsection we report from [138]

Lemma 9.7 *Let $\chi \in C_0^\infty$ be the cut-off-function with support inside of a segment Γ^j of Γ. Let $f \in H^s(\Gamma)$, $s \geq \frac{1}{2}$ and $\psi \in H^{-\frac{1}{2}}(\Gamma)$ be a solution of $V\psi = f$ on Γ. Then we have the a-priori estimate:*

$$\|\chi\psi\|_{H^{s-1}(\Gamma)} \leq C \cdot \left\{ \|f\|_{H^s(\Gamma)} + \|\psi\|_{H^{-\frac{1}{2}}(\Gamma)} \right\} \tag{9.10}$$

Proof We have $\chi V\psi = \chi f$ and will consider the following situation:
Then there holds (Fig. 9.4):

$$\chi V \chi_1 \psi = -\chi V(1-\chi_1)\psi + \chi f =: \chi h \,. \tag{9.11}$$

Here we note that with $\chi(1-\chi_1) \equiv 0$ the integral-kernel of $\chi V(1-\chi_1)$ is a C^∞-function. Thus,

$$\|\chi h\|_{H^s(\Gamma)} \leq C \cdot \left\{ \|f\|_{H^s(\Gamma)} + \|\psi\|_{H^{-\frac{1}{2}}(\Gamma)} \right\} \tag{9.12}$$

Therefore (9.11) may be conceived as an equation on a simply connected C^∞-curve $\widetilde{\Gamma}$, containing Γ^j with $\mathrm{cap}(\widetilde{\Gamma}) \neq 1$. On $\widetilde{\Gamma}$ we will now consider V^{-1} which is a pseudo-differential operator of order 1. We then have

$$V^{-1}\chi V \chi_1 \psi = \chi\chi_1\psi + \underbrace{\left(V^{-1}\chi - \chi V^{-1}\right)}_{\text{pseudo-diff. op. of order 0}} V\chi_1\psi \overset{\chi=\chi\chi_1}{=} V^{-1}\chi h$$

This yields the estimate

$$\|\chi\psi\|_{H^{s-1}(\Gamma)} \leq C \cdot \left\{ \|\chi_1\psi\|_{H^{s-2}(\Gamma)} + \|\chi h\|_{H^s(\Gamma)} \right\} .$$

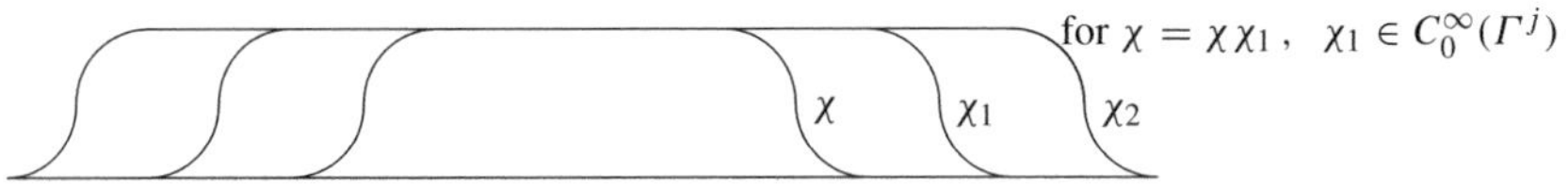

Fig. 9.4 Cut-off functions

Repeating the above arguments gives

$$\|\chi_1\psi\|_{H^{s-2}(\Gamma)} \le C' \cdot \left\{\|\chi_2\psi\|_{H^{s-3}(\Gamma)} + \|\chi_1 h\|_{H^{s-1}(\Gamma)}\right\}.$$

with $\chi_2 \in C_0^\infty(\Gamma^j)$ such that $\chi_1 = \chi_1\chi_2, \quad \chi_1(1-\chi_2) \equiv 0$. After a finite number of applications of these arguments we obtain:

$$\|\chi\psi\|_{H^{s-1}(\Gamma)} \le C \cdot \left\{\|\chi_n\psi\|_{H^{-\frac{1}{2}}(\Gamma)} + \|\chi_n h\|_{H^s(\Gamma)}\right\},$$

which, in combination with (9.12), proves the lemma. For further details see [138]. □

For the regularity at the vertices we will use the Mellin-transformed equations and Cauchy's integral theorem. One may observe that the singularity-functions are given by:

$$\psi \cong (\psi_-, \psi_+) = (c_-, c_+)x^{-i\lambda-1}\chi(x), \tag{9.13}$$

for λ being the zero of the transcendental equation

$$\sinh(2\pi - \omega)\lambda \cdot \sinh(\omega\lambda) = 0 \tag{9.14}$$

and $(c_-, c_+) \in \mathbb{C}^2$ the corresponding eigenvector of the Mellin-symbol of V. In the case that λ is a double zero of (9.14) there holds:

$$\psi \cong \sum_{l=0}^{1}(c_{l_-}, c_{l_+})x^{-i\lambda-1}\log^l(x)\chi(x). \tag{9.15}$$

The local regularity results for the solution of the integral equation may be summed up as follows:

Theorem 9.3 *Let $f \in H^s(\Gamma)$, $s > 1/2$, $s \ne \frac{1}{2} + \Im(\lambda)$ for all zeroes λ of (9.14). Let further $\psi \in H^{-\frac{1}{2}}(\Gamma)$ be a solution of $V\psi = f$ on Γ. Then ψ has the local representation*

$$\chi\psi = \chi\psi^{(s)} + \sum_{0<\Im(\lambda_k)<s-\frac{1}{2}} c_k v_k, \qquad c_k \in \mathbb{R},$$

for v_k being functions of the form of either (9.13) or (9.15) with λ being a zero of (9.14) or $\lambda_k = i \cdot m$, $m = 1, 2, \ldots$ and $\chi\psi^{(s)} \in H^{s-1}(\Gamma^\omega)$. Further there holds

$$\|\chi\psi^{(s)}\|_{H^{s-1}(\Gamma^\omega)} + \sum_{0<\Im(\lambda_k)<s-\frac{1}{2}} |c_k| \le C \cdot \left\{\|f\|_{H^s(\Gamma)} + \|\psi\|_{H^{-\frac{1}{2}}(\Gamma)}\right\}.$$

Proof First we will transform the integral equation $V\psi = f$ on Γ such that Mellin transformation can be applied. To this end we take the cut-off-functions $\chi_1, \chi_2, \chi_3 \in C_0^\infty(\mathbb{C})$, which only depend on $|z|$ with

$$0 \le \chi_j \le 1 \quad \text{and} \quad \chi_j \equiv \begin{cases} 1 & \text{for } |z| \le \alpha_j \\ 0 & \text{for } |z| \ge \beta_j, \end{cases}$$

with $j = 1, \ldots, 3$; $\alpha_j < \beta_j$ and $\beta_1 < \alpha_2$, $\beta_2 < \alpha_3$. Furthermore the support of χ_j contains only the corner at the origin. In the neighbourhood of this vertex we have

$$V\psi = f \iff \chi_2 V \chi_1 \psi = \chi_2 \Big(f - V(1-\chi_1)\psi \Big) =: \chi_2 F \quad \text{on } \Gamma .$$

Obviously, the same will hold on Γ^ω. The right hand side of the above equation satisfies the relation

$$\|\chi_2 F\|_{H^s(\Gamma^\omega)} \le C \cdot \left\{ \|f\|_{H^s(\Gamma)} + \|\psi\|_{H^{-\frac{1}{2}}(\Gamma)} \right\},$$

since

$$\chi_2 V(1-\chi_1)\psi = \chi_2 V(1-\chi_1)\chi_3\psi + \chi_2 V(1-\chi_1)(1-\chi_3)\psi ,$$

where on one hand

$$\|\chi_2 V(1-\chi_1)\chi_3\psi\|_{H^s(\Gamma^\omega)} \le C \cdot \|(1-\chi_1)\chi_3\psi\|_{H^{s-1}(\Gamma)} \le C \cdot \left\{ \|f\|_{H^s(\Gamma)} + \|\psi\|_{H^{-\frac{1}{2}}(\Gamma)} \right\}$$

(by Lemma 9.7 and $\chi = (1-\chi_1)\chi_3$)
and on the other hand we have

$$\|\chi_2 V(1-\chi_1)(1-\chi_3)\psi\|_{H^s(\Gamma^\omega)} \le C \cdot \|\psi\|_{H^{-\frac{1}{2}}(\Gamma)},$$

since the operator has a C^∞-kernel. Thus for the rest of the proof we only need to give a lower estimate for $\|\chi_2 V \chi_1 \psi\|_{H^s(\Gamma^\omega)}$. For that purpose we approximate $\chi_1\psi$ in $H^{-\frac{1}{2}}(\Gamma^\omega)$ by $\phi \in C_0^\infty(\Gamma^\omega)$. Then $\widehat{\phi}(\lambda - i)$ converges to $\widehat{\chi_1\psi}(\lambda - i)$ for $\Im(\lambda) < 0$ and thus $\widehat{V\chi_1\psi}(\lambda) = \widehat{V}(\lambda)\widehat{\chi_1\psi}(\lambda - i)$ for $\Im(\lambda) \in (-1, 0)$.

$$\Longrightarrow \quad V\chi_1\psi = (1-\chi_2)V\chi_1\psi + \chi_2 F =: H \quad \text{on } \Gamma^\omega .$$

Application of the Mellin-transformation now yields:

$$\widehat{V\chi_1\psi}(\lambda) = \widehat{H}(\lambda) \quad \forall \Im(\lambda) \in (-1, 0),$$

which can be continued meromorphically for all λ with $\Im(\lambda) \in (-1, s - \frac{1}{2})$.

The rest of the proof easily follows with the estimates of the last section. □

9.3 A Direct Boundary Element Method for Interface Crack Problems

We present from [424] a Galerkin boundary element method to solve the transmission problem with a crack in the interface for the Helmholtz equation. We investigate the regularity for polygons and improve the rate of convergence by various methods: (i) refinement of the mesh toward the crack tips, (ii) augmenting the finite element space by singular functions, (iii) increasing the order of polynomials for fixed mesh (p-method). For other boundary value problems or elasticity problems one can proceed analogously, applying the convergence analysis in [128] to the general method in [422].

We consider a bounded Lipschitz domain $\Omega_1 \subset \mathbb{R}^2$ and its exterior $\Omega_2 = \mathbb{R}^2 \setminus \overline{\Omega_1}$. The boundary $\partial\Omega_1$ consists of a crack and a transmission part. The two sides of the crack facing Ω_1 and Ω_2 are denoted by Γ_1 and Γ_4,resp., and the two sides of the interface by Γ_2 and Γ_3. The normal vector on Γ_1 and Γ_2 points to Ω_2, the normal vector on Γ_3 and Γ_4 points to Ω_1. The crack tips are denoted by z_1 and z_2. We look for solutions of the following crack transmission problem: Find $u_j \in H^1_{loc}(\Omega_j),\ j = 1, 2$; satisfying

$$\left(\Delta + k_j^2\right) u_j = 0 \text{ in } \Omega_j,\ j = 1, 2 \tag{9.16}$$

$$u_1|_{\Gamma_2} = u_2|_{\Gamma_3} + v_2, \qquad \frac{1}{\rho_1}\frac{\partial u_1}{\partial n}\bigg|_{\Gamma_2} = -\frac{1}{\rho_2}\frac{\partial u_2}{\partial n}\bigg|_{\Gamma_3} + \psi_2 \tag{9.17}$$

$$\frac{1}{\rho_1}\frac{\partial u_1}{\partial n}\bigg|_{\Gamma_1} = \psi_1 \qquad \frac{1}{\rho_2}\frac{\partial u_2}{\partial n}\bigg|_{\Gamma_4} = \psi_4 \tag{9.18}$$

Here we assume for the wave numbers $\operatorname{Im} k_j^2 \geqslant 0,\ k_j \neq 0$ and for the densities $\rho_j > 0$. The it is shown in [422] that for given boundary data $\psi_1, v_2, \psi_2, \psi_4$ in the space

$$Y = \{(\psi_1, v_2, \psi_2, \psi_4) \in H^{-1/2}(\Gamma_1)\times H^{1/2}(\Gamma_2) \times H^{-1/2}(\Gamma_2) \times H^{-1/2}(\Gamma_4) \mid (\psi_1 + \psi_4, \psi_2) \in H^{-1/2}(\partial\Omega_1)\} \tag{9.19}$$

there exists a unique solution u_1, u_2 of (9.16)–(9.18). The unknown boundary data $\left(u_1|_{\Gamma_1}, \rho_1^{-1} \frac{\partial u_1}{\partial n}\Big|_{\Gamma_2}, u_1|_{\Gamma_2}, u_2|_{\Gamma_4}\right)$ are an element of the affine space

$$X =\{(w_1, \phi_2, w_2, w_4) \in H^{1/2}(\Gamma_1) \times H^{-1/2}(\Gamma_2) \times H^{1/2}(\Gamma_2) \times H^{1/2}(\Gamma_4) \mid (w_1, w_2) \in H^{1/2}(\partial\Omega_1), (w_4, w_2 - v_2) \in H^{1/2}(\partial\Omega_2), (\psi_1, \phi_2) \in H^{-1/2}(\partial\Omega_1)\}. \tag{9.20}$$

Following the method in [422] we derive the following boundary integral equation for the unknown boundary data $\zeta = (w_1, \phi_2, w_2, w_4) \in X$

$$H\zeta = g \tag{9.21}$$

where g is determined by the given boundary data $(\psi_1, v_2, \psi_2, \psi_4) \in Y$ and

$$H = \begin{bmatrix} \rho_1^{-1} W_{11} & K_{12}' & \rho_1^{-1} W_{12} & 0 \\ -K_{21} & \rho_1 V_{22} + \rho_2 V_{33} & -K_{22} + K_{33} & K_{34} \\ \rho_1^{-1} W_{21} & K_{22}' - K_{33}' & \rho_1^{-1} W_{22} + \rho_2^{-1} W_{33} & \rho_2^{-1} W_{34} \\ 0 & -K_{43}' & \rho_2^{-1} W_{43} & \rho_2^{-1} W_{44} \end{bmatrix} \tag{9.22}$$

The boundary integral operators in H are the traces and normal derivatives of the single and double layer potential, which are defined with the fundamental solution $G_j(x, y) = \frac{i}{4} H_0^{(1)}(k_j|x - y|)$ of (9.16). Because of the singularities these integrals are understood as finite part integrals: ($r = 1$ for $i, j \leqslant 2$, $r = 2$ for $i, j \geqslant 3$, $x \in \Gamma_i$)

$$V_{i,j}\psi_j(x) = 2\int_{\Gamma_j} G_r(x, y)\psi_j(y)ds_y, \quad K_{ij}\psi_j(x) = 2\frac{\partial}{\partial n_x}\int_{\Gamma_j} G_r(x, y)\psi_j(y)ds_y,$$

$$K_{ij}'v_j(x) = 2\int_{\Gamma_j} \frac{\partial}{\partial n_y} G_r(x, y)v_j(y)ds_y, \quad W_{ij}v_j = -2\frac{\partial}{\partial n_x}\int_{G_j} \frac{\partial}{\partial n_y} G_r(x, y)v_j(y)ds_y.$$

There hold the following results for (9.21) (see [424]):

Theorem 9.4 *The interface crack problem* (9.16)–(9.18) *and the boundary integral equation* (9.21) *are equivalent: Let* $u_j \in H^1_{loc}(\Omega_j)$ *be a weak solution of* (9.16)–(9.18), *then*

$$\zeta = \left(u_1|_{\Gamma_1}, \rho_1^{-1} \frac{\partial u_1}{\partial n}\bigg|_{\Gamma_2}, u_2|_{\Gamma_2}, u_2|_{\Gamma_4} \right)$$

belongs to X and satisfies (9.21). *If on the other hand* $\zeta = (w_1, \phi_2, w_2, w_4)$ *satisfies* (9.21)*, then* w_1, w_2, ψ_1, ϕ_2 *define by Green's representation formula a function* $u_1 \in H^1_{loc}(\Omega_1)$ *and* $w_3 := w_2 - v_2$, w_4, $\phi_3 := -\phi_2 - \psi_2$, ψ_4 *define a function* $u_2 \in H^1_{loc}(\Omega_2)$ *such that* u_1 *and* u_2 *satisfy* (9.16)–(9.18).

Theorem 9.5

(I) *Let* $(\psi_1, v_2, \psi_2, \psi_4) \in Y$ *be given. Then there exists exactly one solution* $\zeta \in X$ *of the integral equation* (9.21).

(II) *There exists a compact operator* $T_0 : \tilde{X} \to Y$ *and* $\gamma > 0$ *such that for all* $\zeta \in \tilde{X}$

$$\Re \big\langle (H + T_0)\, \zeta, \overline{\zeta} \big\rangle_{Y \times \tilde{X}} \geqslant \gamma \, \|\zeta\|^2_{\tilde{X}} \tag{9.23}$$

where $\tilde{X}$ *is the homogeneous space corresponding to X. This is the dual space of Y and* $\langle \cdot, \cdot \rangle_{Y \times \tilde{X}}$ *denotes the natural duality.*

A general Galerkin procedure involves a family of finite dimensional affine subspaces $S_h^0 \subset X$ for $h \in (0, h_0)$ such that $\bigcup\limits_{h>0} S_h^0$ is dense in X, and solving the problem: Find $\zeta_h \in S_h^0$ such that for all $\xi_h \in \tilde{S}_h^0$

$$\langle H\zeta_h - g, \xi_h \rangle = 0 \tag{9.24}$$

where $\tilde{S}_h^0$ denotes the homogeneous space corresponding to S_h^0.

Now Gårding's inequality (9.23) and the invertibility of H together imply the following quasioptimal error estimate for the Galerkin procedure (9.24) by standard arguments [408] (see Theorem 6.1):

Proposition 9.1 *There exists* $h_0 > 0$ *such that for any* $h \in (0, h_0)$ *the Galerkin equations* (9.24) *have a unique solution* ζ_h, *and there exists* $C > 0$ *such that for the exact solution* $\zeta \in X$ *of* (9.21) *holds*

$$\|\zeta - \zeta_h\|_{\tilde{X}} \leqslant C \inf_{\xi_h \in S_h^0} \|\zeta - \xi_h\|_{\tilde{X}} \,. \tag{9.25}$$

We choose $S^{t,k}$*-systems of finite elements for* S_h^0 :

$$S_h := S_h^{t,k}(\Gamma_1) \times S_h^{t-1,k-1}(\Gamma_2) \times S_h^{t,k}(\Gamma_2) \times S_h^{t,k}(\Gamma_4), t \in \mathbb{N},\ k \in \mathbb{N},\ t > k \tag{9.26}$$

and define S_h^0 as the affine subspace of S_h which satisfies the inhomogeneous compatibility conditions in the crack tips:

$$w_1^h(z_i) = w_2^h(z_i),\ \ w_2^h(z_i) = w_4^h(z_i) + v_2(z_i),\ \ i = 1, 2.$$

In order to derive convergence rates in (9.25) *we need to know the regularity of the exact solution ζ of* (9.21)*. Let $\mathscr{H}^s(\Gamma_j) := \prod_{l=1}^{n_j} H^s\left(\Gamma_j^l\right)$ where Γ_j^l, $l = 1, \ldots, n_j$, are the sides of the polygon contained in Γ_j. Then define*

$$Y^s = \left\{(\psi_1, v_2, \psi_2, \psi_4) \in Y|_{v_2} \in H^{1/2}(\Gamma_2), \psi_j \in \mathscr{H}^{-1/2+s}(\Gamma_j),\ j = 1, 2, 4\right\}$$

$$X^s = \left\{(w_1, \phi_2, w_2, w_4) \in X|_{w_j} \in H^{1/2+s}(\Gamma_j),\ j = 1, 2, 4,\ \phi_2 \in \mathscr{H}^{-1/2+s}(\Gamma_j)\right\}$$

Proceeding as in[128],[129] we use the Mellin transformation and obtain, see Sect. 9.1

Theorem 9.6 *Let $(\psi_1, v_2, \psi_2, \psi_4) \in Y^s$ be given. Then the solution ζ of* (9.21) *has the form*

$$\zeta = \sum_{j=1}^{J} \sum_{l=1}^{L_j^*} C_{jl}\zeta_{jl} + \zeta_s \tag{9.27}$$

with $\zeta_s \in X^s$, $s \notin A$ where A is the set of singular exponents described below. Here the singular functions $\zeta_{jl} = \left(w_1^{jl}, \phi_2^{jl}, w_2^{jl}, w_4^{jl}\right)$ are of the form

$$w^{jl}(x) = \left|x - z_j\right|^{\alpha} \log^r \left|x - z_j\right| \chi_j(x), \quad \phi^{jl}(x) = \left|x - z_j\right|^{\alpha-1} \log^r \left|x - z_j\right| \chi_j(x)$$

where z_j, $j = 1 \ldots J$, are the crack tips and the corners of the polygon, $\chi_j(x)$ are cut-off functions. The singular exponents α are either in $\mathbb{N}$ or a zero of $\det \hat{H}_{(j)}(\mathrm{i}\alpha)$*. $H_{(j)}$ is the boundary integral operator describing the problem in the neighbourhood of z_j, and $\hat{H}_{(j)}(\lambda)$ is its Mellin symbol.*

For corners z_j in the crack with angle ω_j one gets the well-known singularities for Neumann problems, namely $\alpha = k\frac{\pi}{\omega_j}$, $k \in \mathbb{N}$ (see [128]). For corners in the interface the results of [129] apply and yield the transcendental equation $(\rho_1 + \rho_2)^2 \sin^2 \pi\alpha = (\rho_1 - \rho_2)^2 \sin^2(\pi - \omega_j)\alpha$. For the singularity in the crack tips we have to calculate the determinant of the Mellin symbol of H in (9.22). This gives the transcendental equation

$$(\rho_1 + \rho_2)^2 \sin^2 \pi 2\alpha = (\rho_1 - \rho_2)^2 \sin^2(\pi - \omega)2\alpha \tag{9.28}$$

Hence the singular exponents are half the exponents of a corner in the interface. For $\omega = \pi$ or $\rho_1 = \rho_2$ we get $\alpha = \frac{1}{2}$, the known singularity of a crack. For $\omega \neq \pi$

let $\omega' = \min\{\omega, 2\pi - \omega\}$, $\alpha_o = \frac{\pi}{2(2\pi-\omega')} > \frac{1}{4}$, $\alpha^o = \min\{\frac{\pi}{2\omega}, \frac{\pi}{(2-\omega')}\} \leqslant \frac{3}{4}$. Then (9.28) has in $(0, \alpha^o)$ exactly two real solutions α_1, α_2 with $\alpha_o < \alpha_1 < \frac{1}{2} < \alpha_2 < \alpha^o$ and $\alpha_1 \to \alpha_o$, $\alpha_2 \to \alpha^o$ as $\frac{\rho_1}{\rho_2}$ tends to zero or to infinity.

Applications of Theorem 4 to the estimate (9.25) yields the convergence rate $\mathcal{O}\left(h^{\alpha_0-\varepsilon}\right)$ where $\alpha_0 = \min A$, $\frac{1}{4} < \alpha_0 \leqslant \frac{1}{2}$. T his low convergence rate can be improved by various methods:

(i) Refinement of the mesh towards the crack tips by a grading paramter $\beta \geqslant 1$: For $\beta > \frac{3}{2\alpha_0}$ we obtain for $t = 2$, $k = 1$ in (9.26) a convergence rate of $\mathcal{O}\left(h^{3/2-\varepsilon}\right)$ (see [422]).

(ii) By augmenting the boundary elements by all singular functions with exponents smaller than some $\alpha_1 \in A$, we get a convergence rate of $\mathcal{O}\left(h^{\alpha_1-\varepsilon}\right)$ if $t > \alpha_1$ in (9.26) (see [128], [129].

(iii) Performing the $p-$method where the degree p of the polynomials on a fixed mesh is increased gives a convergence rate of $\mathcal{O}\left(p^{-2\alpha_0+2\epsilon}\right)$, using results in [405], [404]. This is twice the rate of the $h-$version with uniform mesh.

9.4 Mixed BVP of Potential Theory on Polygons

The following boundary integral operators (and some closely related to them) appear in these applications:

$$Vf(z) = \int_\Gamma f(\zeta)G(z,\zeta)ds_\zeta;$$

$$Kf(z) = \int_\Gamma f(\zeta)\frac{\partial}{\partial n_z}G(z,\zeta)ds_\zeta;$$

$$K'f(z) = \int_\Gamma f(\zeta)\frac{\partial}{\partial n_z}G(z,\zeta)ds_\zeta;$$

$$Wf(z) = -\frac{\partial}{\partial n_z}\int_\Gamma f(\zeta)\frac{\partial}{\partial n_z}G(z,\zeta)ds_\zeta;$$

$$Sf(z) = \frac{1}{i\pi}\int_\Gamma (\zeta - z)^{-1}f(\zeta)d\zeta,$$

where $\frac{1}{2}G(z,\zeta)$ is the fundamental solution $-(1/2\pi)\ln|z-\zeta|$ of the Laplace or $(i/4\pi)H_0^{(l)}(k|z-\zeta|)$ of the Helmholtz equation.

As an example, we consider the mixed Dirichlet-Neumann problem with given Cauchy data $g = u$ on Γ_1 and $h = \frac{\partial u}{\partial n}$ on Γ_2, and unknown Cauchy data $v := u$ on

Γ_2 and $\psi := \frac{\partial u}{\partial n}$ on Γ_1. The representation formula is

$$2u(z) = \int_\Gamma u(\zeta)\frac{\partial}{\partial n_\zeta}G(z,\zeta)ds_\zeta + \int_\Gamma \frac{\partial u}{\partial n_\zeta}(\zeta)G(z,\zeta)ds_\zeta \quad \text{for } z \in \Omega_1,$$

where $\partial u/\partial n$ means the derivative with respect to the normal pointing from Ω_1 to Ω_2.

Taking the limit of $u(z)$ for $z \in \Gamma_2$ and the normal derivative $\partial u(z)/\partial n$ for $z \in \Gamma_1$ in this formula and using the jump relations, one finds the system

$$\begin{pmatrix} W_{22} & K'_{12} \\ -K_{21} & V_{11} \end{pmatrix}\begin{pmatrix} v \\ \psi \end{pmatrix} = \begin{pmatrix} -W_{12} & 1-K'_{22} \\ 1+K_{11} & -V_{21} \end{pmatrix}\begin{pmatrix} g \\ h \end{pmatrix} \tag{9.29}$$

where the subscripts at W_{jk} etc. mean: integration over Γ_j and evaluation on Γ_k.

This is a system of first kind integral equations for v and ψ. The same problem has been studied in [128] using a second kind integral equation for v instead of the first line in (9.29). This was obtained by taking the limit of $u(z)$ for $z \in \Gamma_1$ instead $\partial u/\partial n$. The system then had to be modified in order to satisfy a Gårding inequality, whereas the system (9.29) does this without modification. Another advantage is that the natural bilinear form associated with (9.29) is equivalent to the "energy norm", i.e., the natural norm for the Cauchy data of the weak (or variational) solution of the problem. On the other hand, (9.29) contains the operator W of the normal derivative of the double layer potential having a hypersingular kernel that therefore was (for obvious reasons) used to be avoided in applications.

Local properties of the boundary integral operators are studied using an infinite sector $\{z \in \mathbb{C} | 0 < \arg z < \omega\}$ as a model domain. The boundary is then

$$\Gamma^\omega = \overline{\Gamma}_- \dot{\subset} \overline{\Gamma}_+ \text{ with } \Gamma_- = e^{i\omega}\cdot\mathbb{R}_+$$
$$\text{and } \Gamma_+ = \mathbb{R}_+,$$

and the equations on Γ are reduced to Γ^ω by means of standard localizing techniques, [128], Sect. 9.1 . In the localized forms of the operators appear integral operators on $\mathbb{R}_+$ with kernels on the following form

$$V_\omega(x,y) = -\frac{1}{\pi}\ln|1 - \frac{x}{y}e^{i\omega}|;$$

$$K_\omega(x,y) = \frac{1}{\pi}\Im\left(\frac{1}{xe^{i\omega}-y}\right); \quad K'_\omega(x,y) = \frac{1}{\pi}\Im\left(\frac{e^{i\omega}}{xe^{i\omega}-y}\right)$$

$$W_\omega(x,y) = -\frac{1}{x}\frac{\partial}{\partial\omega}K_\omega(x,y).$$

These kernels are functions positively homogeneous in (x, y) and thus for the corresponding operators the Mellin convolution theorem (Lemma 9.6) can be applied.

If the differential operator contains lower order terms or the boundary Γ is curved near the corners, then the kernels contain in addition an expansion

$$\sum_{l,\ell,n} x^k y^\ell (xe^{i\omega} - y)^{-n} \tag{9.30}$$

into terms of ascending degree of homogeneity. Finally, the localization procedure produces additional smoothing operators thus introducing smooth or compact perturbations of the results obtained by the symbolic calculus.

The convolution theorem (Lemma 9.6) gives for $f \in C_0^\infty(0, \omega)$ the Mellin symbols:

$$\widehat{V_\omega f}(\lambda) = \frac{\cosh(\pi - \omega)\lambda}{\lambda \sinh \pi\lambda} \hat{f}(\lambda - i) =: \hat{V}_\omega(\lambda)\hat{f}(\lambda - i)$$

$$(\Im\lambda \in (0, 1));$$

$$\widehat{K_\omega f}(\lambda) = -\frac{\sinh(\pi - \omega)\lambda}{\sinh \pi\lambda} \hat{f}(\lambda) =: \hat{K}_\omega(\lambda)\hat{f}(\lambda)$$

$$(\Im\lambda \in (-1, 1));$$

$$\widehat{K'_\omega f}(\lambda) = \hat{K}_\omega(\lambda + i)\hat{f}(\lambda) \qquad (\Im\lambda \in (-2, 0));$$

$$\widehat{W_\omega f}(\lambda) = -(\lambda + i)^2 \hat{V}_\omega(\lambda + i)\hat{f}(\lambda + i) \qquad (\Im\lambda \in (-2, 0)).$$

Introducing this into the localized form of the boundary integral equations $AU = F$, and including smoothing parts into the right hand side Γ, one finds the transformed equation

$$\hat{A}(\lambda)\hat{U}(\lambda) = \hat{F}(\lambda) - \sum_{k=l}^{m} \hat{A}^k(\lambda)\hat{U}(\lambda - ki). \tag{9.31}$$

The additional terms on the right side correspond to the lower order terms (9.30) in the kernels [128].

In our example we have locally, at a corner where the boundary condition changes, Γ_- corresponding to Γ_2 and Γ_+ to Γ_1:

$$A = \begin{pmatrix} -W_0 & K'_\omega \\ -K_\omega & V_0 \end{pmatrix}, \tag{9.32}$$

hence

$$\hat{A}(\lambda) = \begin{pmatrix} \lambda^2 \hat{V}_0(\lambda) & \hat{K}'_\omega(\lambda) \\ -\hat{K}_\omega(\lambda) & \hat{V}_0(\lambda) \end{pmatrix};$$
$$\hat{U}(\lambda) = \begin{pmatrix} \hat{V}(\lambda) \\ \hat{\psi}(\lambda - i) \end{pmatrix}; \tag{9.33}$$
$$\hat{F}(\lambda) = \begin{pmatrix} \hat{f}_1(\lambda - i) \\ \hat{f}_2(\lambda) \end{pmatrix}.$$

The right hand side in (9.31) is converted by local Mellin transforms into

$$\hat{F}(\lambda) = \begin{pmatrix} \lambda^2 \hat{V}_\omega(\lambda) & 1 \\ 1 & -\hat{V}_\omega(\lambda) \end{pmatrix} \begin{pmatrix} g(\lambda) \\ \hat{h}(\lambda - i) \end{pmatrix}.$$

Having calculated the Mellin symbols, one uses the Parseval relation and the connection between singular expansion at the origin and poles of the Mellin transform (Lemma 9.5) to deduce from equation (9.31) results of the following kind:

(i) Continuity of the operators in Sobolev spaces $H^s(\Gamma)$, $H^s(\Gamma_j)$ or $\tilde{H}^s(\Gamma_j)$, ($j = 1, 2$), where, as usual, for $s > 0$ $H^s(\Gamma)$ is the space of traces of $H^{s+1/2}(\mathbb{R}^2)$ on Γ, $H^0(\Gamma) = L^2(\Gamma)$, and $H^{-s}(\Gamma$ is the dual space of $H^s(\Gamma)$. Furthermore, $\tilde{H}^s(\Gamma_1 = \left\{u \in H^s(\Gamma) | u|_{\Gamma_2} = 0\right\}$ and $H^s(\Gamma_1) = H^s(\Gamma)/\tilde{H}^s(\Gamma_2)$ have to be used.
(ii) A Gårding inequality on the boundary.
(iii) Regularity of the solution. This includes an expansion into explicitly given singular functions at the corners and a more regular part, and an a-priori estimate for the coefficients of the singular functions and for the regular part in terms of Sobolev norms.

These results are then used to reach the following conclusions: Firstly, Fredholm's alternative holds, yielding existence of a solution of the boundary value problem if uniqueness is known. (Uniqueness, in general, has to be deduced from different principles).

Secondly, standard arguments for approximation schemes yield convergence and quasi-optional asymptotic error estimates for the general Galerkin procedure applied to the integral equations. The regularity results (iii) show that higher convergence rates are obtained by including the singular functions in the approximate solution.

In our example, the above results are obtained in the following way (we always localize at a "mixed" corner, the cases of "Dirichlet" or "Neumann" corners have been treated earlier (see Sect. 9.2):

(i) Here, the symbol $\hat{A}(\lambda)$ has to be estimated from above: For $s \in \left(-\frac{1}{2}, \frac{3}{2}\right)$, one finds for $\Im\lambda = s - \frac{1}{2}$ the qualitative behaviour:

$$\left|\hat{V}_0(\lambda)\right| \approx \left(1 + |\lambda|^2\right)^{-1/2},$$
$$\left|\hat{K}_\omega(\lambda)\right| \approx e^{-\min\{\omega, 2\pi-\delta\}\cdot|\lambda|},$$

hence,

$$\left|\hat{A}(\lambda)\hat{U}(\lambda)\right|^2 \le C\left(\left(1 + |\lambda|^2\right)|\hat{v}(\lambda)|^2 + \left(1 + |\lambda|^2\right)^{-1}|\hat{\psi}(\lambda - i)|^2\right).$$

By Parseval's relation, this means continuity from $H^s(\mathbb{R}_+) \oplus H^{s-1}(\mathbb{R}_+)$ to $H^{s-1}(\mathbb{R}_+) \oplus \tilde{H}^s(\mathbb{R}_+)$. Duality and interpolation extends the range to $s \in \left(-\frac{1}{2}, \frac{3}{2}\right)$, and patching up the local results, one obtains:

Theorem 9.7 *Let* $s \in \left(-\frac{1}{2}, \frac{3}{2}\right)$. *The operator* $B = \begin{pmatrix} W_{22} & K'_{12} \\ -K_{21} & V_{11} \end{pmatrix}$ *maps*

$$\begin{matrix} \tilde{H}^s(\Gamma_2) \\ \oplus \\ \tilde{H}^{s-1}(\Gamma_1) \end{matrix} \quad \textit{continuously into} \quad \begin{matrix} H^{s-1}(\Gamma_2) \\ \oplus \\ H^s(\Gamma_1), \end{matrix}.$$

(ii) For Gårding's inequality, the symbol has to be estimated from below on the appropriate line $\Im\lambda = s - \frac{1}{2}$. In our example $s = \frac{1}{2}$. For real λ, we obtain

$$\begin{aligned}
&\Re\hat{A}(\lambda)\hat{U}(\lambda)\cdot\overline{\hat{U}(\lambda)} \\
&= \Re\left\{(\lambda^2\hat{V}_0(\lambda)\hat{v}(\lambda) + \hat{K}_\omega(\lambda)\hat{\psi}(\lambda - i))\overline{\hat{v}(\lambda)}\right. \\
&\quad \left. + \left(-\hat{K}_\omega(\lambda)\hat{v}(\lambda) + \hat{V}_0(\lambda)\hat{\psi}(\lambda - i)\right)\overline{\hat{\psi}(\lambda - i)}\right\} \\
&= \lambda^2\hat{V}_0(\lambda)|\hat{v}(\lambda)|^2 + \hat{V}_0(\lambda)|\hat{\psi}(\lambda - i)|^2 \\
&\ge \gamma\left((1 + |\lambda|^2)^{1/2}|\hat{v}(\lambda)|^2 + \frac{1 + |\lambda|}{\lambda^2}\left|\hat{\psi}(\lambda - i)\right|^2\right).
\end{aligned}$$

Therefore,

$$\begin{aligned}
\Re\, 2\pi\int_0^\infty AU(x)\cdot\overline{U(x)}dx &= \Re\int_{\Im\lambda=0}\hat{A}\hat{U}(\lambda)\cdot\overline{\hat{U}(\lambda)}d\lambda \\
&\ge \gamma\left(\int_{\Im\lambda=0}(1 + |\lambda|^2)^{1/2}|\hat{v}(\lambda)|^2 d\lambda\right.
\end{aligned}$$

$$+ \int\limits_{\Im\lambda=-1} \frac{1+|\lambda|}{|\lambda+i|^2} |\hat{\psi}(\lambda)|^2 d\lambda\Bigg)$$

$$\geq \gamma \left(\|v\|^2_{\tilde{H}^{1/2}(\mathbb{R}_+)} + \|\psi\|^2_{\tilde{H}^{-1/2}(\mathbb{R}_+)} \right).$$

Note that after applying Parseval's relation for $s = 0$, we shifted the path of integration from $\Im\lambda = -\frac{1}{2}$ to $\Im\lambda = 0$, e.g., :

$$2\pi \int\limits_{\Im\lambda=0} (-W_0 v(x))\overline{v(x)}dx = \int\limits_{\Im\lambda=-\frac{1}{2}} \frac{1}{2}\widehat{-W_0 v}(\lambda)\hat{v}(\lambda)d\lambda$$

$$= \int\limits_{\Im\lambda=-\frac{1}{2}} (\lambda+i)^2 \hat{V}_0(\lambda+i)\hat{v}(\lambda+i)\overline{\hat{v}(\lambda+i)}d\lambda$$

$$= \int\limits_{\Im\lambda=0} \lambda^2 \hat{V}_0(\lambda)|\hat{v}(\lambda)|^2 d\lambda.$$

This shifting is allowed if we assume $v \in C_0^\infty(0,\infty)$ so that $\hat{v}$ is analytic. Moreover, we used that $\tilde{H}^{-1/2} = (H^{1/2})'$, and therefore from

$$\int\limits_{\Im\lambda=0} \frac{|\lambda|^2}{1+|\lambda|} |\hat{u}(\lambda)|^2 d\lambda = C|u|^2_{1/2} \leq C\|u\|^2_{H^{1/2}(\mathbb{R}_+)}$$

follows

$$\|u\|^2_{\tilde{H}^{-1/2}(\mathbb{R}_+)} \leq C \int\limits_{\Im\lambda=-1} \frac{1+|\lambda|}{|\lambda+i|^2} |\hat{u}(\lambda)|^2 d\lambda.$$

Together with the local Gårding inequalities for Dirichlet and Neumann corners shown earlier, we thus obtain a global Gårding inequality:

Theorem 9.8 *There exists a constant $\gamma > 0$ and a compact operator C : $\tilde{H}^{1/2}(\Gamma_2) \oplus \tilde{H}^{-1/2}(\Gamma_2) \rightarrow H^{-1/2}(\Gamma_2) \oplus H^{1/2}(\Gamma_1)$ such that for all $U = (v,\psi) = \tilde{H}^{1/2}(\Gamma_2) \oplus \tilde{H}^{-1/2}(\Gamma_1)$ there holds*

$$\Re\langle (B+C)U, \bar{U}\rangle \geq \gamma(\|v\|^2_{\tilde{H}^{1/2}(\Gamma_2)} + \|\psi\|^2_{\tilde{H}^{-1/2}(\Gamma_1)}.$$

Here $\langle\cdot,\cdot\rangle$ means the natural duality between $\tilde{H}^{1/2}(\Gamma_2) \oplus \tilde{H}^{-1/2}(\Gamma_1)$ and $H^{-1/2}(\Gamma_2) \oplus H^{1/2}(\Gamma_1)$, which is, for smooth functions $U = (v,\psi)$ and

$W = (f_1, f_2)$, *given by*

$$\langle W, U \rangle := \int_{\Gamma_2} f_1 v ds + \int_{\Gamma_1} f_2 \psi ds.$$

(iii) For a singular expansion and an a priori estimate, one has, according to Lemma 9.5 , to find the poles and the residues of the meromorphic function $\hat{U}(\lambda)$. In equation (9.31), one may assume F to be smooth which means that $\hat{F}$ is meromorphic whith poles only at $\lambda \in i\mathbb{Z}$. Then from (9.31) there arise two kinds of poles for $\hat{U}$, namely at the poles of $\hat{A}(\lambda)^{-1}$ and additionally at poles of $\hat{U}(\lambda - ki), k = 1, \ldots, m$, due to the lower order terms $\hat{A}^k(\lambda)$. The latter give in this way, for every pole of $\hat{A}(\lambda)^{-1}$ at $\lambda = \lambda_0$, rise to infinitely many poles at $\lambda = \lambda_0 + i, \lambda_0 + 2i, \ldots$. In our example, no lower order terms are present, so we only have to find the poles of $\hat{A}(\lambda)^{-1}$, i.e., the zeros of

$$\det \hat{A}(\lambda) = \lambda^2 \hat{V}_0(\lambda)^2 + \hat{K}_\omega(\lambda)^2 = \frac{\cosh \omega\lambda \cosh(2\pi - \omega)\lambda}{\sinh^2 \pi\lambda}.$$

For $\cosh \omega\lambda = 0$ we find the poles at $\lambda = i\alpha, \alpha = \frac{2k-1}{2}\frac{\pi}{\omega}, k \in \mathbb{N}$, which give the well known functions $|z - z_j|^\alpha \log |z - z_j|^r$ for $v(z)$ and $|z - z_j|^{\alpha-1} \log |z - z_j|^r$ for $\psi(z)(r = 0, 1)$. Here z_j is the corner point where the boundary conditions change. Taking into account the poles of $\hat{U}(\lambda)$ for $\Im\lambda \in \left(0, s - \frac{1}{2}\right)$ at each corner point $z_j (j = 1, \ldots, J)$ and the corresponding singular functions $v_{j\ell}, \psi_{j\ell} (\ell = 1, \ldots, L_j)$, one finds for the Cauchy data of the weak solution of the mixed boundary value problem the expansion

$$\begin{pmatrix} v \\ \psi \end{pmatrix} = \sum_{j=l}^{J} \sum_{\ell=l}^{L_j} c_{j\ell} \begin{pmatrix} v_{j\ell} \\ \psi_{j\ell} \end{pmatrix} + \begin{pmatrix} v_s \\ \psi_s \end{pmatrix}$$

$$\text{with } (v_s, \psi_s) \in \tilde{H}^s(\Gamma_2) \oplus H^{s-1}(\Gamma_1)$$

and a corresponding a priori estimate.

9.5 Boundary Integral Operators in Countably Normed Spaces

Our regularity investigations below will be based on the weighted Sobolev spaces and the countably normed spaces as introduced in the following.

Let $I = (0, 1)$. By $H_\beta^{m,l}(I)$ ($m \geq l \geq 1$, integers and $0 < \beta < 1$) we denote the completion of the set of all infinitely differentiable functions under the norm

$$\|\phi\|^2_{H_\beta^{m,l}(I)} = \|\phi\|^2_{H^{l-1}(I)} + \sum_{j=l}^{m} |\phi|^2_{H_\beta^{j,l}(I)} \tag{9.34}$$

where

$$|\phi|_{H_\beta^{j,l}(I)} := \|x^{\beta+j-l}\phi^{(j)}\|_{L^2(I)}. \tag{9.35}$$

The countably normed spaces $B_\beta^l(I)$ on I are defined as

$$\begin{aligned} B_\beta^l(I) = \{\phi \in H_\beta^{m,l}(I), m = l, l+1, \ldots ;\ \exists C \geq 0, d \geq 1\ \forall j = l, l+1, \ldots \\ |\phi|_{H_\beta^{j,l}(I)} \leq C d^{(j-l)}(j-l)!\} \quad (l \geq 1,\ \text{integer}). \end{aligned} \tag{9.36}$$

On Γ these spaces are defined as the product spaces

$$H_\beta^{m,l}(\Gamma) = \Pi_{j=1}^J \Pi_{k=1}^2 H_\beta^{m,l}(\Gamma_k^j),$$

$$B_\beta^l(\Gamma) = \Pi_{j=1}^J \Pi_{k=1}^2 B_\beta^l(\Gamma_k^j) \cap H^{l-1}(\Gamma) \tag{9.37}$$

where each boundary piece Γ_k^j has to be mapped onto I such that the vertex t_{j+k-2} falls onto 0 in order to apply the definition (9.36). If we want to emphasize the dependence on the constants C and d we will write $B_{\beta,C,d}^l$ instead of B_β^l.

For technical reasons we need the following representation of the countable normed spaces:

$$B_{\beta,C,d}^l(I) = \bigcap_{L=l}^{\infty} B_{\beta,C,d}^{l,L}(I) \tag{9.38}$$

where

$$B_{\beta,C,d}^{l,L}(I) := \{\phi \in H_\beta^{L,l}(I);\ |\phi|_{H_\beta^{j,l}(I)} \leq C d^{(L-l)}(j-l)!,\ j = l, l+1, \ldots, L\}. \tag{9.39}$$

The spaces $B_\beta^{l,L}(\Gamma)$ are defined accordingly to (9.37). For localization techniques one needs to introduce cut-off functions. These turn out not to be comprised by the general countable normed spaces B_β^l. But, evidently, for the spaces $B_\beta^{l,L}$ there exist partitions of unity. Furthermore, they can be chosen such that the constants C and d of $B_{\beta,C,d}^{l,L}(\Gamma)$ do not depend on the parameter L.

Lemma 9.8 ([241, Lemma 2.1]) *Let $U \subset \Gamma$ be an open set and $U_\delta := \{x \in \Gamma : \operatorname{dist}(x, U) \leq \delta\}$ for $\delta > 0$. Let $\phi \in B^{l,L}_{\beta,C,d}(U_\delta)$ for all $L \geq l$. Then there exists for each $L \geq l$ a cut-off function $\chi_L \in C^\infty(\Gamma)$ such that*

$$\chi_L|_U \equiv 1 \qquad \text{and} \qquad \chi_L|_{\Gamma \setminus U_\delta} \equiv 0$$

and

$$\chi_L \phi \in B^{l,L}_{\beta,\tilde{C},\tilde{d}}(\Gamma)$$

with constants $\tilde{C}$ and $\tilde{d}$ independent of L.

For $\phi \in C_0^\infty(0, \infty)$ the Mellin transformation is defined by

$$\mathscr{M}(\phi)(\lambda) := \hat{\phi}(\lambda) := \int_0^\infty x^{i\lambda - 1} \phi(x) dx. \tag{9.40}$$

The seminorm $|\phi|_{H^{j,l}_\beta(I)}$ can be characterized by using this transformation.

Lemma 9.9 ([241, Lemma 2.4]) *Let $\phi \in C_0^\infty(I)$ and $0 < \beta < 1$. Then*

$$|\phi|^2_{H^{j,l}_\beta(I)} \simeq \int_{\Im(\lambda) = l - 1/2 - \beta} |f_j(\lambda)|^2 |\hat{\phi}(\lambda)|^2 \, d\lambda \quad (j \geq l),$$

where $f_j(\lambda) := i\lambda \cdot (i\lambda + 1) \cdots (i\lambda + j - 1)$. The constants in the mutual estimates do not depend on j.

Now we present our main results concerning the Poincaré–Steklov operator in countable normed spaces from [242] . Before doing so we need to recall the respective results for standard Sobolev spaces.

For Lipschitz domains continuity and regularity of the integral operators (2.18)–(2.21) as mappings between usual Sobolev spaces have been investigated by Costabel, see [114]. Using these estimates and noting that the Poincaré-Steklov operator maps the Dirichlet datum onto the Neumann datum, we obtain the following proposition.

Proposition 9.2 ([114]) *For all $\sigma \in [0, 1/2]$ $S : H^{1/2+\sigma}(\Gamma) \to H^{-1/2+\sigma}(\Gamma)$ is continuous. For $\sigma \in [0, 1/2]$ let $v \in H^{1/2}(\Gamma)$ satisfy $Sv \in H^{-1/2+\sigma}(\Gamma)$. Then $v \in H^{1/2+\sigma}(\Gamma)$, and there holds the a priori estimate*

$$\|v\|_{H^{1/2+\sigma}(\Gamma)} \leq C \left(\|Sv\|_{H^{-1/2+\sigma}(\Gamma)} + \|v\|_{H^{1/2}(\Gamma)} \right).$$

Following Costabel and Stephan in [129] we use the method of Mellin transformation to investigate the Poincaré-Steklov operator acting on countable normed spaces, see also [241]. First we look at the local properties on an infinite angle Γ^ω. In a second step we apply these results to the boundary Γ of a polygonal domain. Let

$\Gamma^\omega = \Gamma^- \cup \{0\} \cup \Gamma^+$ with $\Gamma^- = e^{i\omega}\mathbb{R}_+$ and $\Gamma^+ = \mathbb{R}_+$ ($\omega \in (0, 2\pi)$). A function ϕ on Γ^ω can be identified with the pair (ϕ_-, ϕ_+) of functions on $\mathbb{R}_+$ defined by $\phi_-(x) = \phi(xe^{i\omega})$, $\phi_+(x) = \phi(x)$ $(x > 0)$. We will choose the representation of ϕ by its even and odd parts (in a formal sense) which are defined by

$$\phi^e = \frac{1}{2}(\phi_- + \phi_+), \qquad \phi^o = \frac{1}{2}(\phi_- - \phi_+).$$

This induces for any operator A acting on functions on Γ^ω a representation by a 2×2-matrix of operators acting on functions on $\mathbb{R}_+$:

$$A \,\hat{=}\, \mathscr{A} := \begin{pmatrix} A_{ee} & A_{oe} \\ A_{eo} & A_{oo} \end{pmatrix} \quad \text{where} \quad \begin{aligned} (A\phi)^e &= A_{ee}\phi^e + A_{oe}\phi^o, \\ (A\phi)^o &= A_{eo}\phi^e + A_{oo}\phi^o. \end{aligned}$$

We need the following operators acting on functions on $\mathbb{R}_+$:

$$\begin{aligned}
V_\omega\phi(x) &:= -\frac{1}{\pi}\int_0^\infty \ln\left|1 - \frac{x}{y}e^{-i\omega}\right|\phi(y)\,dy, \qquad V_0 = V_\omega \quad \text{for} \quad \omega = 0, \\
K_\omega\phi(x) &:= \frac{1}{\pi}\int_0^\infty \Im\left(\frac{1}{xe^{i\omega} - y}\right)\phi(y)\,dy, \\
K'_\omega\phi(x) &:= \frac{1}{\pi}\int_0^\infty \Im\left(\frac{e^{i\omega}}{xe^{i\omega} - y}\right)\phi(y)\,dy, \\
W_\omega\phi(x) &:= -\frac{1}{x}\frac{\partial}{\partial\omega}K_\omega\phi(x), \qquad W_0 = \lim_{\omega\to 0} W_\omega.
\end{aligned} \tag{9.41}$$

Then, with the exception of finite dimensional operators which are negligible in our theory, the integral operators (9.41) can be represented by the following matrices (see [129]):

$$V \,\hat{=}\, \mathscr{V} = \begin{pmatrix} V_0 + V_\omega & 0 \\ 0 & V_0 - V_\omega \end{pmatrix},$$

$$W \,\hat{=}\, \mathscr{W} = \begin{pmatrix} W_\omega - W_0 & 0 \\ 0 & -(W_0 + W_\omega) \end{pmatrix},$$

$$K \,\hat{=}\, \mathscr{K} = \begin{pmatrix} K_\omega & 0 \\ 0 & -K_\omega \end{pmatrix}, \qquad K' \,\hat{=}\, \mathscr{K}' = \begin{pmatrix} K'_\omega & 0 \\ 0 & -K'_\omega \end{pmatrix}.$$

Using these representations we also obtain the representation of the Poincaré-Steklov operator acting on even and odd functions on the infinite angle:

$$S \,\hat{=}\, \mathscr{S} = \begin{pmatrix} S_{ee} & S_{oe} \\ S_{eo} & S_{oo} \end{pmatrix}$$

with

$$\begin{aligned} S_{ee} &= W_\omega - W_0 + (I + K'_\omega)(V_0 + V_\omega)^{-1}(I + K_\omega), \\ S_{oo} &= W_0 - W_\omega + (I - K'_\omega)(V_0 - V_\omega)^{-1}(I - K_\omega), \end{aligned} \tag{9.42}$$

and $S_{eo} = S_{oe} = 0$.

The Mellin symbols of all the components are explicitly known (see [129] Sect. 9.1):

$$\mathcal{M}(V_\omega\phi)(\lambda) = \hat{V}_\omega(\lambda)\hat{\phi}(\lambda - i) := \frac{\cosh[(\pi-\omega)\lambda]}{\lambda\sinh\pi\lambda}\hat{\phi}(\lambda - i), \quad \Im(\lambda) \in (0, 1), \tag{9.43}$$

$$\begin{aligned} \mathcal{M}(W_\omega\phi)(\lambda) &= \hat{W}_\omega(\lambda + i)\hat{\phi}(\lambda + i) \\ &:= -(\lambda + i)\frac{\cosh[(\pi-\omega)(\lambda + i)]}{\sinh[\pi(\lambda + i)]}\hat{\phi}(\lambda + i), \quad \Im(\lambda) \in (-2, 0), \end{aligned} \tag{9.44}$$

$$\mathcal{M}(K_\omega\phi)(\lambda) = \hat{K}_\omega(\lambda)\hat{\phi}(\lambda) := -\frac{\sinh[(\pi-\omega)\lambda]}{\sinh\pi\lambda}\hat{\phi}(\lambda), \quad \Im(\lambda) \in (-1, 1), \tag{9.45}$$

$$\mathcal{M}(K'_\omega\phi)(\lambda) = \hat{K}_\omega(\lambda + i)\hat{\phi}(\lambda), \quad \Im(\lambda) \in (-2, 0). \tag{9.46}$$

On an infinite angle the continuity with respect to seminorms is as follows.

Lemma 9.10 *Let $\rho < \beta < 1$ for $\rho := 3/2 - \min\{\frac{\pi}{2\pi-\omega}, \frac{\pi}{\omega}\}$. Then there holds*

$$|S\phi|_{H^{j,1}_\beta(\Gamma^\omega)} \le C|\phi|_{H^{j+1,2}_\beta(\Gamma^\omega)} \quad (j \ge 1)$$

The constant $C > 0$ does not depend on j.

Proof We use the representation (9.42) of the Poincaré-Steklov operator and Lemma 9.9 to handle the norms of the weighted Sobolev spaces. To calculate the seminorms $|\cdot|_{H^{j,l}_\beta(\mathbb{R}_+)}$ it suffices to concentrate on test functions with compact support in $(0,\infty)$: Let $\phi \in C_0^\infty[0,\infty)$ with $\operatorname{supp}(1-\phi) \subset (0,\infty)$. Then we automatically have

$$\operatorname{supp}(\phi^{(j)}) \subset C_0^\infty(0,\infty), \quad j \ge l \ge 1,$$

and the values $\phi(x)$ for x near 0 are not taken into account for calculating the seminorms $|\cdot|_{H^{j,l}_\beta(\mathbb{R}_+)}$.

Therefore we take $\phi \hat{=} (\phi^e, \phi^o) \in C_0^\infty(\mathbb{R}_+)^2$. Then we have

$$\mathcal{S}\phi = (S_{ee}\phi^e + S_{oe}\phi^o, S_{eo}\phi^e + S_{oo}\phi^o)^T = (S_{ee}\phi^e, S_{oo}\phi^o)^T$$

and it remains to estimate the two components of the right hand side by means of the seminorm $|\phi|_{H^{j,l}_\beta(\mathbb{R}_+)}$.

Using the representations (9.43)–(9.46) we obtain

$$\begin{aligned}
&\mathcal{M}(S_{ee}\phi^e)(\lambda - i)\\
&= \mathcal{M}(W_\omega - W_0)(\lambda)\phi^e(\lambda) + \mathcal{M}(I + K'_\omega)(\lambda)\mathcal{M}\big((V_0 + V_\omega)^{-1}\big)(\lambda)\mathcal{M}(I + K_\omega)(\lambda)\hat{\phi}^e(\lambda)\\
&= 2\lambda \frac{\sinh \pi\lambda - \sinh(\pi - \omega)\lambda}{\cosh \pi\lambda + \cosh(\pi - \omega)\lambda}\hat{\phi}^e(\lambda) =: \mathcal{M}(S_{ee})(\lambda)\hat{\phi}^e(\lambda). \qquad (9.47)
\end{aligned}$$

Therefore, by Lemma 9.9,

$$\begin{aligned}
|S_{ee}\phi^e|^2_{H^{j,1}_\beta(\mathbb{R}_+)} &\simeq \int_{\Im(\lambda)=1/2-\beta} |f_j(\lambda)|^2 |\mathcal{M}(S_{ee}\phi^e)(\lambda)|^2\, d\lambda\\
&= 4\int_{\Im(\lambda)=3/2-\beta} |f_{j+1}(\lambda)|^2 \left|\frac{\sinh \pi\lambda - \sinh(\pi - \omega)\lambda}{\cosh \pi\lambda + \cosh(\pi - \omega)\lambda}\right|^2 |\phi^e(\lambda)|^2\, d\lambda.
\end{aligned}$$

Here we used the relation $|f_j(\lambda)(\lambda + i)| = |f_{j+1}(\lambda + i)|$.

Now, $\cosh \pi\lambda + \cosh[(\pi - \omega)\lambda]$ does not vanish if $\Im(\lambda) = 3/2 - \beta < \min\{\frac{\pi}{2\pi-\omega}, \frac{\pi}{\omega}\}$, i.e. if $\beta > 3/2 - \min\{\frac{\pi}{2\pi-\omega}, \frac{\pi}{\omega}\} = \rho$.
Thus, since

$$\left|\frac{\sinh \pi\lambda - \sinh(\pi - \omega)\lambda}{\cosh \pi\lambda + \cosh(\pi - \omega)\lambda}\right|$$

is bounded for $|\Re(\lambda)| \to \infty$ when avoiding the roots of the denominator, there holds

$$\begin{aligned}
|S_{ee}\phi^e|^2_{H^{j,1}_\beta(\mathbb{R}_+)} &\simeq \int_{\Im(\lambda)=3/2-\beta} |f_{j+1}(\lambda)|^2 \left|\frac{\sinh \pi\lambda - \sinh(\pi - \omega)\lambda}{\cosh \pi\lambda + \cosh(\pi - \omega)\lambda}\right|^2 |\phi^e(\lambda)|^2\, d\lambda\\
&\le C\int_{\Im(\lambda)=3/2-\beta} |f_{j+1}(\lambda)|^2 |\phi^e(\lambda)|^2\, d\lambda \simeq |\phi^e|^2_{H^{j+1,2}_\beta(\mathbb{R}_+)}
\end{aligned}$$

for $\beta > \rho$ and $j \ge 1$.

Analogously, we obtain

$$\begin{aligned}
|S_{oo}\phi^o|^2_{H^{j,1}_\beta(\mathbb{R}_+)} &\simeq \int_{\Im(\lambda)=3/2-\beta} |f_{j+1}(\lambda)|^2 \left|\frac{\sinh \pi\lambda + \sinh(\pi - \omega)\lambda}{\cosh \pi\lambda - \cosh(\pi - \omega)\lambda}\right|^2 |\phi^o(\lambda)|^2\, d\lambda\\
&\le C\int_{\Im(\lambda)=3/2-\beta} |f_{j+1}(\lambda)|^2 |\phi^o(\lambda)|^2\, d\lambda \simeq |\phi^o|^2_{H^{j+1,2}_\beta(\mathbb{R}_+)}
\end{aligned}$$

for $\beta > \rho$ and $j \ge 1$.

Therefore

$$
\begin{aligned}
|\mathscr{S}\phi|^2_{H^{j,1}_\beta(\Gamma^\omega)} &\leq C(|\phi^e|^2_{H^{j+1,2}_\beta(\mathbb{R}_+)} + |\phi^o|^2_{H^{j+1,2}_\beta(\mathbb{R}_+)}) \\
&\leq c(|\phi_-|^2_{H^{j+1,2}_\beta(\mathbb{R}_+)} + |\phi_+|^2_{H^{j+1,2}_\beta(\mathbb{R}_+)}) = c|\phi|^2_{H^{j+1,2}_\beta(\Gamma^\omega)}
\end{aligned}
$$

and the assertion of the lemma is proved. □

Using this lemma we obtain the continuity of S within countable normed spaces on the whole polygon Γ.

Theorem 9.9 *For $\rho < \beta < 1$ with $\rho := 3/2 - \min\{\frac{\pi}{2\pi-\omega}, \frac{\pi}{\omega}\}$ let $\phi \in B^2_\beta(\Gamma)$. Then there holds $S\phi \in B^1_\beta(\Gamma)$.*

Proof Let $\phi \in B^2_{\beta,C,d}(\Gamma)$. Due to the definition of the countable normed spaces in (9.36), (9.37) and the respective norms (9.34) we have to consider the global norm $\|S\phi\|_{L^2(\Gamma)}$ and the seminorms $|S\phi|_{H^{j,1}_\beta(\Gamma^i_k)}$ $(j \geq 1)$ for $i = 1, \ldots, J$ and $k = 1, 2$. The continuity with respect to the L^2-norm is proved by Proposition 9.2. To show the continuity with respect to the seminorms (9.35) we need a partition of unity which exists due to Lemma 9.8 for each $B^{2,L}_{\beta,C,d}(\Gamma)$ $(L \geq 2)$, cf. (9.38). Let $\chi_i \in C^\infty(\Gamma)$, $i = 1, \ldots, J$, such that $\operatorname{supp} \chi_i \subset \Gamma_i \cup \{t_i\} \cup \Gamma_{i+1}$ and $\sum_{i=1}^J \chi_i = 1$ and

$$
\chi_i\phi \in B^{2,L}_{\beta,\tilde{C},\tilde{d}}(\Gamma_i \cup \{t_i\} \cup \Gamma_{i+1})
$$

($\chi_i\phi$ is supposed to be extended by 0 outside $\operatorname{supp} \chi_i$). Then for $\beta > \rho$, Lemma 9.10 yields

$$
\begin{aligned}
|S\chi_i\phi|_{H^{j,1}_\beta(\Gamma_i\cup\{t_i\}\cup\Gamma_{i+1})} &\leq c|\chi_i\phi|_{H^{j+1,2}_\beta(\Gamma_i\cup\{t_i\}\cup\Gamma_{i+1})} \\
&\leq c\tilde{C}\tilde{d}^{L-2}(j-1)!, \quad j = 1, \ldots, L-1
\end{aligned}
\tag{9.48}
$$

where the second inequality is caused by the regularity of $\chi_i\phi$ and definition (9.39). Now, the already known boundedness of $\|S\chi_i\phi\|_{L^2(\Gamma)}$ and the estimate (9.48) yield due to (9.39) the local regularity

$$
S\chi_i\phi \in B^{1,L-1}_{\beta,c\tilde{C},\tilde{d}}(\Gamma_i \cup \{t_i\} \cup \Gamma_{i+1}). \tag{9.49}
$$

Again, there exists a partition of unity $\{\zeta_i;\ i = 1, \ldots, J\}$ for $S\chi_i\phi$ and the index $L-1$. Due to (9.49) there holds

$$
\zeta_i S\chi_i\phi \in B^{1,L-1}_{\beta,C',d'}(\Gamma_i \cup \{t_i\} \cup \Gamma_{i+1}) \tag{9.50}
$$

and due to the analyticity of the kernel of the Poincaré-Steklov operator aside the diagonal $x = y$ we also have

$$\zeta_j S\chi_i\phi \in B^{1,L-1}_{\beta,C',d'}(\Gamma \setminus \overline{\Gamma_i \cup \{t_i\} \cup \Gamma_{i+1}}) \quad (j \neq i). \tag{9.51}$$

Putting (9.50) and (9.51) together for each of the corners $\{t_i\}$ we obtain

$$S\phi = \sum_{j=1}^{m}\sum_{i=1}^{m} \zeta_j S\chi_i\phi \in B^{1,L-1}_{\beta,C^\star,d^\star}(\Gamma) \tag{9.52}$$

where $C^\star$ and $d^\star$ are the largest numbers of the different C's and d's, respectively. Since the constants C' and d' do not depend on the parameter L and the partitions of unity $\{\zeta_i\}$ and $\{\chi_i\}$ corresponding to L, eq. (9.52) finally yields together with the representation (9.38)

$$S\phi \in B^1_{\beta,C^\star,d^\star}(\Gamma). \qquad \square$$

We now investigate the inverse of the Poincaré-Steklov operator. First let us formulate the local regularity result which corresponds to the continuity given by Lemma 9.10.

Lemma 9.11 *Let $\rho = 3/2 - \min\{\frac{\pi}{2\pi-\omega}, \frac{\pi}{\omega}\}$. For $\phi \in H^1(\Gamma^\omega)$ such that $S\phi \in H^{j-1,1}_\beta(\Gamma^\omega)$ for $j \geq 2$ there holds $\phi \in H^{j,2}_\beta(\Gamma^\omega)$ for $\rho < \beta < 1$ and*

$$|\phi|_{H^{j,2}_\beta(\Gamma^\omega)} \leq C|S\phi|_{H^{j-1,1}_\beta(\Gamma^\omega)}.$$

The constant C does not depend on j.

Proof Again we use the representation of functions on Γ^ω by their even and odd parts on Γ^- and Γ^+ and the induced matrix representation of the Poincaré-Steklov operator as in the proof of Lemma 9.10, cf. (9.42). With regard to (9.47) there holds

$$\hat{\phi}^e(\lambda) = \frac{\cosh\pi\lambda + \cosh(\pi-\omega)\lambda}{2\lambda\big(\sinh\pi\lambda - \sinh(\pi-\omega)\lambda\big)}\mathscr{M}(S_{ee}\phi^e)(\lambda - i), \quad \Im(\lambda) \in (-1,1).$$

Noting that $|f_j(\lambda)/\lambda| = |f_{j-1}(\lambda - i)|$ we obtain by Lemma 9.9 for $\beta \in (1/2, 1)$

$$|\phi^e|^2_{H^{j,2}_\beta(\mathbb{R}_+)} \simeq \int_{\Im(\lambda)=3/2-\beta} |f_j(\lambda)|^2|\hat{\phi}^e(\lambda)|^2\,d\lambda \tag{9.53}$$
$$= \frac{1}{4}\int\limits_{\Im(\lambda)=1/2-\beta} |f_{j-1}(\lambda)|^2\left|\frac{\cosh\pi(\lambda+i) + \cosh(\pi-\omega)(\lambda+i)}{\sinh\pi(\lambda+i) - \sinh(\pi-\omega)(\lambda+i)}\right|^2 |\mathscr{M}(S_{ee}\phi^e)(\lambda)|^2\,d\lambda.$$

Since

$$\left|\frac{\cosh\pi(\lambda+i)+\cosh(\pi-\omega)(\lambda+i)}{\sinh\pi(\lambda+i)-\sinh(\pi-\omega)(\lambda+i)}\right|$$

is bounded for $\Re(\lambda) \to \pm\infty$ provided the denominator does not vanish at the horizontal strip under consideration, we obtain by (9.53) and Lemma 9.9

$$|\phi^e|^2_{H^{j,2}_\beta(\mathbb{R}_+)} \simeq$$
$$\int\limits_{\Im(\lambda)=1/2-\beta} |f_{j-1}(\lambda)|^2\left|\frac{\cosh\pi(\lambda+i)+\cosh(\pi-\omega)(\lambda+i)}{\sinh\pi(\lambda+i)-\sinh(\pi-\omega)(\lambda+i)}\right|^2 |\mathscr{M}(S_{ee}\phi^e)(\lambda)|^2\,d\lambda$$
$$\leq C\int_{\Im(\lambda)=1/2-\beta} |f_{j-1}(\lambda)|^2|\mathscr{M}(S_{ee}\phi^e)(\lambda)|^2\,d\lambda \simeq |S_{ee}\phi^e|^2_{H^{j-1,1}_\beta(\mathbb{R}_+)}$$

for $\beta > 3/2 - \pi/(2\pi-\omega)$. Analogously, we obtain for the odd part

$$\hat{\phi}^o(\lambda) = \frac{\cosh\pi\lambda - \cosh(\pi-\omega)\lambda}{2\lambda\big(\sinh\pi\lambda+\sinh(\pi-\omega)\lambda\big)}\mathscr{M}(S_{oo}\phi^o)(\lambda-i), \quad \Im(\lambda)\in(-1,1)$$

and therefore,

$$|\phi^o|^2_{H^{j,2}_\beta(\mathbb{R}_+)} \simeq$$
$$\int\limits_{\Im(\lambda)=1/2-\beta} |f_{j-1}(\lambda)|^2\left|\frac{\cosh\pi(\lambda+i)-\cosh(\pi-\omega)(\lambda+i)}{\sinh\pi(\lambda+i)+\sinh(\pi-\omega)(\lambda+i)}\right|^2 |\mathscr{M}(S_{oo}\phi^o)(\lambda)|^2\,d\lambda$$
$$\leq C\int_{\Im(\lambda)=1/2-\beta} |f_{j-1}(\lambda)|^2|\mathscr{M}(S_{oo}\phi^o)(\lambda)|^2\,d\lambda \simeq |S_{oo}\phi^o|^2_{H^{j-1,1}_\beta(\mathbb{R}_+)}$$

for $\beta > 3/2 - \pi/\omega$.

Altogether, since $\phi \in H^1(\Gamma^\omega)$ by assumption, we proved that $\phi^e \in H^{j,2}_\beta(\mathbb{R}_+)$ and $\phi^o \in H^{j,2}_\beta(\mathbb{R}_+)$ for $j \geq 2$ and therefore we have $\phi \in H^{j,2}_\beta(\Gamma^\omega)$ and the proof of the lemma is finished. □

Now we use again the partitions of unity to prove the regularity of the Poincaré-Steklov operator on the whole polygon.

Theorem 9.10 *Let* $\rho = 3/2 - \min\{\frac{\pi}{2\pi-\omega_j}, \frac{\pi}{\omega_j};\ j = 1,\dots,J\}$ *and* $\rho < \beta < 1$. *Then there holds* $\phi \in B^2_\beta(\Gamma)$ *if* $S\phi \in B^1_\beta(\Gamma)$ *with* $\int_\Gamma S\phi\,ds = 0$ *where* ϕ *is unique up to a constant.*

Proof The proof is analogous to the proof of Theorem 9.9. The regularity with respect to the global Sobolev norms is given by Proposition 9.2. The boundedness with respect to the seminorms $|\cdot|_{H_\beta^{j,l}(\Gamma_k^j)}$ follows from Lemma 9.11. Here again, we have to use a partition of unity as in the proof of Theorem 9.9. □

For Mellin convolution equations on an interval (second kind integral equations) stability and exponential convergence in the L_q-norm $1 \leq q \leq \infty$ for Galerkin and Collocation methods are proved for piecewise polynomials on geometrically refined meshes in [154].

For the use of Mellin techniques for integral operators over polyhedral domains see [152, 153, 351, 367].

In [125] Mellin techniques are applied to first kind integral equations for linear elasticity in polygonal domains.

Chapter 10
A-BEM

First in this chapter we give a general framework of adaptive Petrov–Galerkin methods for the solution of operator equations in Banach spaces. This approach is made precise in the application to Symm's integral equation. Then we present more general adaptive BEM. Here we use the residual error estimator and prove reliability and efficiency in 2D. Finally we analyze the hierarchical error estimator and demonstrate its applicability in two-level adaptive BEM for scalar and vector boundary value problems. Special emphasis is given to the 3D case for the weakly singular integral equation (Sect. 10.3) and for the hypersingular integral equation (Sect. 10.4). In Sect. 10.5 we present a two-level adaptive BEM for the weakly singular operator and the h-version on surface pieces. In Sect. 10.6 based on a two-level subspace decomposition for the p-version BEM we give hierarchical error estimators for the hypersingular integral operator on curves. Finally recent developments on the convergence of the adaptive BEM for the h-version are given in Sect. 10.7.

We do not want to miss to mention further work on adaptive boundary elements. Faermann in [170] provides local a posteriori error indicators for the Galerkin discretization of boundary integral equations of the Babuşka-Rheinboldt type. These error indicators are for a wide class of integral operators reliable on special meshes (like $K-$meshes) and efficient for arbitrary meshes. The proof of efficiency is problematic for BEM: Carstensen derives efficiency only for uniform meshes (see Sect. 10.2.2) and Mund et al. show efficiency and reliability only for uniform meshes and under the assumption of a saturation condition (see Sect. 10.5). Faermann also used local double-integral seminorms as estimators in [171, 172] and their relation to weighted residual error indicators and multilevel (hierarchical) error indicators shown in the survey paper [78]. The earliest suggestions for error indicators and adaptive BEM [348, 434] used the concept of an influence index. They are equivalent to Faermann's Babuşka-Rheinboldt error indicators. Super convergence properties and gradient recovering techniques for the purpose of adaptive mesh refinements are used in [375] and [88]. For the adaptive method there are also some

© Springer International Publishing AG, part of Springer Nature 2018
J. Gwinner, E. P. Stephan, *Advanced Boundary Element Methods*,
Springer Series in Computational Mathematics 52,
https://doi.org/10.1007/978-3-319-92001-6_10

different concepts as in [369–371, 389, 390]. In the following we concentrate on estimators of residual type and hierarchical type and present some of the respective analysis.

10.1 General Frame for A Posteriori Error Estimates for Boundary Element Methods

In this section we consider the analogue of adaptive feedback algorithms developed within finite element methods and present their analogue within boundary element methods. This can be generally described as follows (see [92]):

Let $X,\ Y$ be Banach spaces and let the operator $A:\ X \to Y$ be linear, bounded and bijective. We want to approximate the continuous solution $u := A^{-1}f \in X$ of the problem

$$Au = f \qquad \text{for} \quad f \in Y \text{ given}$$

by $u_h \in X_h \subset X$ such that:

$$\langle t, Au_h \rangle = \langle t, f \rangle \iff 0 = \langle t, f - Au_h \rangle \qquad \forall t \in T_h \subset Y^* . \tag{10.1}$$

If we define the residual R by $R := f - Au_h \in Y$, and assume that it can be computed, at least numerically, we have by the Banach's inverse mapping theorem that A^{-1} is bounded and there holds

$$\|u - u_h\|_X = \|A^{-1}R\|_X \leq \|A^{-1}\|_{L(Y;X)} \cdot \|R\|_Y \tag{10.2}$$

Let u_h be obtained by some Galerkin procedure such that there exists a finite dimensional subspace T_h of Y^*, the dual of Y and let $\langle , \rangle_{Y^*\times Y}$ denote the dual pairing between Y^* and Y. Then

$$0 = \langle t_h, R \rangle_{Y^*\times Y} \qquad \forall t_h \in T_h \subset Y^* \tag{10.3}$$

By a consequence of the Hahn-Banach theorem there exists $\tilde{\rho} \in Y^*$ with norm 1 and $\|R\|_Y = \langle \tilde{\rho}, R \rangle_{Y^*\times Y}$. Setting $\rho = \|R\|_Y \tilde{\rho}$ shows the following assertion.

Lemma 10.1 *There exists some* $\rho \in Y^*$ *with*

$$\|\rho\|_{Y^*}^2 = \|R\|_Y^2 = \langle \rho, R \rangle_{Y^*\times Y} .$$

Choosing some ρ as in the lemma, we have with (10.3)

$$\|R\|_Y^2 = \langle \rho, R \rangle_{Y^*\times Y} = \inf_{t_h \in T_h} \langle \rho - t_h, R \rangle_{Y^*\times Y} \leq \|R\|_Y \cdot \inf_{t_h \in T_h} \|\rho - t_h\|_{Y^*} ,$$

yielding

$$\|R\|_Y \;\le\; \inf_{t_h\in T_h} \|\rho - t_h\|_{Y^*}\,. \tag{10.4}$$

Comparing this with (10.2) yields the abstract a posteriori error estimate

$$\|u-u_h\|_X \;\le\; \|A^{-1}\|_{L(Y;X)}\cdot \inf_{t_h\in T_h}\|\rho - t_h\|_{Y^*}\,.$$

Combining interpolation with this estimate gives:

Theorem 10.1 *Let* X, Y_0, Y_θ, Y_1 *be Banach spaces with* $Y_1 \subseteq Y_\theta \subseteq Y_0$ *. Let* c_θ *be some positve constant such that for some* $\theta \in (0,1)$

$$\|y\|_{Y_\theta} \;\le\; c_\theta\cdot \|y\|_{Y_0}^{1-\theta}\cdot\|y\|_{Y_1}^{\theta} \qquad \forall y\in Y_1\,. \tag{10.5}$$

Let $A:\; X\to Y_\theta$ *be linear, bounded and bijective. Let* $f\in Y_\theta,\;u:=A^{-1}f$ *, and let* $u_h\in X$ *with* $R:=f-Au_h\in Y_1$ *satisfying*

$$0=\langle t_h, R\rangle_{Y_0^*\times Y_0} \qquad \forall t_h\in T_h\,, \tag{10.6}$$

where $\langle,\rangle_{Y_0'\times Y_0}$ *denoting the dual pairing between* Y_0^* *and* Y_0*,* $T_h\subseteq Y_0^*$ *. Let further* $\rho\in Y_0^*$ *be defined as in Lemma 10.1 (with* Y_0 *replacing* Y*), i.e. with*

$$\|\rho\|_{Y_0^*}^2 \;=\; \|R\|_{Y_0}^2 = \langle\rho, R\rangle_{Y_0^*\times Y_0}\,. \tag{10.7}$$

Then there holds:

$$\|u-u_h\|_X \;\le\; c_\theta\cdot\|A^{-1}\|_{L(Y_\theta;X)}\cdot\|R\|_{Y_1}^{\theta}\cdot\inf_{t_h\in T_h}\|\rho-t_h\|_{Y_0^*}^{1-\theta}\,.$$

Proof Using (10.2) with Y_θ replacing Y and estimating $\|R\|_{Y_\theta}$ via (10.5) we obtain

$$\|u-u_h\|_X \;\le\; c_\theta\cdot\|A^{-1}\|_{L(Y_\theta;X)}\cdot\|R\|_{Y_1}^{\theta}\cdot\|R\|_{Y_0}^{1-\theta}\,.$$

By Lemma 10.1 we have some ρ satisfying (10.7). Using (10.4) with Y_0 replacing Y concludes the proof. □

We now want to apply the above Theorem to Symm's integral equation and explicitly determine ρ for this example.

10.1.1 Symm's Integral Equation

Given a bounded Lipschitz domain $\Omega \subseteq \mathbb{R}^n$ with boundary Γ the Dirichlet problem for the Laplacian is related with the *Symm's* integral equation for the unknown density ϕ on Γ:

$$V\phi(x) = g(x), \qquad x \in \Gamma \tag{10.8}$$

with the weakly singular operator V.

Let $T_h = S_h^\circ \subseteq L^2(\Gamma)$ denote the vector space of piecewise polynomials with respect to a 'triangulation' of the boundary $\Gamma = \cup_{j=1}^N \Gamma_j$, where the 'elements' $\Gamma_1, \ldots, \Gamma_N$ are either equal or have at most one common point or side, respectively. For $x \in \Gamma_j$ we define the piecewise constant function

$$h : \begin{cases} \Gamma \longrightarrow [0, \infty) \\ x \mapsto |\Gamma_j| =: h_j \ , \end{cases}$$

$|\Gamma_j| > 0$ denoting the element size. Let $\phi_h \in T_h$ denote the Galerkin solution of (10.8), i.e.

$$\langle t_h, R\rangle = 0 \qquad \forall\, t_h \in T_h \tag{10.9}$$

where $R = g - V\phi_h$. Assume that $g \in H^1(\Gamma)$ so that, according to $T_h \subseteq L^2(\Gamma)$, $R = V(\phi - \phi_h) \in H^1(\Gamma)$ due to the mapping properties of the single-layer potential. The following result gives an a posteriori error estimate using the residual R on the right hand side.

Theorem 10.2 ([92])
Under the above assumptions, there holds for $0 \leq s \leq 1$:

$$\|\phi - \phi_h\|_{H^{-s}(\Gamma)} \leq \|V^{-1}\|_{L(H^{1-s}(\Gamma); H^{-s}(\Gamma))} \cdot \|R\|_{H^1(\Gamma)}^{1-s} \cdot \|h \cdot \nabla_\Gamma R\|_{L^2(\Gamma)}^s \, . \tag{10.10}$$

Proof Note that $H^1(\Gamma) \subseteq H^{1-s}(\Gamma) \subseteq L^2(\Gamma) = H^0(\Gamma)$ and (10.5) holds with $c_\theta = 1$, therefore we can apply Theorem 10.1 with $A = V$, $X = H^{-s}(\Gamma)$, $Y_0 = L^2(\Gamma)$, $Y_1 = H^1(\Gamma)$, $\theta = 1 - s$ and $Y_\theta = H^{1-s}(\Gamma)$. Note that $\rho = R$, identifying Y_0^* with Y_0 by the Riesz representation theorem, satisfies (10.7) and there holds

$$\inf_{t_h \in T_N} \|R - t_h\|_{L^2(\Gamma)} \leq \|h \cdot \nabla_\Gamma R\|_{L^2(\Gamma)}$$

by standard interpolation and approximation arguments. Hence Theorem 10.1 yields the assertion. □

We can obtain an adaptive feedback algorithm as follows (see [92]): For a given 'triangulation' of the boundary as above we can compute an approximation of the contribution a_j of one element Γ_j by numerical integration. Defining for Symm's integral equation $a_j := \|R'\|_{L^2(\Gamma_j)}$, we have the error estimate in the energy-norm:

$$\|\phi - \phi_h\|_{H^{-\frac{1}{2}}(\Gamma)} \leq c \cdot \left(\sum_{j=1}^{N} a_j^2 \right)^{\frac{1}{4}} \cdot \left(\sum_{j=1}^{N} h_j^2 a_j^2 \right)^{\frac{1}{4}} .$$

The mesh may be steered be the following, adaptive feedback algorithm where $0 \leq \theta \leq 1$ is a global parameter:

Algorithm *Given some coarse, e.g. uniform mesh, refine it successively by halving some of the elements due to the following rule. For any triangulation define* $a_1, \ldots, a_N$ *as above and refine the element* Γ_j *if and only if*

$$h_j \cdot a_j \geq \theta \cdot \max_{1 \leq k \leq N} h_k \cdot a_k .$$

Note for $\theta = 0$ we have a uniform triangulation whereas the number of refined elements decreases with increasing θ.

Remark 10.1 The above abstract setting allows also to handle *hp*-adaptive boundary element methods. In [82] the following estimates are derived for the hypersingular operator W and the weakly singular operator V, respectively; see [265] for a different approach with a γ-shape regular mesh on Γ and a γ_p-shape regular polynomial degree distribution,

$$\|u - u_N\|_{H^{1/2}(\Gamma)} \leq \|(h/p)^{1/2} R_N\|_{L^2(\Gamma)}$$

for $R_N = g - W u_N$, where $Wu = g$ on Γ,

$$\|\phi - \phi_N\|_{H^{-1/2}(\Gamma)} \leq \|(h/p + 1)^{1/2} \nabla_\Gamma R_N\|_{L^2(\Gamma)}$$

for $R_N = f - V\phi_N$, where $V\phi = f$ on Γ.

10.2 Adaptive Boundary Element Methods

Let $A : \mathcal{H} \to \mathcal{H}^*$ with $\mathcal{H} \subseteq H^m(\Gamma)$, $m \in \mathbb{R}$, $\mathcal{H}^*$ the dual of $\mathcal{H}$, be some pseudodifferential operator which is bounded, linear and positive definite on the closed subspace $\mathcal{H}$ of $H^m(\Gamma)$, i.e., we have a constant $\alpha > 0$ with

$$< Au, u > \geq \alpha \|u\|^2_{H^m(\Gamma)} \quad \text{for all } u \in \mathcal{H} \subseteq H^m(\Gamma).$$

Here, $<,>$ extends the $L^2(\Gamma)$ scalar product to the duality between the Sobolev spaces $\mathscr{H}$ and $\mathscr{H}^*$. Due to the Lax–Milgram lemma we then have a unique solution $u \in \mathscr{H} \subseteq H^m(\Gamma)$ of

$$Au = f \tag{10.11}$$

for any given right hand side $f \in H^{-m}(\Gamma) \subseteq \mathscr{H}^*$.

For the numerical approximation of u let $\Gamma = \overline{\cup_{j=1}^{N} \Gamma_j}$ be partitioned into N pairwise disjoint elements $\Gamma_1, \dots, \Gamma_N$ and let S_h^k denote a finite dimensional subspace of $\mathscr{H} \subseteq H^m(\Gamma)$ of piecewise polynomials, i.e., $v_h|_{\Gamma_j}$ is a polynomial of degree at most k for any $v_h \in S_h^k$ and $j = 1, \dots, N$. Then, the Galerkin equations

$$< Au_h, v_h > = < f, v_h > \quad \text{for all } v_h \in S_h^k \tag{10.12}$$

have a unique solution $u_h \in S_h^k$. Due to the Céa lemma we have

$$\|u - u_h\|_{H^m(\Gamma)} \leq C \cdot \inf_{v_h \in S_h^k} \|u - v_h\|_{H^m(\Gamma)} \tag{10.13}$$

where C denotes a generic positive constant which is independent of the data f and S_h^k. In [93] we prove a posteriori error estimates of the form

$$\|u - u_h\|_{H^m(\Gamma)} \leq C \cdot \sum_{j=1}^{N} a_j \cdot h_j^{1-r} \tag{10.14}$$

via "local interpolation" for the class of K-meshes where the quotient of the mesh sizes of two neighbouring elements is uniformly bounded by some constant $K > 1$.

Let $\Omega \subset \mathbb{R}^2$ be a bounded domain with Lipschitz boundary $\tilde{\Gamma}$ and $\Gamma \subset \tilde{\Gamma}$ be a connected piece of $\tilde{\Gamma}$. The Dirichlet problem for the Laplacian is equivalently related to Symm's integral equation

$$V\phi(x) = g(x) \quad (x \in \Gamma) \tag{10.15}$$

where

$$V\phi(x) := -\frac{1}{\pi} \int_\Gamma \phi(y) \ln|x - y| ds_y \tag{10.16}$$

Note if $cap(\Gamma) < 1$ (see Section 2.3), then $A := V$ is bijective between $\mathscr{H} := \tilde{H}^{-1/2}(\Gamma)$ and $\mathscr{H}^* = H^{1/2}(\Gamma)$. Letting $m = -1/2, k = 0$, the equations (10.11) and (10.12), i.e., (10.15) and

$$< V\phi_h, \psi_h > = < g, \psi_h > \quad \text{for all } \psi_h \in S_h^0, \tag{10.17}$$

S_h^0 being the piecewise constant functions on $\Gamma = \overline{\cup_{j=1}^N \Gamma_j}$, have unique solutions ϕ and ϕ_h, respectively, and and there holds a quasi-optimal convergence estimate (10.13).

Remark 10.2 Note that, for given g in $H^1(\Gamma)$ the residual $R := g - V\phi_h$ lies also in $H^1(\Gamma)$ for $\phi_h \in L^2(\tilde{\Gamma})$. Moreover,

$$< R, \psi_h >= 0 \qquad \text{for all } \psi_h \in S_h^0 \tag{10.18}$$

by the Galerkin procedure.

The Neumann problem for the Laplacian is equivalently related to the hypersingular integral equation

$$Wv(x) = f(x) \quad (x \in \Gamma), \tag{10.19}$$

where

$$Wv(x) := \frac{1}{\pi}\frac{\partial}{\partial n_x}\int_\Gamma v(y)\frac{\partial}{\partial n_y}\ln|x-y|ds_y \tag{10.20}$$

is hypersingular and f can be computed from the given normal derivative of the displacement field u which gives $v = u|_{\tilde{\Gamma}}$ on Γ.

Thus, $A := W$ is bijective between $\mathscr{H} = H^{1/2}(\Gamma)/\mathbb{R}$ and its dual. Letting $m = 1/2$, $k = 1$, the problems (10.11) and (10.12), i.e., (10.19) and

$$< Wv_h, w_h > = < f, w_h > \quad \text{for all } w_h \in S_h^1,$$

have unique solutions v and v_h, respectively. Here, S_h^1 are continuous piecewise linear functions on $\Gamma = \overline{\cup_{j=1}^N \Gamma_j}$ with support in Γ, e.g., the trial functions vanish at the endpoints of Γ if Γ is an open arc. Again we have a quasi-optimal convergence estimate (10.13).

Remark 10.3 Note that, for given f in $L^2(\Gamma) \cap \mathscr{H}^*$ the residual $f - Wv_h$ lies also in $L^2(\Gamma)$ for $v_h \in \tilde{H}^1(\Gamma)$. Moreover,

$$< f - Wv_h, w_h > = 0 \qquad \text{for all } w_h \in S_h^1 \tag{10.21}$$

by the Galerkin procedure.

10.2.1 Reliability of A Posteriori BEM Error Estimates

With the above notations we have the following a posteriori error estimates.

Their proofs are based on "local interpolation" whereas (10.10) is based on "global interpolation" (see [92]).

Let $K \geq 1$ denote the maximum of all quotients of sizes $h_1, \ldots, h_N$ of neighbouring elements $\Gamma_1, \ldots, \Gamma_N$ which describe the discretization within the Galerkin method.

Theorem 10.3 ([93]) *Let* $g \in H^1(\Gamma)$ *and* $cap(\Gamma) < 1$. *For Symm's integral equation we have for* $0 \leq s \leq 1$ *with* $R' = \frac{\partial}{\partial s} R$

$$\|\phi - \phi_h\|_{\tilde{H}^{-s}(\Gamma)} \leq \sqrt{2} K^{1-s} \|V^{-1}\|_{L(H^{1-s}(\Gamma); \tilde{H}^{-s}(\Gamma))} \sum_{j=1}^{N} h_j^s (h_j^2 + 1)^{(1-s)/2} \|R'\|_{L^2(\Gamma_j)} \tag{10.22}$$

Theorem 10.4 ([93]) *Let* $f \in L^2(\Gamma)$ *and* $cap(\Gamma) < 1$. *For the hypersingular integral equation we have for* $0 \leq s \leq 1$

$$\|v - v_h\|_{\overline{\mathscr{H}^s}} \leq \sqrt{2} K^s \|W^{-1}\|_{L(\mathscr{H}^{s-1}; \overline{\mathscr{H}^s})} \cdot \sum_{j=1}^{N} h_j^{1-s} (h_j^2 + 1)^{s/2} \cdot \|f - W v_h\|_{L^2(\Gamma_j)}. \tag{10.23}$$

The proofs of the above theorems use the following proposition where S_h^0 denotes the piecewise constant functions with respect to a partition $\Gamma_1, \ldots, \Gamma_N$ of $\Gamma = \overline{\cup_{j=1}^N \Gamma_j}$.

Proposition 10.1 ([93]) *If* $f \in H^1(\Gamma)$ *satisfies*

$$< f, \psi_h > = 0 \qquad \text{for all } \psi_h \in S_h^0, \tag{10.24}$$

then for $0 \leq \sigma \leq 1$ *there holds*

$$\|f\|_{H^\sigma(\Gamma)} \leq \sqrt{2} \cdot K^\sigma \sum_{j=1}^{N} \|f'\|_{L^2(\Gamma_j)} \cdot h_j^{1-\sigma} (1 + h_j^2)^{\sigma/2}.$$

Proof Since $\int_{\Gamma_j} f\, ds = 0$ we have at least one zero y_j of the continuous function f in the interior of Γ_j, $j = 1, \ldots, N$. Define $\tilde{f}_j \in H^1(\Gamma)$ to be equal to f on the part of Γ between y_j and y_{j+1} and equal to 0 on the remaining part of Γ (note that $\tilde{f}_j$ is continuous, e.g., at y_j and hence absolute continuous on Γ and piecewise in H^1; thus $\tilde{f}_j \in H^1(\Gamma)$). Here we set $y_0 = y_N$ and $M := N$ if Γ is closed; y_0 and y_{N+1} as the starting point and endpoint of Γ, respectively, and $M := N + 1$ if Γ is an open arc. Then, $\tilde{f}_j \in H^\sigma(\Gamma)$ and we conclude from the triangle inequality that

$$\|f\|_{H^\sigma(\Gamma)} \le \sum_{j=0}^{M-1} \|\tilde{f}_j\|_{H^\sigma(\Gamma)}. \tag{10.25}$$

Since $\tilde{\Gamma}_j := supp\ \tilde{f}_j \subseteq \Gamma_j \cup \Gamma_{j+1}$, we have by definition of $\tilde{f}_j$, and interpolation of $H^1(\Gamma)$ and $L^2(\Gamma)$ [44]

$$\|\tilde{f}_j\|_{H^\sigma(\Gamma)} \le \|\tilde{f}_j\|^\sigma_{H^1(\Gamma)} \cdot \|\tilde{f}_j\|^{1-\sigma}_{L^2(\Gamma)} = \|f\|^\sigma_{H^1(\tilde{\Gamma}_j)} \cdot \|f\|^{1-\sigma}_{L^2(\tilde{\Gamma}_j)}. \tag{10.26}$$

Note that in any nonempty $\overline{\Gamma_j \cap \tilde{\Gamma}_k} =: \Gamma_{jk}$ (hence $k = j$ or $k = j - 1$, because everywhere else $\overline{\Gamma_j \cap \tilde{\Gamma}_k} =: \emptyset$) there is at least one zero of f. Using the fundamental theorem of calculus gives:

$$\|f\|_{L^\infty(\Gamma_{jk})} \le \|f'\|_{L^1(\Gamma_{jk})}$$

yielding

$$\|f\|_{L^2(\Gamma_{jk})} \le h_{jk} \cdot \|f'\|_{L^2(\Gamma_{jk})}$$

where $h_{jk} := |\Gamma_j \cap \tilde{\Gamma}_k|$ is the length of $\Gamma_j \cap \tilde{\Gamma}_k$. Hence, (10.26) gives

$$\|\tilde{f}_j\|_{H^\sigma(\Gamma)} \le \|f\|^\sigma_{H^1(\widetilde{\Gamma}_j)} \|f\|^{1-\sigma}_{L^2(\widetilde{\Gamma}_j)}$$

$$\|\tilde{f}_j\|_{H^\sigma(\Gamma)} \le \|f\|^\sigma_{H^1(\widetilde{\Gamma}_j)} \|f\|^{1-\sigma}_{L^2(\widetilde{\Gamma}_j)} \le \Big[(\|f'\|_{L^2(\Gamma_{jj})} + \|f'\|_{L^2(\Gamma_{j+1j})})^2 \tag{10.27}$$

$$+ (h_{jj}\|f'\|_{L^2(\Gamma_{jj})} + h_{j+1j}\|f'\|_{L^2(\Gamma_{j+1j})})^2\Big]^{\sigma/2}$$

$$\cdot \left(h_{jj}\|f'\|_{L^2(\Gamma_{jj})} + h_{j+1j}\|f'\|_{L^2(\Gamma_{j+1j})}\right)^{1-\sigma}.$$

Use the following estimate for positive a, b, α, β: $(a+b)^2 + (\alpha a + \beta b)^2 \le (a\sqrt{1+\alpha^2} + b\sqrt{1+\beta^2})^2$, (10.27) shows with $h_{jk} \le h_j$

$$\|\tilde{f}_j\|_{H^\sigma(\Gamma)} \le \left(\|f'\|_{L^2(\Gamma_{jj})}\sqrt{1+h_j^2} + \|f'\|_{L^2(\Gamma_{j+1j})}\sqrt{1+h_{j+1}^2}\right)^\sigma$$

$$\cdot \left(h_j\|f'\|_{L^2(\Gamma_{jj})} + h_{j+1}\|f'\|_{L^2(\Gamma_{j+1j})}\right)^{1-\sigma}. \tag{10.28}$$

Now

$$(a+b)^{1-s} \cdot (\alpha a + \beta b)^s \le K^{1-s} \cdot (a\alpha^s + b\beta^s) \tag{10.29}$$

for any $a, b, \alpha, \beta \ge 0$ with $1/K \le \alpha/\beta \le K$ and $0 \le s \le 1$.

Now we set $\alpha = h_j/\sqrt{1+h_j^2}$ and $\beta = h_{j+1}/\sqrt{1+h_{j+1}^2}$. Since $1/K \leq h_j/h_{j+1} \leq K$ (by the definition of K as the maximum of all quotients of neighbouring elements)

$$\frac{\alpha}{\beta} = \sqrt{1 + \frac{h_j^2/h_{j+1}^2 - 1}{h_j^2 + 1}}$$

belongs to the interval $[1/K, K]$ as well.

Hence, we may use (10.29) and obtain

$$\|\tilde{f}_j\|_{H^\sigma(\Gamma)} \leq K^\sigma \Big(\|f'\|_{L^2(\Gamma_{jj})} h_j^{1-\sigma} (1+h_j^2)^{\sigma/2} + \|f'\|_{L^2(\Gamma_{j+1j})} h_{j+1}^{1-\sigma} (1+h_{j+1}^2)^{\sigma/2} \Big). \tag{10.30}$$

Estimating each summand in (10.25) with (10.30) leads to

$$\|f\|_{H^\sigma(\Gamma)} \leq K^\sigma \sum_{j=1}^N \Big(\|f'\|_{L^2(\Gamma_{jj})} + \|f'\|_{L^2(\Gamma_{jj-1})} \Big) h_j^{1-\sigma} (1+h_j^2)^{\frac{\sigma}{2}}$$

$$\leq K^\sigma \sqrt{2} \sum_{j=1}^N \|f'\|_{L^2(\Gamma_j)} \cdot h_j^{1-\sigma} (1+h_j^2)^{\sigma/2}$$

which proves the proposition. □

Proof (of Theorem 10.3) Due to Remark 10.2 we have that $R := g - V\phi_h \in H^1(\Gamma)$ satisfies (10.24). Now Proposition 10.1 leads to

$$\|\phi - \phi_h\|_{\tilde{H}^{-s}(\Gamma)} = \|V^{-1}R\|_{\tilde{H}^{-s}(\Gamma)} \leq \|V^{-1}\|_{L(H^{1-s}(\Gamma);\tilde{H}^{-s}(\Gamma))} \cdot \|R\|_{H^{1-s}(\Gamma)}$$
$$\leq \sqrt{2} \cdot K^{1-s} \cdot \|V^{-1}\|_{L(H^{1-s}(\Gamma);\tilde{H}^{-s}(\Gamma))} \cdot \sum_{j=1}^N h_j^s \cdot (h_j^2+1)^{(1-s)/2} \cdot \|R'\|_{L^2(\Gamma_j)}.$$

This concludes the proof. □

Proof (of Theorem 10.4) Due to Remark 10.3 we have that $f - Wv_h \in L^2(\Gamma)$ Hence, an application of Proposition 10.1 leads to (see [93])

$$\|R\|_{H^\sigma(\Gamma)} \leq \sqrt{2} \cdot K^\sigma \sum_{j=1}^N \|f - Wv_h\|_{L^2(\Gamma_j)} \cdot h_j^{1-\sigma} (1+h_j^2)^{\sigma/2}.$$

Finally, we consider $\frac{\partial}{\partial s} : \mathscr{H}^\sigma \to \mathscr{H}^{\sigma-1}$ (i.e., differentiation with respect to the arc length) which is bounded by 1 in the operator norms for $\sigma = 1$ and as well for

$\sigma = 0$ which follows by duality when $\frac{\partial}{\partial s}$ is defined in the distributional sense. By interpolation this shows $\|f - Wv_h\|_{\mathscr{H}^{\sigma-1}} \le \|R\|_{\mathscr{H}^\sigma} = \|R\|_{H^\sigma(\Gamma)}$ for $0 \le \sigma \le 1$.

Since $W : \mathscr{H}^s \to \mathscr{H}^{s-1}$ is linear, bounded, and bijective, one concludes this proof as in the proof of Theorem 10.3. □

10.2.2 Efficiency of A Posteriori BEM Error Estimates (2D)

In this section we report on the paper of Carstensen [74]. Let $\pi = \{\Gamma_1, \ldots, \Gamma_n\}$ be a partition of the polygon Γ in intervals $\Gamma_1, \ldots, \Gamma_n$. Then

$$S_\pi^0 := \{v_h \in L^\infty(\Gamma) : v_h\big|_{\Gamma_j} \in \mathbb{R} \ \forall j = 1, \ldots, N\}$$

denotes the linear space of piecewise constant functions and $h(\pi) \in S_\pi^0$ is defined as the local mesh size, i.e., $h(\pi)\big|_{\Gamma_j} := |\Gamma_j| :=$ length of Γ_j.

In the following we consider for Symm's integral equation the Galerkin method with piecewise constants on quasi uniform meshes on Γ, i.e. meshes for which there exists a global constant c_u such that for all meshes π under consideration,

$$\max h(\pi)/\min h(\pi) = \max_{j \neq k} |\Gamma_j|/|\Gamma_k| \le c_u \tag{10.31}$$

with $\max h(\pi) := \|h(\pi)\|_{L^\infty(\Gamma)}$ and $\min h(\pi) := \min\{h(\pi)(x) : x \in \Gamma\}$.

In [74] Carstensen shows the following efficiency result.

Theorem 10.5 ([74]) *If f is continuous and smooth on each side of Γ, there exist constants $c_0, h_0 > 0$ (depending only on Γ, f and c_u) such that for all partitions π of Γ with* $\max h(\pi) < h_0$ *and (10.31), and for $s \in [0, 1]$, one has*

$$\max h(\pi)^s \cdot \|R_h'\|_{L^2(\Gamma)} \le c_0 \cdot \|\phi - \phi_h\|_{H^{-s}(\Gamma)} \tag{10.32}$$

where ϕ solves Symm's integral equation $V\phi = f$ on Γ and ϕ_h denotes its Galerkin solution; $R_h = f - V\phi_h$.

Let us first introduce an abstract setting and give some results which are used to prove the above theorem. Let $X_1 \subset X_0$ and $Y_1 \subset Y_0$ be real Banach spaces, and let $X_\theta := [X_0, X_1]_\theta$, $Y_\theta := [Y_0, Y_1]_\theta$, $0 \le \theta \le 1$ be defined by interpolation. We assume, that there are positive constants $c_{\theta,X}$ and $c_{\theta,Y}$ such that for all $x \in X_1$ and $y \in Y_1$ (10.5) is satisfied:

$$\|x\|_{X_\theta} \le c_{\theta,X} \cdot \|x\|_{X_0}^{1-\theta} \cdot \|x\|_{X_1}^\theta, \qquad \|y\|_{Y_\theta} \le c_{\theta,Y} \cdot \|y\|_{Y_0}^{1-\theta} \cdot \|y\|_{Y_1}^\theta. \tag{10.33}$$

Let $L(X; Y)$ denote the Banach space of linear bounded mappings between the Banach spaces X and Y, and let $\|\cdot\|_{L(X;Y)}$ be the corresponding operator norm.

Then, for each $A_j \in L(X_j; Y_j)$, $j = 0, 1$ with $A_0\big|_{X_1} = A_1$, the restriction $A_\theta = A_0\big|_{X_\theta}$ belongs to $L(X_\theta; Y_\theta)$ and

$$\|A_\theta\|_{L(X_\theta;Y_\theta)} \leq \|A_\theta\|^{1-\theta}_{L(X_0;Y_0)} \cdot \|A_\theta\|^{\theta}_{L(X_1;Y_1)} \tag{10.34}$$

Let $A = A_\theta$ be such a mapping and assume, in addition, that $A_\theta : X_\theta \mapsto Y_\theta$ is bijective. Then, fix a right-hand side $f \in Y_1$ and the solution $u \in X_\theta$ of

$$Au = f. \tag{10.35}$$

We apply the Galerkin method to approximate u. Let $S_h \subset X_1$ and $T_h \subset Y_0^*$ be finite-dimensional subspaces such that there exists some $u_h \in S_h$ satisfying (10.1). We define the residual $R_h := f - Au_h$ and the error $e_h := u - u_h$. Note that now there holds the assertion of Theorem 10.1 with X_θ instead of X. We consider a family of Galerkin methods described by a family of discrete subspaces $(S_h : h \in I)$ of X_1 and $(T_h : h \in I)$ of Y_0^*, where I is an index set. We assume the three properties: (i) Approximation property ,(ii) Inverse assumption, (iii) Stability, made precise in the following.

(i) Approximation property. Assume that the solution $u \in X_\theta$ of (10.35), also belongs to X_1. Then, for each $h \in I$, let

$$E(u, S_h) := \inf\{\|u - v_h\|_{X_1} : v_h \in S_h\} = \|u - \Pi_h u\|_{X_1}$$

be the best approximation error in the norm of X_1 and let $\Pi_h : X_1 \mapsto S_h$ denote a projection such that $\Pi_h u$ is the best approximation in S_h. Furthermore we define:

$$F(u, S_h) := \frac{\|u - \Pi_h u\|_{X_0}}{\|u - \Pi_h u\|_{X_1}}$$

(ii) Inverse assumption. For each $h \in I$ let

$$G(S_h) := \sup\left\{\frac{\|v_h\|_{X_1}}{\|v_h\|_{X_0}} : v_h \in S_h\backslash\{0\}\right\}$$

(iii) Stability. For each $h \in I$ let $P_h : X_0 \mapsto S_h$ be a projection such that $P_h \in L(X_0, X_0)$ and $P_h\big|_{X_1} \in L(X_1, X_1)$ with norms

$$\|P_h\|_j := \sup\left\{\frac{\|P_h v\|_{X_j}}{\|v\|_{X_j}} : v \in X_j\backslash\{0\}\right\}, \qquad j = 0, 1$$

Then for $h, H \in I$ define

$$\delta(u, S_h, S_H) := \frac{E(u, S_H)}{E(u, S_h)} \cdot \left(1 + c_{\theta,X} \cdot \|P_H\|_1^{\theta}\, [F(u, S_H)G(S_H)\|P_H\|_0]^{1-\theta}\right) \tag{10.36}$$

Theorem 10.6 ([74]) *Let $A \in L(X_1, Y_1), h, H \in I$ with $S_h \subseteq S_H$; consider $u, u_h \in X_1, e_h := u - u_h, R_h := Ae_h \in Y_1$, and assume (i),(ii),(iii). If $\delta(u, S_h, S_H) < 1$, then*

$$\|R_h\|_{Y_1} \leq \frac{\|P_H\|_0^{1-\theta} \cdot \|P_H\|_1^{\theta}}{1 - \delta(u, S_h, S_H)} \cdot \|A\|_{L(X_1,Y_1)} \cdot G(S_H)^{1-\theta} \cdot \|e_h\|_{X_\theta} \tag{10.37}$$

The proof in [74] is based on two further corollaries, below.

In case if S_h are spline function spaces $E(u, S_h)$, $F(u, S_h)$ and $G(S_h)$ can be bounded. Now: Let $(S_h : h \in I)$ be a family of subspaces of X, where the index h is a positive parameter, say, $I \subset (0, 1)$.Suppose $S_h \subset S_H \; \forall h, H \in I$ with $H < h$ and that $\cup_{h \in I} S_h$ is dense in X_1. Suppose that there exist positive constants c_α, c_β, c_p and real constants α, β such that $\forall h \in I$

$$F(u, S_h) \leq c_\alpha \cdot h^\alpha \tag{10.38}$$

$$G(S_h) \leq c_\beta \cdot h^\beta \tag{10.39}$$

$$\|P_h\|_0^{1-\theta} \cdot \|P_h\|_1^{\theta} \leq c_p \tag{10.40}$$

Corollary 10.1 *Assume (10.38)(10.39)(10.40) and $\alpha + \beta \geq 0$. Define*

$$c_1 := 2 + 2c_{\theta,X} \cdot c_p \cdot c_\alpha^{1-\theta} \cdot c_\beta^{1-\theta}, \qquad c_2 := 2c_p \cdot \|A\|_{L(X_1;Y_1)} \cdot c_\beta^{1-\theta}$$

Then, for each $h \in I$, we can find $H \in I$ with

$$E(u, S_H) \leq \frac{1}{c_1} E(u, S_h) \qquad H < h \tag{10.41}$$

and we have $\|R_h\|_{Y_1} \leq c_2 \cdot H^{\beta(1-\theta)} \cdot \|e_h\|_{X_\theta}$

As chosen in Corollary 10.1, H depends highly on h and we need more information on $E(u, S_h)$ to control this in (10.41).

Corollary 10.2 *In addition to the assumption of Corollary 10.1 let there exist constants $\eta, q, h_0, 0 < \eta, q < 1$,such that for all $h \in I$ with $h < h_0$ we have*

$$E(u, S_{\eta \cdot h}) \leq q \cdot E(u, S_h) \qquad \eta \cdot h \in I \tag{10.42}$$

Then, there exists $c_0 > 0$ such that for all $h \in I$ with $h < h_0$

$$\|R_h\|_{Y_1} \leq c_0 \cdot h^{\beta(1-\theta)} \cdot \|e_h\|_{X_\theta} \tag{10.43}$$

Lemma 10.2 *[74] For any $v_h \in S_h, h \in I$, we have*

$$\|v_h\|_{X_1} \leq \|P_h\|_0^{1-\theta} \|P_h\|_1^{\theta} \cdot G(S_h)^{1-\theta} \cdot \|v_h\|_{X_\theta}$$

and

$$\|u - \Pi_h u\|_{X_\theta} \leq c_{\theta,X} \cdot F(u, S_h)^{1-\theta} \cdot \|u - \Pi_h u\|_{X_1}$$

Symm's integral equation for a polygon Γ, $V\phi = f$ has a unique solution $\phi \in L^2(\Gamma)$ for a given right hand side $f \in H^1(\Gamma)$ if $cap(\Gamma) < 1$, which is assumed now. We are interested in its $L^2(\Gamma)$-best approximation error:

$$E(\phi, S_h) = \min\left\{\|\phi - \psi_h\|_{L^2(\Gamma)} \;:\; \psi_h \in S_\pi^0(\Gamma)\right\} = \|\phi - \Pi_h\phi\|_{L^2(\Gamma)} \qquad (10.44)$$

where, Π_h is the orthogonal projection onto $S_\pi^0(\Gamma)$ in $L^2(\Gamma)$.

Proposition 10.2 ([74]) *Provided ϕ is not constant, there exists positive constants γ, h_0, c_γ and c'_γ (depending only on Γ, f and c_u in (10.31)) such that $0 < \gamma \leq 1$ and either*

$$c'_\gamma \leq \max h(\pi)^{-\gamma} \cdot E(\phi, S_h) \leq c_\gamma \qquad (10.45)$$

or $\gamma = 1/2$ and

$$c'_\gamma \leq -\max h(\pi)^{-1/2} \cdot \log^{-1/2}(\max h(\pi)) \cdot E(\phi, S_h) \leq c_\gamma \qquad (10.46)$$

holds for all meshes π with $\max h(\pi) < h_0$ and satisfying (10.31).

Proof The solution ϕ of $V\phi = f$ has the form (see Section 7.1)

$$\phi(x) = \phi_0(x) + \sum_{j=1}^{m} c_j \cdot \phi_j(x), \qquad x \in \Gamma \qquad (10.47)$$

where $\phi_0 \in H^2(\Gamma)$ and the real constants c_j depend on f. The singular functions ϕ_j depend on the corners of the polygon with interior angle ω_j as follows:

$$\phi_j(x) = r^{\beta_j} \cdot \chi_j(x), \qquad \beta_j + 1 = k_j\pi/\omega_j,\ k_j \ \textit{integer} \geq 0$$

or

$$\phi_j(x) = r^{\beta_j} \cdot \log(r) \cdot \chi_j(x), \qquad \beta_j + 1 = k_j\pi/\omega_j \ \textit{integer},\ \beta_j \geq 1$$

The proof in [74] is split into several steps. ϕ is approximated by $S_h^0(0, 1)$ on a quasi-uniform mesh on $(0, 1)$ described by a partition $0 = x_0 < x_1 < x_2 < \cdots < x_n <$

$x_{n+1} = 1$. Define $m_j = h_j^{-1} \cdot \int_{x_j}^{x_{j+1}} \phi(x)dx = \Pi_h\phi\Big|_{(x_j,x_{j+1})}$ and $h_j := x_{j+1} - x_j$ for $j = 0, \dots, n$.

Let $\phi(x) = x^\beta$ for $x \in (0, 1)$, and $-1/2 < \beta < 1/2$, $\beta \neq 0$. Let $0 \leq a < a + h \leq 1$, and consider the error $\|\phi - m\|_{L^2(a,a+h)}$, where ϕ is approximated by the constant $m = h^{-1} \int_a^{a+h} \phi(x)dx$. One obtains:

$$\|\phi - m\|^2_{L^2(a,a+h)} = a^{2\beta+1} \cdot \eta(h/a) \qquad (a > 0) \tag{10.48}$$

$$\|\phi - m\|^2_{L^2(0,h)} = \frac{\beta^2}{(1+\beta^2)(2\beta+1)} \qquad (a = 0), \tag{10.49}$$

where

$$\eta(\delta) := \frac{(1+\delta)^{2\beta+1} - 1}{2\beta+1} - \frac{[(1+\delta^2)^\beta - 1]^2}{\delta(1+\beta)^2} \qquad (\delta > 0)$$

yielding

$$\eta(\delta) = c_1 \cdot \delta^3 + hot(\delta) \tag{10.50}$$

with a positive constant c_1 (depending only on $\beta > -1/2$) and $hot(\delta)$ denoting higher order terms in δ. Moreover, one can conclude from (10.50) and (10.48) that

$$c_2 \leq \eta(\delta) \cdot \delta^{-3} \leq c_3 \qquad \forall \delta \in (0, c_u] \tag{10.51}$$

with constants c_j depending on β and c_u. Since $\Gamma = \cup \Gamma_j$ for $j = 0, \dots, n$ and using (10.48) and (10.49) with $a = x_j$ gives

$$\|\phi - \Pi_h\phi\|^2_{L^2(\Gamma)} = h_0^{2\beta+1} \cdot \frac{\beta^2}{(1+\beta^2)(2\beta+1)} + \sum_{j=1}^n x_j^{2\beta+1} \cdot \eta(h_j/x_j) \tag{10.52}$$

$$= h_0^{2\beta+1} \left(\frac{\beta^2}{(1+\beta^2)(2\beta+1)} + \sum_{j=1}^n \left(\frac{x_j}{h_0}\right)^{2\beta+1} \cdot \eta(h_j/x_j) \right)$$

Estimating with the Riemann's Zeta function ζ gives

$$\sum_{j=1}^n j^{2\beta-2} \leq \zeta(2-2\beta) = \sum_{j=1}^\infty j^{2\beta-2}$$

and yields (10.45) with $\gamma = \beta + 1/2$. Similar arguments hold for $\beta = 1/2$ and for $\phi(x) = x^\beta \log x$, $\beta \geq 1$. See [74] for details. □

Proposition 10.3 *There holds* $F(\phi, S_h) \leq \max h(\pi)$ *with* $X_0 = H^{-1}(\Gamma)$, $X_1 = L^2(\Gamma)$.

Proof Using $\|\eta - \Pi_h \eta\|_{L^2(\Gamma)} \leq \max h(\pi) \cdot \|\eta'\|_{L^2(\Gamma)}$ for all $\eta \in H^1(\Gamma)$ one obtains

$$\|\phi - \Pi_h \phi\|_{H^{-1}(\Gamma)} = \sup_{\eta \in H^1(\Gamma)} \frac{\int_\Gamma (\phi - \Pi_h \phi)(\eta - \Pi_h \eta) ds}{\|\eta\|_{H^1(\Gamma)}} \leq \max h(\pi) \cdot \|\phi - \Pi_h \phi\|_{L^2(\Gamma)}$$

yielding the assertion. □

Next we define

$$S^1_\pi(\Gamma) := \{w_h \in H^1(\Gamma) \, : \, \frac{\partial w_h}{\partial s} \in S^0_\pi(\Gamma)\} \tag{10.53}$$

the linear space of continuous and piecewise linear functions with respect to a mesh π.

The following result shows $G(S_h) \leq C/h$.

Proposition 10.4 *[74] There exists a constant* $c > 0$ *such that for all meshes* π *with* $\max h(\pi) < 1$,

$$\|w_h\|_{H^1(\Gamma)} \leq c \cdot \min h(\pi)^{-1} \cdot \|w_h\|_{L^2(\Gamma)} \qquad \forall \, w_h \in S^1_\pi(\Gamma), \tag{10.54}$$

$$\|\psi_h\|_{L^2(\Gamma)} \leq c \cdot \min h(\pi)^{-1} \cdot \|\psi_h\|_{H^{-1}(\Gamma)} \qquad \forall \, w_h \in S^1_\pi(\Gamma). \tag{10.55}$$

Proof We already know the inverse inequality (10.54), which also can be easily proved by direct calculations on each element. Let $I(f)$ be defined ny integrating f along Γ with respect to the arclength, then $I : S^0_\pi(\Gamma) \cap L^2_0(\Gamma) \mapsto S^1_\pi(\Gamma) \cap L^2_0(\Gamma)$ is an isomorphism and (10.55) follows from (10.54). □

For each $\psi \in H^{-1}(\Gamma)$ define $P_h \psi \in S^0_\pi(\Gamma)$ by

$$P_h \psi := \psi^0 + \frac{\partial}{\partial s} \Pi^1_h I(\psi - \psi^0)$$

for $\psi^0 := |\Gamma|^{-1} \int_\Gamma \psi ds$, where Π^1_h is the L^2-Projection onto $S^1_\pi(\Gamma)$

Proposition 10.5 *[74] The operator* P_h *is a projection onto* $S^0_\pi(\Gamma)$*, which is bounded as a mapping between* $H^{-1}(\Gamma)$ *and* $H^{-1}(\Gamma)$ *or between* $L^2(\Gamma)$ *and* $L^2(\Gamma)$

Now let us prove Theorem 10.5

Proof With the above notation and the Propositions 10.3, 10.4 and 10.5 we get the properties (10.38) , (10.39) and (10.40), where the index parameter h is identified with $\max h(\pi)$ for a mesh π satisfying (10.31). For a given mesh π, let $\eta = 1/k$ for an integer k and define a new mesh by dividing each element Γ_j of π in k pieces of length $|\Gamma_j|/k$. Then, the new mesh also satisfies (10.31). Moreover, according to Proposition 10.2 we obtain (10.42) of Corollary 10.2 with some q which depends on k, c_γ, c'_γ and γ. If we choose k large enough, we can obtain $0 < q < 1$. Note that η depends only on c_γ, c'_γ and γ. Application of Corollary 10.2 concludes the proof. □

A corresponding result holds for the hypersingular integral equation $Wv = g$ on Γ when the Galerkin solution $v_h \in S^1_\pi(\Gamma) \cap H^1_0(\Gamma)$ is computed on a quasi uniform mesh: Assume g is smooth on each side of the polygon. Then there exist constants $c_0, h_0 > 0$ (depending on Γ, g and c_u) such that for all partitions with (10.31) and for $0 < s < 1$ there holds

$$\max h(\pi)^{1-s} \cdot \|R_h\|_{L^2(\Gamma)} \leq c_0 \cdot \|v - v_h\|_{H^s(\Gamma)}. \tag{10.56}$$

10.3 The Weakly Singular Integral Equation in 3D

Here we report from [87]. Let us consider Symm's integral equation which is equivalent to interior or exterior Dirichlet problems for the Laplacian in a bounded Lipschitz domain $\Omega \subset \mathbb{R}^3$ with boundary $\partial\Omega$ or on the open surface $\Gamma \subset \partial\Omega$: *Given f find ψ with*

$$V\psi(x) := \frac{1}{4\pi} \int_\Gamma \frac{\psi(y)}{|x-y|} \, ds_y = f(x) \quad (x \in \Gamma). \tag{10.57}$$

A Galerkin discretisation provides ψ_N and a partition $\mathscr{T} = \{\Gamma_1, \ldots, \Gamma_N\}$ of Γ in elements $\Gamma_1, \ldots, \Gamma_N$ with mesh sizes $h_1, \ldots, h_N$ with the property that the residual R_N,

$$R_N(x) := f(x) - V\psi_N(x) \quad (x \in \Gamma),$$

satisfies a Poincaré inequality on Γ_j, i.e., there holds

$$\|R_N\|_{L^2(\Gamma_j)} \leq C(\Gamma_j) \, \|\nabla R_N\|_{L^2(\Gamma_j)} \tag{10.58}$$

Piecewise constant test functions cause $\int_{\Gamma_j} R_N \, ds = 0$ and so (10.58) follows with $C(\Gamma_j) \leq \operatorname{diam}(\Gamma_j)$. Following [87] we show below that (10.58) is sufficient for an a posteriori error estimate with localized residuals

$$\|\psi - \psi_N\|_{\tilde{H}^{-\alpha}(\Gamma)} \leq c(\alpha, \mathscr{T}) \Big(\sum_{j=1}^{N} h_j^{2\alpha} \|\nabla R_N\|^2_{L^2(\Gamma_j)} \Big)^{1/2}, \tag{10.59}$$

where ∇ is the surface gradient on Γ, $(0 \le \alpha \le 1)$ (see also Theorem 10.3 in Sect. 10.2.1 for the two-dimensional situation). The upper bound consists of a sum of computable residuals $\eta_j := h_j^\alpha \|\nabla R_N\|_{L^2(\Gamma_j)}$, which serve as error indicators in an adaptive mesh refinement algorithm (Algorithm A) in Sect. 10.3.1. The underlying meshes are shape-regular and locally uniform.

Next a localisation is performed by multiplication with functions from a partition of unity with local supports.

Definition 10.1 Let Ω be a bounded Lipschitz domain in $\mathbb{R}^3$. A finite partition of unity of $\partial\Omega$ is a finite sequence $\Phi := (\varphi_1, \ldots, \varphi_M)$ of Lipschitz functions $\varphi_1, \ldots, \varphi_M : \partial\Omega \to \mathbb{R}$ such that on $\partial\Omega$

$$1 = \varphi_1 + \ldots + \varphi_M \text{ and } \varphi_1, \ldots, \varphi_M \ge 0. \tag{10.60}$$

The overlap $K(\Phi)$ is defined by ($card\{S\}$ denotes the number of elements in a set S)

$$K(\Phi) := \max_{j=1,\ldots,M} card\{k \in \{1, \ldots, M\} : \varphi_k \varphi_j \ne 0 \text{ on } \partial\Omega\}.$$

$K(\Phi)$ may be much smaller than M, even bounded, while M is increasing to infinity as the mesh-size tends to zero. Hence, one distributes $\varphi_1, \ldots, \varphi_M$ into a minimal number $K \le K(\Phi)$ of groups to apply the following lemma (Lemma 7.3 in Section 7.4) in order to derive Theorem 10.7, below.

Lemma 10.3 ([405, 423]) *Let $f_1, \ldots, f_n \in H^\alpha(\partial\Omega)$, $0 \le \alpha \le 1$, such that $f_j f_k = 0$ on $\partial\Omega$ whenever $1 \le j < k \le n$. Let $\omega_j :=$ interior(supp f_j) satisfy $\overline{\omega}_j =$ supp f_j. Then*

$$\|\sum_{j=1}^n f_j\|^2_{H^\alpha(\partial\Omega)} \le C_1 \sum_{j=1}^n \|f_j\|^2_{H^\alpha(\omega_j)}.$$

The constant C_1 depends on $\partial\Omega$ but does not depend on f_j or on n.

Furthermore one needs the following result.

Lemma 10.4 ([87]) *Let Φ be a finite partition of unity of $\partial\Omega$ with overlap $K(\Phi)$. Then there exists a partition of $\{1, \ldots, M\}$ into $K \le K(\Phi)$ non-empty subsets $M_1, \ldots, M_K$,*

$$\bigcup_{j=1}^K M_j = \{1, \ldots, M\} \text{ and } M_j \cap M_k = \emptyset \text{ if } j \ne k \quad (j, k = 1, \ldots, K),$$

such that, for all $\ell \in \{1, \ldots, K\}$ and $j, k \in M_\ell$ with $j \ne k$,

$$\varphi_j \varphi_k = 0 \text{ on } \partial\Omega. \tag{10.61}$$

Theorem 10.7 *Let Γ be a connected subpiece of $\partial\Omega$ and let Φ be a finite partition of unity of $\partial\Omega$ with overlap $K(\Phi)$. Then, for any $f \in H^\alpha(\partial\Omega)$, $0 \le \alpha \le 1$, we have*

$$\|f\|^2_{H^\alpha(\Gamma)} \le K(\Phi) \sum_{j=1}^{M} \|f\,\varphi_j\|^2_{H^\alpha(\omega_j)}. \tag{10.62}$$

Let Φ be a finite partition of unity on $\partial\Omega$ with overlap $K(\Phi)$ and let ω_j the interior of $\operatorname{supp}\varphi_j$ and $d_j := \text{width}(\omega_j)$ for each $j \in \{1, \dots, M\}$. Then, for $0 < \alpha < 1$, $\Gamma \subset \partial\Omega$, and $f \in H^1(\partial\Omega)$, we have

$$\|f\|^2_{H^\alpha(\Gamma)} \le K(\Phi) \sum_{j=1}^{M} d_j^{2(1-\alpha)}\,(1+d_j^2)^\alpha\,\|\nabla(\varphi_j\,f)\|^2_{L^2(\omega_j)}. \tag{10.63}$$

Next, let $\psi_N \in L^2(\Gamma)$ such that $R_N := f - V\psi_N$ satisfies $\int_{\Gamma_k} R_N\,ds = 0$ for each element Γ_k with diameter h_k of a partition $\mathscr{T} = \{\Gamma_1, \dots, \Gamma_N\}$ of Γ. Suppose that the supports of hat functions φ_j are matched exactly by a finite number of elements. Then, the Galerkin condition $\int_{\Gamma_k} R_N\,ds = 0$ (for all $k = 1, \dots, N$) implies $\int_{\omega_j} R_N\,ds = 0$ (for all $j = 1, \dots, M$).

Corollary 10.3 *On a locally uniform mesh there exists a constant $C > 0$ that depends only on $0 < \alpha < 1$, Γ, $\partial\Omega$, and the shape (not the size) of the elements and patches such that for any $\psi_N \in L^2(\Gamma)$ and $R_N := f - V\psi_N$ with $\int_{\omega_j} R_N\,ds = 0$ for all $j = 1, \dots, M$, we have*

$$\|\Psi - \Psi_N\|^2_{\tilde{H}^{-\alpha}(\Gamma)} \le C \sum_{j=1}^{N} h_j^{2\alpha}\,\|\nabla R_N\|^2_{L^2(\Gamma_j)}. \tag{10.64}$$

Proof First with $\psi - \psi_N = V^{-1}R_N$ and the boundedness of V^{-1} we observe that (10.63) implies: There exists a constant $C > 0$ that depends only on $0 < \alpha < 1$, Γ, and $\partial\Omega$ such that for any $\psi_N \in L^2(\Gamma)$ and $R_N := f - V\psi_N$ we have

$$\|\Psi - \Psi_N\|^2_{\tilde{H}^{-\alpha}(\Gamma)} \le C \sum_{j=1}^{N} d_j^{2\alpha}\,\|\nabla(\varphi_j R_N)\|^2_{L^2(\omega_j)}. \tag{10.65}$$

Furthermore, a Poincaré inequality gives

$$\|R_N\|^2_{L^2(\omega_j)} \le C(\omega_j)\|\nabla R_N\|^2_{L^2(\omega_j)}.$$

Thus

$$\|\nabla(\varphi_j R_N)\|^2_{L^2(\omega_j)} \leq \|R_N \nabla\varphi_j\|^2_{L^2(\omega_j)} + \|\varphi_j \nabla R_N\|^2_{L^2(\omega_j)}$$
$$\leq Lip(\varphi_j)\|R_N\|^2_{L^2(\omega_j)} + \|\nabla R_N\|^2_{L^2(\omega_j)} \leq C\|\nabla R_N\|^2_{L^2(\omega_j)}.$$

since $Lip(\varphi_j)\,C(\omega_j) \leq C$ for a locally uniform mesh. Here $Lip(\varphi_j)$ denotes the Lipschitz constant of the function φ_j. □

The above corollary motivates the isotropic error indicator

$$\mu_j := h_j^{1/2}\|\nabla R_N\|_{L^2(\omega_j)} \tag{10.66}$$

since $C(\sum_{j=1}^N \mu_j^2)^{1/2}$ is a computable upper error bound with respect to the energy norm.

Open surfaces yield singularities near the edge (see Subsection 7.3.1) which limit the regularity of the exact solution, so in general $\psi \notin L^2(\Gamma)$. Here an anisotropic error indicator such as

$$\mu_{j,k} := h_{j,k}^{1/2}\|\partial R_N/\partial x_k\|_{L^2(\Gamma_j)} \quad (k = 1, 2) \tag{10.67}$$

with an axes parallel rectangle Γ_j with edge-lengthes $h_{j,1}$ and $h_{j,2}$ reflects the singular behaviour better than μ_j.

A two-level ansatz and a saturation assumption are used in [314](see Section 10.5) to see for quasi-uniform meshes, that, up to multiplicative constants $(\sum_{j=1}^N \eta_{j,1}^2 + \eta_{j,2}^2 + \eta_{j,3}^2)^{1/2}$ is a lower and upper error bound, where,

$$\eta_{j,k} := -\frac{\langle R_N, \xi_{j,k}\rangle}{\langle V\xi_{j,k}, \xi_{j,k}\rangle^{1/2}} \quad (k = 1, 2, 3)$$

and the ansatz functions $\xi_{j,k}$ are defined for one rectangle by dividing it into four congruent rectangles, see Fig. 10.4, where $\xi_{j,k}$ is denoted by $\beta_{j+1,\cdot,k}$.

The error indicators μ_j, $\mu_{j,k}$, $\nu_j := \|R_N\|_{H^\alpha(\omega_j)}$ and $\eta_{j,k}$ can used for steering automatic mesh-refinements.

Theorem 10.8 ([87]) *There exist constants $c_1, c_2, c_3 > 0$ which depend on the aspect ratio ($\max_{j\neq l} \frac{|\Gamma_j|}{|\Gamma_l|}$) of the elements $\Gamma_1, \ldots, \Gamma_N$ in $\mathcal{T}$ and on Γ but not on f, $R_N := f - V\psi_N$,or $\psi = V^{-1}f$ and neither on the sizes nor numbers of elements in $\mathcal{T}$. We have*

$$\eta_{j,k} \leq c_1\mu_{j,k} \leq c_1\mu_j \qquad (k = 1, 2;\ j = 1, \ldots, N) \tag{10.68}$$

$$\left(\sum_{\Gamma_l\in\overline{\omega}_j} (\eta_{1,l}^2 + \eta_{2,l}^2)\right)^{1/2} \leq c_2\nu_j \leq c_3\left(\sum_{\Gamma_l\in\overline{\omega}_j} \mu_l^2\right)^{1/2} \qquad (j = 1, \ldots, M) \tag{10.69}$$

Estimates with $\eta_{j,k}$ require $\alpha = 1/2$ while the other holds for all $\alpha \in [0, 1]$.

Proof $\xi_{j,k}$ can be written as the derivative of a hat function $\phi_{j,k}$ with height $h_{j,k}/2$. Then

$$\langle R_N, \xi_{j,k}\rangle = -\langle \partial_{x_k} R_N, \phi_{j,k}\rangle \leq \|\phi_{j,k}\|_{L^2(\Gamma_j)} \|\partial_{x_k} R_N\|_{L^2(\Gamma_j)}$$

Now we note that for each triangulation and hat function φ_z at node z with $h_z :=$ diam(supp(φ_z)) there holds

$$\|\varphi_z\|_{\tilde{H}^s(\Gamma)} \sim h_z^{1-s} \quad \text{for } 0 \leq s \leq 1 \tag{10.70}$$

Therefore taking a hat function φ with supp $(\varphi) \subset \omega$ and $\|\varphi\|_{L^2(\Gamma)} \sim h_\omega :=$ diam (ω) we have with the characteristic function χ of ω

$$\langle V\chi, \chi\rangle \approx \|\chi\|^2_{\tilde{H}^{-1/2}(\Gamma_j)} \geq \sup_{\substack{\eta \in H^{1/2}(\Gamma_j) \\ \eta \neq 0}} \frac{\langle \chi, \eta\rangle^2}{\|\eta\|^2_{H^{1/2}(\Gamma_j)}} \geq \frac{\langle \chi, \varphi\rangle^2}{\|\phi\|^2_{H^{1/2}(\Gamma)}} \approx h_\omega |\omega|$$

Thus

$$\frac{\langle R_N, \xi_{j,k}\rangle}{\langle V\xi_{j,k}, \xi_{j,k}\rangle^{1/2}} \leq c_1 \frac{\|\phi_{j,k}\|_{L^2(\Gamma_j)}}{(h_{j,k}|\Gamma_j|)^{1/2}} \|\partial_{x_k} R_N\|_{L^2(\Gamma_j)} \lesssim \mu_{j,k}$$

This implies the first estimate in (10.68). For (10.69) see [87] . □

10.3.1 Adaptive Algorithms

The error estimators derived above lead to the following three algorithms, Algorithm A without direction control, Algorithm B with a direction control and (hierarchical) Algorithm C.

Let the parameter $0 \leq \theta \leq 1$ and an initial partition π_0 of Γ be given. Let $S_{0,N_0}(\Gamma)$ be the finite dimensional space of piecewise constant functions of π_0.

The adaptive algorithms read as follows:

Algorithm A For $m = 0, 1, 2, \ldots$

(i) Compute the Galerkin solution $\psi_N \in S_{0,N_m}(\Gamma)$ according to the partition $\pi_m = \{\Gamma_1, \ldots, \Gamma_{N_m}\}$.

(ii) For each element $\Gamma_i \in \pi_m$ compute the local error indicator

$$\mu_i := h_i^{1/2} \|\nabla R_N\|_{L^2(\Gamma_i)}$$

where $h_i = \operatorname{diam} \Gamma_i$. Check for a stopping criterion.

(iii) Compute $\mu_{\max} := \max\{\mu_i\,;\, i = 1, \dots, N\}$ and refine the rectangle Γ_i into four equal sized rectangles iff

$$\mu_i \geq \Theta \mu_{\max}$$

This defines the new partition π_{m+1} and the refined N_{m+1}-dimensional space $S_{0,N_{m+1}}(\Gamma) \supset S_{0,N_m}(\Gamma)$.

Algorithm B For $m = 0, 1, 2, \dots$

(i) Compute the Galerkin solution $\psi_N \in S_{0,N_m}(\Gamma)$ according to the partition $\pi_m = \{\Gamma_1, \dots, \Gamma_{N_m}\}$.
(ii) For each element $\Gamma_i \in \pi_m$ compute the local error indicators

$$\mu_{i,k} := h_{i,k}^{1/2} \|\partial_{x_k} R_N\|_{L^2(\Gamma_i)}$$

where $h_{i,k}$ = diameter of Γ_i in x_k-direction ($k = 1, 2$). Check for a stopping criterion.
(iii) Compute $\mu_{\max,k} := \max\{\mu_{i,k}\,;\, i = 1, \dots, N\}$ ($k = 1, 2$) and refine the rectangle Γ_i along the x_k-axis iff

$$\mu_{i,k} \geq \Theta \mu_{\max,k} \quad (k = 1, 2).$$

This defines the new partition π_{m+1} and the refined N_{m+1}-dimensional space $S_{0,N_{m+1}}(\Gamma) \supset S_{0,N_m}(\Gamma)$.

Algorithm C For $m = 0, 1, 2, \dots$

(i) Compute the Galerkin solution $\psi_N \in S_{0,N_m}(\Gamma)$ according to the partition $\pi_m = \{\Gamma_1, \dots, \Gamma_{N_m}\}$.
(ii) For each element $\Gamma_i \in \pi_m$ compute the local error indicators

$$\eta_{i,1}^H := \frac{|\langle f - V\psi_N, \xi_{i,1}\rangle|}{V(\xi_{i,1}, \xi_{i,1})^{1/2}}, \quad \eta_{i,2}^H := \frac{|\langle f - V\psi_N, \xi_{i,2}\rangle|}{V(\xi_{i,2}, \xi_{i,2})^{1/2}}$$

where $\xi_{i,1} = \beta_{i+1,\cdot,1}, \xi_{i,2} = \beta_{i+1,\cdot,2}$ see Fig. 10.4 and the global error indicator

$$\Sigma_n := \left(\sum_{i=N_m+1}^{4N_m} \eta_i^2 \right)^{1/2}.$$

The algorithm stops if $\Sigma_n < \varepsilon_0$ where $\varepsilon_0 > 0$ is a given constant.
(iii) Compute $\eta_{\max}^H := \max\{\eta_{i,1}^H, \eta_{i,2}^H\,;\, i = 1, \dots, N\}$ ($k = 1, 2$) and refine the rectangle Γ_i along the x_k-axis iff

$$\eta_{i,k}^H \geq \Theta \eta_{\max}^H \quad (k = 1, 2).$$

This defines the new partition π_{m+1} and the refined N_{m+1}-dimensional space $S_{0,N_{m+1}}(\Gamma) \supset S_{0,N_m}(\Gamma)$.

10.3.2 Numerical Example

Let Γ be the L-Shape and $f = 1$. The energy norm of the solution ψ of $V\psi = 1$ is known to be $\|\psi\|_{\tilde{H}^{-1/2}} = 2.878293$. We start with a uniform mesh containing 12 elements and apply our algorithms A,B and the algorithm C using the hierarchical error indicator. Table 10.1 gives the results for Algorithm B and Table 10.2 gives the results for the hierarchical error indicator. Figure 10.1 gives the error curves for the adaptive algorithms, the uniform h-version and for algebraically graded meshes with different grading parameters. We see that Algorithm B and Algorithm C (hierarchical error indicator) are giving nearly identical curves with optimal convergence rates. The adaptive schemes generate the same curves for $\Theta \in [0.5, 1]$, only the number of newly generated meshes raises with increasing Θ.

The error indicators of Algorithms A and B have been computed by numerical integration of the analytically computed gradient of the residual by an 4×4-Gaussian quadrature rule. The error indicators of Algorithm C have been calculated analytically using the same algorithms as in the computation of the Galerkin matrix.

All computations have been performed on a Sun Ultrasparc-II (300MHz) at the Institute for Applied Mathematics, University of Hannover, using the program system *maiprogs* [293].

Table 10.1 Residual error indicators and efficiency for $\theta = 0.5$ (Alg. B) [87]

N	Indicator	Error	Effectivity	Rate
12	0.9558277	0.6646427	1.438108	
33	0.7157455	0.4917142	1.455613	0.29789
64	0.5231094	0.3603837	1.451534	0.46911
105	0.3791923	0.2621146	1.446666	0.64310
148	0.2857298	0.1968658	1.451394	0.83396
184	0.2220168	0.1548582	1.433678	1.10237
245	0.1641732	0.1158337	1.417319	1.01408
306	0.1275614	0.0909866	1.401980	1.08598
371	0.0984725	0.0707287	1.392257	1.30757
450	0.0789983	0.0579766	1.362589	1.02987
540	0.0638628	0.0465304	1.372496	1.20629
656	0.0509174	0.0376706	1.351647	1.08548
805	0.0411453	0.0308533	1.333580	0.97535
968	0.0340271	0.0260617	1.305636	0.91532
1243	0.0266476	0.0207189	1.286148	0.91749
1543	0.0217106	0.0173712	1.249805	0.81513
1902	0.0179114	0.0144254	1.241652	0.88835
2418	0.0145247	0.0118803	1.222581	0.80867
3127	0.0116897	0.0097656	1.197027	0.76231
3926	0.0096766	0.0081669	1.184856	0.78566

Table 10.2 Hierarchical error indicators and efficiency for $\theta = 0.5$ (Alg. C)

N	Indicator	Error	Effectivity	Rate
12	0.3460879	0.6646427	0.520713	
33	0.2516634	0.4917142	0.511808	0.29789
64	0.1843478	0.3603837	0.511532	0.46911
105	0.1343572	0.2621146	0.512590	0.64310
158	0.0961964	0.1898212	0.506774	0.78970
194	0.0755879	0.1485726	0.508760	1.19363
255	0.0576133	0.1104080	0.521822	1.08590
324	0.0451620	0.0838526	0.538589	1.14883
382	0.0376194	0.0670436	0.561118	1.35852
469	0.0320552	0.0526239	0.609138	1.18028
606	0.0251860	0.0410832	0.613048	0.96602
759	0.0206936	0.0328849	0.629274	0.98873
977	0.0168162	0.0262459	0.640718	0.89313
1240	0.0138612	0.0212237	0.653100	0.89097
1667	0.0109750	0.0167291	0.656044	0.80418
1670	0.0108687	0.0165264	0.657656	
2487	0.0079173	0.0120928	0.654716	0.78429
3188	0.0063593	0.0096686	0.657722	0.90097
4053	0.0053212	0.0077425	0.687276	0.92541

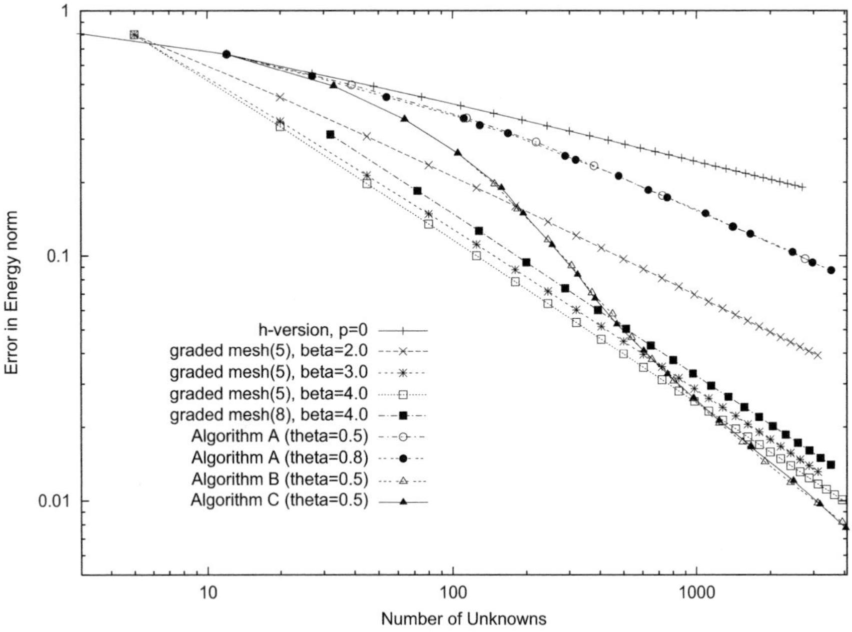

Fig. 10.1 Dirichlet problem on the L-shape in $\mathbb{R}^3$ [87]

Remark 10.4 For uniform and graded meshes, the superiority of a grading with a parameter $(j/J)^4$ for $j = 0, \dots, J$ gives the best result. Algorithm (A) is not competitive, there is a need for anisotropic mesh-refinement. Algorithms (B) and (C) yield similar results and can compete for higher degrees of freedom with optimally graded meshes.

10.4 The Hypersingular Integral Equation in 3D

Here we report from [86]. Each element Γ_j of the triangulation $\mathscr{T}$ is supposed to be a closed (flat) triangle or parallelogram in $\mathbb{R}^3$.

The set of all nodes in the triangulation is denoted by $\mathscr{N}$ and the free nodes by $\mathscr{K} := \mathscr{N} \setminus \gamma$, where γ is the boundary of Γ. The set of all edges in the triangulation is denoted by $\mathscr{E}$ and $\mathscr{E}$ is split into edges on the boundary γ, namely $\mathscr{E}_\gamma := \{E \in \mathscr{E} : E \subseteq \gamma\}$, and interior edges $\mathscr{E}_\Gamma := \{E \in \mathscr{E} : E \cap \gamma = \emptyset\}$.

For each free node $z \in \mathscr{K}$ there is a hat function φ_z which equals zero on each element Γ_k (if $z \notin \Gamma_k$) or one of the nodal local basis functions (if z is a vertex of Γ_k) such that $\varphi_z(z) = 1$ and $\varphi_z(x) = 0$ for all $x \in \mathscr{N} \setminus \{z\}$. The hat functions are Lipschitz continuous and form a partition of unity. Their linear hull

$$\mathscr{S} := \operatorname{span}\{\varphi_z : z \in \mathscr{K}\} \subseteq H_0^1(\Gamma) \tag{10.71}$$

satisfies proper boundary conditions. Since $\{\varphi_z : z \in \mathscr{K}\}$ is, in general, not a partition of unity, we choose a node $\zeta(z) \in \mathscr{K}$ for each $z \in \mathscr{N} \setminus \mathscr{K}$ and define $\zeta(z) = z$ for $z \in \mathscr{K}$. We get a partition of $\mathscr{N}$ into $card\{\mathscr{K}\}$ classes $I(z) := \{\tilde{z} \in \mathscr{N} : \zeta(\tilde{z}) = z\}$, $z \in \mathscr{K}$. For each $z \in \mathscr{K}$ set

$$\psi_z := \sum_{\tilde{z} \in I(z)} \varphi_{\tilde{z}} \tag{10.72}$$

and notice that $\{\psi_z : z \in \mathscr{K}\}$ is a partition of unity. It is required that

$$\Omega_z := \{x \in \Gamma : 0 < \psi_z(x)\} \tag{10.73}$$

is connected and contains only a limited number of elements. We remark that $\psi_z \neq \varphi_z$ implies that $\gamma \cap \partial\Omega_z$ has a positive surface measure. Those definitions follow and adapt [77] in order to employ an approximation operator $(0 \leq s \leq 1)$

$$\mathscr{J} : \tilde{H}^s(\Gamma) \to \mathscr{S} \subseteq \tilde{H}^s(\Gamma). \tag{10.74}$$

For each $g \in L^1(\Gamma)$ let $g_z \in \mathbb{R}$ be $g_z := \dfrac{\int_{\Omega_z} g\psi_z\, ds}{\int_{\Omega_z} \varphi_z\, ds}$ for $z \in \mathcal{K}$ and then set

$$\mathcal{J} g := \sum_{z\in\mathcal{K}} g_z\, \varphi_z \quad \in \mathcal{S}. \tag{10.75}$$

Lemma 10.5 ([77]) *There is a constant $c_4 > 0$ that depends on Γ and the aspect ratio of the elements (but not on their sizes) such that for all $z \in \mathcal{K}$ and $g \in H_0^1(\Gamma)$ $= \widetilde{H}^1(\Gamma)$ we have*

$$\|\psi_z g - \varphi_z g_z\|_{L^2(\Omega_z)} \le c_4 \min\left\{\|g\|_{L^2(\Omega_z)},\ h_z\|\nabla g\|_{L^2(\Omega_z)}\right\}, \tag{10.76}$$

where $h_z = \mathrm{diam}\,(\Omega_z)$.

Suppose that the residual $R := f - Wu_h = W(u - u_h) \in L^2(\Gamma)$ with the hypersingular operator W satisfies the Galerkin conditions

$$\langle R, \varphi_z\rangle = 0 \qquad \text{for all } z \in \mathcal{K}, \tag{10.77}$$

where $\langle\cdot,\cdot\rangle$ denotes the duality pairing on $\tilde{H}^{-s}$ and H^s. Then we have the following residual-based a posteriori error estimate.

Theorem 10.9 *There is a constant $c_5 > 0$ such that for all $R \in L^2(\Gamma)$ with* (10.77) *and $0 \le s \le 1$ there holds*

$$\|R\|_{H^{s-1}(\Gamma)} \le c_5\Big(\sum_{z\in\mathcal{K}} h_z^{2-2s}\|R\|^2_{L^2(\Omega_z)}\Big)^{1/2}. \tag{10.78}$$

Proof Since $(\psi_z : z \in \mathcal{K})$ defines a partition of unity, we have $R = \sum_{z\in\mathcal{K}} \psi_z R$. This combined with (10.77) shows for $g \in H_0^1(\Gamma)$

$$\langle R, g\rangle = \sum_{z\in\mathcal{K}} \langle R, g\psi_z\rangle = \sum_{z\in\mathcal{K}} \langle R, g\psi_z - g_z\varphi_z\rangle \tag{10.79}$$

with $g_z \in \mathbb{R}$ defined as above. Now

$$\|g\psi_z - g_z\varphi_z\|_{L^2(\Omega_z)} \le c_1 h_z^{1-s}\|g\|_{H_D^{1-s}(\Omega_z)} \tag{10.80}$$

and (10.79) imply

$$\langle R, g\rangle \le c_2\Big(\sum_{z\in\mathcal{K}} h_z^{2-2s}\,\|R\|^2_{L^2(\Omega_z)}\Big)^{1/2}\Big(\sum_{z\in\mathcal{K}} \|g\|^2_{H_D^{1-s}(\Omega_z)}\Big)^{1/2} \tag{10.81}$$

A coloring argument (see proof of Lemma 3.1 [87]) completes the proof. □

The following result guarantees reliability of the estimator

$$\eta := \Big(\sum_{T\in\mathscr{T}} \eta_T^2\Big)^{1/2} \quad \text{with } \eta_T := h_T^{1-s}\|R\|_{L^2(T)} \text{ and } h_T := \operatorname{diam}(T) \quad \text{for } T \in \mathscr{T}.$$

Theorem 10.10 *For $0 < s < 1$, there is a constant $c_6 > 0$ that depends on s, Γ, and the aspect ratio of the elements (but not their size) such that*

$$\|u - u_h\|_{\tilde{H}^s(\Gamma)} \le c_6 c_7 \, \|h_{\mathscr{T}}^{1-s} R\|_{L^2(\Gamma)}, \tag{10.82}$$

where the $\mathscr{T}$-piecewise constant $h_{\mathscr{T}} \in L^\infty(\Gamma)$ is defined by $h_{\mathscr{T}}(x) = h_T$ for $x \in T$ and $c_7 := \max\{h_z/h_T : z \in \mathscr{K}, T \in \mathscr{T}$ with $T \subseteq \Omega_z\}$.

Proof $W : \tilde{H}^s(\Gamma) \to H^{s-1}(\Gamma)$ is a continuous, linear bijection and by the open mapping theorem the inverse map W^{-1} is continuous i.e., $\|W^{-1}\| < \infty$. Since $R = W(u - u_h)$ we obtain

$$\|u - u_h\|_{\tilde{H}^s(\Gamma)} \le \|W^{-1}\| \, \|R\|_{H^{s-1}(\Gamma)}. \tag{10.83}$$

Theorem 10.9 and the finite overlap of the $\{\Omega_z : z \in \mathscr{K}\}$ imply

$$\|R\|_{H^{s-1}(\Gamma)} \le c_5 \Big(\sum_{z\in\mathscr{K}} \|h^{1-s} R\|^2_{L^2(\Omega_z)}\Big)^{1/2} \le M c_5 c_7 \, \|h_{\mathscr{T}}^{1-s} R\|_{L^2(\Gamma)} \tag{10.84}$$

with a bounded number M and $h(x) := \max\{h_z : x \in \Omega_z\}$. □

Now we compare η with the multilevel error estimator μ from Section 10.5.1 for pw. linears on triangles and from[313] for pw. bilinears on rectangles.

We need at least two meshes where one triangulation $\mathscr{T}_h$ is a refinement of $\mathscr{T}_H$. The set of free nodes $\mathscr{K}_h$ and $\mathscr{K}_H$ give rise to hat functions $(\varphi_z : z \in \mathscr{K}_h)$ and $(\varphi_z : z \in \mathscr{K}_H)$ with respect to $\mathscr{T}_h$ and $\mathscr{T}_H$, respectively. If $\mathscr{S}_h$ and $\mathscr{S}_H$ denote the respective discrete spaces, $\mathscr{S}_H \subseteq \mathscr{S}_h$ and

$$\mathscr{S}_h = \mathscr{S}_H \oplus \operatorname{span}\{\varphi_z : z \in \mathscr{K}_h \backslash \mathscr{K}_H\}, \tag{10.85}$$

with respective discrete solutions u_h and u_H. For the practical computation, only u_H is required, $\mathscr{S}_h$ plays the role of a fictitious larger space. However, the saturation assumption,

$$\|u - u_h\|_W \le \kappa \|u - u_H\|_W, \tag{10.86}$$

for some fixed $0 \le \kappa < 1$, plays an essential role. In contrast to the finite element context for partial differential equations [84, 143, 144], a proof of (10.86) is unknown for boundary element problems.

Definition 10.2 For each $z \in \mathscr{K}_h$, let $\mu_z := \langle R, \varphi_z \rangle / \|\varphi_z\|_W$ with $R := f - W u_H$.

Theorem 10.11 ([313]) *Under the saturation assumption* (10.86) *we have with mesh independent constants* c_8, c_9

$$c_8 \sum_{z \in \mathscr{K}_h \setminus \mathscr{K}_H} \mu_z^2 \leq \|u - u_H\|_W^2 \leq c_9 \sum_{z \in \mathscr{K}_h \setminus \mathscr{K}_H} \mu_z^2. \tag{10.87}$$

The proof of the above theorem is given in Subsection 10.5.1.

Remark 10.5 The constant c_8 in (10.87) depends on $1/(1-\kappa)$ and (possibly) degenerates as $\kappa \to 1$. The constant c_7 is robust with $0 \leq \kappa < 1$. The two-level error estimator

$$\mu := \Big(\sum_{z \in \mathscr{K}_h \setminus \mathscr{K}_H} \mu_z^2 \Big)^{1/2} \tag{10.88}$$

performs very accurately in practice.

Theorem 10.12 *There is an* $h_{\mathscr{T}}$*-independent constant* c_{10} *such that, for each* $z \in \mathscr{K}_h \setminus \mathscr{K}_H$ *and* supp $\varphi_z \subseteq T_1 \cup T_2$ *with* $T_1, T_2 \in \mathscr{T}_H$*, we have*

$$\mu_z^2 \leq c_{10} \|h_{\mathscr{T}}^{1/2} R\|_{L^2(T_1 \cup T_2)} \leq c_{10}(\eta_{T_1} + \eta_{T_2}). \tag{10.89}$$

Proof Since $\|\varphi_z\|_W^2 \approx \|\varphi_z\|_{\tilde{H}(\Gamma)^{1/2}}^2 \approx h_z$ due to (10.70) we have

$$\mu_z^2 \leq \|R\|_{L^2(T_1 \cap T_2)}^2 \|\varphi_z\|_{L^2(T_1 \cap T_2)}^2 \leq h_z \|R\|_{L^2(T_1 \cap T_2)}^2$$ □

In our numerical experiments we use the estimators

$$\eta_N := \|h_{\mathscr{T}}^{1/2} R\|_{L^2(\Gamma)} \quad \text{for the residual } R := f - W u_N$$

$$\mu_N := \Big(\sum_{z \in \mathscr{K}_h \setminus \mathscr{K}_k} \mu_z^2 \Big)^{1/2} \quad \text{from (10.88)}$$

for the current coarse mesh $\mathscr{T}_H = \mathscr{T}_k$ and one (fictitious) refinement $\mathscr{T}_h$ with the new nodes $\mathscr{K}_h \setminus \mathscr{K}_k$ on all edges. We perform the refinement strategy of Fig. 12.2. In [86] we have performed numerical experiments for the Neumann screen problem of the Laplacian with boundary data $g(x) = 1$ on the L-shaped screen (Example 1) and with $g(x) = \frac{1}{|x-(0.1,0.1)|}$ on $\Gamma = [-1,1]^2$ (Example 2). The numerical experiments are compactly displayed in Figs. 10.2 and 10.3, where we plotted the energy norm E_N versus the number of degrees of freedom N. Both axes are scaled logarithmically. For description of algorithms (A_R) (residual) and (A_H) (hierarchical) see [86]. In both examples the experimental convergence rate of the uniform h-version is $\approx 1/2$. The examples validate the reliability of

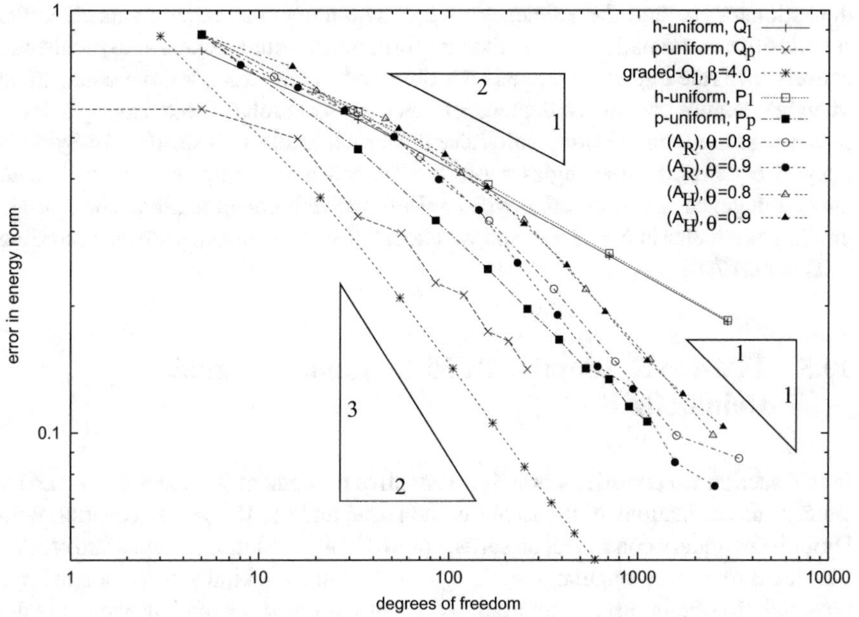

Fig. 10.2 Neumann problem on the L-shape in Example 1 [86]

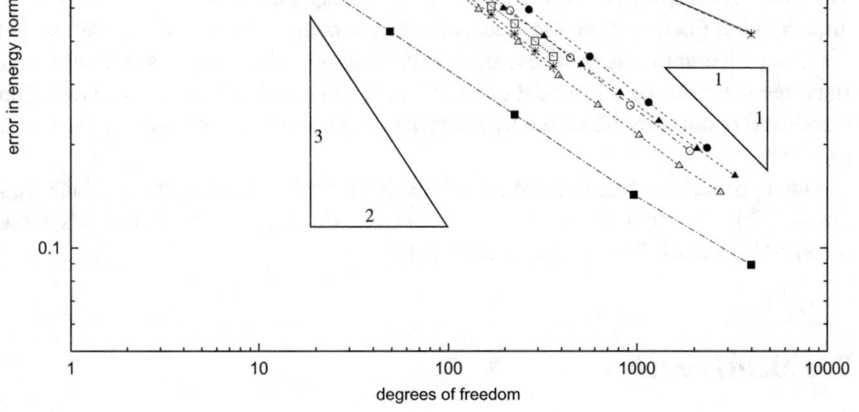

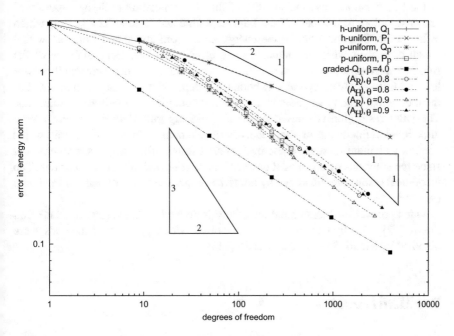

Fig. 10.3 Neumann problem on the unit square in Example 2 [86]

the estimator η_N and the efficiency of μ_N. The numerical experiments show the superiority of automatic adaptive over uniform mesh-refinements. An experimental convergence rate higher than one is not observed in our examples; the conforming triangles require too many degrees of freedom to resolve the anisotropic layer structure as they refine isotropically. The numerical results in Sect. 10.3 for Symm's integral equation on rectangles show that indeed a speed-up of a factor $3/2$ in the convergence rate is possible when anisotropic refinement is used. The optimal grading parameter is $\beta = 3 + \epsilon$ and we choose $\beta = 4$ in the experiments according to Theorem 7.18.

10.5 Two-Level Adaptive BEM for Laplace, Lamé, Helmholtz

In this section we consider hierarchical adaptive refinement strategies for boundary element discretizations of the Laplace, the Lamé and the Helmholtz equation with Dirichlet boundary conditions on surface pieces in $\mathbb{R}^3$ modelling screens and cracks. This leads to weakly singular integral equations of the first kind with the single layer potential. To obtain discrete approximations of the integral equations we apply the h-version Galerkin boundary element method with piecewise constant test and trial functions. We define a two-level decomposition of the discrete spaces for the case of the Laplacian and use the stability of this decomposition to derive a-posteriori error estimators for the case of the Laplace, Lamé and Helmholtz operators which estimate the Galerkin error in the energy norm from above and from below and which consist of easily computable local error indicators. Here, stability of the subspace decomposition means that the condition number of the corresponding two-level additive Schwarz operator is bounded independently of the mesh size h. For the theoretical results we assume a saturation condition and uniform refinements. The latter restriction is relaxed in the adaptive algorithm where we allow local directional refinements of the quadrilateral elements which lead to anisotropic boundary element meshes. These meshes are essential for optimal approximations since the solutions of our model problems have edge and corner singularities. The theoretical results are illustrated by numerical experiments for the case of the Lamé equation.

Our approach to hierarchical adaptivity is related to the results by Bank and Smith [32]. The first results for 3D adaptive boundary element methods were obtained by Mund, Stephan and Weisse [314].

The Model Problem

Let $\Gamma \subset \mathbb{R}^3$ denote an open surface piece. For simplicity we assume that Γ is a right angled polygonal domain in the (x_1, x_2)-plane. In the unbounded domain

$\Omega = \mathbb{R}^3 \setminus \overline{\Gamma}$ we consider the following Dirichlet problem: Given a function $g \in (H^{1/2}(\Gamma))^d$ find $u \in (H^1_{loc}(\Omega))^d$ such that

$$\begin{aligned} Du &= 0 \quad \text{in } \Omega \\ u &= f \quad \text{on } \Gamma \end{aligned}$$

where the differential operator D is either the Laplace operator Δ or the Lamé operator Δ^* or the Helmholtz operator $\Delta + \kappa^2$ $(\kappa > 0)$. If $D = \Delta^*$ we have $d = 3$ and otherwise $d = 1$. Furthermore we assume the radiation condition $u = O(r^{-1})$ and additionally in the case of the Helmholtz equation

$$\frac{\partial u}{\partial r} - iku = o(r^{-1}) \qquad \text{for} \quad r = |x| \to \infty .$$

According to Lions and Magenes [284] we define the Sobolev space $\mathcal{H}^d = (\widetilde{H}^{-1/2}(\Gamma))^d$ as the dual space of $(H^{1/2}(\Gamma))^d$.

The problem can be reformulated as the following boundary integral equation [396] and [398]: Find $\phi_D \in \mathcal{H}^d$ such that

$$V_D \phi_D = g \tag{10.90}$$

where $g = 2f$ and $V_D : \mathcal{H}^d \to (H^{1/2}(\Gamma))^d$ is the single layer potential operator

$$(V_D \psi)(x) = -2 \int_\Gamma G_D(x, y)\, \psi(y)\, d\sigma_y .$$

Here, $G_D(x, y) \in \mathbb{R}^{d \times d}$ denotes the fundamental solution of the differential operator D, i.e.

$$G_D(x, y) = \begin{cases} \frac{1}{4\pi} \frac{\exp(i\kappa|x-y|)}{|x-y|} & \text{if } D = \Delta + \kappa^2 \\ \frac{\lambda+3\mu}{8\pi\mu(\lambda+2\mu)} \left\{ \frac{1}{|x-y|} I_{3\times 3} + \frac{\lambda+\mu}{\lambda+3\mu} \frac{(x-y)(x-y)^\top}{|x-y|^3} \right\} & \text{if } D = \Delta^*. \end{cases}$$

with $\kappa \geq 0$ and the Lamé constants λ and μ satisfy $\mu > 0$ and $2\lambda + \mu > 0$.

The variational formulation of (10.90) is given by: Find $\phi_D \in \mathcal{H}^d$ such that

$$\langle V_D \phi_D, \psi \rangle = \langle g, \psi \rangle \tag{10.91}$$

for all $\psi \in \mathcal{H}^d$ where $\langle \cdot, \cdot \rangle$ denotes the duality pairing of $(H^{1/2}(\Gamma))^d$ and $\mathcal{H}^d$. We define the symmetric bilinear form $V_D(\chi, \psi) = \langle V_D \chi, \psi \rangle$. If $D = \Delta$ or $D = \Delta^*$ then $V_D(\cdot, \cdot)$ is an inner product in $\mathcal{H}^d$ and in this case there is a constant $\eta > 0$ such that

$$V_D(\psi, \psi) \geq \eta\, \|\psi\|^2_{\mathcal{H}^d} \qquad (D \in \{\Delta, \Delta^*\}) \tag{10.92}$$

for all $\psi \in \mathscr{H}^d$. The single layer potential for the Helmholtz operator satisfies

$$|\langle (V_{\Delta+\kappa^2} - V_\Delta)\chi\,,\,\psi\rangle| \;\le\; c\,\|\chi\|_{L^2(\Gamma)}\,\|\psi\|_{\mathscr{H}^d} \tag{10.93}$$

for all $\chi, \psi \in \mathscr{H}^d$ and for the constant $c > 0$.

Now, let τ_0 be a an initial partition of Γ into regular quadrilaterals of size h_0 and let $\{\tau_j\}_{j=1}^{\infty}$ be a sequence of uniform refinements of τ_0 where τ_j is obtained by refining all elements in τ_{j-1} into four equally sized quadrilaterals of size $h_j = h_0 2^{-j}$. The number of elements in τ_j is denoted by n_j. We seek approximations to the exact solution ϕ_D of (10.91) in the spaces

$$S_{j,D} \;=\; \{\psi : \Gamma \to \Delta^d \mid \psi \text{ piecewise constant w.r.t. } \tau_j\} \tag{10.94}$$

where $\Delta = \mathbb{C}$ if $D = \Delta + \kappa^2$ ($\kappa > 0$) and $\Delta = \mathbb{R}$ otherwise.

The Galerkin method at level j reads as follows: Find $\phi_{j,D} \in S_{j,D}$ such that

$$V_D(\phi_{j,D}, \psi) \;=\; \langle g, \psi\rangle \tag{10.95}$$

for all $\psi \in S_{j,D}$.

A Stable Two-Level Decomposition

In this section we consider a two-level decomposition by Mund et al. [314] for the Laplacian Δ. The stability of this decomposition will be used in the next section to derive a-posteriori estimates for the Galerkin error in the cases of the Laplace, Lamé and Helmholtz equations.

For $J \in \mathbb{N}$ we consider hierarchical basis functions $\beta_{j,i,l}$ in $S_{J,\Delta}$ where $j \in \{0, \dots, J\}$ denotes the level, $i \in \{1, \dots, n_j\}$ denotes the number of the element and $l \in \{1, 2, 3\}$. We start with the standard basis of piecewise constant brick functions in $S_{0,\Delta}$ and define the hierarchical basis recursively. Whenever an element $\Gamma_i^j \in \tau_j$ ($j \ge 0$) is divided into the four elements $\Gamma_{i_1}^{j+1}$, $\Gamma_{i_2}^{j+1}$, $\Gamma_{i_3}^{j+1}$ and $\Gamma_{i_4}^{j+1}$ we extend the basis of $S_{j,\Delta}$ by the basis functions $\beta_{j+1,i,1}$, $\beta_{j+1,i,2}$ and $\beta_{j+1,i,3}$ which are defined as

$$\beta_{j+1,i,1}(x) = \begin{cases} 1 \text{ if } x \in \Gamma_{i_1}^{j+1} \cup \Gamma_{i_2}^{j+1} \\ -1 \text{ if } x \in \Gamma_{i_3}^{j+1} \cup \Gamma_{i_4}^{j+1} \\ 0 \text{ otherwise} \end{cases}$$

$$\beta_{j+1,i,2}(x) = \begin{cases} 1 \text{ if } x \in \Gamma_{i_1}^{j+1} \cup \Gamma_{i_4}^{j+1} \\ -1 \text{ if } x \in \Gamma_{i_2}^{j+1} \cup \Gamma_{i_3}^{j+1} \\ 0 \text{ otherwise} \end{cases}$$

$$\beta_{j+1,i,3}(x) = \beta_{j+1,i,1}(x) \cdot \beta_{j+1,i,2}(x) \qquad \text{(see Fig. 10.4)}.$$

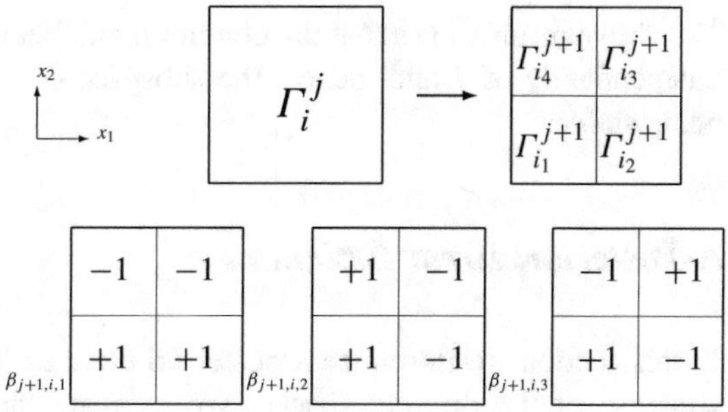

Fig. 10.4 Refinement of Γ_i^j into four new elements and the additional basis functions

We define the one-dimensional spaces $Y_{j+1,i,l} = \text{span}\{\beta_{j+1,i,l}\}$ and note that adding $Y_{j+1,i,1}$ to the discrete space $S_{j,\Delta}$ corresponds to a bisection of Γ_i^j (the i-th element at level j) along the x_1-axis and adding $Y_{j+1,i,2}$ corresponds to a bisection of Γ_i^j along the x_2-axis. Now, we consider the following two-level subspace decomposition of $S_{j+1,\Delta}$:

$$S_{j+1,\Delta} = S_{j,\Delta} \oplus \bigoplus_{i=1}^{n_j} \bigoplus_{l=1}^{3} Y_{j+1,i,l}\,. \tag{10.96}$$

To define the corresponding two-level additive Schwarz operator we need the Galerkin projections $P_j : S_{j+1,\Delta} \to S_{j,\Delta}$ and $P_{j+1,i,l} : S_{j+1,\Delta} \to Y_{j+1,i,l}$ which are defined by

$$V_\Delta(P_j \chi, \psi) = V(\chi, \psi) \qquad \forall \psi \in S_{j,\Delta}$$

and

$$V_\Delta(P_{j+1,i,l}\chi, \psi) = V(\chi, \psi) \qquad \forall \psi \in Y_{j+1,i,l}$$

with $\chi \in S_{j+1,\Delta}$.

The two-level additive Schwarz operator $P_{j+1}^\Delta : S_{j+1,\Delta} \to S_{j+1,\Delta}$ is now defined as

$$P_{j+1}^\Delta = P_j + \sum_{i=1}^{n_j} \sum_{l=1}^{3} P_{j+1,i,l}\,.$$

The following result was proved by Mund et al. [314]:

Theorem 10.13 *There are constants $\mu_1, \mu_2 > 0$ independent of j such that*

$$\mu_1 V_\Delta(\psi, \psi) \le V_\Delta(P_{j+1}^\Delta \psi, \psi) \le \mu_2 V_\Delta(\psi, \psi) \qquad \forall \psi \in S_{j+1,\Delta}\,. \tag{10.97}$$

The above result shows that the condition number of the operator P^{Δ}_{j+1} is bounded independently of j and, hence, the subspace decomposition (10.96) is termed as being *stable*.

A-Posteriori Error Estimates

In this section we introduce a-posteriori error estimates based on the hierarchical structure of the discrete spaces. We estimate the difference between the exact solution ϕ_D of (10.91) and the Galerkin solution $\phi_{j,D}$ of (10.95) in the $\mathscr{H}^d$-norm for $D = \Delta+\kappa^2$ ($\kappa \geq 0$) and $d = 1$ or $D = \Delta^*$ and $d = 3$. The estimate is proved for the case of uniform refinements and under the assumption of the following saturation condition:

Assumption 10.1 (A$_D$) *There exists an integer j_0 and a constant $\varrho < 1$ such that*

$$\|\phi_D - \phi_{j+1,D}\|_{\mathscr{H}^d} \leq \varrho \, \|\phi_D - \phi_{j,D}\|_{\mathscr{H}^d} \qquad \forall\, j \geq j_0 \,. \tag{10.98}$$

This assumption is certainly satisfied if $\|\phi_D - \phi_{j,D}\|_{\mathscr{H}^d} \sim n_j^{-\alpha}$ for some constant $\alpha > 0$ and where n_j denotes again the number of elements in τ_j. From (10.98) we conclude that

$$\frac{1}{1+\rho} \, \|\phi_{j+1,D} - \phi_{j,D}\|_{\mathscr{H}^d} \leq \|\phi_D - \phi_{j,D}\|_{\mathscr{H}^d} \leq \frac{1}{1-\rho} \, \|\phi_{j+1,D} - \phi_{j,D}\|_{\mathscr{H}^d} \tag{10.99}$$

and, hence, it remains to estimate the difference between two successive Galerkin solutions.

For the Laplacian, i.e. $D = \Delta$, we obtain from Theorem 10.13 and from the orthogonality of the Galerkin projections $P_{j+1,i,l}$ that

$$\|\chi\|^2_{\mathscr{H}^d} \sim V_\Delta(P^{\Delta}_{j+1}\chi, \chi) = V_\Delta(P_j\chi, \chi) + \sum_{i=1}^{n_j} \sum_{l=1}^{3} V_\Delta(P_{j+1,i,l}\chi, \chi) \tag{10.100}$$

for all $\chi \in S_{j+1,\Delta}$.

For $\chi = \phi_{j+1,\Delta} - \phi_{j,\Delta}$ the terms on the right hand side of (10.100) can be calculated explicitly: From the Galerkin property of $\phi_{j+1,\Delta}$ and $\phi_{j,\Delta}$ we obtain

$$V_\Delta(P_j(\phi_{j+1,\Delta} - \phi_{j,\Delta}), \, \phi_{j+1,\Delta} - \phi_{j,\Delta}) = 0 \,. \tag{10.101}$$

Since $P_{j+1,i,l}$ projects the space $S_{j+1,\Delta}$ onto the one-dimensional space $Y_{j+1,i,l}$ it is easy to verify that

$$\eta^{\Delta}_{j,i,l} := V_\Delta(P_{j+1,i,l}(\phi_{j+1,\Delta} - \phi_{j,\Delta}), \phi_{j+1,\Delta} - \phi_{j,\Delta}) = \frac{|\langle g - V_\Delta\phi_{j,\Delta}, \beta_{j+1,i,l}\rangle|}{V_\Delta(\beta_{j+1,i,l}, \beta_{j+1,i,l})^{1/2}} \,. \tag{10.102}$$

The following a-posteriori error estimate follows now from (10.99)–(10.102):

Theorem 10.14 *Under Assumption* $\mathbf{A}_\Delta$ *there exist constants* $C_1, C_2 > 0$ *and an integer* j_0 *such that for all* $j \geq j_0$ *the following holds:*

$$C_1 \sum_{i=1}^{n_j} \sum_{l=1}^{3} \left(\eta_{j,i,l}^{\Delta}\right)^2 \leq \|\phi_\Delta - \phi_{j,\Delta}\|_{\mathscr{H}}^2 \leq C_2 \sum_{i=1}^{n_j} \sum_{l=1}^{3} \left(\eta_{j,i,l}^{\Delta}\right)^2 .$$

The local error indicators $\eta_{j,i,l}^{\Delta}$ *are defined in (10.102).*

A corresponding result for $D = \Delta + \kappa^2$ with $\kappa > 0$ was obtained by Maischak et al. [294] by taking advantage of (10.93):

Theorem 10.15 *Under Assumption* $\mathbf{A}_{\Delta+\kappa^2}$ *there exist constants* $C_1, C_2 > 0$ *and an integer* j_0 *such that for all* $j \geq j_0$ *the following holds:*

$$C_1 \sum_{i=1}^{n_j} \sum_{l=1}^{3} \left(\eta_{j,i,l}^{\Delta+\kappa^2}\right)^2 \leq \|\phi_{\Delta+\kappa^2} - \phi_{j,\Delta+\kappa^2}\|_{\mathscr{H}}^2 \leq C_2 \sum_{i=1}^{n_j} \sum_{l=1}^{3} \left(\eta_{j,i,l}^{\Delta+\kappa^2}\right)^2 .$$

The local error indicators are defined as

$$\eta_{j,i,l}^{\Delta+\kappa^2} = \frac{|\langle g - V_{\Delta+\kappa^2}\phi_{j,\Delta+\kappa^2}, \beta_{j+1,i,l}\rangle|}{V_\Delta(\beta_{j+1,i,l}, \beta_{j+1,i,l})^{1/2}} .$$

In the case of the Lamé operator, $D = \Delta^*$, we use the fact that

$$V_{\Delta^*}(\psi, \psi) \sim \sum_{k=1}^{3} V_\Delta(\psi_k, \psi_k) \qquad \forall\, \psi = (\psi_1, \psi_2, \psi_3) \in \mathscr{H}^3$$

which follows from (10.92) and from the continuity of the single layer potential operators V_Δ and V_{Δ^*}. By applying the stability result of Theorem 10.13 to each component of the vector $\phi_{j+1,\Delta^*} - \phi_{j,\Delta^*}$ we obtain local error indicators similar to those in (10.102). Since the spaces spanned by the basis functions $\beta_{j+1,i,l}$ are now three-dimensional we have to solve a 3×3 linear system (see (10.103) below) to obtain the error indicators $\eta_{j,i,l}^{\Delta^*}$.

Theorem 10.16 *Under Assumption* $\mathbf{A}_{\Delta^*}$ *there exist constants* $C_1, C_2 > 0$ *and an integer* j_0 *such that for all* $j \geq j_0$ *the following holds:*

$$C_1 \sum_{i=1}^{n_j} \sum_{l=1}^{3} \left(\eta_{j,i,l}^{\Delta^*}\right)^2 \leq \|\phi_{\Delta^*} - \phi_{j,\Delta^*}\|_{\mathscr{H}}^2 \leq C_2 \sum_{i=1}^{n_j} \sum_{l=1}^{3} \left(\eta_{j,i,l}^{\Delta^*}\right)^2 .$$

The local error indicators are defined as

$$\eta^{\Delta^*}_{j,i,l} = V_{\Delta^*}(e_{j,i,l}, e_{j,i,l})^{1/2}$$

and $e_{j,i,l} \in (\mathrm{span}\{\beta_{j+1,i,l}\})^3$ *is obtained by solving*

$$V_{\Delta^*}(e_{j,i,l}, \psi) = \langle g - V_{\Delta^*}\phi_{j,\Delta^*}, \psi\rangle \tag{10.103}$$

for all $\psi \in (\mathrm{span}\{\beta_{j+1,i,l}\})^3$.

The Adaptive Algorithm

Based on the above a-posteriori estimates we formulate a refinement strategy for h-adaptivity. We relax the regularity assumptions on the meshes τ_j and allow in this section non-uniform and anisotropic refinements. The following adaptive algorithm is closely related to those used by Maischak et al. [294] and Mund et al. [314].

Algorithm 10.1 *Let the parameter* $0 \leq \theta \leq 1$ *and an initial subdivision* τ_0 *of* Γ *be given. Let* $S_{0,D}$ *be the finite dimensional space of piecewise constant functions over* τ_0 (cf. (10.94)) *where D denotes the Laplace, Helmholtz or Lamé operator. For* $j = 0, 1, 2, \ldots$ *we perform the following:*

(1) Compute the Galerkin solution $\phi_j \in S_{j,D}$.
(2) For each element $\Gamma^j_i \in \tau_j$ *compute the local error indicators* $\eta^D_{j,i,1}$ *and* $\eta^D_{j,i,2}$ *where we note that* $\eta^D_{j,i,l}$ *indicates the error along the* x_l*-axis* ($l \in \{1, 2\}$). *Check the stopping criterion.*
(3) Compute $\eta_{\max} := \max_{\Gamma^j_i \in \tau_j}\{\eta^D_{j,i,1}, \eta^D_{j,i,2}\}$ *and refine* Γ^j_i *along the* x_l*-axis iff*

$$\eta^D_{j,i,l} \geq \theta\, \eta_{\max}\,.$$

This defines the subdivision τ_{j+1} *and the refined space* $S_{j+1,D} \supset S_{j,D}$. *Go back to Step 1.*

We stop the algorithm in Step 2 if the criterion

$$\Sigma_j := \left(\sum_{i=1}^{n_j}\sum_{l=1}^{3}(\eta^D_{j,i,l})^2\right)^{1/2} < \varepsilon_0 \tag{10.104}$$

is satisfied, where $\varepsilon_0 > 0$ is a given constant. This stopping criterion is motivated by the a-posteriori error estimates in Theorems 10.14–10.16.

In Step 3 of the algorithm, refinement of Γ_i^j along the x_l-axis means bisection of Γ_i^j ($l \in \{1, 2\}$) and refinement along both axes means subdivision of Γ_i^j into four elements (cf. Fig. 10.4). Note that the third indicator $\eta_{j,i,3}^D$ appears in (10.104) but is not used to decide whether or not to refine the element Γ_i^j since the corresponding basis function $\beta_{j+1,i,3}$ only appears if Γ_i^j is refined along both axes.

To compute the local error indicators $\eta_{j,i,l}^D$ we can use the same subroutines as for the computation of the Galerkin matrix and of the right hand side vector. Hence, an existing boundary element code which allows basis transformations and local mesh refinements can be easily equipped with the above algorithm.

Numerical Results

Let Γ denote the L-shaped surface piece modelling a screen or crack with corners at $(-1,-1,0)$, $(1,-1,0)$, $(1,0,0)$, $(0,0,0)$, $(0,1,0)$ and $(-1,1,0)$. We consider the Dirichlet boundary value problem (10.90) with the Lamé operator $D = \Delta^*$ and right hand side $f = -1/2$ and this leads to the boundary integral equation $V_{\Delta^*}\phi = 1$ on Γ (cf. (10.90)). Let $\phi_j = \phi_{j,\Delta^*} \in S_{j,\Delta^*}$ denote the Galerkin approximation of ϕ at level j (cf. (10.95)). As in the previous section S_{j,Δ^*} is the space of piecewise constant functions on the locally refined mesh τ_j.

We apply Algorithm 10.1 with $\theta = 0.7$ and the initial mesh τ_0 consisting of 12 equally sized squares and we need 10 adaptive refinement steps to reach the given accuracy $\varepsilon_0 = 5 \cdot 10^{-2}$ in (10.104). At each step we compute the Galerkin error $(V_{\Delta^*}(\phi - \phi_j, \phi - \phi_j))^{1/2}$ in energy norm which is equivalent to the norm in $\mathscr{H}^3 = (H_{00}^{-1/2}(\Gamma))^3$. From the Galerkin property of ϕ_j we obtain

$$\mathscr{E}_j := (V_{\Delta^*}(\phi - \phi_j, \phi - \phi_j))^{1/2} = \big(V_{\Delta^*}(\phi, \phi) - V_{\Delta^*}(\phi_j, \phi_j)\big)^{1/2}. \qquad (10.105)$$

To compute (10.105) we replace the quantity $V_{\Delta^*}(\phi, \phi)$ by the value 3.72844, which was obtained from the values $V_{\Delta^*}(\phi_j, \phi_j)$ by extrapolation.

It is well known that the exact solution of our model problem has edge and corner singularities at the boundary of Γ, with the exception of the incoming corner at $(0, 0, 0)$. Hence, we may expect anisotropically refined meshes at the edges of Γ. The optimal convergence rate of our adaptive Galerkin method is given by $\mathscr{E}_j \sim n_j^{-3/4+\epsilon}$ where n_j denotes the number of elements in τ_j and $\epsilon > 0$ is arbitrarily small. However, for uniform refinements it is only $\mathscr{E}_j \sim n_j^{-1/4}$.

The values of $\mathscr{E}_j$ as a function of the number of unknowns $N_j = 3n_j$ ($0 \le j \le 10$) are plotted in Fig. 10.5. For comparison we also plotted the optimal curve $cN_j^{-3/4}$ for some constant $c > 0$. We observe that both curves have approximatly the same slope for N_j sufficiently large and this indicates the reliablity of the adaptive

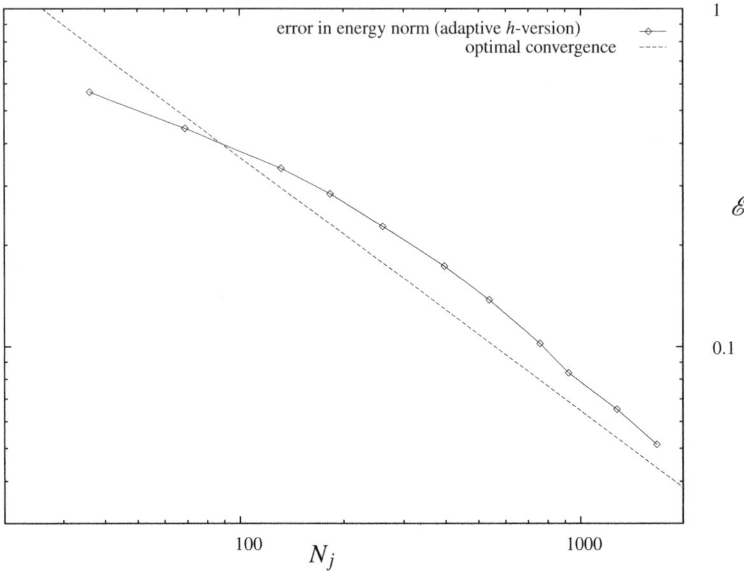

Fig. 10.5 Error $\mathscr{E}_j$ in the energy norm plotted versus the number of unknowns N_j

algorithm and of the a-posteriori error estimate in Theorem 10.16 (for locally refined meshes). The sequence of refined meshes is shown in Fig. 10.6.

10.5.1 A Stable Two-Level Subspace Decomposition for the Hypersingular Operator

Let $S_h^1(\Gamma)$ denote the space of continuous, piecewise linear functions on Γ. For a partition of unity $\{\theta_j;\ j = 1, \ldots, J_\nu\}$ which consists of continuous, piecewise linear functions:

$$\sum_j \theta_j = 1, \operatorname{supp} \theta_j = \overline{\Gamma_j'}, \quad 0 \leq \theta_j \leq 1. \tag{10.106}$$

there holds

$$|\frac{\partial}{\partial_x}\theta_j|, |\frac{\partial}{\partial_y}\theta_j| \leq c/h.$$

where the domain Γ_j' is the union of the elements Γ_i, which are adjacent to the node x_j.

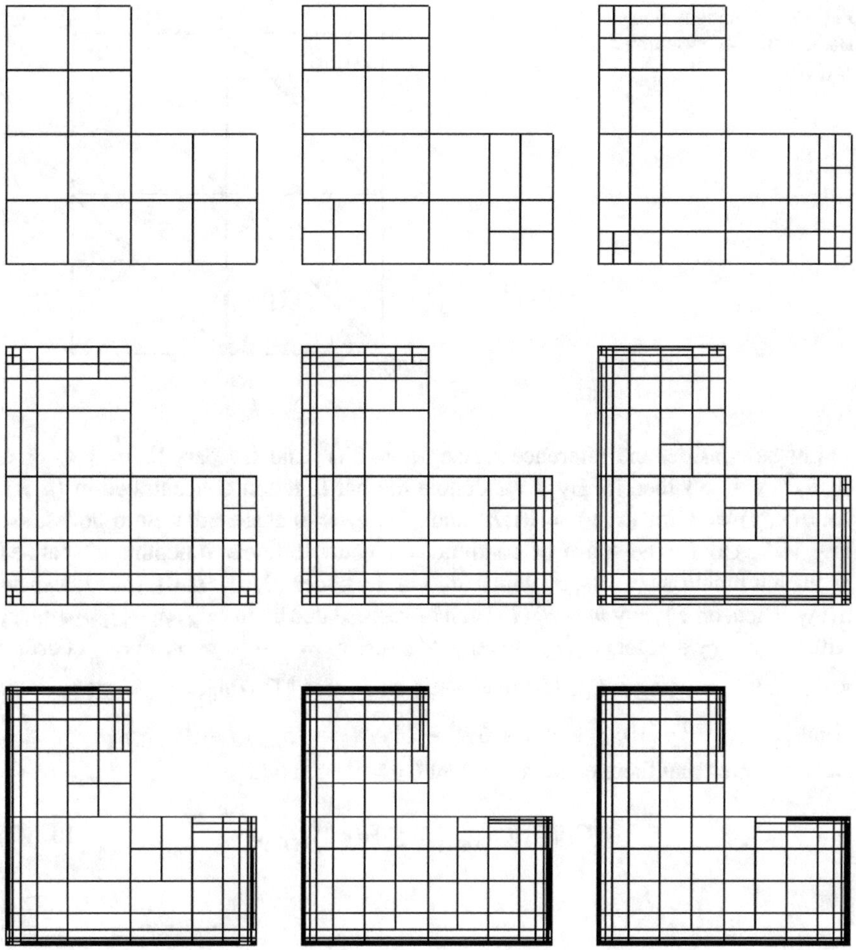

Fig. 10.6 The sequence $\{\tau_j\}_{j=0}^{8}$ of refined meshes

Lemma 10.6 *For any* $w \in S_h^1(\Gamma)$ *there holds for the linear interpolation operator* Π_1 *in the nodal points*

$$\|\Pi_1 \theta_j w\|_{\tilde{H}^{1/2}(\Gamma)} \le C \|\theta_j w\|_{\tilde{H}^{1/2}(\Gamma)}$$

uniformly in h *and for all cut-off functions* θ_j *in (10.106). Here,* C *is an arbitrary constant* ≥ 4.07.

Proof We first prove the continuity of Π_1 with respect to the L^2- and H_0^1-norms and then interpolate these results.

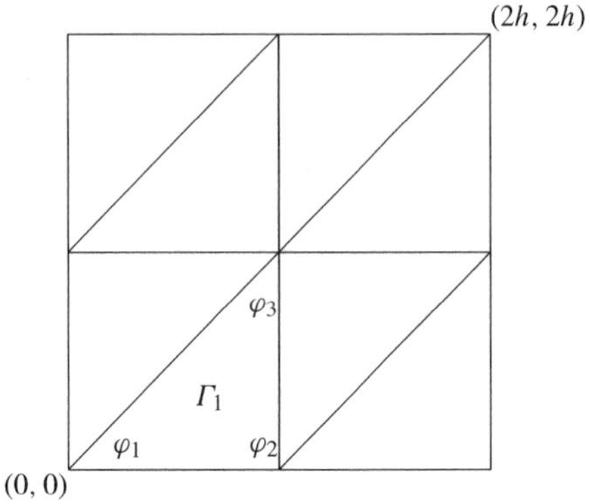

Fig. 10.7 Configuration: Hat function θ concentrated at (h, h)

Let us consider the reference rectangle $(0, 2h)^2$ and the part $\Gamma_1 = \{(x, y) \in (0, h)^2 \mid y < x\}$ therein. By θ we denote the hat function concentrated in (h, h), i.e. $\theta(x, y) = 1$ for $(x, y) = (h, h)$ and $\theta(x, y) = 0$ at the adjacent 6 nodes, see Fig. 10.7. On Γ_1 the space of continuous, piecewise linear functions is spanned by the hat functions $\varphi_1, \varphi_2, \varphi_3$ which are 1 at the nodes $(0, 0)$, $(h, 0)$, (h, h), respectively. Then, on Γ_1, any $w \in S_h^1(\Gamma_1)$ can be represented by $w = \chi_1\varphi_1 + \chi_2\varphi_2 + \chi_3\varphi_3$ with $\chi_1, \chi_2, \chi_3 \in \mathbb{R}$ and $\varphi_1(x, y) = 1 - \frac{x}{h}$, $\varphi_2(x, y) = \frac{x}{h} - \frac{y}{h}$, $\varphi_3(x, y) = \frac{y}{h}$. Further we have $\theta w = \varphi_3 w$ on Γ_1 , $\Pi_1\theta w = \chi_3\varphi_3$ on Γ_1 and $\|\Pi_1\theta w\|^2_{L^2(\Gamma_1)} = \chi_3^2 \frac{h^2}{12}$, $\|\theta w\|^2_{L^2(\Gamma_1)} = \frac{h^2}{180}\left(10\chi_1^2 + 4\chi_2^2 + 6\chi_3^2 - 11\chi_1\chi_2 + 15\chi_1\chi_3 - 9\chi_2\chi_3\right)$.
Next we show that there exists a constant $C_1 > 0$ such that

$$C_1\|\Pi_1\theta w\|^2_{L^2(\Gamma_1)} \le \|\theta w\|^2_{L^2(\Gamma_1)}. \tag{10.107}$$

For

$$f_C(\chi_1, \chi_2, \chi_3) := \frac{180}{h^2}(\|\theta w\|^2_{L^2(\Gamma_1)} - C\|\Pi_1\theta w\|^2_{L^2(\Gamma_1)})$$

we find

$$\nabla f_C(\chi_1, \chi_2, \chi_3) = \begin{pmatrix} 20 & -11 & 15 \\ -11 & 8 & -9 \\ 15 & -9 & 2(6-15C) \end{pmatrix} \begin{pmatrix} \chi_1 \\ \chi_2 \\ \chi_3 \end{pmatrix}.$$

Due to

$$\det(20) = 20 > 0, \quad \det\begin{pmatrix} 20 & -11 \\ -11 & 8 \end{pmatrix} = 39 > 0$$

and

$$\det\begin{pmatrix} 20 & -11 & 15 \\ -11 & 8 & -9 \\ 15 & -9 & 2(6-15C) \end{pmatrix} = 18 - 1170C > 0, \text{ if } C < \frac{1}{65}$$

it is obvious that $f_C(\chi)$ tends to $+\infty$ for $|\chi| \to +\infty$ if $C < \frac{1}{65}$. Since ∇f_C is a multi-linear function the only extreme value of f_C in $\mathbb{R}^3$ is a minimum at 0. Thus

$$f_C(\chi) \geq \inf_{\chi\in\mathbb{R}^3} f_C(\chi) = f_C(0) = 0 \quad (C < \frac{1}{65})$$

and therefore (10.107) holds for $C_1 < \frac{1}{65}$.

Now we show that there exists a constant $C_2 > 0$ such that

$$C_2 \|\Pi_1 \theta w\|^2_{H_0^1(\Gamma_1)} \leq \|\theta w\|^2_{H_0^1(\Gamma_1)}. \tag{10.108}$$

Analogously as before we find $\|\Pi_1\theta w\|^2_{H_0^1(\Gamma_1)} = \frac{1}{2}\chi_3^2$ and

$$\|\theta w\|^2_{H_0^1(\Gamma_1)}$$
$$= \left(\frac{1}{3}\chi_1^2 + \frac{1}{3}\chi_2^2 + \frac{1}{3}\chi_3^2 - \frac{1}{2}\chi_1\chi_2 + \frac{1}{2}\chi_1\chi_3 - \frac{1}{2}\chi_2\chi_3\right)$$

For $g_C(\chi_1, \chi_2, \chi_3) := \|\theta w\|^2_{H_0^1(\Gamma_1)} - C\|\Pi_1\theta w\|^2_{H_0^1(\Gamma_1)}$
we find

$$\nabla g_C(\chi_1, \chi_2, \chi_3) = \begin{pmatrix} 2/3 & -1/2 & 1/2 \\ -1/2 & 2/3 & -1/2 \\ 1/2 & -1/2 & 2/3 - C \end{pmatrix} \begin{pmatrix} \chi_1 \\ \chi_2 \\ \chi_3 \end{pmatrix}.$$

It is obvious that $g_C(\chi)$ tends to $+\infty$ for $|\chi| \to +\infty$ if $C < \frac{5}{21}$. Since ∇g_C is a multi-linear function the only extreme of g_C in $\mathbb{R}^3$ is a minimum at 0. Thus

$$g_C(\chi) \geq \inf_{\chi\in\mathbb{R}^3} g_C(\chi) = g_C(0) = 0 \quad (C < \frac{5}{21})$$

and therefore (10.108) holds for $C_2 < \frac{5}{21}$.

Obviously (10.107) and (10.108) are valid on the whole reference rectangle and we obtain via interpolation

$$\|\Pi_1 \theta_j w\|_{\tilde{H}^{1/2}(\Gamma)} \leq C \|\theta_j w\|_{\tilde{H}^{1/2}(\Gamma)}, \quad j = 1, \ldots, J_\nu,$$

for a constant

$$C \geq (C_1C_2)^{-1/4} > (\frac{1}{65}\frac{5}{21})^{-1/4}$$

which is independent of h.

10.5.1.1 2-Level Method

We decompose the ansatz space as

$$S_h^1 = \text{span}\left\{S_H^1 \cup S_{h,1}^1 \cup \ldots \cup S_{h,J_v}^1\right\}, \tag{10.109}$$

where S_H^1 is the space of continuous linear functions on a coarser mesh with size $H = 2h$ and where

$$S_{h,j}^1 = \text{span}\left\{\phi_{h_j}\right\}$$

Let $P = P_H + \sum_{j=1}^{J_v} P_i$ be the two-level additive Schwarz operator belonging to the subspace decomposition (10.109) and the bilinear form $< W.,. >$, i.e.,

$$\langle W P_H\varphi, \psi\rangle = \langle W\varphi, \psi\rangle \quad \forall\psi \in S_H^1, \varphi \in S_h^1$$

$$\langle W P_h\varphi, \psi\rangle = \langle W\varphi, \psi\rangle \quad \forall\psi \in S_{h,i}^1, \varphi \in S_h^1 .$$

Then there holds

Theorem 10.17 *There exist constants $c_1, c_2 > 0$, independent of h, such that for all $u \in S_h^1$*

$$c_1\langle Wu, u\rangle \leq \langle WPu, u\rangle \leq c_2\langle Wu, u\rangle . \tag{10.110}$$

Proof First we show that for all $v \in S_h^1$ we can find a representation $v = v_H + v_{h,1} + \ldots + v_{h,J_v}$ such that

$$\|v_H\|^2_{\widetilde{H}^{1/2}(\Gamma)} + \sum_{j=1}^{J_v} \|v_{h,j}\|^2_{\widetilde{H}^{1/2}(\Gamma)} \leq c\,\|v\|^2_{\widetilde{H}^{1/2}(\Gamma)}$$

This is done by use of cut-off-functions θ_j and the interpolation operator Π_1.

Let us define the L^2-projection: $\widetilde{H}^{1/2}(\Gamma) \to S_H^1$ by $v_H = Q_H v \in S_1^1(\Gamma_H)$ and $v_{h,j} = \Pi_1(\theta_j w_h)$ with $w_h = v - v_H$. Then $\text{supp}\,\Pi_1\left(\theta_j w_h\right) = \overline{\Gamma_j'}$, and

$\Pi_1\left(\theta_j w_h\right) \in C^0$ is piecewise linear. Thus $v_{h,j} \in S^1_{h,j}(\Gamma'_j)$. Furthermore

$$\sum_{j=1}^{J_\nu} v_{h,j} = \Pi_1\left(\sum_{j=1}^{J_\nu} \Theta_j w_h\right) = w_h$$

since w_h and $\sum_{j=1}^{J_\nu} \theta_j w_h$ have the same nodal values. Next we define the localization operator by

$$\Lambda : S^1_{h,1}(\Gamma) \to S^1_H \times \prod_{j=1} S^1_{h,j}\left(\Gamma'_j\right)$$

$$v \mapsto (\Lambda v)_{j=0}^J = \left(Q_H v, \Pi_1\theta_1 w_h, \ldots, \Pi_1\theta_{J_{nu}} w_h\right)$$

Then Lemma 10.6 implies

$$\sum_{j=1}^{J_\nu} \|(\Lambda v)_j\|^2_{\widetilde{H}^{1/2}(\Gamma)} \le c\,\|v\|^2_{\widetilde{H}^{1/2}(\Gamma)}$$

yielding the lower bound in (10.110).

To estimate the maximal eigenvalue of P we show for all $v \in S^1_h(\Gamma)$ and for all representations $v = v_H + v_{h,1} + \ldots + v_{h,J_\nu}$:

$$\|v\|^2_{\widetilde{H}^{1/2}(\Gamma)} \le c\left(\|v_H\|^2_{\widetilde{H}^{1/2}(\Gamma)} + \sum_{j=1}^{J_\nu} \|v_{h,j}\|^2_{\widetilde{H}^{1/2}(\Gamma)}\right)$$

Using the triangle inequality together with a colouring argument we obtain

$$\begin{aligned}
\|v\|^2_{\widetilde{H}^{1/2}(\Gamma)} &\le 2\left(\|v_H\|^2_{\widetilde{H}^{1/2}(\Gamma)} + \left\|\sum_{j=1}^{J_\nu} v_{h,j}\right\|^2_{\widetilde{H}^{1/2}(\Gamma)}\right) \\
&\le 2\left(\|v_H\|^2_{\widetilde{H}^{1/2}(\Gamma)} + c\sum_{j=1}^{J_\nu} \|v_{h,j}\|^2_{\widetilde{H}^{1/2}(\Gamma'_j)}\right) \\
&\le 2\left(\|v_H\|^2_{\widetilde{H}^{1/2}(\Gamma)} + c\sum_{j=1}^{J_\nu} \|v_{h,j}\|^2_{\widetilde{H}^{1/2}(\Gamma)}\right)
\end{aligned}$$

This completes the proof. □

Next we prove Theorem 10.11 for the above setting of piecewise linear functions on a uniform partition γ_h of Γ into triangles (Fig. 10.7).

Proof (of Theorem 10.11) The saturation assumption (10.86) yields the equivalence of norms

$$\|u_h - u_H\|_{1/2} \sim \|u - u_H\|_{1/2}\,.$$

Due to Theorem 10.17 we have

$$\begin{aligned} &c_1\|u_h - u_H\|^2_{1/2} \\ &\leq \|P_H(u_h - u_H)\|^2_{1/2} + \sum_{i=1}^{J_\nu} \|P_i(u_h - u_H)\|^2_{1/2} \leq c_2\|u_h - u_H\|^2_{1/2}\,. \end{aligned}$$

First, we observe that since u_H and u_h satisfiy the Galerkin equation there holds for any $w \in S^1_H$

$$\langle W P_H u_h, w\rangle = \langle W u_h, w\rangle = \langle f, w\rangle = \langle W u_H, w\rangle = \langle W P_H u_H, w\rangle\,.$$

Hence

$$\|P_H(u_h - u_H)\|^2_{1/2} = 0\,.$$

The error indicator μ_j is obtained by solving a linear problem in the space $S^1_{h,j}$. The function $v_{h,j} = P_j(u_h - u_H) \in S^1_{h,j}$ solves for any $v \in S^1_{h,j}$:

$$\langle W v_{h,j}, v\rangle = \langle f - W u_H, v\rangle\,. \tag{10.111}$$

Hence, firstly one solves (10.111) for $0 \leq j \leq J_\nu$ and then one computes the terms $\mu_j = \langle W v_{h,j}, v_{h,j}\rangle^{1/2}$. Since $S^1_{h,j} = \text{span}\{\phi_{h_j}\}$ is a one-dimensional space, we have $v_{h,j} = c\phi_{h_j}$ with coefficient

$$c = \frac{\langle f - W u_H, \phi_{h_j}\rangle}{\langle W\phi_{h_j}, \phi_{h_j}\rangle}\,.$$

Hence

$$\mu_j = |c|\langle W\phi_{h_j}, \phi_{h_j}\rangle^{1/2}$$

and there holds with constants c_1, c_2 independent of h,

$$c_1 \sum_{j=1}^{J_\nu} \mu_j^2 \leq \|u - u_H\|^2_{1/2} \leq c_2 \sum_{j=1}^{J_\nu} \mu_j^2\,,$$

which is Theorem 10.11. □

10.6 Two-Level Subspace Decomposition for the p-Version BEM

Here we consider the p-version of the boundary element method for the hypersingular integral equation in two dimensions. We present from [238] an a-posteriori error estimate that is based on a stable two-level subspace decomposition of the enriched ansatz space. The Galerkin error is estimated by inverting local projection operators that are defined on small subspaces of the second level. We consider an enriched space on $\Gamma = \cup_{j=1}^{J} \Gamma_j$

$$\tilde{H}_N := S_{h,p}(\Gamma) + Z_1 + \cdots + Z_J, \tag{10.112}$$

where

$$S_{h,p}(\Gamma) = \{\phi \in C^0(\Gamma);\ \phi|_{\partial\Gamma} = 0, \phi|_{\Gamma_j} \in \mathscr{P}_{p_j}(\Gamma_j), j = 1, \ldots, J\}.$$

Here, the local enrichment is given by adding bubble functions on the elements

$$Z_j = \operatorname{span}(\psi_{p_j+1} \circ T_j^{-1}), \qquad j = 1, \ldots, J,$$

with the affine map $T_j : I = (-1, 1) \to \Gamma_j$ and

$$\psi_j(x) = \sqrt{\frac{2j-1}{2}} \int_{-1}^{x} L_{j-1}(t)\, dt \quad (2 \le j \le p_j),$$

where L_{j-1} is the Legendre polynomial of degree $j-1$ and

$$\psi_0(x) = \frac{1-x}{2}, \quad \psi_1(x) = \frac{1+x}{2}.$$

We use the notation $\tilde{\phi}_j = \phi_j \circ T_j$ for functions ϕ_j defined on Γ_j.

Now for the hypersingular integral equation on a given curve Γ we consider the problem: *given $f \in H^{-1/2}(\Gamma)$, find $u \in \tilde{H}^{1/2}(\Gamma)$ such that*

$$\langle Wu, v\rangle_{L^2(\Gamma)} = \langle f, v\rangle_{L^2(\Gamma)} \quad \forall v \in \tilde{H}^{1/2}(\Gamma), \tag{10.113}$$

where $\|u\|_W^2 = \langle Wu, u\rangle_{L^2(\Gamma)}$.

Lemma 10.7 *There exist positive constants c_1 and c_2, independent of the mesh and p, such that*

$$c_1(1 + \log p_{\max})^{-2} \sum_{j=1}^{J} \|\phi_j\|_W^2 \le \|\phi_1^*\|_W^2 \le c_2 \sum_{j=1}^{J} \|\phi_j\|_W^2$$

for all $\phi_1^* = \sum_{j=1}^{J} \phi_j \in Z$, *where* $\phi_j \in Z_j$, $j = 1, \dots, J$. *Here,* $p_{\max} = \max\{p_1, \dots, p_J\}$.

Proof First, we observe that

$$\|\phi_1^*\|_W^2 \le C\, \|\phi_1^*\|_{\tilde{H}^{1/2}(\Gamma)}^2 \le C \sum_{j=1}^{J} \|\phi_j\|_{\tilde{H}^{1/2}(\Gamma_j)}^2 \le C \sum_{j=1}^{J} \|\phi_j\|_W^2.$$

On the other hand there holds

$$\sum_{j=1}^{J} \|\phi_j\|_{\tilde{H}^{1/2}(\Gamma_j)}^2 \le C \sum_{j=1}^{J} \|\tilde{\phi}_j\|_{\tilde{H}^{1/2}(I)}^2$$

$$\le C \sum_{j=1}^{J} (1 + \log(p_j + 1))^2 \|\tilde{\phi}_j\|_{H^{1/2}(I)}^2 \le C(1 + \log p_{\max})^2 \sum_{j=1}^{J} \|\tilde{\phi}_j\|_{H^{1/2}(I)}^2. \tag{10.114}$$

For the first inequality we used that the $\tilde{H}^{1/2}$-norm scales with a constant that is independent of the element size under affine transformations, see Lemma 3.1 in [405] and Lemma 2 in [233]. The second inequality is Theorem 6.5 in [15]. Note that $\tilde{\phi}_j(-1) = \tilde{\phi}_j(+1) = 0$. Therefore, we can apply the Poincaré–Friedrichs inequality and obtain

$$\sum_{j=1}^{J} \|\tilde{\phi}_j\|_{H^{1/2}(I)}^2 \le C \sum_{j=1}^{J} |\tilde{\phi}_j|_{H^{1/2}(I)}^2 \le C \sum_{j=1}^{J} |\phi_j|_{H^{1/2}(\Gamma_j)}^2 \le C |\phi_1^*|_{H^{1/2}(\Gamma)}^2. \tag{10.115}$$

The last inequality is a direct application of the definition of the Sobolev–Slobo--deckij seminorm $|\cdot|_{H^{1/2}(\Gamma)}$. Finally, by combining (10.114) and (10.115) and by using the fact that W is positive definite on $\tilde{H}^{1/2}(\Gamma)$, the proof is finished. □

Next we present a two-level subspace decomposition.

Lemma 10.8 (subspace decomposition) *There exist positive constants* c_3 *and* c_4, *independent of the mesh and* p, *such that*

$$c_3(1 + \log p_{\max})^{-2} \sum_{j=0}^{J} \langle W\phi_j, \phi_j \rangle \le \langle W\phi, \phi \rangle \le c_4 \sum_{j=0}^{J} \langle W\phi_j, \phi_j \rangle$$

for all $\phi = \sum_{j=0}^{J} \phi_j \in \tilde{H}_N$, *where* $\phi_0 \in H_N$ *and* $\phi_j \in Z_j$, $j = 1, \dots, J$.

Proof With Lemma 10.7 we note that there holds

$$\langle W\phi, \phi \rangle = \|\phi_0 + \phi_1^*\|_W^2 \le 2\,\{\|\phi_0\|_W^2 + \|\phi_1^*\|_W^2\} \le C \sum_{j=0}^{J} \langle W\phi_j, \phi_j \rangle,$$

which is the right inequality of the assertion. To prove the left inequality we show that there exists $C > 0$ such that

$$C\,(\|\phi_0\|_W^2 + \|\phi_1^*\|_W^2) \le \|\phi\|_W^2.$$

It suffices to show $\|\phi_0\|_W \le C\,\|\phi\|_W$ since this implies with $\phi_1^* = \phi - \phi_0$ that $\|\phi_1^*\|_W \le C\,\|\phi\|_W$ by the triangle inequality. To show $|\phi_0|_{H^1(\Gamma)} \le C\,|\phi|_{H^1(\Gamma)}$ we prove $|\phi_0|^2_{H^1(\Gamma_j)} \le C\,|\phi|^2_{H^1(\Gamma_j)}$ and then sum over j.

On $I = (-1, 1)$:

$$\tilde{\phi}(x) = \left(\phi|_{\Gamma_j} \circ T_j\right)(x) = \sum_{i=0}^{p+1} c_i \psi_i(x)$$

$$\tilde{\phi}_0(x) := \left(\phi_0|_{\Gamma_j} \circ T_j\right)(x) = c_0\psi_0(x) + c_1\psi_1(x) + \sum_{i=2}^{p} c_i \psi_i(x)$$

and

$$\tilde{\phi}_1^*(x) := \left(\phi_1^*|_{\Gamma_j} \circ T_j\right)(x) = c_{p+1}\psi_{p+1}(x).$$

$$\frac{d\tilde{\phi}_0}{dx} = -\frac{1}{2}c_0 + \frac{1}{2}c_1 + \sum_{i=2}^{p} c_i \sqrt{\frac{2i-1}{2}} L_{i-1}(x)$$

and

$$\frac{d\tilde{\phi}_1^*}{dx} = c_{p+1}\sqrt{\frac{2p+1}{2}} L_p(x)$$

Note $\frac{d\tilde{\phi}_0}{dx}$ and $\frac{d\tilde{\phi}_1^*}{dx}$ are orthogonal in $L^2(I)$. Therefore

$$|\phi|^2_{H^1(\Gamma_j)} = |\phi_0|^2_{H^1(\Gamma_j)} + |\phi_1^*|^2_{H^1(\Gamma_j)},$$

which shows that $|\phi_0|^2_{H^1(\Gamma_j)} \le C\,|\phi|^2_{H^1(\Gamma_j)}$ and thus, $|\phi_0|_{H^1(\Gamma)} \le C\,|\phi|_{H^1(\Gamma)}$.

We consider the normalized shape functions $\psi_i^* := \psi_i / \|\psi_i\|_{L^2(I)}$, $i = 0, \ldots, p+1$. Then, we can represent

$$\tilde{\phi}_0(x) := \left(\phi_0|_{\Gamma_j} \circ T_j\right)(x) = v_0\psi_0^*(x) + v_1\psi_1^*(x) + \sum_{i=2}^{p} v_i \psi_i^*(x)$$

and

$$\tilde{\phi}(x) := \left(\phi|_{\Gamma_j} \circ T_j\right)(x) = v_0\psi_0^*(x) + v_1\psi_1^*(x) + \sum_{i=2}^{p+1} v_i\psi_i^*(x).$$

Note $\|\tilde{\phi}_0\|^2_{L^2(I)} = v^T Av$, where A is a matrix of dimension $(p+1)\times(p+1)$ with entries given by

$$a_{ij} = \int_{-1}^{1} \psi_i^*(x)\psi_j^*(x)\,dx \quad i, j = 0, \ldots, p$$

and $v^T = (v_0, \ldots, v_p)$. We get

$$A = \begin{bmatrix} 1 & c_0 & b_0 & -b_1 & & & & \\ c_0 & 1 & b_0 & b_1 & & & & \\ b_0 & b_0 & 1 & 0 & b_2 & & & \\ -b_1 & b_1 & 0 & 1 & 0 & b_3 & & \\ & & b_2 & 0 & 1 & \ldots & \ldots & \\ & & & \ldots & \ldots & \ldots & \ldots & b_{p-2} \\ & & & & \ldots & \ldots & 1 & 0 \\ & & & & & b_{p-2} & 0 & 1 \end{bmatrix}$$

where $c_0 = \frac{1}{2}$, $b_0 = -\sqrt{\frac{5}{8}}$, $b_1 = -\sqrt{\frac{7}{40}}$ and

$$b_j = -\frac{1}{2}\sqrt{\frac{(2j-3)(2j+5)}{(2j-1)(2j+3)}} \quad \forall\, j \geq 2.$$

It follows that

$$\|\tilde{\phi}\|^2_{L^2(I)} = \tilde{u}^T \begin{bmatrix} A & b^T \\ b & 1 \end{bmatrix} \tilde{u},$$

where $\tilde{u}^T = (v_0, \ldots, v_p, v_{p+1})$ and $b = (0, 0, \ldots, b_{p-1}, 0)$. To show that $\|\phi_0\|_{L^2(\Gamma_j)} \leq C\,\|\phi\|_{L^2(\Gamma_j)}$, we need to bound the maximum eigenvalue of

$$\begin{bmatrix} A & 0 \\ 0 & 0 \end{bmatrix}\begin{bmatrix} v \\ w \end{bmatrix} = \lambda \begin{bmatrix} A & b \\ b^T & 1 \end{bmatrix}\begin{bmatrix} v \\ w \end{bmatrix}.$$

Following Pavarino [337], we deduce that $|\lambda| < C < \infty$ which shows $\|\phi_0\|^2_{L^2(\Gamma_j)} \leq C\,\|\phi\|^2_{L^2(\Gamma_j)}$. This gives $\|\phi_0\|^2_{L^2(\Gamma)} \leq C\,\|\phi\|^2_{L^2(\Gamma)}$. The assertion now follows by interpolation. □

Application of Lemma(10.8) yields an a posteriori error estimate for the Galerkin solution with hierarchical error indicators when the following saturation assumption holds: There exists $\sigma \in [0, 1)$ such that

$$\|u - \tilde{u}_N\|_W \leq \sigma \|u - u_N\|_W$$

with the solution u of (10.113) and the corresponding Galerkin solutions $u_N \in S_{h,p}(\Gamma)$, $\tilde{u}_N \in \tilde{H}_N$ of (10.112).

Theorem 10.18 *Assume that the mesh $\mathcal{T}_h$ is locally quasi-uniform. Then there exist positive constants c_1 and c_2 such that for the error estimator based upon the decomposition (10.112) there holds*

$$c_1 \sum_{j=1}^{J} \theta_j^2 \leq \|u - u_N\|_W^2 \leq \frac{c_2}{1-\sigma^2}(1 + \log p_{\max})^2 \sum_{j=1}^{J} \theta_j^2.$$

Here, $u_N \in S_{h,p}(\Gamma)$ is the Galerkin approximation of the solution u of (10.113) and σ is the saturation parameter. $\theta_j = \|P_j(\tilde{u}_N - u_N)\|_W$, $\tilde{u}_N \in \tilde{H}_N$ is the Galerkin solution of the enriched space $\tilde{H}_N$ and $P_j : \tilde{H}_N \to Z_j$ is definied by

$$\langle W P_j \phi, \psi \rangle = \langle W\phi, \psi \rangle \qquad \psi \in Z_j.$$

The corresponding result for the single layer potential operator using Legendre polynomials as bubble functions can also be found in [238] together with numerical experiments for p- and hp- adaptive algorithms. In [239] a p-adaptive algorithm with bubble functions is investigated for the BEM with the hypersingular operator on the plane screen.

10.7 Convergence of Adaptive BEM for Estimators Without *h*-Weighting Factor

In this section we present some results from [176].

A posteriori error estimation and related adaptive mesh-refining algorithms are one important basis of modern scientific computing. Starting from an initial mesh $\mathcal{T}_0$ and based on a computable a posteriori error estimator, such algorithms iterate the loop

$$\boxed{\texttt{solve}} \quad \rightarrow \quad \boxed{\texttt{estimate}} \quad \rightarrow \quad \boxed{\texttt{mark}} \quad \rightarrow \quad \boxed{\texttt{refine}} \tag{10.116}$$

to create a sequence of successive locally refined meshes $\mathcal{T}_\ell$, corresponding discrete solutions U_ℓ, as well as a posteriori error estimators μ_ℓ. We consider the frame of conforming Galerkin discretizations, where $\mathscr{T}_\ell$ is linked to a finite-dimensional subspace $\mathscr{X}_\ell$ of a Hilbert space $\mathscr{H}$ with corresponding Galerkin solution $U_\ell \in \mathscr{X}_\ell$, where successive refinement guarantees nestedness $\mathscr{X}_\ell \subseteq \mathscr{X}_{\ell+1} \subset \mathscr{H}$ for all $\ell \in \mathbb{N}_0$.

Convergence of this type of adaptive algorithm in the sense of

$$\lim_{\ell\to\infty} \|u - U_\ell\|_{\mathscr{H}} = 0 \tag{10.117}$$

has first been addressed in [26] for 1D FEM and [143] for 2D FEM. We note that already the pioneering work [26] observed that validity of some Céa-type quasi-optimality and nestedness $\mathscr{X}_\ell \subseteq \mathscr{X}_{\ell+1}$ for all $\ell \in \mathbb{N}_0$ imply a priori convergence

$$\lim_{\ell\to\infty} \|U_\infty - U_\ell\|_{\mathscr{H}} = 0, \tag{10.118}$$

where U_∞ is the unique Galerkin solution in $\mathscr{X}_\infty := \overline{\bigcup_{\ell\in\mathbb{N}_0} \mathscr{X}_\ell}$. From a conceptual point of view, it thus only remained to identify the limit $u = U_\infty$. Based on such an a priori convergence result (10.118), a general theory of convergence of adaptive FEM is devised in [308, 377], where the analytical focus is on *estimator convergence*

$$\lim_{\ell\to\infty} \mu_\ell = 0. \tag{10.119}$$

Moreover, the recent work [79] gives an analytical frame to guarantee convergence with optimal convergence rates; see also the overview article [174] for the current state of the art of adaptive BEM. Throughout, it is however implicitly assumed that the local contributions $\mu_\ell(T)$ of the error estimator μ_ℓ are weighted with the local mesh-size, i.e., $|T|^\alpha$ for some appropriate $\alpha > 0$, or that μ_ℓ is *locally* equivalent to a mesh-size weighted error estimator.

Our analysis in [176] covers the two-level error estimators for BEM considered Sections 10.4, 10.5, 10.6 and in [164, 234, 238, 294, 313, 314] or the adaptive FEM-BEM coupling considered in Subsection 12.3.3 and in [12, 198, 274, 312]. The local contributions are projections of the computable error between two Galerkin solutions onto one-dimensional spaces, spanned by hierarchical basis functions. These estimators are known to be efficient. On the other hand, reliability is only proven under an appropriate saturation assumption which is even equivalent to reliability for the symmetric BEM operators [11, 161, 162]. However, such a saturation assumption is formally equivalent to asymptotic convergence of the adaptive algorithm [178] which cannot be guaranteed mathematically in general and is expected to fail on coarse meshes.

Next we take an abstract setting as follows. Let $\mathscr{H}$ be a Hilbert space with dual space $\mathscr{H}^\star$ and $A : \mathscr{H} \mapsto \mathscr{H}^\star$ be a bi-Lipschitz continuous, not necessarily linear operator, i.e.

$$C_{\text{cont}}^{-1}\|w - v\|_{\mathscr{H}} \le \|Aw - Av\|_{\mathscr{H}^\star} \le C_{\text{cont}}\|w - v\|_{\mathscr{H}} \text{ for all } v, w \in \mathscr{H} \tag{10.120}$$

Here, $\|\cdot\|_{\mathscr{H}^\star}$, denotes the operator norm on $\mathscr{H}^\star$,

$$\|F\|_{\mathscr{H}^\star} = \sup_{v\in\mathscr{H}\setminus\{0\}} \frac{|\langle F, v\rangle|}{\|v\|_{\mathscr{H}}} \text{ for all } F \in \mathscr{H}^\star \tag{10.121}$$

Suppose that there exists some subspace $\mathscr{X}_{00} \subseteq \mathscr{H}$ such that for any given closed subspace $\mathscr{X}_{00} \subseteq \mathscr{X}_\star \subseteq \mathscr{H}$ and any continuous linear functional $F \in \mathscr{H}^\star$ on $\mathscr{H}$ the Galerkin formulation

$$\langle AU_\star, V_\star\rangle = \langle F, V_\star\rangle \text{ for all } V_\star \in \mathscr{X}_\star \tag{10.122}$$

admits a unique solution $U_\star \in \mathscr{X}_\star$, where $\langle\cdot,\cdot\rangle$ denote the duality bracket between $\mathscr{H}$ and $\mathscr{H}^\star$. This implies the existence of a unique solution $u \in \mathscr{H}$ of

$$Au = F \tag{10.123}$$

We shall assume that $\mathscr{X}_\ell$ is a finite-dimensional subspace of $\mathscr{H}$ related to some triangulation $\mathscr{T}_\ell$ and that $U_\ell(F) \in \mathscr{X}_\ell$ is the corresponding Galerkin solution (10.122) for $\mathscr{X}_\star = \mathscr{X}_\ell$. Starting from an initial mesh $\mathscr{T}_0$, the triangulations $\mathscr{T}_\ell$ are successively refined by means of the following realization of (10.116), where for all $\mathscr{E}_\ell \subseteq \mathscr{T}_\ell$

$$\mu_\ell(F;\mathscr{E}_\ell) := \Big(\sum_{T\in\mathscr{E}_\ell} \mu_\ell(F;T)^2\Big)^{1/2} < \infty \quad \text{and} \quad \mu_\ell(F) := \mu_\ell(F;\mathscr{T}_\ell) \tag{10.124}$$

is a computable a posteriori error estimator. Its local contributions $\mu_\ell(F;T) \geq 0$ measure, at least heuristically, the error $u(F) - U_\ell(F)$ locally on each element $T \in \mathscr{T}_\ell$.

Algorithm 10.2 INPUT: *Right-hand side* $F \in \mathscr{H}^*$, *initial mesh* $\mathscr{T}_0$ *with* $\mathscr{X}_0 \supseteq \mathscr{X}_{00}$, *and parameter* $0 < \theta \leq 1$.

For $\ell = 0, 1, 2, \ldots$ *iterate the following:*

(i) *Compute Galerkin solution* $U_\ell(F) \in \mathscr{X}_\ell$.

(ii) *Compute refinement indicators* $\mu_\ell(F;T)$ *for all* $T \in \mathscr{T}_\ell$.

(iii) *Determine some set* $\mathscr{M}_\ell \subseteq \mathscr{T}_\ell$ *of marked elements which satisfies*

$$\theta\,\mu_\ell(F)^2 \leq \mu_\ell(F;\mathscr{M}_\ell)^2. \tag{10.125}$$

(iv) *Generate a new mesh* $\mathscr{T}_{\ell+1}$ *and hence an enriched space* $\mathscr{X}_{\ell+1}$ *by refinement of at least all marked elements* $T \in \mathscr{M}_\ell$.

OUTPUT: *Sequence of successively refined triangulations* $\mathscr{T}_\ell$ *as well as corresponding Galerkin solutions* $U_\ell(F) \in \mathscr{X}_\ell$ *and error estimators* $\mu_\ell(F)$, *for* $\ell \in \mathbb{N}_0$.

The convergence results of Propositions 2.4 and 2.5 in [176] require an auxiliary error estimator

$$\rho_\ell(F) := \rho_\ell(F; \mathscr{T}_\ell) \text{ with } \rho_\ell(F; \mathscr{E}_\ell) := \Big(\sum_{T \in \mathscr{E}_\ell} \rho_\ell(F; T)^2\Big)^{1/2} < \infty \quad \text{for all } \mathscr{E}_\ell \subseteq \mathscr{T}_\ell \tag{10.126}$$

with local contributions $\rho_\ell(F; T) \geq 0$. For all $\ell \in \mathbb{N}_0$, we suppose that there exists some set $\mathscr{R}_\ell \subseteq \mathscr{T}_\ell$ with $\mathscr{M}_\ell \subseteq \mathscr{R}_\ell$ which satisfies the following three assumptions (A1)–(A3):

(A1). $\mu_\ell(F)$ is a local lower bound of $\rho_\ell(F)$: There is a constant $C_1 > 0$ such that for all $\ell \in \mathbb{N}_0$ holds

$$\mu_\ell(F; \mathscr{M}_\ell) \leq C_1 \, \rho_\ell(F; \mathbb{R}_\ell). \tag{10.127}$$

(A2). $\rho_\ell(F)$ is contractive on $\mathbb{R}_\ell$: There is a constant $C_2 > 0$ such that for all $\ell, m \in \mathbb{N}_0$ and all $\delta > 0$ holds

$$\begin{aligned} C_2^{-1} \, \rho_\ell(F; \mathbb{R}_\ell)^2 \leq \rho_\ell(F)^2 &- \frac{1}{1+\delta} \, \rho_{\ell+m}(F)^2 \\ &+ (1+\delta^{-1}) C_2 \, \|U_{\ell+m}(F) - U_\ell(F)\|_{\mathscr{H}}^2. \end{aligned} \tag{10.128}$$

The constants $C_1, C_2 > 0$ may depend on F, but are independent of the level $\ell \in \mathbb{N}_0$, i.e., in particular independent of the discrete spaces $\mathscr{X}_\ell$ and the corresponding Galerkin solutions $U_\ell(F)$. If $\rho_\ell(F)$ is not well-defined for all $F \in \mathscr{H}^*$, but only on a dense subset $D \subseteq \mathscr{H}^*$, we require the following additional assumption:

(A3). $\mu_\ell(\cdot)$ is stable on $\mathscr{M}_\ell$ with respect to F: There is a constant $C_3 > 0$ such that for all $\ell \in \mathbb{N}_0$ and $F' \in \mathscr{H}^*$ holds

$$|\mu_\ell(F; \mathscr{M}_\ell) - \mu_\ell(F'; \mathscr{M}_\ell)| \leq C_3 \|F - F'\|_{\mathscr{H}}^*. \tag{10.129}$$

Some remarks are in order to relate the abstract assumptions (A1)–(A3) to the applications we have in mind.

Choice of ρ_ℓ $\mu_\ell(F)$ being the two-level error estimator for BEM considered in Sections 10.4, 10.5, 10.6 , see also [11, 161, 162, 164, 234, 238, 294, 313, 314] and for the FEM-BEM coupling in Subsection 12.3.3, see also [12, 198, 312]. $\rho_\ell(F)$ denotes some weighted-residual error estimator, see [76, 86, 87, 92, 93] and Sects. 10.1–10.4 for BEM and [8, 91, 198] and Subsection 12.3.2 for the FEM-BEM coupling.

Necessity of (A3) In these cases, the weighted-residual error estimator ρ_ℓ imposes additional regularity assumptions on the given right-hand side F. For instance, the weighted-residual error estimator for the weakly singular integral equation [76, 87,

92, 93] requires $F \in H^1(\Gamma)$, while the natural space for the residual is $H^{1/2}(\Gamma)$. Convergence (10.119) of Algorithm 10.2 for arbitrary $F \in H^{1/2}(\Gamma)$ then follows by means of stability (A3).

Verification of (A1)–(A2) For two-level estimators, (A1) has first been observed in [86] for BEM and follows essentially from scaling arguments for the hierarchical basis functions. Finally, the novel observation (A2) follows from an appropriately constructed mesh-size function and refinement of marked elements as well as appropriate inverse-type estimates, where we shall build on the recent developments of [9]; see e.g. the proof of Theorem 10.19 in [176].

Verification of (A3) Suppose that the operator A is linear and $\mu_\ell(\cdot)$ is efficient

$$\mu_\ell(F) \leq C_{\text{eff}} \, \|u(F) - U_\ell(F)\|_{\mathscr{H}} \quad \text{for all } F \in \mathscr{H}^*. \tag{10.130}$$

Provided $\mu_\ell(\cdot)$ has a semi-norm structure, the corresponding triangle inequality yields

$$\begin{aligned} \mu_\ell(F) \leq \mu_\ell(F') + \mu_\ell(F - F') &\leq \mu_\ell(F') + C_{\text{eff}} \, \|u(F - F') - U_\ell(F - F')\|_{\mathscr{H}} \\ &\leq \mu_\ell(F') + C_{\text{eff}} C_{\text{cea}} \, \|u(F - F')\|_{\mathscr{H}} \\ &\leq \mu_\ell(F') + C_{\text{eff}} C_{\text{cea}} \, \|A^{-1}\| \, \|F - F'\|_{\mathscr{H}^*}, \end{aligned} \tag{10.131}$$

where $\|A^{-1}\|$ denotes the operator norm of A^{-1}, and the (bounded) inverse exists due to (10.120). This proves stability (A3) with $C_3 = C_{\text{eff}} C_{\text{cea}} \, \|A^{-1}\|$.

As a model problem we consider the weakly singular integral equation

$$Au(x) = -2 \int_\Gamma G(x - y)\, u(y)\, d\Gamma(y) = F(x) \quad \text{for all } x \in \Gamma \tag{10.132}$$

on a relatively open, polygonal part $\Gamma \subseteq \partial\Omega$ of the boundary of a bounded, polyhedral Lipschitz domain $\Omega \subset \mathbb{R}^d$, $d = 2, 3$. For $d = 3$, we assume that the boundary of Γ (a polygonal curve) is Lipschitz itself. Here,

$$G(z) = \frac{1}{2\pi} \ln |z| \quad \text{resp.} \quad G(z) = -\frac{1}{4\pi} |z|^{-1} \tag{10.133}$$

denotes the fundamental solution of the Laplacian in $d = 2, 3$. The reader is referred to Chapter 2, Sections 2.3 and 2.4for proofs of and details on the following facts: The singe layer integral operator $A : \mathscr{H} \to \mathscr{H}^*$ is a continuous linear operator between the fractional-order Sobolev space $\mathscr{H} = \widetilde{H}^{-1/2}(\Gamma)$ and its dual $\mathscr{H}^* = H^{1/2}(\Gamma) := \{\widehat{v}|_\Gamma : \widehat{v} \in H^1(\Omega)\}$. Duality is understood with respect to the extended $L^2(\Gamma)$-scalar product $\langle \cdot, \cdot \rangle$. In 2D, we additionally assume diam$\,(\Omega) < 1$ which can always be achieved by scaling. Then, the single layer integral operator is also elliptic, i. e.

$$\langle v, Av \rangle \geq C_{\text{ell}} \, \|v\|^2_{\widetilde{H}^{-1/2}(\Gamma)} \quad \text{for all } v \in \mathscr{H} = \widetilde{H}^{-1/2}(\Gamma) \tag{10.134}$$

with some constant $C_{\text{ell}} > 0$ which depends only on Γ. Thus, A meets all assumptions of the abstract setting at the beginning of this section, and $\|v\|_A^2 := \langle Av, v\rangle$ even defines an equivalent Hilbert norm on $\mathcal{H}$.

Next we introduce the discretization. Let $\mathcal{T}_\star$ be a γ-shape regular triangulation of Γ into affine line segments for $d = 2$ resp. plane surface triangles for $d = 3$. For $d = 3$, γ-shape regularity means

$$\sup_{T\in\mathcal{T}_\star} \frac{\operatorname{diam}(T)^2}{|T|} \leq \gamma < \infty \tag{10.135a}$$

with $|\cdot|$ being the two-dimensional surface measure, whereas for $d = 2$, we impose uniform boundedness of the local mesh-ratio

$$\frac{\operatorname{diam}(T)}{\operatorname{diam}(T')} \leq \gamma < \infty \quad \text{for all } T, T' \in \mathcal{T}_\star \text{ with } T \cap T' \neq \emptyset. \tag{10.135b}$$

To abbreviate notation, we shall write $|T| := \operatorname{diam}(T)$ for $d = 2$. In addition, we assume that $\mathcal{T}_\star$ is regular in the sense of Ciarlet for $d = 3$, i.e., there are no hanging nodes.

With $\mathcal{X}_\star = \mathcal{P}^0(\mathcal{T}_\star)$ being the space of $\mathcal{T}_\star$-piecewise constant functions, we now consider the Galerkin formulation (10.122).

A weighted residual error estimator (see Sections 10.1–10.4) is our next concern. According to the Galerkin formulation (10.122), the residual $F - AU_\star(F) \in H^{1/2}(\Gamma)$ has $\mathcal{T}_\star$-piecewise integral mean zero, i.e.,

$$\int_T (F - AU_\star(F))\, d\Gamma = 0 \quad \text{for all } T \in \mathcal{T}_\star. \tag{10.136}$$

Suppose for the moment that the right-hand side has additional regularity $F \in H^1(\Gamma) \subset H^{1/2}(\Gamma)$. Since $A : \widetilde{H}^{-1/2}(\Gamma) \to H^{1/2}(\Gamma)$ is an isomorphism with additional stability $A : \widetilde{H}^{-1/2+s}(\Gamma) \to H^{1/2+s}(\Gamma)$ for all $-1/2 \leq s \leq 1/2$ (We note that A is *not* isomorphic for $s = \pm 1$ and $\Gamma \subsetneqq \partial\Omega$.), a Poincaré-type inequality in $H^{1/2}(\Gamma)$ shows

$$\begin{aligned} \|u(F) - U_\star(F)\|\widetilde{H}^{-1/2}(\Gamma) &\simeq \|F - AU_\star(F)\|_{H^{1/2}(\Gamma)} \\ &\lesssim \|h_\star^{1/2}\nabla_\Gamma(F - AU_\star(F))\|_{L^2(\Gamma)} =: \eta_\star(F), \end{aligned} \tag{10.137}$$

see [76, 87, 92, 93]. Here, $\nabla_\Gamma(\cdot)$ denotes the surface gradient, and $h_\star \in \mathcal{P}^0(\mathcal{T}_\star)$ is the local mesh-width function defined pointwise almost everywhere by $h_\star|_T := \operatorname{diam}(T)$ for all $T \in \mathcal{T}_\star$. Overall, this proves the reliability estimate

$$\|u(F) - U_\star(F)\|_{\widetilde{H}^{-1/2}(\Gamma)} \leq \widetilde{C}_{\text{rel}}\, \eta_\star(F), \tag{10.138}$$

and the constant $\widetilde{C}_{\mathrm{rel}} > 0$ depends only on Γ and the γ-shape regularity (10.135) of $\mathscr{T}_\star$; see [87]. In 2D, it holds that $\widetilde{C}_{\mathrm{rel}} = C \ln^{1/2}(1+\gamma)$, where $C > 0$ depends only on Γ; see [76]. In particular, the weighted-residual error estimator can be localized via

$$\eta_\star(F) = \Big(\sum_{T\in\mathscr{T}_\star} \eta_\star(F;T)^2\Big)^{1/2} \qquad (10.139)$$
$$\text{with } \eta_\star(F;T) = \operatorname{diam}(T)^{1/2}\|\nabla_\Gamma(F - AU_\star(F))\|_{L^2(T)}.$$

Recently, convergence of Algorithm 10.2 has been shown even with quasi-optimal rates, if $\eta_\ell(F) = \mu_\ell(F)$ is used for marking (10.125); see [175, 177]. We stress that our approach with $\eta_\ell(F) = \rho_\ell(F) = \mu_\ell(F)$ would also give convergence $\eta_\ell(F) \to 0$ as $\ell \to \infty$. Since this is, however, a much weaker result than that of [177], we omit the details.

Unlike reliability (10.138) of $\eta_\star(F)$ which is proved for general $F \in H^1(\Gamma)$, efficiency $\eta_\star(F) \lesssim \|u(F) - U_\star(F)\|_{\widetilde{H}^{-1/2}(\Gamma)}$ is only known for special right-hand sides $F \in H^1(\Gamma)$ which guarantee equivalence of the weakly singular integral equation (10.132) to some 2D Laplace problem

$$-\Delta U = 0 \text{ in } \Omega \subset \mathbb{R}^2 \text{ subject to } U = g \text{ on } \Gamma = \partial\Omega$$

with smooth Dirichlet data g; see [74] for quasi-uniform meshes and the very recent work [10] for the generalization to locally refined meshes which are γ-shape regular (10.135b), see also Sect. 10.2.2.

Next we consider a two-level error estimator. In the frame of weakly singular integral equations (10.132), the two-level error estimator was introduced in [314], Sect. 10.5. Let $\widehat{\mathscr{T}}_\star$ denote the uniform refinement of $\mathscr{T}_\star$. For each element $T \in \mathscr{T}_\star$, let $\widehat{\mathscr{T}}_\star|_T := \{T' \in \widehat{\mathscr{T}}_\star : T' \subset T\}$ denote the set of sons of T. Let $\{\chi_T, \varphi_{T,1}, \ldots, \varphi_{T,D}\}$ be a basis of $\mathscr{P}^0(\widehat{\mathscr{T}}_\star|_T)$ with fine-mesh functions $\varphi_{T,j}$ which satisfy $\operatorname{supp}(\varphi_{T,j}) \subseteq T$ and $\int_T \varphi_{T,j}\, d\Gamma = 0$. We note that usually $D = 1$ for $d = 2$ and $D = 3$ for $d = 3$. Typical choices are Fig. 10.4 and Fig. 10.8. Then, the local contributions of the two-level error estimator from [161, 164, 238, 294, 314] read

$$\mu_\star(F;T)^2 = \sum_{j=1}^{D} \mu_{\star,j}(F;T)^2 \quad \text{with} \quad \mu_{\star,j}(F;T) = \frac{\langle F - AU_\star(F), \varphi_{T,j}\rangle}{\langle A\varphi_{T,j}, \varphi_{T,j}\rangle^{1/2}}. \qquad (10.140)$$

Put differently, we test the residual $F - AU_\star(F) \in H^{1/2}(\Gamma)$ with the additional basis functions from $\mathscr{P}^0(\widehat{\mathscr{T}}_\star)\backslash\mathscr{P}^0(\mathscr{T}_\star)$. This quantity is appropriately scaled by the corresponding energy norm $\|\varphi\|_{\widetilde{H}^{-1/2}(\Gamma)} \simeq \langle A\varphi, \varphi\rangle^{1/2} = \|\varphi\|_A$.

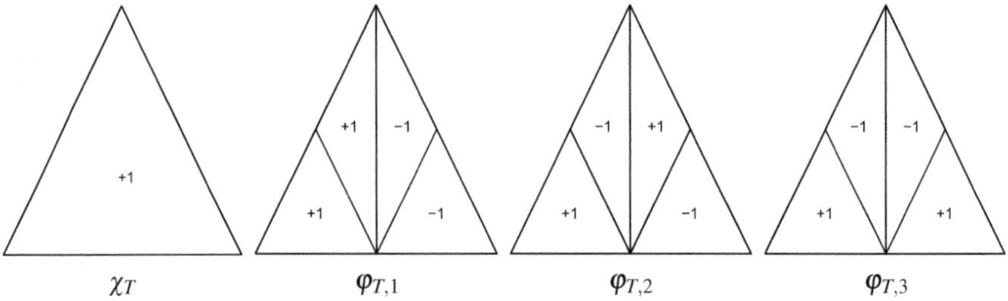

Fig. 10.8 For $d = 3$, uniform bisection-based mesh-refinement usually splits a coarse mesh element $T \in \mathscr{T}_\ell$ (left) into four sons $T' \in \widehat{\mathscr{T}}_\ell$ (right) so that $|T|/4 = |T'|$. Typical hierarchical basis functions $\varphi_{T,j}$ are indicated by their piecewise constant values ± 1 on the son elements T'

Theorem 10.19 ([176]) *Suppose that the two-level error estimator* (10.140) *is used for marking* (10.125)*. Suppose that the mesh-refinement guarantees uniform γ-shape regularity* (10.135) *of the meshes $\mathscr{T}_\ell$ generated, as well as that all marked elements $T \in \mathscr{M}_\ell$ are refined into sons $T' \in \mathscr{T}_{\ell+1}$ with $|T'| \leq \kappa\, |T|$ with some uniform constant $0 < \kappa < 1$. Then, Algorithm 10.2 guarantees*

$$\mu_\ell(F) \to 0 \quad as\ \ell \to \infty \tag{10.141}$$

for all $F \in H^{1/2}(\Gamma)$.

For the corresponding results on the convergence of adaptive schemes for the hypersingular operator and the Johnson-Nedelec coupling and the symmetric coupling of Mund and Stephan in [312] see [176].

For further reading we suggest [89] and [13].

Chapter 11
BEM for Contact Problems

In literature we find various finite element discretization schemes that tackle variational inequalities that arise from scalar unilateral Signorini problems and from contact problems without and with friction in solid mechanics, see e.g. [199, 249, 266]. Each scheme has to overcome several challenges, mainly the discretization of a cone, a primal one in variational inequalities or a dual one in mixed methods, the non-differentiability of the friction functional in the classical sense and the reduced regularity of the solution at the a priori unknown free boundary/interface from contact to non-contact and from stick to slip.

In many cases the insufficient resolution of these interfaces is the dominant source of error. As they lie on the boundary only, it seems to be favorable to reduce these nonlinear boundary value problems to boundary variational inequalities as shown in Chap. 5 and use the boundary element method. Thereby one only requires a boundary mesh.

This chapter starts with the h-BEM for scalar Signorini problems with a first convergence result. Then more advanced hp-versions of the BEM for boundary variational inequalities arising from frictional contact are treated. The chapter concludes with the study of h-BEM and hp-BEM for nonmonotone contact problems from delamination of adhesively bonded interfaces in material science.

Mixed methods are typically studied for low order finite element schemes where the discrete Lagrange multiplier is a piecewise constant or linear discontinuous function which allows a conforming discretization of the sign and box constraints of the Lagrange multiplier. For higher polynomial degrees these conditions are no longer easy to realize and are only satisfied in a discrete sense. Therefore it is important that the missing conformity is captured in the a posteriori error estimates. Now as shown in Sect. 11.4.2 the a posteriori error estimate can be localized (approximately) and therefore used to recover/improve the convergence rate within an adaptive scheme compared to a uniform scheme which has reduced convergence rate due to the reduced regularity from the contact conditions.

© Springer International Publishing AG, part of Springer Nature 2018

J. Gwinner, E. P. Stephan, *Advanced Boundary Element Methods*,
Springer Series in Computational Mathematics 52,
https://doi.org/10.1007/978-3-319-92001-6_11

In adaptive schemes a sequence of solutions must be computed. Hence it is important that the convergence rate of the adaptive scheme is significantly higher than that of the uniform scheme. Here, hp-adaptive schemes show their superiority over pure h-adaptive schemes.

The a posteriori error estimate strategy of Theorem 11.4 gives an upper and lower bound for the discretization error under the saturation assumption, but a global variational inequality over a p-enriched boundary element space needs to be solved for the evaluation of the error estimate. Whereas the a posteriori error estimate strategy of Theorem 11.10 gives a much easier to evaluate upper bound for the discretization error, a proof of a corresponding lower bound seems possible only for lowest order discretizations.

11.1 h-BEM for the Signorini Problem

11.1.1 Discretization of the Boundary Variational Inequality

In the following we start from the variational formulation of the scalar Signorini problem as a boundary variational inequality given in the Sect. 5.1. We suppose that for simplicity $\Omega \subset \mathbb{R}^2$ is polygonal, but not necessarily convex. Let Γ be represented by

$$x_i = X_i(s)\,, \quad 0 \le s \le L \quad (i = 1, 2)$$

with $X_i(0) = X_i(L)$ $(i = 1, 2)$. We partition Γ into finitely many segments by the points $P_j = (X_1(s_j), X_2(s_j))$, $j = 1, \ldots, J$, where the endpoints of $\overline{\Gamma}_D$ and $\overline{\Gamma}_S$ are included and where $s_1 = 0$, $s_{J+1} = L$. The partitioning of Γ is characterized by the mesh size

$$h := \max_{j=1,\ldots,J} |s_{j+1} - s_j|\,.$$

Note that the boundary variational inequality (5.13) splits into a variational equality in $H^{-1/2}(\Gamma)$, which can be discretized in a standard way, and a novel variational inequality in the convex cone $K \subset H^{1/2}(\Gamma)$. As an important issue we want to treat not only piecewise linear, but also piecewise quadratic and piecewise cubic approximations of K. To this end we introduce the space $\mathscr{P}^\kappa$ of polynomials of degree less than or equal to κ $(\kappa = 1, 2, 3)$ and the subsequent finite point sets:

$$\Sigma_1^h := \{s_j : j = 1, \ldots, J\}$$

$$\Sigma_2^h := \{s \in (0, L) : s \text{ is a midpoint of an interval } (s_j, s_{j+1}) \text{ for some } j = 1, \ldots, J\}$$

$$\Sigma_3^h := \{s \in (0, L) : s \text{ divides an interval } (s_j, s_{j+1}) \text{ by the ratio 1:2}\}$$

Moreover,

$$\Pi_1^h := \{P_j : j = 1, \ldots, J\} \cap \overline{\Gamma}_S ,$$

where with appropriate $1 \leq j_0 \leq j_1 < J$,

$$\Pi_1^h = \{P_j : j = j_0, \ldots, j_1 + 1\} ,$$

and for $\kappa = 2, 3$ we set

$$\Pi_\kappa^h := \left\{ P = (X_1(s), X_2(s)) : s \in \Sigma_1^h \cup \Sigma_\kappa^h \right\} \cap \overline{\Gamma}_S .$$

Recall

$$H^1_{\Gamma_D,0}(\Omega) = \{v \in H^1(\Omega) : v|\Gamma_D = 0\} ,$$

introduce the trace space

$$H^{1/2}_{\Gamma_D,0}(\Gamma) = \{v \in H^{1/2}(\Gamma) : v|\Gamma_D = 0\}$$

which can be approximated by the finite dimensional subspace

$$U^h_{\kappa,\mu} := \Big\{ v^h \in C^\mu(\Gamma) : v^h \circ X \mid (s_j, s_{j+1}) \in \mathscr{P}^\kappa \\ (j = 1, \ldots, J) \, ; \; v^h|_{\Gamma_D} = 0\} .$$

Further approximate the convex cone K by the convex cone

$$K^h_{\kappa,\mu} := \{v^h \in U^h_{\kappa,\mu} : v^h(P) \leq 0 \ \ (\forall P \in \Pi^h_\kappa)\} ,$$

imposing only finitely many inequality constraints, and $H^{-1/2}(\Gamma)$ by the finite dimensional subspace

$$\Phi^h_{\kappa-1,\mu-1} := \Big\{ \psi^h \in C^{\mu-1}(\Gamma) : \psi^h \circ X \mid (s_j, s_{j+1}) \in \mathscr{P}^{\kappa-1} \\ (j = 1, \ldots, J) \Big\} .$$

Here $\mu \in \mathbb{N}_0$ with $\mu \leq \kappa - 1$ and $C^{-1}(\Gamma)$ denotes the space of discontinuous functions. Note that $K^h_{1,0} \subset K$ holds for all $h > 0$.

Thus we are led to the following discretized variational problem:
Find $[u^h, \varphi^h] \in K^h_{\kappa,\mu} \times \Phi^h_{\kappa-1,\mu-1}$ *such that*

$$(\pi^h_\kappa) \quad \begin{cases} a\Big(\dfrac{du^h}{ds}, \dfrac{dv^h}{ds} - \dfrac{du^h}{ds}\Big) + b(\varphi^h, v^h - u^h) \geq \ell(v^h - u^h) \quad \forall v^h \in K^h_{\kappa,\mu}, \\ a(\psi^h, \varphi^h) = b(\psi^h, u^h) \quad \forall \psi^h \in \Phi^h_{\kappa-1,\mu-1}; \end{cases}$$

or equivalently

$$A([u^h, \varphi^h], [v^h, \psi^h] - [u^h, \varphi^h]) \geq \ell(v^h - u^h)$$
$$\forall [v^h, \psi^h] \in K^h_{\kappa,\mu} \times \Phi^h_{\kappa-1,\mu-1}. \tag{11.1}$$

Let us remark that the condition (C.28) in the Appendix guarantees the existence and uniqueness of not only the solution $[u, \varphi]$ of the problem (π), but also of the solution $[u^h, \varphi^h]$ of the approximate problems (π^h_κ) because our discretization does not affect the linear form ℓ.

11.1.2 The Convergence Result

Theorem 11.1 *Let solutions* $[u, \varphi]$ *to* (π) *and* $[u^h_\kappa, \varphi^h_\kappa]$ *to* (π^h_κ) $(h > 0)$ *exist. Assume that the solution* $[u, \varphi]$ *is unique. Then for* $\kappa = 1, 2, 3$

$$\lim_{h\to 0} \| [u^h_\kappa, \varphi^h_\kappa] - [u, \varphi] \|_{H^{1/2}(\Gamma)\times H^{-1/2}(\Gamma)} = 0$$

Proof In virtue of Lemma 5.2, the bilinear form $A(\cdot, \cdot)$ satisfies the Gårding inequality (5.16). Therefore the convergence theorem C.7 in the Sect. C.3.2 in Appendix C applies and requires the following hypotheses:

H1 If $\{v^h\}_{h>0}$ weakly converges to v, where $v^h \in K^h := K^h_{\kappa,\mu}$, then $v \in K$.

H2 There exist a subset $M \subset H^{1/2}(\Gamma)$ such that $\overline{M} = K$ and mappings $\rho^h : M \to U^h = U^h_{\kappa,\mu}$ with the property that, for each $w \in M$, $\rho^h w$ strongly converges to w (as $h \to 0+$) and $\rho^h w \in K^h$ for all $0 < h < h_0(w)$.

We note that the analogous hypotheses for the approximation of $\psi \in H^{-1/2}(\Gamma)$ by $\psi^h \in \Phi^h$ are trivally satisfied in view of $\Phi^h \subset H^{-1/2}(\Gamma)$ and well-known density and approximation properties.

Verification of (H1). Since $K^h_{1,\mu}$ is contained in the weakly closed set K for all $h > 0$, we have only to consider the cases $\kappa = 2$ and $\kappa = 3$ with $\mu \in \mathbb{N}_0$ such that $\mu \leq \kappa - 1$.

Let the polygonal boundary part $\overline{\Gamma}_S$ be partitioned by

$$\overline{\Gamma}_S = \bigcup_{j=j_0}^{j_1} [P_j, P_{j+1}],$$

where the closed line segment $[P_j, P_{j+1}]$ has the intermediate point $P_{j+\frac{1}{2}} \in \Pi_2^h$, respectively the two intermediate points $P_{j+\frac{1}{3}}, P_{j+\frac{2}{3}} \in \Pi_3^h$. For any $\psi \in C^0(\overline{\Gamma}_S)$ with $\psi \geq 0$ we define

$$\psi^h = \sum_{j=j_0}^{j_1} \psi(P_{j+\frac{1}{2}})\, \chi_{j+\frac{1}{2}},$$

where $\chi_{j+\frac{1}{2}}$ denotes the characteristic function of the open segment $]P_j, P_{j+1}[$. Then $\psi^h \geq 0$ on Γ_S $(\kappa = 2, 3)$ and by the uniform continuity of ψ on $\overline{\Gamma}_S$

$$\lim_{h \to 0} \|\psi^h - \psi\|_{L^\infty(\Gamma_S)} = 0. \tag{11.2}$$

Now let $\{v^h\}_{j>0}$ be a family weakly convergent to $v \in H^{1/2}_{\Gamma_D,0}(\Gamma)$, where $v^h \in K^h_{\kappa,\mu}$ $(h > 0;\ \kappa = 2$ or $\kappa = 3)$. Since the embedding $H^{1/2}(\Gamma) \subset L^1(\Gamma_S)$ is weakly continuous, the functions v^h converge weakly to v in $L^1(\Gamma_S)$ and are norm bounded. Therefore by the estimate

$$\left|\int_{\Gamma_S} (v^h \psi^h - v\psi) ds\right| \leq \|v^h\|_{L^1(\Gamma_S)}\, \|\psi^h - \psi\|_{L^\infty(\Gamma_S)} + \left|\int_{\Gamma_S} (v^h - v)\psi\, ds\right|,$$

using (11.2) and $\psi \in L^\infty(\Gamma_S) = (L^1(\Gamma_S))^*$, we obtain that

$$\lim_{h \to 0} \int_{\Gamma_S} v^h \psi^h\, ds = \int_{\Gamma_S} v\psi\, ds. \tag{11.3}$$

From Simpson's rule it follows for $v^h \in K^h_{2,\mu}$ and all $\psi \in C^0(\Gamma)$ with $\psi \geq 0$ that

$$\begin{aligned}
\int_{\Gamma_S} v^h \psi^h\, ds &= \sum_{j=j_0}^{j_1} \int_{s_j}^{s_{j+1}} \psi(P_{j+\frac{1}{2}}) \sum_{i=1}^{2} (v^h \circ X)(s)\, ds \\
&= \frac{1}{6} \sum_{j=j_0}^{j_1} \psi(P_{j+\frac{1}{2}})(s_{j+1} - s_j) \left[v^h(P_j) + 4v^h(P_{j+\frac{1}{2}}) + v^h(P_{j+1})\right] \\
&\leq 0,
\end{aligned} \tag{11.4}$$

whereas from Newton's pulcherrima quadrature rule [226, §7.1.5]) for $v^h \in K^h_{3,\mu}$

$$\int_{\Gamma_S} v^h \psi^h \, ds = \frac{1}{8} \sum_{j=j_0}^{j_1} \psi(P_{j+\frac{1}{2}})\,(s_{j+1} - s_j) \Big[v^h(P_j) + 3v^h(P_{j+\frac{1}{3}})$$

$$+3v^h(P_{j+\frac{2}{3}}) + v^h(P_{j+1}) \Big] \leq 0 \,. \tag{11.5}$$

Combining (11.3) and (11.4), respectively (11.5) we obtain that for all $\psi \in C^0(\overline{\Gamma}_S)$ with $\psi \geq 0$

$$\int_{\Gamma_S} v\psi \, ds \leq 0 \,,$$

hence $v \leq 0$ almost everywhere on Γ_S or $v \in K$. This proves (H1).

Verification of (H2). In virtue of (5.19), we can take $M = K \cap C^\infty(\Gamma)$. Now we define $\rho^h_\kappa : H^{1/2}(\Gamma) \cap C^\infty(\Gamma) \to U^h_{\kappa,\kappa-1} \subseteq U^h_{\kappa,\mu}$ by L-periodic spline interpolation subordinated to the partitioning of Γ. Thus in particular

$$\rho^h_\kappa \, w(P) = w(P) \,, \quad \forall P \in \Pi^h_\kappa \quad (\kappa = 1, 2, 3) \,.$$

Hence $\rho^h_\kappa w$ belongs to $K^h_{\kappa,\kappa-1} \subseteq K^h_{\kappa,\mu}$ for any $w \in M$, since $\mu \leq \kappa - 1$. Moreover by spline interpolation theory, $U_{\kappa,\kappa-1}$ is a regular family of finite elements in the sense of Babuška and Aziz [14] and therefore we have

$$\|w - \rho^h_\kappa w\|_{H^{1/2}(\Gamma)} \leq c h^{\kappa - 1/2} \, \|w\|_{H^\kappa(\Gamma)} \quad (\kappa = 1, 2, 3)$$

with $c > 0$ independent of h and w. Hence we conclude that

$$\lim_{h \to 0} \|w - \rho^h_\kappa w\|_{H^{1/2}(\Gamma)} = 0 \,, \quad \forall w \in M \,;\ \kappa = 1, 2, 3 \,.$$

□

Remark 11.1 By the proof above (see especially the estimates (11.4) and (11.5)) we have shown that boundary element convergence holds true for arbitrary piecewise polynomial approximations as long as the corresponding Newton-Cotes quadrature formula has positive weights. This is a reasonable restriction for practical computations and is satisfied for the Newton-Cotes formulae up to the order $\kappa = 8$ [160, §6.2.1].

In this section we considered the simplest elliptic equation. However the method presented can be extended to the more general problems of unilateral contact involving the Navier - Lame -system of linear elasticity, see [222]. A priori error estimates with linear boundary elements are provided in [292, 386, 387]. Nitsche type error estimates for variational inequalities are derived in [393].

11.2 *hp*-BEM with Hierarchical Error Estimators for Scalar Signorini Problems

In this section we report from [297] a priori and a posteriori error estimates for the $hp-$ discretization of a boundary integral formulation of the Signorini problem of the Laplacian. We present $hp-$ convergence results for the BEM Galerkin solution in the energy norm. The a priori error estimate shows $O\left(h^{\frac{1}{4}}p^{-\frac{1}{4}}\right)$ convergence rate and corresponds to the FEM result (originally derived by Falk in [173] for the h version). The presented a posteriori error estimate is efficient and reliable. The hierarchical error estimators used are computed by enriching the boundary element spaces by bubble functions on each element. This enrichment defines two-level subspace decompositions with corresponding additive Schwarz operators where the latter have condition numbers which depend only logarithmically on the polynomial degrees. For ease of reading, we present most of the proofs from [297]. Numerical experiments in [297] show that a three-step adaptive algorithm (steered by the hierarchical error estimators) leads to appropriate mesh refinement and reasonable polynomial degree distribution. For extension to friction problems see [103, 222].

Let $\Omega \subset \mathbb{R}^n$, $n \geq 2$ be a bounded domain with Lipschitz boundary $\Gamma = \partial\Omega$ which is a disjoint union of Γ_D, Γ_N and $\Gamma_S \neq \emptyset$. We consider the following problem.

Given $h \in H^{-1/2}(\Gamma_N \cup \Gamma_S)$, $g \in H^{1/2}(\Gamma_D \cup \Gamma_S) \cap C^0(\bar{\Gamma}_D \cup \bar{\Gamma}_S)$, find $\hat{u} \in H^1(\Omega)$ such that

$$\begin{aligned} \Delta\hat{u} &= 0 && \text{in } \Omega, \\ \hat{u} &= g && \text{on } \Gamma_D, \\ \tfrac{\partial\hat{u}}{\partial n} &= h && \text{on } \Gamma_N, \\ \hat{u} \leq g,\ \tfrac{\partial\hat{u}}{\partial n} \leq h,\ (\hat{u}-g)(\tfrac{\partial\hat{u}}{\partial n} - h) &= 0 && \text{on } \Gamma_S \end{aligned} \tag{11.6}$$

As we have seen in Sect. 5.1 this problem can be formulated equivalently as a variational inequality over Ω with the convex subset $\hat{K} := \{\hat{v} \in H^1(\Omega) : \hat{v}|_{\Gamma_D} = g|_{\Gamma_D}, \hat{v}|_{\Gamma_S} \leq g|_{\Gamma_S}\} \subset H^1(\Omega)$: Find $\hat{u} \in K$ such that

$$\int_\Omega \nabla\hat{u} \cdot \nabla(\hat{v} - \hat{u})\, dx \geq \int_{\Gamma_N \cup \Gamma_S} h(\hat{v} - \hat{u})\, ds \quad \forall\hat{v} \in \hat{K}. \tag{11.7}$$

The variational inequality (11.7) has a unique solution if $\Gamma_D \neq \emptyset$ or $\int_{\Gamma_N \cup \Gamma_S} h\, ds < 0$ (cf. [386] and Appendix C.3.1). We reformulate (11.6) as an equivalent variational inequality involving only boundary integral operators on Γ.

With the integral operators of the single layer, double layer potential and their normal derivatives V, K, K', W respectively we have the problem (L):

Find $(u, \varphi) \in K^\Gamma \times H^{-1/2}(\Gamma)$ such that

$$\begin{aligned} \langle u, W(v-u)\rangle + \langle\varphi, (I+K)(v-u)\rangle &\geq 2l(v-u) \\ \langle\psi, V\varphi\rangle - \langle\psi, (I+K)u\rangle &= 0 \qquad \forall(v,\psi) \in K_\Gamma \times H^{-1/2}(\Gamma) \end{aligned} \tag{11.8}$$

where

$$K^{\Gamma} := \{v \in H^{1/2}(\Gamma) \,:\, v|_{\Gamma_D} = g|_{\Gamma_D}, v|_{\Gamma_S} \le g|_{\Gamma_S}\} \tag{11.9}$$

and

$$l(v) = \int_{\Gamma_N \cup \Gamma_S} hv\, ds.$$

System (L) can be rewritten with the coercive and non-symmetric bilinear form

$$B(u, \varphi; v, \psi) := \langle Wu, v\rangle + \langle (I + K)^t \varphi, v\rangle + \langle V\varphi, \psi\rangle - \langle (I + K)u, \psi\rangle$$

as: Find $(u, \varphi) \in K^{\Gamma} \times H^{-1/2}(\Gamma)$ such that

$$B(u, \varphi; v - u, \psi) \ge \mathscr{L}(v - u, \psi) \quad \forall (v, \psi) \in K^{\Gamma} \times H^{-1/2}(\Gamma) \tag{11.10}$$

where

$$\mathscr{L}(v, \psi) := 2 \int_{\Gamma_N \cup \Gamma_S} hv\, ds. \tag{11.11}$$

On the other hand, eliminating φ in (L) leads to the equivalent problem (S): Find $u \in K^{\Gamma}$ such that

$$\langle Su, v - u\rangle \ge l(v - u) \quad \forall v \in K^{\Gamma} \tag{11.12}$$

with the symmetric Poincaré–Steklov operator S for the interior problem

$$S := \frac{1}{2}(W + (K' + I)V^{-1}(K + I)) \,:\, H^{1/2}(\Gamma) \to H^{-1/2}(\Gamma) \tag{11.13}$$

which is positive definite on $H^{1/2}(\Gamma)/\mathbb{R}$. Existence and uniqueness of the solution of problems (S) and (L), respectively, have been shown by Houde Han [227]; for the corresponding elasticity problem see Gwinner and Stephan [222].

Let ω_h, γ_h be two not necessarily identical regular partitions of Γ, such that all corners of Γ and all "end points" $\bar{\Gamma}_S \cap \bar{\Gamma}_N$, $\bar{\Gamma}_N \cap \bar{\Gamma}_D$, $\bar{\Gamma}_D \cap \bar{\Gamma}_S$ are nodes of ω_h, γ_h.

Let $\mathbf{p} = (p_e)_{e\in\omega_h}$, or $\mathbf{q} = (q_e)_{e\in\gamma_h}$ be degree vectors which associate each element of ω_h or γ_h with a polynomial degree $p_e \ge 1$ or $q_e \ge 0$.

On the interval $[-1, 1]$ we choose $N + 1$ Gauss-Lobatto quadrature points, i.e. the points ξ_j^{N+1}, $0 \le j \le N$, that are the zeros of $(1 - \xi^2)L_N'(\xi)$, where L_N denotes the Legendre polynomial of degree N. It is known (cf. [46, Prop. 2.2, (2.3)]) that there exist positive weight factors $\varrho_j^{N+1} := \frac{1}{N(N+1)L_N^2(\xi_j^{N+1})}$ such that $\forall \phi \in \mathscr{P}_{2N-1}([-1, 1]) \,:\, \sum_{j=0}^N \phi(\xi_j)\varrho_j^{N+1} = \int_{-1}^1 \phi(\xi)\, d\xi$.

By an affine transformation we define the set of Gauss-Lobatto points $G_{e,hp}$ on each element e of the partition ω_h of Γ, corresponding to the polynomial degree p_e and set $G_{hp} := \bigcup_{e\in\omega_h} G_{e,hp}$.

On the partition ω_h of Γ we introduce σ_{hp} as the space of continuous polynomials with $\sigma_{hp}|_e \subseteq \mathscr{P}_{p_e}(e)$, $\forall e \in \omega_h$. A suitable basis of σ_{hp} is given by the Lagrange interpolation polynomials on the set of Gauss-Lobatto points of each element, see Fig. 11.1, On the partition γ_h of Γ we introduce τ_{hp} as the space of piecewise polynomials with $\tau_{hp}|_e \subseteq \mathscr{P}_{q_e}(e)$, $\forall e \in \gamma_h$; here a suitable basis is given by the Legendre polynomials.

Therefore, our test and trial spaces are defined as

$$\sigma_{hp} := \{v_{hp} \in C^0(\Gamma;\mathbb{R}) \, : \, v_{hp}|_e \in \mathscr{P}_{p_e}(e), \forall e \in \omega_h\} \tag{11.14}$$

$$\tau_{hp} := \{\psi_{hp} \in L^2(\Gamma;\mathbb{R}) \, : \, \psi_{hp}|_e \in \mathscr{P}_{q_e}(e), \forall e \in \gamma_h\}. \tag{11.15}$$

Now, we chose

$$K^{\Gamma}_{hp} := \{v \in \sigma_{hp} \mid \forall x \in G_{hp} \cap \Gamma_S \, : \, v(x) \le g(x) \text{ and } \forall x \in G_{hp} \cap \Gamma_D \, : \, v(x) = g(x)\}, \tag{11.16}$$

which is a convex, closed subset of σ_{hp}. Note that $K^{\Gamma}_{hp} \not\subseteq K^{\Gamma}$ for $p \ge 2$ (as Fig. 11.1 illustrates) or for non-concave gap function g.

Based on the local set of Gauss-Lobatto points $G_{e,hp}$ we define the local interpolation operator $i_{e,p_e} : C^0(\bar{e}) \to \mathscr{P}_{p_e}(e)$ by

$$(i_{e,p_e}\psi)(x) = \psi(x) \quad \forall x \in G_{e,hp}\,, \forall \psi \in C^0(\bar{e})$$

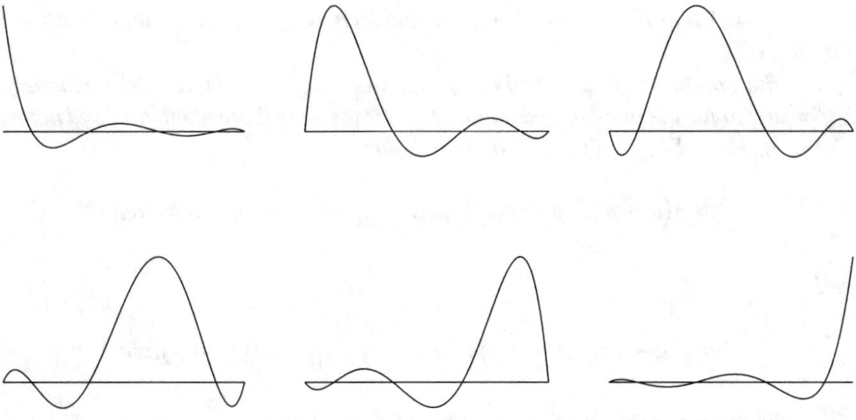

Fig. 11.1 Lagrange Polynomials , $N = 5$

and the global interpolation operator $i_{\omega_h,\mathbf{p}} : C^0(\Gamma) \to \sigma_{hp}$ by

$$i_{\omega_h,\mathbf{p}}\psi := \sum_{e\in\omega_h} \chi_e i_{e,p_e}\psi|_{\bar{e}} \qquad \forall\psi \in C^0(\Gamma) \tag{11.17}$$

where χ_e is the characteristic function of $e \in \omega_h$.

With the above choices, the hp-version of (L) is given by the discrete problem (L_{hp}):

Find $(u_{hp}, \varphi_{hp}) \in K^{\Gamma}_{hp} \times \tau_{hp}$ such that

$$B(u_{hp}, \varphi_{hp}; v - u_{hp}, \psi) \geq 2l(v - u_{hp}) \quad \forall(v, \psi) \in K^{\Gamma}_{hp} \times \tau_{hp}. \tag{11.18}$$

Via the canonical imbeddings $j_{hp} : \sigma_{hp} \hookrightarrow H^{1/2}(\Gamma)$ and $k_{hp} : \tau_{hp} \hookrightarrow H^{-1/2}(\Gamma)$ and their duals j^*_{hp}, and k^*_{hp} the discrete Poincaré-Steklov operator $S_{hp} : \sigma_{hp} \to \sigma^*_{hp}$

$$S_{hp} := \frac{1}{2}(j^*_{hp} W j_{hp} + j^*_{hp}(I + K')k_{hp}(k^*_{hp} V k_{hp})^{-1} k^*_{hp}(I + K) j_{hp}) \tag{11.19}$$

is well defined (see [73]). Now the hp-version of (S) given by the discrete problem (S_{hp}) for the general hp-version (11.12) reads: Find $u_{hp} \in K^{\Gamma}_{hp}$ such that

$$\langle S_{hp} u_{hp}, v_{hp} - u_{hp}\rangle \geq l(j_{hp} v_{hp} - j_{hp} u_{hp}) \quad \forall v_{hp} \in K^{\Gamma}_{hp}. \tag{11.20}$$

There holds the following convergence result for the Galerkin solution of (L_{hp}) in the energy norm without any regularity assumptions.

Theorem 11.2 ([297]) *Let $(\omega_h, \gamma_h)_{h\in I}$ be a family of quasi uniform meshes, such that $h := \max\{|e|, e \in \omega_h or\ e \in \gamma_h\}$, where $I \subset (0, \infty)$ with $0 \in I$. Let $\mathbf{p} = (p_e)_{e\in\omega_h}$, such that $p_e = p$ for all $e \in \omega_h$, and let $\mathbf{q} = (q_e)_{e\in\gamma_h}$, such that $q_e = p-1$ for all $e \in \gamma_h$.*

Let the solutions (u, φ) of (11.10) and (u_{hp}, φ_{hp}) of (11.18) exist uniquely. Suppose that for the polygonal domain Ω, there are only finite number of end points $\bar{\Gamma}_S \cap \bar{\Gamma}_D$, $\bar{\Gamma}_D \cap \bar{\Gamma}_N$, $\bar{\Gamma}_N \cap \bar{\Gamma}_S$. Then there holds

$$\lim_{p\to\infty} \|(u - u_{hp}, \varphi - \varphi_{hp})\|_{H^{1/2}(\Gamma)\times H^{-1/2}(\Gamma)} = 0, \quad \text{if } h \text{ fixed}$$

and

$$\lim_{h\to 0} \|(u - u_{hp}, \varphi - \varphi_{hp})\|_{H^{1/2}(\Gamma)\times H^{-1/2}(\Gamma)} = 0, \quad \text{if } p \text{ fixed}$$

For its proof see the proof of Theorem 11.6 in the subsequent section that treats the discretization of a friction-type functional in addition.

If we assume higher regularity of the solution (u, φ) of (L) and of the contact functon g in (11.6), i.e $u, g \in H^{3/2}(\Gamma), \varphi \in H^{1/2}(\Gamma)$ we obtain the following a priori error estimate which proposes a convergence rate of $O(h^{1/4}p^{-1/4})$.

Theorem 11.3 ([297]) *Let $(\omega_h, \gamma_h)_h$ be a family of quasi uniform meshes, such that $h := \max\{|e|, e \in \omega_h or \in \gamma_h\}$ and let $\mathbf{p} = (p_e)_{e\in\omega_h}$, $p_e = p$, $\mathbf{q} = (q_e)_{e\in\gamma_h}$, $q_e = p - 1$. Let $(u, \varphi) \in K^\Gamma \times H^{1/2}(\Gamma)$ be a solution of problem (L) and let $(u_{hp}, \varphi_{hp}) \in K^\Gamma_{hp} \times \tau_{hp}$ be solution of problem (L_{hp}) and assume $u, g \in H^{3/2}(\Gamma)$ and $h^* - Su \in L^2(\Gamma)$, with $h^*|_{\Gamma_N\cup\Gamma_S} = h$ and $h^*|_{\Gamma_D} = 0$. Then there holds*

$$\|(u - u_{hp}, \varphi - \varphi_{hp})\|_{H^{1/2}(\Gamma)\times H^{-1/2}(\Gamma)} \leq Ch^{1/4}p^{-1/4}\|u\|_{H^{3/2}(\Gamma)}$$

for $p \to \infty$ or $h \to 0$.

For its proof see the proof of Theorem 11.7 in the subsequent section that treats the discretization of a friction-type functional in addition.

Next we consider a hierarchical boundary element method, where we extend the finite dimensional spaces σ_{hp} and τ_{hp} by bubble functions on each element in ω_h and γ_h.

Each element $e \in \omega_h$ is associated with a polynomial degree p_e and the affine mapping F_e with $F_e(\xi) = x(\xi) \in e$ for $\xi \in [-1, 1]$.

With the Legendre polynomial L_j of degree j we set

$$\psi_0(\xi) := \frac{1-\xi}{2}, \quad \psi_1(\xi) := \frac{1-\xi}{2}, \quad \psi_j(\xi) := \sqrt{\frac{2j-1}{2}} \int_{-1}^{\xi} L_{j-1}(t)\,dt, \quad 2 \leq j,$$

and take

$$\sigma_e = \text{span}\{\psi_{e,p_e+1}\}, \qquad \psi_{e,j}(x) := \psi_j(F_e^{-1}(x)). \tag{11.21}$$

On the other hand, each element $e \in \gamma_h$ is associated with a polynomial degree q_e. Setting

$$\phi_0(t) = \frac{1}{2}, \quad \phi_j(t) := \sqrt{\frac{2j+1}{2}} L_j(t), \quad 1 \leq j,$$

we take

$$\tau_e = \text{span}\{\phi_{e,q_e+1}\}, \qquad \phi_{e,j}(x) := \phi_j(F_e^{-1}(x)). \tag{11.22}$$

In this way we obtain the subspace decompositions

$$\sigma_{h,p+1} := \sigma_{hp} \oplus L_p, \qquad L_p := \sum_{e\in\omega_h} \sigma_e, \tag{11.23}$$

$$\tau_{h,p+1} := \tau_{hp} \oplus \lambda_p, \qquad \lambda_p := \sum_{e \in \gamma_h} \tau_e. \tag{11.24}$$

Hence the polynomial degree vector associated with $\sigma_{h,p+1}$ is $(p_e + 1)_{e \in \omega_h}$ and the polynomial degree vector associated with $\tau_{h,p+1}$ is $(q_e + 1)_{e \in \gamma_h}$.

Let $P_{hp} : \sigma_{h,p+1} \to \sigma_{hp}$, $P_{hp,e} : \sigma_{h,p+1} \to \sigma_e$, $p_{hp} : \tau_{h,p+1} \to \tau_{hp}$, $p_{hp,e} : \tau_{h,p+1} \to \tau_e$ be the Galerkin projections with respect to the bilinear forms $\langle W\cdot, \cdot\rangle$ and $\langle V\cdot, \cdot\rangle$. For all $u \in \sigma_{h,p+1}$ we define P_{hp} and $P_{hp,e}$ by

$$\langle W P_{hp} u, v\rangle = \langle Wu, v\rangle \qquad \forall v \in \sigma_{hp} \tag{11.25}$$

$$\langle W P_{hp,e} u, v\rangle = \langle Wu, v\rangle \qquad \forall v \in \sigma_e,\, e \in \omega_h \tag{11.26}$$

and for all $\varphi \in \tau_{h,p+1}$ we define p_{hp} and $p_{hp,e}$ by

$$\langle V p_{hp} \varphi, \phi\rangle = \langle V\varphi, \phi\rangle \qquad \forall \phi \in \tau_{hp} \tag{11.27}$$

$$\langle V p_{hp,e} \varphi, \phi\rangle = \langle V\varphi, \phi\rangle \qquad \forall \phi \in \tau_e,\, e \in \gamma_h. \tag{11.28}$$

Finally, we define the two-level additive Schwarz operators

$$P_\sigma := P_{hp} + \sum_{e \in \omega_h} P_{hp,e} \quad \text{and} \quad p_\tau := p_{hp} + \sum_{e \in \gamma_h} p_{hp,e}. \tag{11.29}$$

For all $v \in \sigma_{h,p+1}$, $\psi \in \tau_{h,p+1}$ there holds with positive constants c_1, c_2, C_1, C_2 independently of h, p and q

$$C_1\,(1 + \log p_{max})^{-2}\, \|v\|_W^2 \le \|P_{hp} v\|_W^2 + \sum_{e \in \omega_h} \|P_{hp,e} v\|_W^2 \le C_2 \|v\|_W^2 \tag{11.30}$$

$$c_1\,(1 + \log q_{max})^{-2}\, \|\psi\|_V^2 \le \|p_{hp} \psi\|_V^2 + \sum_{e \in \gamma_h} \|p_{hp,e} \psi\|_V^2 \le c_2 \|\psi\|_V^2 \tag{11.31}$$

with $p_{max} := \max\{p_e, e \in \omega_h\}$, $q_{max} := \max\{q_e, e \in \gamma_h\}$ (see [239]).

For all $(u, \varphi) \in H^{1/2}(\Gamma)/\mathbb{R} \times H^{-1/2}(\Gamma)$ we define the norm $\|(u, \varphi)\|_H = (\|u\|_W^2 + \|\varphi\|_V^2)^{1/2}$, which is equivalent to $(\|u\|^2_{H^{1/2}(\Gamma)/\mathbb{R}} + \|\varphi\|^2_{H^{-1/2}(\Gamma)})^{1/2}$. The norm $\|(u, \varphi)\|_H$ is generated by the bilinear form

$$a(u, \varphi; v, \psi) := \langle Wu, v\rangle + \langle V\varphi, \psi\rangle.$$

Let (u, φ) be the solution of the variational inequality (11.10) and let $(u_{hp}, \varphi_{hp}) \in \sigma_{hp} \times \tau_{hp}$, $(u_{h,p+1}, \varphi_{h,p+1}) \in \sigma_{h,p+1} \times \tau_{h,p+1}$ be the solutions of the corresponding discrete problems.

As for finite element problems (see, e.g. [32]) we make the saturation assumption: There exists a parameter $0 \le \kappa < 1$ such that for all discrete spaces holds:

$$\|(u - u_{h,p+1}, \varphi - \varphi_{h,p+1})\|_H \le \kappa \|(u - u_{hp}, \varphi - \varphi_{hp})\|_H. \tag{11.32}$$

Now the saturation assumption implies with the triangle inequality:

$$\begin{aligned}(1-\kappa)\|(u - u_{hp}, \varphi - \varphi_{hp})\|_H &\le \|(u_{h,p+1} - u_{hp}, \varphi_{h,p+1} - \varphi_{hp}\|_H \\ &\le (1+\kappa)\|(u - u_{hp}, \varphi - \varphi_{hp})\|_H.\end{aligned}$$

Theorem 11.4 *We assume that (11.32) holds for the solution* (u, φ) *of the variational inequality (11.10) and the solutions* (u_{hp}, φ_{hp}) , $(u_{h,p+1}, \varphi_{h,p+1})$ *of the corresponding discrete problems. Then there are constants* $\zeta_1, \zeta_2 > 0$ *such that*

$$\zeta_1 \eta_{hp} \le \|(u - u_{hp}, \varphi - \varphi_{hp})\|_H \le \zeta_2 (1 + \log \max\{p_{max}, q_{max}\}) \eta_{hp} \tag{11.33}$$

where

$$\eta_{hp} := \left(\Theta_{hp}^2 + \eta_{u,hp}^2 + \eta_{\varphi,hp}^2\right)^{1/2}, \qquad \Theta_{hp} := \|P_{hp} e_{h,p+1}\|_W, \tag{11.34}$$

$$\eta_{u,hp} := \Big(\sum_{e \in \omega_h} \Theta_{hp,e}^2\Big)^{1/2}, \qquad \Theta_{hp,e} := \|P_e e_{h,p+1}\|_W, \tag{11.35}$$

$$\eta_{\varphi,hp} := \Big(\sum_{e \in \gamma_h} \theta_{hp,e}^2\Big)^{1/2}, \qquad \theta_{hp,e} := \frac{|B(u_{hp}, \varphi_{hp}; 0, \psi_{e,q_e+1})|}{\|\psi_{e,q_e+1}\|_V}, \tag{11.36}$$

and $e_{h,p+1} \in K_{u_{hp}}$ *is the solution of the variational inequality*

$$\langle W e_{h,p+1}, v - e_{h,p+1}\rangle \ge \mathscr{L}(v - e_{h,p+1}, 0) - B(u_{hp}, \varphi_{hp}; v - e_{h,p+1}, 0) \quad \forall v \in K_{u_{hp}} \tag{11.37}$$

with $K_{u_{hp}} := \{v - u_{hp} \mid v \in K_{h,p+1}^{\Gamma}\} = K_{h,p+1}^{\Gamma} - u_{hp}$.

Proof For the defect $(\tilde{e}_{h,p+1}, \tilde{\varepsilon}_{h,p+1}) := (u_{h,p+1} - u_{hp}, \varphi_{h,p+1} - \varphi_{hp})$ of the solution to

$$B(u_{h,p+1}, \varphi_{h,p+1}; v - u_{h,p+1}, \psi) \ge \mathscr{L}(v - u_{h,p+1}, \psi) \quad \forall (v, \psi) \in K_{h,p+1}^{\Gamma} \times \tau_{h,p+1} \tag{11.38}$$

we can write

$$B(\tilde{e}_{h,p+1}, \tilde{\varepsilon}_{h,p+1}; v - \tilde{e}_{h,p+1}, \psi) \geq \mathscr{L}(v - \tilde{e}_{h,p+1}, \psi) - B(u_{hp}, \varphi_{hp}; v - \tilde{e}_{h,p+1}, \psi) \tag{11.39}$$

for all $(v, \psi) \in K_{u_{hp}} \times \tau_{h,p+1}$.

We define $(e_{h,p+1}, \varepsilon_{h,p+1}) \in K_{u_{hp}} \times \tau_{h,p+1}$ by

$$a(e_{h,p+1}, \varepsilon_{h,p+1}; v - e_{h,p+1}, \psi) \geq \mathscr{L}(v - e_{h,p+1}, \psi) - B(u_{hp}, \varphi_{hp}; v - e_{h,p+1}, \psi) \tag{11.40}$$

for all $(v, \psi) \in K_{u_{hp}} \times \tau_{h,p+1}$.

This variational inequality can be separated into an inequality and an equality which are independent, i.e. (11.40) is equivalent to

$$\langle W e_{h,p+1}, v - e_{h,p+1} \rangle \geq \mathscr{L}(v - e_{h,p+1}, 0) - B(u_{hp}, \varphi_{hp}; v - e_{h,p+1}, 0) \quad \forall v \in K_{u_{hp}}, \tag{11.41}$$

$$\langle V \varepsilon_{h,p+1}, \psi \rangle \qquad = \mathscr{L}(0, \psi) - B(u_{hp}, \varphi_{hp}; 0, \psi) \quad \forall \psi \in \tau_{h,p+1}. \tag{11.42}$$

Now, application of (11.30) gives:

$$\begin{aligned} C_1(1 + \log p_{max})^{-2} \|e_{h,p+1}\|_W^2 + c_1(1 + \log q_{max})^{-2} \|\varepsilon_{h,p+1}\|_V^2 \\ \leq \|(P_{hp} e_{h,p+1}, p_{hp} \varepsilon_{h,p+1})\|_H^2 + \sum_{e \in \omega_h} \|P_e e_{h,p+1}\|_W^2 + \sum_{e \in \gamma_h} \|p_e \varepsilon_{h,p+1}\|_V^2 \\ \leq C_2 \|e_{h,p+1}\|_W^2 + c_2 \|\varepsilon_{h,p+1}\|_V^2 \end{aligned}$$

By definition of p_{hp} and the Galerkin orthogonality we have

$$\|p_{hp} \varepsilon_{h,p+1}\|_V^2 = \langle V \varepsilon_{h,p+1}, p_{hp} \varepsilon_{h,p+1} \rangle = 0 \tag{11.43}$$

yielding

$$\|(P_{hp} e_{h,p+1}, p_{hp} \varepsilon_{h,p+1})\|_H^2 = \|P_{hp} e_{h,p+1}\|_W^2. \tag{11.44}$$

Furthermore

$$
\begin{aligned}
p_e\varepsilon_{h,p+1} &= \frac{\langle V\varepsilon_{h,p+1}, \phi_{e,q_e+1}\rangle}{\langle V\phi_{e,q_e+1}, \phi_{e,q_e+1}\rangle}\phi_{e,q_e+1} = \frac{\langle V\varepsilon_{h,p+1}, \phi_{e,q_e+1}\rangle}{\|\phi_{e,q_e+1}\|_V^2}\phi_{e,q_e+1} \\
&= \frac{\mathscr{L}(0, \phi_{e,q_e+1}) - B(u_{hp}, \varphi_{hp}; 0, \phi_{e,q_e+1})}{\|\phi_{e,q_e+1}\|_V^2}\phi_{e,q_e+1} \\
&= \frac{-B(u_{hp}, \varphi_{hp}; 0, \phi_{e,q_e+1})}{\|\phi_{e,q_e+1}\|_V^2}\phi_{e,q_e+1}
\end{aligned}
$$

and hence

$$
\|p_e\varepsilon_{h,p+1}\|_V = \frac{|B(u_{hp}, \varphi_{hp}; 0, \phi_{e,q_e+1})|}{\|\phi_{e,q_e+1}\|_V} =: \theta_{hp,e}. \tag{11.45}
$$

Collecting together we get

$$
\begin{aligned}
c_3\|(e_{h,p+1}, \varepsilon_{h,p+1})\|_H^2 &\le \|P_{hp}e_{h,p+1}\|_W^2 + \sum_{e\in\omega_h} \|P_e e_{h,p+1}\|_W^2 + \sum_{e\in\gamma_h} \theta_{hp,e}^2 \simeq \eta_{hp} \\
&\le c_4\|(e_{h,p+1}, \varepsilon_{h,p+1})\|_H^2
\end{aligned}
$$

with $c_3 := \min\{C_1(1 + \log p_{max})^{-2}, c_1(1 + \log q_{max})^{-2}\}$, $c_4 = \max\{C_1, C_2\}$.

But (see [Lemma 16 ,[297])

$$
\|(e_{h,p+1}, \varepsilon_{h,p+1})\|_H \simeq \|(\tilde{e}_{h,p+1}, \tilde{\varepsilon}_{h,p+1})\|_H = \|(u_{h,p+1} - u_{hp}, \varphi_{h,p+1} - \varphi_{hp})\|_H \tag{11.46}
$$

and therefore

$$
\frac{1}{\sqrt{c_4}}\eta_{hp} \le \|(u_{h,p+1} - u_{hp}, \varphi_{h,p+1} - \varphi_{hp})\|_H \le \frac{1}{\sqrt{c_3}}\eta_{hp}\,.
$$

Application of the saturation assumption yields the assertion of the theorem with $\zeta_1 = \frac{1}{(1+\kappa)\sqrt{\max\{C_2,c_2\}}}$ and $\zeta_2 = \frac{1}{(1-\kappa)\sqrt{\min\{C_1,c_1\}}}$. □

11.3 *hp*-BEM for a Variational Inequality of the Second Kind Modelling Unilateral Contact and Friction

This section continues Sect. 5.2 and is based on [217, 218, 297]. Here we give a further contribution to the analysis of the *hp*-version of the BEM for nonsmooth boundary value problems.

We recall the primal boundary variational inequality (5.22), which is equivalent to (5.20) and (5.21): Find $u \in K^\Gamma$ such that for all $v \in K^\Gamma$

$$\langle Su, v - u \rangle + j(v) - j(u) \geq l(v - u)$$

with the convex closed set

$$K^\Gamma := \left\{ v \in H^{1/2}(\Gamma) : v|\Gamma_D = \chi|\Gamma_D \text{ a.e. and } v|\Gamma_C \leq \chi|\Gamma_C \text{ a.e.} \right\},$$

the symmetric Steklov-Poincaré operator S for the interior problem,

$$S := \frac{1}{2}\left[W + \left(K' + I\right) V^{-1} \left(K + I\right) \right] : H^{1/2}(\Gamma) \to H^{-1/2}(\Gamma),$$

the continuous, positively homogeneous and sublinear, hence convex friction-type functional

$$j(v) := \int_{\Gamma_C} g|v|\, ds\,,$$

and the continuous linear form

$$l(v) := \int_{\Gamma_N \cup \Gamma_C} f v\, ds\,,$$

where the data $f \in H^{-1/2}(\Gamma_N \cup \Gamma_C)$, $\chi \in H^{1/2}(\Gamma_C \cup \Gamma_D) \cap C^0[\overline{\Gamma}_C \cup \overline{\Gamma}_D]$, $g \in L^\infty(\Gamma_C)$ with $g \geq 0$ are given. Here $\Omega \subset \mathbb{R}^d$ $(d \geq 2)$ is a bounded Lipschitz domain with its boundary Γ and mutually disjoint parts Γ_D, Γ_N, and Γ_C such that $\Gamma = \overline{\Gamma}_D \cup \overline{\Gamma}_N \cup \overline{\Gamma}_C$ and meas $(\Gamma_C) > 0$.

In contrast to the preceding section taken from [297] and to a related paper of Guediri [205] on a boundary variational inequality of the second kind modelling friction, we take the quadrature error of the friction-type functional into account of the error analysis. At first without any regularity assumptions, we prove convergence of the hp-BEM Galerkin approximation in the energy norm. Then under mild regularity assumptions, we establish an a priori error estimate that is based on a Céa-Falk lemma for abstract variational inequalities of the second kind. This lemma permits to split the total discretization error into three different parts: the approximation error due to the approximation of the Steklov-Poincaré operator by its discrete counterpart, the distance of the continuous solution to the convex set of approximations in the trial space, and the consistency error caused by the nonconforming approximation. Here as in [297] we apply the well-known approximation theory of spectral methods [47], extend the approximation analysis of Falk [173], and use interpolation arguments to obtain estimates in Sobolev norms of fractional order. Moreover, we exploit the special structure of the friction functional. Thus for our more general variational problem we arrive under mild regularity assumptions at an a priori error estimate of the same convergence order as in [297] which is suboptimal because of the appearance of the consistency error in the nonconforming approximation scheme and because of the well-known regularity threshold in unilateral problems [267].

11.3.1 The hp-Version Galerkin Boundary Element Scheme

For simplicity let Ω be a polygonal, planar domain and let g be a piecewise constant function on Γ_C. These are no restrictions of generality. In fact, the approximation of nonpolygonal domains/curves by polygons is well-understood. The analysis to follow can be extended to higher dimensional domains by tensor product approximation.

Let $\mathscr{S}_N$ ($N \in \mathbb{N}$) be a sequence of regular partitions of Γ such that all corners of Γ and all "end points" $\overline{\Gamma}_C \cap \overline{\Gamma}_N, \overline{\Gamma}_N \cap \overline{\Gamma}_D, \overline{\Gamma}_D \cap \overline{\Gamma}_C$ are nodes of $\mathscr{S}_N$ and that moreover, g is constant on each edge e of $\mathscr{S}_N$ contained in Γ_C. We introduce the set of edges on the contact boundary,

$$\mathscr{E}_{c,N} = \{e : e \subset \Gamma_C \text{ is an edge of } \mathscr{S}_N\}\,.$$

Further we denote by $p_{N,e} \geq 1$ a polynomial degree for each edge $e \in \mathscr{S}_N$. We assume that neighboring elements have comparable polynomial degrees, i.e. there exists a constant $c > 0$ such that for edges $e, e' \in \mathscr{S}_N$ with $\overline{e} \cap \overline{e'} \mathrm{N} e \emptyset$ there holds

$$c^{-1}\, p_{N,e} \leq p_{N,e'} \leq c\, p_{N,e}\,.$$

As in [297] and in the preceding section we employ Gauss-Lobatto integration in the discretization procedure. To this end we introduce for $p \geq 1$ on the reference interval $[-1, 1]$ the $p+1$ Gauss-Lobatto points, i.e., the zeros ξ_j^{p+1} $(0 \leq j \leq p)$ of $(1 - \xi^2)L_p'(\xi)$, where L_p denotes the Legendre polynomial of degree $p \geq 1$. It is known (see [47], chapter I, section 4) that there exist positive weights

$$\omega_j^{p+1} := \frac{1}{p(p+1)L_N^2(\xi_j^{p+1})}$$

such that the quadrature formula

$$\int_{-1}^{1} \phi(\xi)\, d\xi = \sum_{j=0}^{p} \omega_j^{p+1} \phi(\xi_j^{p+1})$$

is exact for all polynomials ϕ up to degree $2p-1$. By affine transformation $F_e : [-1,1] \to \overline{e}$ we define the set of Gauss-Lobatto points $G_{e,N}$ for each element e of $\mathscr{S}_N$ and set $G_N := \bigcup\{G_{e,N} : e \in \mathscr{S}_N\}$. Note that $\xi_0^{p+1} = -1$ and $\xi_N^{p+1} = 1$ are the end points of the reference interval. Therefore we can introduce the space Σ_N of continuous functions on Γ that are piecewise polynomial up to degree $p_{N,e}$ on each $e \in \mathscr{S}_N$ as our ansatz space for u,

$$\Sigma_N := \left\{ v_N \in C^0(\Gamma) : v_N|e \in \mathscr{P}^{p_{N,e}},\ \forall\ e \in \mathscr{S}_N \right\}.$$

A suitable basis of Σ_N is given by the Lagrange interpolation polynomials on the set of Gauss-Lobatto points of each element, since because of the essential boundary condition we have to evaluate the ansatz functions in the sets $G_{e,N}$ for each e in the Dirichlet boundary. To approximate K^Γ we choose the Gauss-Lobatto points as control points of the unilateral constraint and define

$$K_N^\Gamma := \{v_N \in \Sigma_N \ :\ v_N = \gamma \text{ on } G_N \cap \overline{\Gamma}_D,\ v_N \leq \chi \text{ on } G_N \cap \overline{\Gamma}_C\}.$$

Clearly, K_N^Γ is a convex closed subset of Σ_N. However, K_N^Γ is generally not contained in K^Γ for polynomial degree ≥ 2 or for a non-concave obstacle χ.

We also approximate the nonlinear nonsmooth functional j using the above quadrature rule by

$$j_N(v) = \sum_{e \in \mathscr{E}_{c,N}} g_e \sum_{j=0}^{p_{N,e}} \omega_j^{p_{N,e}+1} \left| v \circ F_e(\xi_j^{p_{N,e}+1}) \right|,$$

where g_e denotes the constant value of the function g on e. Note that for the piecewise polygonal boundary Γ, $v \circ F_e$ is piecewise polynomial of the same degree as v.

Associated to the Gauss-Lobatto points $G_{e,N}$ we have the local interpolation operator $i_{e,N} := i_{e,p_{N,e}} : C^0(\overline{e}) \to \mathscr{P}^{p_{N,e}}$ given by

$$(i_{e,N}\eta)(x) = \eta(x),\ \forall\ x \in G_{e,N},\ \eta \in C^0(\overline{e})$$

and the global interpolation operator $i_N : C^0(\Gamma) \to \Sigma_N$ by

$$i_N\eta = \sum_{e \in \mathscr{S}_N} 1_e\, i_{e,N}\eta|\overline{e},\ \forall\ \eta \in C^0(\Gamma),$$

where 1_e denotes the $\{0,1\}$-valued characteristic function of $e \in \mathscr{S}_N$.

For later use we recall from [47, Theorem 13.49] the following result on the polynomial interpolation error in the reference interval $\Lambda = (-1, 1)$.

Theorem 11.5 *For any real numbers r and s satisfying $s > (1+r)/2$ and $0 \leq r \leq 1$, there exists a positive constant c depending only on s such that for any function*

$\eta \in H^s(\Lambda)$ *the following estimate holds:*

$$\|\eta - i_{\Lambda,p}\eta\|_{H^r(\Lambda)} \le c\, p^{r-s}\, \|\eta\|_{H^s(\Lambda)}\,.$$

To approximate the dual variable $\varphi \in H^{-1/2}(\Gamma)$, we take as ansatz space

$$\Phi_N := \left\{\psi_N \in L^2(\Gamma) : \psi_N|e \in \mathscr{P}^{p_{N,e}-1}\ \forall\, e \in \mathscr{S}_N\right\}.$$

Here a suitable basis is given by the Legendre polynomials.

Thus we arrive at the following discrete variational problem (π_N) as approximation to our variational problem (π) given in (5.21): Find $(u_N, \varphi_N) \in K_N^\Gamma \times \Phi_N$ such that for all $(v, \psi) \in K_N^\Gamma \times \Phi_N$,

$$\frac{1}{2}\, B(u_N, \varphi_N; v - u_N, \psi) + j_N(v) - j_N(u_N) \ge l(v - u_N)\,. \tag{11.47}$$

Note that we only replaced the nonlinear functional j by its approximate j_N. In most computations, however, also B and l have to be replaced by some approximations that take into account e.g. numerical integration, matrix compression, or approximation of a curved boundary. Since such approximations are well documented in the literature of numerical analysis of linear elliptic boundary value problems, we omit this aspect here.

Instead we take care of the inconsistency when approximating the Steklov-Poincarè operator S by its discrete counterpart, here $S_N : \Sigma_N \to \Sigma_N^*$ taking $S_N = S_{hp}$ as in the previous section and in [297]:

$$S_N := \frac{1}{2}(\iota_N^* W \iota_N + \iota_N^* \left(K' + I\right) \kappa_N (\kappa_N^* V \kappa_N)^{-1} \kappa_N^* \left(K + I\right) \iota_N)\,,$$

where $\iota_N : \Sigma_N \hookrightarrow H^{1/2}(\Gamma)$, $\kappa_N : \Phi_N \hookrightarrow H^{-1/2}(\Gamma)$ denote the canonical embeddings and ι_N^*, κ_N^* their duals. Thus analogously to the primal formulation (5.22) of our variational problem (π), the discrete problem (π_N) reads: Find $u_N \in K_N^\Gamma$ such that for all $v_N \in K_N^\Gamma$,

$$\langle S_N u_N, v_N - u_N\rangle + j_N(v_N) - j_N(u_N) \ge (l \circ \iota_N)(v_N - u_N)\,. \tag{11.48}$$

Recall that the primal variable u and the dual variable φ, respectively their *hp*-approximates u_N and φ_N are related by

$$V\varphi = (I + K)u\,; \quad \kappa_N^* V \kappa_N \varphi_N = \kappa_N^*(I + K)\iota_N u_N\,.$$

For later use we provide the following a priori estimate.

Lemma 11.1 *Let $E_N = \iota_N^* S \iota_N - S_N$. Then there holds*

$$\|\varphi - \kappa_N \varphi_N\|^2_{H^{-1/2}(\Gamma)} \lesssim \langle E_N u, u \rangle + \|u - \iota_N u_N\|^2_{H^{1/2}(\Gamma)} .$$

Proof Define $\hat{\varphi}_N \in \Phi_N$ as the solution of

$$\kappa_N^* V \kappa_N \hat{\varphi}_N = \kappa_N^* (I + K) u .$$

Then by Galerkin orthogonality and the boundedness of the Galerkin projection,

$$\begin{aligned}
\|\varphi - \kappa_N \varphi_N\|^2_{H^{-1/2}(\Gamma)} &\simeq \langle V(\varphi - \kappa_N \varphi_N), \varphi - \kappa_N \varphi_N \rangle = \|\varphi - \kappa_N \varphi_N\|^2_V \\
&\le 2\|\varphi - \kappa_N \hat{\varphi}_N\|^2_V + 2\|\kappa_N \hat{\varphi}_N - \kappa_N \varphi_N\|^2_V \\
&= 4\langle E_N u, u \rangle + 2\langle (I + K)(u - \iota_N u_N), \kappa_N (\kappa_N^* V \kappa_N)^{-1} \kappa_N^* (I + K)(u - \iota_N u_N) \rangle \\
&\lesssim \langle E_N u, u \rangle + \|u - \iota_N u_N\|^2_{H^{1/2}(\Gamma)} .
\end{aligned}$$

□

Without any regularity assumption for the solution (u, φ) of (π) we can show the following convergence result for the Galerkin BEM solution of (π_N) in the energy norm.

Theorem 11.6 *Let the solution (u, φ) of (π) given by (5.21) exist uniquely. Suppose that for the polygonal domain Ω there are only a finite number of "end points" $\overline{\Gamma}_C \cap \overline{\Gamma}_D, \overline{\Gamma}_D \cap \overline{\Gamma}_N, \overline{\Gamma}_N \cap \overline{\Gamma}_C$ and the gap function $\chi|\Gamma_C$ belongs to $H^{1/2+\varepsilon}(\Gamma_C)$ for some $\varepsilon > 0$. Then (u_N, φ_N) given by (11.47) are bounded in $H^{1/2}(\Gamma) \times H^{-\frac{1}{2}}(\Gamma)$ independently of N. Moreover for $N \to \infty$ with $\max_{e \in \mathscr{S}_N} h_{N,e} \to 0$ and with $p_{N,e} = p$ fixed for all $e \in \mathscr{S}_N$ or with $\min_{e \in \mathscr{S}_N} p_{N,e} \to \infty$ and with $h_{N,e} = h_e$ fixed for $e \in \mathscr{S}_N$ there holds $(u_N, \varphi_N) \to (u, \varphi)$ in $H^{1/2}(\Gamma) \times H^{-1/2}(\Gamma)$.*

Proof Since

$$\begin{aligned}
0 \le j_N(w_N) &\le \|g\|_{L^\infty(\Gamma_C)} \sum_{e \in \mathscr{E}_{c,N}} \sum_{j=0}^{p_{N,e}} \omega_j^{p_{N,e}+1} |w_N \circ F_E(\xi_j^{p_{N,e}+1})| \\
&\le 3\|g\|_{L^\infty(\Gamma_C)} \operatorname{meas}(\Gamma_C)^{1/2} \|w_N\|_{L^2(\Gamma_C)} \\
&\lesssim \|w_N\|_{H^{1/2}(\Gamma)} \quad \forall w_N \in \mathscr{S}_N ,
\end{aligned}$$

by the L^2 stability of Gauss Lobatto quadrature, see [48, Lemma 2.2], the indirect argument for the a priori bound of approximate solutions of a variational inequality of the first kind, which is given in the proof of Theorem C.7 in Sect. C.3.2 of the Appendix, extends to the boundary variational inequality (11.47) of the second kind.

Thus the assumed uniqueness of the solution (u, φ) of (π) implies the boundedness of (u_N, φ_N) in $H^{1/2}(\Gamma) \times H^{-\frac{1}{2}}(\Gamma)$.

To prove the claimed convergence assertions we adapt the discretization theory of Glowinski [199] to more general semicoercive variational inequalities of the second kind over a convex subset instead over the whole space, see Theorem C.7 in Sect. C.3.2 in the Appendix for variational inequalities of the first kind and also [221] for variational inequalities of the second kind. Thus we have to show the following hypotheses:

H1 If $v_N \rightharpoonup v$ in $H^{1/2}(\Gamma)$ for $N \to \infty$ with $v_N \in K_N^\Gamma$, then $v \in K$ and

$$\liminf_{N\to\infty} j_N(v_N) \geq j(v).$$

H2 There exists a subset $M \subset K^\Gamma$ dense in K^Γ and mappings $\varrho_N : M \to \Sigma_N$ such that, for each $w \in M$, $\varrho_N(w) \to w$ for $N \to \infty$,

$$\lim_{N\to\infty} j_N\big(\varrho_N(w)\big) = j(w),$$

and $\varrho_N(w) \in K_N^\Gamma$ for all $N \geq N_0(w)$ for some $N_0(w) \in \mathbb{N}$.

Note that the analogous hypotheses for the approximation of ψ by ψ_N are trivially satisfied in view of the inclusion $\Phi_N \subset H^{-1/2}(\Gamma)$.

The convergence of h-BEM, respectively h-FEM for the variational problem under study has been treated in [221], respectively in [215], where Newton-Cotes formulas in numerical quadrature are used instead of Gauss-Lobatto quadrature. Inspecting the proof of the respective convergence theorem shows that the norm convergence for a fixed polynomial and quadrature order hinges on the positiveness of the quadrature weights, what is satisfied for all quadrature orders with Gauss-Lobatto quadrature. For the proof of norm convergence (in the case $j = 0$) for $h \to 0$ with $p_{N,e} = p$ fixed for all $e \in \mathscr{S}_N$ we can also refer to [297]. Therefore in the following we can focus to the case where for all $e \in \mathscr{S}_N$, $h_{N,e} = h_e$ is fixed for all N and $\min_{e\in\mathscr{S}_N} p_{N,e} \to \infty$.

To verify H1 it is enough to show that for any $\lambda \in C^0(\Gamma)$ with $\lambda|\Gamma_C \geq 0$,

$$\int_{\Gamma_C} (v - \chi)\lambda \, ds \leq 0, \tag{11.49}$$

and to show that for any $\mu \in C^0(\Gamma)$ with $|\mu| \leq 1$ on Γ_C there holds

$$\int_{\Gamma_C} g \, v\mu \, ds \leq \lim_{N\to\infty} j_N(v_N), \tag{11.50}$$

since by duality with respect to (L^1, L^∞), see [151, chapter 4.3], and density

$$j(v) = \sup\Big\{ \int_{\Gamma_C} g\, v\mu\, ds | \mu \in C^0(\Gamma), |\mu| \le 1 \Big\}.$$

Moreover, since the partition $\mathscr{S}_N$ is here independent of N, we can simply consider the above integrals on any fixed edge $e \in \mathscr{S}_N$ such that $e \subset \Gamma_C$ with $g_e = g|e$ and $p := p_{N,e}$ instead. Thus fix $\lambda, \mu \in C^0[\overline{e}]$ with $\lambda \ge 0$, $|\mu| \le 1$ and as [297] approximate these functions by a combination of Bernstein polynomials B_N with the local mapping $F_e : [-1, 1] \to \overline{e}$ to define $\lambda_N := B_N\lambda \circ F_e$, $\mu_N := B_N\mu \circ F_e$ via

$$\lambda_N(t) = (B_N\lambda \circ F_e)(t) := \sum_{k=0}^{p} \binom{p}{k} (\frac{1+t}{2})^k (\frac{1-t}{2})^{p-k} (\lambda \circ F_e) \left(\frac{2k}{p} - 1\right).$$

Since the Bernstein operators are monotone, $\lambda_N \ge 0$ and $|\mu_N| \le 1$. By [142, Chapter 1, Theorem 2.3],

$$\lim_{p\to\infty} \|\lambda_N - \lambda\|_{L^\infty(e)} = \lim_{p\to\infty} \|\mu_N - \mu\|_{L^\infty(e)} = 0\,. \tag{11.51}$$

For the obstacle function $\chi \in H^{1/2+\varepsilon}(\Gamma_C)$ we use the interpolate $\chi_N := i_{e,p}\chi$ as approximation. By Theorem 11.5 with $r = 0$, $s = \frac{1}{2} + \varepsilon$

$$\lim_{N\to\infty} \|\chi_N - \chi\|_{L^2(e)} = 0. \tag{11.52}$$

Since the embedding $H^{1/2}(\Gamma) \hookrightarrow L^1(e)$ is weakly continuous, $v_N \rightharpoonup v$ in $L^1(e)$ and $\|v_N\|_{L^1(e)}$ is bounded. Therefore from

$$\Big| \int_e \big[(v_N - \chi_N)\lambda_{p_{N,e}-1} - (v-\chi)\lambda\big]\, dt \Big|$$

$$\le \|v_N - \chi_N\|_{L^1(e)} \|\lambda_{p-1} - \lambda\|_{L^\infty(e)} + \Big| \int_e \big[(v_N - \chi_N) - (v - \chi)\big]\lambda\, dt \Big|;$$

$$\Big| \int_e [v_N\mu_{p_{N,e}-1} - v\mu]\, dt \Big| \le \|v_N\|_{L^1(e)} \|\mu_{p-1} - \mu\|_{L^\infty(e)} + \Big| \int_e [v_N - v]\mu\, dt \Big|,$$

(11.51), (11.52) and using $\lambda, \mu \in L^\infty(e) = \big(L_1(e)\big)^*$, we conclude

$$\lim_{N\to\infty} \int_e (v_N - \chi_N)\lambda_{p_{N,e}-1}\,dt = \int_e (v - \chi)\lambda\,dt\,, \tag{11.53}$$

$$\lim_{N\to\infty} \int_e v_N \mu_{p_{N,e}-1} dt = \int_e v\mu\,dt\,. \tag{11.54}$$

On the other hand, $(v_N - \chi_N)\lambda_{p-1}$ and $v_N\mu_{p-1}$ are polynomials of degree $2p-1$. Hence the above integrals can be evaluated exactly by the Gauss-Lobatto quadrature formula to obtain

$$\int_e (v_N - \chi_N)\lambda_{p-1}\,dt = \sum_{j=0}^{p} \omega_j^{p+1}\big[(v_N - \chi_N)\lambda_{p-1}\big] \circ F_e(\xi_j^{p+1})\,,$$

$$\int_e v_N\mu_{p-1}\,dt = \sum_{j=0}^{p} \omega_j^{p+1}(v_N\mu_{p-1}) \circ F_e(\xi_j^{p+1})\,.$$

Since the weights $\omega_j^{p+1} > 0$, $\lambda_{p-1} \geq 0$, $(v_N - \chi_N) \circ F_e(\xi_j^{p+1}) \leq 0$ by $v_N \in K_N^\Gamma$, respectively $|\mu_{p-1}| \leq 1$, $g_e \geq 0$ we arrive at

$$\int_e (v_N - \chi_N)\lambda_{p-1}\,dt \leq 0,$$

$$g_e \int_e v_N\mu_{p-1}\,dt \leq g_e \sum_{j=0}^{p} \omega_j^{p+1}|v_N \circ F_e(\xi_j^{p+1})| =: j_{e,N}(v_N)\,,$$

$$\sum_{e\in\mathscr{E}_{c,N}} j_{e,N}(v_N) = j_N(v_N)\,.$$

In view of (11.53) (11.54) this proves our claim (11.49),(11.50).

In the last step let us prove H2. By the finiteness assumption we have due to [222, Lemma 3.3] the density relation

$$\overline{K^\Gamma \cap C^\infty(\Gamma)} = K^\Gamma\,.$$

Therefore we can take $M = K^\Gamma \cap C^\infty(\Gamma)$ and define $\varrho_N : M \to \Sigma_N$ by $\varrho_N := i_N$. Moreover, since $w \in M$ satisfies the constraints in K^Γ pointwise, $\varrho_N w \in K_N^\Gamma$ for all $w \in M$. By Theorem 11.5, $\varrho_N w \to w$ in $H^{1/2}(\Gamma)$. Finally by $j_N(w) = j(\varrho_N w)$,

$$\begin{aligned} |j(w) - j_N(\varrho_N w)| &\leq |j(w) - j(\varrho_N w)| + |j_N(w) - j_N(\varrho_N w)| \\ &\leq \|g\|_{L^\infty(\Gamma_C)}[\|w - \varrho_N w\|_{L^1(\Gamma_C)} + \|w - \varrho_N w\|_{L^\infty(\Gamma_C)}] \to 0 \quad (N \to \infty). \end{aligned}$$ □

11.3.2 A Céa-Falk Lemma for Variational Inequalities of the Second Kind

In this subsection we provide from [217] an abstract Céa-Falk approximation lemma for variational inequalities of the second kind, so that the above boundary variational inequality (5.22) is included as a special case.

Consider real normal vector spaces $(E, \|.\|_E)$, $(G, \|.\|_G)$ and their duals E^*, G^* such that $E \subset G$ continuously. Let E_N be a subspace of E $(N \in \mathbb{N})$ with the embedding $\iota_N \in \mathscr{L}(E_N, E)$. Let $C \subseteq E$ and $C_N \subseteq E_N$ be convex sets. Let $f^* \in E^*$ and for simplicity, $f_N^* = \iota_N^* f^*$. Let $B \in \mathscr{L}(E, E^*)$, $B_N \in \mathscr{L}(E_N, E_N^*)$ uniformly bounded and positive definite with respect to $\|, \|_E$; i.e. there exist some $\underline{c}_B, \bar{c}_B > 0$ (independent of E_N) such that

$$\begin{aligned} \underline{c}_B \|v\|_E^2 &\le \langle Bv, v\rangle &&\le \bar{c}_B \|v\|_E^2 \; (\forall\, v \in E), \\ \underline{c}_B \|v_N\|_E^2 &\le \langle B_N v_N, v_N\rangle &&\le \bar{c}_B \|v_N\|_E^2 \; (\forall\, v_N \in E_N). \end{aligned}$$

In addition following the analysis in [73], we assume an estimate of $\iota_N^* B \iota_N - B_N$ in the form

$$\langle (\iota_N^* B \iota_N - B_N) v_N, w_N\rangle \le c_0 \|w_N\|_E \Big(e_N(u) + \|u - v_N\|_E \Big) \tag{11.55}$$

for any $u \in E$; $v_N, w_N \in E_N$, where $c_0 > 0$ is independent of N and e_N is some function that describes the inconsistency and satisfies $e_N(u) \to 0$ as $N \to \infty$.

Lemma 11.2 *Let the preceding assumptions of this subsection be satisfied; let $u \in K$ and $u_N \in K_N$ such that for all $v \in K$, respectively for all $v_N \in K_N$,*

$$\langle Bu, v - u\rangle + j(v) - j(u) \ge \langle f^*, v - u\rangle\,, \tag{11.56}$$

$$\langle B_N u_N, v_N - u_N\rangle + j_N(v_N) - j_N(u_N) \ge \langle f_N^*, v_N - u_N\rangle\,. \tag{11.57}$$

Assume $Bu - f^* \in G^*$. Then there exists a constant $c > 0$ which depends on $\underline{c}_B, \bar{c}_B, c_0$ but not on N such that

$$\begin{aligned} c\|u - u_N\|_E^2 \le{}& \big[e_N(u)\big]^2 \\ &+ \inf_{v \in K} \Big\{ \|Bu - f^*\|_{G^*} \|u_N - v\|_G + |j_N(u_N) - j(v)| \Big\} \\ &+ \inf_{v_N \in K_N} \Big\{ \|u - v_N\|_E^2 + |j(u) - j_N(v_N)| \\ &+ \|Bu - f^*\|_{G^*} \|u - v_N\|_G \Big\}. \end{aligned}$$

Proof Estimate similar to [297][p.432/433]

$$\underline{c}_B \, \|\iota_N(u_N - v_N)\|_E^2 \leq \langle B_N(u_N - v_N), u_N - v_N \rangle$$

using $f_N^* = \iota_N^* f$,(11.56) , and (11.57) for any $v \in K$ and $v_N \in K_N$

$$\begin{aligned}
&\leq \langle Bu, v - \iota_N u_N \rangle + \langle Bu, \iota_N v_N - u \rangle \\
&\quad + \langle Bu, \iota_N u_N - \iota_N v_N \rangle + \langle B_N v_N, v_N - u_N \rangle \\
&\quad - \langle f^*, v - u \rangle - \langle f^*, \iota_N(v_N - u_N) \rangle \\
&\quad + j(v) - j(u) + j_N(v_N) - j_N(u_N) \\
&\leq \langle Bu - f^*, v - \iota_N u_N \rangle + \langle Bu - f^*, \iota_N v_N - u \rangle \\
&\quad + \langle B(u - \iota_N v_N), \iota_N u_N - \iota_N v_N \rangle \\
&\quad + \langle (\iota_N^* B \iota_N - B_N) v_N, u_N - v_N \rangle + J,
\end{aligned}$$

where we abbreviate

$$J := \Big| j(u) - j_N(v_N) | + | j_N(u_N) - j(v) \Big| .$$

Hence using

$$\langle (\iota_N^* B \iota_N - B_N) v_N, u_N - v_N \rangle \leq c_0 \Big\| \iota_N(u_N - v_N) \Big\|_E (e_N(u) + \|u - \iota_N v_N\|_E)$$

with arbitrary $\varepsilon > 0$

$$\begin{aligned}
&\underline{c}_B \, \Big\| \iota_N(u_N - v_N) \|_E^2 \\
&\leq \frac{1}{2} \overline{c}_B \left(\varepsilon \, \|\iota_N u_N - \iota_N v_N\|_E^2 + \frac{1}{\varepsilon} \, \|u - \iota_N v_N\|_E^2 \right) \\
&\quad + c_0 \, \varepsilon \, \|\iota_N u_N - \iota_N v_N\|_E^2 + \frac{c_0}{2} \frac{1}{\varepsilon} \Big[e_N(u) \Big]^2 \\
&\quad + \frac{c_0}{2} \frac{1}{\varepsilon} \, \|u - \iota_N v_N\|_E^2 + J \\
&\quad + \|Bu - f^*\|_{G^*} \Big(\|v - \iota_N u_N\|_G + \|\iota_N v_N - u\|_G \Big) .
\end{aligned}$$

Choose a suitable $\varepsilon > 0$ and again by Cauchy's inequality conclude for some $c = c(\underline{c}_B, \overline{c}_B, c_0) > 0$

$$\begin{aligned}
c \, \|u - \iota_N u_N\|_E^2 \leq \|u - \iota_N v_N\|_E^2 + \Big[e_N(u) \Big]^2 + J \\
+ \|Bu - f^*\|_{G^*} \Big(\|u - \iota_N v_N\|_G + \|v - \iota_N u_N\|_G \Big) .
\end{aligned}$$

□

11.3.3 A Priori Error Estimate for hp-Approximation

In this section we apply the Lemma 11.2 to obtain an a priori error estimate for the hp-approximate (u_N, φ_N) of the variational problem (π) under the realistic regularity assumptions of [297], in particular assuming $H^{3/2}(\Gamma)$ regularity of the solution u.

Theorem 11.7 *Let $(u, \varphi) \in K^\Gamma \times H^{-1/2}(\Gamma)$ be the unique solution of the problem (π) given by (5.21) with meas $(\Gamma_D) > 0$ and let $(u_N, \varphi_N) \in K_N^\Gamma \times \Phi_N$ be the solution of (11.47). Assume $u \in H^{3/2}(\Gamma)$, $\chi|\Gamma_C \in H^{3/2}(\Gamma_C)$, $\chi|\Gamma_D \in H^{3/2}(\Gamma_D)$, $f^* - Su \in L^2(\Gamma)$, where $f^*|\Gamma_N \cup \Gamma_C = f$ and $f^*|\Gamma_D = 0$. Then there exists $c = c(u, \chi, f^*, g) > 0$, independent of N such that for $N \to \infty$*

$$\left\|(u - u_N, \varphi - \varphi_N)\right\|_{H^{1/2}(\Gamma)\times H^{-1/2}(\Gamma)} \leq c \max_{e\in\mathscr{S}_N} h_{N,e}^{1/4}\, p_{N,e}^{-1/4} .$$

Proof Let us fix $N \in \mathbb{N}$ and write $h = \max_{e\in\mathscr{S}_N} h_{N,e}$, $p = \min_{e\in\mathscr{S}_N} p_{N,e}$ for short. First we prove the claimed error estimate for $\|u - u_N\|$ in $H^{1/2}(\Gamma)$. Since $u \in K^\Gamma$, respectively $u_N \in K_N^\Gamma$ satisfies

$$\langle Su, v - u\rangle + j(v) - j(u) \geq l(v - u) \equiv \int_\Gamma f^*(v - u)\, ds,\ \forall\, v \in K^\Gamma;$$

$$\langle S_N u_N, v_N - u_N\rangle + j_N(v_N) - j_N(u_N) \geq (l \circ \iota_N)\,(v_N - u_N),\ \forall\, v_N \in K_N^\Gamma,$$

respectively, we can apply the Lemma 11.2 to the boundary variational inequality (11.48) in the setting

$$E := H^{1/2}(\Gamma) \subset G := L^2(\Gamma),\ K := K^\Gamma,\ K_N := K_N^\Gamma,\ B := S,\ B_N := S_N .$$

Moreover, the nondifferentiable convex functionals j and j_N are given as previously by

$$j(v) = \int_{\Gamma_C} g|v|\, dt = \sum_{\substack{e\in\mathscr{S}_h\\ e\subset\Gamma_C}} g_e \int_e |v|\, dt ,$$

$$j_N(v) = \sum_{\substack{e\in\mathscr{S}_h\\ e\subset\Gamma_C}} g_e \sum_{j=0}^{p} \omega_j^{p+1} \left|v \circ F_e(\xi_j^{p+1})\right| .$$

Thus the Lemma 11.2 splits the error under study into three different error terms:

$$
\begin{aligned}
c\|u-u_N\|^2_{H^{1/2}(\Gamma)} \leq [e_N(u)]^2 + & \qquad\qquad (11.58)\\
\inf_{v\in K^\Gamma} & \Big\{\|Su-f^*\|_{L^2(\Gamma)}\|u_N-v\|_{L^2(\Gamma)} + |j_N(u_N)-j(v)|\Big\}\\
+ \inf_{v_N\in K_N^\Gamma} & \Big\{\|u-v_N\|^2_{H^{1/2}(\Gamma)} + |j(u)-j_N(v_N)|\\
+ \|Su-f^*\|_{L^2(\Gamma)}\|u-v_N\|_{L^2(\Gamma)}\Big\}.
\end{aligned}
$$

First we verify the inconsistency estimate (11.55) for $E_N := \iota_N^* S \iota_N - S_N$,

$$
e_N(u) := \operatorname{dist}_{H^{-1/2}(\Gamma)}(\varphi, \Phi_N) \inf_{\psi\in\Phi_N} = \|V^{-1}(K+I)u - \psi\|_{H^{-1/2}(\Gamma)},
$$

and bound the first term in (11.58). By the subsequent Lemma 11.3 we obtain

$$
\begin{aligned}
\langle E_N v_N, w_N\rangle &\leq \|E_N v_N\|_{H^{-1/2}(\Gamma)}\,\|w_N\|_{H^{1/2}(\Gamma)}\\
&\leq \big(c_0\, e_N(u) + \|E_N\|\,\|u-v_N\|_{H^{1/2}(\Gamma)}\big)\,\|w_N\|_{H^{1/2}(\Gamma)},
\end{aligned}
$$

where E_N is bounded independently of N. Moreover by [47, Theorem 6.1, Remark 6.1], [63, Theorem 4.4.20]

$$
e_N(u) \leq c_I \left(\frac{h}{p}\right)^{s-\frac{1}{2}} \|u\|_{H^s(\Gamma)}, \quad 1/2 < s \leq 3/2. \tag{11.59}
$$

To bound next the approximation error $\inf\{\ldots | v_N \in K_N^\Gamma\}$, which is the third term in (11.58), take $v_N = u_N^* := i_N u \in K_N^\Gamma$, the interpolate of $u \in H^{3/2}(\Gamma) \subset C^0(\Gamma)$. By Theorem 11.5, [63, Theorem 4.4.20] there are constants $c_1, c_2 > 0$ independent of u and N such that

$$
\begin{aligned}
\|u-u_N^*\|_{L^2(\Gamma)} &\leq c_1 \left(\frac{h}{p}\right)^{s} \|u\|_{H^s(\Gamma)}, \quad 1/2 < s \leq 3/2,\\
\|u-u_N^*\|_{H^1(\Gamma)} &\leq c_2 \left(\frac{h}{p}\right)^{s-1} \|u\|_{H^s(\Gamma)}, \quad 3/4 < s \leq 3/2,
\end{aligned}
$$

hence by real interpolation, see e.g. [373],

$$
\|u-u_N^*\|_{H^{1/2}(\Gamma)} \leq c_3 \frac{h}{p} \|u\|_{H^{3/2}(\Gamma)},
$$

Further, by construction,

$$\begin{aligned} j_N(u_N^*) &= \sum_{e\subset\Gamma_C} g_e \sum_{j=0}^{p} \omega_j^{p+1} \left|(u_N^* \circ F_e)(\xi_j^{p+1})\right| \\ &= \sum_{e\subset\Gamma_C} g_e \sum_{j=0}^{p} \omega_j^{p+1} \left|(u \circ F_e)(\xi_j^{p+1})\right| = j_N(u) \,. \end{aligned}$$

Hence using the interpolation operators $i_{e,p}$, i_N and the exactness of Gauss-Lobatto quadrature

$$\begin{aligned} \left|j(u) - j_N(u_N^*)\right| &\le \sum_{e\subset\Gamma_C} g_e \left| \int_e |u| ds - \int_e i_{e,p}\big(|u|\big) ds \right| \\ &\le \|g\|_{L^\infty(\Gamma_C)} \int_{\Gamma_C} \Big| |u| - i_N\big(|u|\big) \Big| ds \\ &\le \text{meas}\,(\Gamma_C)^{1/2} \|g\|_{L^\infty(\Gamma_C)} \Big\| |u| - i_N\big(|u|\big) \Big\|_{L^2(\Gamma_C)} \,. \end{aligned} \tag{11.60}$$

Since $|u|$ can only be guaranteed to lie in $H^1(\Gamma)$ (as the max of the two absolutely continuous functions u, $-u$, see also [267, Corollary A.6]) we can use the regularity assumption to derive only

$$\big\| |u| \big\|_{H^t(\Gamma)} \le \|u\|_{H^t(\Gamma)} \,, \quad 1/2 < t \le 1 \,. \tag{11.61}$$

what for $t < 1$ follows easily from the triangle inequality in the Sobolev-Slobodetskii norm. Therefore we can conclude by Theorem 11.5,[63, Theorem 4.4.20]

$$\begin{aligned} \left|j(u) - j_N(u_N^*)\right| &\le \tilde{c} \left(\frac{h}{p}\right)^t \big\| |u| \big\|_{H^t(\Gamma)} \\ &\le \tilde{c} \left(\frac{h}{p}\right)^t \|u\|_{H^t(\Gamma)} \,, \quad 1/2 < t \le 1 \,. \end{aligned}$$

Thus for the approximation error,

$$\inf\left\{ \ldots | v_N \in K_N^\Gamma \right\} \le c_{II}(u, f^*, g)\, \frac{h}{p}\, \|u\|_{H^{3/2}(\Gamma)} \,. \tag{11.62}$$

To bound the consistency error $\inf\left\{ \ldots | v \in K^\Gamma \right\}$, which is the second term in (11.58), let $\chi_N := i_N \chi$ and take (with min defined pointwise a.e.) $v^* = \min(u_N - \chi_N, 0) + \chi$ on Γ_C, $v^* = \gamma$ on Γ_D, $v^* = u_N$ on Γ_N. Clearly, $v^* \in K^\Gamma$. To show that

$v^* \in H^1(\Gamma)$ we use the arguments of [297]: For any $e \subset \overline{\Gamma_C \cup \Gamma_D}$, the polynomial $u_N - \chi_N \mid E$, which is of degree $p_{N,e} \le p$, has at most p zeros on e. Hence, the level set $\{x \in \Gamma_C \cup \Gamma_D \,:\, (u_N - \chi_N)(x) < 0\}$ is the finite union of open subintervals and $\min(u_N - \chi_N, 0)$ is continuous and piecewise a polynomial on $\Gamma_C \cup \Gamma_D$. Therefore $v^* \in H^1(\Gamma)$ as claimed.

Moreover, following [297] write $v^* - u_N = \min(u_N - \chi_N, 0) + \chi - u_N = \chi - \chi_N + \min(0, \chi_N - u_N) - 0$ on $\Gamma_C \cup \Gamma_D$. Hence

$$\|v^* - u_N\|_{L^2(\Gamma_C \cup \Gamma_D)} \le \|\chi - \chi_N\|_{L^2(\Gamma_C \cup \Gamma_D)} + \|\min(0, \chi_N - u_N) - 0\|_{L^2(\Gamma_C \cup \Gamma_D)} .$$

For the first term there holds

$$\|\chi - \chi_N\|_{L^2(\Gamma_C \cup \Gamma_D)} \le C_1 h^{\frac{1}{2}} p^{-\frac{1}{2}} \|\chi\|_{H^{\frac{1}{2}}(\Gamma_C \cup \Gamma_D)} .$$

By $u_N \in K_N^{\Gamma}$, $i_N \min(0, \chi_N - u_N) = 0$ and hence by Theorem 11.5 for the second term there holds

$$\begin{aligned}
&\|\min(0, \chi_N - u_N) - 0\|_{L^2(\Gamma_C \cup \Gamma_D)} \le \|\chi_N - u_N\|_{L^2(\Gamma_C \cup \Gamma_D)} \\
&\|\min(0, \chi_N - u_N) - 0\|_{L^2(\Gamma_C \cup \Gamma_D)} \le C_2 h^1 p^{-1} \|\min(0, \chi_N - u_N)\|_{H^1(\Gamma_C \cup \Gamma_D)} \\
&\le C_2 h^1 p^{-1} \|\chi_N - u_N\|_{H^1(\Gamma_C \cup \Gamma_D)} ,
\end{aligned}$$

thus by real interpolation,

$$\|\min(0, \chi_N - u_N) - 0\|_{L^2(\Gamma_C \cup \Gamma_D)} \le C h^{\frac{1}{2}} p^{-\frac{1}{2}} \|\chi_N - u_N\|_{H^{\frac{1}{2}}(\Gamma_C \cup \Gamma_D)} .$$

By Theorem 11.6 for some $C' > 0$, independent of h and p,

$$\|u_N\|_{H^{\frac{1}{2}}(\Gamma_C \cup \Gamma_D)} \le C' \|u\|_{H^{\frac{1}{2}}(\Gamma_C \cup \Gamma_D)} .$$

By definition of v^*, $\|v^* - u_N\|_{L^2(\Gamma_N)} = 0$. Hence we arrive at the estimate

$$\|v^* - u_N\|_{L^2(\Gamma)} \le C h^{1/2} p^{-1/2} \big(\|\chi\|_{H^{1/2}(\Gamma)} + \|u\|_{H^{1/2}(\Gamma)}\big) . \tag{11.63}$$

Further by construction,

$$\begin{aligned}
&v^*\big(F_e(\xi_j^{p+1})\big) = (u_N - \chi_N + \chi)\big(F_e(\xi_j^{p+1})\big) = u_N\big(F_e(\xi_j^{p+1})\big) , \\
&j_N(v^*) = j_N(u_N) .
\end{aligned}$$

On the other hand, by Theorem 11.5,

$$\begin{aligned}\|v^*\|_{H^1(\Gamma_C)} &\le \|u_N - \chi_N\|_{H^1(\Gamma_C)} + \|\chi\|_{H^1(\Gamma_C)}\\ &\le \|u_N\|_{H^1(\Gamma)} + 2\|\chi\|_{H^1(\Gamma_C)} + cp^{-1/2}\|\chi\|_{H^{3/2}(\Gamma_C)},\\ \|v^*\|_{L^2(\Gamma_C)} &\le \|u_N - \chi_N\|_{L^2(\Gamma_C)} + \|\chi\|_{L^2(\Gamma_C)}\\ &\le \|u_N\|_{L^2(\Gamma)} + 2\|\chi\|_{L^2(\Gamma_C)} + c' p^{-3/2}\|\chi\|_{H^{3/2}(\Gamma_C)}.\end{aligned}$$

Hence by real interpolation of the nonlocal H^t-norm and by the boundedness of $\|u_N\|$ in virtue of Theorem 11.6, v^* is bounded, too, in $H^{1/2}(\Gamma_C)$. Thus with similar arguments as above, see (11.60) and (11.61), we can conclude

$$\begin{aligned}\left|j(v^*) - j_N(u_N)\right| &\le \sum_{e\in\mathscr{E}_{c,N}} g_e\left|\int_e |v^*|ds - \int_e i_{e,p_{N,e}}\big(|v^*|\big)ds\right|\\ &\le \text{meas}\,(\Gamma_C)^{1/2}\|g\|_{L^\infty(\Gamma_C)}\Big\||v^*| - i_{E,p_{N,E}}\big(|v^*|\big)\Big\|_{L^2(\Gamma_C)}\\ &\le c\,h\,p^{-1}\|v^*\|_{H^1(\Gamma_C)}.\end{aligned}$$

To show the analogue estimate with respect to the L^2 norm we estimate separately:

$$j(v^*) = \sum_{e\in\mathscr{E}_{c,N}} g_e\int_e |v^*|ds \le \text{meas}\,(\Gamma_C)^{1/2}\|g\|_{L^\infty(\Gamma_C)}\|v^*\|_{L^2(\Gamma_C)}$$

and with Cauchy-Schwarz inequality

$$\begin{aligned}j_N(v^*) &= \sum_{e\in\mathscr{E}_{c,N}} g_e\sum_{j=0}^{p_{N,e}}\omega_j^{p_{N,E}+1}\left|u_N\circ F_E(\xi_j^{p_{N,e}+1})\right|\\ &\le 3\|g\|_{L^\infty(\Gamma_C)}\,\text{meas}\,(\Gamma_C)^{1/2}\,\|u_N\|_{(L^2(\Gamma_C))^2},\end{aligned}$$

where we use the L^2 stability of Gauss Lobatto quadrature, see [48, Lemma 2.2]. Therefore by real interpolation we arrive at

$$\left|j(v^*) - j_N(u_N)\right| \le c\,h^{1/2}\,p^{-1/2}\|v^*\|_{H^{1/2}(\Gamma_C)}.$$

Hence for the consistency error,

$$\inf\left\{\ldots|v\in K^\Gamma\right\} \le c_{III}(u,f^*,\chi,g)\,h^{1/2}\,p^{-1/2}. \tag{11.64}$$

Altogether, (11.58), (11.59), (11.62), and (11.64) yield the claimed estimate of the error $\|u - u_N\|$.

It remains to prove the claimed estimate of the error $\|\varphi - \varphi_N\|$ in $H^{-1/2}(\Gamma)$. This follows immediately from the proved estimate of the error $\|u - u_N\|$ in virtue of Lemma 11.1 and Lemma 11.3 using (11.59). □

Remark 11.2 The preceding proof is given in the case of dimension $d = 2$. It extends to $d > 2$ by tensor product approximation on a quadrilateral mesh.

In the proof above we used the following lemma, due to Maischak and Stephan, see [297, Lemma 15].

Lemma 11.3 *Let $E_N = \iota_N^* S \iota_N - S_N$. Then E_N is bounded independently of N and there holds*

$$\|E_N u\|_{H^{-1/2}(\Gamma)} \lesssim e_N(u) = \inf_{\psi \in \Phi_N} \|V^{-1}(K+I)u - \psi\|_{H^{-1/2}(\Gamma)},$$

$$0 \leq \langle E_N(u), u\rangle \lesssim e_N(u)^2 .$$

Proof Let $y := (I + K)u \in H^{1/2}(\Gamma)$. Since V is positive definite, there exists a solution $z \in H^{\frac{1}{2}}(\Gamma)$ of $Vz = y$ and a solution $z_N \in \Phi_N$ of $\kappa_N^* V \kappa_N z_N = \kappa_N^* y$, i.e., z_N is the Galerkin approximation of $z \in H^{\frac{1}{2}}(\Gamma)$ in Φ_N and we have

$$\|z\|_{H^{-1/2}(\Gamma)} \lesssim \|y\|_{H^{1/2}(\Gamma)} \text{ and } \|z_N\|_{H^{-1/2}(\Gamma)} \lesssim \|y\|_{H^{1/2}(\Gamma)} .$$

Then it results

$$\begin{aligned}\langle E_N(u), v\rangle &= \frac{1}{2}\langle z - \kappa_N z_N, (I+K)v\rangle \\ &\lesssim \|y\|_{H^{1/2}(\Gamma)}\|(I+K)v\|_{H^{1/2}(\Gamma)} \lesssim \|u\|_{H^{1/2}(\Gamma)}\|v\|_{H^{1/2}(\Gamma)},\end{aligned}$$

i.e., E_N is bounded independently of N. By Galerkin orthogonality we have

$$\begin{aligned}&2\langle E_N(u), v\rangle = \langle y, V^{-1}y\rangle - \langle y, \kappa_N(\kappa_N^* V \kappa_N)^{-1}\kappa_N^* y\rangle \\ &= \langle z, Vz\rangle - \langle \kappa_N z_N, V\kappa_N z_N\rangle = \|z\|_V^2 - \|\kappa_N z_N\|_V^2 = \|z - \kappa_N z_N\|_V^2 \geq 0,\end{aligned}$$

i.e., due to the quasi-optimal error estimate we have

$$0 \leq \langle E_N(u), u\rangle = \frac{1}{2}\|z - \kappa_N z_N\|_V^2 \lesssim e_N(u)^2 .$$

Using the boundedness of $(I + K)$ and the quasi-optimal error estimate we have

$$\|E_N u\|_{H^{-1/2}(\Gamma)} \leq \frac{1}{2}\|(I+K)\|\|z - \kappa_N z_N\|_{H^{-1/2}(\Gamma)} \lesssim e_N(u) .$$

Therefore, the assertion of the Lemma follows. □

The approach using the Steklov-Poincaré operator and the treatment of the resulting consistency error in our analysis opens the way to attack contact friction problems with nonlinear material behaviour by more efficient $hp-$ methods. For such problems $hp-$finite element methods can be employed in a small limited subdomain such that a coupling procedure permits to use the advantages of the boundary element method, specifically in the reduction of dimension.

Thus we anticipate that many of the details of our numerical analysis will be applicable to various interesting unilateral contact problems and other free boundary value problems.

11.4 Mixed hp-BEM for Frictional Contact Problems

11.4.1 Boundary Integral Formulation for Contact Problem

Now we report from [38] and consider the unilateral contact problem with Tresca friction. For a given gap function g, friction threshold $\mathscr{F} \geq 0$, elasticity tensor $\mathscr{C}$ and Neumann data f, find u such that

$$-\operatorname{div}\sigma(u) = 0 \qquad \text{in } \Omega \subset \mathbb{R}^d,\ d = 2 \text{ or } 3 \tag{11.65a}$$

$$\sigma(u) = \mathscr{C} : \epsilon(u) \qquad \text{in } \Omega \tag{11.65b}$$

$$u = 0 \qquad \text{on } \Gamma_D \tag{11.65c}$$

$$\sigma(u) \cdot n = f \qquad \text{on } \Gamma_N \tag{11.65d}$$

$$\sigma_n \leq 0,\ u_n \leq g,\ \sigma_n(u_n - g) = 0 \qquad \text{on } \Gamma_C \tag{11.65e}$$

$$|\sigma_t| \leq \mathscr{F}, \quad \sigma_t u_t + \mathscr{F}|u_t| = 0 \qquad \text{on } \Gamma_C \tag{11.65f}$$

Here, n, t denote the outer unit normal, tangential, respectively, and $\sigma_n = (\sigma(u)n)n \in \mathbb{R}$, $\sigma_t = (\sigma(u)n)t \in \mathbb{R}^{d-1}$, $u_n = u \cdot n \in \mathbb{R}$, $u_t = u \cdot t \in \mathbb{R}^{d-1}$ the normal, tangential component of $\sigma(u)n$ or of u. The friction constraint (11.65f) is equivalent to

$$|\sigma_t| \leq \mathscr{F}, \quad |\sigma_t| < \mathscr{F} \Rightarrow u_t = 0, \quad |\sigma_t| = \mathscr{F} \Rightarrow \exists\, \nu \geq 0 : u_t = -\nu\sigma_t$$

where we use $|\cdot|$ to abbreviate the Euclidean norm. We denote $\Gamma_\Sigma := \bar{\Gamma}_N \cup \bar{\Gamma}_C$ where $\partial\Omega = \bar{\Gamma}_N \cup \bar{\Gamma}_C \cup \bar{\Gamma}_D$ and $\bar{\Gamma}_C \cap \bar{\Gamma}_D = \emptyset$ for simplicity of the notation. The space dimension d may be two or three. Since the main source of difficulties, i.e. the Signorini (11.65e) and friction (11.65f) condition, lie on the boundary it maybe favorable to choose a boundary integral formulation with the symmetric Steklov-Poincaré operator $S : H^{\frac{1}{2}}(\Gamma) \to H^{-\frac{1}{2}}(\Gamma)$.

$$S := \frac{1}{2}\{W + (K' + I)V^{-1}(K + I)\}$$

Here, for $x \in \Gamma$

$$V\mu(x) = 2\int_\Gamma G(x,y)\mu(y)ds_y, \qquad Kv(x) = 2\int_\Gamma \left(T_y G(x,y)\right)^\top v(y)ds_y \tag{11.66}$$

$$K^\top \mu(x) = 2T_x\int_\Gamma G(x,y)\mu(y)ds_y, \qquad Wv(x) = -2T_x\int_\Gamma \left(T_y G(x,y)\right)^\top v(y)ds_y \tag{11.67}$$

with $T_n(u) := \sigma(u)|_\Gamma \cdot n$ denote the single layer potential, the double layer potential, its adjoint operator and the hypersingular operator with the fundamental solution of the Lamé equation

$$G(x,y) = \begin{cases} \frac{\lambda+3\mu}{4\pi\mu(\lambda+2\mu)}\left(\ln\frac{1}{|x-y|}I + \frac{\lambda+\mu}{\lambda+3\mu}\frac{(x-y)(x-y)^\top}{|x-y|^2}\right), & \text{if d=2} \\ \frac{\lambda+3\mu}{8\pi\mu(\lambda+2\mu)}\left(|x-y|^{-1}I + \frac{\lambda+\mu}{\lambda+3\mu}\frac{(x-y)(x-y)^\top}{|x-y|^3}\right), & \text{if d=3} \end{cases}$$

In [37] we consider the following variational inequality formulation on the boundary only, which is a reformulation of (11.65): Find $u \in K$ s.t.

$$\langle Su, v-u\rangle_{\Gamma_\Sigma} + j(v) - j(u) \geq \langle f, v-u\rangle_{\Gamma_N} \quad \forall\, v \in K \tag{11.68}$$

with friction functional $j(v) := \int_{\Gamma_C} \mathscr{F}\,|v_t|\; ds$ and convex cone of admissible functions $K := \left\{v \in \tilde{H}^{\frac{1}{2}}(\Gamma_\Sigma) : v_n \leq g \text{ a.e. on } \Gamma_C\right\}$. Equivalently, there holds a mixed formulation on the boundary only, in which the Lagrange multiplier $\lambda = -\sigma n|_{\Gamma_C}$ is sought as an additional unknown: Find $(u,\lambda) \in \tilde{H}^{\frac{1}{2}}(\Gamma_\Sigma) \times M^+(\mathscr{F})$ s.t.

$$\langle Su, v\rangle_{\Gamma_\Sigma} + \langle \lambda, v\rangle_{\Gamma_C} = \langle f, v\rangle_{\Gamma_N} \qquad \forall\, v \in \tilde{H}^{\frac{1}{2}}(\Gamma_\Sigma) \tag{11.69a}$$

$$\langle u, \mu-\lambda\rangle_{\Gamma_C} \leq \langle g, \mu_n - \lambda_n\rangle_{\Gamma_C} \quad \forall\, \mu \in M^+(\mathscr{F}) \tag{11.69b}$$

where

$$M^+(\mathscr{F}) := \left\{\mu \in \tilde{H}^{-\frac{1}{2}}(\Gamma_C) : \langle \mu, v\rangle_{\Gamma_C} \leq \langle \mathscr{F}, |v_t|\rangle_{\Gamma_C}\, \forall v \in H^{\frac{1}{2}}(\Gamma_C),\ v_n \leq 0\right\}$$

is the set of admissible Lagrange multipliers. The connection between these three formulations is as follows:

Theorem 11.8 ([37]) *The problems* (11.68) *and* (11.69) *are equivalent and are also equivalent to* (11.65) *in a distributional sense. Furthermore, there exists exactly one solution to* (11.68) *and* (11.69). *This means:*

(i) Any solution of (11.69) *is also a solution of* (11.68).
(ii) For the solution $u \in K$ *of* (11.68) *there exists a* $\lambda \in M^+(\mathscr{F})$ *such that* (u, λ) *is a solution of* (11.69).
(iii) There exists a unique solution to (11.69).

11.4.2 hp-Boundary Element Procedure with Lagrange Multiplier and Fast Solver

To set up the discrete problem we firstly introduce biorthogonal *hp*-BE-spaces, namely: The approximation of the primal variable u is sought in

$$V_{hp} = \Big\{ v_{hp} \in \tilde{H}^{\frac{1}{2}}(\Gamma_\Sigma) \cap C^0(\Gamma_\Sigma) : v_{hp}|_E \circ \Psi_E \in \Big[\mathbb{P}_{p_E}([-1,1]^{d-1})\Big]^d \;\; \forall E \in \mathscr{T}_h \cap \Gamma_\Sigma,$$
$$v_{hp} = 0 \text{ in the endpoints of } \Gamma_\Sigma \Big\}$$

for a given boundary mesh $\mathscr{T}_h$ of Γ, consisting of line segments in 2D and quadrilaterals in 3D, in which the nodes, edges coincide with the boundary of the boundary parts Γ_D, Γ_N and Γ_C, and with polynomial degree distribution $p : \mathscr{T}_h \to \mathbb{N}$. Here, $\Psi_E : [-1,1]^{d-1} \to E \in \mathscr{T}_h$ is the affine mapping onto the physical element. The dual, Lagrange multiplier variable λ_{hp} is sought in

$$M^+_{hp}(\mathscr{F}) = \left\{ \mu_{hp} = \sum_{i=1}^{N_C} \boldsymbol{\mu}_i \psi_i : \langle \mu_{hp}, v_{hp} \rangle_{\Gamma_C} \le \big\langle \mathscr{F}, |(v_{hp})_t|_h \big\rangle \; \forall v_{hp} \in V_{hp}, (\mathbf{v}_i)_n \le 0 \right\}$$

Here, we take $|(v_{hp})_t|_h := \sum_i |(\mathbf{v}_i)_t| \phi_i$ where $v_{hp} = \sum_{i=1}^{N} \mathbf{v}_i \phi_i$, $\mathbf{v}_i \in \mathbb{R}^d$ and where ϕ_i denotes a scalar Gauss-Lobatto-Lagrange basis function (for u_{hp}) and ψ_j a biorthogonal basis function (for λ_{hp}), see Fig. 11.2, i.e.

$$\int_{\Gamma_C} \phi_i \psi_j \, ds = \delta_{ij} \int_{\Gamma_C} \phi_i \, ds \quad 1 \le i, j \le N_C.$$

In particular, we assume that the first N_C basis functions are associated with a Gauss-Lobatto point which lies on Γ_C. Furthermore, $(\mathbf{v}_i)_n := \mathbf{v}_i^\top \mathbf{n}$, $(\boldsymbol{\mu}_i)_n := \boldsymbol{\mu}_i^\top \mathbf{n}$ and $(\mathbf{v}_i)_t := \mathbf{v}_i - (\mathbf{v}_i)_n \mathbf{n}$, $(\boldsymbol{\mu}_i)_t := \boldsymbol{\mu}_i - (\boldsymbol{\mu}_i)_n \mathbf{n}$ are the normal, tangential components of the expansion coefficients $\boldsymbol{v}$, $\boldsymbol{\mu}$. Note that we use the same mesh and polynomial degree distribution for $u_{hp}|_{\Gamma_C}$ and for λ_{hp}. Due to the biorthogonality, the discrete inf-sup condition is still satisfied [36].

(a)

(b)

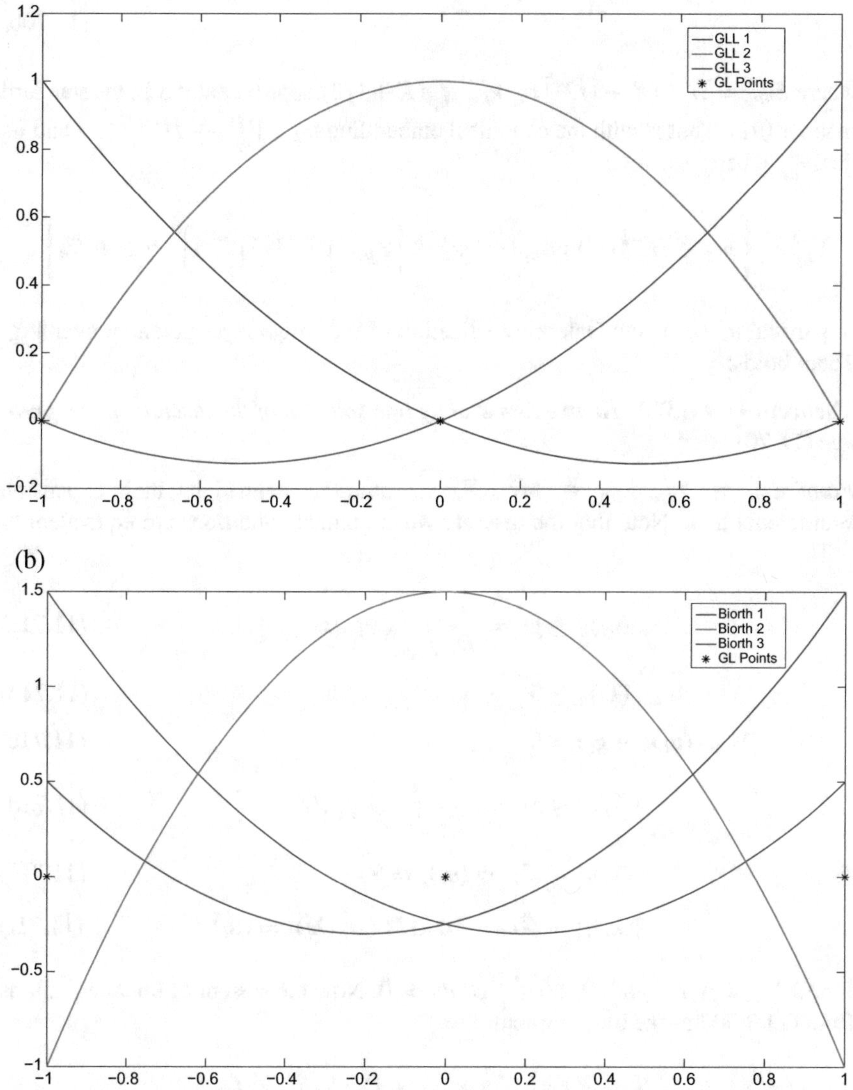

Fig. 11.2 Basisfunctions for $p = 2$ [37]. (**a**) Gauss-Lobatto-Lagrange. (**b**) Biorthogonal

In [37] we analyse the following mixed *hp*-boundary element formulation: Find $(u_{hp}, \lambda_{hp}) \in V_{hp} \times M^+_{hp}(\mathscr{F})$ such that

$$\langle S_{hp} u_{hp}, v_{hp} \rangle_{\Gamma_\Sigma} + \big\langle \lambda_{hp}, v_{hp} \big\rangle_{\Gamma_C} = \langle f, v_{hp} \rangle_{\Gamma_N} \qquad \forall v_{hp} \in V_{hp} \tag{11.70a}$$

$$\langle u_{hp}, \mu_{hp} - \lambda_{hp}\rangle_{\Gamma_C} \leq \langle g, (\mu_{hp})_n - (\lambda_{hp})_n\rangle_{\Gamma_C} \ \forall\, \mu_{hp} \in M_{hp}^{+}(\mathscr{F}) \tag{11.70b}$$

where $S_{hp} := W + (K+1/2)^\top i_{hp} \mathscr{V}_{hp}^{-1} i_{hp}^*(K+1/2)$ approximates S in the standard manner [91]. That is with the canonical embedding $i_{hp} : V_{hp}^D \mapsto H^{-1/2}(\Gamma)$ and its dual i_{hp}^* where

$$V_{hp}^D = \left\{ \phi_{hp} \in H^{-\frac{1}{2}}(\Gamma) : \phi_{hp}|_E \circ \Psi_E \in \left[\mathbb{P}_{p_E-1}([-1,1]^{d-1})\right]^d \ \forall\, E \in \mathscr{T}_h \right\}.$$

In particular, $\mathscr{V}_{hp}$ is the Galerkin realization of the single layer potential over V_{hp}^D. There holds:

Theorem 11.9 ([37]) *There exists exactly one solution of the discrete mixed problem* (11.70).

Proof $u_{hp} \in \mathscr{V}_{hp}$, $\lambda_{hp} \in M_{hp}^{+}(\mathscr{F})$ are uniquely defined by their expansion coefficients $\boldsymbol{u}$, $\boldsymbol{\lambda}$. Note that the discrete weak contact conditions are equivalent to

$$(\mathbf{u}_i)_n \leq g_i := \frac{1}{D_i} \int_{\Gamma_C} g\psi_i \, ds \tag{11.71a}$$

$$(\boldsymbol{\lambda}_i)_n \geq 0 \tag{11.71b}$$

$$(\boldsymbol{\lambda}_i)_n \left((\mathbf{u}_i)_n - g_i\right) = 0 \tag{11.71c}$$

$$|(\boldsymbol{\lambda}_i)_t| \leq \mathscr{F}_i := \frac{1}{D_i} \int_{\Gamma_C} \mathscr{F}\phi_i \, ds \tag{11.71d}$$

$$|(\boldsymbol{\lambda}_i)_t| < \mathscr{F}_i \Rightarrow (\mathbf{u}_i)_t = 0 \tag{11.71e}$$

$$|(\boldsymbol{\lambda}_i)_t| = \mathscr{F}_i \Rightarrow \exists\, \alpha \in \mathbb{R} : \alpha^2 (\boldsymbol{\lambda}_i)_t = (\mathbf{u}_i)_t \tag{11.71f}$$

for all $1 \leq i \leq N_C$ and $D_i := \int_{\Gamma_C} \phi_i \, ds > 0$. Now the system of linear equations from (11.70a) has the block structure

$$\begin{pmatrix} \boldsymbol{S}_{\mathscr{C}\mathscr{C}} & \boldsymbol{S}_{\mathscr{C}\mathscr{R}} \\ \boldsymbol{S}_{\mathscr{R}\mathscr{C}} & \boldsymbol{S}_{\mathscr{R}\mathscr{R}} \end{pmatrix} \begin{pmatrix} \boldsymbol{u}_{\mathscr{C}} \\ \boldsymbol{u}_{\mathscr{R}} \end{pmatrix} + \begin{pmatrix} \boldsymbol{D}_{\mathscr{C}} \\ \boldsymbol{0} \end{pmatrix} \boldsymbol{\lambda} = \begin{pmatrix} \boldsymbol{f}_{\mathscr{C}} \\ \boldsymbol{f}_{\mathscr{R}} \end{pmatrix}.$$

Let $\mathscr{P}_{\geq 0}$, $\mathscr{P}_{B(\mathscr{F}_i)}$ be closest point projections onto $\mathbb{R}_{\geq 0}$, $B(\mathscr{F}_i)$, a ball with center zero and radius $\mathscr{F}_i$, respectively. Let

$$T(\boldsymbol{\lambda}) := \left(\mathscr{P}_{\geq 0}\left((\boldsymbol{\lambda}_i)_n + r((\mathbf{u}_i)_n - g_i)\right), \mathscr{P}_{B(\mathscr{F}_i)}\left((\boldsymbol{\lambda}_i)_t + r(\mathbf{u}_i)_t\right)\right)_{i=1}^{N_C}.$$

Then

$$\|T(\boldsymbol{\lambda}_1) - T(\boldsymbol{\lambda}_2)\|_2^2 \leq \|\boldsymbol{\lambda}_1 - \boldsymbol{\lambda}_2\|_2^2 \left(1 - 2\alpha r\gamma^2 + r^2\gamma^2\right) < \|\boldsymbol{\lambda}_1 - \boldsymbol{\lambda}_2\|_2^2$$

for $r = \alpha$ where $\alpha = \lambda_{\min}\left(\boldsymbol{D}_{\mathscr{C}}^{-1}\left(\boldsymbol{S}_{\mathscr{C}\mathscr{C}} - \boldsymbol{S}_{\mathscr{C}\mathscr{R}}\boldsymbol{S}_{\mathscr{R}\mathscr{R}}^{-1}\boldsymbol{S}_{\mathscr{R}\mathscr{C}}\right)\right) > 0$ is the minimal eigenvalue, $\gamma := \|\delta\boldsymbol{u}_{\mathscr{C}}\|_2 \,/\, \|\delta\boldsymbol{\lambda}\|_2$. Thus T is a strict contraction and Banach's fixed point theorem yields the assertion for λ_{hp}. Since S_{hp} is $\tilde{H}^{1/2}(\Gamma_\Sigma)$-coercive, u_{hp} exists and is unique as well. □

An important benefit of this approach with biorthogonal basis functions is the componentwise decoupling of the weak contact conditions [37]. For the non-linear complementarity functions (NCF)

$$\begin{aligned}\varphi_\eta : \mathbb{R}^{d\cdot N} \times \mathbb{R}^{d\cdot N_C} \to \mathbb{R}^{N_C} \text{ with } \left(\varphi_\eta(\boldsymbol{u}, \boldsymbol{\lambda})\right)_i &= \eta\Big((\boldsymbol{\lambda}_i)_n + (g_i - (\mathbf{u}_i)_n) \\ &\quad - \sqrt{(\boldsymbol{\lambda}_i)_n^2 + (g_i - (\mathbf{u}_i)_n)^2}\Big) + (1-\eta)\,(\boldsymbol{\lambda}_i)_n^+\,(g_i - (\mathbf{u}_i)_n)^+\end{aligned}$$

for the Signorini condition and for $c_t > 0$

$$\begin{aligned}C_T : \mathbb{R}^{d\cdot N}\times\mathbb{R}^{d\cdot N_C} \to \mathbb{R}^{(d-1)\cdot N_C} \text{with } (C_T(\boldsymbol{u}, \boldsymbol{\lambda}))_i &= \max\{\mathscr{F}_i, |(\boldsymbol{\lambda}_i)_t + c_t(\mathbf{u}_i)_t|\}(\boldsymbol{\lambda}_i)_t \\ &\quad - \mathscr{F}_i\cdot((\boldsymbol{\lambda}_i)_t + c_t(\mathbf{u}_i)_t)\,,\ 1 \leq i \leq N_C\end{aligned}$$

for the Tresca condition, there holds: $\varphi_\eta(\boldsymbol{u}, \boldsymbol{\lambda}) = 0$ if and only if (11.71a)–(11.71c) hold, and $C_T(\boldsymbol{u}, \boldsymbol{\lambda}) = 0$ if and only if (11.71d)–(11.71f) hold. Therewith, the discrete mixed problem (11.70) is equivalent to

$$0 = F(\boldsymbol{u}, \boldsymbol{\lambda}) = \begin{pmatrix} \boldsymbol{S}\boldsymbol{u} + \boldsymbol{D}\boldsymbol{\lambda} - \boldsymbol{f} \\ \varphi_\eta(\boldsymbol{u}, \boldsymbol{\lambda}) \\ C_T(\boldsymbol{u}, \boldsymbol{\lambda}) \end{pmatrix} \tag{11.72}$$

where $\boldsymbol{S}\boldsymbol{u}+\boldsymbol{D}\boldsymbol{\lambda}-\boldsymbol{f} = 0$ is matrix representation of the variational equality (11.70a). As shown in [37] a fast solver for (11.72) is the semi-smooth Newton (SSN) method which converges locally super-linearly and even locally quadratically in the frictionless case, i.e, when $\mathscr{F} \equiv 0$.

11.4.3 Error Controlled hp-Adaptive Schemes

In [37] we derive the following a posteriori error estimate with residual type error indicators.

Theorem 11.10 *Let* (u, λ), (u_{hp}, λ_{hp}) *be the solution of* (11.69), (11.70) *respectively. Then there exists a constant* $C > 0$, *independent of h and p s.t.*

$$\left\|\lambda_{hp} - \lambda\right\|^2_{\tilde{H}^{-\frac{1}{2}}(\Gamma_C)} + \left\|u - u_{hp}\right\|^2_{\tilde{H}^{\frac{1}{2}}(\Gamma_\Sigma)} \leq C \sum_{E \in \mathscr{T}_h} \eta^2_{res}(E) + \textit{contact terms}$$

with $\psi_{hp} := i_{hp} V_{hp}^{-1} i^*_{hp}(K + \frac{1}{2}) u_{hp}$

$$\eta^2_{res}(E) := \frac{h_E}{p_E} \left\|f - S_{hp} u_{hp}\right\|^2_{L^2(E \cap \Gamma_N)} + \frac{h_E}{p_E} \left\|\lambda_{hp} + S_{hp} u_{hp}\right\|^2_{L^2(E \cap \Gamma_C)}$$
$$+ h_E \left\| \frac{\partial}{\partial s}(V \psi_{hp} - (K + I) u_{hp}) \right\|^2_{L^2(E)}$$

where the contact terms (resulting from the violation of the contact condition by the discrete solution of (11.70)*) are*

$$\left\|(\lambda_{hp})^-_n\right\|^2_{\tilde{H}^{-\frac{1}{2}}(\Gamma_C)} + \left\langle(\lambda_{hp})^+_n, (g - (u_{hp})_n)^+\right\rangle_{\Gamma_C} + \left\|(g - (u_{hp})_n)^-\right\|^2_{H^{\frac{1}{2}}(\Gamma_C)}$$
$$+ \left\|(\left|(\lambda_{hp})_t\right| - \mathscr{F})^+\right\|^2_{\tilde{H}^{-\frac{1}{2}}(\Gamma_C)} - \int_{\Gamma_C} (\left|(\lambda_{hp})_t\right| - \mathscr{F})^- \left|(u_{hp})_t\right|$$
$$+ (\lambda_{hp})_t (u_{hp})_t - \left|(\lambda_{hp})_t\right| \left|(u_{hp})_t\right| \; ds \;\; \textit{with } v^+ = \max\{v, 0\}, \;\; v^- = \min\{v, 0\}.$$

These contact terms can be interpreted as: violation of the consistency, complementarity and non-penetration condition with respect to the normal component of the solution, and violation of the consistency, violation of the stick condition and of having the same the sign in slip condition wrt. the tangential component of the solution.

Proof The starting point is to consider the auxiliary problem (Braess' trick [59]): Let $z \in \tilde{H}^{1/2}(\Gamma_\Sigma)$ such that

$$\langle Sz, v\rangle_{\Gamma_\Sigma} = \langle f, v\rangle_{\Gamma_N} - \left\langle\lambda_{hp}, v\right\rangle_{\Gamma_C} \quad \forall v \in \tilde{H}^{1/2}(\Gamma_\Sigma) \tag{11.73}$$

for which u_{hp} is the Galerkin approximation. The ellipticity and continuity of S gives

$$\alpha \left\|u_{hp} - u\right\|^2_{\tilde{H}^{1/2}(\Gamma_\Sigma)} \leq C \left\|u_{hp} - z\right\|_{\tilde{H}^{1/2}(\Gamma_\Sigma)} \left\|u_{hp} - u\right\|_{\tilde{H}^{1/2}(\Gamma_\Sigma)}$$
$$+ \left\langle\lambda_n - \left(\lambda_{hp}\right)_n, \left(u_{hp}\right)_n - u_n\right\rangle_{\Gamma_C} + \left\langle\lambda_t - \left(\lambda_{hp}\right)_t, \left(u_{hp}\right)_t - u_t\right\rangle_{\Gamma_C}.$$

Furthermore

$$
\begin{aligned}
&\langle \lambda_n - (\lambda_{hp})_n , (u_{hp})_n - u_n \rangle_{\Gamma_C} \\
&\leq \left\langle (\lambda_{hp})_n^+ , (g - (u_{hp})_n)^+ \right\rangle_{\Gamma_C} + \left\| (\lambda_{hp})_n - \lambda_n \right\|_{\tilde{H}^{-1/2}(\Gamma_C)} \left\| (g - (u_{hp})_n)^- \right\|_{H^{1/2}(\Gamma_C)} \\
&\quad + \left\| (\lambda_{hp})_n^- \right\|_{\tilde{H}^{-1/2}(\Gamma_C)} \left\| (u_{hp})_n - u_n \right\|_{H^{1/2}(\Gamma_C)}
\end{aligned}
$$

whereas

$$
\begin{aligned}
&\langle \lambda_t - (\lambda_{hp})_t , (u_{hp})_t - u_t \rangle_{\Gamma_C} \\
&\leq \left\| \left(\|(\lambda_{hp})_t\|_2 - \mathscr{F} \right)^+ \right\|_{\tilde{H}^{-1/2}(\Gamma_C)} \left\| \|u_t - (u_{hp})_t\|_2 \right\|_{H^{1/2}(\Gamma_C)} \\
&\quad - \left\langle \left(\|(\lambda_{hp})_t\|_2 - \mathscr{F} \right)^- , \|(u_{hp})_t\|_2 \right\rangle_{\Gamma_C} \\
&\quad - \langle (\lambda_{hp})_t , (u_{hp})_t \rangle_{\Gamma_C} + \left\langle \|(\lambda_{hp})_t\|_2 , \|(u_{hp})_t\|_2 \right\rangle_{\Gamma_C} .
\end{aligned}
$$

From the continuous inf-sup condition it follows

$$
\|\lambda_{hp} - \lambda\|^2_{\tilde{H}^{-1/2}(\Gamma_C)} \leq \frac{2C^2}{\beta^2} \|u - u_{hp}\|^2_{\tilde{H}^{1/2}(\Gamma_\Sigma)} + \frac{2C^2}{\beta^2} \|u_{hp} - z\|^2_{\tilde{H}^{1/2}(\Gamma_\Sigma)}
$$

yielding

$$
\begin{aligned}
&\|\lambda_{hp} - \lambda\|^2_{\tilde{H}^{-1/2}(\Gamma_C)} + \|u_{hp} - u\|^2_{\tilde{H}^{1/2}(\Gamma_\Sigma)} \\
&\leq C \Bigg(\|u_{hp} - z\|^2_{\tilde{H}^{1/2}(\Gamma_\Sigma)} + \left\| (\lambda_{hp})_n^- \right\|^2_{\tilde{H}^{-1/2}(\Gamma_C)} + \left\| \left(\left\|(\lambda_{hp})_t\right\|_2 - \mathscr{F} \right)^+ \right\|^2_{\tilde{H}^{-1/2}(\Gamma_C)} \\
&\quad - \left\langle \left(\left\|(\lambda_{hp})_t\right\|_2 - \mathscr{F} \right)^- , \left\|(u_{hp})_t\right\|_2 \right\rangle_{\Gamma_C} - \left\langle (\lambda_{hp})_t , (u_{hp})_t \right\rangle_{\Gamma_C} \\
&\quad + \left\langle \left\|(\lambda_{hp})_t\right\|_2 , \left\|(u_{hp})_t\right\|_2 \right\rangle_{\Gamma_C} + \left\langle (\lambda_{hp})_n^+ , \left(g - (u_{hp})_n \right)^+ \right\rangle_{\Gamma_C} \\
&\quad + \left\| \left(g - (u_{hp})_n \right)^- \right\|^2_{H^{1/2}(\Gamma_C)} \Bigg) .
\end{aligned}
$$

It remains to estimate $\|u_{hp} - z\|^2_{\tilde{H}^{1/2}(\Gamma_\Sigma)} \leq C \sum_{E \in \mathscr{T}_h} \eta^2_{res}(E)$ with the local contributions

$$
\eta^2_{res}(E) := \frac{h_E}{p_E} \|\tilde{t} - S_{hp} u_{hp}\|^2_{L^2(E \cap \Gamma_\Sigma)} + h_E \left\| \frac{\partial}{\partial s} (V \psi_{hp} - (K + I) u_{hp}) \right\|^2_{L^2(E)}
$$

where $\tilde{t}|_{\Gamma_N} = f$ and $\tilde{t}|_{\Gamma_C} = -\lambda_{hp}$. $\psi_{hp} := i_{hp} V_{hp}^{-1} i_{hp}^{*}(K + I)u_{hp}$ is a natural side product when estimating the error induced by the approximation of V^{-1}. □

In [37] we replace $\mathscr{F}$ with $\mathscr{F}\lambda_n$, $\mathscr{F}(\lambda_{hp})_n$ in the continuous and in the discrete cases, respectively, as modifications for Coulomb friction. There we show that the decoupling of contact constraints still holds with $\mathscr{F}_i := \frac{1}{D_i}\int_{\Gamma_C} \mathscr{F}(\lambda_{hp})_n \phi_i ds$ and $\mathscr{F}_i = \mathscr{F}(\lambda_i)_n$ for constant $\mathscr{F}$. Furthermore, if $\|\mathscr{F}\|_{L^\infty(\Gamma_C)}$ is sufficiently small, the frictional part of the contact terms in Theorem 11.10 changes to

$$\left\|\left(\left|\left(\lambda_{hp}\right)_t\right| - \mathscr{F}(\lambda_{hp})_n\right)^+\right\|_{\tilde{H}^{-1/2}(\Gamma_C)} - \left\langle\left(\left|\left(\lambda_{hp}\right)_t\right| - \mathscr{F}(\lambda_{hp})_n\right)^-, \left|\left(u_{hp}\right)_t\right|\right\rangle_{\Gamma_C}$$
$$- \left\langle\left(\lambda_{hp}\right)_t, \left(u_{hp}\right)_t\right\rangle_{\Gamma_C} + \left\langle\left|\left(\lambda_{hp}\right)_t\right|, \left|\left(u_{hp}\right)_t\right|\right\rangle_{\Gamma_C}$$

Alternatively, we can take a bubble error estimate instead of a residual estimate for the variational equality part, but then the saturation assumption

$$\kappa \in (0,1): \ \left\|u_{hp+1} - z\right\|_W^2 + \left\|\psi - \psi_{hp+1}\right\|_V^2 \leq \kappa^2 \left\|u_{hp} - z\right\|_W^2 + \left\|\psi - \psi_{hp}\right\|_V^2$$

must hold. For details see [37] where the following algorithm is performed.

Algorithm 11.1 *(Solve-mark-refine algorithm for hp-adaptivity)*

(i) Choose initial discretization $\mathscr{T}_h$ and p, steering parameters $\theta \in (0, 1)$ and $\delta \in (0, 1)$.

(ii) For $k = 0, 1, 2, \ldots$ do

a. solve discrete mixed problem (11.70).

b. compute local indicators Ξ^2 to current solution.

c. mark all elements $E \in \mathscr{N} := \operatorname{argmin}\left\{\left|\left\{\hat{\mathscr{N}} \subset \mathscr{T}_h : \sum_{E\in\hat{\mathscr{N}}} \Xi^2(E) \geq \theta \sum_{E\in\mathscr{T}_h} \Xi^2(E)\right\}\right|\right\}$ for refinement

d. estimate local analyticity [254], i.e. compute Legendre coefficients of

$$u_{hp}|_E(\Psi_E(x)) = \sum_{j=0}^{p} a_i L_i(x), \quad a_i = \frac{2i+1}{2}\int_{-1}^{1} u_{hp}|_E(\Psi_E(x))L_i(x)\,dx$$

Use a least square approach to compute the slop m of $|\ln|a_i|| = mi + b$. If $e^{-m} \leq \delta$ then p-refine, else h-refine marked element E. (This can be done analogously in higher dimensions [36]).

e. refine marked elements based on the decision in 2(d).

Example 11.1 We take a disc with diameter one, i.e. $\Gamma = \left\{x \in \mathbb{R}^2 : |x| = \frac{1}{2}\right\}$. The boundary is split into $\Gamma_N = \Gamma \cap \left\{x \in \mathbb{R}^2 : x_2 \geq 0\right\}$ and $\Gamma_C = \Gamma \setminus \Gamma_N$. The rigid

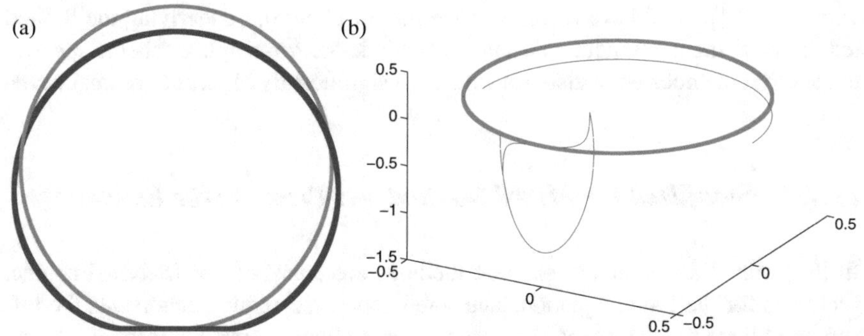

Fig. 11.3 Solution of the Hertz problem with Coulomb-friction, uniform mesh 2048 elements, $p = 2$, [37]. (**a**) Reference (gray), deformed (blue). (**b**) λ_y and t_y (blue), λ_x (red)

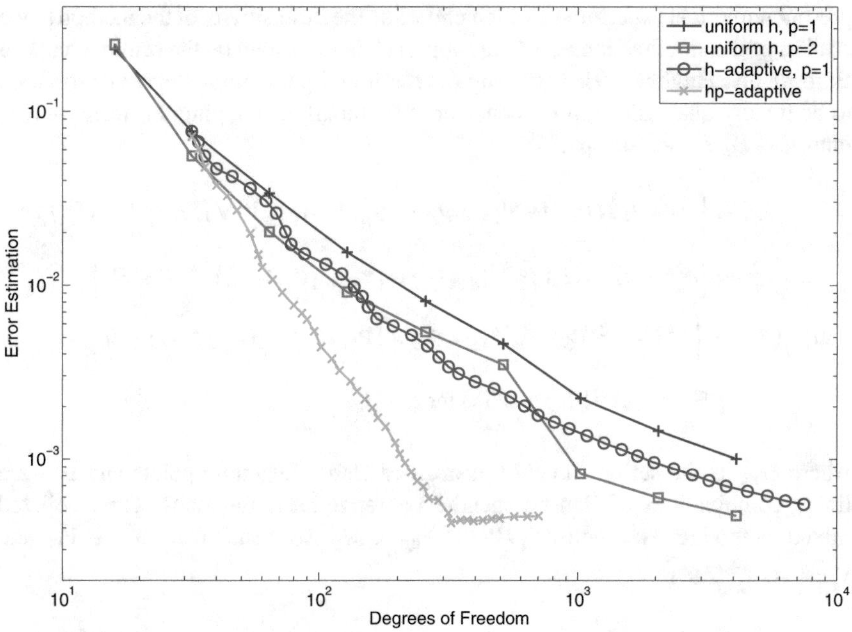

Fig. 11.4 Bubble error estimates for different families of discrete solutions (Hertz) [37]

body motions are set to zero by the iterative solver to obtain a unique solution. The Young's modul is $E = 5$ and the Poisson's ratio is $\nu = 0.45$. The Coulomb friction coefficient is $\mathscr{F} \equiv 1$, the Neumann force is $f = (0, -0.2)^\top$ and the gap is $\mathrm{dist}(\Gamma_C, -0.5)$. Figure 11.3a visualizes the reference configuration in grey and the deformed state in blue. The corresponding Neumann (λ on Γ_C) data are depicted in Fig. 11.3b. The reduction of the bubble error indicator for different families of discrete solutions is displayed in Fig. 11.4. In particular, the uniform h-versions with $p = 1$, $p = 2$ have a convergence rate of about 0.5 underlying the limited regularity of the continuous solution. Not only is the hp-adaptive scheme superior to the other methods but it also has with 2.4 a significantly higher convergence rate.

11.4.4 Stabilized hp-Mixed Method—A Priori Error Estimate

In [33] mixed hp-boundary element methods are analyzed for frictional contact problems for the Lame equation. The stabilization technique circumvents the inf-sup conditions for the mixed problem and thus allows to use the same mesh and polynomial degree for primal and dual variables. A priori estimates are given for Tresca friction using Gauss-Legendre-Lagrange polynomials as test and trial functions for the Lagrange multiplier. In [33] we review about Coulomb friction and present numerical experiments which underline the insensitivity of the method to the scaling of the stabilization term. This approach is motivated by the seminal work of Barbosa and Hughes [39]. Assuming the mesh and polynomial degree distribution to be locally quasi-uniform we consider with the affine mapping Ψ_E form $[-1, 1]$ onto $E \in \mathscr{T}_h$ the ansatz spaces

$$V_{hp} = \left\{v^{hp} \in \tilde{H}^{\frac{1}{2}}(\Gamma_\Sigma) : v^{hp}|_E \circ \Psi_E \in \left[\mathbb{P}_{p_E}([-1, 1])\right]^2 \ \forall\, E \in \mathscr{T}_h\right\} \subset C^0(\Gamma_\Sigma)$$

$$V_{hp}^D = \left\{\phi^{hp} \in H^{-\frac{1}{2}}(\Gamma) : \phi^{hp}|_E \circ \Psi_E \in \left[\mathbb{P}_{p_E-1}([-1, 1])\right]^2 \ \forall\, E \in \mathscr{T}_h\right\},$$

$$\tilde{M}^+_{k,q}(\mathscr{F}) := \Big\{\mu^{kq} \in L^2(\Gamma_C) : \ \mu^{kq}|_E \circ \Psi_E \in \left[\mathbb{P}_{q_E}([-1, 1])\right]^2, \mu^{kq}(x) \geq 0,$$
$$- \mathscr{F}(x) \leq \mu^{kq}|_t(x) \leq \mathscr{F}(x) \text{ for } x \in G_{kq}\Big\}$$

where G_{kq} is the set of affinely transformed Gauss-Legendre points and μ^{kq} are linear combinations of Gauss-Legendre-Lagrange basis functions. The stabilized mixed method reads: Find $(u^{hp}, \lambda^{kq}) \in V_{hp} \times \tilde{M}^+_{k,q}(\mathscr{F})$ such that $\forall v^{hp} \in V_{hp}$ and $\forall \mu^{kq} \in \tilde{M}^+_{k,q}(\mathscr{F})$

$$\begin{aligned} \langle S_{hp}u^{hp}, v^{hp}\rangle_{\Gamma_\Sigma} + \langle \lambda^{kq}, v^{hp}\rangle_{\Gamma_C} - \langle \gamma(\lambda^{kq} + S_{hp}u^{hp}), S_{hp}v^{hp}\rangle_{\Gamma_C} &= \langle f, v^{hp}\rangle_{\Gamma_N} \\ \langle \mu^{kq} - \lambda^{kq}, u^{hp}\rangle_{\Gamma_C} - \langle \gamma(\mu^{kq} - \lambda^{kq}), \lambda^{kq} + S_{hp}u^{hp}\rangle_{\Gamma_C} &\leq \langle g, \mu_n^{kq} - \lambda_n^{kq}\rangle_{\Gamma_C} \end{aligned} \tag{11.74}$$

Here γ is a piecewise constant function on Γ_C such that $\gamma|_E = \gamma_0 h_E^{1+\beta} p_E^{-2-\eta}$ with constants $\gamma_0 > 0$, $\beta, \eta \geq 0$ for all elements $E \in \mathscr{T}_h|_{\Gamma_C}$.

In [33] it is shown that the solution $(u, \lambda) \in \tilde{H}^{1/2}(\Gamma_\Sigma) \times M^+(\mathscr{F})$ of (11.69) is approximated by the solution $(u^{hp}, \lambda^{kq}) \in V_{hp} \times \tilde{M}^+_{k,q}(\mathscr{F})$ of (11.74) with $h = k$

sufficiently small, $p = q, 0 \leq \alpha < 1/2$ satisfying

$$\|u - u^{hp}\|^2_{\tilde{H}^{1/2}(\Gamma_\Sigma)} + \|\gamma^{1/2}(\lambda - \lambda^{kq})\|^2_{L^2(\Gamma_C)} \leq ch^{\alpha/2} p^{-\alpha/3}$$

(see in [33] Theorem 16 and Remark 17). For improved stabilization see [35].

11.4.5 A Priori Error Estimates for hp-Penalty-BEM for Contact Problems in Elasticity

From [102] we report an a priori error analysis for $hp-$version of the penalty Galerkin BEM for frictionless contact problems.

Let us consider an elastic body in two dimensions under small strain assumption. We associate the body with a bounded two-dimensional polygonal domain Ω with boundary $\Gamma = \partial\Omega$. Assume that some part of the boundary $\Gamma_D \subset \Gamma$ is fixed, which prevents the rigid body motions and therefore provides uniqueness of the solution. Further, we denote by Γ_N the boundary part with prescribed boundary tractions. Finally, we call $\Gamma_C \subset \Gamma$ the boundary part which potentially can come into contact with a rigid obstacle. The three parts of the boundary are assumed to be disjoint and satisfy $\overline{\Gamma} = \overline{\Gamma}_D \cup \overline{\Gamma}_N \cup \overline{\Gamma}_C$.

As we have seen in Sect. 5.1 the problem can be rewritten in a weak sense as a variational inequality. Due to the non-penetration conditions on the contact boundary the space of the admissible solutions is restricted by an inequality constraint and forms a convex cone $\mathcal{K}$. The main difficulty in deriving the discrete Galerkin formulation of the original problem is the discretization of $\mathcal{K}$.

One possibility will be to impose the inequality constraint only in the Gauss-Lobatto points, and then solve the resulting constrained optimization problem with e.g. generalized conjugate gradient method, cf. [297] (see also Sect. 11.2). This approach is nonconforming since the set of discrete solutions $\mathcal{K}_{hp}$ is not a subset of the continuous cone: $\mathcal{K}_{hp} \nsubseteq \mathcal{K}$. Unfortunately, this allows only to prove the a priori error estimates with reduced rate of convergence [297, Theorem 3] (see also Sect. 11.3)

$$||u - U||_{H^{1/2}(\Gamma)} \leq C\,(h/p)^{1/4}\,||u||_{H^{3/2}(\Gamma)},$$

where $u \in \mathcal{K}$ solves the variational inequality and $U \in \mathcal{K}_{hp}$ solves its discrete version.

The other way to solve the problem is to use the penalty method, [90, 149, 266]. In this approach, we approximate the variational inequality by introducing a penalty parameter $\epsilon > 0$ that connects the normal displacement u_n, the normal boundary stress (traction) σ_n and the distance g to the rigid obstacle, requiring $-\sigma_n := \epsilon^{-1}(u_n - g)^+$, where $(\cdot)^+$ denotes the positive part of the function (11.83). The penetration of the body into the obstacle is now allowed, but if the penalty parameter is very small, it will cause a large outer pressure, which pushes the body back and prevents large penetrations. The space of the admissible displacements is now unconstrained and, therefore, can be discretized in a conforming way.

The total error consists now of two parts. The reduction of the element size and increasing of the polynomial degree with fixed penalty parameter will not lead to the convergence of the method, the same as the decreasing of the penalty parameter alone with fixed discretization parameters. Only combined changing of ϵ, h and p provides convergence to the exact solution. We carry out the corresponding a priori error analysis and show (Theorem 11.14) that the convergence rate $\mathscr{O}((h/p)^{1-\epsilon})$ is achieved, if $\epsilon = \tilde{C}(h/p)^{1-\epsilon}$ for some fixed $\epsilon \in (0, 1)$ and $\tilde{C} > 0$.

Now we consider (11.65) with $\sigma_t = 0$ on Γ_C. Further, we introduce the functional spaces and sets required for the forthcoming analysis

$$\mathscr{V} := \tilde{H}^{1/2}(\Sigma), \tag{11.75}$$

$$\mathscr{W} := H^{-1/2}(\Gamma), \tag{11.76}$$

$$\mathscr{K} := \left\{ v \in \mathscr{V} : (v_n - g)|_{\Gamma_C} \leq 0 \right\}, \tag{11.77}$$

$$\Lambda := \left\{ \lambda \in \tilde{H}^{-1/2}(\Gamma_C) : \quad \forall v \in \mathscr{V},\ v_n|_{\Gamma_C} \leq 0,\ \int_{\Gamma_C} \lambda v_n \, ds \geq 0 \right\}. \tag{11.78}$$

The classical problem (11.65) can be reformulated in a weak form with the Poincaré–Steklov operator S as a variational inequality (see [299] or Sect. 11.2):

Find $u \in \mathscr{K}$:

$$\langle Su, v - u \rangle \geq L(v - u) \qquad \forall v \in \mathscr{K}, \tag{11.79}$$

or equivalently as the saddle point formulation with Lagrange multiplier (cf. [266], Appendix C.1.3): Find $u \in \mathscr{V}, \lambda \in \Lambda$:

$$\begin{aligned} \langle Su, v \rangle - \langle \lambda, v_n \rangle &= L(v) \ \forall v \in \mathscr{V}, \\ \langle \mu - \lambda, u_n - g \rangle &\geq 0 \qquad \forall \mu \in \Lambda. \end{aligned} \tag{11.80}$$

The existence and uniqueness of the solution of the variational inequality is a well known result of convex analysis (e.g. [222]). Note that both formulations (11.79) and (11.80) include the inequality constraints, which is very inconvenient for construction of the discrete formulation, error estimation and implementation. The penalty formulation allows to avoid such inequality constraints in the set of admissible solutions and to obtain a variational equation. The penalty formulation is given as follows, [149]: Find $u^\epsilon \in \mathscr{V}$:

$$\langle Su^\epsilon, v \rangle - \langle p^\epsilon, v_n \rangle = L(v) \qquad \forall v \in \mathscr{V}, \tag{11.81}$$

$$p^\epsilon := -\frac{1}{\epsilon}(u_n^\epsilon - g)^+. \tag{11.82}$$

Here the penalty parameter $\epsilon > 0$ must be chosen in advance. Furthermore we denote the positive and negative part of a scalar-valued function f by

$$f^+ := (|f| + f)/2 \geq 0, \qquad f^- := (|f| - f)/2 \geq 0, \tag{11.83}$$

which provides $f = f^+ - f^-$.

Next we investigate, how good the solution u^ϵ of the penalty formulation (11.81) and the function p^ϵ approximate the solution (u, λ) of the saddle point formulation (11.80) depending on the penalty parameter ϵ.

Now, we derive an upper bound for the energy norm of the error, caused by the approximation of the solution of the saddle point problem by the solution of the penalty formulation.

Theorem 11.11 *Let $u \in \mathcal{V}$, $\lambda \in \Lambda \cap H^{1/2}(\Gamma_C)$ solve the Lagrange multiplier formulation (11.80), let $u^\epsilon \in \mathcal{V}$ solve the penalty formulation (11.81), and let p^ϵ be defined with (11.82). Then there holds*

$$||u - u^\epsilon||_{\tilde{H}^{1/2}(\Sigma)} \leq \frac{C_S}{c_S \alpha} ||\epsilon\lambda||_{H^{1/2}(\Gamma_C)}, \tag{11.84}$$

$$||\lambda - p^\epsilon||_{\tilde{H}^{-1/2}(\Gamma_C)} \leq \frac{C_S^2}{c_S \alpha^2} ||\epsilon\lambda||_{H^{1/2}(\Gamma_C)}. \tag{11.85}$$

Proof Since u and u^ϵ solve (11.80) and (11.81) respectively, there holds

$$\langle Su, v\rangle - \langle \lambda, v_n\rangle = L(v) \qquad \forall v \in \mathcal{V},$$

$$\langle Su^\epsilon, v\rangle - \langle p^\epsilon, v_n\rangle = L(v) \qquad \forall v \in \mathcal{V}.$$

Subtracting those variational equations and choosing $v := u - u^\epsilon \in \mathcal{V}$ we obtain

$$\langle S(u - u^\epsilon), u - u^\epsilon\rangle = \langle \lambda - p^\epsilon, u_n - g\rangle + \langle p^\epsilon - \lambda, u_n^\epsilon - g\rangle$$

Now $\langle \lambda - p^\epsilon, u_n - g\rangle \leq 0$, and $\langle p^\epsilon - \lambda, (u_n^\epsilon - g)^-\rangle \geq 0$. Thus,

$$\begin{aligned}\langle S(u - u^\epsilon), u - u^\epsilon\rangle &\leq \langle p^\epsilon - \lambda, u_n^\epsilon - g\rangle \\ &\leq \langle p^\epsilon - \lambda, (u_n^\epsilon - g)^+\rangle.\end{aligned}$$

Definition (11.82) provides $\langle p^\epsilon - \lambda, (u_n^\epsilon - g)^+\rangle = \langle p^\epsilon - \lambda, -\epsilon p^\epsilon\rangle$. Further, since $\langle p^\epsilon - \lambda, \epsilon(p^\epsilon - \lambda)\rangle \geq 0$, we have

$$\begin{aligned}\langle S(u - u^\epsilon), u - u^\epsilon\rangle &\leq \langle p^\epsilon - \lambda, -\epsilon p^\epsilon\rangle \\ &\leq \langle p^\epsilon - \lambda, -\epsilon p^\epsilon\rangle + \langle p^\epsilon - \lambda, \epsilon(p^\epsilon - \lambda)\rangle\end{aligned}$$

$$
\begin{aligned}
&= \langle \lambda - p^{\epsilon}, \epsilon\lambda \rangle \\
&\le ||\lambda - p^{\epsilon}||_{\tilde{H}^{-1/2}(\Gamma_C)} ||\epsilon\lambda||_{H^{1/2}(\Gamma_C)} \\
&\le \frac{C_S}{\alpha} ||u - u^{\epsilon}||_{\tilde{H}^{1/2}(\Sigma)} ||\epsilon\lambda||_{H^{1/2}(\Gamma_C)},
\end{aligned}
\tag{11.86}
$$

Now, ellipticity of the Steklov-Poincaré operator S yields the assertion. □

Next we introduce the discrete penalty formulation as follows: Find $U^{\epsilon} \in \mathcal{V}_{hp}$:

$$
\langle S_{hp} U^{\epsilon}, v \rangle - \langle P^{\epsilon}, v_n \rangle = L(v) \qquad \forall v \in \mathcal{V}_{hp}, \tag{11.87}
$$

where

$$
P^{\epsilon} := -\frac{1}{\epsilon}(U_n^{\epsilon} - g)^+, \qquad \epsilon > 0. \tag{11.88}
$$

Furthermore, for $u^{\epsilon} \in \mathcal{V}$ and $U^{\epsilon} \in \mathcal{V}_{hp}$ we define the traction-like functions

$$
\begin{aligned}
\psi &:= V^{-1}(K + I)u^{\epsilon}, \\
\Psi^* &:= V^{-1}(K + I)U^{\epsilon}, \\
\Psi &:= i_{hp} V_{hp}^{-1} i_{hp}^{*} (K + I) U^{\epsilon}.
\end{aligned}
\tag{11.89}
$$

Lemma 11.4 *Let $u^{\epsilon} \in \mathcal{V}$, $U^{\epsilon} \in \mathcal{V}_{hp}$ and traction-like functions defined by (11.89). Then the following identity holds*

$$
||u^{\epsilon} - U^{\epsilon}||_W^2 + ||\psi - \Psi||_V^2 = 2\langle Su^{\epsilon} - S_{hp}U^{\epsilon}, u^{\epsilon} - U^{\epsilon} \rangle + \langle V(\Psi^* - \Psi), \psi - \Psi \rangle,
$$

where

$$
\begin{aligned}
||u^{\epsilon} - U^{\epsilon}||_W &:= \langle W(u^{\epsilon} - U^{\epsilon}), u^{\epsilon} - U^{\epsilon} \rangle^{1/2}, \\
||\psi - \Psi||_V &:= \langle V(\psi - \Psi), \psi - \Psi \rangle^{1/2}.
\end{aligned}
$$

Lemma 11.5 *For Ψ^*, Ψ defined in (11.89) there holds*

$$
\langle V(\Psi^* - \Psi), \Phi \rangle = 0, \qquad \forall \Phi \in \mathcal{W}_{hp}.
$$

Theorem 11.12 *Let u^{ϵ} solve the continuous penalty problem (11.81), let U^{ϵ} solve the discrete penalty problem (11.87). Let ψ, Ψ be defined by (11.89). Then there exists $C > 0$ independent of h, p, ϵ such that for $\forall w \in \mathcal{V}_{hp}$, $\forall \Phi \in \mathcal{W}_{hp}$ there holds*

$$
\begin{aligned}
&||u^{\epsilon} - U^{\epsilon}||_{\tilde{H}^{1/2}(\Sigma)} + ||\psi - \Psi||_{H^{-1/2}(\Gamma)} + ||\epsilon^{1/2}(p^{\epsilon} - P^{\epsilon})||_{L_2(\Gamma_C)} \\
&\quad \le C(||u^{\epsilon} - w||_{\tilde{H}^{1/2}(\Sigma)} + ||\psi - \Phi||_{H^{-1/2}(\Gamma)} + ||\epsilon^{-1/2}(w_n - u_n^{\epsilon})||_{L_2(\Gamma_C)}).
\end{aligned}
$$

Assume that $u^\epsilon \in \tilde{H}^{3/2}(\Sigma)$ and $\psi \in H^{1/2}(\Gamma)$. According to [45, 405] the following approximation properties hold

$$\inf_{w \in \mathscr{V}_{hp}} ||u^\epsilon - w||_{\tilde{H}^{1/2}(\Sigma)} \leq C\frac{h}{p}||u^\epsilon||_{\tilde{H}^{3/2}(\Sigma)}, \tag{11.90}$$

$$\inf_{\Phi \in \mathscr{W}_{hp}} ||\psi - \Phi||_{H^{-1/2}(\Gamma)} \leq C\frac{h}{p}||\psi||_{H^{1/2}(\Gamma)}, \tag{11.91}$$

$$\inf_{w \in \mathscr{V}_{hp}} ||\epsilon^{-1/2}(w_n - u_n^\epsilon)||_{L_2(\Gamma_C)} \leq C\left(\frac{h}{p}\right)^{3/2} ||\epsilon^{-1/2}u_n^\epsilon||_{H^{3/2}(\Gamma_C)}. \tag{11.92}$$

Let T be the Dirichlet-to-Neumann operator, which maps the function $u \in \tilde{H}^{1/2}(\Gamma)$ to the function $Tu \in H^{-1/2}(\Gamma)$, such that the prolongation of the Cauchy data u, Tu into the domain Ω satisfies the homogeneous Lamé equation (the first equation in (11.65)), see e.g. [391]. Employing the well-known jump conditions it can be shown that T can be written in the non-symmetric form as

$$T = V^{-1}(K + I). \tag{11.93}$$

From (11.89) we have $\psi = Tu^\epsilon$.

The approximation properties (11.90)–(11.92) combined with Theorem 11.12 yield the following a priori error estimate for the solution of the penalty formulation (11.81).

Theorem 11.13 *Let $u^\epsilon \in \tilde{H}^{3/2}(\Sigma)$ be a solution of (11.81) and $Tu^\epsilon \in H^{1/2}(\Gamma)$. Let $U^\epsilon \in \mathscr{V}_{hp}$ be a solution of (11.87). Then there exists a constant $C > 0$ independent of h, p, ϵ, such that*

$$||u^\epsilon - U^\epsilon||_{\tilde{H}^{1/2}(\Sigma)} + ||Tu^\epsilon - \Psi||_{H^{-1/2}(\Gamma)} + ||\epsilon^{1/2}(p^\epsilon - P^\epsilon)||_{L_2(\Gamma_C)}$$
$$\leq C\left(\frac{h}{p}||u^\epsilon||_{\tilde{H}^{3/2}(\Sigma)} + \frac{h}{p}||Tu^\epsilon||_{H^{1/2}(\Gamma)} + \left(\frac{h}{p}\right)^{3/2} ||\epsilon^{-1/2}u_n^\epsilon||_{H^{3/2}(\Gamma_C)}\right).$$

In order to obtain an a priori error estimate for the total error between the solutions of problems (11.80) and (11.87) in terms of the solution of the variational inequality (11.81) we need to combine the results of Theorem 11.11 and Theorem 11.12.

Theorem 11.14 *Let $u \in \tilde{H}^{3/2}(\Sigma)$, $\lambda \in \Lambda \cap H^{1/2}(\Gamma)$ be a solution of (11.80) and let $Tu \in H^{1/2}(\Gamma)$, where T is defined by (11.93). Let $U^\epsilon \in \mathscr{V}_{hp}$ solve (11.87). Assume that $\epsilon = \tilde{C}(h/p)^{1-\tilde{\epsilon}}$ for arbitrary $\tilde{\epsilon} \in (0; 1)$ and $\tilde{C} > 0$. Then there exists a constant $C > 0$ independent of h, p, ϵ such that*

$$||u - U^\epsilon||_{\tilde{H}^{1/2}(\Sigma)} \leq C\left(\frac{h}{p}||u||_{\tilde{H}^{3/2}(\Sigma)} + \left(\epsilon + \frac{h}{p}\right)||Tu||_{H^{1/2}(\Gamma)}\right). \tag{11.94}$$

11.5 *h*-Version BEM for a Nonmonotone Contact Problem from Delamination

This section continues 5.3 and based on [333], it presents the *h*-version BEM for the considered nonmonotone contact problem.

Now let $\Omega \subset \mathbb{R}^d$, $d = 2, 3$, be a polygonal domain with the boundary Γ. We start from a triangulation $\mathcal{T}_h$ of edges in the 2D case and triangles in the 3D case on Γ that is consistent with the decomposition of Γ into Γ_0 and Γ_D. For the discretization of the displacement $\mathbf{u}$ we choose continuous piecewise linear functions on $\mathcal{T}_h$ and define

$$\mathcal{V}_h = \{\mathbf{v_h} \in \mathbf{C}(\Gamma) \,:\, \mathbf{v_h}|_E \in [\mathcal{P}_1]^{d-1}\, \forall E \in \mathcal{T}_h,\ \mathbf{v_h} = 0 \text{ on } \bar{\Gamma}_D\} \subset \mathcal{V} = \mathbf{H}_D^{1/2}(\Gamma),$$

$$\mathcal{K}_h^{\Gamma} = \{\mathbf{v_h} \in \mathcal{V}_h \,:\, (\mathbf{v_h} \cdot \mathbf{n})(P_i) \leq 0 \quad \forall P_i \in \Sigma_h,\ P_i \in \bar{\Gamma}_C \backslash \bar{\Gamma}_D\},$$

where Σ_h is the set of all nodes of $\mathcal{T}_h$.

To discretize the stresses we use the space of piecewise constant functions on $\mathcal{T}_h$:

$$\mathcal{W}_h = \{\psi \in \mathbf{L}^2(\Gamma) \,:\, \psi|_E \in [\mathcal{P}_0]^{d-1} \quad \forall E \in \mathcal{T}_h\} \subset \mathbf{H}^{-1/2}(\Gamma).$$

Let $\{\varphi_i\}_{i=1}^{N_D}$ and $\{\psi_j\}_{j=1}^{N_N}$ be the nodal bases in $\mathcal{V}_h$ and $\mathcal{W}_h$, respectively. Then the boundary element matrices associated to the boundary integral operators V, K, K', W are given by

$$\begin{aligned} V_h &= \{\langle V\psi_i, \psi_j\rangle\}_{i,j=1}^{N_N,N_D} & K_h &= \{\langle K\phi_i, \psi_j\rangle\}_{i,j=1}^{N_D,N_N} \\ K_h' &= \{\langle K'\psi_i, \phi_j\rangle\}_{i,j=1}^{N_N,N_D} & W_h &= \{\langle W\phi_i, \phi_j\rangle\}_{i,j=1}^{N_D,N_N} \end{aligned}$$

The matrix V_h is symmetric and positive definite, so it can be inverted by Cholesky decomposition. This gives the Schur complement matrix

$$S_h = \frac{1}{2}\left(W_h + \left(K_h' + I_h\right) V_h^{-1} \left(K_h + I_h\right)\right).$$

With the canonical embeddings

$$\begin{aligned} k_h &: \mathcal{W}_h \hookrightarrow \mathbf{H}^{-1/2}(\Gamma) \\ i_h &: \mathcal{V}_h \hookrightarrow \mathbf{H}^{1/2}(\Gamma) \end{aligned}$$

and their duals k_h^* and i_h^*, we obtain the discrete Poincaré-Steklov operator S_h : $\mathcal{V}_h \to \mathcal{V}_h^*$ represented by

$$S_h = \frac{1}{2}\left(i_h^* W i_h + i_h^* \left(K' + I\right) k_h (k_h^* V k_h)^{-1} k_h^* (K + I) i_h\right).$$

Due to [73], this operator is well-defined and satisfies

$$\langle S_h \mathbf{u}_h, \mathbf{u}_h \rangle \geq c \| i_h \mathbf{u}_h \|^2_{\mathbf{H}^{1/2}(\Gamma)}. \tag{11.95}$$

Further, we define the operator $E_h : \mathbf{H}^{1/2}(\Gamma) \to \mathbf{H}^{-1/2}(\Gamma)$, reflecting the consistency error in the discretization of the Poincaré-Steklov operator S (see also the previous Sect. 11.3, in particular the proof of Theorem 11.7), by

$$E_h := S - S_h = \frac{1}{2}(I + K')(V^{-1} - i_h(i_h^* V i_h)^{-1} i_h^*)(I + K).$$

Due to [73, 297], the operator E_h is bounded and satisfies

$$\|E_h(\mathbf{u})\|_{\mathbf{H}^{-1/2}(\Gamma)} \leq c \inf_{\mathbf{w} \in \mathscr{W}_h} \|V^{-1}(I + K)\mathbf{u} - \mathbf{w}\|_{\mathbf{H}^{-1/2}(\Gamma)} \quad \forall \mathbf{u} \in \mathbf{H}^{1/2}(\Gamma), \tag{11.96}$$

hence by periodic polynomial spline approximation theory, see Theorem 6.12,

$$\|E_h(\mathbf{u})\|_{\mathbf{H}^{-1/2}(\Gamma)} \leq c \, \|\mathbf{u}\|_{\mathbf{H}^{1/2}(\Gamma)} .$$

Lemma 11.6

(i) If $\mathbf{u}_h \rightharpoonup \mathbf{u}$ *(weak convergence) and* $\mathbf{v}_h \to \mathbf{v}$ *in* $\mathbf{H}^{1/2}(\Gamma)$*, then* $\lim_{h \to 0} \langle S_h \mathbf{u}_h, \mathbf{v}_h \rangle = \langle S\mathbf{u}, \mathbf{v} \rangle$.

(ii) If $\mathbf{u}_h \to \mathbf{u}$ *and* $\mathbf{v}_h \rightharpoonup \mathbf{v}$ *in* $\mathbf{H}^{1/2}(\Gamma)$*, then* $\lim_{h \to 0} \langle S_h \mathbf{u}_h, \mathbf{v}_h \rangle = \langle S\mathbf{u}, \mathbf{v} \rangle$.

Proof The part (i) follows immediately from the estimate below. Indeed, there exists a constant c_0 such that

$$\langle S_h \mathbf{v}_h - i_h^* S \mathbf{v}, \mathbf{w}_h \rangle_{\mathscr{V}_h} \leq c_0 \|\mathbf{w}_h\|_{\mathbf{H}^{1/2}(\Gamma)} \left(e_h(\mathbf{v}) + \|\mathbf{v}_h - \mathbf{v}\|_{\mathbf{H}^{1/2}(\Gamma)} \right)$$

for any $\mathbf{v} \in \mathbf{H}^{1/2}(\Gamma)$ and for any $\mathbf{v}_h, \mathbf{w}_h \in \mathscr{V}_h$, where $e_h(\mathbf{v})$ satisfies $e_h(\mathbf{v}) \to 0$ as $h \to 0$.

Hence, using the symmetry of S and S_h, we obtain

$$\begin{aligned} \langle S_h \mathbf{u}_h, \mathbf{v}_h \rangle - \langle S\mathbf{u}, \mathbf{v} \rangle &= \langle S_h \mathbf{v}_h - k_h^* S \mathbf{v}, \mathbf{u}_h \rangle + \langle S\mathbf{v}, k_h \mathbf{u}_h - \mathbf{u} \rangle \\ &\leq c_0 \|\mathbf{u}_h\|_{\mathbf{H}^{1/2}(\Gamma)} \left(e_h(\mathbf{v}) + \|\mathbf{v}_h - \mathbf{v}\|_{\mathbf{H}^{1/2}(\Gamma)} \right) + \langle S\mathbf{v}, \mathbf{u}_h - \mathbf{u} \rangle \end{aligned}$$

and thus, (i) is satisfied.

The proof of (ii) follows in the same way. □

Now, we turn to the discretization of the regularized problem (5.29). To this end, we define $\Pi : \mathbf{H}^{1/2}(\Gamma) \to L^2(\Gamma_C)$ by

$$\Pi \mathbf{u}_h = \mathbf{u}_h \cdot \mathbf{n} \quad \text{on} \quad \Gamma_C .$$

The mapping Π is linear continuous from $\mathbf{H}^{1/2}(\Gamma)$ into $L^2(\Gamma_C)$, i.e.

$$\exists c > 0 \ : \ \|\Pi \mathbf{v}\|_{L^2(\Gamma_C)} \leq c\|\mathbf{v}\|_{\mathbf{H}^{1/2}(\Gamma)} \quad \forall v \in \mathbf{H}^{1/2}(\Gamma). \tag{11.97}$$

Further, we denote by $\widetilde{\mathscr{V}_h}$ the image of $\mathscr{V}_h$ with respect to Π, i.e.

$$\widetilde{\mathscr{V}_h} = \{w_h \in C(\overline{\Gamma}_C) \ : \ w_h|_E \in P_1(E) \quad \forall E \in \mathscr{T}_h|_{\Gamma_C}, \ w_h = 0 \text{ on } \bar{\Gamma}_D\},$$

where $\mathscr{T}_h|_{\Gamma_C}$ denotes the partition of Γ_C induced by $\mathscr{T}_h$.

Let $\{P_i\}_{i=0}^m$ be the set of all nodes of $\mathscr{T}_h$ lying on Γ_C. To approximate the Gâteaux derivative $\langle DJ_\varepsilon(\cdot), \cdot\rangle$ we use Kepler's trapezoidal rule for numerical integration and define

$$\langle DJ_{\varepsilon,h}(\mathbf{u}_h), \mathbf{v}_h\rangle :=$$

$$\frac{1}{2}\sum_{i=0}^{m-1} |P_i P_{i+1}| \Big[\frac{\partial \tilde{S}}{\partial x}(\Pi \mathbf{u}_h(P_i), \varepsilon)\Pi \mathbf{v}_h(P_i) + \frac{\partial \tilde{S}}{\partial x}(\Pi \mathbf{u}_h(P_{i+1}), \varepsilon)\Pi \mathbf{v}_h(P_{i+1})\Big].$$

Herewith the discretization of the regularized problem (5.29) reads as follows:

Problem $(\mathscr{P}_{\varepsilon,h})$ Find $u_{\varepsilon,h} \in \mathscr{K}_h^\Gamma$ such that for all $\mathbf{v}_h \in \mathscr{K}_h^\Gamma$

$$\langle \mathbf{v}_h - \mathbf{u}_{\varepsilon,h}, S_h \mathbf{u}_{\varepsilon,h}\rangle + \langle DJ_{\varepsilon,h}(\mathbf{u}_{\varepsilon,h}), \mathbf{v}_h - \mathbf{u}_{\varepsilon,h}\rangle \geq \int_{\Gamma_N} \mathbf{t} \cdot (\mathbf{v}_h - \mathbf{u}_{\varepsilon,h})\, ds. \tag{11.98}$$

Let $\mathscr{D}_h$ be another partitioning of Γ_C consisting of elements K_i joining the midpoints $P_{i-1/2}$, $P_{i+1/2}$ of the edges $E \in \mathscr{T}_h$ lying on Γ_C sharing P_i as a common point. If P_i is a vertex of $\partial\Omega$ then K_i is half of the edge. Moreover, if the element K_i is linked to the boundary node P_i of Γ_D, it will be added to its neighbour element K_{i+1}, see Fig. 11.5. Further, on $\mathscr{D}_h$ we introduce the space $\mathscr{Y}_h$ of piecewise constant functions by

$$\mathscr{Y}_h = \{\mu_h \in L^\infty(\Gamma_C) \ : \ \mu_h|_K \in P_0(K) \quad \forall K \in \mathscr{D}_h\}$$

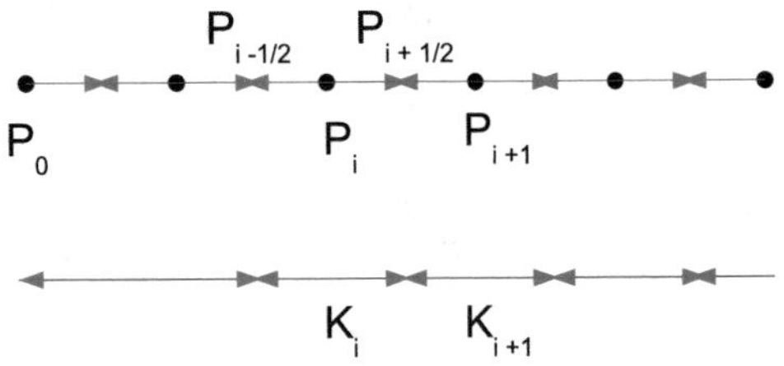

Fig. 11.5 Discretization on Γ_C; P_0 is a boundary point for Γ_D [333]

and define the piecewise constant Lagrange interpolation operator $L_h : \widetilde{\mathscr{V}}_h \to \mathscr{Y}_h$ by

$$L_h(w_h)(x) = \sum_{P_i \in \Gamma_C \cap \Sigma_h} w_h(P_i)\, \chi_{K_i}(x),$$

where χ_{K_i} is the characteristic function of the interior of K_i in Γ_C.

It holds that

$$\langle DJ_{\varepsilon,h}(\mathbf{u}_h), \mathbf{v}_h \rangle_{\Gamma_C} = \int_{\Gamma_C} S_x(\varepsilon, L_h(\Pi \mathbf{u}_h)) L_h(\Pi \mathbf{v}_h)\, ds. \tag{11.99}$$

Moreover, the operator $DJ_{\varepsilon,h} : \mathscr{V}_h \to \mathscr{V}_h^*$ is strongly continuous, and there exists a constant $C > 0$ independent of ε and h such that

$$\exists C > 0 : \langle DJ_{\varepsilon,h}(\mathbf{u}_h), \mathbf{u}_h \rangle_{\Gamma_C} \geq -C \|\mathbf{u}_h\|_{\mathbf{H}^{1/2}(\Gamma)}, \tag{11.100}$$

for the proofs see [332].

From [200], we know that

$$\|L_h(\mathbf{v}_h \cdot \mathbf{n})\|_{\mathbf{L}^2(\Gamma)} \leq 2\, \|\mathbf{v}_h \cdot \mathbf{n}\|_{\mathbf{L}^2(\Gamma)} \tag{11.101}$$

and therefore,

$$\|\mathbf{v}_h \cdot \mathbf{n} - L_h(\mathbf{v}_h \cdot \mathbf{n})\|_{\mathbf{L}^2(\Gamma)} \leq 3\, \|\mathbf{v}_h \cdot \mathbf{n}\|_{\mathbf{L}^2(\Gamma)}. \tag{11.102}$$

Let now $\mathbf{H}^s(\Gamma^j)$, $s \geq 0$, be the standard Sobolev space from [204, 284] defined on the open straight pieces Γ^j by

$$\mathbf{H}^s(\Gamma^j) = \{u|_{\Gamma^j} : u \in \mathbf{H}^s(\Gamma)\}.$$

According to Grisvard [204], $\mathbf{H}^s(\Gamma) \subset \prod_{j=1}^{J} \mathbf{H}^s(\Gamma^j)$ for $s \in [1/2, 3/2)$ and

$$\sum_{j=1}^{J} \|\mathbf{u}\|^2_{\mathbf{H}^s(\Gamma^j)} \leq C \|\mathbf{u}\|^2_{\mathbf{H}^s(\Gamma)}. \tag{11.103}$$

Again from [200],

$$\|\mathbf{v}_h \cdot \mathbf{n} - L_h(\mathbf{v}_h \cdot \mathbf{n})\|^2_{L^2(\Gamma^j)} \leq Ch^2 \|\mathbf{v}_h \cdot \mathbf{n}\|^2_{H^1(\Gamma^j)} \leq Ch^2 \|\mathbf{v}_h\|^2_{H^1(\Gamma^j)}. \tag{11.104}$$

Summing over all j and using thereafter

$$\|\mathbf{v}_h \cdot \mathbf{n} - L_h(\mathbf{v}_h \cdot \mathbf{n})\|_{L^2(\Gamma)} \leq Ch \|\mathbf{v}_h\|_{H^1(\Gamma)}. \tag{11.105}$$

By interpolation between $L^2(\Gamma)$ and $H^1(\Gamma)$ we deduce from (11.102) and (11.105) that

$$\|\mathbf{v}_h \cdot \mathbf{n} - L_h(\mathbf{v}_h \cdot \mathbf{n})\|_{L^2(\Gamma)} \le Ch^{1/2}\|\mathbf{v}_h\|_{H^{1/2}(\Gamma)}. \tag{11.106}$$

By the compactness of $H^{s_1}(\Gamma) \subset H^{s_2}(\Gamma)$ for $0 \le s_2 < s_1$ ($\Omega \subset \mathbb{R}^2$) (see Theorem 3.1), this gives

$$\mathbf{v}_h \rightharpoonup \mathbf{v} \text{ in } \mathbf{H}^{1/2}(\Gamma) \Rightarrow \|L_h(\mathbf{v}_h \cdot \mathbf{n}) - \mathbf{v} \cdot \mathbf{n}\|_{\mathbf{L}^2(\Gamma)} \to 0. \tag{11.107}$$

Further, we introduce the functional $\varphi_{\varepsilon,h} : \mathscr{V}_h \times \mathscr{V}_h \to \mathbb{R}$ by

$$\varphi_{\varepsilon,h}(\mathbf{u}_h, \mathbf{v}_h) := \langle DJ_{\varepsilon,h}(\mathbf{u}_h), \mathbf{u}_h \rangle. \tag{11.108}$$

Due to [332], see also Sect. C.3.5 in Appendix C, this functional is pseudomonotone and upper semicontinuous with respect to the first argument. Moreover by arguments similar as in [332] one can show the following assertions:

(i) If $\{\mathbf{v}_h\}$ weakly converges to $\mathbf{v}$ in $\widetilde{\mathbf{H}}^{1/2}(\Gamma_0)$, $\mathbf{v}_h \in \mathscr{K}_h^\Gamma$, then $\mathbf{v} \in \mathscr{K}^\Gamma$.
(ii) For any $\mathbf{v} \in \mathscr{K}^\Gamma$ there exists $\{\mathbf{v}_h\}$ such that $\mathbf{v}_h \in \mathscr{K}_h^\Gamma$ and $\mathbf{v}_h \to \mathbf{v}$ in $\widetilde{\mathbf{H}}^{1/2}(\Gamma_0)$.
(iii) For any $\{\mathbf{u}_h\}$ and $\{\mathbf{v}_h\}$ such that $\mathbf{u}_h \in \mathscr{K}_h^\Gamma$, $\mathbf{v}_h \in \mathscr{K}_h^\Gamma$, $\mathbf{u}_h \rightharpoonup \mathbf{u}$ and $\mathbf{v}_h \to \mathbf{v}$ in $\mathscr{V}$ we have

$$\limsup \varphi_{\varepsilon,h}(\mathbf{u}_h, \mathbf{v}_h) \le \varphi(\mathbf{u}, \mathbf{v}).$$

(iv) There exist constants $c > 0$, $d, d_0 \in \mathbb{R}$ and $\alpha > 1$ such that for some $\mathbf{v}_h \in \mathscr{K}_h^\Gamma$ with $\mathbf{v}_h \to \mathbf{v}$ there holds

$$-\varphi_{\varepsilon,h}(\mathbf{u}_h, \mathbf{v}_h) \ge c\|\mathbf{u}_h\|_V^\alpha + d\|\mathbf{u}_h\|_V + d_0 \quad \forall \mathbf{u}_h \in \mathscr{K}_h^\Gamma.$$

Based on these assertions, the general approximation result [220, Theorem 3.1], here Theorem C.9 in Appendix C, applies to arrive at the following convergence result.

Theorem 11.15 *The problem ($\mathscr{P}_{\varepsilon,h}$) has at least one solution $\mathbf{u}_{\varepsilon,h}$. Moreover, the family $\{\mathbf{u}_{\varepsilon,h}\}$ of solutions is uniformly bounded in $\mathscr{V} = \widetilde{\mathbf{H}}^{1/2}(\Gamma_0)$ and any weak accumulation point of $\{\mathbf{u}_{\varepsilon,h}\}$ is a solution to the problem ($\mathscr{P}$).*

Here we show the uniform boundedness of $\{\mathbf{u}_{\varepsilon,h}\}$. Indeed, the choice $\mathbf{v}_h = 0$ in (11.98), and the estimates (11.95) and (11.100) lead to

$$\begin{aligned} c\|\mathbf{u}_{\varepsilon,h}\|_{\mathscr{V}}^2 &\le \langle S_h\mathbf{u}_{\varepsilon,h}, \mathbf{u}_{\varepsilon,h}\rangle \le \|\mathbf{t}\|_{\mathscr{V}^*}\|\mathbf{u}_{\varepsilon,h}\|_{\mathscr{V}} + \varphi_{\varepsilon,h}(\mathbf{u}_{\varepsilon,h}, 0) \\ &= \|\mathbf{t}\|_{\mathscr{V}^*}\|\mathbf{u}_{\varepsilon,h}\|_{\mathscr{V}} + \langle DJ_{\varepsilon,h}(\mathbf{u}_{\varepsilon,h}), -\mathbf{u}_{\varepsilon,h}\rangle \\ &\le \|\mathbf{t}\|_{\mathscr{V}^*}\|\mathbf{u}_{\varepsilon,h}\|_{\mathscr{V}} + c\|\mathbf{u}_{\varepsilon,h}\|_{\mathscr{V}}. \end{aligned}$$

Further, in case of uniqueness we can improve the convergence result of Theorem 11.15 and show that the weak convergence can be replaced by the strong one.

Theorem 11.16 *Let the solutions* $\mathbf{u}$ *to* $(\mathscr{P})$ *and* $\mathbf{u}_{\varepsilon,h}$ *to* $(\mathscr{P}_{\varepsilon,h})$ *exist uniquely. Then*

$$\lim_{\varepsilon\to 0, h\to 0} \|\mathbf{u}_{\varepsilon,h} - \mathbf{u}\|_{\widetilde{\mathbf{H}}^{1/2}(\Gamma_0)} = 0.$$

Proof Let $\{h_n\}$ and $\{\varepsilon_n\}$ be arbitrary sequences such that $h_n \to 0^+$ and $\varepsilon_n \to 0^+$ as $n \to \infty$. In view of (**ii**), there exists a sequence $\{\bar{\mathbf{u}}_{\varepsilon_n,h_n}\}$ such that $\bar{\mathbf{u}}_{\varepsilon_n,h_n} \in \mathscr{K}_{h_n}^{\Gamma}$ and $\bar{\mathbf{u}}_{\varepsilon_n,h_n} \to \mathbf{u}$ in $\mathscr{V} := \widetilde{\mathbf{H}}^{1/2}(\Gamma_0)$.

Using (11.95), we obtain

$$\begin{aligned} &c\|\bar{\mathbf{u}}_{\varepsilon_n,h_n} - \mathbf{u}_{\varepsilon_n,h_n}\|_{\mathscr{V}}^2 \\ &\le \langle S_h(\bar{\mathbf{u}}_{\varepsilon_n,h_n} - \mathbf{u}_{\varepsilon_n,h_n}), \bar{\mathbf{u}}_{\varepsilon_n,h_n} - \mathbf{u}_{\varepsilon_n,h_n}\rangle \\ &= \langle S_h\bar{\mathbf{u}}_{\varepsilon_n,h_n}, \bar{\mathbf{u}}_{\varepsilon_n,h_n} - \mathbf{u}_{\varepsilon_n,h_n}\rangle - \langle S_h\mathbf{u}_{\varepsilon_n,h_n}, \bar{\mathbf{u}}_{\varepsilon_n,h_n} - \mathbf{u}_{\varepsilon_n,h_n}\rangle. \end{aligned} \tag{11.109}$$

Since $\bar{\mathbf{u}}_{\varepsilon_n,h_n} \to \mathbf{u}$ in $\mathscr{V}$ and $\mathbf{u}_{\varepsilon_n,h_n} \rightharpoonup \mathbf{u}$ in $\mathscr{V}$, it follows from Lemma 11.6 (ii) that the first term on the right-hand side of (11.109) tends to zero.

Using the definition of (P_{ε_n,h_n}), inequality (11.98), the second term can be estimated as follows:

$$\begin{aligned} &|\langle S_h\mathbf{u}_{\varepsilon_n,h_n}, \mathbf{u}_{\varepsilon_n,h_n} - \bar{\mathbf{u}}_{\varepsilon_n,h_n}\rangle| \\ &\le |\langle \mathbf{g}, \mathbf{u}_{\varepsilon_n,h_n} - \bar{\mathbf{u}}_{\varepsilon_n,h_n}\rangle| + |\langle DJ_{\varepsilon_n,h_n}(\mathbf{u}_{\varepsilon_n,h_n}), \bar{\mathbf{u}}_{\varepsilon_n,h_n} - \mathbf{u}_{\varepsilon_n,h_n}\rangle|, \end{aligned} \tag{11.110}$$

where

$$\begin{aligned} &|\langle DJ_{\varepsilon_n,h_n}(\mathbf{u}_{\varepsilon_n,h_n}), \bar{\mathbf{u}}_{n,h_n} - \mathbf{u}_{n,h_n}\rangle| \\ &= \left|\int_{\Gamma_C} \frac{\partial\tilde{S}}{\partial x}(L_{h_n}(\Pi\mathbf{u}_{\varepsilon_n,h_n}), \varepsilon_n) L_{h_n}(\Pi(\bar{\mathbf{u}}_{\varepsilon_n,h_n} - \mathbf{u}_{\varepsilon_n,h_n}))\, ds\right| \\ &\le \|\frac{\partial\tilde{S}}{\partial x}(L_{h_n}(\Pi\mathbf{u}_{\varepsilon_n,h_n}), \varepsilon_n)\|_{L^2(\Gamma_C)} \|L_{h_n}(\Pi(\bar{\mathbf{u}}_{\varepsilon_n,h_n} - \mathbf{u}_{\varepsilon_n,h_n}))\|_{L^2(\Gamma_C)} \end{aligned}$$

converges to zero, as follows from the boundedness of $\{\mathbf{u}_{\varepsilon_n,h_n}\}$ in $\widetilde{\mathbf{H}}^{1/2}(\Gamma_0)$, (11.107) and the boundedness of $\{\frac{\partial\tilde{S}}{\partial x}(L_{h_n}(\Pi\mathbf{u}_{\varepsilon_n,h_n}), \varepsilon_n)\}$ in $L^2(\Gamma)$, what we show next.

From

$$\exists c > 0 : \left|\frac{\partial\tilde{S}}{\partial x}(\varepsilon, x)\right| \le c(1 + |x|) \quad \forall x \in \mathbb{R},$$

the elementary inequality $(a+b)^2 \leq 2(a^2+b^2)$ and integration over Γ we obtain

$$\int_{\Gamma_C} \left| \frac{\partial \tilde{S}}{\partial x}(\varepsilon, L_h(\Pi \mathbf{u}_h(s)) \right|^2 ds \leq 2c^2 \operatorname{meas}(\Gamma_C) + 2c^2 \|L_h(\Pi \mathbf{u}_h)\|^2_{L^2(\Gamma_C)}.$$

Hence,

$$\begin{aligned} \left\| \frac{\partial \tilde{S}}{\partial x}(L_h(\Pi \mathbf{u}_h), \varepsilon) \right\|_{L^2(\Gamma_C)} &\leq \left(2c^2 \operatorname{meas}(\Gamma_C) + 2c^2 \|L_h(\Pi \mathbf{u}_h)\|^2_{L^2(\Gamma_C)}\right)^{1/2} \\ &\leq \sqrt{2}c \left((\operatorname{meas}(\Gamma_C))^{1/2} + \|L_h(\Pi \mathbf{u}_h)\|_{L^2(\Gamma_C)}\right) \\ &\leq \tilde{c}(1 + \|\mathbf{u}_h\|_{\mathbf{H}^{1/2}(\Gamma_C)}), \end{aligned} \tag{11.111}$$

where we have used (11.102) and (11.97), and the elementary inequality $\sqrt{a^2+b^2} \leq |a| + |b|$.

Passing now to the limit superior in (11.111), we get

$$\limsup_{n\to\infty} \langle S_h \mathbf{u}_{\varepsilon_n,h_n}, \mathbf{u}_{\varepsilon_n,h_n} - \bar{\mathbf{u}}_{\varepsilon_n,h_n} \rangle \leq 0.$$

Hence, (11.109) entails in the limit

$$\|\bar{\mathbf{u}}_{\varepsilon_n,h_n} - \mathbf{u}_{\varepsilon_n,h_n}\|_{\mathscr{V}} \to 0.$$

Finally, from the triangle inequality

$$\|\mathbf{u}_{\varepsilon_n,h_n} - \mathbf{u}\|_{\mathscr{V}} \leq \|\mathbf{u}_{\varepsilon_n,h_n} - \bar{\mathbf{u}}_{\varepsilon_n,h_n}\|_{\mathscr{V}} + \|\bar{\mathbf{u}}_{\varepsilon_n,h_n} - u\|_{\mathscr{V}},$$

we get the strong convergence of $\{\mathbf{u}_{\varepsilon_n,h_n}\}$ to $\mathbf{u}$ in $\mathscr{V}$. □

As an advantage of the combination of regularization methods of nondifferentiable optimization with the h-BEM we arrive at smooth optimization problems at the discrete level which can be solved by standard optimization methods, like trust region methods [111]. For numerical experiments using h-FEM instead of h-BEM we can refer to [220].

Similar nonmontone contact problems from adhesion have been treated by the h-BEM directly in [326]. Then special nonsmooth optimization solver, like bundle methods, have to be employed at the discrete level. For the convergenc analysis of the h-BEM and numerical results for a similar benchmark problem along this latter approach we refer to [326].

For further reading we refer to [27, 147, 149, 150] where boundary integral equations and boundary element methods for related contact problems are treated, especially in [148] different adaptive methods are presented.

11.6 *hp*-BEM for Delamination Problems

To avoid domain approximation, let $\Omega \subset \mathbb{R}^d$, $d = 2, 3$, be a polygonal domain. Let $\mathscr{T}_h$ be a sufficiently fine finite element mesh of the boundary Γ respecting the decomposition of Γ into Γ_D, Γ_N and Γ_C, $p = (p_T)_{T\in\mathscr{T}_h}$ a polynomial degree distribution over $\mathscr{T}_h$, $\mathbb{P}_{p_T}(\hat{T})$ the space of polynomials of order p_T on the reference element $\hat{T}$, and $\Psi_T : \hat{T} \to T \in \mathscr{T}_h$ a bijective, (bi)-linear transformation. In 2D, $\hat{T}$ is the interval $[-1, 1]$, whereas in 3D it is the reference square $[-1, 1]^2$. Let $\Sigma_{T,hp}$ be the set of all $(p_T + 1)^{d-1}$ affinely transformed (tensor product based) Gauss-Lobatto nodes on the element T of the partition $\mathscr{T}_h$ of Γ, and set $\Sigma_{hp} := \bigcup_{T\in\mathscr{T}_h|\Gamma_C} \Sigma_{T,hp}$, see [218, 275, 297]. Furthermore, we assume in this section that $g \in C^0(\overline{\Gamma}_C)$ to allow point evaluation.

For the discretization of the displacement $\mathbf{u}$ we use

$$\mathscr{V}_{hp} = \{\mathbf{v_{hp}} \in \mathbf{C}^0(\Gamma) : \mathbf{v_{hp}}|_T \circ \Psi_T \in [\mathbb{P}_{p_T}(\hat{T})]^d \quad \forall T \in \mathscr{T}_h,\ \mathbf{v_{hp}} = 0 \text{ on } \overline{\Gamma}_D\},$$

$$\mathscr{K}_{hp}^{\Gamma} = \{\mathbf{v_{hp}} \in \mathscr{V}_{hp} : (\mathbf{v_{hp}} \cdot \mathbf{n})(P_i) \le g(P_i) \quad \forall P_i \in \Sigma_{hp}\}.$$

In general $\mathscr{K}_{hp}^{\Gamma} \not\subseteq \mathscr{K}^{\Gamma}$. For the approximation S_{hp} of the Poincaré-Steklov operator, we need the space

$$\mathscr{W}_{hp} = \{\psi_{hp} \in \mathbf{L}^2(\Gamma) : \psi_{hp}|_T \circ \Psi_T \in [\mathbb{P}_{p_T-1}(\hat{T})]^d \quad \forall T \in \mathscr{T}_h\} \subset \mathbf{H}^{-1/2}(\Gamma).$$

Now, we turn to the discretization of the regularized problem $(\mathscr{P}_\varepsilon)$, see (5.29) in Sect. 5.3. The discretized regularized problem $(\mathscr{S}_{\varepsilon,hp})$ is: Find $\mathbf{u}_{hp}^{\varepsilon} \in \mathscr{K}_{hp}^{\Gamma}$ such that for all $\mathbf{v}_{hp} \in \mathscr{K}_{hp}^{\Gamma}$

$$\langle S_{hp}\mathbf{u}_{hp}^{\varepsilon}, \mathbf{v}_{hp} - \mathbf{u}_{hp}^{\varepsilon}\rangle_{\Gamma_0} + \langle DJ_\varepsilon(\mathbf{u}_{hp}^{\varepsilon}), \mathbf{v}_{hp} - \mathbf{u}_{hp}^{\varepsilon}\rangle_{\Gamma_C} \ge \langle \mathbf{F}, \mathbf{v}_{hp} - \mathbf{u}_{hp}^{\varepsilon}\rangle_{\Gamma_0}. \qquad (11.112)$$

Lemma 11.7 *Let $\mathbf{u}_\varepsilon \in \mathscr{K}^{\Gamma}$ be the solution of the problem $(\mathscr{P}_\varepsilon)$ and let $\mathbf{u}_{hp}^{\varepsilon} \in \mathscr{K}_{hp}^{\Gamma}$ be the solution of the problem $(\mathscr{P}_{\varepsilon,hp})$. Assume that $\alpha_0 < c_S$ in (5.30), where c_S is the coerciveness constant of S, further $\mathbf{u}_\varepsilon \in \mathbf{H}^{3/2}(\Gamma)$, $g \in H^{3/2}(\Gamma_C)$ and $S\mathbf{u}_\varepsilon - \mathbf{F} \in \mathbf{L}^2(\Gamma)$. Then there exists a constant $c = c(\mathbf{u}_\varepsilon, g, \mathbf{F}) > 0$, but independent of h and p such that*

$$\begin{aligned}
c\|\mathbf{u}_\varepsilon - \mathbf{u}_{hp}^{\varepsilon}\|_{\mathbf{H}^{1/2}(\Gamma)}^2 &\le \|E_{hp}(\mathbf{u}_\varepsilon)\|_{\mathbf{H}^{-1/2}(\Gamma)}^2 \\
&+ \inf_{v\in\mathscr{K}^{\Gamma}} \{\|S\mathbf{u}_\varepsilon - \mathbf{F}\|_{\mathbf{L}^2(\Gamma)}\|\mathbf{u}_{hp}^{\varepsilon} - \mathbf{v}\|_{\mathbf{L}^2(\Gamma)} + \langle DJ_\varepsilon(\mathbf{u}_\varepsilon), \mathbf{v} - \mathbf{u}_{hp}^{\varepsilon}\rangle_{\Gamma_C}\} \\
&+ \inf_{v_{hp}\in\mathscr{K}_{hp}^{\Gamma}} \Big\{\|\mathbf{u}_\varepsilon - \mathbf{v}_{hp}\|_{\mathbf{H}^{1/2}(\Gamma)}^2 + \|S\mathbf{u}_\varepsilon - \mathbf{F}\|_{\mathbf{L}^2(\Gamma)}\|\mathbf{u}_\varepsilon - \mathbf{v}_{hp}\|_{\mathbf{L}^2(\Gamma)} \\
&+ \langle DJ_\varepsilon(\mathbf{u}_{hp}^{\varepsilon}), \mathbf{v}_{hp} - \mathbf{u}_\varepsilon\rangle_{\Gamma_C}\Big\}.
\end{aligned}$$

Proof Using the definitions of $(\mathscr{P}_\varepsilon)$ and $(\mathscr{P}_{\varepsilon,hp})$, and estimates similar to [297, Theorem 3], we obtain for all $v \in \mathscr{K}^\Gamma$, $v_{hp} \in \mathscr{K}^\Gamma_{hp}$

$$\begin{aligned} c_P \|\mathbf{u}_\varepsilon - \mathbf{u}^\varepsilon_{hp}\|^2_{\mathbf{H}^{1/2}(\Gamma)} &\le \|E_{hp}(\mathbf{u}_\varepsilon)\|^2_{\mathbf{H}^{-1/2}(\Gamma)} + \|\mathbf{u}_\varepsilon - \mathbf{v}_{hp}\|^2_{\mathbf{H}^{1/2}(\Gamma)} \\ &+ \|S\mathbf{u}_\varepsilon - \mathbf{F}\|_{\mathbf{L}^2(\Gamma)} \left(\|\mathbf{u}_\varepsilon - \mathbf{v}_{hp}\|_{\mathbf{L}^2(\Gamma)} + \|\mathbf{u}^\varepsilon_{hp} - \mathbf{v}\|_{\mathbf{L}^2(\Gamma)} \right) + \mathrm{D}, \end{aligned}$$

where we abbreviate

$$\mathrm{D} = \langle DJ_\varepsilon(\mathbf{u}_\varepsilon), \mathbf{v} - \mathbf{u}_\varepsilon \rangle_{\Gamma_C} + \langle DJ_\varepsilon(\mathbf{u}^\varepsilon_{hp}), \mathbf{v}_{hp} - \mathbf{u}^\varepsilon_{hp} \rangle_{\Gamma_C}.$$

To bound the term D, we use (5.30) and estimate as follows:

$$\begin{aligned} \mathrm{D} &= \langle DJ_\varepsilon(\mathbf{u}_\varepsilon), \mathbf{v} - \mathbf{u}^\varepsilon_{hp} \rangle_{\Gamma_C} + \langle DJ_\varepsilon(\mathbf{u}^\varepsilon_{hp}), \mathbf{v}_{hp} - \mathbf{u}_\varepsilon \rangle_{\Gamma_C} \\ &\quad + \langle DJ_\varepsilon(\mathbf{u}_\varepsilon) - DJ_\varepsilon(\mathbf{u}^\varepsilon_{hp}), \mathbf{u}^\varepsilon_{hp} - \mathbf{u}_\varepsilon \rangle_{\Gamma_C} \\ &\le \langle DJ_\varepsilon(\mathbf{u}_\varepsilon), \mathbf{v} - \mathbf{u}^\varepsilon_{hp} \rangle_{\Gamma_C} + \langle DJ_\varepsilon(\mathbf{u}^\varepsilon_{hp}), \mathbf{v}_{hp} - \mathbf{u}_\varepsilon \rangle_{\Gamma_C} + \alpha_0 \|\mathbf{u}_\varepsilon - \mathbf{u}^\varepsilon_{hp}\|^2_{\mathscr{V}}. \end{aligned}$$

Therefore, since $\alpha_0 < c_S$ by assumption, we obtain the assertion. □

Theorem 11.17 *Let $\mathbf{u}_\varepsilon \in \mathscr{K}^\Gamma$ be the solution of the problem $(\mathscr{P}_\varepsilon)$ and let $\mathbf{u}^\varepsilon_{hp} \in \mathscr{K}^\Gamma_{hp}$ be the solution of the problem $(\mathscr{P}_{\varepsilon,hp})$. Assume that $\alpha_0 < c_S$ in (5.30), $\mathbf{u}_\varepsilon \in \mathbf{H}^{3/2}(\Gamma)$, $g \in H^{3/2}(\Gamma_C)$ and $S\mathbf{u}_\varepsilon - \mathbf{F} \in \mathbf{L}^2(\Gamma)$. Then there exists a constant $c = c(\mathbf{u}_\varepsilon, g, \mathbf{F}) > 0$, but independent of h and p such that*

$$\|\mathbf{u}_\varepsilon - \mathbf{u}^\varepsilon_{hp}\|_{\mathbf{H}^{1/2}(\Gamma)} \le c h^{1/4} p^{-1/4}. \tag{11.113}$$

Proof Taking into account the estimates obtained by Maischak and Stephan in their Theorem 3 in [297] for the consistency error, the approximation error, and for $\|E_{hp}\mathbf{u}\|_{\mathbf{H}^{-1/2}(\Gamma)}$, we only need to estimate

$$\langle DJ_\varepsilon(\mathbf{u}_\varepsilon), \mathbf{v}^* - \mathbf{u}^\varepsilon_{hp} \rangle_{\Gamma_C} \tag{11.114}$$

and

$$\langle DJ_\varepsilon(\mathbf{u}^\varepsilon_{hp}), \mathbf{v}_{hp} - \mathbf{u}_\varepsilon \rangle_{\Gamma_C}. \tag{11.115}$$

To estimate (11.114)–(11.115) we must consider the same test functions $\mathbf{v}^*$ and $\mathbf{v}_{hp}$ as in [297, Theorem 3] used to estimate the standard error terms. Let $\mathbf{v}^* \in \mathscr{K}^\Gamma \cap \mathbf{H}^1(\Gamma)$ be defined by

$$\mathbf{v}^* := \begin{cases} \mathbf{u}^\varepsilon_{hp,t} + [g + \inf\{u^\varepsilon_{hp,n} - g_{hp}, 0\}]\mathbf{n} & \text{on } \Gamma_C \\ 0 & \text{on } \Gamma_D \\ \gamma_N \mathbf{u}^\varepsilon_{hp} & \text{on } \Gamma_N, \end{cases}$$

where $g_{hp} := I_{hp} g$ is the interpolate of the gap function g, and γ_N is the trace map onto Γ_N.

As shown in [334, Lemma 2]

$$\begin{aligned}\langle DJ_\varepsilon(\mathbf{u}_\varepsilon), \mathbf{v}^* - \mathbf{u}^\varepsilon_{hp}\rangle_{\Gamma_C} &= \int_{\Gamma_C} \frac{\partial}{\partial x}\tilde{S}(u_{\varepsilon,n}, \varepsilon)(v_n^* - u^\varepsilon_{hp,n})\, ds \\ &\leq c\left(1 + \|u_{\varepsilon,n}\|_{L^2(\Gamma_C)}\right) \|v_n^* - u^\varepsilon_{hp,n}\|_{L^2(\Gamma_C)}. \end{aligned} \tag{11.116}$$

The elaborate analysis in [297], see the proof of Theorem 3 there, gives

$$\|v_n^* - u^\varepsilon_{hp,n}\|_{L^2(\Gamma_C)} \leq C_2 h^{1/2} p^{-1/2} \left(\|g\|_{H^{1/2}(\Gamma_C)} + \|\mathbf{u}^\varepsilon_{hp}\|_{H^{1/2}(\Gamma_C)}\right). \tag{11.117}$$

Further, let $\mathbf{v}_{hp} := I_{hp}\mathbf{u}_\varepsilon \in \mathscr{K}^\Gamma_{hp}$ be the interpolate of $\mathbf{u}_\varepsilon \in \mathbf{H}^{3/2}(\Gamma) \subset C^0(\Gamma)$. Analogously to (11.116), we have

$$\begin{aligned}\langle DJ_\varepsilon(\mathbf{u}^\varepsilon_{hp}), \mathbf{v}_{hp} - \mathbf{u}_\varepsilon\rangle_{\Gamma_C} &= \int_{\Gamma_C} \frac{\partial}{\partial x}\tilde{S}(u^\varepsilon_{hp,n}, \varepsilon)(v_{hp,n} - u_{\varepsilon,n})\, ds \\ &\leq c\left(1 + \|u^\varepsilon_{hp,n}\|_{L^2(\Gamma_C)}\right) \|u_{\varepsilon,n} - v_{hp,n}\|_{L^2(\Gamma_C)}. \end{aligned} \tag{11.118}$$

By [47, Theorems 4.2 and 4.5] and by the real interpolation between $H^1(\Gamma)$ and $L^2(\Gamma)$ there exists a constant $C_1 > 0$ such that

$$\|u_{\varepsilon,n} - v_{hp,n}\|_{H^{1/2}(\Gamma)} \leq C_1 h^1 p^{-1} \|\mathbf{u}_\varepsilon\|_{H^{3/2}(\Gamma)}. \tag{11.119}$$

Finally, combining the error estimates for the interpolation (11.119) and the consistency (11.117) with (11.118) and (11.116), respectively, and taking into account the boundedness of $\|u^\varepsilon_{hp,n}\|$ in $H^{1/2}(\Gamma_C)$, we prove the asserted bound for (11.114) and (11.115). □

To be able to split the approximation error into the discretization error of a simpler variational equation and contributions arising from the constraints on Γ_C we introduce the mixed regularized formulation (11.120a)–(11.120b), which is equivalent to the regularized problem $(\mathscr{P}_\varepsilon)$.

Find $(\mathbf{u}^\varepsilon, \lambda^\varepsilon) \in \mathscr{V} \times M(\mathbf{u}^\varepsilon)$ such that

$$\left\langle S\mathbf{u}^\varepsilon, \mathbf{v}\right\rangle_{\Gamma_0} + \left\langle \lambda^\varepsilon, v_n\right\rangle_{\Gamma_C} = \langle \mathbf{F}, \mathbf{v}\rangle_{\Gamma_0} \quad \forall \mathbf{v} \in \mathscr{V} \tag{11.120a}$$

$$\left\langle \mu - \lambda^\varepsilon, u_n^\varepsilon - g\right\rangle_{\Gamma_C} \leq 0 \qquad \forall \mu \in M(\mathbf{u}^\varepsilon) \tag{11.120b}$$

with the set of admissible Lagrange multipliers

$$M(\mathbf{u}^\varepsilon) := \left\{\mu \in X^* : \langle \mu, \eta\rangle_{\Gamma_C} \geq \left\langle DJ_\varepsilon(\mathbf{u}^\varepsilon), \eta\right\rangle_{\Gamma_C} \ \forall \eta \in X,\ \eta \geq 0 \text{ a.e. on } \Gamma_C\right\}$$

where $X = \{w \mid \exists \mathbf{v} \in \mathscr{V},\ v_n|_{\Gamma_C} = w\} \subseteq H^{1/2}(\Gamma_C)$ and X^* its dual space.

Lemma 11.8

(i) Let $\mathbf{u}^\varepsilon$ *solve the regularized problem* $(\mathscr{P}_\varepsilon)$*, then there exists a* $\lambda^\varepsilon \in M(\mathbf{u}^\varepsilon)$ *such that* $(\mathbf{u}^\varepsilon, \lambda^\varepsilon)$ *solves* (11.120).
(ii) Let $(\mathbf{u}^\varepsilon, \lambda^\varepsilon)$ *solve* (11.120)*, then* $\mathbf{u}^\varepsilon$ *solves* $(\mathscr{P}_\varepsilon)$.

Given the discrete solution $\mathbf{u}^\varepsilon_{hp} \in \mathscr{K}^\Gamma_{hp}$ to $(\mathscr{P}_{\varepsilon,hp})$, we reconstruct $\lambda^\varepsilon_{hp} \in \operatorname{span}\{\psi_i\}_{i=1}^M$ such that

$$\left\langle \lambda^\varepsilon_{hp}, v_n \right\rangle_{\Gamma_C} = \langle \mathbf{F}, \mathbf{v}\rangle_{\Gamma_0} - \left\langle S_{hp}\mathbf{u}^\varepsilon_{hp}, \mathbf{v}\right\rangle_{\Gamma_0} \quad \forall\, \mathbf{v} \in \mathscr{V}_{hp} \tag{11.121}$$

by solving a potentially over-constrained system of linear equations for an arbitrary choice of basis $\{\psi\}$.

Following the Braess trick [59] as e.g. in [37], we define the auxiliary problem

$$\mathbf{z} \in \mathscr{V}: \quad \langle S\mathbf{z}, \mathbf{v}\rangle_{\Gamma_0} = \langle \mathbf{F}, \mathbf{v}\rangle_{\Gamma_0} - \left\langle \lambda^\varepsilon_{hp}, v_n\right\rangle_{\Gamma_C} \quad \forall\, \mathbf{v} \in \mathscr{V}. \tag{11.122}$$

Subtracting (11.120a) and (11.122) yields

$$\left\langle S(\mathbf{u}^\varepsilon - \mathbf{z}), \mathbf{v}\right\rangle_{\Gamma_0} = \left\langle \lambda^\varepsilon_{hp} - \lambda^\varepsilon, v_n\right\rangle_{\Gamma_C} \quad \forall\, \mathbf{v} \in \mathscr{V} \tag{11.123}$$

and additionally with the continuous inf-sup condition [101, Theorem 3.2.1] this yields (see [37])

$$\left\| \lambda^\varepsilon_{hp} - \lambda^\varepsilon \right\|_{X^*} \le \frac{C}{\beta} \left\| \mathbf{u}^\varepsilon - \mathbf{z} \right\|_{\mathscr{V}} \le \frac{C}{\beta} \left\| \mathbf{u}^\varepsilon - \mathbf{u}^\varepsilon_{hp} \right\|_{\mathscr{V}} + \frac{C}{\beta} \left\| \mathbf{u}^\varepsilon_{hp} - \mathbf{z} \right\|_{\mathscr{V}} \tag{11.124}$$

with inf-sup constant $\beta > 0$. See [101, Theorem 3.2.1] for a proof of the inf-sup condition for the difficult case when $\bar{\Gamma}_C \cap \bar{\Gamma}_D = \emptyset$, i.e. $X^* = \tilde{H}^{-1/2}(\Gamma_C)$.

Theorem 11.18 *Under the assumption* (5.32) *and if* $\frac{\partial}{\partial x}\tilde{S}(\cdot, \varepsilon)$ *is Lipschitz continuous, then there exists a constant* C *independent of* h *and* p *such that for* $0 < \varsigma < \frac{c_P - \alpha_0}{4}$ *arbitrary*

$$\begin{aligned}(c_P - \alpha_0 - 4\varsigma)\, \|\mathbf{u}^\varepsilon - \mathbf{u}^\varepsilon_{hp}\|^2_{\mathscr{V}} \le{}& \left(\frac{C}{\varsigma} + 1\right) \|\mathbf{z} - \mathbf{u}^\varepsilon_{hp}\|^2_{\mathscr{V}} + \frac{1}{4\varsigma} \left\| (\lambda^\varepsilon_{hp} - DJ_\varepsilon(\mathbf{u}^\varepsilon_{hp}))^- \right\|^2_{X^*} \\ &+ C\left(\frac{1}{\varsigma} + \frac{1}{\beta^2} + \frac{1}{\varsigma\beta^2}\right) \left\| (u^\varepsilon_{hp,n} - g)^+ \right\|^2_X \\ &- \left\langle (\lambda^\varepsilon_{hp} - DJ_\varepsilon(\mathbf{u}^\varepsilon_{hp}))^+, (u^\varepsilon_{hp,n} - g)^- \right\rangle_{\Gamma_C}\end{aligned}$$

with $(\mathbf{u}^\varepsilon_{hp}, \lambda^\varepsilon_{hp})$ *satisfying* (11.121)*,* $\mathbf{z}$ *solving* (11.122) *and* $\mathbf{u}^\varepsilon$ *solving* $(\mathscr{P}_\varepsilon)$.

The a-posteriori error estimate decomposes into the discretization error of a variational equality $\|\mathbf{z} - \mathbf{u}^{\varepsilon}_{hp}\|^2_{\mathcal{V}}$, which can be further estimated by e.g. residual error estimates [75] or bubble error estimates, e.g. [37], and violation of the consistency condition $\left\|(\lambda^{\varepsilon}_{hp} - DJ_{\varepsilon}(\mathbf{u}^{\varepsilon}_{hp}))^-\right\|^2_{X^*}$, violation of the non-penetration condition $\left\|(u^{\varepsilon}_{hp,n} - g)^+\right\|^2_{X}$ and violation of the complementarity condition $-\left\langle(\lambda^{\varepsilon}_{hp} - DJ_{\varepsilon}(\mathbf{u}^{\varepsilon}_{hp}))^+, (u^{\varepsilon}_{hp,n} - g)^-\right\rangle_{\Gamma_C}$. Localizing an approximation of the global a-posteriori error estimate gives rise to the following solve-mark-refine algorithm for *hp*-adaptivity.

For the numerical experiments we choose $\Omega = (0, 1/2)^2$, $\Gamma_D = \{0\} \times [0, 1/2]$, $\Gamma_C = (0, 1/2] \times \{0\}$, $\Gamma_N = \partial\Omega \setminus (\Gamma_D \cup \Gamma_C)$. The material parameters are $E = 5$, $\nu = 0.45$, $\mathbf{f} \equiv 0$, $\mathbf{t} = 0.25$ on $[1/4, 1/2] \times \{1\}$ and zero elsewhere, $g = 0$. The delamination law is given via

$$\begin{aligned} f(u_n(x)) &= \min\{g_1(g(x) - u_n(x)), g_2(g(x) - u_n(x)), g_3(g(x) - u_n(x))\} \\ &= -\max\{-g_1(-u_n(x)), -g_2(-u_n(x)), -g_3(-u_n(x))\} \end{aligned}$$

with

$$g_1(y) = \frac{A_1}{2t_1}y^2, \qquad g_2(y) = b_2(y^2 - t_1^2) + d_2, \qquad g_3(y) = d_3$$

and parameters

$$A_1 = 0.05, \quad A_2 = 0.03, \quad t_1 = 0.02, \quad t_2 = 0.04,$$

$$b_2 = \frac{A_2}{2t_2}, \quad d_2 = A_1\frac{t_1}{2}, \quad d_3 = b_2(t_2^2 - t_1^2) + d_2.$$

The regularized delamination law S_x with regularization parameter $\varepsilon = 10^{-4}$ is plotted in Fig. 11.6. The characteristic saw tooth shape is already present, but the absolute value in the tips and the slope approximating the jump are still noticeable coarse approximated.

The discrete Lagrange multiplier $\lambda^{\varepsilon}_{hp}$ is obtain by solving (11.121) where ψ_i are discontinuous, piecewise polynomials on Γ_C on a one time coarsened mesh ($H = 2h$) with polynomial degree reduced by one ($q = p - 1$) compared to the mesh and polynomial degree distribution of $\mathbf{u}^{\varepsilon}_{hp}$. Figure 11.7 displays the deformation of the rectangle and the normal stresses on Γ_C obtained from the lowest order uniform h-method with 16384 elements and regularization parameter $\varepsilon = 10^{-4}$. The normal stress on Γ_C, Fig. 11.7b, reflects the delamination law from Fig. 11.6 well.

Figure 11.8 displays the reduction of the error in $(\mathbf{u}, \lambda)$ and of the error estimate. Since the exact solution is not known, we compute the error approximately by $\|\mathbf{u}_{fine} - \mathbf{u}_{hp}\|_S$ and $\|\lambda_{fine} - \lambda_{hp}\|_V$, with norms induced by the Poincaré-Steklov

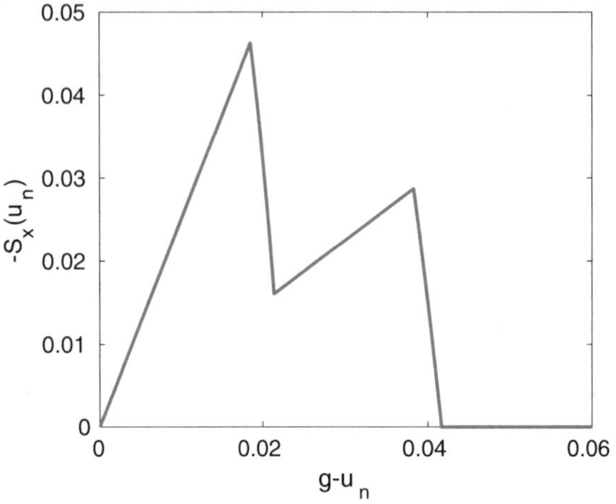

Fig. 11.6 Regularized delamination law S_x for $\varepsilon = 10^{-4}$

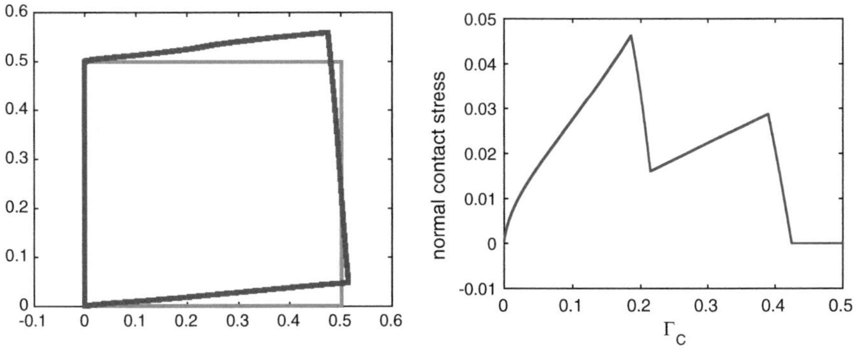

(a) Deformation

(b) Contact stresses

Fig. 11.7 Uniform h-version, p=1, 16384 elements with 4096 elements on Γ_C. (**a**) Deformation. (**b**) Contact stresses

operator S and the single layer potential V acting on Γ_C, respectively. The pair $(\mathbf{u}_{fine}, \lambda_{fine})$ is a very fine (last) approximation for each sequence of discretization.

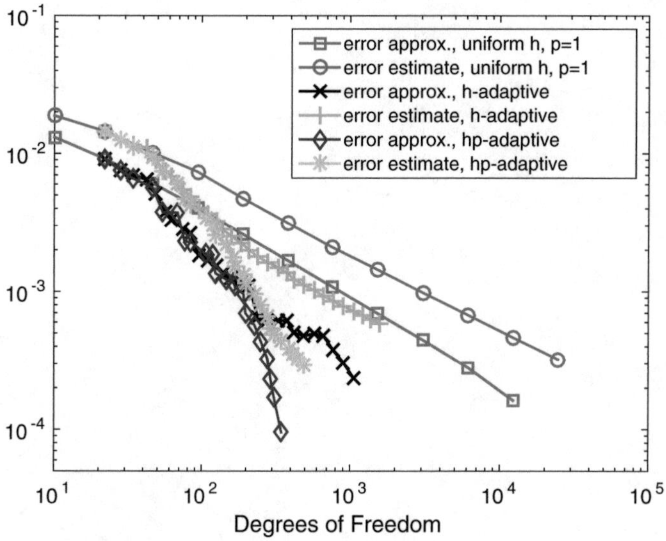

Fig. 11.8 $\sqrt{\|\mathbf{u}_{fine} - \mathbf{u}_{hp}\|_P^2 + \|\lambda_{fine} - \lambda_{hp}\|_V^2}$ and error estimate for different families of discrete solutions, $\varepsilon = 10^{-4}$ [334]

Chapter 12
FEM-BEM Coupling

The BEM is well established for the solution of linear elliptic boundary value problems. Its essential feature is the reduction of the partial differential equation in the domain to an integral equation on the surface. Then, for the numerical treatment, only the surface has to be discretized. This leads to a comparatively small number of unknowns. It is possible to solve problems in unbounded domains. In contrast, the FEM requires a discretization of the domain. However, when dealing with nonlinear problems, the latter method is more versatile. Typical examples for which the coupling of both methods is advantageous are rubber sealings and bearings that are located between construction elements made of steel, concrete, or glass. For these elements, linear elasticity often is a sufficient model, and the BEM is favorable.In contrast, for sealings and bearings the nonlinear material behavior imply that the FEM is preferable. Moreover, for rubberlike materials the incompressibility has to be taken into account, which requires mixed finite elements. Thus to combine the advantages of both discretization methods we are led to study FEM-BEM coupling, "marriage a‘ la mode" [439]. Here we focus first to symmetric coupling and consider two alternative approaches: (i) the abstract setting of saddle point problems introduced by Costabel and Stephan in [136] is reported in Sect. 12.1 with application to an elastic interface problem in Sect. 12.2.1, (ii) the use of the Poincare-Steklov operator in the variational formulation as given by Carstensen and Stephan in [91] is reported in Sect. 12.3 together with adaptive coupling versions using residual type error indicators in Sect. 12.3.2 and hierarchical type error indicators in Sect. 12.3.3. In Sect. 12.3.5 we report on other coupling methods like Johnson-Nedelec coupling and Bielak-MacCamy coupling. Other topics are least squares FEM-BEM couplings in Sect. 12.4 and FEM/BEM coupling for Signorini contact problems in Sect. 12.5 with a primal method in Sect. 12.5.1 and dual methods in Sect. 12.5.2. In Sect. 12.6 we consider a primal mixed FEM-BEM coupling for plane elasticity. An elliptic interface problem with a strongly nonlinear differential operator is considered in Sect. 12.7. In Sect. 12.8 the time-harmonic

© Springer International Publishing AG, part of Springer Nature 2018
J. Gwinner, E. P. Stephan, *Advanced Boundary Element Methods*,
Springer Series in Computational Mathematics 52,
https://doi.org/10.1007/978-3-319-92001-6_12

eddy-current problem is treated with Hiptmair's symmetric coupling method and a posteriori error estimates of residual type as well as of p-hierarchical type are presented. Section 12.9 presents a discontinuous (in time) Galerkin method for a parabolic-elliptic interface problem. For further reading we refer to the seminal works of M.Costabel [115] and H.Han [228]; see also [121, 132, 133, 439].

12.1 Abstract Framework of Some Saddle Point Problems

In this section we report from [136]. As investigated there the direct boundary integral equation method for the interface problem leads to an operator of the form $\begin{pmatrix} P & Q \\ Q' & -A \end{pmatrix}$ where P and A are strongly elliptic operators and Q' is the adjoint of Q. The same structure is shared by the equations that arise from the symmetric coupling method. This typical form leads to critical points of saddle point type for functionals that are strictly convex in one direction and strictly concave in the other direction. As we will see, this situation can be reduced to the study of the minimum of a strictly convex functional. Due to the strict monotonicity of both operators P and A, we do not need any analogue of the Babuška-Brezzi stability condition.

Let X, Y be reflexive Banach spaces with duals X', Y'. Let $P\colon X \to X'$ be a nonlinear operator. Let $Q\colon Y \to X'$ be a continuous, linear operator. Let $A\colon Y \to Y'$ be a continuous, linear, bijective operator. We define $P_1 : X \times Y \to X' \times Y'$ by

$$P_1 \begin{pmatrix} u \\ \phi \end{pmatrix} := \begin{bmatrix} P(u) + Q\phi \\ Q'u - A\phi \end{bmatrix}.$$

We identify P_1 with the matrix

$$P_1 = \begin{pmatrix} P & Q \\ Q' & -A \end{pmatrix}.$$

In addition, we define the operator $P_2\colon X \to X'$ by

$$P_2(u) := P(u) + QA^{-1}Q'u$$

Then, if $1_{X'}$ and $1_{Y'}$ denote the respective identity mappings, there holds the following relation:

$$\begin{pmatrix} 1_{X'} & QA^{-1} \\ 0 & 1_{Y'} \end{pmatrix} \cdot P_1 = \begin{pmatrix} P_2 & 0 \\ Q' & -A \end{pmatrix} \tag{12.1}$$

This relation is the key to the proof of the following equivalence theorem.

Theorem 12.1

(i) P_1 *is surjective if and only if* P_2 *is surjective.*
(ii) P_1 *is injective if and only if* P_2 *is injective.*
(iii) P_1 *has a bounded inverse if and only if* P_2 *has a bounded inverse.*

Proof The operator

$$\begin{pmatrix} 1_{X'} & QA^{-1} \\ 0 & 1_{Y'} \end{pmatrix}$$

on $X' \times Y'$ is a bounded, linear, bijective operator with bounded inverse

$$\begin{pmatrix} 1_{X'} & -QA^{-1} \\ 0 & 1_{Y'} \end{pmatrix}.$$

Therefore (12.1) shows that the statements of the theorem are equivalent to the following.

The operator $P_2 \colon X \to X'$ is surjective, injective, or boundedly invertible if and only if the operator

$$\begin{pmatrix} P_2 & 0 \\ Q' & -A \end{pmatrix} : X \times Y \to X' \times Y'$$

has the respective properties. The claimed equivalence is now clear, because A was assumed to be bijective. It can be seen from the equality

$$\begin{pmatrix} P_2 & 0 \\ Q' & -A \end{pmatrix}^{-1} = \begin{pmatrix} P_2^{-1} & 0 \\ A^{-1} Q' P_2^{-1} & -A^{-1} \end{pmatrix}$$ □

Next we assume that the problem is given in variational form, i.e. the operators are given as derivatives of functionals. As an example we present in Sect. 12.2.1 the elastoplastic interface problem for material obeying the Hencky-von Mises stress-strain relation. We introduce

- $J \colon X \to \mathbb{R}$ a functional defined everywhere.
- $Q \colon Y \to X'$ linear, continuous, $q \colon X \times Y \to \mathbb{R}$ bilinear form with $q(u, \phi) := \langle Q\phi, u \rangle = \langle Q'u, \phi \rangle$.
- $A \colon Y \to Y'$ linear, continuous, bijective, self-adjoint, $a \colon Y \to \mathbb{R}$, $a(\phi) := \frac{1}{2}\langle A\phi, \phi \rangle$.
- $J_1 \colon X \times Y \to \mathbb{R}$, $J_1(u, \phi) := J(u) + q(u, \phi) - a(\phi)$.
- $J_2 \colon X \to \mathbb{R}$, $J_2(u) := J(u) + \frac{1}{2} q(u, A^{-1}Q'u) = J(u) + \frac{1}{2}\langle QA^{-1}Q'u, u \rangle$.

If J is Gateaux differentiable, we denote its Gateaux derivative by $DJ(u) \in X'$ and its Gateaux derivative in direction w by $DJ(u, w) \in \mathbb{R}$.

Theorem 12.2 *Let J be differentiable on X. Then*

(i) J_1 and J_2 are differentiable
(ii) For $(u, \phi) \in X \times Y$ there holds

$$\begin{pmatrix} 1_{X'} & QA^{-1} \\ 0 & 1_{Y'} \end{pmatrix} \cdot DJ_1(u, \phi) = \begin{pmatrix} DJ_2(u) \\ Q'u - A\phi \end{pmatrix} \in X' \times Y' \tag{12.2}$$

(iii) For $u \in X$ there holds $DJ_2(u) = 0$ if and only if there exists $\phi \in Y$ with $DJ_1(u, \phi) = 0$. If this is satisfied, then

$$\phi = A^{-1}Q'u$$

Proof Let $u, w \in X$. Then from the definition of J_2 and the symmetry of A^{-1} there follows

$$DJ_2(u, w) = DJ(u, w) + \langle QA^{-1}Q'u, w\rangle,$$

hence $DJ_2(u) = DJ(u) + QA^{-1}Q'u$.

Let $\psi \in Y$. Then the definition of J_1 implies

$$DJ_1(u, \phi; w, \psi) = DJ(u; w) + \langle Q\phi, w\rangle + \langle Q\psi, u\rangle - \langle A\phi, \psi\rangle$$

hence

$$DJ_1(u, \phi) = \begin{pmatrix} DJ(u) + Q\phi \\ Q'u - A\phi \end{pmatrix};$$

thus J_1 and J_2 are differentiable and (12.2) is already verified, see (12.1) with $P = DJ$. In order to show *(iii)*, assume that $DJ_2(u) = 0$ is satisfied and define $\phi \in Y$ by $\phi := A^{-1}Q'u$. Then the right hand side in (12.2) vanishes, and therefore $DJ_1(u, \phi) = 0$. Conversely, (12.2) also shows that $DJ_1(u, \phi) = 0$ implies $DJ_2(u) = 0$ and $\phi = A^{-1}Q'u$. □

Lemma 12.1 *Assume J is twice continuously differentiable and there exist constants $\lambda, \Lambda > 0$ such that $\forall u, w \in X$:*

$$\lambda\|w\|_X^2 \leqslant D^2J(u; w, w) \leqslant \Lambda\|w\|_X^2. \tag{12.3}$$

Assume further that

$$a(\phi) \geqslant 0 \quad \forall\phi \in Y. \tag{12.4}$$

Then there is a $\Lambda' \geqslant \Lambda$ *such that for all* $u, w \in X$

$$\lambda \|w\|_X^2 \leqslant D^2 J_2(u; w, w) \leqslant \Lambda' \|w\|_X^2.$$

Proof The assertion follows from

$$D^2 J_2(u; w, w) = D^2 J(u; w, w) + \langle Q A^{-1} Q' w, w \rangle = D^2 J(u; w, w) + 2a(A^{-1} Q' w)$$

□

In the following we assume (12.3) and (9.19) to hold. As a corollary, we obtain an existence and uniqueness result for critical points of J_1.

Theorem 12.3

(i) J_2 *has exactly one critical point* $u \in X$. *This is a minimum.*
(ii) J_1 *has exactly one critical point* $(u, \phi) \in X \times Y$. *This is a saddle point:*

$$J_1(u, \phi + \psi) \leqslant J_1(u, \phi) \quad \forall \psi \in Y$$
$$J_1(u + w, \phi) \geqslant J_1(u, \phi) \quad \forall w \in X$$

Proof From Lemma 12.1 it is clear that J_2 has precisely one critical point u which is a minimum, because J_2 is coercive , lower semicontinuous, and strictly convex. From Theorem 12.2 we see that J_1 has a unique critical point $(u, \phi) \in X \times Y$ and that $Q'u = A\phi$ holds. Thus for $\psi \in Y$ we obtain with (9.19)

$$\begin{aligned} J_1(u, \phi + \psi) &= J(u) + \langle Q'u, \phi + \psi \rangle - \frac{1}{2}\langle A(\phi + \psi), \phi + \psi \rangle \\ &= J(u) - a(\phi) + \langle Q'u, \phi \rangle - a(\psi) = J_1(u, \phi) - a(\psi) \leqslant J_1(u, \phi). \end{aligned}$$

In order to show the second saddle point inequality we define a functional J_1^ϕ by $J_1^\phi(w) := J_1(w, \phi)$. It satisfies $D^2 J_1^\phi(w) = D^2 J(w)$ and it is strictly convex due to (12.3). Hence its critical point u is a minimum. □

12.2 Galerkin Approximation of Saddle Point Problems

Suppose (12.3) and (12.4) hold, and J_1 has a unique critical point $(u, \phi) \in X \times Y$. Let $X_N \subset X$, $Y_N \subset Y$ be closed subspaces of finite dimension. Let $d_X(w, X_N) := \inf\{\|w - v\|_X \mid v \in X_N\}$, $d_Y(\psi, Y_N)$ denote the distances to X_N, Y_N. The restriction of J_1 to $X_N \times Y_N$ inherits all relevant properties from J_1. Thus due to Theorem 12.3 it has exactly one critical point $(u_N, \phi_N) \in X_N \times Y_N$. There holds

Theorem 12.4 *There exist exactly one* $(u_N, \phi_N) \in X_N \times Y_N$ *such that* $DJ_1(u_N, \phi_N; w, \psi) = 0$ *for all* $(w, \psi) \in X_N \times Y_N$.

There exists $C > 0$ *independent of* X_N, Y_N *such that*

$$\|u - u_N\|_X + \|\phi - \phi_N\|_Y \leqslant C\,(d_X(u, X_N) + d_Y(\phi, Y_N))$$

To prove Theorem 12.4 we need some elementary consequences of the strong ellipticity assumption (12.3).

Lemma 12.2 *Let* $\Phi : X \to \mathbb{R}$ *be twice continuously differentiable and assume there exist* $\lambda, \Lambda > 0$ *such that* $\lambda \|w\|_X^2 \leqslant D^2\Phi(v; w, w) \leqslant \Lambda \|w\|_X^2$ $\forall v, w \in X$. *Then*

$$\lambda \|v - w\|_X^2 \leqslant D\Phi(v; v - w) - D\Phi(w; v - w) \quad \forall v, w \in X \tag{12.5}$$

and

$$\frac{\lambda}{2}\|v - w\|_X^2 \leqslant \Phi(v) - \Phi(w) - D\Phi(w; v - w) \leqslant \frac{\Lambda}{2}\|v - w\|_X^2, \quad \forall v, w \in X \tag{12.6}$$

The functional Φ *has a unique minimum* $u \in X$. *The restriction of* Φ *to* X_N *has a unique minimum* $u_N^* \in X_N$ *and there holds*

$$\frac{\lambda}{2}\|u - u_N^*\|_X^2 \leqslant \Phi(u_N^*) - \Phi(u) \leqslant \frac{\Lambda}{2} d_X(u, X_N)^2 \tag{12.7}$$

Proof (12.5) follows from

$$\lambda \|v - w\|_X^2 \leqslant \int_0^1 D^2\Phi(w + t(v - w); v - w, v - w)dt = D\Phi(v; v - w) - D\Phi(w; v - w). \tag{12.8}$$

similarly (12.6) follows from

$$\begin{aligned} \frac{\lambda}{2}\|v - w\|_X^2 &\leqslant \int_0^1 \int_0^1 D^2\Phi(w + t\tau(v - w); v - w, v - w)t\,dt\,d\tau \\ &= \Phi(v) - \Phi(w) - D\Phi(w; v - w) \\ &\leqslant \frac{\Lambda}{2}\|v - w\|_X^2 \end{aligned}$$

Finally for (12.7) we choose an arbitrary $w \in X_N$ and obtain from $D\Phi(u; v) = 0 \forall v \in X$, $\Phi(w) \geqslant \Phi(u_N^*)$ and (12.6)

$$\frac{\lambda}{2}\|u - u_N^*\|_X^2 \leqslant \Phi(u_N^*) - \Phi(u) \leqslant \Phi(w) - \Phi(u) \leqslant \frac{\Lambda}{2}\|w - u\|_X^2$$

□

Now let J_{1N} be the restriction of J_1 to $X_N \times Y_N$. Then we have for $(w, \psi) \in X_N \times Y_N$

$$\begin{aligned} J_{1N}(w, \psi) &= J(w) + q(w, \psi) - a(\psi) \\ &= J(w) + \langle Q_N\psi, w\rangle - \frac{1}{2}\langle A_N\psi, \psi\rangle \end{aligned}$$

where the operators $Q_N : Y_N \to X'_N$, $A_N : Y_N \to Y'_N$ are defined by the relations:

$$\begin{aligned} \langle Q_N\psi, w\rangle &= \langle Q\psi, w\rangle \quad \forall (w, \psi) \in X_N \times Y_N \\ \langle A_N\psi, \chi\rangle &= \langle A\psi, \chi\rangle \quad \forall (\psi, \chi) \in Y_N \times Y_N. \end{aligned}$$

Note that $A_N : Y_N \to Y'_N$ is invertible, see Theorem 1.1 i).

Lemma 12.3 *Let $t \in Y'$ be given and $\psi := A^{-1}t \in Y$, $\psi_N := A_N^{-1}P_N t \in Y_N$ where $P_N : Y' \to Y'_N$ is the natural projection, i.e. $\langle P_N t, \chi\rangle = \langle t, \chi\rangle \quad \forall \chi \in Y_N$. Then*

$$\|\psi - \psi_N\|_Y \leqslant C \cdot d_Y(\psi, Y_N) \tag{12.9}$$

Proof ψ_N solves the Galerkin equation

$$\langle A\psi_N, \chi\rangle = \langle t, \chi\rangle \quad \forall \chi \in Y_N.$$

Hence (12.9) is the quasi-optimality of the Galerkin error for the self-adjoint operator A, see Theorem 1.1 iii) in the more special situation of a Hilbert space Y. □

Proof (of Theorem 12.4) Define

$$J_{2N}(w) := J(w) + \frac{1}{2}\langle Q_N A_N^{-1} Q'_N w, w\rangle \quad \forall w \in X_N$$

Note J_{2N} does <u>not</u> coincide with the restriction of J_2 to X_N. There are unique critical points u_N, u_N^* satisfying

$$\begin{aligned} DJ_{2N}(u_N; w) &= DJ(u_N; w) + \langle Q_N A_N^{-1} Q'_N u_N, w\rangle = 0 \quad \forall w \in X_N \\ DJ_2(u_N^*; w) &= DJ(u_N^*, w) + \langle Q A^{-1} Q' u_N^*, w\rangle = 0 \qquad \forall w \in X_N \end{aligned} \tag{12.10}$$

Note

$$DJ_2(u; w) = DJ(u; w) + \langle QA^{-1}Q'u, w\rangle = 0 \quad \forall w \in X.$$

Now by Lemma 12.2, (12.7)

$$\|u - u_N^*\|_X \leqslant \sqrt{\frac{\Lambda}{\lambda}}\, d_X(u, X_N). \tag{12.11}$$

Thus, in order to obtain an error estimate for $u - u_N$, we estimate $u_N^* - u_N$ as follows: With $\Phi = J_{2N}$ on X_N we get

$$\begin{aligned}\lambda\|u_N^* - u_N\|_N^2 &\leqslant DJ_{2N}(u_N^*, u_N^* - u_N)\\ &= DJ(u_N^*, u_N^* - u_N) + \langle Q_N A_N^{-1} Q_N' u_N^*, u_N^* - u_N\rangle \end{aligned}\tag{12.12}$$

With (12.10) we can rewrite this as

$$\begin{aligned}&DJ_2(u_N^*; u_N^* - u_N) + \langle Q_N A_N^{-1} Q_N' u_N^*, u_N^* - u_N\rangle - \langle QA^{-1}Q'u_N^*, u_N^* - u_N\rangle\\ &= \langle\left(A_N^{-1}Q_N' - A^{-1}Q'\right)u_N^*, Q'\left(u_N^* - u_N\right)\rangle\\ &= \langle\left(A_N^{-1}P_N - A^{-1}\right)Q'u_N^*, Q'\left(u_N^* - u_N\right)\rangle\\ &\leq \left\{\|\left(A_N^{-1}P_N - A^{-1}\right)Q'(u_N^* - u)\|_Y + \|\left(A_N^{-1}P_N - A^{-1}\right)Q'u\|_Y\right\}\|Q'(u_N^* - u_N)\|_{Y'}\\ &\leqslant C\cdot\left\{\|u_N^* - u\|_X + d_Y\left(A^{-1}Q'u, Y_N\right)\right\}\cdot\|u_N^* - u_N\|_X\end{aligned}$$

Here we used

$$\langle\psi, Q_N'w\rangle = \langle Q_N\psi, w\rangle = \langle Q\psi, w\rangle = \langle\psi, Q'w\rangle = \langle\psi, P_N Q'w\rangle \quad \forall(w,\psi) \in X_N \times Y_N$$

and the stability of the Galerkin scheme for the operator A and (12.9).

Next, note $DJ_1(u, \phi) = 0$ with $\phi = A^{-1}Q'u$ by Theorem 12.2. Then (12.11), (12.12) yield

$$\|u_N^* - u_N\|_X \leqslant C\left(\|u_N^* - u\|_X + d_Y(\phi, Y_N)\right) \leqslant C\left(d_X(u, X_N) + d_Y(\phi, Y_N)\right)$$

and hence

$$\|u - u_N\|_X \leqslant C\left(d_X(u, Y_N) + d_Y(\phi, Y_N)\right). \tag{12.13}$$

In order to estimate $\|\phi_N - \phi\|_Y$ we write $\phi_N - \phi = \phi_N - \phi_N^* + \phi_N^* - \phi$ with $\phi_N^* := A_N^{-1}P_N Q'u$, where $\phi_N = A_N^{-1}Q_N' u_N = A_N^{-1}P_N Q' u_N$.

Therefore

$$\|\phi_N - \phi_N^*\|_Y = \|A_N^{-1} P_N Q'(u_N - u)\|_Y \leqslant C\|u_N - u\|_X$$

and again by (12.9,

$$\|\phi_N^* - \phi\|_Y = \| \left(A_N^{-1} P_N - A^{-1}\right) Q' u\|_Y \leqslant C \cdot d_Y(\phi, Y_N).$$

Together with (12.13) this gives

$$\|\phi_N - \phi\|_Y \leqslant C\left(d_X(u, X_N) + d_Y(\phi, Y_N)\right). \qquad \square$$

12.2.1 Symmetric FE/BE Coupling for a Nonlinear Interface Problem

Let us consider the interface problem **(TMP)**: For given F, u_0 find u_1, u_2 satisfying $P_1(u_1) = F$ in Ω_1 , $P_2(u_2) := \Delta_2^* u_2 = 0$ in Ω_2, $u_1 = u_2$, $t_1 = t_2$ on Γ_c, $u_1 = u_0$ on Γ_1 with the geometry as in Fig. 12.1, the Lamé operator Δ_2^* and $P_1(u_1)_l := \frac{\partial}{\partial x_l}(k - 2/3\mu(\Gamma(u_1))) \operatorname{div} u_1 + \sum_{j=1}^{3} 2\frac{\partial}{\partial x_l}\mu(\Gamma(u))e_{ij}(u_1), l = 1, 2, 3.$

Here the nonlinear material is described by the Hencky - von Mises stress-strain relation. We set

$$\Phi_1(u, w) := \int_{\Omega_1} \{k - 2/3\mu(\Gamma(u)) \operatorname{div} u \operatorname{div} w + \sum_{j=1}^{3} 2\mu(\Gamma(u))e_{ij}(u)e_{ij}(w)\}dx$$

with the bulk modulus k, the Lamé function $\mu(\Gamma)$ for the nonlinear material and the strain tensor e_{ij}. Then the weak formulation of (TMP) reads: Find

$$u \in H^1_{\Gamma_1}(\Omega_1) = \left\{u \in \left(H^1(\Omega_1)\right)^3, u = 0 \text{ on } \Gamma_1\right\}, \ \phi \in H^{-1/2}(\Gamma_c)$$

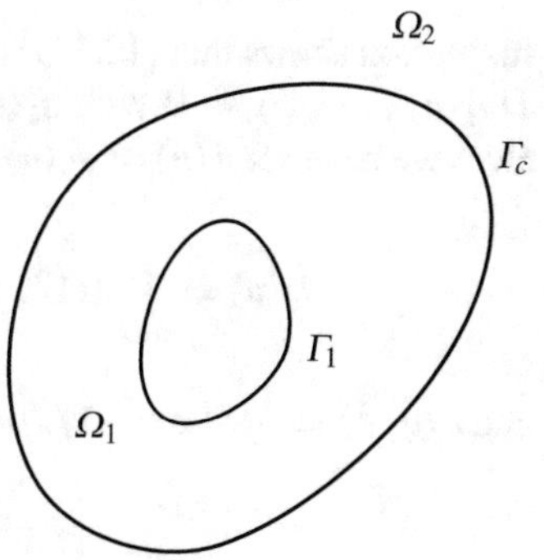

Fig. 12.1 Geometrical setting

such that

$$b(u, \phi; w; \psi) = l(w, \psi) \quad \forall w \in H^1_{\Gamma_1}(\Omega_1), \psi \in H^{-1/2}(\Gamma_c). \tag{12.14}$$

Here

$$b(u, \phi; w, \psi) = \Phi_1(u, w) - \int\limits_{\Gamma_c} \phi \cdot w ds + \left(\begin{pmatrix} z \\ -\psi \end{pmatrix}, (1 - \mathscr{C}_2) \begin{pmatrix} v \\ \phi \end{pmatrix} \right) \tag{12.15}$$

where

$$z = w|_{\Gamma_c}, \quad , v = u|_{\Gamma_c} \quad , \left(\begin{pmatrix} z \\ \psi \end{pmatrix}, \begin{pmatrix} v \\ \phi \end{pmatrix} \right) := \langle z, \phi \rangle + \langle v, \psi \rangle$$

and

$$\mathscr{C}_2 = \begin{pmatrix} 1/2 + \Lambda_2 & -V_2 \\ -W_2 & 1/2 - \Lambda_2' \end{pmatrix}, \; l(w, \psi) := \int\limits_{\Omega_1} F \cdot w dx \text{ for } F \in (L^2(\Omega_1))^3$$

Explicitly (12.15) becomes

$$b(u, \phi; w, \psi) = \Phi_1(u, w) + \langle z, W_2 v \rangle - \langle z, \left(\frac{1}{2} - \Lambda_2' \right) \phi \rangle - \langle \left(\frac{1}{2} - \Lambda_2 \right) v, \psi \rangle - \langle \psi, V_2 \phi \rangle$$

Here $V_2, \Lambda_2, \Lambda_2', W_2$ are the boundary integral operators of single layer, double layer, and it's adjoint and it's traction for the Lamé operator.

We use $\mathscr{L}_1, \mathscr{L}_2$.

$$\mathscr{L}_1 = \left\{ u_1 \in \left(H^1(\Omega_1) \right)^3, P_1 u_1 = F \in \left(L^2(\Omega_1) \right)^3, u = 0 \text{ on } \Gamma_1 \right\}$$

$$\mathscr{L}_2 = \left\{ u_2 \in \left(H^1_{loc}(\overline{\Omega}_2) \right)^3, P_2 u_2 = 0 \text{ on } \Omega_2 \text{ and } u_2 = \mathscr{O}(\frac{1}{|x|}) \text{ as } |x| \to \infty \right\}$$

Inspection shows that (12.15) is a weak form of the Euler equation $DJ_1(u, \phi; w, \varphi) = 0$ with $J_1(u, \phi) := J(u) + q(u, \phi) - a(\phi)$ where $v = u|_{\Gamma_c}$. Here we have set $J(u) = J_0(u) + 1/2\langle v, W_2 v \rangle$ with

$$J_0(u) = \int_{\Omega_1} \{(1/2)k|\operatorname{div} u|^2 + \int_0^{\Gamma(u)} \mu(t) dt - F \cdot u\} dx,$$

and $q(u, \phi) = \langle \phi, (\Lambda_2 - 1/2) v \rangle$, $a(\phi) = 1/2\langle \phi, V_2 \phi \rangle$.

Then there holds

Theorem 12.5 *Let $F \in (L^2(\Omega_1))^3$. Then, if $u_j \in \mathcal{L}_j$ $(j = 1, 2)$ solve (TMP), then $u = u_1$ and $\phi = T_1(u_1)|_{\Gamma_c}$ solve* (12.14)*. Conversely, if $(u, \phi) \in H^1_{\Gamma_1}(\Omega_1) \times H^{-1/2}(\Gamma_c)$ satisfy* (12.14) *and if u_2 is defined by the representation formula*

$$u_2(x) = \int_{\Gamma_c} \{T_2(x, y)v_2(y) - G_2(x, y)\phi_2(y)\}\, ds_y \tag{12.16}$$

with $v_2 = v = u|_{\Gamma_c}$ and $\phi_2 = \phi$ on Γ_c, then $u_1 = u \in \mathcal{L}_1$ and $u_2 \in \mathcal{L}_2$ solve (TMP). Here

$$G_2(x, y) = \frac{\lambda_2 + 3\mu_2}{8\pi\mu_2(\lambda_2 + 2\mu_2)} \left\{ \frac{1}{|x - y|} I + \frac{\lambda_2 + \mu_2}{\lambda_2 + 3\mu_2} \frac{(x - y)(x - y)^T}{|x - y|^3} \right\},$$

$$T_2(x, y) := T_{2,y}(G_2(x, y))^T.$$

Proof Let (u, ϕ) satisfy(12.14) and define $u_1 = u \in H^1_{\Gamma_1}(\Omega_1)$,and $u_2 \in \mathcal{L}_2$ via (12.16) with $v_2 = v = u|_{\Gamma_c}$ and $\phi_2 = \phi$. Then from the definition of $\mathcal{C}_2$ we have

$$u_2|_{\Gamma_c} = \left(\frac{1}{2} + \Lambda_2\right) v - V_2\phi \tag{12.17}$$

and

$$T_2(u_2)|_{\Gamma_c} = -W_2 v + \left(\frac{1}{2} - \Lambda_2'\right) \phi \tag{12.18}$$

Setting $w = 0$ in (12.14), and taking $\psi \in H^{-1/2}(\Gamma_c)$ arbitrarily gives with (12.16) the relation

$$\left(\frac{1}{2} - \Lambda_2\right) v + V_2\phi = 0$$

Hence with (12.17) $u_2|_{\Gamma_c} = v$ and thus $u_1 = u_2$ on Γ_c.

Next, take $\psi = 0$ in (12.14) and $w \in H^1_{\Gamma_1}(\Omega_1)$ arbitrarily. Then comparison with

$$\int_{\Omega_1} P_1 u \cdot w \partial x = \Phi_1(u, w) - \langle T_1 u, w \rangle$$

shows that

$$P_1 u_1 = F \text{ in } \Omega_1, \quad T_1(u_1) = \left(\frac{1}{2} - \Lambda_2'\right)\phi - W_2 v \text{ on } \Gamma_c.$$

Therefore together with (12.18) we get $T_1(u_1) = T_2(u_2)$ on Γ_c. □

Lemma 12.4

(i) $\exists \gamma_1 > 0\ \forall \phi \in H^{-1/2}(\Gamma_c) \quad \langle \phi, V_2\phi\rangle \geqslant \gamma_1 \|\phi\|^2_{H^{-1/2}(\Gamma_c)}$

(ii) $\forall v \in H^{1/2}(\Gamma_c) \quad \langle v, W_2 v\rangle \geqslant 0$

Proof Let $v \in H^{1/2}(\Gamma_c), \phi \in H^{-1/2}(\Gamma_c)$ be given. From the definition of the Calderon projector $\mathscr{C}_2$ we have

$$\langle \phi, V_2\phi\rangle + \langle v, W_2 v\rangle = \left(\begin{pmatrix} v \\ \phi \end{pmatrix}, \left(\frac{1}{2} - \mathscr{C}_2\right)\begin{pmatrix} v \\ \phi \end{pmatrix}\right) \tag{12.19}$$

Let $\Omega_2^c := \mathbb{R}^3 \setminus \overline{\Omega_2}$ be the complementary domain of Ω_2. Define u_2^c in Ω_2^c and u_2 in Ω_2 by the representation formula (12.16) with v replacing v_2 and ϕ replacing ϕ_2. Let

$$\begin{pmatrix} v_2 \\ \phi_2 \end{pmatrix} := \begin{pmatrix} u_2 \\ T_2(u_2) \end{pmatrix}\Bigg|_{\Gamma_c} \text{ and } \begin{pmatrix} v_2^c \\ \phi_2^c \end{pmatrix} := \begin{pmatrix} u_2^c \\ T_2(u_2^c) \end{pmatrix}\Bigg|_{\Gamma_c}$$

denote the respective Cauchy data. Then the jump relations yield

$$\begin{pmatrix} v_2 \\ \phi_2 \end{pmatrix} - \begin{pmatrix} v_2^c \\ \phi_2^c \end{pmatrix} = \begin{pmatrix} v \\ \phi \end{pmatrix}$$

and

$$\begin{pmatrix} v_2 \\ \phi_2 \end{pmatrix} + \begin{pmatrix} v_2^c \\ \phi_2^c \end{pmatrix} = 2\left(\mathscr{C}_2 - 1/2\right)\begin{pmatrix} v \\ \phi \end{pmatrix}.$$

Therefore we can rewrite (12.19) as follows,

$$\begin{aligned}
\langle \phi, V_2\phi\rangle + \langle v, W_2 v\rangle &= \frac{1}{2}\left(\begin{pmatrix} v_2^c - v_2 \\ \phi_2^c - \phi_2 \end{pmatrix}, \begin{pmatrix} v_2^c + v_2 \\ \phi_2^c + \phi_2 \end{pmatrix}\right) \\
&= \frac{1}{2}\left\{\langle v_2^c - v_2, \phi_2^c + \phi_2\rangle + \langle v_2^c + v_2, \phi_2^c - \phi_2\rangle\right\} \\
&= \langle v_2^c, \phi_2^c\rangle - \langle v_2, \phi_2\rangle
\end{aligned}$$

Next, we need the first Green formula for P_2 in both Ω_2 and Ω_2^c. Then

$$\Phi_2(u_2, u_2) + \langle \phi_2, v_2 \rangle = 0$$

and

$$\Phi_2^c(u_2^c, u_2^c) - \langle \phi_2^c, v_2^c \rangle = 0$$

where

$$\Phi_2^c(u, v) = \int_{\Omega_2^c} \sum_{i,h,k,l=1}^{3} a_{ihkl}^2 \epsilon_{kl}(u)\epsilon_{ih}(v)dx \ , \quad a_{ihkl}^2 = \lambda_2 \delta_{ih}\delta_{kl} + \mu_2(\delta_{ik}\delta_{hl} + \delta_{il}\delta_{hk})$$

with $\delta_{ik} = 1$ for $i = k$ and $\delta_{ik} = 0$ for i different from k. Hence

$$\langle \phi, V_2\phi \rangle + \langle v, W_2 v \rangle = \Phi_2(u_2, u_2) + \Phi_2^c(u_2^c, u_2^c) \tag{12.20}$$

Now, the right hand side is nonnegative, which proves the assertion *(ii)*. Furthermore, Korn's inequality shows that the right hand side satisfies a Gårding inequality in the H^1 norm. This implies that V_2 and W_2 satisfy Gårding inequalities in the $H^{-1/2}(\Gamma_c)$ and $H^{1/2}(\Gamma_c)$ norms, respectively. Thus V_2 is positive up to a compact perturbation. It remains to show that $\langle \phi, V_2\phi \rangle > 0$ holds for $\phi \neq 0$. Thus let $v = 0$. Then $v_2^c = v_2$ which shows that $u_2 \in H^1_{loc}(\mathbb{R}^3)$ if we define $u_2 = u_2^c$ in Ω_2^c. Therefore (12.20) yields

$$\langle \phi, V_2\phi \rangle = \int_{\mathbb{R}^3} \sum_{i,h,k,l=1}^{3} a_{i,h,k,l}^2 \epsilon_{kl}(u_2)\epsilon_{ih}(u_2)dx > 0$$

unless u_2 is a rigid body motion, then implying $\phi = \phi_2 - \phi_2^c = 0$. Thus for $\phi \neq 0$ there holds $\langle \phi, V_2\phi \rangle > 0$. □

As an application of Theorem 12.3 we obtain the following existence and uniqueness result for (TMP) if we assume that P_1 is strongly monotone (see [136]).

Theorem 12.6 *Let $F \in (L^2(\Omega_1))^3$ be given. Then there exists exactly one solution $(u, \phi) \in H^1_{\Gamma_1}(\Omega_1) \times H^{-1/2}(\Gamma_c)$ of* (12.14) *yielding precisely one solution u_1 in $\mathcal{L}_1$, $u_2 \in \mathcal{L}_2$ of (TMP).*

Next, we choose finite-dimensional subspaces X_N, Y_N of $X = H^1_{\Gamma_1}(\Omega_1)$, $Y = H^{-1/2}(\Gamma_c)$ with

$$d_X(w, X_N) \to 0, \ d_Y(\psi, Y_N) \to 0 \text{ as } N \to \infty \ \forall w \in H^1_{\Gamma_1}(\Omega_1), \ \psi \in H^{-1/2}(\Gamma_c).$$

The Galerkin scheme reads: Find $(u_N, \phi_N) \in X_N \times Y_N$ such that

$$b(u_N, \phi_N; w, \psi) = l(w, \psi) \quad \forall w \in X_N, \ \psi \in Y_N \tag{12.21}$$

Theorem 12.7 *For all $N \in \mathbb{N}$ there exists exactly one solution $(u_N, \phi_N) \in X_N \times Y_N$ solution of* (12.21).

Furthermore there exists $C > 0$ independent of N such that

$$\|u - u_N\|_{H^1(\Omega_1)} + \|\phi - \phi_N\|_{H^{-1/2}(\Gamma_c)} \leqslant C\{d_X(u, X_N) + d_Y(\phi, Y_N)\}$$

where $u \in H^1_{\Gamma_1}(\Omega_1)$ and $\phi) \in H^{-1/2}(\Gamma_c)$ solve (12.14).

Proof Direct consequence of Theorem 12.4. □

12.3 Symmetric FE/BE Coupling—Revisited

Firstly, we introduce the symmetric coupling method of Costabel from [113, 115] and show its equivalence with the original transmission problem (IP) (Theorem 12.8). Secondly we report from [91] an h-adaptive procedure for the symmetric coupling of FEM and BEM. The theoretical results are obtained via the Poincaré-Steklov operator. For an alternative approach via saddle points see [136] and the foregoing Sect. 12.1 . An a posteriori error estimate is presented which guarantees a given bound for the energy norm.

Let $\Omega_1 := \Omega \subset \mathbb{R}^d$, $d \geq 2$ be a bounded domain with Lipschitz boundary $\Gamma = \partial\Omega_1$, and $\Omega_2 := \mathbb{R}^d \setminus \bar{\Omega}_1$ with normal n on Γ pointing into Ω_2. In the case $d = 2$ we always assume $cap(\Gamma) < 1$ in the following. This can always be achieved by scaling. For given $f \in L^2(\Omega_1)$, $u_0 \in H^{1/2}(\Gamma)$, $t_0 \in H^{-1/2}(\Gamma)$ we consider the model interface problem (IP): Find $u_1 \in H^1(\Omega_1)$, $u_2 \in H^1_{\text{loc}}(\Omega_2)$ such that

$$-\operatorname{div} A(\nabla u_1) = f \text{ in } \Omega_1 \tag{12.22}$$

$$\Delta u_2 = 0 \text{ in } \Omega_2 \tag{12.23}$$

$$u_1 = u_2 + u_0 \text{ on } \Gamma \tag{12.24}$$

$$A(\nabla u_1) \cdot n = \frac{\partial u_2}{\partial n} + t_0 \text{ on } \Gamma \tag{12.25}$$

$$u_2(x) = \begin{cases} b \ln|x| + o(1), & d = 2 \\ \mathcal{O}(|x|^{2-d}), & d \geq 3 \end{cases}, \ |x| \to \infty \tag{12.26}$$

where $b \in \mathbb{R}$ is a constant (depending on u_2). The operator A is assumed to be uniformly monotone and Lipschitz continuous, i.e. there exist positive constants α and C such that for all $\eta, \tau \in L^2(\Omega)^d$

$$\int_\Omega (A(\eta) - A(\tau)) \cdot (\eta - \tau)\, dx \geq \alpha \|\eta - \tau\|_{0,\Omega}^2 \tag{12.27}$$

$$\|A(\eta) - A(\tau)\|_{0,\Omega} \leq C\|\eta - \tau\|_{0,\Omega}. \tag{12.28}$$

Here $\|.\|_{0,\Omega}$ denotes the norm in $L^2(\Omega)^d$. Examples for operators of this type can be found in [399] and for models of nonlinear elasticity in Section 62 of [437]. The definition of the Sobolev spaces is as usual:

$$H^s(\Omega) = \{\phi|_\Omega\, ;\ \phi \in H^s(\mathbb{R}^d)\} \quad (s \in \mathbb{R}),$$

$$H^s(\Gamma) = \begin{cases} \{\phi|_\Gamma\, ;\ \phi \in H^{s+1/2}(\mathbb{R}^d)\} & (s > 0) \\ L^2(\Gamma) & (s = 0) \\ (H^{-s}(\Gamma))' \quad \text{(dual space)} & (s < 0) \end{cases}$$

In the following we often write $\|.\|_{s,B}$ for the Sobolev norm $\|\cdot\|_{H^s(B)}$ with $B = \Omega$ or Γ. We now derive the symmetric coupling method as discussed in detail in [121]. Green's formula together with the decaying condition (12.26) leads to the representation formula for the solution in the exterior domain u_2 of (12.23).

$$u_2(x) = \int_\Gamma \left\{ \frac{\partial}{\partial n(y)} G(x, y) u_2(y) - G(x, y) \frac{\partial u_2}{\partial n(y)} \right\} ds_y, \quad x \in \Omega_2 \tag{12.29}$$

with the fundamental solution of the Laplacian given by

$$G(x, y) = \begin{cases} -\frac{1}{\omega_2} \ln |x - y|, & d = 2 \\ \frac{1}{\omega_d} |x - y|^{2-d}, & d \geq 3 \end{cases} \tag{12.30}$$

where we have $\omega_2 = 2\pi$, $\omega_3 = 4\pi$.

By using the boundary integral operators

$$V\psi(x) := 2 \int_\Gamma G(x, y)\psi(y)\, ds_y, \quad x \in \Gamma \tag{12.31}$$

$$K\psi(x) := 2 \int_\Gamma \frac{\partial}{\partial n_y} G(x, y)\psi(y)\, ds_y, \quad x \in \Gamma \tag{12.32}$$

$$K'\psi(x) := 2 \frac{\partial}{\partial n_x} \int_\Gamma G(x, y)\psi(y)\, ds_y, \quad x \in \Gamma \tag{12.33}$$

$$W\psi(x) := -2 \frac{\partial}{\partial n_x} \int_\Gamma \frac{\partial}{\partial n_y} G(x, y)\psi(y)\, ds_y, \quad x \in \Gamma \tag{12.34}$$

together with their jump conditions (see Chap. 2) we obtain from (12.29) the following integral equations

$$2\frac{\partial u_2}{\partial n} = -Wu_2 + (I - K')\frac{\partial u_2}{\partial n} \tag{12.35}$$

$$0 = (I - K)u_2 + V\frac{\partial u_2}{\partial n}. \tag{12.36}$$

One observes that (12.36) gives one part of the weak formulation of problem (IP)

$$-\langle u_1, \psi\rangle - \langle V\frac{\partial u_2}{\partial n}, \psi\rangle + \langle Ku_1, \psi\rangle = -\langle u_0, \psi\rangle + \langle Ku_0, \psi\rangle \quad \forall \psi \in H^{-1/2}(\Gamma). \tag{12.37}$$

The second part of the weak formulation has to couple the exterior problem (12.23) and the interior problem (12.22). To this end we use

$$a(u_1, v) := \int_{\Omega_1} A(\nabla u_1)\cdot\nabla v\, dx = \int_\Gamma (A(\nabla u_1)\cdot n)v\, ds + \int_{\Omega_1} f v\, dx \quad \forall v \in H^1(\Omega_1). \tag{12.38}$$

Taking the integral equation (12.35) and substituting (12.24) and (12.25) into (12.38) one obtains for all $v \in H^1(\Omega_1)$

$$2a(u_1, v) - \langle\frac{\partial u_2}{\partial n}, v\rangle + \langle K'\frac{\partial u_2}{\partial n}, v\rangle + \langle Wu_1, v\rangle = 2(f, v) + 2\langle t_0, v\rangle + \langle Wu_0, v\rangle, \tag{12.39}$$

where $(f, v) = \int_{\Omega_1} f v\, dx$.

Note that in this way we obtain the following variational formulation $(\widetilde{P})$:

Given $(f, u_0, t_0) \in L^2(\Omega) \times H^{1/2}(\Gamma) \times H^{-1/2}(\Gamma)$ find $u := u_1 \in H^1(\Omega_1)$ and $\phi := \frac{\partial u_2}{\partial n} \in H^{-1/2}(\Gamma)$ such that for all $v \in H^1(\Omega)$ and $\psi \in H^{-1/2}(\Gamma)$

$$\begin{aligned} 2a(u, v) + \langle (K' - I)\phi, v\rangle + \langle Wu, v\rangle &= 2\langle t_0, v\rangle + \langle Wu_0, v\rangle + 2(f, v) \\ \langle (K - I)u, \psi\rangle - \langle V\phi, \psi\rangle &= \langle (K - I)u_0, \psi\rangle. \end{aligned} \tag{12.40}$$

There holds the following equivalence:

Theorem 12.8 *The problems (IP) and $(\widetilde{P})$ are equivalent in the following sense. If $(u, v) \in H^1(\Omega_1)\times H^1_{loc}(\Omega_2)$ is a solution of (IP) then $(u, \phi) \in H^1(\Omega_1)\times H^{-1/2}(\Gamma)$ solves $(\widetilde{P})$ with $\phi := \frac{\partial v}{\partial n}|_\Gamma$. If, conversely, (u, ϕ) is a solution of problem $(\widetilde{P})$ then*

(u, v) solves (IP) with $v \in H^1_{loc}(\Omega_2)$ defined by

$$v(z) = -\frac{1}{2\pi}\int_\Gamma \phi(\zeta)\cdot G(z,\zeta)\,ds_\zeta + \frac{1}{2\pi}\int_\Gamma (u-u_0)(\zeta)\cdot\frac{\partial}{\partial n_\zeta}G(z,\zeta)\,ds_\zeta \quad (z\in\Omega_2). \tag{12.41}$$

Proof By deriving the coupling formulation we have already shown that if (u, v) solves (IP) then (u, ϕ) solves $(\widetilde{P})$.

Conversely, let (u, ϕ) solve $(\widetilde{P})$ and define v by (12.41). Then, according to [129], v satisfies (12.23), (12.26) and with the jump relations we have

$$\begin{pmatrix} v|_\Gamma \\ \frac{\partial v}{\partial n}|_\Gamma \end{pmatrix} = \frac{1}{2}(Id - H)\begin{pmatrix} u|_\Gamma - u_0 \\ \phi \end{pmatrix} \text{ with } H := \begin{pmatrix} -K & V \\ W & K' \end{pmatrix}. \tag{12.42}$$

The first component of (12.42) together with (12.37) yields $u|_\Gamma = v|_\Gamma + u_0$. From the second identity in (12.42) we then have

$$\frac{\partial v}{\partial n}|_\Gamma = -\frac{1}{2}\{W(u|_\Gamma - u_0) + (K' - 1)V\phi\}.$$

Using this in (12.38) gives, by Green's formula

$$\int_\Omega (\operatorname{div}(A\operatorname{grad} u) + f)\eta\, d\Omega = \langle (A\operatorname{grad} u)\cdot n|_\Gamma - \frac{\partial v}{\partial n}|_\Gamma - t_0, \eta|_\Gamma\rangle$$

for all $\eta \in H^1(\Omega)$. Choosing $\eta \in H^1_0(\Omega)$, the completion of $C_0^\infty(\Omega)$ in the H^1-norm, we get the weak form of (12.22). Hence using (12.22) we get (12.25). □

Remark 12.1 We note that

$$W1 = 0 \quad \text{and} \quad K1 = -1 \tag{12.43}$$

with 1 being the constant function with the value one. The identities (12.43) follow from $H\binom{1}{0} = \binom{1}{0}$ (cf. [129, Lemma 3.5]).

For the Galerkin scheme we choose finite dimensional subspaces $X_M \subset H^1(\Omega)$ and $Y_N \subset H^{-1/2}(\Gamma)$ and define the Galerkin solution $(u_M, \phi_N) \in X_M \times Y_N$ by

$$\begin{aligned} 2a(u_M, v) + \langle (K' - I)\phi_N, v\rangle + \langle W u_M, v\rangle &= 2\langle t_0, v\rangle + \langle W u_0, v\rangle + 2(f, v) \\ \langle (K - I)u_M, \psi\rangle - \langle V\phi_N, \psi\rangle &= \langle (K - I)u_0, \psi\rangle \end{aligned} \tag{12.44}$$

for all $v \in X_M$ and $\psi \in Y_N$.

There holds the following convergence result as application of Theorem 12.4. In the next section we will present a different proof via the Poincare-Steklov operator, cf. Corollary 12.1.

Theorem 12.9 ([136]) *Every Galerkin scheme (12.44) with approximating finite dimensional spaces $X_M \subset H^1(\Omega)$ and $Y_N \subset H^{-1/2}(\Gamma)$ converges with optimal order, i.e. with the exact solution (u, ϕ) of (12.40) and the Galerkin solution (u_M, ϕ_N) of (12.44) there holds the estimate*

$$\|u - u_M\|_{1,\Omega} + \|\phi - \phi_N\|_{-1/2,\Gamma} \le C\{\inf_{\hat{u}\in X_M} \|u - \hat{u}\|_{1,\Omega} + \inf_{\hat{\phi}\in Y_N} \|\phi - \hat{\phi}\|_{-1/2,\Gamma}\} \tag{12.45}$$

where the constant C is independent of M, N, u and ϕ.

12.3.1 Convergence Analysis

In this section we prove existence and uniqueness of the weak (variational) solution of the interface problem (IP) in Sect. 12.3 and show convergence of the Galerkin solution proving Theorem 12.9 above, now following [91] and using heavily the strong coerciveness of the Poincaré-Steklov operator (for the exterior problem) and of its discrete analogue.

Firstly, we note that the weak formulation (12.40) is

Problem (P): Find $(u, \phi) \in H^1(\Omega_1) \times H^{-1/2}(\Gamma)$ with

$$B(\begin{pmatrix} u \\ \phi \end{pmatrix}, \begin{pmatrix} v \\ \psi \end{pmatrix}) = L\begin{pmatrix} v \\ \psi \end{pmatrix} \qquad \forall (v, \psi) \in H^1(\Omega_1) \times H^{-1/2}(\Gamma). \tag{12.46}$$

Here the continuous mapping $B : (H^1(\Omega) \times H^{-1/2}(\Gamma))^2 \to \mathbb{R}$ and the linear form $L : H^1(\Omega) \times H^{-1/2}(\Gamma) \to \mathbb{R}$ are defined by

$$B(\begin{pmatrix} u \\ \phi \end{pmatrix}, \begin{pmatrix} v \\ \psi \end{pmatrix}) := \int_{\Omega_1} A(\nabla u) \cdot \nabla v \, dx + \frac{1}{2}\langle Wu|_\Gamma + (K' - I)\phi, v|_\Gamma\rangle$$

$$+ \frac{1}{2}\langle \psi, V\phi + (I - K)u|_\Gamma\rangle \tag{12.47}$$

$$L\begin{pmatrix} v \\ \psi \end{pmatrix} := \int_{\Omega_1} f \cdot v \, dx + \frac{1}{2}\langle \psi, (I - K)u_0\rangle + \langle t_0 + \frac{1}{2}Wu_0, v|_\Gamma\rangle \tag{12.48}$$

for any $(u, \phi), (v, \psi) \in H^1(\Omega_1) \times H^{-1/2}(\Gamma)$.

Note that (12.36) is equivalent to

$$\phi = -V^{-1}(I - K)(u_1 - u_0) \tag{12.49}$$

which we use to eliminate $\phi = \frac{\partial u_2}{\partial n}$ in (12.40). Thus we arrive at the following equivalent formulation: Find $u \in H^1(\Omega_1)$ with

$$A'(u)(\eta) := 2\int_{\Omega_1} A(\nabla u)\cdot\nabla\eta\,dx + \langle Su|_\Gamma, \eta|_\Gamma\rangle \tag{12.50}$$

$$= L'(\eta) := 2\int_{\Omega_1} f v\,dx + \langle 2t_0 + Su_0, \eta|_\Gamma\rangle \quad (\eta \in H^1(\Omega))$$

with the Poincaré-Steklov operator for the exterior problem

$$S := W + (I - K')V^{-1}(I - K) : H^{1/2}(\Gamma) \longrightarrow H^{-1/2}(\Gamma) \tag{12.51}$$

Lemma 12.5 ([91]) *(Suppose* $cap(\Gamma) < 1$ *in 2D). The operator* $S := W + (1 - K')V^{-1}(1 - K) : H^{1/2}(\Gamma) \to H^{-1/2}(\Gamma)$ *is linear, bounded, symmetric, and positive definite.*

Proof Due to the above mentioned properties of W, K, V, K', the operator S is linear, bounded, symmetric, positive semidefinite, and a Fredholm operator of index zero. Thus, it suffices to prove that the kernel $ker\ S$ is trivial in order to conclude that S is positive definite. Let $u \in ker\ S$, then $0 = \langle Su, u\rangle$. On the other hand $\langle Su, u\rangle \geq \langle Wu, u\rangle \geq 0$, so that $\langle Wu, u\rangle = 0$. By Theorem 2.5, u is constant. Therefore $0 = \langle V^{-1}(1-K)u, (1-K)u\rangle$. By Theorem 2.5, V^{-1} is positive definite so that $(1 - K)u = 0$. Using (12.43), this implies that the constant u is equal to zero. Thus, $ker\ S = \{0\}$. The arguments can be extended to cover also the 3D case making use of Lemma 12.4. □

Lemma 12.6 *There exists a constant* $\beta > 0$ *such that for all* $(u, \phi), (v, \psi) \in H^1(\Omega) \times H^{-1/2}(\Gamma)$ *we have*

$$\beta \cdot \left\| \begin{pmatrix} u - v \\ \phi - \psi \end{pmatrix} \right\|_{H^1(\Omega)\times H^{-1/2}(\Gamma)} \cdot \left\| \begin{pmatrix} u - v \\ \eta - \delta \end{pmatrix} \right\|_{H^1(\Omega)\times H^{-1/2}(\Gamma)}$$
$$\leq B\left(\begin{pmatrix} u \\ \phi \end{pmatrix}, \begin{pmatrix} u - v \\ \eta - \delta \end{pmatrix}\right) - B\left(\begin{pmatrix} v \\ \psi \end{pmatrix}, \begin{pmatrix} u - v \\ \eta - \delta \end{pmatrix}\right) \tag{12.52}$$

with $2\eta := \phi + V^{-1}(I - K)u|_\Gamma$, $2\delta := \psi + V^{-1}(I - K)v|_\Gamma \in H^{-1/2}(\Gamma)$.

Proof Some calculations show

$$B(\begin{pmatrix} u \\ \phi \end{pmatrix}, \begin{pmatrix} u - v \\ \eta - \delta \end{pmatrix}) - B(\begin{pmatrix} v \\ \psi \end{pmatrix}, \begin{pmatrix} u - v \\ \eta - \delta \end{pmatrix})$$
$$= \int_\Omega \Big((A\nabla u) - (A\nabla v)\Big)\cdot\nabla(u - v)\,dx$$
$$+ \frac{1}{4}\langle W(u - v), u - v\rangle + \frac{1}{4}\langle S(u - v), u - v\rangle + \frac{1}{4}\langle V(\phi - \psi), \phi - \psi\rangle.$$

Since A is uniformly monotone, W is positive semi-definite, S and V are positive definite, the right hand side in (12.52) is bounded below by $\tilde{c}\|\begin{pmatrix} u-v \\ \phi-\psi \end{pmatrix}\|^2_{H^1(\Omega)\times H^{-1/2}(\Gamma)}$ with a suitable constant $\tilde{c}$.

On the other hand, by definition of η, δ, we have with a constant c'

$$\|\eta - \delta\|_{H^{-1/2}(\Gamma)} \leq c'\|\begin{pmatrix} u-v \\ \phi-\psi \end{pmatrix}\|_{H^1(\Omega)\times H^{-1/2}(\Gamma)},$$

yielding (12.52). □

Theorem 12.10 *The interface problem (IP) and the problem (P) have unique solutions.*

Proof The operator A' on the left hand side in (12.50) maps $H^1(\Omega_1)$ into its dual; it is continuous, bounded, uniformly monotone and therefore bijective. This yields the existence of u satisfying (12.50). Letting ϕ as in (12.49) we have that (u, ϕ) solves problem (P). Uniqueness of the solution follows from Lemma 12.6 yielding also the unique solvability of the equivalent interface problem (IP). □

Next we treat the discretization of problem (P) in the 2D case.

Let $(H_h \times H_h^{-1/2} : h \in I)$ for $I \subseteq (0,1)$ with $0 \in \bar{I}$ be a family of finite dimensional subspaces of $H^1(\Omega)\times H^{-1/2}(\Gamma)$. Then, the coupling of finite elements and boundary elements consists in the following Galerkin procedure.

Definition 12.1 (Problem (P_h)) For $h \in I$ find $(u_h, \phi_h) \in H_h \times H_h^{-1/2}$ such that

$$B(\begin{pmatrix} u_h \\ \phi_h \end{pmatrix}, \begin{pmatrix} v_h \\ \psi_h \end{pmatrix}) = L\begin{pmatrix} v_h \\ \psi_h \end{pmatrix} \tag{12.53}$$

for all $(v_h, \psi_h) \in H_h \times H_h^{-1/2}$.

In order to prove a discrete Babuška–Brezzi condition if A is linear, we need some notations and the positive definiteness of the discrete Poincaré-Steklov operator.

Assumption 12.1 *For any $h \in I$ let $H_h \times H_h^{-1/2} \subseteq H^1(\Omega) \times H^{-1/2}(\Gamma)$. Suppose $1 \in H_h^{-1/2}$ for any $h \in I$, where 1 denotes the constant function with value 1.*

Let $i_h : H_h \hookrightarrow H^1(\Omega)$ and $j_h : H_h^{-1/2}(\Gamma) \hookrightarrow H^{-1/2}(\Gamma)$ denote the canonical injections with their duals $i_h^* : H^1(\Omega)^* \to H_h^*$ and $j_h^* : H^{1/2}(\Gamma) \to H_h^{-1/2})(\Gamma)^*$ being projections. Let $\gamma : H^1(\Omega) \to H^{1/2}(\Gamma)$ denote the trace operator, $\gamma u = u|_\Gamma$ for all $u \in H^1(\Omega)$, with the dual γ^*. Then, define

$$V_h := j_h^* V j_h, \quad K_h := j_h^* K \gamma i_h, \quad W_h := i_h^* \gamma^* W \gamma i_h, \quad K_h' := i_h^* \gamma^* K' j_h \tag{12.54}$$

and, since V_h is positive definite as well as its continuous analogue V,

$$S_h := W_h + (I_h^* - K_h')V_h^{-1}(I_h - K_h) : H_h \to H_h^* \tag{12.55}$$

with $I_h := j_h^* \gamma i_h$ and its dual I_h^*.

A key role is played by the following coerciveness of the discrete version of the Poincaré-Steklov operator, due to [91].

Lemma 12.7 *There exist constants $c_0 > 0$ and $h_0 > 0$ such that for any $h \in I$ with $h < h_0$ we have*

$$\langle S_h u_h, u_h \rangle \geq c_0 \cdot \|u_h|_\Gamma\|^2_{H^{1/2}(\Gamma)} \quad \textit{for all } u_h \in H_h.$$

Proof Assume that the assertion is false. Then one can construct a sequence of functions $(u_{h_n})_{n=1,2,3,\ldots}$ in $H^1(\Omega)$ with

$$u_{h_n} \in H_{h_n},\ \|u_{h_n}|_\Gamma\|_{H^{1/2}(\Gamma)} = 1,\ \langle S_{h_n} u_{h_n}, u_{h_n} \rangle \leq \frac{1}{n}\ (n = 1, 2, 3, \ldots)$$

and $\lim_{n\to\infty} h_n = 0$. Due to the Banach–Alaoglu theorem $(u_{h_n}|_\Gamma)_{n=1,2,3,\ldots}$ converges weakly towards some $w \in H^{1/2}(\Gamma)$ in $H^{1/2}(\Gamma)$ (a subsequence at least).

Then, by definition of S_h ,first we conclude that $\langle W u_{h_n}|_\Gamma, u_{h_n}|_\Gamma \rangle$ tends towards zero so that $\langle Ww, w \rangle = 0$, i.e. $w|_\Gamma$ is constant. A decomposition of $u_{h_n}|_\Gamma = v_n + w_n$ with $v_n \in H_0^{1/2}(\Gamma) = \{v \in H^{1/2}(\Gamma), \langle v, 1 \rangle = 0\}$ and $w_n \in \mathbb{R}$ shows additionally that $(v_n)_{n=1,2,3,\ldots}$ tends towards zero strongly in $H^{1/2}(\Gamma)$, since W is positive definite on $H_0^{1/2}(\Gamma)$. Hence we have also strong convergence of $(u_{h_n}|_\Gamma)_{n=1,2,3,\ldots}$ towards the constant $w \in \mathbb{R}$ in $H^{1/2}(\Gamma)$.
On the other hand we have $0 = \lim_{n\to\infty} \langle V z_n, z_n \rangle$ with $z_n := V_{h_n}^{-1}(\phi_n) \in H_{h_n}^{-1/2} \subseteq H^{-1/2}(\Gamma)$, $\phi_n := j_{h_n}^* y_n \in (H_{h_n}^{-1/2})^*$, $y_n := u_{h_n} - K u_{h_n} \in H^{1/2}(\Gamma)$.

Thus, $0 = \lim_{n\to\infty} \|z_n\|_{H^{-1/2}(\Gamma)}$ whence $0 = \lim_{n\to\infty} \|\phi_n\|_{(H_{h_n}^{-1/2})^*}$.

Because of $(u_{h_n}|_\Gamma)_{n=1,2,3,\ldots} \to w$ we get $(y_n)_{n=1,2,3,\ldots} \to 2w$ (strongly) in $H^{1/2}(\Gamma)$ (by (12.43) and $w \in \mathbb{R}$). Hence,

$$2w\langle 1, 1 \rangle = \lim_{n\to\infty} \langle 1, y_n \rangle = \lim_{n\to\infty} \langle j_{h_n} 1, y_n \rangle = \lim_{n\to\infty} \langle 1, \phi_n \rangle = 0,$$

i.e. $w = 0$. This contradicts $\|w\|_{H^{1/2}(\Gamma)} = \lim_{n\to\infty} \|u_{h_n}|_\Gamma\|_{H^{1/2}(\Gamma)} = 1$. □

In the above Lemma it is assumed that the initial boundary mesh is sufficiently fine. This assumption has first been proved to be unnecessary in [13], where the original argument of the above proof is refined.

Lemma 12.8 *There exist constants $\beta_0 > 0$ and $h_0 > 0$ such that for any $h \in I$ with $h < h_0$ we have that for any $(u_h, \phi_h), (v_h, \psi_h) \in H_h \times H_h^{-1/2}$*

$$\beta_0 \| \begin{pmatrix} u_h - v_h \\ \phi_h - \psi_h \end{pmatrix} \|_{H^1(\Omega)\times H^{-1/2}(\Gamma)} \cdot \| \begin{pmatrix} u_h - v_h \\ \eta_h - \delta_h \end{pmatrix} \|_{H^1(\Omega)\times H^{-1/2}(\Gamma)} \tag{12.56}$$

$$\leq B(\begin{pmatrix} u_h \\ \phi_h \end{pmatrix}, \begin{pmatrix} u_h - v_h \\ \eta_h - \delta_h \end{pmatrix}) - B(\begin{pmatrix} v_h \\ \psi_h \end{pmatrix}, \begin{pmatrix} u_h - v_h \\ \eta_h - \delta_h \end{pmatrix}) \tag{12.57}$$

with $2\eta_h := \phi_h + V_h^{-1}(I_h - K_h)u_h$, $2\delta_h := \psi_h + V_h^{-1}(I_h - K_h)v_h \in H_h^{-1/2}$.

Proof The proof is analogous to that of Lemma 12.6.Due to Lemma 12.7 the constants are independent of h as well so that β_0 does not depend on $h < h_0$, h_0 chosen in Lemma 12.7. This concludes the proof. □

Corollary 12.1 *There exist constants $c_0 > 0$ and $h_0 > 0$ such that for any $h \in I$ with $h < h_0$ the problem (P_h) has a unique solution (u_h, ϕ_h) and with the solution (u, ϕ) of (P), there holds*

$$\| \begin{pmatrix} u - u_h \\ \phi - \phi_h \end{pmatrix} \|_{H^1(\Omega)\times H^{-1/2}(\Gamma)} \leq c_0 \cdot \inf_{\binom{v_h}{\psi_h} \in H_h \times H_h^{-1/2}} \| \begin{pmatrix} u - v_h \\ \phi - \psi_h \end{pmatrix} \|_{H^1(\Omega)\times H^{-1/2}(\Gamma)}.$$

Proof The existence and uniqueness of the discrete solutions follows as in the proof of Theorem 12.10. Let $(U_h, \Phi_h) \in H^h \times H_h^{-1/2}$ be the orthogonal projections onto $H^h \times H_h^{-1/2}$ of the solution (u, ϕ) of Problem (P) in $H^1(\Omega) \times H^{-1/2}(\Gamma)$. From Lemma 12.8 we conclude with appropriate $(\eta_h, \delta_h) \in H^h \times H_h^{-1/2}$ that

$$\beta_0 \cdot \| \begin{pmatrix} U_h - u_h \\ \Phi_h - \phi_h \end{pmatrix} \|_{H^1(\Omega)\times H^{-1/2}(\Gamma)} \cdot \| \begin{pmatrix} U_h - u_h \\ \eta_h - \delta_h \end{pmatrix} \|_{H^1(\Omega)\times H^{-1/2}(\Gamma)}$$

$$\leq B(\begin{pmatrix} U_h \\ \Phi_h \end{pmatrix}, \begin{pmatrix} U_h - u_h \\ \eta_h - \delta_h \end{pmatrix}) - B(\begin{pmatrix} u_h \\ \phi_h \end{pmatrix}, \begin{pmatrix} U_h - u_h \\ \eta_h - \delta_h \end{pmatrix}).$$

Using the Galerkin equations and the Lipschitz continuity of B with constant L, the right hand side is bounded by

$$L \cdot \| \begin{pmatrix} U_h - u_h \\ \eta_h - \delta_h \end{pmatrix} \|_{H^1(\Omega)\times H^{-1/2}(\Gamma)} \cdot \| \begin{pmatrix} U_h - u \\ \Phi_h - \phi \end{pmatrix} \|_{H^1(\Omega)\times H^{-1/2}(\Gamma)},$$

what gives the assertion. □

12.3.2 Adaptive FE/BE Coupling: Residual Based Error Indicators

In this section we present a posteriori error estimates for the h-version of the symmetric coupling method from [91].

For simplicity, we restrict ourselves to linear ansatz functions on triangles as finite elements in H_h and to piecewise constant functions in $H_h^{-1/2}$.

Assumption 12.2 *Let Ω be a two-dimensional domain with polygonal boundary Γ on which we consider a family $\mathcal{T} := (\mathcal{T}_h : h \in I)$ of decompositions $\mathcal{T}_h = \{\Delta_1, \ldots, \Delta_N\}$ of Ω in closed triangles $\Delta_1, \ldots, \Delta_N$ such that $\bar{\Omega} = \cup_{i=1}^N \Delta_i$, and two different triangles are disjoint or have a side or a vertex in common. Let $\mathcal{S}_h$ denote the sides, i.e. $\mathcal{S}_h = \{\partial T_i \cap \partial T_j : i \neq j \text{ with } \partial T_i \cap \partial T_j \text{ is a common side}\}$, ∂T_j being the boundary of T_j. Let $\mathcal{G}_h = \{E : E \in \mathcal{S}_h \text{ with } E \subseteq \Gamma\}$ be the set of "boundary sides" and let $\mathcal{S}_h^0 = \mathcal{S}_h \setminus \mathcal{G}_h$ be the set of "interior sides". We assume that all the angles of some $\Delta \in \mathcal{T}_h \in \mathcal{T}$ are $\geq \Theta$ for some fixed $\Theta > 0$ which does not depend on Δ or $\mathcal{T}_h$.*

Then, define

$$H_h := \{\eta_h \in C(\Omega) : \eta_h|_\Delta \in P_1 \text{ for any } \Delta \in \mathcal{T}_h\} \tag{12.58}$$

$$H_h^{-1/2} := \{\eta_h \in L^\infty(\Gamma) : \eta_h|_E \in P_0 \text{ for any } E \in \mathcal{G}_h\} \tag{12.59}$$

where P_j denotes the polynomials with degree $\leq j$.

For fixed $\mathcal{T}_h$ let h be the piecewise constant function defined such that the constants $h|_\Delta$ and $h|_E$ equal the element sizes $\operatorname{diam}(\Delta)$ of $\Delta \in \mathcal{T}_h$ and $\operatorname{diam}(E)$ of $E \in \mathcal{S}_h$. We assume that $A(\nabla v_h) \in C^1(\Delta)$ for any $\Delta \in \mathcal{T}_h \in \mathcal{T}$ and any trial function $v_h \in H_h$. Finally, let $f \in L^2(\Omega)$, $u_0 \in H^1(\Gamma)$, and $t_0 \in L^2(\Gamma)$.

Let n be the exterior normal on Γ and on any element boundary $\partial\Delta$, let n have a fixed orientation so that $[A(\nabla u_h) \cdot n]|_E \in L^2(E)$ denotes the jump of the discrete tractions $A(\nabla u_h) \cdot n$ over the side $E \in \mathcal{S}_h^0$. Define

$$R_1^2 := \sum_{\Delta \in \mathcal{T}_h} \operatorname{diam}(\Delta)^2 \cdot \int_\Delta |f + \operatorname{div} A(\nabla u_h)|^2 \, dx \tag{12.60}$$

$$R_2^2 := \sum_{E \in \mathcal{S}_h^0} \operatorname{diam}(E) \cdot \int_E |[A(\nabla u_h) \cdot n]|^2 \, ds \tag{12.61}$$

$$R_3 := \|\sqrt{h} \cdot \Big(t_0 - A(\nabla u_h) \cdot n + \frac{1}{2} W(u_0 - u_h|_\Gamma) - \frac{1}{2}(K' - I)\phi_h\Big)\|_{L^2(\Gamma)} \tag{12.62}$$

$$R_4 := \sum_{E \in \mathcal{G}_h} \operatorname{diam}(E)^{1/2} \cdot \|\frac{\partial}{\partial s}\{(I - K)(u_0 - u_h|_\Gamma) - V\phi_h\}\|_{L^2(E)}. \tag{12.63}$$

Under the above assumptions there holds the following a posteriori estimate where (u, ϕ) and (u_h, ϕ_h) solve problem (P) and (P_h).

Theorem 12.11 *There exists some constant $c > 0$ such that for any $h \in I$ with $h < h_0$ (h_0 from Lemma 12.7) we have*

$$\|\begin{pmatrix} u - u_h \\ \phi - \phi_h . \end{pmatrix}\|_{H^1(\Omega)\times H^{-1/2}(\Gamma)} \le c \cdot (R_1 + R_2 + R_3 + R_4).$$

The proof of Theorem 12.11 is divided into several lemmas. We set

$$e := u - u_h, \quad \epsilon := \phi - \phi_h, \quad \delta := \frac{1}{2}(\epsilon + V^{-1}(1 - K)e|_\Gamma).$$

Lemma 12.9 *We have*

$$\beta \cdot \|\begin{pmatrix} e \\ \epsilon \end{pmatrix}\|_{H^1(\Omega)\times H^{-1/2}(\Gamma)} \cdot \|\begin{pmatrix} e \\ \delta \end{pmatrix}\|_{H^1(\Omega)\times H^{-1/2}(\Gamma)} \le T_1 + T_2 + T_3 + T_4$$

where, for any $(e_h, \delta_h) \in H_h \times H_h^{-1/2}$,

$$T_1 := \sum_{\Delta \in \mathcal{T}_h} \int_\Delta (f + \operatorname{div}(A \operatorname{grad} u_h))(e - e_h)\, d\Omega$$

$$T_2 := -\sum_{E \in \mathcal{S}_h^0} \int_E [(A \operatorname{grad} u_h) \cdot n](e - e_h)|_E\, ds$$

$$T_3 := \langle t_0 - (A \operatorname{grad} u_h) \cdot n + \frac{1}{2} W(u_0 - u_h|_\Gamma) - \frac{1}{2}(K' - 1)\phi_h, (e - e_h)|_\Gamma \rangle$$

$$T_4 := \frac{1}{2}\langle \delta - \delta_h, (1 - K)(u_0 - u_h|_\Gamma) - V\phi_h \rangle.$$

Proof Due to the arguments of the proof of Lemma 12.6 we have

$$\begin{aligned} &\beta \cdot \|\begin{pmatrix} e \\ \epsilon \end{pmatrix}\|_{H^1(\Omega)\times H^{-1/2}(\Gamma)} \cdot \|\begin{pmatrix} e \\ \delta \end{pmatrix}\|_{H^1(\Omega)\times H^{-1/2}(\Gamma)} \\ &\le B(\begin{pmatrix} u \\ \phi \end{pmatrix}, \begin{pmatrix} e \\ \delta \end{pmatrix}) - B(\begin{pmatrix} u_h \\ \phi_h \end{pmatrix}, \begin{pmatrix} e \\ \delta \end{pmatrix}) \\ &= L\begin{pmatrix} e - e_h \\ \delta - \delta_h \end{pmatrix} - B(\begin{pmatrix} u_h \\ \phi_h \end{pmatrix}, \begin{pmatrix} e - e_h \\ \delta - \delta_h \end{pmatrix}) \end{aligned}$$

using (12.53) and (12.46). By definition of B and L, the last term equals

$$\int_{\Omega} (f(e - e_h) - A \operatorname{grad} u_h \operatorname{grad}(e - e_h))\, d\Omega$$

$$+ \langle t_0 + \frac{1}{2} W(u_0 - u_h|_\Gamma) - \frac{1}{2}(K' - 1)\phi_h, (e - e_h)|_\Gamma \rangle$$

$$+ \frac{1}{2} \langle \delta - \delta_h, (1 - K)(u_0 - u_h|_\Gamma) - V\phi_h \rangle.$$

Using Green's formula on all elements $\Delta \in \mathcal{T}_h$ we obtain

$$- \int_{\Omega} A \operatorname{grad} u_h \operatorname{grad}(e - e_h)\, d\Omega$$

$$= \sum_{\Delta \in \mathcal{T}_h} \int_{\Delta} \operatorname{div}(A \operatorname{grad} u_h)(e - e_h)\, d\Omega$$

$$- \sum_{E \in \mathcal{S}_h^0} \int_E [(A \operatorname{grad} u_h) \cdot n](e - e_h)|_E\, ds$$

$$- \langle (A \operatorname{grad} u_h) \cdot n, (e - e_h)|_\Gamma \rangle.$$

yielding the assertion. □

We note that under the Assumption 12.2 Clement interpolation can be applied and gives the following lemma where $c > 0$ is a generic constant and depends only on $\mathcal{T}$ but not on h, Δ, N, u, etc.

Lemma 12.10 *There exists a family of interpolation operators $(I_h : H^1(\Omega) \to H_h : h \in I)$ and a constant $c > 0$ such that the following holds. For any $\Delta \in \mathcal{T}_h \in \mathcal{T}$ and integers k, q with $0 \le k \le q \le 2$ and with $N := \cup\{\Delta' \in \mathcal{T}_h : \Delta' \cap \Delta \neq \emptyset\}$, the union of all neighbor elements of Δ, and for all $u \in H^q(N)$,*

$$|I_h u - u|^2_{H^k(\Delta)} \le c \cdot \operatorname{diam}(T)^{2(q-k)} \cdot |u|^2_{H^q(N)}.$$

Furthermore, choosing $e_h := I_h e$ we have $T_i \le c \cdot |e|_{H^1(\Omega)} \cdot R_i, \quad i = 1, 2, 3.$

Additionally, there exists a constant $c > 0$ such that for any E, E is one side of $\Delta \in \mathcal{T}_h \in \mathcal{T}$, and any $u \in H^1(\Delta)$ there holds

$$\operatorname{diam}(\Delta)\|u\|^2_{L^2(E)} \le c \cdot \left(\|u\|^2_{L^2(\Delta)} + \operatorname{diam}(\Delta)^2 \cdot |u|^2_{H^1(\Delta)}\right).$$

Lemma 12.11 *For $\psi := (1 - K)(u_0 - u_h|_\Gamma) - V\phi_h$ we have with a constant c*

$$\|\psi\|_{H^{1/2}(\Gamma)} \le c \cdot \|\sqrt{h} \cdot \psi'\|_{L^2(\Gamma)}.$$

Proof (of Theorem 12.11) The assertion follows from Lemmas 12.10, and 12.11 to estimate T_1, T_2, T_3, and T_4 (with $\delta_h = 0$) in Lemma 12.9, respectively. Then, division by $\|\binom{e}{\delta}\|_{H^1(\Omega)\times H^{-1/2}(\Gamma)}$ proves the theorem. □

12.3.2.1 Adaptive Feedback Procedure

For a given triangulation $\mathcal{T}_h = \{\Delta_1, \ldots, \Delta_N\}$ of Ω and the related partition $\{\Gamma_1, \ldots, \Gamma_M\} = \mathcal{G}_h$ of the boundary Γ we can consider one element $\Delta_j \in \mathcal{T}_h$ and compute its contributions a_j, b_k to the right hand side of the a posteriori error estimate in Theorem 12.11

$$\begin{aligned} a_j^2 := diam(\Delta_j)^2 \cdot \int_{\Delta_j} |f + \operatorname{div}(A \operatorname{grad} u_h)|^2 \, d\Omega \\ + \sum_{E\in\mathcal{S}_h^0, E\subseteq\partial\Delta_j} \operatorname{diam}(E) \cdot \int_E |[(A \operatorname{grad} u_h)\cdot n]_E|^2 \, ds \\ + \operatorname{diam}(\Gamma\cap\partial\Delta_j) \cdot \|t_0 - (A \operatorname{grad} u_h)\cdot n + \frac{1}{2} W(u_0 - u_h|_\Gamma) \\ - \frac{1}{2}(K' - 1)\phi_h\|^2_{L^2(\Gamma\cap\partial\Delta_j)} \\ b_k := \operatorname{diam}(\Gamma_k)^{1/2} \cdot \|\frac{\partial}{\partial s}\{(1-K)(u_0 - u_h|_\Gamma) - V\phi_h\}\|_{L^2(\Gamma_k)}. \end{aligned}$$

If we neglect the constant $c > 0$ in Theorem 12.11, the error in the energy norm is bounded by

$$\sqrt{\sum_{j=1}^N a_j^2 + \sum_{k=1}^M b_k}. \tag{12.64}$$

Note that the different nature of the coefficients a_j and b_k is, in general, caused by two different discretizations: a_j is related to a finite element, b_k is related to a boundary element. Because of a simple storage organization and a simple computation of the stiffness matrices, it is convenient to use only one mesh, i.e. to take the boundary element discretization induced by the finite element triangulation. Therefore, we consider this case in the sequel. For any element Δ_j let

$$c_j := a_j + \sum_{k=1, \Gamma_k\subseteq\overline{\Delta_j}}^N b_k$$

where the sum may be zero or consists of one or two summands.

The meshes in our numerical examples are steered by the following algorithm where $0 \leq \theta \leq 1$ is a global parameter:

Algorithm 12.1 ((A)) *Given some coarse e.g. uniform mesh refine it successively by halving some of the elements due to the following rule. For any triangulation define $a_1, \ldots, a_N$ as above and divide some element Γ_j by halving the largest side if*

$$c_j \geq \theta \cdot \max_{k=1,\ldots,N} c_k.$$

In a subsequent step all hanging nodes are avoided by further refinement in order to obtain a regular mesh.

Remark 12.2

(i) Note that in Algorithm (A) $\theta = 0$ gives a uniform refinement and with increasing θ the number of refined elements in the present step decreases.
(ii) By observing (12.64) we have some error control which, in some sense, yields a *reliable* algorithm. In particular, the relative improvement of (12.64) may be used as a reasonable termination criterion.
(iii) If in some step of Algorithm (A), (12.64) does not become smaller then we may add some uniform refinement steps ($\theta = 0$). It can be proved that in this case (12.64) decreases and tends towards zero. If we allow this modification we get *convergence* of the adaptive algorithm.

In [91] we consider (IP) with $p = 1$, $f = 0$, Ω the L-shape region with vertices $(0, 0)$, $(1, 0)$, $(1, 1)$, $(-1, 1)$, $(-1, -1)$, $(0, -1)$ and take

$$u = r^{2/3} \cdot \sin(\frac{2}{3}\alpha) \quad \text{and} \quad v = \frac{1}{2} \ln((x + \frac{1}{2})^2 + (y - \frac{1}{2})^2)$$

such that u_0, t_0 are given by (12.24), (12.25).

In Table 12.1 we have the numerical results for the uniform mesh ($\theta = 0$) and for the meshes generated by Algorithm (A) for $\theta = 0.2$, 0.4, and 0.6. Here, we show only the number of degrees of freedom N for the finite element method (chosen by the algorithm; a new row corresponds to a new refinement step in the adaptive algorithm), and the corresponding relative error of the displacements e_N in the $H^1(\Omega)$-norm.

Let γ_N be the error in energy norm divided by (12.64). Hence, by Theorem 12.11, γ_N is bounded which can be observed from Table 12.1 Moreover, γ_N is bounded below which indicates efficiency of the estimate and hence of the adaptive scheme. For further experiments see [91].

Table 12.1 Numerical results for the linear transmission problem [91]

Uniform mesh			(A) for $\theta = 0.4$			(A) for $\theta = 0.6$		
N	e_N	γ_N	N	e_N	γ_N	N	e_N	γ_N
8	0.20434	.152	8	0.20434	.152	8	0.20434	.152
11	0.18587	.173	11	0.18587	.173	10	0.20467	.173
21	0.14485	.164	15	0.17074	.176	13	0.17286	.176
33	0.12564	.185	21	0.14520	.182	17	0.14848	.185
65	0.09563	.149	26	0.12197	.188	21	0.13954	.193
113	0.08027	.159	31	0.11007	.201	26	0.11594	.196
225	0.06230	.148	40	0.09420	.168	33	0.10579	.209
(A) for $\theta = 0.2$			48	0.08544	.177	38	0.09402	.214
N	e_N	γ_N	55	0.07824	.180	50	0.08328	.181
8	0.20434	.152	71	0.06837	.182	55	0.07744	.181
11	0.18587	.173	80	0.06260	.184	69	0.06742	.183
19	0.14621	.163	101	0.05633	.187	78	0.06448	.185
27	0.12844	.182	134	0.04959	.187	97	0.05639	.189
41	0.10297	.155	157	0.04510	.184	08	0.05448	.189
52	0.09020	.166	201	0.03904	.183	49	0.04533	.189
66	0.07554	.162	226	0.03656	.184	64	0.04367	.185
75	0.06900	.172				211	0.03783	.184
102	0.05947	.174				239	0.03562	.185
134	0.05128	.176						
156	0.04646	.175						
201	0.04004	.177						
235	0.03604	.177						

12.3.3 *Adaptive FE/BE Coupling with a Schur Complement Error Indicator*

Recently, the use of adaptive hierarchical methods has becoming increasingly popular. Using the discretization of the Poincaré-Steklov operator we present from [274] for the symmetric FE/BE coupling method an a posteriori error estimate with 'local' error indicators; for an alternative method which uses the full coupling formulation see [312]. By using stable hierarchical basis decompositions for finite elements we have two-level subspace decompositions for locally refined meshes. Assuming a saturation condition to hold an adaptive algorithm is formulated to compute the finite element solution on a sequence of refined meshes in the interior domain and on the interface boundary. At the end of this subsection we present numerical experiments which show efficiency and reliability of the error indicators.

Let $\rho \in \mathscr{C}^1(\mathbb{R}_+)$ satisfy the conditions

$$\rho_0 \le \rho(t) \le \rho_1 \quad \text{and} \quad \rho_2 \le \rho(t) + t\rho'(t) \le \rho_3 \tag{12.65}$$

for some global constants $\rho_0, \rho_1, \rho_2, \rho_3 > 0$. We consider the following nonlinear interface problem (NP) (cf. [274]) in $\mathbb{R}^2$:

Problem (NP): Given the functions $f : \Omega_1 \to \mathbb{R}$ and $u_0, t_0 : \Gamma \to \mathbb{R}$ find $u_i : \Omega_i \to \mathbb{R}$, $i = 1, 2$, and $b \in \mathbb{R}$ such that

$$-\operatorname{div}(\rho(|\nabla u_1|)\, \nabla u_1) = f \qquad \text{in } \Omega_1 \tag{12.66a}$$

$$-\Delta u_2 = 0 \qquad \text{in } \Omega_2 \tag{12.66b}$$

$$u_1 - u_2 = u_0 \qquad \text{on } \Gamma \tag{12.66c}$$

$$\rho(|\nabla u_1|)\, \frac{\partial u_1}{\partial n} - \frac{\partial u_2}{\partial n} = t_0 \qquad \text{on } \Gamma \tag{12.66d}$$

$$u_2(x) = b \ln |x| + o(1) \qquad \text{for } |x| \to \infty \tag{12.66e}$$

where $\frac{\partial v}{\partial n}$ is the normal derivative of v pointing from Ω_1 into Ω_2.

By the symmetric coupling method the problem (12.66) is transformed into the following variational problem (cf. [91]):

Given $f \in (H^1(\Omega_1))'$, $u_0 \in H^{1/2}(\Gamma)$ and $t_0 \in H^{-1/2}(\Gamma)$ find $u \in H^1(\Omega_1)$ and $\phi \in H^{-1/2}(\Gamma)$ such that

$$a(u, v) + B(u, \phi; v, \psi) \;=\; \mathscr{L}(v, \psi) \tag{12.67}$$

for all $v \in H^1(\Omega_1)$ and $\psi \in H^{-1/2}(\Gamma)$ where the form $a(\cdot;\cdot)$ is defined as

$$a(u, v) := 2 \int_{\Omega_1} \rho(|\nabla u|)\, \nabla u \cdot \nabla v \, dx \,,$$

the bilinear form $B(\cdot;\cdot)$ is defined as

$$B(u, \phi; v, \psi) \;:=\; \langle W u_{|\Gamma} + (K' - I)\phi \,,\, v_{|\Gamma}\rangle - \langle \psi \,,\, (K - I)u_{|\Gamma} - V\phi\rangle \,,$$

and the linear form $\mathscr{L}(\cdot)$ is defined as

$$\mathscr{L}(v, \psi) \;:=\; 2(f, v) + \langle 2t_0 + W u_0 \,,\, v_{|\Gamma}\rangle - \langle \psi \,,\, (K - I)u_0\rangle \,.$$

Here, $(\cdot,\cdot)$ and $\langle\cdot,\cdot\rangle$ denote the duality pairings between $(H^1(\Omega_1))'$ and $H^1(\Omega_1)$ and between $H^{-1/2}(\Gamma)$ and $H^{1/2}(\Gamma)$, respectively. The unknowns in (12.66) satisfy $u_1 = u$ and $\frac{\partial u_2}{\partial n} = \phi$ and u_2 can be obtained via a representation formula (see the foregoing section).

Lemma 12.12 *The following problem is equivalent to* (12.67):
Find $u \in H^1(\Omega_1)$ *such that*

$$a(u, v) + \langle Su_{|\Gamma}, v\rangle = F(v) \quad \forall v \in H^1(\Omega_1)$$
$$\text{where} \qquad F(v) := 2\int_{\Omega_1} f v dx + \langle 2t_0 + Su_0, v_{|\Gamma}\rangle \,, \tag{12.68}$$

$a(\cdot,\cdot)$ *as in* (12.67)*, and the Poincaré-Steklov operator for the exterior domain represented by* $S := W + (K' - I)V^{-1}(K - I)$ *is a continuous map from* $H^{1/2}(\Gamma)$ *into* $H^{-1/2}(\Gamma)$ *and coercive on* $H^{1/2}(\Gamma)$ *for* $cap(\Gamma) < 1$.

Next, we describe the coupling of the finite element method (FEM) and the boundary element method (BEM) to compute approximations to the solution (u, ϕ) of (12.67). We consider regular triangulations ω_H of Ω_1 and partitions γ_H of Γ. Our test and trial spaces are defined as

$$T_H := \{v_H : \Omega_1 \to \mathbb{R}\,;\ v_H \text{ p.w. linear on } \omega_H\,,\ v_H \in \mathscr{C}^0(\Omega_1)\}\,, \tag{12.69}$$
$$\tau_H := \{\psi_H : \Gamma \to \mathbb{R}\,;\ \psi_H \text{ p.w. constant on } \gamma_H\}\,. \tag{12.70}$$

For simplicity, we assume that the mesh for the discretization of the boundary element part γ_H is induced by that of the finite element part. This yields the following discretization of problem (12.67):
Find $(u_H, \phi_H) \in T_H \times \tau_H$ such that

$$2\int_{\Omega_1} \rho(|\nabla u_H|)\,\nabla u_H \cdot \nabla v\, dx + B(u_H, \phi_H; v, \psi) \;=\; \mathscr{L}(v, \psi) \tag{12.71}$$

for all $(v, \psi) \in T_H \times \tau_H$.
Application of Newton's method to (12.71) yields a sequence of linear systems to be solved. Given an initial guess $(u_H^{(0)}, \phi_H^{(0)})$ we compute

$$(u_H^{(l)}, \phi_H^{(l)}) = (u_H^{(l-1)}, \phi_H^{(l-1)}) + (d_H^{(l)}, \delta_H^{(l)}) \qquad (l = 1, 2, \dots)$$

such that

$$a_{u_H^{(l-1)}}(d_H^{(l)}, v) + B(d_H^{(l)}, \delta_H^{(l)}; v, \psi) = \mathscr{L}(v, \psi) - a(u_H^{(l-1)}, v) - B(u_H^{(l-1)}, \phi_H^{(l-1)}; v, \psi) \tag{12.72}$$

for all $(v, \psi) \in T_H \times \tau_H$ with $a(\cdot,\cdot)$, $B(\cdot,\cdot;\cdot,\cdot)$, and $\mathscr{L}(\cdot,\cdot)$ as in (12.67). The bilinear form $a_w(\cdot,\cdot)$ is defined by

$$a_w(u, v) := 2\int_{\Omega_1} \tilde{\rho}(\nabla w)\,\nabla u \cdot \nabla v\, dx\,, \tag{12.73}$$

and $\tilde{\rho} \in \mathbb{R}^2$ is the Jacobian of $x \to \rho(|x|)x$, i.e.

$$\tilde{\rho} = \rho(|x|)I_{2\times 2} + \rho'(|x|)\frac{x \cdot x^T}{|x|} \quad (x \in \mathbb{R}^2). \tag{12.74}$$

From the assumptions on ρ in (12.65) it follows that there exist constants $\nu, \mu > 0$ such that

$$a_w(u, v) \leq \nu \|u\|_{H^1(\Omega_1)}\|v\|_{H^1(\Omega_1)} \quad \text{and} \quad \mu \|u\|^2_{H^1(\Omega_1)} \leq a_w(u, u) \tag{12.75}$$

for all $w, u, v \in H^1(\Omega_1)$.

By the assumptions on ρ in (12.65) the energy functional of (12.67) is strictly convex, and hence, Newton's method converges locally.

For the implementation of (12.72) we define the piecewise linear basis functions b_i via

$$b_i(\nu_j) := \delta_{i,j} \qquad (1 \leq j \leq n_{\text{in}}, \quad n_{\text{in}} + 1 \leq j \leq n_T)$$

where $\nu_i \in \Omega_1 \backslash \Gamma$ $(1 \leq i \leq n_{\text{in}})$ are the inner nodes of ω_H and $\nu_i \in \Gamma$

$(n_{\text{in}} + 1 \leq i \leq n_H := \dim T_H)$ are the boundary nodes of ω_H counted along the closed curve Γ.

The above result obviously holds for subspaces of T_h. That means that ω_h does not necessarily has to be a uniform refinement of ω_H. Thus (12.77) holds also for locally refined meshes.On the boundary the following basis of τ_H is introduced: Let $\mu_i \in \gamma_H$ be the boundary element induced by the nodes $\nu_{n_{\text{in}}+i}, \nu_{n_{\text{in}}+i+1}$ $(1 \leq i \leq n_\tau - 1, n_\tau := \dim \tau_H)$ and μ_{n_τ} by the nodes $\nu_{n_T}, \nu_{n_{\text{in}}+1}$. With each μ_i we associate the basis function

$$\beta_i(x) := \begin{cases} 1 & \text{if} \quad x \in \mu_i \\ 0 & \text{if} \quad x \in \Gamma \backslash \mu_i \end{cases} .$$

With the basis functions b_i and β_i (12.72) yields a linear system which may be solved with the hybrid modified conjugate residual (HMCR) scheme together with efficient preconditioners [236].

In [312] an adaptive algorithm is given based on a posteriori error estimates of the solution (u_H, ϕ_H) of (12.71). Here we apply a Schur complement method based on a Galerkin discretization of the variational formulation (12.68) eliminating the unknown vector ϕ. In this way we also obtain a discretization of the Poincaré-Steklov operator which will be used to develop an a posteriori error indicator which needs only a refinement of the mesh defining T_H and does not need a finer discretization as τ_H.

Next, we introduce **hierarchical two-level decompositions** for the finite element space T_h on ω_h (cf. (12.69)) where we get ω_h by the refinement shown in Fig. 12.2.

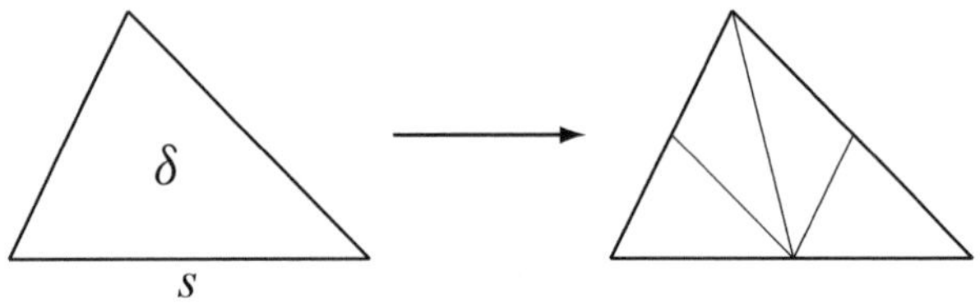

Fig. 12.2 Refinement of $\delta \in \omega_H$. The longest edge of δ is denoted by s. The new nodes are the midpoints of the edges of δ [312]

These decompositions will be used to derive an a posteriori error estimate for the Galerkin solution to (12.67) which is obtained by applying a Schur complement to (12.71).

We take the hierarchical two-level subspace decomposition

$$T_h := T_H \oplus L_h\,, \quad L_h := T_1 \oplus T_2 \oplus \ldots \oplus T_n$$

with $T_i := \text{span}\{\hat{b}_i\}$ where $\hat{b}_i$ denote the piecewise linear basis functions in the new n node-points ν_i of the fine grid, [312, 436]. Let $P_H : T_h \longrightarrow T_H$, $P_i : T_h \longrightarrow T_i$ be the Galerkin projections with respect to the bilinear form $b(\cdot, \cdot)$ which is defined as

$$b(u, v) := \int_{\Omega_1} (\nabla u \cdot \nabla v + uv)\, dx\,. \tag{12.76}$$

For all $u \in T_h$ we define P_H and P_i by

$$b(P_H u, v) = b(u, v) \quad \forall v \in T_H$$
$$b(P_i u, v) = b(u, v) \quad \forall v \in T_i.$$

The following result states that the two-level additive Schwarz operator $P := P_H + \sum_{i=1}^m P_i$ has bounded condition number: There are constants $c_1, c_2 > 0$ which depend only on the smallest angle of the triangles in ω_H and on the diameter of Ω_1 such that

$$c_1\|v\|^2_{H^1(\Omega_1)} \leq \|P_H v\|^2_{H^1(\Omega_1)} + \sum_{i=1}^m \|P_i v\|^2_{H^1(\Omega_1)} \leq c_2\|v\|^2_{H^1(\Omega_1)} \qquad \forall v \in T_h. \tag{12.77}$$

Now, we introduce the approximate Poincaré-Steklov operator on fine mesh functions

$$\tilde{S}_h := W_h + (K^*_{H,h} - I^*_{H,h})V_H^{-1}(K_{h,H} - I_{h,H}) \tag{12.78}$$

where for $u, v \in T_h$ and $\phi, \psi \in \tau_H$

$$\begin{aligned}\langle W_h u_{|\Gamma}, v_{|\Gamma}\rangle &= \langle W u_{|\Gamma}, v_{|\Gamma}\rangle\\ \langle (K_{h,H} - I_{h,H}) u_{|\Gamma}, \psi|_\Gamma\rangle &= \langle (K - I) u_{|\Gamma}, \psi|_\Gamma\rangle\\ \langle V_H \phi|_\Gamma, \psi|_\Gamma\rangle &= \langle V\phi|_\Gamma, \psi|_\Gamma\rangle\\ \langle (K^*_{H,h} - I^*_{H,h})\phi|_\Gamma, v_{|\Gamma}\rangle &= \langle (K' - I)\phi|_\Gamma, v_{|\Gamma}\rangle\,.\end{aligned}$$

Furthermore we consider the discrete Poincaré-Steklov operator

$$S_H := W_H + (K^*_{H,H} - I^*_{H,H}) V_H^{-1} (K_{H,H} - I_{H,H}) \tag{12.79}$$

on coarse mesh functions where the operators are defined as above by substituting T_H for T_h.

With the discrete Poincaré-Steklov operators $\tilde{S}_h$ and S_H we formulate discrete problems to (12.68):

Find $u_H \in T_H$ such that

$$a(u_H, v) + \langle S_H u_{H|\Gamma}, v_{|\Gamma}\rangle = F_H(v) \quad \forall v \in T_H, \tag{12.80}$$

and

find $\tilde{u}_h \in T_h$ such that

$$a(\tilde{u}_h, v) + \langle \tilde{S}_h \tilde{u}_{h|\Gamma}, v_{|\Gamma}\rangle = \tilde{F}_h(v) \quad \forall v \in T_h \tag{12.81}$$

where $F_H(\cdot)$ and $\tilde{F}_h(\cdot)$ are obtained by substituting S_H for S in F of (12.68) and $\tilde{S}_h$, respectively.

For our analysis to derive an a posteriori error estimate (Theorem 12.12) we have to make the following *saturation assumption*.

Assumption 12.3 *Let $u, u_H, \tilde{u}_h$ be defined as in* (12.68), (12.80) *and* (12.81). *There exists a constant $\kappa \in (0, 1)$ independent of H, h such that*

$$\|u - \tilde{u}_h\|_{H^1(\Omega_1)} \le \kappa \|u - u_H\|_{H^1(\Omega_1)}$$

The foregoing assumption immediately implies

$$(1-\kappa)\|u - u_H\|_{H^1(\Omega_1)} \le \|\tilde{u}_h - u_H\|_{H^1(\Omega_1)} \le (1+\kappa)\|u - u_H\|_{H^1(\Omega_1)}. \tag{12.82}$$

The following a posteriori error estimate is proved in[274] (see also [312]).

Theorem 12.12 *Assume Assumption 12.3 holds. Let $T_0 \subset T_1 \subset T_2 \subset \ldots$ be a sequence of hierarchical subspaces where T_0 is an initial FEM space (cf. (12.69)). The refinement of all triangles defining T_k according to Fig. 12.2 gives us $T_{h,k}$. Let k denote the number of the refinement level and u_k the corresponding Galerkin* *solution of* (12.80) *and u the exact solution of* (12.68)*, then there are constants* $\zeta_1, \zeta_2 > 0$, $k_0 \in \mathbb{N}_0$*, such that for all* $k \geq k_0$

$$\zeta_1 \Big(\sum_{i=1}^{n} \theta_{i,k}^2 \Big)^{1/2} \leq \| u - u_k \|_{H^1(\Omega_1)} \leq \zeta_2 \Big(\sum_{i=1}^{n} \theta_{i,k}^2 \Big)^{1/2} \tag{12.83}$$

where the local error indicators

$$\theta_{i,k} := \frac{|2\vartheta_\Omega(b_{i,k}) + \vartheta_\Gamma(b_{i,k})|}{\|b_{i,k}\|_{H^1(\Omega_1)}} \tag{12.84}$$

are obtained via basis functions $b_{i,k} \in T_{h,k} \backslash T_k$ *by a domain part*

$$\vartheta_\Omega(b_{i,k}) := \int_{\Omega_1} f\, b_{i,k} dx - \int_{\Omega_1} \rho(|\nabla u_k|)\, \nabla u_k \cdot \nabla b_{i,k}\, dx \tag{12.85}$$

and a boundary part

$$\vartheta_\Gamma(b_{i,k}) := \big\langle 2t_0 + \tilde{S}_{h,k} u_0, b_{i,k|\Gamma} \big\rangle - \big\langle \tilde{S}_{h,k} u_{k|\Gamma}, b_{i,k|\Gamma} \big\rangle \tag{12.86}$$

with $\tilde{S}_{h,k}$ *defined as in (12.78) with respect to* $T_{h,k}$*,* T_k *instead of* T_h*,* T_H*.*

Proof We define the form

$$Q(w, u_k, v) = 2 \int_{\Omega_1} \big[\rho(|\nabla w|)\, \nabla w - \rho(|\nabla u_k|)\, \nabla u_k - \tilde{\rho}(\nabla u_k)\, \nabla(w - u_k) \big] \cdot \nabla v\, dx$$

for all $w, v \in T_{h,k}$. Note that

$$Q(w, u_k, v) = a(w, v) - a(u_k, v) - a_{u_k}(w - u_k, v) \tag{12.87}$$

for all $w, v \in T_{h,k}$. The bilinear form a_w is defined in (12.73). Since the function

$$G : \begin{cases} \mathbb{R}^2 & \longrightarrow \mathbb{R}^2 \\ x & \longmapsto 2\rho(|x|)x \end{cases}$$

is differentiable and $2\tilde{\rho}$ (see (12.74)) is the Jacobian of G we obtain

$$\delta(k) := \frac{\|G(\nabla \tilde{u}_{h,k}) - G(\nabla u_k) - 2\tilde{\rho}(\nabla u_k)(\nabla(\tilde{u}_{h,k} - u_k))\|_{[L^2(\Omega_1)]^2}}{\|\nabla(\tilde{u}_{h,k} - u_k)\|_{[L^2(\Omega_1)]^2}} \longrightarrow 0$$

for

$$\|\tilde{u}_{h,k} - u_k\|_{H^1(\Omega_1)} \to 0 \tag{12.88}$$

and

$$|Q(\tilde{u}_{h,k}, u_k, v)| \le \delta(k)\|\tilde{u}_{h,k} - u_k\|_{H^1(\Omega_1)}\|v\|_{H^1(\Omega_1)} \tag{12.89}$$

where $\tilde{u}_{h,k}$ denotes the Galerkin solution of (12.81) substituting T_h by $T_{h,k}$. (12.88) is obvious for $k \to \infty$ with (12.82).

We define $e_k \in T_{h,k}$ by

$$b(e_k, v) = \tilde{F}_k(v) - a(u_k, v) - \langle \tilde{S}_k u_k, v\rangle \quad \forall v \in T_{h,k} \tag{12.90}$$

where the bilinear form $b(\cdot,\cdot)$ is given in (12.76), $a(\cdot,\cdot)$ in (12.67), and $\tilde{F}_k$, $\tilde{S}_k$ in (12.81) taking T_k for T_H. Next, we show that there are constants $\mu_0, \nu_0 > 0$, which are independent of k, such that

$$\mu_0\|\tilde{u}_{h,k} - u_k\|_{H^1(\Omega_1)} \le \|e_k\|_{H^1(\Omega_1)} \le \nu_0\|\tilde{u}_{h,k} - u_k\|_{H^1(\Omega_1)}\,. \tag{12.91}$$

By (12.75), (12.87), (12.81), and (12.90) we obtain

$$\begin{aligned}
\mu\|\tilde{u}_{h,k} - u_k\|_{H^1(\Omega_1)} \le\; & \frac{a_{u_k}(\tilde{u}_{h,k} - u_k, \tilde{u}_{h,k} - u_k)}{\|\tilde{u}_{h,k} - u_k\|_{H^1(\Omega_1)}} \\
\le\; & \sup_{v\in M}\Big(a_{u_k}(\tilde{u}_{h,k} - u_k, v) + \langle \tilde{S}_k(\tilde{u}_{h,k} - u_k), v\rangle\Big) \\
\le\; & \sup_{v\in M}\Big(a(\tilde{u}_{h,k}, v) - a(u_k, v) + \langle \tilde{S}_k(\tilde{u}_{h,k} - u_k), v\rangle - Q(\tilde{u}_{h,k}, u_k, v)\Big) \\
\le\; & \sup_{v\in M}\Big(\tilde{F}_k(v) - a(u_k, v) - \langle \tilde{S}_k u_k, v\rangle - Q(\tilde{u}_{h,k}, u_k, v)\Big) \\
\le\; & \sup_{v\in M}\Big(b(e_k, v) - Q(\tilde{u}_{h,k}, u_k, v)\Big) \\
\le\; & \|e_k\|_{H^1(\Omega_1)} + \delta(k)\|\tilde{u}_{h,k} - u_k\|_{H^1(\Omega_1)}
\end{aligned}$$

where $M := \{v \in T_{h,k} \mid \|v\|_{H^1(\Omega_1)} = 1\}$. The second inequality follows by the positive definiteness of $\tilde{S}_k$, the last by (12.89) and the Cauchy-Schwarz inequality. Furthermore, we obtain by (12.90), (12.81), (12.87), and (12.75) that

$$\begin{aligned}
\|e_k\|^2_{H^1(\Omega_1)} =\; & b(e_k, e_k) = a(\tilde{u}_{h,k}, e_k) - a(u_k, e_k) + \langle \tilde{S}_k(\tilde{u}_{h,k} - u_k), e_k\rangle \\
=\; & a_{u_k}(\tilde{u}_{h,k} - u_k, e_k) + Q(\tilde{u}_{h,k}, u_k, e_k) + \langle \tilde{S}_k(\tilde{u}_{h,k} - u_k), e_k\rangle \\
\le\; & (\nu + \delta(k))\|\tilde{u}_{h,k} - u_k\|_{H^1(\Omega_1)}\|e_k\|_{H^1(\Omega_1)} \\
& + \|\tilde{S}_k(\tilde{u}_{h,k} - u_k)\|_{H^{-1/2}(\Gamma)}\|e_k\|_{H^{1/2}(\Gamma)} \\
\le\; & (\nu + \delta(k) + \nu_S)\|\tilde{u}_{h,k} - u_k\|_{H^1(\Omega_1)}\|e_k\|_{H^1(\Omega_1)}
\end{aligned}$$

where the last inequalities follow, again, by (12.89) and the existence of a constant ν_S since $\tilde{S}_k$ is uniformly bounded for all $k > k_0$, $k_0 \in \mathbb{N}_0$ constant. Here, the uniform boundedness of $\tilde{S}_k$ follows by the approximation properties of the discrete BE spaces defining $\tilde{S}_k$ and the boundedness of S.

If k_0 is sufficiently large (such that $\delta(k) \le \delta_0 < \mu$ for all $k \ge k_0$) then (12.91) follows with $\mu_0 = \mu - \delta_0$ and $\nu_0 = \nu + \delta_0 + \nu_S$.

Since

$$\|e_k\|^2_{H^1(\Omega_1)} = b(e_k, e_k)$$

where $b(\cdot,\cdot)$ is defined in (12.76), we can apply (12.77) to obtain

$$c_1\|e_k\|^2_{H^1(\Omega_1)} \le \|P^{(k)}e_k\|^2_{H^1(\Omega_1)} + \sum_{i=1}^{m_k}\|P_{i,k}e_k\|^2_{H^1(\Omega_1)} \le c_2\|e_k\|^2_{H^1(\Omega_1)} \,. \tag{12.92}$$

Here $P^{(k)} : T_{h,k} \to T_k$ and $P_{i,k} : T_{h,k} \to \operatorname{span}\{b_{i,k}\}$ are the Galerkin projections with respect to the bilinear form $b(\cdot,\cdot)$. With the notations of (12.77) $P^{(k)} = P_H$.

By definition of $P^{(k)}$ and $P_{i,k}$, by (12.90), (12.81), and (12.80) it follows that

$$\|P^{(k)}e_k\|^2_{H^1(\Omega_1)} = b(e_k, P^{(k)}e_k) = \tilde{F}_k(P^{(k)}e_k) - \tilde{F}_k(P^{(k)}e_k) = 0 \tag{12.93}$$

and

$$P_{i,k}e_k = \frac{b(e_k, b_{i,k})}{b(b_{i,k}, b_{i,k})}b_{i,k} = \frac{\tilde{F}_k(b_{i,k}) - a(u_k, b_{i,k}) - \langle \tilde{S}_k u_k, b_{i,k}\rangle}{\|b_{i,k}\|^2_{H^1(\Omega_1)}}b_{i,k} \,.$$

Hence, we have

$$\|P_{i,k}e_k\|_{H^1(\Omega_1)} = \frac{|\tilde{F}_k(b_{i,k}) - a(u_k, b_{i,k}) - \langle \tilde{S}_k u_k, b_{i,k}\rangle|}{\|b_{i,k}\|_{H^1(\Omega_1)}} = \theta_{i,k} \,. \tag{12.94}$$

By (12.77), (12.93) and (12.94) we obtain

$$c_1\|e_k\|^2_{H^1(\Omega_1)} \le \sum_{i=1}^{m_k}\theta^2_{i,k} \le c_2\|e_k\|^2_{H^1(\Omega_1)}.$$

This yields together with (12.91) that

$$\frac{1}{\nu_0\sqrt{c_2}}\left(\sum_{i=1}^{m_k}\theta^2_{i,k}\right)^{1/2} \le \|\tilde{u}_{h,k} - u_k\|_{H^1(\Omega_1)} \le \frac{1}{\mu_0\sqrt{c_1}}\left(\sum_{i=1}^{m_k}\theta^2_{i,k}\right)^{1/2}$$

Table 12.2 Results for adaptive algorithm based on Theorem 12.12 for (NP) with u_1, u_2 from (12.95), $\zeta = 0.15$

L	n_k	dim T_k	dim τ_k	E_k	η_k	η_k/E_k	κ_k	α_k
0	37	21	16	0.10608	0.13067	1.232	–	–
1	55	37	18	0.07596	0.08283	1.090	0.716	0.842
2	78	58	20	0.05511	0.06495	1.179	0.725	0.919
3	109	85	24	0.04510	0.05596	1.241	0.818	0.599
4	163	129	34	0.03626	0.04373	1.206	0.804	0.542
5	454	396	58	0.02063	0.02419	1.172	0.569	0.550
6	677	595	82	0.01654	0.01936	1.171	0.802	0.554

and, finally, we obtain the assertion of the theorem by (12.82) with

$$\zeta_1 = \frac{1}{(1+\kappa)\nu_0\sqrt{c_2}} \text{ and } \zeta_2 = \frac{1}{(1-\kappa)\mu_0\sqrt{c_1}} . \qquad \square$$

In Table 12.2 , we list the numerical experiment for (NP) with $\varrho \equiv 1$ (for $\varrho(t) = \frac{1}{6}\left(1 + \frac{5}{1+5t}\right)$ see [91]) and choose Ω_1 to be the L-shaped domain with corners at $(0,0)$, $(0,\frac{1}{4})$, $(-\frac{1}{4},\frac{1}{4})$, $(-\frac{1}{4},-\frac{1}{4})$, $(\frac{1}{4},-\frac{1}{4})$, $(\frac{1}{4},0)$. The exact solution of the model problem (NP) is given by

$$u_1(r,\alpha) = r^{2/3}\sin\tfrac{2}{3}(\alpha - \tfrac{\pi}{2}), \quad u_2(x_1,x_2) = \ln\sqrt{(x_1+\tfrac{1}{8})^2 + (x_2+\tfrac{1}{8})^2}. \tag{12.95}$$

The functions u_0, t_0, f are chosen to yield the exact solution. The quantities in Table 12.2 are given as follows: With k we denote the refinement level, with n_k the total number of unknowns and with N_k the total number of triangles defining T_k. The error E_k is defined as

$$E_k := \|u - u_k\|_{1,\Omega_1} .$$

The global error indicator η_k is defined by

$$\eta_k = \left(\sum_{i=1}^{N_k}\eta_{i,k}^2\right)^{1/2}, \qquad \eta_{i,k} := \left(\theta_{i_1,k}^2 + \theta_{i_2,k}^2 + \theta_{i_3,k}^2\right)^{1/2} \quad (i = 1,\dots,N_k).$$

Here i_1, i_2, i_3 denote the three edges and the corresponding new base functions for every element of the old mesh. The values of the quotient η_k/E_k, the efficiency index, indicate the efficiency of the error indicator η_k and confirm Theorem 12.12. The quantity

$$\kappa_k := \frac{\|u - u_k\|_{1,\Omega_1}}{\|u - u_{k+1}\|_{1,\Omega_1}}$$

estimates the saturation constant κ. Since κ_k is bounded by a constant less than 1 the saturation condition (Assumption 12.3) is satisfied for the sequence of meshes which is generated by our adaptive algorithm. The experimental convergences rates α_k are given by

$$\alpha_k = \frac{\ln(E_k/E_{k-1})}{\ln(n_{k-1}/n_k)}.$$

From Table 12.2 we see that α_k approaches $1/2$, which is the convergence rate in case of a smooth solution. This shows the quality of the adaptive algorithm. For uniform meshes one obtains the non-optimal convergence rate $\alpha = 1/3$ The above hierarchical method is easily implemented since for the computation of the error indicators one can use the same routine as for the computation of the entries of the Galerkin matrix.

12.3.4 Convergence of Adaptive FEM-BEM Couplings

Let ζ_l denote an a posteriori estimator, e.g. the residual estimator $R_1 + R_2 + R_3 + R_4$ of Sect. 12.3.2. We assume $\zeta_l^2 = \zeta_l(I_l)^2 := \sum_{\Delta \in I_l} \zeta_l(\Delta)^2$, where $I_l = T_l^{\Omega} \cup \mathscr{E}_l^{\Omega} \cup T_l^{\Gamma}$ with $T_l^{\Omega} = \{\Delta_1, \ldots, \Delta_N\}$, $\mathscr{E}_l^{\Omega}$ the set of interior edges, T_l^{Γ} the set boundary edges. $\zeta_l(\Delta)$ denotes the local refinement indicator for $\Delta \in I_l$.

Algorithm 12.2 (Adaptive) *Input: Initial mesh T_0, $l_i = 0$, $0 < \theta \le 1$:*

i) Compute discrete solution $\mathbf{u}_l \in H_l = X_{N_l} \times Y_{N_l}$
ii) Compute refinement indicators $\zeta_l(\Delta)\ \forall\, \Delta \in I_l$
iii) Determine $\mathscr{M}_l \subseteq I_l$ such that Dörfler marking

$$\theta \zeta_l^2 \le \sum_{\Delta \in \mathscr{M}_l} \zeta_l(T)^2$$

holds
iv) Compute new triangulation T_{l+1}, where at least all marked elements $\Delta \in \mathscr{M}_l$ are refined.
v) Increase counter l and go to i)

Output: Sequence of Galerkin solutions $\{\mathbf{u}_l\}_{l=0}^L$, sequence of error estimators $\{\zeta_l\}_{l=0}^L$ and sequence of triangulations $\{T_l\}_{l=0}^L$.

For the symmetric coupling method the reliability of the corresponding error estimators follows from Theorem 12.11 in Sect. 12.3.2. Corresponding residual type error estimators and a priori estimates hold for the Johnson-Nedelec and the Bielak-MacCamy couplings. As shown in [[8] for Bielak-MacCamy coupling], in [[9] for the symmetric coupling], in [[181] for the Johnson-Nedelec coupling] the above adaptive algorithm, (steered by the residual error estimator) converges. The proof crucially needs the following inverse estimates of the boundary integral operators.

Lemma 12.13 ([264]) *Let T_l^Γ be a regular triangulation of Γ. Let $h_l \in \mathbb{P}^0(T_l^\Gamma)$ with $h_l\Big|_\Delta = |\Delta|^{1/(d-1)}$, where d is the dimension of Ω. Then there exist constants $C_{inv}^K, C_{inv}^V > 0$ with*

$$\|h_l^{1/2} \nabla_\Gamma K v_l\|_{L^2(\Gamma)} \le C_{inv}^K \|v_l\|_{H^{1/2}(\Gamma)}$$

$$\|h_l^{1/2} W v_l\|_{L^2(\Gamma)} \le C_{inv}^K \|v_l\|_{H^{1/2}(\Gamma)}$$

$$\|h_l^{1/2} \nabla_\Gamma V \psi_l\|_{L^2(\Gamma)} \le C_{inv}^V \|\psi_l\|_{H^{-1/2}(\Gamma)}$$

$$\|h_l^{1/2} K' v_l\|_{L^2(\Gamma)} \le C_{inv}^V \|\psi_l\|_{H^{-1/2}(\Gamma)}$$

for all $v_l \in S^p(T_l^\Gamma)$, $\psi_l \in \mathbb{P}^p(T_l^\Gamma)$ (continous, respectively discontinous polynomials of degree p). The constants C_{inv}^K, C_{inv}^V only depend on the γ-regularity of T_l^Γ, the boundary mesh, and the polynomial degree p.

A further ingredient of the proof is the newest vertex bijection (NVB) as refinement strategy. See [181] for further details. In [176] we prove convergence of the adaptive algorithm with the hierarchical two-level estimator, considered in Sect. 12.3.3. There we show that the usual adaptive algorithm (with the weighted-residual error estimator terms in Theorem 12.11) drives the hierarchical estimator to zero.

12.3.5 Other Coupling Methods

In this section we consider the Johnson-Nedelec coupling [263, 439] which is often called the direct one-equation coupling, since only one equation of the Calderon projector is used. The first stability results rely on the compactness of the double layer operator K ([263]). This has the disadvantage that Γ needs to be smooth (then K is compact in the Laplace case), this is for standard FEM resp. BEM not optimal. Here the work by ([360]) turned out to be the breakthrough showing (for the first time for the Laplace transmission problem (IP)) that the Johnson-Nedelec coupling is well-defined on polygonal domains. The proof shows stability of the adjoint problem and was applied to some problems in linear elastostatics in [186]. A different approach was developed in [392], where an explicit stabilisation is introduced which leads to an equivalent problem in the continuous case. Steinbach shows in [392] that this equivalent problem (with a linear operator A in Ω_1) is elliptic under the assumption $c_{mon} > 1/4$. Here c_{mon} is the smallest eigenvalue of A. This condition was improved to $c_{mon} > c_K/4$ where $c_K \in [1/2, 1)$ is the contraction constant of the double layer potential K [330]. Unfortunately with this stabilization an additional boundary integral equation must be solved on each discrete level whereas for the implicit stabilization, given in [181], the equivalence to (IP) holds on the discrete level. Thus Johnson-Nedelec coupling is obtained by taking (12.38) and testing (12.36) in $H^{-1/2}(\Gamma)$, i.e.:

Find $(u, \phi) \in H^1(\Omega_1) \times H^{-1/2}(\Gamma)$ such that for all $(v, \psi) \in H^1(\Omega_1) \times H^{-1/2}(\Gamma)$ there holds

$$\langle B(u, \phi), (v, \psi)\rangle = F((v, \psi)) \tag{12.96}$$

where

$$\begin{aligned}\langle B(u, \phi), (v, \psi)\rangle &:= (A\nabla u, \nabla v) - \langle \phi, v\rangle + \langle \psi, (I - K)u + V\phi\rangle \\ F((v, \psi)) &:= (f, v) + \langle t_0, v\rangle + \langle \psi, (I - K)u_0\rangle\end{aligned}$$

The implicit stabilization reads: Assume there exists $\xi \in Y_N$ with $\langle \xi, 1\rangle \neq 0$ there holds

$$\langle \tilde{B}(u, \phi), (v, \psi)\rangle = \tilde{F}((v, \psi)) \tag{12.97}$$

where

$$\begin{aligned}\langle \tilde{B}(u, \phi), (v, \psi)\rangle &:= \langle B(u, \phi), (v, \psi)\rangle + \langle \xi, (I - K)u + V\phi\rangle\langle \xi, (I - K)v + V\psi\rangle \\ \tilde{F}((v, \psi)) &:= F((v, \psi)) + \langle \xi, (I - K)u_0\rangle\langle \xi, (I - K)v + V\psi\rangle\end{aligned}$$

Now $(u_h, \phi_h) \in X_N \times Y_N$ solves (12.96) $\forall (v_h, \psi_h) \in X_N \times Y_N$ if and only if it solves (12.97) for all $\forall (v_h, \psi_h) \in X_N \times Y_N$. One obtains under the assumption (12.27), (12.28) quasi-optimality of the Galerkin scheme for (12.97) (see [181] for details).

Our model problem (IP) can also be reformulated with the Bielak-MacCamy coupling [54]. This method is also called indirect one-equation coupling, since an indirect ansatz (which does not use the Calderon system) is applied to solve the exterior problem. The Bielak-MacCamy coupling reads:

Find $(u, \phi) \in H^1(\Omega) \times H^{-1/2}(\Gamma)$ such that for all $(v, \psi) \in H^1(\Omega) \times H^{-1/2}(\Gamma)$ there holds

$$\langle B(u, \phi), (v, \psi)\rangle = F((v, \psi)) \tag{12.98}$$

where

$$\begin{aligned}\langle B(u, \phi), (v, \psi)\rangle &:= (A\nabla u, \nabla v) - \langle (I - K')\phi, v\rangle + \langle \psi, V\phi - u\rangle \\ F((v, \psi) &:= (f, v) + \langle t_0, v\rangle + \langle \psi, u_0\rangle\end{aligned}$$

Now the implicit stabilization reads with ξ as above:

$$\langle \tilde{B}(u, \phi), (v, \psi) \rangle = \tilde{F}((v, \psi)) \tag{12.99}$$

where

$$\langle \tilde{B}(u, \phi), (v, \psi) \rangle := \langle B(u, \phi), (v, \psi) \rangle + \langle \xi, V\phi - u \rangle \langle \xi, V\psi - v \rangle$$
$$\tilde{F}((v, \psi)) := F((v, \psi)) - \langle \xi, u_0 \rangle \langle \xi, V\psi - v \rangle$$

Analogously $(u_h, \phi_h) \in X_N \times Y_N$ solves 12.98 if and only if (u_h, ϕ_h) solves 12.99 on $X_N \times Y_N$ and again the Galerkin solution (u_h, ϕ_h) converges quasioptimally in the energy norm. These results hold true for polygonal Γ and coupling problems in elastostatics (see again [181] for details).

12.4 Least Squares FEM/BEM Coupling for Transmission Problems

Here we report from [296] a least squares formulation for the numerical solution of second-order linear transmission problems, where in a bounded domain the second order partial differential equation is rewritten as first-order system. The least squares functional is given in terms of Sobolev norms of order -1 and of order $1/2$ and uses boundary integral operators. In [296] these norms are computed by approximating the corresponding inner product using multilevel preconditioners (multigrid and BPX) for the differential operator and weakly singular integral operator.

Let $\Omega_1 := \Omega \subset \mathbb{R}^d$, $d \geq 2$ be a bounded domain with Lipschitz boundary $\Gamma = \partial\Omega_1$, and $\Omega_2 := \mathbb{R}^d \backslash \bar{\Omega}_1$ with normal n on Γ pointing into Ω_2. Let $f \in L^2(\Omega_1)$, $u_0 \in H^{1/2}(\Gamma)$, $t_0 \in H^{-1/2}(\Gamma)$. We consider the model transmission problem of finding $u_1 \in H^1(\Omega_1)$, $u_2 \in H^1_{\text{loc}}(\Omega_2)$ such that

$$-\operatorname{div}(a\nabla u_1) = f \text{ in } \Omega_1 \tag{12.100}$$

$$\Delta u_2 = 0 \text{ in } \Omega_2 \tag{12.101}$$

$$u_1 = u_2 + u_0 \text{ on } \Gamma \tag{12.102}$$

$$(a\nabla u_1) \cdot n = \frac{\partial u_2}{\partial n} + t_0 \text{ on } \Gamma \tag{12.103}$$

$$u_2(x) = \begin{cases} A \ln|x| + o(1), & d = 2 \\ \mathcal{O}(|x|^{2-d}), & d \geq 3 \end{cases}, \ |x| \to \infty \tag{12.104}$$

Let $a_{ij} \in L^\infty(\Omega_1)$ such that $a = (a_{ij})$ satisfies for some $\alpha > 0$

$$\alpha \|z\|^2 \leq z^T a(x) z \quad \forall z \in \mathbb{R}^d \text{ and for almost all } x \in \Omega_1.$$

In the following, we will apply the boundary integral equation method in Ω_2 and reduce the original problem to a nonlocal transmission problem on the bounded domain Ω. The fundamental solution of the Laplacian is given by

$$G(x,y) = \begin{cases} -\frac{1}{\omega_2} \ln|x-y|, \; d=2 \\ \frac{1}{\omega_d}|x-y|^{2-d}, \;\; d \geq 3 \end{cases}$$

where we have $\omega_2 = 2\pi$, $\omega_3 = 4\pi$. For all $x \in \Omega_2$

$$u_2(x) = \int_\Gamma \left\{ \frac{\partial}{\partial n(y)} G(x,y)u(y) - G(x,y)\frac{\partial u}{\partial n(y)} \right\} ds_y$$

satisfies the Laplace equation (12.101) and the radiation condition (12.104). By using the boundary integral operators

$$V\psi(x) := 2\int_\Gamma G(x,y)\psi(y)\, ds_y, \quad x \in \Gamma \tag{12.105}$$

$$K\psi(x) := 2\int_\Gamma \frac{\partial}{\partial n_y} G(x,y)\psi(y)\, ds_y, \quad x \in \Gamma \tag{12.106}$$

$$K'\psi(x) := 2\frac{\partial}{\partial n_x}\int_\Gamma G(x,y)\psi(y)\, ds_y, \quad x \in \Gamma \tag{12.107}$$

$$W\psi(x) := -2\frac{\partial}{\partial n_x}\int_\Gamma \frac{\partial}{\partial n_y} G(x,y)\psi(y)\, ds_y, \quad x \in \Gamma \tag{12.108}$$

together with their well known-jump conditions we obtain the following integral equations

$$2\frac{\partial u_2}{\partial n} = -Wu_2 + (I-K')\frac{\partial u_2}{\partial n} \tag{12.109}$$

$$0 = (I-K)u_2 + V\frac{\partial u_2}{\partial n}. \tag{12.110}$$

In this way, the original transmission problem (12.100) — (12.104) reduces to the following non-local boundary value problem in Ω. Find $(u, \sigma) \in H^1(\Omega) \times H^{-1/2}(\Gamma)$ such that

$$-\operatorname{div}(a\nabla u) = f \quad \text{in } \Omega \tag{12.111}$$

$$\sigma = (a\nabla u)\cdot n \quad \text{on } \Gamma \tag{12.112}$$

$$2(\sigma - t_0) = -W(u-u_0) + (I-K')(\sigma - t_0) \quad \text{on } \Gamma \tag{12.113}$$

$$0 = (I-K)(u-u_0) + V(\sigma - t_0) \quad \text{on } \Gamma \tag{12.114}$$

Note that the flux variable $\theta := a\nabla u$ belongs to the Hilbert space.

$$H(\operatorname{div};\Omega) = \{\theta \in [L^2(\Omega)]^d \; : \; \|\theta\|^2_{[L^2(\Omega)]^d} + \|\operatorname{div}\theta\|^2_{L^2(\Omega)} < \infty\}.$$

with the inner product

$$(\theta, \zeta)_{H(\mathrm{div};\Omega)} = (\theta, \zeta)_{[L^2(\Omega)]^d} + (\operatorname{div}\theta, \operatorname{div}\zeta)_{L^2(\Omega)} \ .$$

Moreover, for all $\zeta \in H(\mathrm{div};\Omega)$ there holds $\zeta \cdot n \in H^{-1/2}(\Gamma)$ and $\|\zeta \cdot n\|_{H^{-1/2}(\Gamma)} \leq \|\zeta\|_{H(\mathrm{div};\Omega)}$ (see [196]).

With the interface conditions we can rewrite the transmission problem as follows with a first order system on Ω:

Find $(\theta, u, \sigma) \in H(\mathrm{div};\Omega) \times H^1(\Omega) \times H^{-1/2}(\Gamma)$ such that

$$\theta = a\nabla u \text{ in } \Omega \tag{12.115}$$

$$-\operatorname{div}\theta = f \text{ in } \Omega \tag{12.116}$$

$$\sigma = \theta \cdot n \text{ on } \Gamma \tag{12.117}$$

$$2(\sigma - t_0) = -W(u - u_0) + (I - K')(\sigma - t_0) \text{ on } \Gamma \tag{12.118}$$

$$0 = (I - K)(u - u_0) + V(\sigma - t_0) \text{ on } \Gamma \tag{12.119}$$

Let $\tilde{H}^{-1}(\Omega)$ denote the dual space of $H^1(\Omega)$, equipped with the dual norm $\|w\|_{\tilde{H}^{-1}(\Omega)} = \sup_{v \in H^1(\Omega)} \frac{(w,v)_{L^2(\Omega)}}{\|v\|_{H^1(\Omega)}}$. Then the solution of (12.115)—(12.119) is a solution of the following minimization problem:

Find $(\theta, u, \sigma) \in X := [L^2(\Omega)]^d \times H^1(\Omega) \times H^{-1/2}(\Gamma)$ such that

$$J(\theta, u, \sigma) = \min_{(\zeta, v, \tau) \in X} J(\zeta, v, \tau) \tag{12.120}$$

where J is the quadratic functional defined by

$$\begin{aligned} J(\zeta, v, \tau) &= \|a\nabla v - \zeta\|^2_{[L^2(\Omega)]^d} + \|(I - K)(v - u_0) + V(\tau - t_0)\|^2_{H^{1/2}(\Gamma)} \\ &+ \|\operatorname{div}\zeta + f - \frac{1}{2}\delta_\Gamma \otimes (W(v - u_0) + 2\zeta \cdot n - 2t_0 - (I - K')(\tau - t_0))\|^2_{\tilde{H}^{-1}(\Omega)} \\ &= \|a\nabla v - \zeta\|^2_{[L^2(\Omega)]^d} + \|(I - K)v + V\tau - (I - K)u_0 - Vt_0\|^2_{H^{1/2}(\Gamma)} \\ &+ \|\operatorname{div}\zeta - \frac{1}{2}\delta_\Gamma \otimes (Wv + 2\zeta \cdot n - (I - K')\tau) \\ &+ f + \frac{1}{2}\delta_\Gamma \otimes (Wu_0 + 2t_0 - (I - K')t_0)\|^2_{\tilde{H}^{-1}(\Omega)}. \end{aligned} \tag{12.121}$$

Here $\delta_\Gamma \otimes \tau$ denotes the distribution in $\tilde{H}^{-1}(\Omega)$ for $\tau \in H^{-1/2}(\Gamma)$ defined by

$$\langle \delta_\Gamma \otimes \tau, \varphi \rangle_{\tilde{H}^{-1}(\Omega) \times H^1(\Omega)} = (\tau, \varphi|_\Gamma)_{H^{-1/2}(\Gamma) \times H^{1/2}(\Gamma)} \quad \forall \varphi \in H^1(\Omega).$$

Due to coercivity and continuity of the corresponding variational problem the authors obtain in [296] uniqueness of (12.120) and equivalence between (12.115)—(12.119) and (12.120).

Defining $g(\zeta, v, \tau) := \operatorname{div} \zeta - \frac{1}{2}\delta_\Gamma \otimes (Wv + 2\zeta \cdot n - (I - K')\tau)$ we introduce with

$$\begin{aligned} B((\theta, u, \sigma), (\zeta, v, \tau)) = &(a\nabla u - \theta, a\nabla v - \zeta)_{L^2(\Omega)} \\ &+((I - K)u + V\sigma, (I - K)v + V\tau)_{H^{1/2}(\Gamma)} \\ &+(g(\theta, u, \sigma), g(\zeta, v, \tau))_{\tilde{H}^{-1}(\Omega)} \end{aligned} \tag{12.122}$$

and

$$\begin{aligned} G(\zeta, v, \tau) = &((I - K)v + V\tau, (I - K)u_0 + Vt_0)_{H^{1/2}(\Gamma)} \\ &-(g(\zeta, v, \tau), f + \frac{1}{2}\delta_\Gamma \otimes (Wu_0 + 2t_0 - (I - K')t_0))_{\tilde{H}^{-1}(\Omega)} \end{aligned} \tag{12.123}$$

the variational formulation for (12.120) as:

Find $(\theta, u, \sigma) \in X = [L^2(\Omega)]^d \times H^1(\Omega) \times H^{-1/2}(\Gamma)$ such that

$$B((\theta, u, \sigma), (\zeta, v, \tau)) = G(\zeta, v, \tau) \quad \forall (\zeta, v, \tau) \in X. \tag{12.124}$$

Theorem 12.13 *The bilinear form $B(\cdot, \cdot)$ is strongly coercive in X, i.e. there holds*

$$B((\zeta, v, \tau), (\zeta, v, \tau)) \gtrsim \|(\zeta, v, \tau)\|_X^2, \qquad \forall (\zeta, v, \tau) \in X. \tag{12.125}$$

Proof Let $(\zeta, v, \tau) \in X = [L^2(\Omega)]^d \times H^1(\Omega) \times H^{-1/2}(\Gamma)$.

We can estimate $\|\zeta\|_{[L^2(\Omega)]^d}$ by

$$\|\zeta\|_{[L^2(\Omega)]^d} \le \|\zeta - a\nabla v\|_{[L^2(\Omega)]^d} + \|a\nabla v\|_{[L^2(\Omega)]^d} \lesssim \|\zeta - a\nabla v\|_{[L^2(\Omega)]^d} + \|v\|_{H^1(\Omega)}. \tag{12.126}$$

Using the boundedness of V^{-1} (as a mapping from $H^{1/2}(\Gamma)$ into $H^{-1/2}(\Gamma)$) and $I - K$ we can estimate

$$\begin{aligned} \|\tau\|_{H^{-1/2}(\Gamma)} &\lesssim \|V\tau\|_{H^{1/2}(\Gamma)} \\ &\lesssim \|V\tau + (I - K)v\|_{H^{1/2}(\Gamma)} + \|(I - K)v\|_{H^{1/2}(\Gamma)} \\ &\lesssim \|V\tau + (I - K)v\|_{H^{1/2}(\Gamma)} + \|v\|_{H^{1/2}(\Gamma)} \\ &\lesssim \|V\tau + (I - K)v\|_{H^{1/2}(\Gamma)} + \|v\|_{H^1(\Omega)}. \end{aligned} \tag{12.127}$$

Now we use the Poincaré-Steklov operator $S : H^{1/2}(\Gamma) \mapsto H^{-1/2}(\Gamma)$ for the exterior domain, given by

$$S := W + (I - K')V^{-1}(I - K).$$

From [91, Lemma 4] we know that with the L^2 inner products $(\cdot,\cdot)$ and $\langle\cdot,\cdot\rangle$ on Ω and Γ, respectively, there holds

$$\|v\|^2_{H^1(\Omega)} \lesssim (a\nabla v, \nabla v) + \frac{1}{2}\langle Sv, v\rangle \quad \forall v \in H^1(\Omega),$$

yielding

$$\|v\|_{H^1(\Omega)} \lesssim \sup_{w\in H^1(\Omega)} \frac{(a\nabla v, \nabla w) + \frac{1}{2}\langle Sv, w\rangle}{\|w\|_{H^1(\Omega)}}.$$

We can expand the expression by

$$\begin{aligned}&(a\nabla v, \nabla w) + \frac{1}{2}\langle Sv, w\rangle\\ &= (a\nabla v - \zeta, \nabla w) - (\operatorname{div}\zeta, w) + \langle \zeta\cdot n, w\rangle + \frac{1}{2}\langle Sv, w\rangle\\ &= (a\nabla v - \zeta, \nabla w) - (\operatorname{div}\zeta - \delta_\Gamma \otimes [\zeta\cdot n + \frac{1}{2}Sv], w)\end{aligned}$$

and obtain

$$\begin{aligned}\|v\|_{H^1(\Omega)} &\lesssim \sup_{w\in H^1(\Omega)} \frac{(a\nabla v - \zeta, \nabla w)}{\|w\|_{H^1(\Omega)}} + \sup_{w\in H^1(\Omega)} \frac{(\operatorname{div}\zeta - \delta_\Gamma \otimes [\zeta\cdot n + \frac{1}{2}Sv], w)}{\|w\|_{H^1(\Omega)}}\\ &\le \|a\nabla v - \zeta\|_{[L^2(\Omega)]^d} + \|\operatorname{div}\zeta - \delta_\Gamma \otimes [\zeta\cdot n + \frac{1}{2}Sv]\|_{\tilde H^{-1}(\Omega)}.\end{aligned} \qquad (12.128)$$

Finally, writing

$$Sv = Wv - (I - K')\tau + (I - K')V^{-1}(V\tau + (I - K)v)$$

we can estimate

$$\begin{aligned}&\|\operatorname{div}\zeta - \delta_\Gamma \otimes [\zeta\cdot n + \frac{1}{2}Sv]\|_{\tilde H^{-1}(\Omega)} \lesssim\\ &\|\operatorname{div}\zeta - \frac{1}{2}\delta_\Gamma \otimes [2\zeta\cdot n + Wv - (I - K')\tau]\|_{\tilde H^{-1}(\Omega)} + \|V\tau + (I - K)v\|_{H^{1/2}(\Gamma)}.\end{aligned} \qquad (12.129)$$

Collecting the bounds (12.126) for $\|\zeta\|_{[L^2(\Omega)]^d}$, (12.127) for $\|\tau\|_{H^{-1/2}(\Gamma)}$, (12.128) for $\|v\|_{H^1(\Omega)}$ and (12.129), we obtain (12.125). □

Theorem 12.14 *The bilinear form $B(\cdot,\cdot)$ is continuous in $X \times X$ and the linear form $G(\cdot)$ is continuous on X.*

Proof Following the definition of $B(\cdot,\cdot)$ we obtain first

$$\begin{aligned}B((\theta,u,\sigma),(\zeta,v,\tau)) \le\ & \|a\nabla u-\theta\|_{[L^2(\Omega)]^d}\cdot\|a\nabla v-\zeta\|_{[L^2(\Omega)]^d}\\ &+\|(I-K)u+V\sigma\|_{H^{1/2}(\Gamma)}\cdot\|(I-K)v+V\tau\|_{H^{1/2}(\Gamma)}\\ &+\|g(\theta,u,\sigma)\|_{\tilde H^{-1}(\Omega)}\cdot\|g(\zeta,v,\tau)\|_{\tilde H^{-1}(\Omega)}.\end{aligned}$$

Using the triangle inequality, the mapping properties and the trace theorem we have

$$\|a\nabla u-\theta\|_{[L^2(\Omega)]^d}\le\|a\|_{L^\infty(\Omega)}\|\nabla u\|_{[L^2(\Omega)]^d}+\|\theta\|_{[L^2(\Omega)]^d}\lesssim\|u\|_{H^1(\Omega)}+\|\theta\|_{[L^2(\Omega)]^d}$$

and

$$\|(I-K)u+V\sigma\|_{H^{1/2}(\Gamma)}\lesssim\|u\|_{H^{1/2}(\Gamma)}+\|\sigma\|_{H^{-1/2}(\Gamma)}\lesssim\|u\|_{H^1(\Omega)}+\|\sigma\|_{H^{-1/2}(\Gamma)}.$$

Finally, there holds

$$\|g(\theta,u,\sigma)\|_{\tilde H^{-1}(\Omega)}\le\|\operatorname{div}\theta-\delta_\Gamma\otimes\theta\cdot n\|_{\tilde H^{-1}(\Omega)}+\frac{1}{2}\|\delta_\Gamma\otimes(Wu-(I-K')\sigma)\|_{\tilde H^{-1}(\Omega)}$$

and we obtain

$$\begin{aligned}\|\operatorname{div}\theta-\delta_\Gamma\otimes\theta\cdot n\|_{\tilde H^{-1}(\Omega)} &= \sup_{v\in H^1(\Omega)}\frac{(\operatorname{div}\theta-\delta_\Gamma\otimes\theta\cdot n,v)}{\|v\|_{H^1(\Omega)}}\\ &=\sup_{v\in H^1(\Omega)}\frac{(\operatorname{div}\theta,v)-\langle\theta\cdot n,v\rangle}{\|v\|_{H^1(\Omega)}}=\sup_{v\in H^1(\Omega)}\frac{(\theta,\nabla v)}{\|v\|_{H^1(\Omega)}}\le\|\theta\|_{[L^2(\Omega)]^d}\end{aligned}$$

and, analogously,

$$\begin{aligned}\|\delta_\Gamma\otimes(Wu-(I-K')\sigma)\|_{\tilde H^{-1}(\Omega)} &= \sup_{v\in H^1(\Omega)}\frac{(\delta_\Gamma\otimes(Wu-(I-K')\sigma),v)}{\|v\|_{H^1(\Omega)}}\\ &=\sup_{v\in H^1(\Omega)}\frac{\langle Wu-(I-K')\sigma,v\rangle}{\|v\|_{H^1(\Omega)}}\\ &\le\|Wu-(I-K')\sigma\|_{H^{-1/2}(\Gamma)}\lesssim\|u\|_{H^1(\Omega)}+\|\sigma\|_{H^{-1/2}(\Gamma)}.\end{aligned}$$

Collecting the individual terms, the continuity of $B(\cdot,\cdot)$ follows. The continuity of $G(\cdot)$ can be seen analogously. □

Now application of the Lax-Milgram lemma gives the following result (see [296])

Theorem 12.15 *There exists a unique solution of the variational least-squares formulation (12.124), which is also a solution of (12.115)—(12.119).*

12.4.1 The Discretized Least Squares Formulation

Following [62] we give an alternative representation for the norm in $\tilde{H}^{-1}(\Omega)$ which will be discretized later. Let $T : \tilde{H}^{-1}(\Omega) \mapsto H^1(\Omega)$ be defined by $Tf := w$ where $w \in H^1(\Omega)$ is the unique function satisfying

$$(\nabla w, \nabla v) + (w, v) = (f, v) \quad \forall v \in H^1(\Omega).$$

As observed in [62, Lemma 2.1], there holds

$$\|v\|^2_{\tilde{H}^{-1}(\Omega)} = \sup_{\theta \in H^1(\Omega)} \frac{(v, \theta)^2}{\|\theta\|^2_{H^1(\Omega)}} = \|Tv\|^2_{H^1(\Omega)} = (v, Tv).$$

Therefore, the inner product on $\tilde{H}^{-1}(\Omega) \times \tilde{H}^{-1}(\Omega)$ is given by (v, Tw), for $v, w \in \tilde{H}^{-1}(\Omega)$.

Let $V_h \subset H^1(\Omega)$. Then let $T_h : \tilde{H}^{-1}(\Omega) \mapsto V_h$ be defined by $T_h f := w$ where $w \in V_h$ is the unique function satisfying

$$(\nabla w, \nabla v) + (w, v) = (f, v) \quad \forall v \in V_h.$$

In case of the space $H^{1/2}(\Gamma)$ we proceed analogously: Let $R : H^{1/2}(\Gamma) \mapsto H^{-1/2}(\Gamma)$ be defined by $Rf := w$ where $w \in H^{-1/2}(\Gamma)$ is the unique function satisfying

$$\langle Vw, v\rangle = \langle f, v\rangle \quad \forall v \in H^{-1/2}(\Gamma).$$

Then there holds

$$\|v\|^2_{H^{1/2}(\Gamma)} = \sup_{\theta \in H^{-1/2}(\Gamma)} \frac{\langle v, \theta\rangle^2}{\|\theta\|^2_{H^{-1/2}(\Gamma)}} \sim \sup_{\theta \in H^{-1/2}(\Gamma)} \frac{\langle v, \theta\rangle^2}{\langle V\theta, \theta\rangle} = \langle v, Rv\rangle.$$

where $\sim$ denotes norm equivalence. Let $S_h \subset H^{-1/2}(\Gamma)$. Then let $R_h : H^{1/2}(\Gamma) \mapsto S_h$ be defined by $R_h f := w$ where $w \in S_H$ is the unique function satisfying

$$\langle Vw, v\rangle = \langle f, v\rangle \quad \forall v \in S_h.$$

For the numerical efficiency of the proposed scheme we replace in [296] T_h by the preconditioner B_h and R_h by the preconditioner C_h such that there holds $(T_h\cdot, \cdot) \sim (B_h\cdot, \cdot)$ and $\langle R_h\cdot, \cdot\rangle \sim \langle C_h\cdot, \cdot\rangle$. B_h and C_h are chosen in such a way that their evaluation is much cheaper than the computation of $T_h v_h$ or $R_h \tau_h$.

For the discretization we assume that there exists projection operators which are bounded independently of h

$$P_h : H^1(\Omega) \to V_h \subset H^1(\Omega) \tag{12.130}$$

$$Q_h : H^{-1/2}(\Gamma) \to S_h \subset H^{-1/2}(\Gamma). \tag{12.131}$$

As a consequence also their adjoints are bounded

$$P_h^* : \tilde{H}^{-1}(\Omega) \to V_h^* \subset \tilde{H}^{-1}(\Omega) \tag{12.132}$$

$$Q_h^* : H^{1/2}(\Gamma) \to S_h^* \subset H^{1/2}(\Gamma). \tag{12.133}$$

Replacing T in the representation of the $\tilde{H}^{-1}(\Omega)$ inner product by the preconditioner B_h and R in the representation of the $H^{1/2}(\Gamma)$ inner product by the preconditioner C_h we obtain the discretized formulation:

Find $(\theta_h, u_h, \sigma_h) \in X^h$ such that

$$B^{(h)}((\theta_h, u_h, \sigma_h), (\zeta_h, v_h, \tau_h)) = G^{(h)}(\zeta_h, v_h, \tau_h) \qquad \forall (\zeta_h, v_h, \tau_h) \in X^h, \tag{12.134}$$

where $X^h = H_h \times V_h \times S_h$, $H_h \subset [L^2(\Omega)]^d$ with

$$\begin{aligned} B^{(h)}((\theta, u, \sigma), (\zeta, v, \tau)) &= (a\nabla u - \theta, a\nabla v - \zeta)_{L^2(\Omega)} \\ &\quad + \langle C_h Q_h^*((I-K)u + V\sigma), Q_h^*((I-K)v + V\tau)\rangle_{L^2(\Gamma)} \\ &\quad + (B_h P_h^* g(\theta, u, \sigma), P_h^* g(\zeta, v, \tau))_{L^2(\Omega)} \end{aligned} \tag{12.135}$$

$$\begin{aligned} G^{(h)}(\zeta, v, \tau) &= \langle C_h Q_h^*((I-K)v + V\tau), Q_h^*((I-K)u_0 + Vt_0)\rangle_{L^2(\Gamma)} \\ &\quad - (B_h P_h^* g(\zeta, v, \tau), P_h^*(f + \frac{1}{2}\delta_\Gamma \otimes (Wu_0 + 2t_0 - (I-K')t_0)))_{L^2(\Omega)} \end{aligned} \tag{12.136}$$

for all $(\theta, u, \sigma), (\zeta, v, \tau) \in X$. Analogously to the proofs of Theorem 12.13 and Theorem 12.14 the authors show in [296] the following result:

Theorem 12.16 *For arbitrary functions $(\zeta_h, v_h, \tau_h) \in X^h$ the following a-priori estimate holds*

$$\begin{aligned} &\|v_h\|^2_{H^1(\Omega)} + \|\zeta_h\|^2_{[L^2(\Omega)]^d} + \|\tau_h\|^2_{H^{-1/2}(\Gamma)} \lesssim B^{(h)}((\zeta_h, v_h, \tau_h), (\zeta_h, v_h, \tau_h)) \\ &\sim \|a\nabla v_h - \zeta_h\|^2_{[L^2(\Omega)]^d} \\ &\quad + \|B_h^{1/2} P_h^*(\operatorname{div} \zeta_h - \frac{1}{2}\delta_\Gamma \otimes [Wv_h + 2\zeta_h \cdot n - (I-K')\tau_h])\|^2_{L^2(\Omega)} \\ &\quad + \|C_h^{1/2} Q_h^*[(I-K)v_h + V\tau_h]\|^2_{L^2(\Gamma)} \end{aligned} \tag{12.137}$$

Furthermore for arbitrary functions $(\zeta, v, \tau) \in X$ *the discretized bilinear form* $B^{(h)}(\cdot,\cdot)$ *and the discretized linear form* $G^{(h)}(\cdot)$ *are continuous, i.e. there holds* $B^{(h)}((\theta, u, \sigma), (\zeta, v, \tau)) \lesssim \|(\theta, u, \sigma)\|_X \cdot \|(\zeta, v, \tau)\|_X$, *and* $G^{(h)}((\zeta, v, \tau)) \lesssim \|(\zeta, v, \tau)\|_X$ *for all* $(\theta, u, \sigma), (\zeta, v, \tau) \in X$ *with constants independent of h.*

For finite dimensional subspaces $X^h := H_h \times V_h \times S_h \subset X$ we assume the usual approximation properties, e.g. for the space V_h of continuous, piecewise linear/bilinear functions on a regular triangulation, for the space H_h of either piecewise constant functions or continuous, piecewise linear/bilinear functions or $H(\mathrm{div}; \Omega)$-conforming Raviart-Thomas elements of lowest order, and for the space S_h of piecewise constant functions on the boundary (see [105, 352]):

There exists $r > 1$ such that for all $u \in H^r(\Omega)$

$$\inf_{v_h \in V_h} \|u - v_h\|_{H^1(\Omega)} \lesssim h^{r-1} \|u\|_{H^r(\Omega)},$$

$$\inf_{\tau_h \in S_h} \|\sigma - \tau_h\|_{H^{-1/2}(\Gamma)} \lesssim h^{r-1} \|\sigma\|_{H^{r-3/2}(\Gamma)} \lesssim h^{r-1} \|u\|_{H^r(\Omega)},$$

$$\inf_{\zeta_h \in H_h} \|\theta - \zeta_h\|_{[L^2(\Omega)]^d} \lesssim h^{r-1} \|\theta\|_{[H^{r-1}(\Omega)]^d} \lesssim h^{r-1} \|u\|_{H^r(\Omega)}.$$

Now, application of Theorem 12.16, the Lax-Milgram lemma and the Second Strang lemma gives the following result (see [296] for details):

Theorem 12.17 *The unique solution* $(\theta_h, u_h, \sigma_h) \in X^h$ *of the discretized formulation (12.134) exists and there holds the following convergence estimate*

$$\|u - u_h\|_{H^1(\Omega)} + \|\theta - \theta_h\|_{[L^2(\Omega)]^d} + \|\sigma - \sigma_h\|_{H^{-1/2}(\Gamma)} \lesssim h^{r-1} \|u\|_{H^r(\Omega)}.$$

For numerical experiments see [296].

12.5 FE/BE Coupling for Interface Problems with Signorini Contact

12.5.1 Primal Method

Here we report from [298] a FEM-BEM coupling procedure which is based on reducing the given nonlinear interface problem with contact to a boundary / domain variational inequality. In [298] also Coulomb friction is considered. For the ease of the reader here we restrict our presentation to the simpler case of Signorini contact (see also [292]).

Let $\Omega \subset \mathbb{R}^d$, $d \geq 2$, be a bounded domain with Lipschitz boundary Γ. Let $\Gamma = \overline{\Gamma_t \cup \Gamma_s}$ where Γ_t and Γ_s are nonempty, disjoint and open in Γ. In the interior part we consider a nonlinear partial differential equation modeling nonlinear material behavior in elasticity, whereas in the exterior part we consider the Laplace equation and impose a radiation condition:

$$-\operatorname{div}(\varrho(|\nabla u|)\cdot\nabla u) = f \quad \text{in} \quad \Omega \tag{12.138}$$

$$-\Delta u = 0 \quad \text{in} \quad \Omega_c = \mathbb{R}^d \backslash \bar{\Omega} \tag{12.139}$$

$$\left.\begin{array}{ll} u(x) = a + \frac{b}{2\pi}\log|x| + o(1) & \text{if } d = 2, \\ u(x) = \mathcal{O}(|x|^{2-d}) & \text{for } d \geq 3, \end{array}\right\} \quad (|x| \to \infty), \tag{12.140}$$

where a, b are real constants (constant for any u but varying with u).

Further, $\varrho : [0,\infty) \to [0,\infty)$ is a $C^1[0,\infty)$ function with $t \cdot \varrho(t)$ being monotonously increasing with t, $\varrho(t) \leq \varrho_0$, $(t\cdot\varrho(t))' \leq \varrho_1$ and further $\varrho(t) + t\cdot\min\{0, \varrho(t)\} \geq \alpha > 0$. With $u_1 := u|_\Omega$ and $u_2 := u|_{\Omega_c}$, the tractions on Γ are given by $\varrho(|\nabla u_1|)\frac{\partial u_1}{\partial n}$ and $-\frac{\partial u_2}{\partial n}$ with normal n pointing into Ω_c.

We consider transmission conditions on Γ_t

$$u_1|_{\Gamma_t} - u_2|_{\Gamma_t} = u_0|_{\Gamma_t} \quad \text{and} \quad \varrho(|\nabla u_1|)\frac{\partial u_1}{\partial n}|_{\Gamma_t} - \frac{\partial u_2}{\partial n}|_{\Gamma_t} = t_0|_{\Gamma_t}, \tag{12.141}$$

and Signorini conditions on Γ_s

$$\begin{aligned} u_1|_{\Gamma_s} - u_2|_{\Gamma_s} &\leq u_0|_{\Gamma_s} \\ \varrho(|\nabla u_1|)\frac{\partial u_1}{\partial n}|_{\Gamma_s} &= \frac{\partial u_2}{\partial n}|_{\Gamma_s} + t_0|_{\Gamma_s} \leq 0 \\ 0 = \varrho(|\nabla u_1|)&\frac{\partial u_1}{\partial n}|_{\Gamma_s}\cdot(u_2 + u_0 - u_1)|_{\Gamma_s}. \end{aligned} \tag{12.142}$$

Given data $f \in L^2(\Omega)$, $u_0 \in H^{1/2}(\Gamma)$, and $t_0 \in H^{-1/2}(\Gamma)$ (with $(f,1)_{L^2(\Omega)} + \langle t_0, 1\rangle = 0$ if $d = 2$) we look for $u_1 \in H^1(\Omega)$ and $u_2 \in H^1_{\text{loc}}(\Omega_c)$ satisfying (12.138)–(12.142) in a weak form.

Setting

$$g(t) = \int_0^t s\cdot\varrho(s)\,ds$$

the assumptions on ϱ yield that

$$G(u) = 2\int_\Omega g(|\nabla u|)\,dx$$

is finite for any $u \in H^1(\Omega)$ and its Fréchet derivative

$$DG(u; v) = 2\int_\Omega \varrho(|\nabla u|)(\nabla u)^T\cdot\nabla v\,dx \quad \forall u, v \in H^1(\Omega) \tag{12.143}$$

is uniformly monotone, i.e., there exists a constant $\gamma > 0$ such that

$$\gamma |u - v|^2_{H^1(\Omega)} \leq DG(u; u - v) - DG(v; u - v) \quad \forall u, v \in H^1(\Omega), \tag{12.144}$$

Let $E := H^1(\Omega) \times \widetilde{H}^{1/2}(\Gamma_s)$ where $\widetilde{H}^{1/2}(\Gamma_s) := \{w \in H^{1/2}(\Gamma) : \operatorname{supp} w \subseteq \Gamma_s\}$ and set

$$D := \{(u, v) \in E \,:\, v \geq 0 \text{ a.e. on } \Gamma_s \text{ and } \langle S1, u|_\Gamma + v - u_0 \rangle = 0 \text{ if } d = 2\},$$

where S denotes the Poincaré-Steklov operator for the exterior problem:

$$S = 1/2 \left(W + (K' - I)V^{-1}(K - I) \right)$$

Then the primal formulation of (12.138)–(12.142), called problem (SP), consists in finding $(\hat{u}, \hat{v})$ in D such that

$$\Psi(\hat{u}, \hat{v}) = \inf_{(u,v)\in D} \Psi(u, v).$$

where

$$\Psi(u, v) := 2 \int_\Omega g(|\nabla u|)\, dx + \frac{1}{2} \langle S(u|_\Gamma + v), u|_\Gamma + v \rangle - \lambda(u, v),$$

and $\lambda \in E^*$, the dual of E, is given by

$$\lambda(u, v) := L(u, u|_\Gamma + v) + \langle S u_0, u|_\Gamma + v \rangle$$

with

$$L(u, v) := 2 \int_\Omega f \cdot u\, dx + 2 \int_\Gamma t_0 \cdot v\, ds$$

for any $(u, v) \in E$.

Due to [85] there exists exactly one solution $(\hat{u}, \hat{v}) \in D$ of problem (SP), which is the variational solution of the transmission problem (12.138)—(12.142). Moreover, $(\hat{u}, \hat{v}) \in D$ is the unique solution of the variational inequality

$$\mathscr{A}(\hat{u}, \hat{v})(u - \hat{u}, v - \hat{v}) \geq \lambda(u - \hat{u}, v - \hat{v}) \tag{12.145}$$

for all $(u, v) \in D$, with

$$\mathscr{A}(u, v)(r, s) := DG(u, r) + \langle S(u|_\Gamma + v), r|_\Gamma + s \rangle. \tag{12.146}$$

For the discretization we take nested regular quasi-uniform meshes $(\mathcal{T}_h)_h$ consisting of triangles or quadrilaterals. Then, let H_h^1 denote the related continuous and piecewise affine-linear trial functions on the triangulation $\mathcal{T}_h$. The mesh on Ω induces a mesh on the boundary, so that we may consider $H_h^{-1/2}$ as the piecewise constant trial functions. Assuming that the partition of the boundary leads also to a partition of Γ_s, $\widetilde{H}_h^{1/2}$ is then the subspace of continuous and piecewise affine-linear functions on the partition of Γ_s which vanish at intersection points in $\bar{\Gamma}_s \cap \bar{\Gamma}_t$. Then we have $H_h^1 \times \widetilde{H}_h^{1/2} \times H_h^{-1/2} \subset H^1(\Omega) \times \widetilde{H}^{1/2}(\Gamma_s) \times H^{-1/2}(\Gamma)$. Now, D_h is given by

$$D_h := \{(u_h, v_h) \in H_h^1 \times \widetilde{H}_h^{1/2} \,:\, v(x_i) \geq 0, \forall x_i \text{ node of the partition of } \Gamma_s, \\ \text{and } \langle S1, u_h|_\Gamma + v_h - u_0 \rangle = 0 \text{ if } d = 2\}. \tag{12.147}$$

Note that $v_h \geq 0$, once the nodal values of v_h are ≥ 0. Therefore we have $D_h \subset D$. With the approximation S_h as in (12.55) of S the primal FE-BE coupling method (SP$_h$) reads: Find $(\hat{u}_h, \hat{v}_h) \in D_h$ such that

$$\mathscr{A}_h(\hat{u}_h, \hat{v}_h)(u_h - \hat{u}_h, v_h - \hat{v}_h) \geq \lambda_h(u_h - \hat{u}_h, v_h - \hat{v}_h) \tag{12.148}$$

for all $(u_h, v_h) \in D_h$, where

$$\mathscr{A}_h(u_h, v_h)(r_h, s_h) := DG(u_h, r_h) + \langle S_h(u_h|_\Gamma + v_h), r_h|_\Gamma + s_h \rangle \tag{12.149}$$

and

$$\lambda_h(u_h, v_h) := L(u_h, u_h|_\Gamma + v_h) + \langle S_h u_0, u_h|_\Gamma + v_h \rangle. \tag{12.150}$$

with the discrete Steklov-Poincaré operator S_h (12.79).

There holds the following a priori error estimate for the solutions $(\hat{u}, \hat{v})$ of (12.145) and $(\hat{u_h}, \hat{v_h})$ of (12.148) with a positive constant C, independent of h, for $h < h_0$, for some $h_0 > 0$,

$$\begin{aligned} \|\hat{u} - \hat{u_h}, \hat{v} - \hat{v_h}\|^2_{H^1(\Omega)\times\tilde{H}^{1/2}(\Gamma_s)} \leq C\{ & \inf_{u_h \in H_h^1} \|\hat{u} - u_h\|^2_{H^1(\Omega)} \\ & + \inf_{v_h \in \tilde{H}_h^{1/2}} \left(\|\hat{v} - v_h\|^2_{\tilde{H}^{1/2}(\Gamma_s)} + \|\hat{v} - v_h\|_{L^2(\Gamma)} \right) \\ & + \operatorname*{dist}_{H^{-1/2}(\Gamma)} \left(V^{-1}(I - K)(\hat{u} + \hat{v} - u_0), H_h^{-1/2}) \right)^2 \} \end{aligned}$$

This error estimate shows that the solution $(\hat{u}_h, \hat{v}_h) \in D_h$ of (SP$_h$) converges for $h \to 0$ towards the solution $(\hat{u}, \hat{v}) \in D$ of (SP).

In [189] we investigate an adaptive FE/BE procedure for scalar nonlinear interface problems involving friction, where the nonlinear uniformly monotone operator such as the p-Laplacian is coupled to the linear Laplace equation on the exterior domain. The procedure is again to reduce the contact problem to a boundary/domain variational inequality.

12.5.2 Dual Mixed Method

Now we consider again the Signorini problem (12.138)—(12.142) with $\varrho \equiv 1$ and present from [188] a dual mixed variational formulation in terms of a convex minimization problem and an associated variational inequality.

In [188] a coupling method is proposed and analyzed for dual mixed finite elements and boundary elements for (12.138)–(12.142) using the inverse Steklov-Poincaré operator R, the Neumann-to-Dirichlet (NtD) map, given by

$$R := S^{-1} = -\frac{1}{2}[V + (I + K)W^{-1}(I + K')] : H^{-1/2}(\Gamma)] \to H^{1/2}(\Gamma)\,. \tag{12.151}$$

Define $\tilde{\Psi} : H(\mathrm{div}; \Omega) \to \mathbb{R} \cup \{\infty\}$ by

$$\tilde{\Psi}(q) := \frac{1}{2}\, \|q\|^2_{[L^2(\Omega)]^d} + \frac{1}{4}\langle q \cdot n, R(q \cdot n)\rangle - \frac{1}{2}\langle q \cdot n, R(t_0) + 2u_0\rangle\,, \tag{12.152}$$

and the subset of admissible functions by

$$\tilde{D} := \{q \in H(\mathrm{div}; \Omega) : q \cdot n \leq 0 \text{ on } \Gamma_s, \quad -\mathrm{div}\, q = f \quad \text{in} \quad \Omega\}\,.$$

Then the uniquely solvable dual formulation $(\widetilde{SP})$ consists in finding $q^D \in \tilde{D}$ such that

$$\tilde{\Psi}(q^D) = \min_{q \in \tilde{D}} \tilde{\Psi}(q)\,. \tag{12.153}$$

As shown in [188] problem $(\widetilde{SP})$ is equivalent to the original Signorini contact problem (12.138) — (12.142) with $\varrho \equiv 1$.

Next a saddle point formulation (M) of $(\widetilde{SP})$ is given with the help of $\mathscr{H}$: $H(\mathrm{div}; \Omega) \times L^2(\Omega) \times \tilde{H}^{1/2}(\Gamma_s) \to \mathbb{R} \cup \{\infty\}$ defined as

$$\mathscr{H}(p, v, \mu) := \tilde{\Psi}(p) + \int_\Omega v\, \mathrm{div}\, p\, dx + \int_\Omega f v\, dx + \langle p \cdot n, \mu\rangle_{\Gamma_s} \tag{12.154}$$

for all $(p, v, \mu) \in H(\mathrm{div}; \Omega) \times L^2(\Omega) \times \tilde{H}^{1/2}(\Gamma_s)$, and consider the subset of admissible functions

$$\tilde{H}^{1/2}_+(\Gamma_s) := \{\mu \in \tilde{H}^{1/2}(\Gamma_s) : \quad \mu \geq 0\}\,. \tag{12.155}$$

The saddle point problem (M) reads:
Find $(\hat{q}, \hat{u}, \hat{\lambda}) \in H(\mathrm{div}; \Omega) \times L^2(\Omega) \times \tilde{H}_+^{1/2}(\Gamma_s)$ such that

$$\mathscr{H}(\hat{q}, u, \lambda) \le \mathscr{H}(\hat{q}, \hat{u}, \hat{\lambda}) \le \mathscr{H}(q, \hat{u}, \hat{\lambda}) \;\; \forall (q, u, \lambda) \in H(\mathrm{div}; \Omega) \times L^2(\Omega) \times \tilde{H}_+^{1/2}(\Gamma_s), \tag{12.156}$$

which is equivalent to finding a solution $(\hat{q}, \hat{u}, \hat{\lambda}) \in H(\mathrm{div}; \Omega) \times L^2(\Omega) \times \tilde{H}_+^{1/2}(\Gamma_s)$ of the variational inequality:

$$a(\hat{q}, q) + b(q, \hat{u}) + d(q, \hat{\lambda}) = \langle q \cdot n, r\rangle \quad \forall q \in H(\mathrm{div}; \Omega), \tag{12.157}$$

$$b(\hat{q}, u) = -\int_\Omega f u \, dx \quad \forall u \in L^2(\Omega), \tag{12.158}$$

$$d(\hat{q}, \lambda - \hat{\lambda}) \le 0 \qquad \forall \lambda \in \tilde{H}_+^{1/2}(\Gamma_s), \tag{12.159}$$

where $r = R(t_0) + 2u_0$ and

$$a(p, q) = 2\int_\Omega p \cdot q \, dx + \langle q \cdot n, R(p \cdot n)\rangle \qquad \forall p, q \in H(\mathrm{div}; \Omega), \tag{12.160}$$

$$b(q, u) = \int_\Omega u \,\mathrm{div}\, q \, dx \qquad \forall (q, u) \in H(\mathrm{div}; \Omega) \times L^2(\Omega), \tag{12.161}$$

$$d(q, \lambda) = \langle q \cdot n, \lambda\rangle_{\Gamma_s} \qquad \forall (q, \lambda) \in H(\mathrm{div}; \Omega) \times \tilde{H}^{1/2}(\Gamma_s), \tag{12.162}$$

The connection between the dual problem $(\widetilde{SP})$ and the saddle point problem (M) is as follows.

Theorem 12.18 ([188]) *The dual problem $(\widetilde{SP})$ is equivalent to the mixed dual variational inequality (M). More precisely:*

(i) If $(\hat{q}, \hat{u}, \hat{\lambda}) \in H(\mathrm{div}; \Omega) \times L^2(\Omega) \times \tilde{H}_+^{1/2}(\Gamma_s)$ is a saddle point of $\mathscr{H}$ in $H(\mathrm{div}; \Omega) \times L^2(\Omega) \times \tilde{H}_+^{1/2}(\Gamma_s)$, then $\hat{q} = \nabla\hat{u}$, $\hat{u} = \frac{1}{2}R(t_0 - \hat{q} \cdot n) + u_0$ on Γ_t, $\hat{\lambda} = -\frac{1}{2}R(\hat{q} \cdot n - t_0) + u_0 - \hat{u}$ on Γ_s, and $\hat{q} \in \tilde{D}$ is the solution of problem $(\widetilde{SP})$.

(ii) Let $q^D \in \tilde{D}$ be the solution of $(\widetilde{SP})$, and define $\hat{\lambda} := -\frac{1}{2}R(q^D \cdot n - t_0) + u_0 - \hat{u}$ on Γ, where $\hat{u} \in H^1(\Omega)$ is the unique solution of the Neumann problem: $-\Delta\hat{u} = f$ in Ω, $\dfrac{\partial \hat{u}}{\partial n} = q^D \cdot n$ on Γ, such that $\langle \mu, \hat{u} + \frac{1}{2}R(q^D \cdot n - t_0) - u_0\rangle \ge 0$ for all $\mu \in H^{-1/2}(\Gamma)$ with $\mu \le -q^D \cdot n$ on Γ_s. Then, $(q^D, \hat{u}, \hat{\lambda})$ is a saddle point of $\mathscr{H}$ in $H(\mathrm{div}; \Omega) \times L^2(\Omega) \times \tilde{H}_+^{1/2}(\Gamma_s)$.

In [188], the problem $(\widetilde{SP})$ is solved approximately by using mixed finite elements in Ω and boundary elements on Γ choosing finite-dimensional subspaces $L_h \times H_h \times H_h^{-1/2} \times H_h^{1/2} \times H_{s,\tilde{h}}^{1/2}$ of $L^2(\Omega) \times H(\mathrm{div}; \Omega) \times H^{-1/2}(\Gamma) \times H^{1/2}(\Gamma)/\mathbb{R} \times \tilde{H}^{1/2}(\Gamma_s)$.

The subspaces $(L_h, H^{1/2}_{s,\tilde{h}})$ and H_h are supposed to verify the usual discrete Babuška-Brezzi condition, which means that there exists $\beta^* > 0$ such that

$$\inf_{\substack{(u_h,\lambda_{\tilde{h}})\in L_h\times H^{1/2}_{s,\tilde{h}} \\ (u_h,\lambda_{\tilde{h}})\neq 0}} \sup_{\substack{q_h\in H_h \\ q_h\neq 0}} \frac{B(q_h,(u_h,\lambda_{\tilde{h}}))}{\|q_h\|_{H(\mathrm{div};\Omega)}\|(u_h,\lambda_{\tilde{h}})\|_{L^2(\Omega)\times\tilde{H}^{1/2}(\Gamma_s)}} \geq \beta^* . \tag{12.163}$$

where $B(q,(u,\lambda)) = b(q,u) + d(q,\lambda)$.

Now, for $h, \tilde{h} \in I$ let $j_h : H_h \hookrightarrow H(\mathrm{div};\Omega)$, $k_h : H_h^{-1/2} \hookrightarrow H^{-1/2}(\Gamma)$ and $l_h : H_h^{1/2} \hookrightarrow H^{1/2}(\Gamma)/\mathbb{R}$ denote the canonical imbeddings with their corresponding duals j_h^*, k_h^* and l_h^*.

In order to approximate R define the discrete operators

$$R_h := j_h^*\gamma^* R\gamma j_h \quad , \quad \tilde{R}_h := j_h^*\gamma^* V\gamma j_h + j_h^*\gamma^*(I+K)l_h(l_h^* W l_h)^{-1} l_h^*(I+K')\gamma j_h ,$$

where $\gamma : H(\mathrm{div};\Omega) \to H^{-1/2}(\Gamma)$ is the trace operator yielding the normal component of functions in $H(\mathrm{div};\Omega)$.

Note that the computation of $\tilde{R}_h$ requires the numerical solution of a linear system with a symmetric positive definite matrix $W_h := l_h^* W l_h$. In general, there holds $\tilde{R}_h \neq R_h$ because $\tilde{R}_h$ is a Schur complement of matrices from discretization while R_h is a discretized Schur complement of operators.

In order to approximate the solution of problem (M), the authors consider in [188] the nonconforming Galerkin scheme (M_h):

Find $(\hat{q}_h, \hat{u}_h, \hat{\lambda}_{\tilde{h}}) \in H_h \times L_h \times H^{1/2}_{s,+,\tilde{h}}$ such that

$$a_h(\hat{q}_h, q_h) + b(q_h, \hat{u}_h) + d(q_h, \hat{\lambda}_{\tilde{h}}) = \langle q_h \cdot n, r_h\rangle \qquad \forall q_h \in H_h , \tag{12.164}$$

$$b(\hat{q}_h, u_h) = -\int_\Omega f u_h \, dx \qquad \forall u_h \in L_h , \tag{12.165}$$

$$d(\hat{q}_h, \lambda_{\tilde{h}} - \hat{\lambda}_{\tilde{h}}) \leq 0 \qquad \forall \lambda_{\tilde{h}} \in H^{1/2}_{s,+,\tilde{h}} , \tag{12.166}$$

where

$$H^{1/2}_{s,+,\tilde{h}} := \{\mu \in H^{1/2}_{s,\tilde{h}} : \quad \mu \geq 0\} , \tag{12.167}$$

$$a_h(p,q) = 2\int_\Omega p \cdot q \, dx + \langle q \cdot n, \tilde{R}_h(p \cdot n)\rangle \quad \forall p, q \in H_h , \tag{12.168}$$

$$b(q,u) = \int_\Omega u \,\mathrm{div}\, q \, dx \qquad \forall (q,u) \in H_h \times L_h , \tag{12.169}$$

$$d(q,\lambda) = \langle q \cdot n, \lambda\rangle_{\Gamma_s} \qquad \forall (q,\lambda) \in H_h \times H^{1/2}_{s,+,\tilde{h}} , \tag{12.170}$$

and

$$r_h := k_h^*((V + (I + K)l_h(l_h^* W l_h)^{-1} l_h^*(I + K'))t_0 + 2u_0) .$$

Note that the nonconformity of problem (M_h) arises from the bilinear form $a_h(\cdot, \cdot)$ approximating $a(\cdot, \cdot)$.

There holds the following a priori error estimate (see [188]) yielding convergence for the solution of the nonconforming Galerkin scheme (M_h) to the weak solution of (M) and therefore to the weak solution of the original Signorini contact problem due to the equivalence result of Theorem 12.18.

Theorem 12.19 ([188]) *Let $(\hat{q}, \hat{u}, \hat{\lambda})$ and $(\hat{q}_h, \hat{u}_h, \hat{\lambda}_{\tilde{h}})$ be the solutions of problems (M) and (M_h), respectively. Define $\hat{\phi} := W^{-1}(I + K')(\hat{q} \cdot n)$ and $\phi_0 := W^{-1}(I + K')t_0$. Then there exists $c > 0$, independent of h and $\tilde{h}$, such that the following Cea type estimate holds*

$$\begin{aligned}
&\|\hat{q} - \hat{q}_h\|_{H(\mathrm{div};\Omega)} + \|\hat{u} - \hat{u}_h\|_{L^2(\Omega)} + \|\hat{\lambda} - \hat{\lambda}_{\tilde{h}}\|_{\tilde{H}^{1/2}(\Gamma_s)} \\
&\leq c \left\{ \inf_{q_h \in H_h} \|\hat{q} - q_h\|_{H(\mathrm{div};\Omega)} + \inf_{u_h \in L_h} \|\hat{u} - u_h\|_{L^2(\Omega)} + \inf_{\lambda_{\tilde{h}} \in H^{1/2}_{s,+,\tilde{h}}} \|\hat{\lambda} - \lambda_{\tilde{h}}\|^{1/2}_{\tilde{H}^{1/2}(\Gamma_s)} \right. \\
&\qquad \left. + \inf_{\phi_h \in H_h^{1/2}} \|\hat{\phi} - \phi_h\|_{H^{1/2}(\Gamma)/R} + \inf_{\phi_h \in H_h^{1/2}} \|\phi_0 - \phi_h\|_{H^{1/2}(\Gamma)/R} \right\} . \qquad (12.171)
\end{aligned}$$

A suitable choice for finite element and boundary element spaces are L_h the set of piecewise constant functions, H_h the space of $H(\mathrm{div}; \Omega)$ conforming Raviart-Thomas elements of order zero and $H^{1/2}_{s,+,\tilde{h}}$ the set of continuous piecewise linear, nonnegative functions of the partition $\tau_{\tilde{h}}$ of Γ_s (see [188] for details).

12.6 Coupling of Primal-Mixed FEM and BEM for Plane Elasticity

Here we report on the solution procedure in [66], where a Stokes-type mixed finite element method with the pressure as the secondary unknown is employed (with the displacement as the primary unknown). In the BEM domain linear elasticity is considered. In the FEM domain an incompressible nonlinear elastic material (governed by a uniformly monotone operator) is assumed. We present from [66] the proofs of existence and uniqueness of the solution and the quasi optimal convergence of a Galerkin method. Finally, we cite from [66] an a posteriori error estimator of explicit residual type.

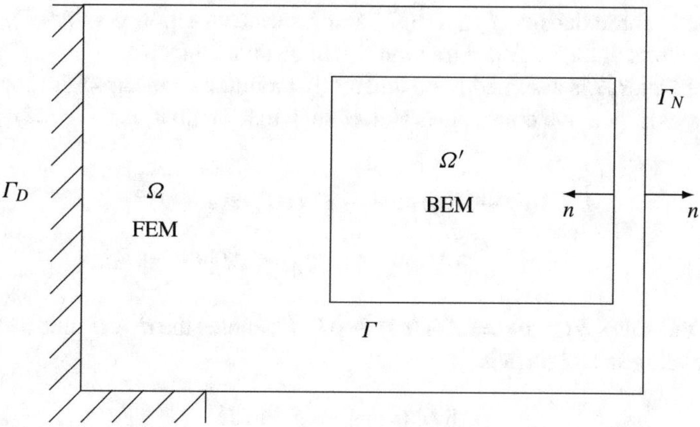

Fig. 12.3 Geometrical setting [66]

Let $\tilde{\Omega} \subset \mathbb{R}^d$, $d = 2, 3$, be a domain with Lipschitz continuous boundary. $\tilde{\Omega}$ is partitioned as $\tilde{\Omega} = \Omega' \cup \Gamma \cup \Omega$, $\Omega' \cap \Omega = \emptyset$ (Fig. 12.3). On the bounded subdomain Ω we will use a finite element method whereas for Ω' we will exploit boundary integral equations such that only the boundary of Ω' will be discretized. For simplicity we assume that all of the boundary $\partial\Omega'$ belongs to the coupling interface Γ, which is also assumed to be Lipschitz continuous.

In Ω the unknowns are the displacement

$$u \in H^1_D(\Omega) := \{v \in [H^1(\Omega)]^d : v|_{\Gamma_D} = 0\}$$

and the pressure

$$p \in L^2(\Omega).$$

$H^1(\Omega)$ is the usual Sobolev space with the norm $\|v\|_{1,\Omega} = (\|v\|^2_{0,\Omega} + \|\text{grad } v\|^2_{0,\Omega})^{1/2}$ and $\|\cdot\|_{0,\Omega}$ denotes the norm in $L^2(\Omega)$.

We seek $(u, p) \in H^1_D(\Omega) \times L^2(\Omega)$ such that

$$\begin{aligned} \int_\Omega \{A(\epsilon(u)) : \epsilon(v) + p \text{ div } v\} dx &= \int_\Gamma \phi \cdot v ds + L(v) \quad \forall v \in H^1_D(\Omega) \\ \int_\Omega q \text{ div } u dx &= 0 \quad \forall q \in L^2(\Omega). \end{aligned} \tag{12.172}$$

The linearized strain is $\epsilon(v) := \frac{1}{2}(\text{grad } v + (\text{grad } v)^T)$, and we use the notation $\sigma : \epsilon = \sum_{ij} \sigma_{ij}\epsilon_{ij}$. The possibly nonlinear operator $A : [L^2(\Omega)]^{d\times d}_{\text{sym}} \to [L^2(\Omega)]^{d\times d}_{\text{sym}}$ yields a symmetric tensor field. The exterior loads are

$$L(v) := \int_\Omega f \cdot v dx + \int_{\Gamma_N} g \cdot v ds.$$

with a body force density $f \in L^2(\Omega)$ and a surface traction $g \in L^2(\Gamma_N)$. In the coupling method, the interface traction ϕ will be an unknown.

The operator A is assumed to be uniformly monotone and Lipschitz continuous, i.e., there exist positive constants α and M such that for all $\epsilon, \eta \in [L^2(\Omega)]^{d\times d}_{\mathrm{sym}}$

$$\int_\Omega (A(\epsilon) - A(\eta)) : (\epsilon - \eta)dx \geq \alpha \|\epsilon - \eta\|^2_{L^2}$$

$$\|A(\epsilon) - A(\eta)\|_{L^2} \leq M\|\epsilon - \eta\|_{L^2}\,.$$

With the stress $\sigma(u, p) := A(\epsilon(u)) + pI$ (I denotes the $d \times d$ unit matrix), the corresponding strong form is

$$\begin{aligned} -\operatorname{div}\sigma(u,p) &= f \quad \text{in } \Omega \\ \operatorname{div} u &= 0 \quad \text{in } \Omega \\ \sigma(u,p)n &= g \quad \text{on } \Gamma_N \\ \sigma(u,p)n &= \phi \quad \text{on } \Gamma \\ u &= 0 \quad \text{on } \Gamma_D. \end{aligned} \tag{12.173}$$

In the BEM-domain we consider linear elasticity. The strong form is

$$\begin{aligned} -\operatorname{div}(2\mu\epsilon(u) + \lambda[\operatorname{div} u]I) &= 0 \quad \text{in } \Omega' \\ (2\mu\epsilon(u) + \lambda[\operatorname{div} u]I)n &= -\phi \quad \text{on } \Gamma. \end{aligned} \tag{12.174}$$

We assume that the Lamé coefficients λ and μ are constant on Ω'. For $d = 3$ we allow Ω' to be unbounded (Fig. 12.4) and in this case require the decay condition

$$u(x) = O(1/|x|) \quad \text{and} \quad \frac{\partial u(x)}{\partial x_j} = O(1/|x|^2), \; j = 1,\ldots,d, \qquad \text{for } |x| \to \infty. \tag{12.175}$$

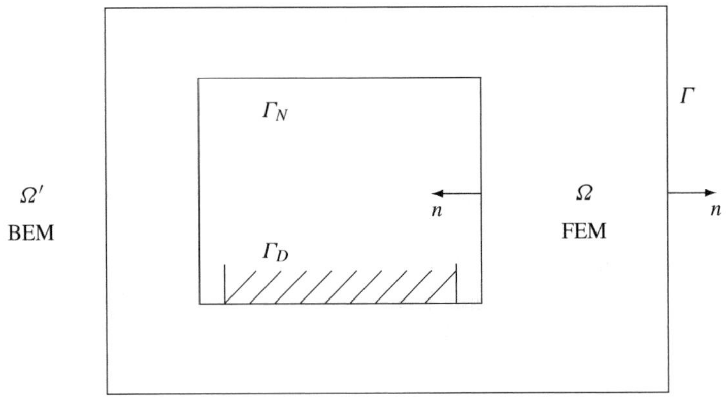

Fig. 12.4 Notation for coupling with unbounded exterior domain Ω' [66]

At any point $x \in \Omega'$, the displacement field can be represented by the Betti formula

$$u(x) = -\int_\Gamma G(x,y)\,T_y u(y)\,ds_y + \int_\Gamma \left(T_y G(x,y)\right)^T u(y)\,ds_y\,.$$

Here $T_y u(y) = (2\mu\epsilon(u(y)) + \lambda[\operatorname{div} u(y)]I)\,n(y)$ is the traction corresponding to u at a point $y \in \Gamma$, and the columns of $T_y G(x,y)$ are the tractions of $G(x,y)$ at y. $G(x,y)$ is the fundamental solution and equals

$$\frac{\lambda+3\mu}{4\pi\mu(\lambda+2\mu)}\left\{\ln\frac{1}{|x-y|}I + \frac{\lambda+\mu}{\lambda+3\mu}\frac{(x-y)(x-y)^T}{|x-y|^2}\right\} \text{ if } d=2,$$

$$\frac{\lambda+3\mu}{8\pi\mu(\lambda+2\mu)}\left\{\frac{1}{|x-y|}I + \frac{\lambda+\mu}{\lambda+3\mu}\frac{(x-y)(x-y)^T}{|x-y|^3}\right\} \text{ if } d=3.$$

Letting $x \to \Gamma$ we obtain with the classical jump relations the boundary integral equation

$$u = -V\phi + Ku \tag{12.176}$$

with $\phi(y) = T_y u(y)$ and the integral operators

$$(V\phi)(x) = 2\int_\Gamma G(x,y)\,\phi(y)\,ds_y, \quad x \in \Gamma,$$

$$(Ku)(x) = 2\int_\Gamma \left(T_y G(x,y)\right)^T u(y)\,ds_y, \quad x \in \Gamma\,.$$

Applying the traction operator T_x we get another boundary integral equation

$$\phi = -K'\phi - Wu \tag{12.177}$$

where

$$(K'\phi)(x) = 2\int_\Gamma T_x G(x,y)\,\phi(y)\,ds_y, \quad x \in \Gamma\,,$$

$$(Wu)(x) = -2T_x\int_\Gamma \left(T_y G(x,y)\right)^T u(y)\,ds_y, \quad x \in \Gamma\,.$$

For $d=3$, V is positive definite, i.e. there is a constant $C > 0$ such that

$$\langle\phi, V\phi\rangle \geq C\|\phi\|^2_{-1/2,\Gamma} \qquad \forall\phi \in H^{-1/2}(\Gamma)$$

For $d=2$, V is positive definite when restricted to

$$H_0^{-1/2}(\Gamma) := \{\phi \in H^{-1/2}(\Gamma) : \int_\Gamma \phi\,ds = 0\}.$$

The operator W has the kernel $\ker W = \ker \epsilon|_\Gamma$, i.e., the kernel consists of the (linearized) rigid body motions. W is positive definite on $H^{1/2}(\Gamma)/\ker\epsilon$. For proofs of these properties we refer to [136]. In the sequel we will use the notation

$$H^{-1/2} := \begin{cases} H^{-1/2}(\Gamma) & \text{if } d = 3 \\ H_0^{-1/2}(\Gamma) & \text{if } d = 2. \end{cases}$$

To obtain the coupled formulation we rewrite the integral equation (12.177) as

$$\phi = (I - K')\phi - Wu$$

and insert this into the right-hand side of (12.172). The first integral equation (12.176) is weighted by a function $\psi \in H^{-1/2}$. Further we assume u to be continuous across the interface Γ. Hence our method reads: Find $(u, \phi, p) \in H_D^1(\Omega) \times H^{-1/2} \times L^2(\Omega)$ such that for all $(v, \psi, q) \in H_D^1(\Omega) \times H^{-1/2} \times L^2(\Omega)$

$$\begin{aligned} \tfrac{1}{2}\mathscr{A}(u, \phi; v, \psi) + b(p, v) &= L(v) \\ b(q, u) \qquad &= \ 0, \end{aligned} \tag{12.178}$$

where

$$\begin{aligned} \mathscr{A}(u, \phi; v, \psi) &:= 2\int_\Omega A(\epsilon(u)) : \epsilon(v)\, dx + \langle v, Wu\rangle - \langle v, (I - K')\phi\rangle \\ &\quad - \langle \psi, V\phi\rangle - \langle \psi, (I - K)u\rangle \\ b(p, v) &:= \int_\Omega p \operatorname{div} v\, dx. \end{aligned}$$

It is well known that the divergence operator $\operatorname{div} : H_D^1(\Omega) \to L^2(\Omega)$ is surjective. The proof can be performed similarly to [196] taking into account that $\partial\Omega \setminus \Gamma_D$ is of positive surface measure. The surjectivity is equivalent to the inf–sup condition: There is a constant $\beta > 0$ such that

$$\inf_{q \in L^2(\Omega)} \sup_{v \in H_D^1(\Omega)} \frac{b(q, v)}{\|q\|_{0,\Omega}\|v\|_{1,\Omega}} \geq \beta. \tag{12.179}$$

Theorem 12.20 *Let $\Gamma_D \neq \emptyset$. Let A be uniformly monotone and Lipschitz continuous. Then* (12.178) *has a unique solution.*

Proof Let us consider the following problem: Find $u \in \ker B := \{v \in H_D^1(\Omega) : b(q, v) = 0\ \forall q \in L^2(\Omega)\}$ and $\phi \in H^{-1/2}$ such that

$$\mathscr{A}(u, \phi; v, \psi) = L(v) \quad \forall (v, \psi) \in \ker B \times H^{-1/2}. \tag{12.180}$$

For all $(w, \chi), (v, \psi) \in \ker B \times H^{-1/2}$ there holds

$$\begin{aligned}
&\mathscr{A}(w, \chi; w - v, -(\chi - \psi)) - \mathscr{A}(v, \psi; w - v, -(\chi - \psi)) \\
&= \int_{\Omega} 2\{A(\epsilon(w)) - A(\epsilon(v))\} : \epsilon(w - v) dx \\
&\quad + \langle w - v, W(w - v)\rangle + \langle \chi - \psi, V(\chi - \psi)\rangle \\
&\geq \alpha \|w - v\|^2_{1,\Omega} + C\|\chi - \psi\|^2_{-1/2,\Gamma} \,. \qquad (12.181)
\end{aligned}$$

Thus $\mathscr{A}$ corresponds to a nonlinear operator which maps $\ker B \times H^{-1/2}$ onto its dual and is uniformly monotone and Lipschitz continuous. Hence the main theorem on monotone operators[437] implies that (12.180) has a unique solution (u, ϕ).

Next we show existence of p. It suffices to find a $p \in L^2(\Omega)$ such that

$$b(p, v) = L(v) - \mathscr{A}(u, \phi; v, 0) \quad \forall v \in H^1_D(\Omega). \qquad (12.182)$$

The right-hand side is a continuous linear functional in v that due to (12.180) lies in $(\ker B)^0$, i.e., it vanishes on $\ker B$. Concerning the left-hand side, note that the operator

$$B' : \begin{cases} L^2(\Omega) \to [H^1_D(\Omega)]' \\ \quad p \quad \mapsto \quad b(p, .) \end{cases}$$

is the adjoint of the divergence operator. Since $\Im B' = (\ker B)^0$, we conclude that (12.182) has a solution. (12.182) implies that the first equation of (12.178) is satisfied with (u, ϕ) being the solution of (12.180). To show uniqueness, let (u, ϕ, p) and $(\tilde{u}, \tilde{\phi}, \tilde{p})$ be solutions of (12.178). Then for all $(v, \psi, q) \in H^1_D(\Omega) \times H^{-1/2} \times L^2(\Omega)$

$$\mathscr{A}(u, \phi; v, \psi) - \mathscr{A}(\tilde{u}, \tilde{\phi}; v, \psi) + b(p - \tilde{p}, v) + b(q, u - \tilde{u}) = 0.$$

Choosing $v = u - \tilde{u}$, $\psi = \phi - \tilde{\phi}$ and $q = -(p - \tilde{p})$ and exploiting uniform monotonicity we conclude $u = \tilde{u}$ and $\phi = \tilde{\phi}$. Now we have $\ker B' = \{0\}$ and thus p is unique. □

We will use a Galerkin method with finite-dimensional subspaces $H^1_h \subset H^1_D(\Omega)$, $H^{-1/2}_h \subset H^{-1/2}$ and $H^0_h \subset L^2(\Omega)$: Find $(u_h, \phi_h, p_h) \in H^1_h \times H^{-1/2}_h \times H^0_h$ such that for all $(v_h, \psi_h, q_h) \in H^1_h \times H^{-1/2}_h \times H^0_h$

$$\begin{aligned}
\mathscr{A}(u_h, \phi_h; v_h, \psi_h) + b(p_h, v_h) &= L(v_h) \\
b(q_h, u_h) &= 0.
\end{aligned} \qquad (12.183)$$

Theorem 12.21 *Let $\Gamma_D \neq \emptyset$. Let A be uniformly monotone and Lipschitz continuous. Let the discrete spaces H_h^1 and H_h^0 satisfy the inf–sup condition: There is an h-independent constant $\beta > 0$ such that*

$$\inf_{q_h \in H_h^0} \sup_{v_h \in H_h^1} \frac{b(q_h, v_h)}{\|q_h\|_{0,\Omega} \|v_h\|_{1,\Omega}} \geq \beta. \tag{12.184}$$

Then (12.183) *has a unique solution which converges quasioptimally:*

$$\begin{aligned} &\|u - u_h\|_{1,\Omega} + \|\phi - \phi_h\|_{-1/2,\Gamma} + \|p - p_h\|_{0,\Omega} \\ &\leq C \left\{ \inf_{v_h \in H_h^1} \|u - v_h\|_{1,\Omega} + \inf_{\psi_h \in H_h^{-1/2}} \|\phi - \psi_h\|_{-1/2,\Gamma} + \inf_{q_h \in H_h^0} \|p - q_h\|_{0,\Omega} \right\}. \end{aligned} \tag{12.185}$$

Proof Theorem 12.20 also holds for finite-dimensional subspaces. This establishes unique solvability of (12.183).

We follow the general theory in [65]. Let $\ker B_h := \{v_h \in H_h^1 : b(q_h, v_h) = 0 \ \forall q_h \in H_h^0\}$. Clearly $u_h \in \ker B_h$. As in (12.181), it can be shown that for all $(w_h, \chi_h) \in \ker B_h \times H_h^{-1/2}$

$$\begin{aligned} &\alpha \|w - v\|_{1,\Omega}^2 + C\|\chi - \psi\|_{-1/2,\Gamma}^2 \\ &\leq \mathscr{A}(w_h, \chi_h; w_h - u_h, \phi_h - \chi_h) - \mathscr{A}(u_h, \phi_h; w_h - u_h, \phi_h - \chi_h). \end{aligned} \tag{12.186}$$

For the last term on the right-hand side we exploit the Galerkin orthogonality

$$\mathscr{A}(u_h, \phi_h; w_h - u_h, \phi_h - \chi_h) = \mathscr{A}(u, \phi; w_h - u_h, \phi_h - \chi_h) + b(p - p_h, w_h - u_h).$$

Since $w_h - u_h \in \ker B_h$ we have

$$b(p - p_h, w_h - u_h) = b(p - q_h, w_h - u_h) \quad \forall q_h \in H_h^0.$$

Now we apply the Cauchy–Schwarz inequality, the Lipschitz continuity and the trace theorem to bound (12.186) further by

$$C(\|w_h - u_h\|_{1,\Omega} + \|\chi_h - \phi_h\|_{-1/2,\Gamma})(\|w_h - u\|_{1,\Omega} + \|\chi_h - \phi\|_{-1/2,\Gamma} + \|p - q_h\|_{0,\Omega}).$$

The triangle inequality yields

$$\|u - u_h\|_{1,\Omega} + \|\phi - \phi_h\|_{-1/2,\Gamma} \leq C\{\|u - w_h\|_{1,\Omega} + \|\phi - \chi_h\|_{-1/2,\Gamma} + \|p - q_h\|_{0,\Omega}\} \tag{12.187}$$

for all $w_h \in \ker B_h$, $\chi_h \in H_h^{-1/2}$ and $q_h \in H_h^0$.

The inf–sup condition (12.184) is equivalent to[65]

$$\sup_{q_h \in H_h^0} \frac{b(q_h, v_h)}{\|q_h\|_{0,\Omega}} \geq \beta \inf_{w_h \in \ker B_h} \|v_h - w_h\|_{1,\Omega} = \beta \|v_h\|_{H_h^1/\ker B_h}$$

for all $v_h \in H_h^1$. The equation $b(q, u) = 0 \ \forall q \in L^2(\Omega)$ implies that for every $v_h \in H_h^1$ there is a $w_h \in \ker B_h$ such that

$$\beta \|v_h - w_h\|_{1,\Omega} \leq \sup_{q_h \in H_h^0} \frac{b(q_h, v_h - u)}{\|q_h\|_{0,\Omega}} \leq C\|u - v_h\|_{1,\Omega},$$

and thus

$$\|u - w_h\|_{1,\Omega} \leq \|u - v_h\|_{1,\Omega} + \|v_h - w_h\|_{1,\Omega} \leq C\|u - v_h\|_{1,\Omega}.$$

Therefore

$$\inf_{w_h \in \ker B_h} \|u - w_h\|_{1,\Omega} \leq C \inf_{v_h \in H_h^1} \|u - v_h\|_{1,\Omega}.$$

Together with (12.187) this implies the bound on $\|u - u_h\|_{1,\Omega} + \|\phi - \phi_h\|_{-1/2,\Gamma}$ as claimed in the theorem.

For arbitrary $q_h \in H_h^0$ the inf–sup condition (12.184) implies

$$\beta \|q_h - p_h\|_{0,\Omega} \leq \sup_{v_h \in H_h^1} \frac{b(q_h - p_h, v_h)}{\|v_h\|_{1,\Omega}}.$$

Exploiting the Galerkin orthogonality

$$b(q_h - p_h, v_h) = -\int_\Omega 2\{A(\epsilon(u)) - A(\epsilon(u_h))\} : \epsilon(v_h)\, dx - \langle v_h, W(u - u_h)\rangle$$
$$+ \langle v_h, (I - K')(\phi - \phi_h)\rangle + b(q_h - p, v_h) \qquad \forall v_h \in H_h^1$$

and using the Cauchy–Schwarz inequality, the Lipschitz continuity and the trace theorem we obtain

$$\beta \|q_h - p_h\|_{0,\Omega} \leq C\{\|u - u_h\|_{1,\Omega} + \|\phi - \phi_h\|_{-1/2,\Gamma} + \|p - q_h\|_{0,\Omega}\}.$$

Now the triangle inequality yields the bound on $\|p - p_h\|_{0,\Omega}$. □

Remark 12.3 Finite element spaces that satisfy (12.184) are well known[65, 196]. A convenient choice is the quadrilateral or hexahedral Q_k/P_{k-1} element: u_h is continuous and a polynomial of degree k in each coordinate whereas p_h is discontinuous and a polynomial of total degree $k - 1$. For $k \geq 2$, (12.184) is fulfilled.

In this section we make a slightly stronger hypothesis on the material law by assuming hyperelasticity:

$$A(\epsilon(u))(x) := \frac{\partial \Psi(\epsilon(u))}{\partial \epsilon}(x) \quad \forall x \in \Omega \tag{12.188}$$

with some stored energy function Ψ. We require that the functional

$$\tilde{\Pi}(v) := \int_{\Omega} \Psi(\epsilon(v))\, dx \tag{12.189}$$

has continuous second-order Gâteaux derivatives and there exist positive constants α and C such that

$$\alpha \|v\|_{1,\Omega}^2 \leq D^2\tilde{\Pi}(u)(v,v) \leq C\|v\|_{1,\Omega}^2 \qquad \forall u, v \in H_D^1(\Omega). \tag{12.190}$$

This implies uniform monotonicity and Lipschitz continuity of the Fréchet derivative

$$D\tilde{\Pi}(u) = \int_{\Omega} A(\epsilon(u))\, dx.$$

For the discretization we assume that the spaces H_h^1 and H_h^0 consist of piecewise polynomial functions on a triangulation $\bar{\Omega} = \cup\{T : T \in \mathscr{T}_h\}$ and $H_h^{-1/2}$ consists of piecewise polynomial functions on a partition of Γ. The elements T typically are closed triangles, quadrilaterals or (in $\mathbb{R}^3$) tetrahedra. For $T \neq T'$, $T \cap T'$ is either empty or a common vertex or edge or side. (In $\mathbb{R}^2$ sides coincide with edges.)

We assume the following approximation property to hold. *Given an element $T \in \mathscr{T}_h$ with diameter h_T, let $\tilde{T} := \cup\{T' \in \mathscr{T}_h : T' \cap T \neq \emptyset\}$. Let S be a side of T with diameter h_S. Then for every $v \in H_D^1(\Omega)$ there exists a $v_h \in H_h^1$ such that*

$$\|v - v_h\|_{0,T} \leq C h_T \|v\|_{1,\tilde{T}} \tag{12.191}$$

$$\|v - v_h\|_{0,S} \leq C h_S^{1/2} \|v\|_{1,\tilde{T}} \tag{12.192}$$

with the constant C being independent of T.

For families of partitions into triangles with interior angles being uniformly bounded from below, (12.191) and (12.192) follow from the analysis in [110].

For each element T with exterior unit normal n_T we define the jump in the computed traction across the side $S \subset \partial T$ as

$$[\![[\![\sigma_h n_T]\!] := \begin{cases} \frac{1}{2}\{\sigma(u_h, p_h)|_{T\cap S} - \sigma(u_h, p_h)|_{T'\cap S}\}n_T & \text{if } S = T \cap T' \\ \sigma(u_h, p_h)n_T - g & \text{if } S \subset \Gamma_N \\ \sigma(u_h, p_h)n_T + Wu_h - (\frac{1}{2}I - K')\phi_h & \text{if } S \subset \Gamma \end{cases}.$$

Note that $[\![\sigma_h n_T]\!]$ is well defined on Γ since $u_h|_\Gamma \in H^1(\Gamma)$ and $\phi_h \in L^2(\Gamma)$ and thus $Wu_h - (\frac{1}{2}I - K')\phi_h \in L^2(\Gamma)$.

For the estimator we need the following residual quantities:

$$\begin{aligned}
R_T^{(1)} &:= h_T \|f + \operatorname{div} \sigma(u_h, p_h)\|_{0,T} \\
R_T^{(2)} &:= \|\operatorname{div} u_h\|_{0,T} \\
R_S^{(3)} &:= h_S^{1/2} \|[\![\sigma_h n]\!]\|_{0,S} \\
R^{(4)} &:= \|V\phi_h + (\frac{1}{2}I - K)u_h\|_{1/2,\Gamma}
\end{aligned}$$

In [66] the following result is shown.

Theorem 12.22 *Let* (12.190) *be satisfied. Let* (u_h, ϕ_h, p_h) *be the solution of* (12.183). *Then there holds the* a posteriori *estimate*

$$\|u - u_h\|_{1,\Omega} + \|\phi - \phi_h\|_{-1/2,\Gamma} + \|p - p_h\|_{0,\Omega}$$

$$\leq C \left\{ \sum_{T \in \mathcal{T}_h} \left\{ R_T^{(1)} + R_T^{(2)} + \sum_{S \subset \partial T \setminus \Gamma_D} R_S^{(3)} \right\} + R^{(4)} \right\}$$

where C is independent of u_h if the mesh is sufficiently fine.

The estimator of Theorem 12.22 is similar to that of [83] for the standard FE/BE coupling. Here the additional term $R_T^{(2)}$ provides for the incompressibility constraint. For the coupling of dual-mixed finite elements and boundary elements applied to elasticity see [187].

For further reading see [81] and [306] for coupling of mixed finite elements and boundary element as well as see [80] for coupling with nonconforming finite elements. In recent year a strong research on coupling with discontinuous (DG) elements has been developped , see e.g. [237].

12.7 Adaptive FE/BE Coupling for Strongly Nonlinear Interface Problems with Tresca Friction

Now we consider again equations (12.138)–(12.141) and write $u_1 = u|_\Omega$ and $u_1 = u|_{\Omega^c}$. For given $g \in L^2(\Gamma_s)$ we set with the contact conditions on the contact boundary Γ_s

$$\begin{aligned}
-\varrho(|\nabla u_1|)\partial_\nu u_1(u_0 + u_2 - u_1) + g|(u_0 + u_2 - u_1)| &= 0, \qquad (12.193)\\
|\varrho(|\nabla u_1|)\partial_\nu u_1| &\leq g \quad \text{on } \Gamma_s
\end{aligned}$$

with interface $\partial\Omega = \overline{\Gamma_s \cup \Gamma_t}$ with disjoint components Γ_s and $\Gamma_t \neq \emptyset$. Here $\varrho(t)$ denotes a function $\varrho \in C(0,\infty)$ satisfying for $p \geq 2$

$$0 \leq \varrho(t) \leq \varrho^*[t^\delta(1+t)^{1-\delta}]^{p-2},$$

$$|\varrho(t)t - \varrho(s)s| \leq \varrho^*[(t+s)^\delta(1+t+s)^{1-\delta}]^{p-2}|t-s| \tag{12.194}$$

and

$$\varrho(t)t - \varrho(s)s \geq \varrho_*[(t+s)^\delta(1+t+s)^{1-\delta}]^{p-2}(t-s)$$

for all $t \geq s > 0$ uniformly in $x \in \Omega$ (with fixed $\delta \in [0,1]$, $\varrho_*, \varrho^* > 0$).

In [189] the above nonlinear interface problemis reformulated as the variational inequality: Find $(\hat{u},\hat{v}) \in X^p = W^{1,p}(\Omega) \times \widetilde{W}^{\frac{1}{2},2}(\Gamma_s)$, $\widetilde{W}^{\frac{1}{2},2}(\Gamma_s) = \{u \in H^{\frac{1}{2}}(\partial\Omega) : \operatorname{supp} u \subset \bar{\Gamma}_s\}$, such that

$$\langle G'\hat{u}, u-\hat{u}\rangle + \langle S(\hat{u}|_{\partial\Omega}+\hat{v}), (u-\hat{u})|_{\partial\Omega} + v - \hat{v}\rangle + j(v) - j(\hat{v}) \geq \lambda(u-\hat{u}, v-\hat{v}) \tag{12.195}$$

for all $(u,v) \in X^p$ with the Steklov-Poincaré operator $S : W^{\frac{1}{2},2}(\partial\Omega) \to W^{-\frac{1}{2},2}(\partial\Omega)$ from (12.51). Here $\lambda(u,v) = \langle t_0 + Su_0, u|_{\partial\Omega} + v\rangle + \int_\Omega fu$, $j(v) = \int_{\Gamma_s} g|v|$ for $v \in L^1(\Gamma_s)$, $\langle G'u, v\rangle = \int_\Omega \varrho(|\nabla u|)\nabla u \cdot \nabla v$ for $u, v \in W^{1,p}(\Omega)$. G is strictly convex and $G' : W^{1,p}(\Omega) \to \left(W^{1,p}(\Omega)\right)'$ bounded and uniformly monotone, hence coercive, with respect to the seminorm $|\cdot|_{1,p}$. The variational inequality (12.195) is uniquely solvable and equivalent to the original problem (12.193).

In order to avoid using $S = \mathscr{W} + (1-\mathscr{K}')\mathscr{V}^{-1}(1-\mathscr{K})$ explicitly, the numerical implementation involves a variant of the variational inequality (12.195) in terms of the layer potentials: Find $(\hat{u},\hat{v},\hat{\phi}) \in X^p \times W^{-\frac{1}{2},2}(\partial\Omega) =: Y^p$, such that

$$\begin{aligned}\langle G'\hat{u}, u-\hat{u}\rangle &+ \langle \mathscr{W}(\hat{u}|_{\partial\Omega}+\hat{v}) + (\mathscr{K}'-1)\hat{\phi}, (u-\hat{u})|_{\partial\Omega} + v - \hat{v}\rangle \\ &+ j(v) - j(\hat{v}) \;\geq\; \langle t_0 + \mathscr{W}u_0, (u-\hat{u})|_{\partial\Omega} + v - \hat{v}\rangle + \int_\Omega f(u-\hat{u}), \\ \langle \phi, \mathscr{V}\hat{\phi} &+ (1-\mathscr{K})(\hat{u}|_{\partial\Omega}+\hat{v})\rangle = \langle \phi, (1-\mathscr{K})u_0\rangle\end{aligned}$$

for all $(u,v,\phi) \in Y^p$. In short,

$$B(\hat{u},\hat{v},\hat{\phi}; u-\hat{u}, v-\hat{v}, \phi-\hat{\phi}) + j(v) - j(\hat{v}) \geq \Lambda(u-\hat{u}, v-\hat{v}, \phi-\hat{\phi})$$

with

$$B(u,v,\phi;\bar{u},\bar{v},\bar{\phi}) = \langle G'u,\bar{u}\rangle + \langle \mathscr{W}(u|_{\partial\Omega}+v) + (\mathscr{K}'-1)\phi, \bar{u}|_{\partial\Omega}+\bar{v}\rangle \\ +\langle \bar{\phi}, \mathscr{V}\phi + (1-\mathscr{K})(u|_{\partial\Omega}+v)\rangle,$$

$$\Lambda(u,v,\phi) = \langle t_0 + \mathscr{W}u_0, u|_{\partial\Omega}+v\rangle + \int_\Omega fu + \langle \phi, (1-\mathscr{K})u_0\rangle.$$

Let $\{\mathscr{T}_h\}_{h\in I}$ a regular triangulation of Ω. Let $W_h^{1,p}(\Omega) \subset W^{1,p}(\Omega)$ the space of functions whose restrictions to any $K \in \mathscr{T}_h$ are linear. $W_h^{\frac{1}{2},2}(\partial\Omega)$ denotes the corresponding space of piecewise linear functions, and $\widetilde{W}_h^{\frac{1}{2},2}(\Gamma_s)$ the subspace of those supported on Γ_s. Finally, $W_h^{-\frac{1}{2},2}(\partial\Omega) \subset W^{-\frac{1}{2},2}(\partial\Omega)$ the space of piecewise constant functions on the boundary mesh.

Then the discretized variational inequality reads with $X_h^p = W_h^{1,p}(\Omega) \times \widetilde{W}_h^{\frac{1}{2},2}(\Gamma_s)$: Find $(\hat{u}_h,\hat{v}_h,\hat{\phi}_h) \in Y_h^p = X_h^p \times W_h^{-\frac{1}{2},2}(\partial\Omega)$ such that for all $(u_h,v_h,\phi_h) \in Y_h^p$:

$$B(\hat{u}_h,\hat{v}_h,\hat{\phi}_h; u_h-\hat{u}_h, v_h-\hat{v}_h, \phi_h-\hat{\phi}_h) + j(v_h) - j(\hat{v}_h) \geq \Lambda(u_h-\hat{u}_h, v_h-\hat{v}_h, \phi_h-\hat{\phi}_h).$$

There holds a Céa type a priori error estimate for the solutions $(\hat{u},\hat{v},\hat{\phi}) \in Y^p$, $(\hat{u}_h,\hat{v}_h,\hat{\phi}_h) \in Y_h^p$ be the solutions of the continuous resp. discretized variational problem, uniformly in $h < h_0$:

$$\|\hat{u}-\hat{u}_h, \hat{v}-\hat{v}_h, \hat{\phi}-\hat{\phi}_h\|_{Y^p}^p \lesssim \inf_{(u_h,v_h,\phi_h)\in Y_h^p} \|\hat{u}-u_h, \hat{v}-v_h, \hat{\phi}-\phi_h\|_{Y^p}^2 + \|\hat{v}-v_h\|_{L^2(\Gamma_s)}.$$

For adaptive error control in [189] a gradient recovery scheme in the interior with a residual type error estimator on the boundary is given: There holds the following a posteriori error estimate where $(e,\tilde{e},\epsilon)$ denotes the error between the Galerkin solution $(\hat{u}_h,\hat{v}_h,\hat{\phi}_h) \in Y_h^p$ and the true solution $(\hat{u},\hat{v},\hat{\phi}) \in Y^p$ $(f \in W^{1,p'}(\Omega))$:

$$\|e,\tilde{e},\epsilon\|_{Y^p}^p \lesssim \eta_{gr}^2 + \eta_f^2 + \eta_S^2 + \eta_\partial^2 + \eta_g^2,$$

where

$$\eta_{gr}^2 = \sum_{K\in\mathscr{T}_h} \int_K G_{p,\delta}(\nabla\hat{u}_h, \nabla\hat{u}_h - G_h\hat{u}_h),$$

$$\eta_f^2 = \sum_{K\in\mathscr{T}_h} \int_K G_{p',1}(|\nabla\hat{u}_h|^{p-1}, h_K(f-f_K)),$$

$$\eta_S^2 = \operatorname{dist}_{W^{-\frac{1}{2},2}(\partial\Omega)}\left(V^{-1}(1-K)(\hat{u}+\hat{v}-u_0), W_h^{-\frac{1}{2},2}(\partial\Omega)\right)^2$$

$$\eta_\partial^2 = \|\nu \cdot A'(\nabla \hat{u}_h) + S(\hat{u}_h|_{\partial\Omega} + \hat{v}_h - u_0) - t_0\|^{p'}_{W^{-1+\frac{1}{p},p'}}$$

$$\eta_g^2 = \|(|\sigma(\hat{u}_h)| - g)_+\|^{p'}_{\widetilde{W}^{-\frac{1}{2},2}(\Gamma_s)} + \int_{\Gamma_s} |(|\sigma(\hat{u}_h)| - g)_-||\hat{v}_h| + \int_{\Gamma_s} (\sigma(\hat{u}_h)\hat{v}_h)_+ .$$

where $G_{p,\delta}(x, y) = |y|^2\omega(x, y)^{p-2} = |y|^2[(|x| + |y|)^\delta(1 + |x| + |y|)^{1-\delta}]^{p-2}$ whenever $|x| + |y| > 0$ and 0 otherwise.

In [189] numerical experiments are presented for (12.138)–(12.141) with (12.193) with the L-shape domain Ω as in the previous section. We set $\varrho(t) = (\varepsilon + t)^{p-2}$, with $p = 3$ and $\varepsilon = 0.00001$, $f = 0$, $u_0 = r^{2/3} \sin \frac{2}{3}(\varphi - \frac{\pi}{2})$, $t_0 = \partial_\nu u_0|_{\partial\Omega}$. The friction parameter is $g = 0.5$, leading to slip conditions on the interface. To solve the variational inequality we apply the following Uzawa algorithm with the damping parameter $\rho = 25$.

Algorithm 12.3 (Uzawa)

(i) Choose $\sigma_h^0 \in \Lambda_h = \{\sigma_h \in \widetilde{W}_h^{-\frac{1}{2},2}(\Gamma_s) : |\sigma_h(x)| \le 1$ *a.e. on* $\Gamma_s\}$.
(ii) For $n = 0, 1, 2, \ldots$ *find* $(u_h^n, v_h^n) \in X_h^p$ *such that*

$$\langle G' u_h^n, u_h\rangle + \langle S_h(u_h^n|_{\partial\Omega} + v_h^n), u_h|_{\partial\Omega} + v_h\rangle + \int_{\Gamma_s} g\sigma_h^n v_h \, ds = \lambda_h(u_h, v_h)$$

for all $(u_h, v_h) \in X_h^p$.
(iii) Set

$$\sigma_h^{n+1} = P_\Lambda(\sigma_h^n + \rho g v_h^n),$$

where for every nodal point of the mesh $\mathcal{T}_h|_{\Gamma_s}$ *there holds* $\delta \mapsto P_\Lambda(\delta) = \sup\{-1, \inf(1, \delta)\}$.
(iv) Repeat with 2. until a convergence criterion is satisfied.

The nonlinear variational problem in the Uzawa algorithm is solved by Newton's method in every Uzawa-iteration step (Table 12.3).

Table 12.3 Convergence rates and Uzawa steps for uniform meshes [189]

DOF	$J_h(\hat{u}_h, \hat{v}_h)$	δJ	α_J	It_{Uzawa}	$\tau(s)$
28	−0.511609	0.017249	–	2	0.190
80	−0.517938	0.010920	−0.435	2	0.640
256	−0.521857	0.007001	−0.382	2	2.440
896	−0.524293	0.004566	−0.341	2	11.05
3328	−0.525841	0.003017	−0.316	2	61.85
12800	−0.526865	0.001993	−0.308	2	437.5
50176	−0.527571	0.001287	−0.320	2	4218.0

Here the terms arizing in the above table have the following meanings:

$$J(\hat{u}, \hat{v}) := G(\hat{u}) + \frac{1}{2}\langle S(\hat{u}|_{\partial\Omega} + \hat{v}), (\hat{u}|_{\partial\Omega} + \hat{v})\rangle - \lambda(\hat{u}, \hat{v})$$

with its approximation $J_h(\hat{u}_h, \hat{v}_h)$ and $\delta J = J_h(\hat{u}_h, \hat{v}_h) - J(\hat{u}, \hat{v})$ where

$$G(u) = \int_\Omega q(|\nabla u|), \qquad q(t) = \int_0^t s\rho(s)ds.$$

Further α_J, It_{Uzawa} and $\tau(s)$ denote the convergence rate , the number of Uzawa iterations and the computation time, respectively (see [189] for details).

12.8 Adaptive FE-BE Coupling for the Eddy-Current Problem in $\mathbb{R}^3$

In this section we present from [282] a reliable and efficient residual based a posteriori error estimator for the following time-harmonic eddy current problem in $\mathbb{R}^3$ and furthermore we give a p-hierarchical error estimator from [283]. The problem is discretized by edge elements inside the conductor and the exterior region is taken into account by means of a suitable boundary integral coupling. Given a conductor and a monochromatic exciting current, the task in eddy current computations is to compute the resulting magnetic and electric fields, in the conductor Ω as well as in the exterior domain Ω_E, which represents air. Let $\Omega \subset \mathbb{R}^3$ be a bounded, simply connected open Lipschitz polyhedron with boundary $\Gamma = \partial\Omega$, and further set $\Omega_E = \mathbb{R}^3 \setminus \bar{\Omega}$. The conductor has conductivity $\sigma \in L^\infty(\mathbb{R}^3)$, $\sigma_1 \geq \sigma(\mathbf{x}) \geq \sigma_0 > 0$ and magnetic permeability $\mu \in L^\infty(\mathbb{R}^3)$, $\mu_1 \geq \mu(\mathbf{x}) \geq \mu_0 > 0$ with positive constants $\sigma_0, \sigma_1, \mu_0, \mu_1$. In Ω_E, we set $\sigma \equiv 0$ and by scaling $\mu \equiv 1$. The elementwise regularity of the material parameters reflects the fact that Ω can consist of different conducting materials, i.e. the conductivity and permeability can jump from one material to another. We assume a source current $\mathbf{J}_0 \in \mathbf{H}(\text{div}, \mathbb{R}^3)$ with $\text{supp}(\mathbf{J}_0) \subset \bar{\Omega}$. Hence $\mathbf{J} \cdot \mathbf{n} = 0$ on Γ (there is no flow of $\mathbf{J}$ through Γ).

A mathematical model of the resulting time-harmonic eddy current problem for low frequencies (cf. Ammari, Buffa & Nédélec [1]) consists of Maxwell's equations

$$\textbf{curl}\,\mathbf{E} = -i\omega\mu\mathbf{H}, \quad \textbf{curl}\,\mathbf{H} = \sigma\mathbf{E} + \mathbf{J}_0 \qquad \text{in } \mathbb{R}^3, \tag{12.196}$$

the Coulomb gauge condition $\text{div}\,\mathbf{E} = 0$ in Ω_E together with the transmission conditions

$$[\mathbf{E} \times \mathbf{n}]_\Gamma = 0, \ [\mathbf{H} \times \mathbf{n}]_\Gamma = 0, \tag{12.197}$$

and the Silver-Müller radiation conditions

$$\mathbf{E}(\mathbf{x}) = O\left(\frac{1}{|\mathbf{x}|}\right), \quad \mathbf{H}(\mathbf{x}) = O\left(\frac{1}{|\mathbf{x}|}\right) \text{ uniformly for } |\mathbf{x}| \to \infty. \tag{12.198}$$

The equations in (12.196) are just the time-harmonic Maxwell equations with neglected displacement currents (formally setting $\omega\epsilon = 0$, where ϵ denotes the electric permittivity). This approximation is justified in view of low frequencies ω. Note that the second equation in (12.196) reduces to $\mathbf{curl}\,\mathbf{H} = 0$ in the exterior domain Ω_E. Therefore $\mathbf{E}$ cannot be uniquely determined in Ω_E and requires the Coulomb gauge condition. The transmission conditions (12.197) result from requiring $\mathbf{curl}\,\mathbf{E}$ and $\mathbf{curl}\,\mathbf{H}$ to be in $\mathbf{L}^2_{\text{loc}}(\mathbb{R}^3)$.

In [246], Hiptmair derives an $\mathbf{E}$-based coupling method for solving the problem (12.196)–(12.198) which is based on Costabel's symmetric coupling method [113].The use of boundary elements for the exterior eddy current problem is not new, we mention the early work of MacCamy & Stephan [285–288] and Nédélec [322, 325] (see also Bossavit [58] for the eddy current problem). The unknowns of Hiptmair's coupled formulation, considered in this section, are $\mathbf{u}$, the electrical field $\mathbf{E}$ in Ω, and λ, the twisted tangential trace of the magnetic field on Γ. The natural Sobolev space for $\mathbf{u}$ is $\mathbf{H}(\mathbf{curl}, \Omega)$, the space of $\mathbf{L}^2$-fields in Ω with rotation in $\mathbf{L}^2(\Omega)$, and the space for λ turns out to be a trace space of $\mathbf{H}(\mathbf{curl}, \Omega)$. The discretization of $\mathbf{u}$ uses the lowest order $\mathbf{H}(\mathbf{curl}, \Omega)$-conforming finite element space of Nédélec [321]. It is then obvious to use the corresponding trace space for discretizing λ, which is just a generalization of the lowest order finite element space of Raviart-Thomas on Γ. Let Ω be a simply connected polyhedron, starlike with respect to a ball and denote the planar boundary faces by Γ_i, $i = 1, \ldots, N_\Gamma$ such that $\partial\Omega = \Gamma = \bigcup_{i=1}^{N_\Gamma} \Gamma_i$.

The complex duality pairings in Ω and on Γ will be denoted by $(\cdot,\cdot)_\Omega$ and $\langle\cdot,\cdot\rangle_\Gamma$. We use the usual Sobolev spaces $H^s(\Omega)$ for scalar functions and $\mathbf{H}^s(\Omega)$ for vector fields of order $s \in \mathbb{R}$. Furthermore we use the spaces

$$\begin{aligned}
\mathbf{H}(\mathbf{curl}, \Omega) &:= \{\mathbf{v} \in \mathbf{L}^2(\Omega) \,:\, \mathbf{curl}\,\mathbf{v} \in \mathbf{L}^2(\Omega)\},\\
\mathbf{H}(\text{div}, \Omega) &:= \{\mathbf{v} \in \mathbf{L}^2(\Omega) \,:\, \text{div}\,\mathbf{v} \in L^2(\Omega)\},\\
\mathbf{X}(\Omega) &:= \mathbf{H}(\mathbf{curl}, \Omega) \cap \mathbf{H}(\text{div}, \Omega),\\
\mathbf{H}_0(\text{div}, \Omega) &:= \{\mathbf{v} \in \mathbf{H}(\text{div}, \Omega) \,:\, \mathbf{v}\cdot\mathbf{n} = 0 \text{ on } \partial\Omega\},\\
\mathbf{H}_{\parallel}^{-1/2}(\text{div}_\Gamma, \Gamma) &:= \{\boldsymbol{\zeta} \in \mathbf{H}_{\parallel}^{-1/2}(\Gamma) \,:\, \text{div}_\Gamma\,\boldsymbol{\zeta} \in H^{-1/2}(\Gamma)\},\\
\mathbf{H}_{\parallel}^{-1/2}(\text{div}_\Gamma 0, \Gamma) &:= \{\boldsymbol{\zeta} \in \mathbf{H}_{\parallel}^{-1/2}(\text{div}_\Gamma, \Gamma) \,:\, \text{div}_\Gamma\,\boldsymbol{\zeta} = 0, \boldsymbol{\zeta} \in H^{-1/2}(\Gamma)\},\\
\mathbf{H}_{\perp}^{-1/2}(\text{curl}_\Gamma, \Gamma) &:= \{\boldsymbol{\zeta} \in \mathbf{H}_{\perp}^{-1/2}(\Gamma) \,:\, \text{curl}_\Gamma\,\boldsymbol{\zeta} \in H^{-1/2}(\Gamma)\},
\end{aligned}$$

with the surface divergence operator $\text{div}_\Gamma\,\mathbf{u} := -\,\text{curl}_\Gamma(\mathbf{u}\times\mathbf{n})$ and the surface curl operator $\text{curl}_\Gamma\,\mathbf{u} := \mathbf{curl}\,\mathbf{u}\cdot\mathbf{n}$, where $\gamma_t^{\times}\mathbf{u} := \mathbf{u}\times\mathbf{n}$. see also [69, 70, 246]. We furthermore need the vectorial surface rotation for a scalar function ϕ defined by $\mathbf{curl}_\Gamma\,\phi := \gamma_t^\times(\mathbf{grad}\,\phi)$. The spaces of distributional tangential fields $\mathbf{H}_\parallel^{-1/2}(\Gamma)$ and $\mathbf{H}_\perp^{-1/2}(\Gamma)$ are introduced in [69] by duality.

In the coupling formulation we use integral operators to represent the exterior problem. These operators are defined for $\mathbf{x} \in \Gamma$ as follows (for their properties see e.g. [246] see also Chapter 4).

$$\mathscr{V}(\lambda)(\mathbf{x}) := \gamma_D \mathbf{V}(\lambda)(\mathbf{x}) = \gamma_D \int_\Gamma \Phi(\mathbf{x},\mathbf{y})\lambda(\mathbf{y})\,ds(\mathbf{y}),$$

$$\mathscr{K}(\lambda)(\mathbf{x}) := \gamma_D \mathbf{K}(\lambda)(\mathbf{x}) = \gamma_D\,\mathbf{curl}_\mathbf{x} \int_\Gamma \Phi(\mathbf{x},\mathbf{y})(\mathbf{n}\times\lambda)(\mathbf{y})\,ds(\mathbf{y}),$$

$$\widetilde{\mathscr{K}}(\lambda)(\mathbf{x}) := \gamma_N \mathbf{V}(\lambda)(\mathbf{x}) = (\gamma_t^\times)\mathbf{K}(\lambda\times\mathbf{n})(\mathbf{x}) = \gamma_N \int_\Gamma \Phi(\mathbf{x},\mathbf{y})\lambda(\mathbf{y})\,ds(\mathbf{y}),$$

$$\mathscr{W}(\lambda)(\mathbf{x}) := \gamma_N \mathbf{K}(\lambda)(\mathbf{x}) = (\gamma_t^\times)\mathbf{W}(\lambda)(\mathbf{x}) = \gamma_N\,\mathbf{curl}_\mathbf{x} \int_\Gamma \Phi(\mathbf{x},\mathbf{y})(\mathbf{n}\times\lambda)(\mathbf{y})\,ds(\mathbf{y})$$

with Laplace kernel $\Phi(\mathbf{x},\mathbf{y}) = \frac{1}{4\pi|\mathbf{x}-\mathbf{y}|}$ and the limits γ_D and γ_N from Ω_E onto Γ of the traces $\gamma_D\mathbf{u} := \mathbf{n}\times(\mathbf{u}\times\mathbf{n}) =: \mathbf{u}_\Gamma$ and $\gamma_N\mathbf{u} := \gamma_t^\times(\mathbf{curl}\,\mathbf{u})$. Furthermore we need $\gamma_n\mathbf{u} := \mathbf{u}\cdot\mathbf{n}$.

Following Buffa *et al* [67, 68, 71, 72] we introduce for $0 < s < 1$ the trace spaces

$$\mathbf{H}_\perp^s(\Gamma) := \gamma_t^\times(\mathbf{H}^{s+1/2}(\Omega)), \quad \mathbf{H}_\parallel^s(\Gamma) := \gamma_D(\mathbf{H}^{s+1/2}(\Omega)).$$

The spaces $\mathbf{H}_\perp^{-s}(\Gamma)$ and $\mathbf{H}_\parallel^{-s}(\Gamma)$, $0 < s < 1$, are then defined as the dual spaces of $\mathbf{H}_\perp^s(\Gamma)$ and $\mathbf{H}_\parallel^s(\Gamma)$, resp., with $\mathbf{L}_t^2(\Gamma) := \{\mathbf{u}\in\mathbf{L}^2(\Gamma)\ :\ \mathbf{u}\cdot\mathbf{n} = 0 \text{ a.e. on } \Gamma\}$ as pivot space. For any $s > \frac{1}{2}$ we define $\mathbf{H}_-^s(\Gamma) := \{\mathbf{u}\in\mathbf{L}_t^2(\Gamma)\ :\ \mathbf{u}_{|\Gamma_j} \in \mathbf{H}_t^s(\Gamma_j),\ j = 1,\ldots,N_\Gamma\}$, furthermore

$$\mathbf{H}_\parallel^s(\mathrm{div}_\Gamma,\Gamma) := \begin{cases} \mathbf{H}_\parallel^{-1/2}(\mathrm{div}_\Gamma,\Gamma), & s = -\frac{1}{2}, \\ \{\lambda\in\mathbf{H}_\parallel^s(\Gamma),\ \mathrm{div}_\Gamma\,\lambda\in H^s(\Gamma)\}, & -\frac{1}{2} < s < \frac{1}{2}, \\ \{\lambda\in\mathbf{H}_\parallel^s(\Gamma),\ \mathrm{div}_\Gamma\,\lambda\in H_-^s(\Gamma)\}, & s > \frac{1}{2}, \end{cases}$$

$$\mathbf{H}_\perp^s(\mathrm{curl}_\Gamma,\Gamma) := \begin{cases} \mathbf{H}_\perp^{-1/2}(\mathrm{curl}_\Gamma,\Gamma), & s = -\frac{1}{2}, \\ \{\lambda\in\mathbf{H}_\perp^s(\Gamma),\ \mathrm{curl}_\Gamma\,\lambda\in H^s(\Gamma)\}, & -\frac{1}{2} < s < \frac{1}{2}, \\ \{\lambda\in\mathbf{H}_\perp^s(\Gamma),\ \mathrm{curl}_\Gamma\,\lambda\in H_-^s(\Gamma)\}, & s > \frac{1}{2}. \end{cases}$$

The trace mappings γ_D and $\gamma_t^\times$ can be extended to continuous mappings

$$\gamma_D : \mathbf{H}^s(\mathbf{curl},\Omega) \to \mathbf{H}_\perp^{s-1/2}(\mathrm{curl}_\Gamma,\Gamma), \qquad \gamma_t^\times : \mathbf{H}^s(\mathbf{curl},\Omega) \to \mathbf{H}_\parallel^{s-1/2}(\mathrm{div}_\Gamma,\Gamma) \tag{12.199}$$

for all $0 \le s < 1$, where $\mathbf{H}^s(\mathbf{curl}, \Omega) := \{\mathbf{u} \in \mathbf{H}^s(\Omega) : \mathbf{curl}\,\mathbf{u} \in \mathbf{H}^s(\Omega)\}$ see [68, 72].

After having collected the operators and spaces needed the coupled variational problem for the eddy current problem introduced by Hiptmair reads as ([246].

Find $\mathbf{u} \in \mathbf{H}(\mathbf{curl}, \Omega)$, $\lambda \in \mathbf{H}_{\|}^{-1/2}(\mathrm{div}_\Gamma 0, \Gamma)$ *such that for all* $\mathbf{v} \in \mathbf{H}(\mathbf{curl}, \Omega)$, $\boldsymbol{\zeta} \in \mathbf{H}_{\|}^{-1/2}(\mathrm{div}_\Gamma 0, \Gamma)$

$$
\begin{aligned}
(\mu^{-1}\,\mathbf{curl}\,\mathbf{u}, \mathbf{curl}\,\mathbf{v})_\Omega + i\omega(\sigma\mathbf{u}, \mathbf{v})_\Omega - \langle \mathscr{W}\mathbf{u}_\Gamma, \mathbf{v}_\Gamma\rangle_\Gamma + \langle \widetilde{\mathscr{K}}\lambda, \mathbf{v}_\Gamma\rangle_\Gamma &= -i\omega(\mathbf{J}_0, \mathbf{v})_\Omega,\\
\langle (I - \mathscr{K})\mathbf{u}_\Gamma, \boldsymbol{\zeta}\rangle_\Gamma + \langle \mathscr{V}\lambda, \boldsymbol{\zeta}\rangle_\Gamma &= 0.
\end{aligned}
\tag{12.200}
$$

For brevity write (12.200) as

$$
\mathscr{A}(\mathbf{u}, \lambda; \mathbf{v}, \boldsymbol{\zeta}) = \mathscr{L}(\mathbf{v}, \boldsymbol{\zeta}).
$$

The above formulation is obtained by using Green's formula in Ω and a Stratton-Chu representation formula for $\mathbf{E}$ in Ω_E. The unknown $\mathbf{u}$ corresponds to $\mathbf{E}_{|\Omega}$, and the unknown λ on the boundary corresponds to $\gamma_N\mathbf{E} = -i\omega\mathbf{H}_{|\Omega_E} \times \mathbf{n}$, which can indeed be seen to be surface divergence free. Due to the transmission conditions there holds $\lambda = \gamma_N\mathbf{u}$. Note that the formulation (12.200) is block skew-symmetric. As observed by Hiptmair [246], the sesquilinear form $\mathscr{A}$ is continuous and elliptic on $(\mathbf{H}(\mathbf{curl}, \Omega) \times \mathbf{H}_{\|}^{-1/2}(\mathrm{div}_\Gamma 0, \Gamma))^2$. Thus, the variational formulation (12.200) admits a unique solution. Setting $\mathbf{E}_{|\Omega} := \mathbf{u}$, $\mathbf{E}_{|\Omega_E} := \mathbf{curl}\,\mathbf{V}(\mathbf{n} \times \gamma_D\mathbf{E}) - \mathbf{V}(\lambda)$ with the single layer potential $\mathbf{V}$ with Laplace kernel and $\mathbf{H} := \frac{1}{i\omega\mu}\,\mathbf{curl}\,\mathbf{E}$ gives a solution to the original problem (12.196)–(12.198).

Next the eddy current problem is discretized by edge elements inside the conductor and the exterior region is taken into account by means of a suitable boundary integral coupling. Let $\mathscr{T}_h$ be a regular triangulation (with tetrahedral or hexahedral elements) of Ω and $K_h = \{T \cap \Gamma : T \in \mathscr{T}_h\}$ the induced triangulation on Γ. For the Galerkin method we use the finite element spaces suggested in [246] , namely the well known $\mathbf{H}(\mathbf{curl}, \Omega)$-conforming finite element space $ND_1(\mathscr{T}_h)$ of first kind Nédélec elements of first order [321] for discretization of the unknown $\mathbf{u} \in \mathbf{H}(\mathbf{curl}, \Omega)$ and $RT_1^0(K_h) := \{\lambda_h \in RT_1(K_h), \mathrm{div}_\Gamma\lambda_h = 0\}$ for the boundary unknown $\lambda \in \mathbf{H}_{\|}^{-1/2}(\mathrm{div}_\Gamma 0, \Gamma)$, where $RT_1(K_h)$ denotes the lowest order $\mathbf{H}_{\|}^{-1/2}(\mathrm{div}_\Gamma, \Gamma)$-conforming finite element space of Raviart-Thomas, which can be obtained as the image of $ND_1(\mathscr{T}_h)$ under the mapping $\gamma_t^\times$. Thus the Galerkin method reads:

Find $\mathbf{u}_h \in ND_1(\mathscr{T}_h)$, $\lambda_h \in RT_1^0(K_h)$ such that $\forall \mathbf{v}_h \in ND_1(\mathscr{T}_h)$, $\boldsymbol{\zeta}_h \in RT_1^0(K_h)$

$$
\begin{aligned}
&(\mu^{-1}\,\mathbf{curl}\,\mathbf{u}_h, \mathbf{curl}\,\mathbf{v}_h)_\Omega + i\omega(\sigma\mathbf{u}_h, \mathbf{v}_h)_\Omega\\
&\qquad - \langle \mathscr{W}\gamma_D\mathbf{u}_h, \gamma_D\mathbf{v}_h\rangle_\Gamma + \langle \widetilde{\mathscr{K}}\lambda_h, \gamma_D\mathbf{v}_h\rangle_\Gamma = -i\omega(\mathbf{J}_0, \mathbf{v}_h)_\Omega,\\
&\qquad \langle (I - \mathscr{K})\gamma_D\mathbf{u}_h, \boldsymbol{\zeta}_h\rangle_\Gamma + \langle \mathscr{V}\lambda_h, \boldsymbol{\zeta}_h\rangle_\Gamma = 0.
\end{aligned}
\tag{12.201}
$$

Now the conformity of the discrete spaces and the strong ellipticity of $\mathscr{A}(\cdot,\cdot)$ imply that the Galerkin formulation (12.201) has a unique solution $(\mathbf{u}_h, \lambda_h) \in ND_1(\mathscr{T}_h) \times RT_1^0(K_h)$.

For simplicity, let σ and μ be piecewise C^∞. Besides the set of elements of the interior mesh $\mathscr{T}_h$, we need the set of faces $\mathscr{F}_h$, the set of exterior faces $\mathscr{F}_h^\Gamma = \{F \in \mathscr{F}_h : F \subset \Gamma\}$ (which coincides with the induced boundary triangulation K_h) and the set of interior faces $\mathscr{F}_h^\Omega = \mathscr{F}_h \setminus \mathscr{F}_h^\Gamma$. Further let h_T denote the maximal diameter of an element $T \in \mathscr{T}_h$ and h_F the maximal diameter of a face $F \in \mathscr{F}_h$. We assume shape regularity of the mesh, which in particular means $h_{T'} \lesssim h_T \quad \forall T, T' \in \mathscr{T}_h,\ T \cap T' \neq \emptyset$ and $h_F \lesssim h_T \quad \forall F \in \mathscr{F}_h(T)$, where $\mathscr{F}_h(T)$ is the set of faces of the element $T \in \mathscr{T}_h$. For $F \in \mathscr{F}_h^\Omega$ a common face of two elements T_1, T_2 and the normal $\mathbf{n}(\mathbf{x})$ pointing into T_2 we define the jump $[\mathbf{n}\cdot\mathbf{q}]_F := \mathbf{n}\cdot\mathbf{q}_{|F\subset T_1} - \mathbf{n}\cdot\mathbf{q}_{|F\subset T_2}$. For $F \in \mathscr{F}_h^\Gamma$ we define $[\mathbf{n}\cdot\mathbf{q}]_F := \mathbf{n}\cdot\mathbf{q}_{|F}$. Analogously we define the jumps $[\mathbf{n}\times\mathbf{q}]_F$. We assumed Γ to be simply connected. Therefore we have $RT_1^0(K_h) = \mathbf{curl}_\Gamma \widetilde{\mathscr{S}_1}(K_h)$, where $\mathscr{S}_1(K_h)$ denotes the finite element space of scalar, continuous piecewise linear functions. Thus we now seek a function $\varphi_h \in \widetilde{\mathscr{S}_1}(K_h) := \{\psi \in \mathscr{S}_1(K_h) : \int_\Gamma \psi\, ds_x = 0\}$ and then set $\lambda_h := \mathbf{curl}_\Gamma(\varphi_h)$. We will use the notations $\mathscr{X} := \mathbf{H}(\mathbf{curl}, \Omega) \times \mathbf{H}_\|^{-1/2}(\mathrm{div}_\Gamma 0, \Gamma)$ for the continuous space of our variational problem (12.200) and $\mathscr{X}_h := ND_1(\mathscr{T}_h) \times \mathbf{curl}_\Gamma \widetilde{\mathscr{S}_1}(K_h)$ for the discrete space of the Galerkin formulation (12.201) and we define the energy norm $\|(\mathbf{v}, \boldsymbol{\zeta})\|_{\mathscr{X}}^2 := \|\mathbf{v}\|_{\mathfrak{E}}^2 + \|\boldsymbol{\zeta}\|_{\mathfrak{e}}^2$ via

$$\|\mathbf{v}\|_{\mathfrak{E}}^2 := (\mu^{-1}\,\mathbf{curl}\,\mathbf{v}, \mathbf{curl}\,\mathbf{v})_\Omega + \omega(\sigma\mathbf{v}, \mathbf{v})_\Omega \simeq \|\mathbf{v}\|_{\mathbf{H}(\mathbf{curl},\Omega)}^2\,, \tag{12.202}$$

$$\|\boldsymbol{\zeta}\|_{\mathfrak{e}}^2 := \langle \mathscr{V}\boldsymbol{\zeta}, \boldsymbol{\zeta}\rangle_\Gamma \simeq \|\boldsymbol{\zeta}\|_{H^{-1/2}(\Gamma)}^2 \tag{12.203}$$

on $\mathbf{H}(\mathbf{curl}, \Omega) \times \mathbf{H}_\|^{-1/2}(\mathrm{div}_\Gamma 0, \Gamma)$.

The following theorem gives a residual-based reliable a posteriori error estimator for the FE-BE coupling method (12.201). Here σ_A and μ_A denote the average of σ and μ on a face F, e.g. $\sigma_A := 0.5(\sigma_{T_1} + \sigma_{T_2})$ with $T_1 \cap T_2 = F$. We assume that σ and μ grow only mildly on neighbouring elements.

Theorem 12.23 *Let* $(\mathbf{u}, \lambda) \in \mathscr{X}$ *and* $(\mathbf{u}_h, \lambda_h) \in \mathscr{X}_h$ *denote the solutions of the continuous resp. the discrete formulation* (12.200) *resp.* (12.201) *and let* $(\mathbf{e}, \varepsilon)$ *be the Galerkin error, i.e.* $\mathbf{e} := \mathbf{u} - \mathbf{u}_h$ *and* $\varepsilon := \lambda - \lambda_h$. *There holds the a posteriori error estimate*

$$\|(\mathbf{e}, \varepsilon)\|_{\mathscr{X}} \lesssim \underbrace{\Big((\eta_0^{\mathscr{T}})^2 + (\eta_1^{\mathscr{T}})^2 + (\eta_0^{\mathscr{F},C})^2 + (\eta_1^{\mathscr{F},C})^2 + (\eta_0^{\mathscr{F},\Gamma})^2 + (\eta_1^{\mathscr{F},\Gamma})^2 + (\eta_2^{\mathscr{F},\Gamma})^2\Big)^{1/2}}_{=:\eta} \tag{12.204}$$

with

$$\eta_j^{\mathscr{T}} := \Big(\sum_{T\in\mathscr{T}_h} (\eta_j^T)^2\Big)^{1/2} \quad (j=0,1), \quad \eta_j^{\mathscr{F},C} := \Big(\sum_{F\in\mathscr{F}_h^{\Omega}} (\eta_j^{F,C})^2\Big)^{1/2} \quad (j=0,1),$$

$$\eta_j^{\mathscr{F},\Gamma} := \Big(\sum_{F\in\mathscr{F}_h^{\Gamma}} (\eta_j^{F,\Gamma})^2\Big)^{1/2} \quad (j=0,1,2) \tag{12.205}$$

and

$$\begin{aligned}
\eta_0^T &:= h_T\sqrt{\omega}\,\|\sqrt{\sigma}^{-1}(\operatorname{div}\mathbf{J}_0 + \operatorname{div}\sigma\mathbf{u}_h)\|_{0,T},\\
\eta_1^T &:= h_T\,\|i\sqrt{\mu}\,\omega\mathbf{J}_0 + i\sqrt{\mu}\,\omega\sigma\mathbf{u}_h + \sqrt{\mu}\,\mathbf{curl}(\mu^{-1}\,\mathbf{curl}\,\mathbf{u}_h)\|_{0,T},\\
\eta_0^{F,C} &:= \sqrt{h_F}\sqrt{\omega}\,\|\sqrt{\sigma_A}^{-1}\,[\sigma\mathbf{u}_h\cdot\mathbf{n}]_F\|_{0,F},\\
\eta_1^{F,C} &:= \sqrt{h_F}\,\|\sqrt{\mu_A}\,[\mu^{-1}\,\mathbf{curl}\,\mathbf{u}_h\times\mathbf{n}]_F\|_{0,F},\\
\eta_0^{F,\Gamma} &:= \sqrt{h_F}\sqrt{\omega}\,\|\sqrt{\sigma}\mathbf{u}_h\cdot\mathbf{n}\|_{0,F},\\
\eta_1^{F,\Gamma} &:= \sqrt{h_F}\,\|\sqrt{\mu}^{-1}\,\mathbf{curl}\,\mathbf{u}_h\times\mathbf{n} - \sqrt{\mu}\,\mathscr{W}\gamma_D\mathbf{u}_h + \sqrt{\mu}\,\widetilde{\mathscr{K}}\lambda_h\|_{0,F},\\
\eta_2^{F,\Gamma} &:= \sqrt{h_F}\,\|\operatorname{curl}_\Gamma\mathbf{u}_h - \operatorname{curl}_\Gamma\mathscr{K}\gamma_D\mathbf{u}_h + \operatorname{curl}_\Gamma\mathscr{V}\lambda_h\|_{0,F}.
\end{aligned}$$

If σ, μ are constant on an element T (or on two elements with common face F), the error estimators can be simplified to

$$\begin{aligned}
\eta_0^T &= h_T\sqrt{\omega\sigma^{-1}}\,\|\operatorname{div}\mathbf{J}_0 + \sigma\operatorname{div}\mathbf{u}_h\|_{0,T},\\
\eta_1^T &= h_T\sqrt{\mu}\,\|i\omega\mathbf{J}_0 + i\omega\sigma\mathbf{u}_h + \mu^{-1}\,\mathbf{curl}\,\mathbf{curl}\,\mathbf{u}_h\|_{0,T},\\
\eta_0^{F,C} &= \sqrt{h_F}\sqrt{\omega\sigma_A^{-1}}\,\|[\sigma\mathbf{u}_h\cdot\mathbf{n}]_F\|_{0,F},\\
\eta_1^{F,C} &= \sqrt{h_F}\sqrt{\mu_A}\,\|[\mu^{-1}\,\mathbf{curl}\,\mathbf{u}_h\times\mathbf{n}]_F\|_{0,F},\\
\eta_0^{F,\Gamma} &= \sqrt{h_F}\sqrt{\omega\sigma}\,\|\sigma\mathbf{u}_h\cdot\mathbf{n}\|_{0,F},\\
\eta_1^{F,\Gamma} &= \sqrt{h_F}\sqrt{\mu}\,\|\mu^{-1}\,\mathbf{curl}\,\mathbf{u}_h\times\mathbf{n} - \mathscr{W}\gamma_D\mathbf{u}_h + \widetilde{\mathscr{K}}\lambda_h\|_{0,F}.
\end{aligned}$$

Using lowest order Nédélec elements, we even obtain:

$$\eta_0^T = h_T\sqrt{\omega\sigma^{-1}}\,\|\operatorname{div}\mathbf{J}_0\|_{0,T},\ \eta_1^T = h_T\omega\sqrt{\mu}\,\|\mathbf{J}_0 + \sigma\mathbf{u}_h\|_{0,T}.$$

Proof Setting $\mathbf{e} := \mathbf{u} - \mathbf{u}_h$, $\varepsilon := \lambda - \lambda_h$. The ellipticity of $\mathscr{A}$ yields

$$\|\mathbf{e}\|_{\mathfrak{E}}^2 + \|\varepsilon\|_{\mathfrak{e}}^2 \lesssim |\mathscr{A}(\mathbf{e},\varepsilon;\mathbf{e},\varepsilon)| = |\mathscr{L}(\mathbf{e},\varepsilon) - \mathscr{A}(\mathbf{u}_h,\lambda_h;\mathbf{e},\varepsilon)|.$$

But for arbitrary $(\mathbf{e}_h, \varepsilon_h) \in \mathscr{X}_h$ we have the equation

$$\mathscr{A}(\mathbf{u}_h, \lambda_h; \mathbf{e}_h, \varepsilon_h) = \mathscr{L}(\mathbf{e}_h, \varepsilon_h),$$

which we can insert in the above equation to obtain

$$\begin{aligned}
&\|\mathbf{e}\|_{\mathfrak{E}}^2 + \|\varepsilon\|_{\mathfrak{e}}^2 \lesssim |\mathscr{L}(\mathbf{e} - \mathbf{e}_h, \varepsilon - \varepsilon_h) - \mathscr{A}(\mathbf{u}_h, \lambda_h; \mathbf{e} - \mathbf{e}_h, \varepsilon - \varepsilon_h)| \\
&= \Big| -i\omega(\mathbf{J}_0 + \sigma\mathbf{u}_h, \mathbf{e} - \mathbf{e}_h)_\Omega - (\mu^{-1}\,\mathbf{curl}\,\mathbf{u}_h, \mathbf{curl}(\mathbf{e} - \mathbf{e}_h))_\Omega \\
&\quad + \langle \mathscr{W}\gamma_D\mathbf{u}_h - \widetilde{\mathscr{K}}\lambda_h, \gamma_D\mathbf{e} - \gamma_D\mathbf{e}_h\rangle_\Gamma + \langle (\mathscr{K} - I)\gamma_D\mathbf{u}_h - \mathscr{V}\lambda_h, \varepsilon - \varepsilon_h\rangle_\Gamma \Big| .
\end{aligned} \tag{12.206}$$

In order to analyze the error $\mathbf{e}$, we decompose it into a weakly solenoidal and an irrotational part. Using the regularized Poincaré map, as investigated in the work by Costabel and McIntosh [123], one has for bounded domains, which are starlike with respect to a ball, the decomposition

$$\mathbf{H}(\mathbf{curl}, \Omega) = \mathbf{H}^1(\Omega) + \mathbf{grad}\ H^1(\Omega). \tag{12.207}$$

For this decomposition see also the proof of Lemma 5.8 in [57]. Therefore for any $\mathbf{v} \in \mathbf{H}(\mathbf{curl}, \Omega)$ there exist functions $\mathbf{v}^\perp \in \mathbf{H}^1(\Omega)$ and $\psi \in H^1(\Omega)/\mathbb{C}$ with $\mathbf{v} = \mathbf{v}^\perp + \mathbf{grad}\ \psi$ such that there holds

$$|\mathbf{v}^\perp|_{\mathbf{H}^1(\Omega)} \lesssim \|\mathbf{curl}\,\mathbf{v}\|_{\mathbf{L}^2(\Omega)}, \tag{12.208}$$

$$\|\mathbf{grad}\ \psi\|_{\mathbf{L}^2(\Omega)} \leq \|\mathbf{v}\|_{\mathbf{H}(\mathbf{curl},\Omega)}. \tag{12.209}$$

We thus split the error term $\mathbf{e}$ into the two parts

$$\mathbf{e} = \mathbf{e}^\perp + \mathbf{grad}\ \psi, \quad \mathbf{e}^\perp \in \mathbf{H}^1(\Omega),\ \psi \in H^1(\Omega)/\mathbb{C} \tag{12.210}$$

using (12.207). For the boundary error term ε we remark that $\varepsilon = \mathbf{curl}_\Gamma\phi$ for some $\phi \in H^{1/2}(\Gamma)/\mathbb{C}$, since $\varepsilon \in \mathbf{H}_\parallel^{-1/2}(\mathrm{div}_\Gamma 0, \Gamma)$ (cf. [69, 70]). Next we define the discrete functions $\mathbf{e}_h$ and ε_h. We choose

$$\mathbf{e}_h = \mathfrak{P}_h^1\mathbf{e}^\perp + \mathbf{grad}\ P_h^1\psi \in ND_1(\mathscr{T}_h)$$

with $\mathbf{e}^\perp$, ψ from (12.210) and the interpolation operators $\mathfrak{P}_h^1 : \mathbf{H}^1(\Omega) \mapsto ND_1(\mathscr{T}_h)$ and $P_h^1 : H^1(\Omega) \mapsto \mathscr{S}_1(\mathscr{T}_h)$ where $\mathscr{S}_1$ denotes the space of continuous and piecewise trilinear functions. On the boundary we choose $\phi_h = p_h^1\phi$ with $p_h^1 : H^{1/2}(\Gamma) \mapsto \mathscr{S}_1(K_h)$ and then

$$\varepsilon_h = \mathbf{curl}_\Gamma\phi_h \in RT_1^0(K_h).$$

With (12.210) and the above definitions of $\mathbf{e}_h$ and ε_h we obtain with (12.206) the residual estimate

$$\|\mathbf{e}\|_{\mathfrak{E}}^2 + \|\varepsilon\|_{\mathfrak{e}}^2 \lesssim \Big| -i\omega(\mathbf{J}_0+\sigma\mathbf{u}_h, \mathbf{e}^\perp - \mathfrak{P}_h^1\mathbf{e}^\perp)_\Omega - (\mu^{-1}\mathbf{curl}\,\mathbf{u}_h, \mathbf{curl}(\mathbf{e}^\perp - \mathfrak{P}_h^1\mathbf{e}^\perp))_\Omega$$
$$+ \langle \mathscr{W}\gamma_D\mathbf{u}_h - \widetilde{\mathscr{K}}\lambda_h, \gamma_D\mathbf{e}^\perp - \gamma_D\mathfrak{P}_h^1\mathbf{e}^\perp\rangle_\Gamma \Big|$$
$$+ \Big| -i\omega(\mathbf{J}_0+\sigma\mathbf{u}_h, \mathbf{grad}\,(\psi - P_h^1\psi))_\Omega + \langle \mathscr{W}\gamma_D\mathbf{u}_h - \widetilde{\mathscr{K}}\lambda_h, \mathbf{grad}_\Gamma(\psi - P_h^1\psi)\rangle_\Gamma \Big|$$
$$+ \Big| \langle (\mathscr{K} - I)\gamma_D\mathbf{u}_h - \mathscr{V}\lambda_h, \mathbf{curl}_\Gamma(\phi - p_h^1\phi)\rangle_\Gamma \Big|. \tag{12.211}$$

Now integration by parts gives

$$(\mu^{-1}\,\mathbf{curl}\,\mathbf{u}_h, \mathbf{curl}(\mathbf{e}^\perp - \mathfrak{P}_h^1\mathbf{e}^\perp))_\Omega = \sum_{T\in\mathscr{T}_h} (\mu^{-1}\,\mathbf{curl}\,\mathbf{u}_h, \mathbf{curl}(\mathbf{e}^\perp - \mathfrak{P}_h^1\mathbf{e}^\perp))_T$$
$$= \sum_{T\in\mathscr{T}_h} \Big((\mathbf{curl}(\mu^{-1}\,\mathbf{curl}\,\mathbf{u}_h), \mathbf{e}^\perp - \mathfrak{P}_h^1\mathbf{e}^\perp)_T + \langle \mu^{-1}\gamma_N\mathbf{u}_h, \gamma_D\mathbf{e}^\perp - \gamma_D\mathfrak{P}_h^1\mathbf{e}^\perp\rangle_{\partial T} \Big)$$
$$= (\mathbf{curl}(\mu^{-1}\mathbf{curl}\,\mathbf{u}_h), \mathbf{e}^\perp - \mathfrak{P}_h^1\mathbf{e}^\perp)_\Omega + \sum_{F\in\mathscr{F}_h} \langle [\mu^{-1}\mathbf{curl}\,\mathbf{u}_h \times \mathbf{n}]_F, \gamma_D\mathbf{e}^\perp - \gamma_D\mathfrak{P}_h^1\mathbf{e}^\perp\rangle_F. \tag{12.212}$$

We have used the fact that the terms $\mu^{-1}\,\mathbf{curl}\,\mathbf{u}_h \times \mathbf{n}$ and $\gamma_D\mathbf{e}^\perp - \gamma_D\mathfrak{P}_h^1\mathbf{e}^\perp$ are in $\mathbf{L}^2(\partial T)$ (since $\mathbf{u}_{h|T}$ is a polynomial and $\mathbf{e}^\perp, \mathfrak{P}_h^1\mathbf{e}^\perp \in \mathbf{H}^1(T)$), such that we can consider the $\mathbf{H}_\parallel^{-1/2}(\mathrm{div}_\Gamma, \partial T) - \mathbf{H}_\perp^{-1/2}(\mathrm{curl}_\Gamma, \partial T)$-duality $\langle\cdot,\cdot\rangle_{\partial T}$ as a $\mathbf{L}^2(\partial T)$-duality. Furthermore we can write $\mathbf{curl}\,\mathbf{u}_h \times \mathbf{n}$ for $\gamma_N\mathbf{u}_h$ due to the regularity of $\mathbf{u}_h$. Since elementwise $u_h \in \mathbf{H}(\mathrm{div})$ we obtain similarly

$$(\mathbf{J}_0+\sigma\mathbf{u}_h, \mathbf{grad}\,\psi - \mathbf{grad}\,P_h^1\psi)_\Omega = -(\mathrm{div}\,\mathbf{J}_0 + \mathrm{div}\,\sigma\mathbf{u}_h, \psi - P_h^1\psi)_\Omega$$
$$+ \sum_{F\in\mathscr{F}_h} \langle [\sigma\mathbf{u}_h\cdot\mathbf{n}]_F, \psi - P_h^1\psi\rangle_F. \tag{12.213}$$

Next, we regard the term $\langle \mathscr{W}\gamma_D\mathbf{u}_h - \widetilde{\mathscr{K}}\lambda_h, \mathbf{grad}_\Gamma\psi - \mathbf{grad}_\Gamma P_h^1\psi\rangle_\Gamma$ from (12.211), which constitutes a $\mathbf{H}_\parallel^{-1/2}(\mathrm{div}_\Gamma, \Gamma) - \mathbf{H}_\perp^{-1/2}(\mathrm{curl}_\Gamma, \Gamma)$-duality pairing (the left hand side is in $\mathbf{H}_\parallel^{-1/2}(\mathrm{div}_\Gamma, \Gamma)$, the right hand side is in $\mathbf{H}_\perp^{-1/2}(\mathrm{curl}_\Gamma, \Gamma)$). With the integration by parts formula given in [69] we obtain

$$\langle \mathscr{W}\gamma_D\mathbf{u}_h - \widetilde{\mathscr{K}}\lambda_h, \mathbf{grad}_\Gamma(\psi - P_h^1\psi)\rangle_\Gamma = -\langle \mathrm{div}_\Gamma\mathscr{W}\gamma_D\mathbf{u}_h - \mathrm{div}_\Gamma\widetilde{\mathscr{K}}\lambda_h, \psi - P_h^1\psi\rangle_\Gamma. \tag{12.214}$$

Next note that for $\mathbf{u} \in \mathbf{H}(\mathbf{curl}, \Omega_E)$, $\lambda \in \mathbf{H}_{\|}^{-1/2}(\mathrm{div}_\Gamma 0, \Gamma)$ there holds

$$\mathrm{div}_\Gamma \widetilde{\mathscr{K}} \lambda = 0 \text{ in } H^{-1/2}(\Gamma), \quad \mathrm{div}_\Gamma \mathscr{W} \gamma_D \mathbf{u} = 0 \text{ in } H^{-1/2}(\Gamma).$$

Thus (12.214) yields

$$\langle \mathscr{W} \gamma_D \mathbf{u}_h - \widetilde{\mathscr{K}} \lambda_h, \mathbf{grad}_{\,\Gamma}(\psi - P_h^1 \psi)\rangle_\Gamma = 0.$$

The last term from (12.211) to consider is $\langle\langle(\mathscr{K}-I)\gamma_D \mathbf{u}_h - \mathscr{V}\lambda_h, \mathbf{curl}_\Gamma \phi - \mathbf{curl}_\Gamma p_h^1 \phi\rangle\rangle_\Gamma$, which is again a duality pairing between $\mathbf{H}_\perp^{-1/2}(\mathrm{curl}_\Gamma, \Gamma)$ and $\mathbf{H}_{\|}^{-1/2}(\mathrm{div}_\Gamma, \Gamma)$. Using again the integration by parts formula from [69] we obtain

$$\langle\langle(\mathscr{K}-I)\gamma_D \mathbf{u}_h - \mathscr{V}\lambda_h, \mathbf{curl}_\Gamma(\phi - p_h^1 \phi)\rangle\rangle_\Gamma = \langle \mathrm{curl}_\Gamma(\mathscr{K}-I)\gamma_D \mathbf{u}_h - \mathrm{curl}_\Gamma \mathscr{V}\lambda_h, \phi - p_h^1\phi\rangle_\Gamma. \tag{12.215}$$

Altogether we have

$$\begin{aligned}
\|\mathbf{e}\|_{\mathfrak{E}}^2 + \|\varepsilon\|_{\mathfrak{e}}^2 \lesssim & \sum_{T \in \mathscr{T}_h} |(-i\omega \mathbf{J}_0 - i\omega\sigma \mathbf{u}_h - \mathbf{curl}(\mu^{-1}\,\mathbf{curl}\,\mathbf{u}_h), \mathbf{e}^\perp - \mathfrak{P}_h^1 \mathbf{e}^\perp)_T| \\
& + \sum_{F \in \mathscr{F}_h^\Omega} |\langle [\mu^{-1}\,\mathbf{curl}\,\mathbf{u}_h \times \mathbf{n}]_F, \gamma_D \mathbf{e}^\perp - \gamma_D \mathfrak{P}_h^1 \mathbf{e}^\perp\rangle_F| \\
& + \sum_{F \in \mathscr{F}_h^\Gamma} |\langle \mu^{-1}\,\mathbf{curl}\,\mathbf{u}_h \times \mathbf{n} - \mathscr{W}\gamma_D \mathbf{u}_h + \widetilde{\mathscr{K}}\lambda_h, \gamma_D \mathbf{e}^\perp - \gamma_D \mathfrak{P}_h^1 \mathbf{e}^\perp\rangle_F| \\
& + \sum_{T \in \mathscr{T}_h} |\omega\,\mathrm{div}\,\mathbf{J}_0 + \omega\,\mathrm{div}\,\sigma \mathbf{u}_h, \psi - P_h^1 \psi)_T| \\
& + \sum_{F \in \mathscr{F}_h^\Omega} |\omega\langle[\sigma \mathbf{u}_h \cdot \mathbf{n}]_F, \psi - P_h^1\psi\rangle_F| + \sum_{F \in \mathscr{F}_h^\Gamma} |\omega\langle \sigma \mathbf{u}_h \cdot \mathbf{n}, \psi - P_h^1 \psi\rangle_F| \\
& + \sum_{F \in \mathscr{F}_h^\Gamma} |\langle \mathrm{curl}_\Gamma(I - \mathscr{K})\gamma_D \mathbf{u}_h + \mathrm{curl}_\Gamma \mathscr{V}\lambda_h, \phi - p_h^1 \phi\rangle_F|.
\end{aligned}$$

Applying the Cauchy-Schwarz inequality and standard approximation properties for $\mathfrak{P}_h^1$, P_h^1 and p_h^1 yields the Theorem. □

Now we present the efficiency of the residual error estimator for FE-BE coupling of the eddy current problem (for details see [281]). We assume that the volume mesh $\mathscr{T}_h$ is shape-regular and that the induced boundary mesh $\mathscr{F}_h^\Gamma = \mathscr{T}_h|_\Gamma$ is quasi-uniform. We assume there holds

$$1 \le \frac{h_{\Gamma,\max}}{h_{\Gamma,\min}} \le \mathscr{Q}(\mathscr{T}_h|_\Gamma) \tag{12.216}$$

for a certain quasi-uniformity constant $\mathcal{Q}(\mathcal{T}_h|_\Gamma)$, independent of the mesh, where $h_{\Gamma,\max} := \max\{h_F,\ F \in \mathcal{F}_h^\Gamma\}$ and $h_{\Gamma,\min} := \min\{h_F,\ F \in \mathcal{F}_h^\Gamma\}$.

Theorem 12.24 ([282]) *Let* $(\mathbf{u}, \lambda) \in \mathcal{X}$ *and* $(\mathbf{u}_h, \lambda_h) \in \mathcal{X}_h$ *denote the solutions of the continuous resp. the discrete formulation* (12.200) *resp.* (12.201) *and let* $(\mathbf{e}, \varepsilon)$ *be the Galerkin error, i.e.* $\mathbf{e} := \mathbf{u} - \mathbf{u}_h$ *and* $\varepsilon := \lambda - \lambda_h$. *Then there exists a constant* $C > 0$, *depending on the quasi-uniformity constant of the boundary element mesh* $\mathcal{Q}(\mathcal{T}_h^\Gamma)$ *of* (12.216) *and on the shape regularity constant of* $\mathcal{T}_h$, *such that there holds for the error estimator* η *in* (12.204) *and* $\delta > 0$

$$\begin{aligned}\eta^2 \le C \Big\{ &\|(\mathbf{e}, \varepsilon)\|_{\mathcal{X}}^2 + \sum_{T \in \mathcal{T}_h} \left(\|\mathbf{e}^0\|_{L^2(T)}^2 + (osc_1^T)^2\right) + \sum_{T \in \mathcal{T}_\Gamma} \|\mathbf{u} - \mathbf{u}_h\|_{\mathfrak{E},T}^2 \\ &+ h_{\Gamma,\max} \|\mathbf{u} - \mathbf{u}_E\|_{H^{1/2}(\mathbf{curl},\Omega)}^2 + h_{\Gamma,\max} \frac{h^{1+2\delta}}{h_{\min}} \|\mathbf{u} - \mathbf{u}_E\|_{\mathbf{H}^{1/2+\delta}(\mathbf{curl},\Omega)}^2 \\ &+ h_{\Gamma,\max} \|\lambda - \lambda_E\|_{\mathbf{H}_{\|}^0(\mathrm{div}_\Gamma,\Gamma)}^2 \Big\}\end{aligned} \tag{12.217}$$

with the interpolant $\mathbf{u}_E := \Pi_1^h \mathbf{u} \in ND_1(\mathcal{T}_h)$, $\lambda_E \in RT_1(K_h)$ *the orthogonal projection of* λ *with respect to the* $\mathbf{H}_{\|}^{-1/2}(\mathrm{div}_\Gamma, \Gamma)$ *inner product. Here* $osc_1^T := h_T \|\sqrt{\mu}\omega(\mathbf{J}_0 - \Pi_1^h \mathbf{J}_0)\|_{\mathbf{L}^2(T)}$ *denotes the oscillation term where* Π_1^h *is an interpolation operator into* $ND_1(\mathcal{T}_h)$. *Furthermore,* $\mathcal{T}_\Gamma$ *denotes the set of elements which have at least one face on the boundary.*

Example 12.1 The geometry in this example is the L-block $\Omega := [-1, 1]^3 \setminus \left([0, 1]^3 \cup [0, 1]^2 \times [-1, 0]\right)$. Here, we consider a singularity function as given current.

$$\mathbf{J}_0 := \operatorname{grad} \left(r^{2/3} \sin(\tfrac{2}{3}\phi)\right) \quad \text{in the L-block,}$$

where r and ϕ are cylindrical coordinates. Hence, one expects an adaptive refinement towards the re-entrant edge.

The energy norm of the unknown exact solution is extrapolated by the energy norms on the sequence of uniform meshes. We perform an adaptive refinement (10% of elements) using hanging nodes. The resulting meshes can be found in Fig. 12.5 and the error in Fig. 12.6. Due of the 2/3-singularity in the interior domain we expect a convergence rate of $\alpha = \frac{2}{3}$ with respect to the mesh size h and a convergence rate of $\alpha = \frac{2}{9}$ with respect to the degrees of freedom. This correspondents to the results in Table 12.4. For the adaptive refinement using the residual error indicators we get a better convergence rate of about 0.4. The effectivity indices are quite constant which underlines the reliability and efficiency of the error estimator.

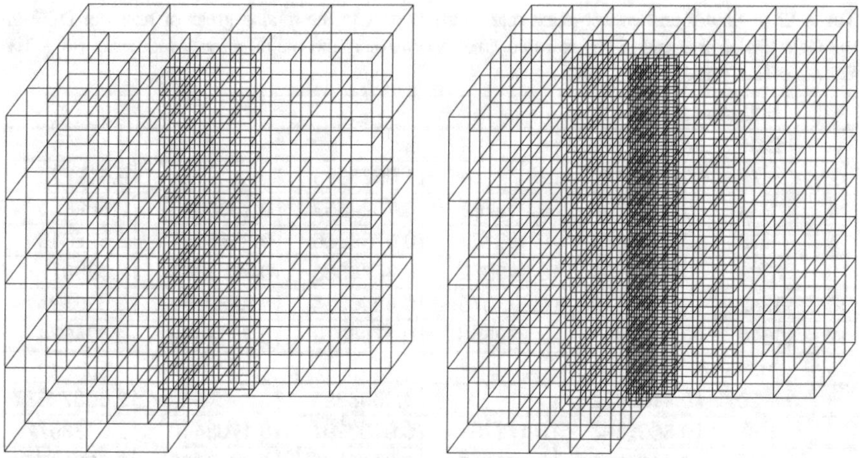

Fig. 12.5 The adaptive meshes (levels of refinement 7 and 9) for Example 12.1 using the residual error estimator [282]

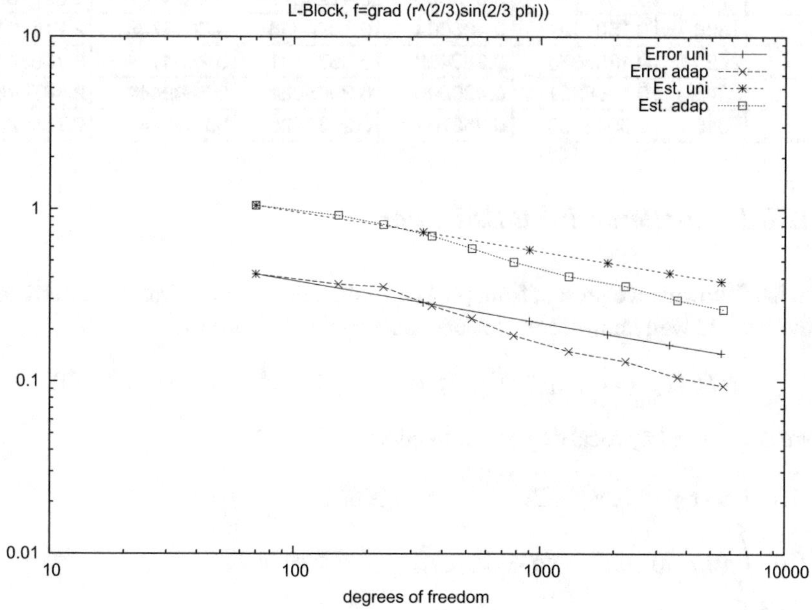

Fig. 12.6 Energy norm e of the Galerkin error and the residual error estimator η of Example 12.1 (L-block) [282]

Table 12.4 Values and convergence rates with respect to the total degrees of freedom DOF of the Galerkin error e and of the residual error estimator η and the effectivity indices $q := \frac{\eta}{e}$ for Example 12.1 (the L-block) [282]

	Uniform refinement					
n	DOF	e	α	η	α	q
2	70	0.4186472		1.0506895		2.509725
4	334	0.2869302	0.241762	0.7324853	0.2308640	2.552834
6	902	0.2246235	0.246421	0.5789936	0.2366966	2.577618
8	1882	0.1881433	0.240962	0.4870686	0.2350675	2.588817
10	3382	0.1638018	0.236375	0.4248354	0.2332292	2.593594
12	5510	0.1462253	0.232553	0.3794070	0.2317002	2.594674
	Adaptive refinement					
	70	0.4186472		1.0506895		2.50972537
	152	0.3661693	0.172731	0.9203291	0.1708448	2.51339776
	231	0.3528255	0.088695	0.8116177	0.3003362	2.30033742
	362	0.2749754	0.554936	0.6961152	0.3417302	2.53155446
	526	0.2319625	0.455246	0.5897867	0.4435999	2.54259503
	778	0.1853135	0.573613	0.4921377	0.4624187	2.65570344
	1306	0.1501191	0.406604	0.4074256	0.3646738	2.71401574
	2229	0.1306073	0.260452	0.3577131	0.2434174	2.73884461
	3648	0.1056281	0.430896	0.2965062	0.3809444	2.80707690
	5615	0.0943108	0.262784	0.2627426	0.2803241	2.78592272

12.8.1 p-Hierarchical Estimator

In the following we present from [401] a p-hiarchical error estimator for tetrahedral meshes. As well known, Nédélec elements on a tetrahedron T

$$ND_k(T) := (\mathbb{P}_{k-1}(T))^3 + \{\mathbf{p} \in (\mathbb{P}_k(T))^3 : \mathbf{p}^T \cdot \mathbf{x} = 0\} \subset (\mathbb{P}_k(T))^3,$$

are determined by local degrees of freedom:

(i) $\displaystyle\int_e \mathbf{u} \cdot \mathbf{t} q \qquad \forall q \in \mathbb{P}_{k-1}, \qquad e \text{ edge of } T,$

(ii) $\displaystyle\int_F (\mathbf{u} \times \mathbf{n}) \cdot \mathbf{q} \qquad \forall \mathbf{q} \in (\mathbb{P}_{k-2})^2, \qquad F \text{ face of } T,$

(iii) $\displaystyle\int_T \mathbf{u} \cdot \mathbf{q} \qquad \forall \mathbf{q} \in (\mathbb{P}_{k-3})^3.$

For $k = 2$ we apply the stable decomposition of Nédélec edge elements from [40]

$$ND_2(\mathcal{T}_h) = ND_1(\mathcal{T}_h) \oplus \mathbf{grad}\, \widetilde{\mathcal{S}}_2(\mathcal{T}_h) \oplus \widetilde{ND}_2^{\perp}(\mathcal{T}_h) \tag{12.218}$$

$$\widetilde{ND}_2^{\perp}(\mathcal{T}_h) := \{\mathbf{u}_h \in ND_2(\mathcal{T}_h) : \langle \mathbf{u}_h, \mathbf{t} \rangle_e = 0, \forall e \text{ edge of } \mathcal{T}_h\}$$

$\widetilde{\mathcal{S}}_2 = \mathcal{S}_2 \setminus \mathcal{S}_1$ (hierarchical surplus), $\mathcal{S}_k = \{w \in C^0 : w|_T \in \mathbb{P}_k\}$.

Let $(\mathbf{u}_h, \boldsymbol{\lambda}_h) \in ND_1(\mathcal{T}_h) \times \mathbf{curl}_\Gamma \mathcal{S}_1(K_h) =: \mathcal{X}_h$, and $(\mathbf{u}_2, \boldsymbol{\lambda}_2) \in ND_2(\mathcal{T}_h) \times \mathbf{curl}_\Gamma \widetilde{\mathcal{S}}_2(K_h) =: \mathcal{X}_2$ (satisfying $\operatorname{div}_\Gamma \boldsymbol{\lambda}_2 = 0$) be Galerkin solutions of (12.201) with exact solution $(\mathbf{u}, \boldsymbol{\lambda}) \in \mathcal{X}$. We make the saturation assumption:

There exists a $\delta \in (0, 1)$ such that there holds

$$\|(\mathbf{u} - \mathbf{u}_2, \boldsymbol{\lambda} - \boldsymbol{\lambda}_2)\|_{\mathcal{X}} \le \delta \|(\mathbf{u} - \mathbf{u}_h, \boldsymbol{\lambda} - \boldsymbol{\lambda}_h)\|_{\mathcal{X}}. \tag{12.219}$$

Theorem 12.25 ([283]) *Assuming (12.219), then on a tetrahedral grid $\mathcal{T}_h$ there holds*

$$\eta \lesssim \|(\mathbf{u} - \mathbf{u}_h, \boldsymbol{\lambda} - \boldsymbol{\lambda}_h)\|_{\mathcal{X}} \lesssim \frac{1}{1-\delta}\eta \tag{12.220}$$

with the local a posteriori estimator

$$\eta^2 := \sum_{i=1}^{M} \left(\Theta^{(e_i)}\right)^2 + \sum_{j=1}^{N} \left(\Theta^{(F_j)}\right)^2 + \sum_{i=1}^{m} \left(\vartheta^{(e_i)}\right)^2,$$

where

$$\vartheta^{(e)} := \frac{|\mathscr{A}(\mathbf{u}_h, \boldsymbol{\lambda}_h; 0, \mathbf{curl}_\Gamma \varphi^{(e)})|}{\|\mathbf{curl}_\Gamma \varphi^{(e)}\|_{\mathfrak{e}}} \qquad \textit{for } \varphi^{(e)} \in \widetilde{\mathcal{S}}_2(K_h),$$

$$\Theta^{(e)} := \frac{|\mathscr{L}(\mathbf{grad}\, \phi^{(e)}, 0) - \mathscr{A}(\mathbf{u}_h, \boldsymbol{\lambda}_h; \mathbf{grad}\, \phi^{(e)}, 0)|}{\|\mathbf{grad}\, \phi^{(e)}\|_{\mathfrak{E}}} \qquad \textit{for } \phi^{(e)} \in \widetilde{\mathcal{S}}_2(\mathcal{T}_h)$$

and

$$\Theta^{(F)} := \|\kappa_1 \mathbf{b}_1^{(F)} + \kappa_2 \mathbf{b}_2^{(F)}\|_{\mathfrak{E}} \qquad \textit{for } \mathbf{b}_i^{(F)} \in \widetilde{ND}_2^{\perp}(\mathcal{T}_h) \quad (i = 1, 2),$$

where $(\kappa_1, \kappa_2)^T$ is the solution of the linear system

$$a(\kappa_1 \mathbf{b}_1^{(F)} + \kappa_2 \mathbf{b}_2^{(F)}, \mathbf{b}_i^{(F)}) = \mathscr{L}(\mathbf{b}_i^{(F)}, 0) - \mathscr{A}(\mathbf{b}_h, \boldsymbol{\lambda}_h; \mathbf{b}_i^{(F)}, 0) \quad (i = 1, 2).$$

and

$$a(\mathbf{u}, \mathbf{v}) := (\mu^{-1}\, \mathbf{curl}\, \mathbf{u}, \mathbf{curl}\, \mathbf{v})_\Omega + i\omega(\sigma \mathbf{u}, \mathbf{v})_\Omega \tag{12.221}$$

Here, M (N) denotes the number of edges (faces) in $\mathcal{T}_h$, $m < M$ the number of edges in K_h (those on Γ).

Example 12.2 We compute the Galerkin solution (12.201) with $\Omega = [-1, 1]^3$ on a series of uniform meshes. Setting $\mu = \sigma = \omega = 1$ we choose the exact solution of (12.196)–(12.198)

$$\mathbf{u}(\mathbf{x}) := \mathbf{curl} \int_\Omega \frac{1}{\|\mathbf{x} - \mathbf{y}\|} \boldsymbol{\rho}(\mathbf{y})\, d\mathbf{y}, \quad \boldsymbol{\lambda} = \mathbf{curl}\,\mathbf{u} \times \mathbf{n}$$

with

$$\boldsymbol{\rho}(\mathbf{x}) = \left((1 - x_1^2)(1 - x_2^2)(1 - x_3^2)\right)^2 x_1 x_2 x_3 \begin{pmatrix} 1 \\ 1 \\ 1 \end{pmatrix} \quad \text{in } \Omega$$

and right hand side

$$\mathbf{J}_0 = -\mathbf{u} + 4\pi i\, \mathbf{curl}\, \boldsymbol{\rho} \quad \text{in } \Omega \text{ and } \mathbf{J}_0 = 0 \text{ in } \Omega_E.$$

Figure 12.7 shows convergence of the FE/BE Galerkin coupling method (12.201) for hexahedral and tetrahedral elements. Furthermore Fig. 12.7 shows that both residual and hierarchical error estimators are reliable and efficient.

For further reading we recommend [247] and for the electric field integral equation [53, 248].

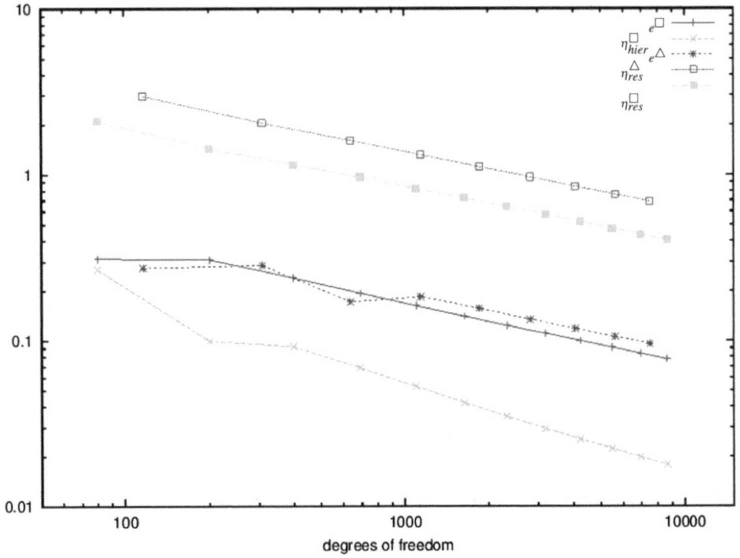

Fig. 12.7 Energy norm e of the Galerkin error $(\mathbf{u} - \mathbf{u}_h, \boldsymbol{\lambda} - \boldsymbol{\lambda}_h)$, residual error estimator η_{res} and reduced hierarchical error estimator η_{hier} for the Example. The superscript $\square$ corresponds to hexahedral elements, $\triangle$ to tetrahedral elements

12.9 Parabolic-Elliptic Interface Problems

In [311] the authors consider the following linear parabolic - elliptic interface problem which describes two-dimensional eddy currents in electro dynamics [289]: Let $\Omega_1 \subset \mathbb{R}^2$ be a bounded simply connected polygonal domain with boundary Γ and complement $\Omega_2 := \mathbb{R}^2 \setminus \bar{\Omega}_1$.

Given $T > 0$, $\lambda > 0$ and the functions $f : \Omega_1 \times [0, T] \to \mathbb{R}$, $v_0, \psi_0 : \Gamma \times [0, T] \to \mathbb{R}$ and $q : \Omega_1 \to \mathbb{R}$ find $A : [0, T] \to \mathbb{R}$ and $u_i : \Omega_i \times [0, T] \to \mathbb{R}$, $i = 1, 2$, such that

$$
\begin{aligned}
\dot{u}_1 - \Delta u_1 + \lambda u_1 &= f && \text{in } \Omega_1 \times (0, T) \\
-\Delta u_2 &= 0 && \text{in } \Omega_2 \times (0, T) \\
u_1 - u_2 &= v_0 && \text{on } \Gamma \times (0, T) \\
\frac{\partial u_1}{\partial n} - \frac{\partial u_2}{\partial n} &= \psi_0 && \text{on } \Gamma \times (0, T) \\
u_1(x, 0) &= q(x) && \forall x \in \Omega_1 \\
u_2(x, t) &= A(t) log|x| + O(\tfrac{1}{|x|}) && \text{for } |x| \to \infty\,.
\end{aligned}
\tag{12.222}
$$

where $\frac{\partial v}{\partial n}$ is the normal derivative of v pointing from Ω_1 into Ω_2.

The symmetric coupling method yields in [311] the following formulation: For $u, w \in H^1(\Omega_1)$ and $\phi, \psi \in H^{-1/2}(\Gamma)$ define the bilinear form

$$
B(u, \phi; w, \psi) = 2\int_{\Omega_1} (\nabla u \cdot \nabla w + \lambda u w)\, dx + \langle Wu + (K' - 1)\phi, w\rangle + \langle \psi, (K - 1)u - V\phi\rangle
$$

and the linear form

$$
L(w, \psi) = \langle 2\psi_0 + W v_0, w\rangle + \langle \psi, (K - 1)v_0\rangle + (2w, f)
$$

where, here, $\langle ., .\rangle$ denotes the duality pairing between $H^{-1/2}(\Gamma)$ and $H^{1/2}(\Gamma)$. The bilinear form $B(.; .)$ satisfies the following Babuska-Brezzi condition:

There exists a constant $\beta > 0$ such that for all $(v, \psi) \in \mathscr{H} := H^1(\Omega_1) \times H^{-1/2}(\Gamma)$:

$$
B(v, \psi; v, -\psi) \geq \beta\, \|(v, \psi)\|^2_{\mathscr{H}} \tag{12.223}
$$

where $\|(v, \psi)\|^2_{\mathscr{H}} = \|v\|^2_{H^1(\Omega_1)} + \|\psi\|^2_{H^{-1/2}(\Gamma)}$.

The problem (12.222) is now transformed into the following variational problem (cf. [120]):

Given $f \in L^2(0,T;H^1(\Omega_1)')$, $v_0 \in L^2(0,T;H^{1/2}(\Gamma))$, $\psi_0 \in L^2(0,T;H^{-1/2}(\Gamma))$ and $q \in H^1(\Omega_1)$ find $u \in Q_T$ and $\phi \in B_T$ such that

$$(\dot{u}, w) + B(u, \phi; w, \psi) = L(w, \psi) \tag{12.224}$$

for all $w \in L^2(0,T;H^1(\Omega_1))$ and $\psi \in L^2(0,T;H^{-1/2}(\Gamma))$ where

$$\begin{aligned} Q_T &= \{u \in L^2(0,T;H^1(\Omega_1));\ \dot{u} \in L^2(0,T;H^1(\Omega_1)'),\ u|_{t=0} = q\} \quad \text{and} \\ B_T &= L^2(0,T;H^{-1/2}(\Gamma)). \end{aligned}$$

For the definition of the Sobolev-spaces $L^2(0,T;X)$ we refer to [284]. The existence of a unique solution (u,ϕ) of (12.224) is proved in [120].

Now, let $(u,\phi) \in Q_T \times B_T$ be the unique solution of (12.224). To obtain a Galerkin aprroximation of (u,ϕ) we have to define suitable finite dimensional subspaces of $L^2(0,T;H^1(\Omega_1))$ and $L^2(0,T;H^{-1/2}(\Gamma))$ as follows: Let $I \subset (0,1)$ be an indexed subset of $(0,1)$ with $0 \in \bar{I}$ and let $\{\Delta^h_{\Omega_1};\ h \in I\}$ be a family of regular triangulations of Ω_1 with corresponding partitions $\{\Delta^h_\Gamma,\ h \in I\}$ of Γ. A triangulation Δ_{Ω_1} of Ω_1 is called *regular* if

- the intersection of any two triangles of Δ_{Ω_1} is either a common side or a common node or empty,
- there exists a constant $\rho > 0$ such that $\rho h_\Delta^2 \le \int_\Delta dx$ for all triangles $\Delta \in \Delta_{\Omega_1}$ with diameter h_Δ,
- there holds $h_\Delta < 1$ for all triangles $\Delta \in \Delta_{\Omega_1}$.

For the discretization of the space variable we need the following finite dimensional subspaces of $H^1(\Omega_1)$ and $H^{-1/2}(\Gamma)$:

$$\begin{aligned} S_h &= \left\{v_h : \Omega_1 \to \mathbb{R};\ v_h \text{ piecewise linear on } \Delta^h_{\Omega_1},\ v_h \in C^0(\Omega_1)\right\}, \\ \tilde{S}_h &= \left\{\psi_h : \Gamma \to \mathbb{R};\ \psi_h \text{ piecewise constant on } \Delta^h_\Gamma\right\}. \end{aligned}$$

Let q_h be the orthogonal L^2-projection of q into the space S_h, i.e. $(q_h, w_h) = (q, w_h)$ for all $w_h \in S_h$. Now, the semi-discrete Galerkin scheme reads as follows:

Find $u_h \in C^1([0,T];S_h)$ and $\phi_h \in C^0([0,T];\tilde{S}_h)$ such that $u_h(0) = q_h$ and

$$(\dot{u}_h, w) + B(u_h, \phi_h; w, \psi) = L(w, \psi) \tag{12.225}$$

for all $(w,\psi) \in S_h \times \tilde{S}_h$.

Choosing a finite element basis $\{w_i\}_{i=1}^M$ of S_h and a boundary element basis $\{\psi_i\}_{i=1}^m$ of $\tilde{S}_h$ we define $U_h = U_h(t) = [u_i(t)]_{i=1,\dots,M}$ and $\Phi_h = \Phi_h(t) = [\phi_i(t)]_{i=1,\dots,m}$ where $u_h(x,t) = \sum_{i=1}^M u_i(t) w_i(x)$ and $\phi_h(x,t) =$

$\sum_{i=1}^{m} \phi_i(t)\psi_i(x)$. Hence, the Galerkin equations (12.225) are equivalent to the following system of ordinary differential equations:

$$\begin{aligned} \mathscr{M}\dot{U}_h + 2(\mathscr{K} + \lambda\mathscr{M})U_h + \mathscr{D}U_h + \mathscr{B}\Phi_h &= F \\ \mathscr{S}\Phi_h - \mathscr{B}^\top U_h &= G \end{aligned} \tag{12.226}$$

where

$$\mathscr{M} = [(w_i, w_j)]_{j=1,\dots,M}^{i=1,\dots,M} \qquad \mathscr{K} = [(\nabla w_i, \nabla w_j)]_{j=1,\dots,M}^{i=1,\dots,M}$$

$$\mathscr{D} = [\langle w_i, W w_j\rangle]_{j=1,\dots,M}^{i=1,\dots,M} \qquad \mathscr{S} = [\langle \psi_i, V\psi_j\rangle]_{j=1,\dots,m}^{i=1,\dots,m}$$

$$\mathscr{B} = [\langle (K'-1)psi_j, w_i\rangle]_{j=1,\dots,m}^{i=1,\dots,M} \qquad \mathscr{B}^\top = [\langle \psi_i, (K-1)w_j\rangle]_{j=1,\dots,M}^{i=1,\dots,m}$$

$$F = [\langle 2\psi_0 + W v_0, w_i\rangle + (2w_i, f)]_{i=1,\dots,M} \text{ and } \quad G = [\langle \psi_i, (1-K)v_0\rangle]_{i=1,\dots,m}$$

For a full discretization of (12.224) we consider partitions $0 = t_0 < t_1 < \ldots < t_n = T$ of the time interval $[0, T]$ in subintervals $I_n = (t_{n-1}, t_n]$ of length $k_n = t_n - t_{n-1}$. With each time interval I_n we associate a regular triangulation $\Delta_{\Omega_1}^n = \Delta_{\Omega_1}^{h_n}$ of Ω_1 and a corresponding partition Δ_Γ^n of Γ. We define $S_h^n := S_{h_n}$ and $\tilde{S}_h^n := \tilde{S}_{h_n}$. To obtain a fully discrete scheme we use the discontinuous (in time) Galerkin method with piecewise linear test and trial functions . Therefore we define the following finite dimensional spaces:

$$V_h^n = \left\{v : I_n \to S_h^n\,;\ v(t) = \chi_0 + t\chi_1\,,\ \chi_0, \chi_1 \in S_h^n,\ t \in I_n\right\},$$

$$\tilde{V}_h^n = \left\{\psi : I_n \to \tilde{S}_h^n\,;\ \psi(t) = \xi_0 + t\xi_1\,,\ \xi_0, \xi_1 \in \tilde{S}_h^n,\ t \in I_n\right\}.$$

and

$$W_{hk} = \{v\,;\ v|_{I_n} \in V_h^n \text{ for } n = 1, \ldots, N\},$$

$$\widetilde{W}_{hk} = \{\psi\,;\ \psi|_{I_n} \in \widetilde{V}_h^n \text{ for } n = 1, \ldots, N\}.$$

With the following notations

$$v_n^+ := \lim_{t\to 0+} v(t_n + t), \quad v_n^- := \lim_{t\to 0-} v(t_n + t) \quad \text{and} \quad [v]_n := v_n^+ - v_n^-$$

the discontinuous Galerkin method for symmetric coupling of FEM/BEM reads as follows: Find $(U, \Phi) \in W_{hk} \times \widetilde{W}_{hk}$ such that for all $(w, \psi) \in W_{hk} \times \widetilde{W}_{hk}$:

$$\int_0^T \{(\dot{U}, w) + B(U, \Phi; w, \psi)\}dt + \sum_{n=2}^{N}([U]_{n-1}, w_{n-1}^+) + (U_0^+, w_0^+) = (q, w_0^+) + \int_0^T L(w, \psi)dt. \tag{12.227}$$

This is equivalent to:

For $n = 1, \ldots, N$ find $(U_n, \Phi_n) \in V_h^n \times \widetilde{V}_h^n$ such that for all $(w, \psi) \in V_h^n \times \widetilde{V}_h^n$:

$$\int_{I_n} \left\{ (\dot{U}_n, w) + B(U_n, \Phi_n; w, \psi) \right\} dt + (U_{n-1}^+, w_{n-1}^+) = (U_{n-1}^-, w_{n-1}^+) + \int_{I_n} L(w, \psi) dt. \tag{12.228}$$

Let (U, Φ) be the corresponding (unique) solution of (12.227) and let (u, ϕ) be the exact solution of (12.224). If $\Delta_{\Omega_1}^h$ is a uniform triangulation of mesh size h, if the time steps are of length k and if $u \in C^1([0, T]; H^r(\Omega_1)) \cap C^3([0, T]; L^2(\Omega_1))$ and $\phi \in C^0([0, T]; H^{r-3/2}(\Gamma))$, $r \in [\frac{3}{2}, 2]$, the following convergence result [311] holds:

$$\|(u - U, \phi - \Phi)\|_{L^2(0,T;\mathscr{H})} = O(h^{r-1} + k^2).$$

A similar result has been proved in [120] for time discretization by the Crank-Nicolson method. An error controlled adaptive scheme is given in [311].

For the time-dependent eddy current problem a FE/BE coupling with the discontinuous Galerkin method in time is established in [341] with a priori error estimates ; a posteriori error control and an adaptive algorithm are given in [342].

Chapter 13
Time-Domain BEM

Time-domain Galerkin boundary elements provide an efficient tool for numerical solution of boundary value problems for the homogeneous wave equation. In Sect. 13.1 we present from [193] a time-domain Galerkin BEM for the wave equation outside a Lipschitz obstacle in an absorbing half-space.A priori error estimates from [193] and a posteriori error estimates from [194] are given in Sect. 13.2.

Efficient and accurate computational methods to simulate sound emission in space and time are of interest from modeling environmental noise to the acoustics of concert halls [34, 94, 197]. This chapter reviews a time-domain Galerkin boundary element method for acoustic wave problems [193, 194]. Our approach proves to be stable and accurate in long-time computations and is competitive with frequency domain methods for realistic problems from the sound emission of tyres.

Computations in time-domain are of particular interest for problems beyond the reach of frequency-domain methods, such as the simulation of transient dynamics, moving sound sources or nonlinear and dynamical contact problems. They can also be applied to obtain results in frequency-domain, for all frequencies in one computation, with the help of the Fast Fourier Transform to translate between time and frequency. This approach proves competitive if a broad band of frequencies is of interest.

As an alternative to time-domain boundary elements, the past few years have seen rapid progress for convolution quadrature methods [30, 31, 361]. These exploit the convolution structure in time for integral equations to approximate them through the frequency domain by an inverse Laplace transform. Given a frequency domain solver, their implementation does not struggle with the careful, accurate computation of distributional integrals like time domain boundary elements do. However, for long time simulations and certain nonlinear problems with constraints, such as dynamic contact and friction problems, the variational nature of Galerkin time domain methods maybe advantageous.

© Springer International Publishing AG, part of Springer Nature 2018

J. Gwinner, E. P. Stephan, *Advanced Boundary Element Methods*,
Springer Series in Computational Mathematics 52,
https://doi.org/10.1007/978-3-319-92001-6_13

13.1 Integral Equations and Anisotropic Space-Time Sobolev Spaces

Let $d = 2$ or 3 and $\Omega^i \subset \mathbb{R}^d$ be a bounded polygonal domain. For simplicity, we assume that the exterior domain $\Omega^e = \mathbb{R}^d \backslash \overline{\Omega^i}$ is connected and that the boundary $\Gamma = \partial\Omega$ is a Lipschitz manifold. Our emphasis will be on the case $d = 3$.

We aim to find a weak solution to an initial-boundary problem for the wave equation in Ω^e:

$$\begin{aligned} \frac{\partial^2 u}{\partial t^2} - \Delta u = 0 \quad & \text{in } \mathbb{R}^+ \times \Omega^e \\ u(0,x) = \frac{\partial u}{\partial t}(0,x) = 0 \quad & \text{in } \Omega^e , \end{aligned} \tag{13.1}$$

with either Dirichlet boundary conditions $u = g$, Neumann boundary conditions $\frac{\partial u}{\partial n} = g$ or more generally acoustic boundary conditions

$$\frac{\partial u}{\partial n} - \alpha \frac{\partial u}{\partial t} = g \quad \text{on } \mathbb{R}^+ \times \Gamma .$$

Here n denotes the inward unit normal vector to $\partial\Omega^e$, g lies in a suitable Sobolev space, $\alpha \in L^\infty(\Gamma)$. In the case of an incoming wave u^{inc} scattered by Ω^i, the right hand side is $g = -\frac{\partial u^{inc}}{\partial n} + \alpha \frac{\partial u^{inc}}{\partial t}$. In order for (13.1) to be well-posed, α should have nonnegative real part, so that waves are not amplified at reflection. We also consider the simpler Dirichlet problem on Γ, for which instead of the absorbing boundary condition, $u|_{\mathbb{R}^+ \times \Gamma}$ is given.

This section reduces the acoustic and Dirichlet boundary problems to time-dependent integral equations on $\mathbb{R}^+ \times \Gamma$ and studies a Galerkin time-domain boundary element method for their approximation. It presents from [193, 194] an a priori and an a posteriori error analysis for methods based on integral formulations of the first kind.

Time-dependent Galerkin boundary element methods for wave problems were introduced by Bamberger and Ha-Duong [29]. Some relevant works on the numerical implementation of the resulting marching-in-on-time scheme include the Ph.D. thesis of Terrasse and [224]. For a survey see Costabel's article [117]

In the special case of the half-space, our work is motivated by the recent explicit formulas for the fundamental solutions obtained by Ochmann [329], which include acoustic boundary conditions on the surface of the street.

Similar to elliptic problems, the initial-boundary value problem (13.1) for the wave equation can be formulated as an integral equation of either the first or second kind on the boundary. Using an appropriate Green's function for the absorbing half-space an equation on the subset Γ of the boundary is obtained.

We introduce the single layer potential in time domain as

$$S\varphi(t,x) = \int_{\mathbb{R}^+\times\Gamma} G(t-\tau,x,y)\,\varphi(\tau,y)\,d\tau\,ds_y\,,$$

where

$$G(t-\tau,x,y) = \frac{1}{2\pi}\frac{H(t-\tau-|x-y|)}{\sqrt{(t-\tau)^2-|x-y|^2}} \quad (2D)$$

$$G(t-\tau,x,y) = \frac{1}{4\pi}\frac{\delta(t-\tau-|x-y|)}{|x-y|} \quad \text{3D}$$

is a fundamental solution to the wave equation with the Heaviside function H and the delta-distribution δ. Specifically in 3 dimensions, it is given by

$$S\varphi(t,x) = \frac{1}{4\pi}\int_\Gamma \frac{\varphi(t-|x-y|,y)}{|x-y|}\,ds_y\,.$$

We similarly define the double layer potential in time domain as

$$D\varphi(t,x) = \int_{\mathbb{R}^+\times\Gamma} \frac{\partial G}{\partial n_y}(t-\tau,x,y)\,\varphi(\tau,y)\,d\tau\,ds_y\,.$$

For acoustic boundary conditions we require the single-layer operator V, its normal derivative K', the double-layer operator K and hypersingular operator W for $x\in\Gamma$, $t>0$:

$$V\varphi(t,x) = 2\int_{\mathbb{R}^+\times\Gamma} G(t-\tau,x,y)\,\varphi(\tau,y)\,d\tau\,ds_y\,,$$

$$K\varphi(t,x) = 2\int_{\mathbb{R}^+\times\Gamma} \frac{\partial G}{\partial n_y}(t-\tau,x,y)\,\varphi(\tau,y)\,d\tau\,ds_y,$$

$$K'\varphi(t,x) = 2\int_{\mathbb{R}^+\times\Gamma} \frac{\partial G}{\partial n_x}(t-\tau,x,y)\,\varphi(\tau,y)\,d\tau\,ds_y\,,$$

$$W\varphi(t,x) = -2\int_{\mathbb{R}^+\times\Gamma} \frac{\partial^2 G}{\partial n_x\partial n_y}(t-\tau,x,y)\,\varphi(\tau,y)\,d\tau\,ds_y\,.$$

The boundary integral operators are considered between space-time anisotropic Sobolev spaces $H^s_\sigma(\mathbb{R}^+,\widetilde{H}^r(\Gamma))$. To define them, if $\partial\Gamma\neq\emptyset$, first extend Γ to a closed, orientable Lipschitz manifold $\widetilde{\Gamma}$.

On Γ one defines the usual Sobolev spaces of supported distributions:

$$\widetilde{H}^r(\Gamma) = \{u\in H^r(\widetilde{\Gamma}) : \operatorname{supp} u\subset\overline{\Gamma}\}\,,\qquad r\in\mathbb{R}\,.$$

Furthermore, $H^r(\Gamma)$ is the quotient space $H^r(\widetilde{\Gamma})/\widetilde{H}^r(\widetilde{\Gamma} \setminus \overline{\Gamma})$.

To write down an explicit family of Sobolev norms, introduce a partition of unity α_i subordinate to a covering of $\widetilde{\Gamma}$ by open sets B_i. For diffeomorphisms φ_i mapping each B_i into the unit cube $\subset \mathbb{R}^d$, a family of Sobolev norms is induced from $\mathbb{R}^d$:

$$||u||_{r,\omega,\widetilde{\Gamma}} = \left(\sum_{i=1}^{p} \int_{\mathbb{R}^d} (|\omega|^2 + |\xi|^2)^r |\mathscr{F} \left\{ (\alpha_i u) \circ \varphi_i^{-1} \right\} (\xi)|^2 d\xi \right)^{\frac{1}{2}} .$$

The norms for different $\omega \in \mathbb{C} \setminus \{0\}$ are equivalent, and $\mathscr{F}$ denotes the Fourier transform. They induce norms on $H^r(\Gamma)$, $||u||_{r,\omega,\Gamma} = \inf_{v \in \widetilde{H}^r(\widetilde{\Gamma} \setminus \overline{\Gamma})} ||u + v||_{r,\omega,\widetilde{\Gamma}}$, and on $\widetilde{H}^r(\Gamma)$, $||u||_{r,\omega,\Gamma,*} = ||e_+ u||_{r,\omega,\widetilde{\Gamma}}$. e_+ extends the distribution u by 0 from Γ to $\widetilde{\Gamma}$. It is stronger than $||u||_{r,\omega,\Gamma}$ whenever $r \in \frac{1}{2} + \mathbb{Z}$.

We now define a class of space-time anisotropic Sobolev spaces:

Definition 13.1 For $s, r \in \mathbb{R}$ define

$$H^s_\sigma(\mathbb{R}^+, H^r(\Gamma)) = \{u \in \mathscr{D}'_+(H^r(\Gamma)) : e^{-\sigma t} u \in \mathscr{S}'_+(H^r(\Gamma)) \text{ and } ||u||_{s,r,\Gamma} < \infty\} ,$$

$$H^s_\sigma(\mathbb{R}^+, \widetilde{H}^r(\Gamma)) = \{u \in \mathscr{D}'_+(\widetilde{H}^r(\Gamma)) : e^{-\sigma t} u \in \mathscr{S}'_+(\widetilde{H}^r(\Gamma)) \text{ and } ||u||_{s,r,\Gamma,*} < \infty\} .$$

$\mathscr{D}'_+(E)$ resp. $\mathscr{S}'_+(E)$ denote the spaces of distributions, resp. tempered distributions, on $\mathbb{R}$ with support in $[0, \infty)$, taking values in $E = H^r(\Gamma), \widetilde{H}^r(\Gamma)$. The relevant norms are given by

$$\|u\|_{s,r,\sigma} := \|u\|_{s,r,\Gamma} = \left(\int_{-\infty+i\sigma}^{+\infty+i\sigma} |\omega|^{2s} \, \|\hat{u}(\omega)\|^2_{r,\omega,\Gamma} \, d\omega \right)^{\frac{1}{2}} ,$$

$$\|u\|_{s,r,\sigma,*} := \|u\|_{s,r,\Gamma,*} = \left(\int_{-\infty+i\sigma}^{+\infty+i\sigma} |\omega|^{2s} \, \|\hat{u}(\omega)\|^2_{r,\omega,\Gamma,*} \, d\omega \right)^{\frac{1}{2}} .$$

For $|r| \leq 1$ the spaces are independent of the choice of α_i and φ_i. See [193, 223] for a more detailed discussion.

The representation formula uses S and D to express a solution to the wave equation in terms of its Dirichlet and Neumann data on Γ:

Theorem 13.1 *Let $u \in L^2(\mathbb{R}^+, H^1(\Omega)) \cap H^1_0(\mathbb{R}^+, L^2(\Omega))$ be the solution of* (13.1) *for a Lipschitz boundary Γ. Then*

$$u(t, x) = S\varphi(t, x) - Dp(t, x) ,$$

where $\varphi = [u]$ is the jump of u across Γ and $p = [\frac{\partial u}{\partial n}]$ is the jump of the normal flux.

The initial boundary value problem (13.1) with acoustic boundary conditions is then equivalent to a system of integral equations of the first kind,

$$\begin{cases} K'p - W\varphi + \alpha\frac{\partial\varphi}{\partial t} = F \\ p + \alpha(V\partial_t p + K\partial_t\varphi) = G\,. \end{cases} \tag{13.2}$$

Here, $\varphi = [u]$ and $p = [\frac{\partial u}{\partial n}]$ as above, and for an incoming wave u^{inc} scattered by Ω^i, we have $F = -2\frac{\partial u^{inc}}{\partial n}$ and $G = -2\alpha\frac{\partial u^{inc}}{\partial t}$. If $\alpha^{-1} \in L^\infty(\Gamma)$, pairing these equations with test functions $\partial_t\psi$ respectively $\frac{q}{\alpha}$, we obtain the following space-time variational formulation:

Find $\Phi = (\varphi, p) \in H^1_\sigma(\mathbb{R}^+, \widetilde{H}^{\frac{1}{2}}(\Gamma)) \times H^1_\sigma(\mathbb{R}^+, L^2(\Gamma))$ such that for all $\Psi = (\psi, q) \in H^1_\sigma(\mathbb{R}^+, \widetilde{H}^{\frac{1}{2}}(\Gamma)) \times H^1_\sigma(\mathbb{R}^+, L^2(\Gamma))$:

$$a(\Phi, \Psi) = l(\Psi)\,. \tag{13.3}$$

where

$$l(\Psi) = \int_0^\infty \int_\Gamma F\partial_t\psi\, ds_x\, d_\sigma t + \int_0^\infty \int_\Gamma \frac{Gq}{\alpha}\, ds_x\, d_\sigma t \tag{13.4}$$

and $a(\Phi, \Psi)$ is given by

$$\iint_{0\,\Gamma}^{\infty} \left(\alpha(\partial_t\varphi)(\partial_t\psi) + \frac{1}{\alpha}pq + K'p(\partial_t\psi) - W\varphi(\partial_t\psi) + V(\partial_t p)q + K(\partial_t\varphi)q\right) ds_x\, d_\sigma t. \tag{13.5}$$

Here $d_\sigma t = e^{-2\sigma t}dt$, $\sigma > 0$, and $\langle u, v\rangle := \int_0^\infty \int_\Gamma u\bar{v}ds_x d_\sigma t$.

The variational formulation of the Dirichlet problem, $V\partial_t\phi = \partial_t f$, similarly reads: Find $\phi \in H^1_\sigma(\mathbb{R}^+, \widetilde{H}^{-\frac{1}{2}}(\Gamma))$ such that

$$b(\phi, \psi) = \langle \partial_t f, \psi\rangle \qquad \forall\psi \in H^1_\sigma(\mathbb{R}^+, \widetilde{H}^{-\frac{1}{2}}(\Gamma))\,, \tag{13.6}$$

where

$$b(\phi, \psi) = \int_0^\infty \int_\Gamma (V\partial_t\phi(t, x))\psi(t, x) ds_x\, d_\sigma t\,,$$

$$\langle \partial_t f, \psi\rangle = \int_0^\infty \int_\Gamma (\partial_t f(t, x))\psi(t, x) ds_x\, d_\sigma t\,.$$

Adapting fundamental observations in [29] and [223] to our situation, the bilinear forms $a(\Phi, \Psi)$ and $b(\phi, \psi)$ are continous and, in a weak sense, coercive. They are related to the physical energy of the system. As a consequence, both the acoustic and the Dirichlet problem admit unique solutions for sufficiently smooth data. See [193] for details.

The Neumann problem, corresponding to $\alpha = 0$, may be discussed similarly [34, 195]. In addition to the variational formulations as integral equations of the first kind, related to the energy, for computations an integral equation of the second kind will prove useful. We will only state the Neumann case, $\alpha = 0$: Find $\varphi(t,x) \in H_\sigma^{\frac{1}{2}}([0,\infty), H^{-\frac{1}{2}}(\Gamma))$ such that for all test functions $\psi(t,x) \in H_\sigma^{\frac{1}{2}}([0,\infty), H^{-\frac{1}{2}}(\Gamma))$ there holds:

$$\int_0^\infty \int_\Gamma \left(-I + K'\right) \varphi(t,x)\psi(t,x)\, ds_x\, d_\sigma t = 2\int_0^\infty \int_\Gamma g(t,x)\psi(t,x)\, ds_x\, d_\sigma t. \tag{13.7}$$

As it is equivalent to the original initial boundary value problem, also this formulation admits a unique solution for smooth right hand sides, though it is not known to be coercive.

For applications to traffic noise, also the wave equation in the half-space $\mathbb{R}^3_+$ is of interest [34]. By choosing an appropriate, modified Green's function G which satisfies the boundary conditions, the formulation may be reduced to the surface Γ of the scatterer. Partially absorbing, acoustic boundary conditions on the road $\partial\mathbb{R}^3_+$, using a modification of K' have been discussed in [193, 329].

We now discuss the discrete spaces used for the numerical approximation of the weak formulations (13.3), (13.6) and (13.7). If Γ is not polygonal we approximate it by a piecewise polygonal curve resp. surface and write Γ again for the approximation. For simplicity, when $d = 3$ we will use here a surface composed of N triangular facets Γ_i such that $\Gamma = \cup_{i=1}^N \Gamma_i$. When $d = 2$, we assume $\Gamma = \cup_{i=1}^N \Gamma_i$ is composed of line segments Γ_i. In each case, the elements Γ_i are closed with $int(\Gamma_i) \neq \varnothing$, and for distinct $\Gamma_i,\ \Gamma_j \subset \Gamma$ the intersection $int(\Gamma_i) \cap int(\Gamma_j) = \varnothing$.

For the time discretisation we consider a uniform decomposition of the time interval $[0,\infty)$ into subintervals $I_n = [t_{n-1}, t_n)$ with time step $|I_n| = \Delta t$, such that $t_n = n\Delta t$ $(n = 0, 1, \dots)$.

We choose a basis $\varphi_1^p, \cdots, \varphi_{N_s}^p$ of the space V_h^p of piecewise polynomial functions of degree p in space (continuous and vanishing at $\partial\Gamma$ if $p \geq 1$) and a basis $\beta^{1,q}, \cdots, \beta^{N_t,q}$ of the space $V_{\Delta t}^q$ of piecewise polynomial functions of degree of q in time (continuous and vanishing at $t = 0$ if $q \geq 1$).

Let $\mathcal{T}_S = \{T_1, \cdots, T_{N_s}\}$ be the spatial mesh for Γ and $\mathcal{T}_T = \{[0, t_1), [t_1, t_2), \cdots, [t_{N_t-1}, T)\}$ the time mesh for a finite subinterval $[0, T)$.

We consider the tensor product of the approximation spaces in space and time, V_h^p and $V_{\Delta t}^q$, associated to the space-time mesh $\mathcal{T}_{S,T} = \mathcal{T}_S \times \mathcal{T}_T$, and we write

$$V_{\Delta t,h}^{p,q} = V_h^p \otimes V_{\Delta t}^q\,.$$

13.2 A Priori and A Posteriori Error Estimates

The approximation spaces lead to Galerkin formulations for the acoustic and Dirichlet problems (13.3), (13.6) and (13.7). E.g. the Galerkin formulation of (13.6) reads: Find $\phi_{\Delta t,h} \in V^{p,q}_{\Delta t,h}$ such that

$$b(\phi_{\Delta t,h}, \psi_{\Delta t,h}) = \langle (\partial_t f)_{\Delta t,h}, \psi_{\Delta t,h} \rangle \qquad \forall \psi_{\Delta t,h} \in V^{p,q}_{\Delta t,h}\,. \tag{13.8}$$

The well-posedness of the continuous and discretized problems is a basic consequence of the continuity and weak coercivity of the bilinear form b:

Corollary 13.1 *Let $f \in H^1_\sigma(\mathbb{R}^+, H^{\frac{1}{2}}(\Gamma))$. Then the Dirichlet problem* (13.6) *and its discretization* (13.8) *admit unique solutions $\phi, \phi_{\Delta t,h} \in H^0_\sigma(\mathbb{R}^+, H^{-\frac{1}{2}}(\Gamma))$ and $\|\phi\|_{0,-\frac{1}{2},\Gamma,*}, \|\phi_{\Delta t,h}\|_{0,-\frac{1}{2},\Gamma,*} \lesssim \|f\|_{1,\frac{1}{2},\Gamma}$.*

In [193], we discuss a priori error estimates and the convergence of Galerkin approximations for (13.3) and (13.6). For the Dirichlet problem the basic estimate is the following:

Theorem 13.2 ([193]) *For the solutions $\phi \in H^1_\sigma(\mathbb{R}^+, \widetilde{H}^{-\frac{1}{2}}(\Gamma))$ of* (13.6), *$\phi_{\Delta t,h} \in \widetilde{V}^{p,q}_{\Delta t,h}$ of* (13.8) *there holds:*

$$\|\phi - \phi_{\Delta t,h}\|_{0,-\frac{1}{2},\Gamma,*} \lesssim \|(\partial_t f)_{\Delta t,h} - \partial_t f\|_{0,\frac{1}{2},\Gamma}$$
$$+ \inf_{\psi_{\Delta t,h} \in \widetilde{V}^{p,q}_{\Delta t,h}} \left\{ (1 + \frac{1}{\Delta t}) \|\phi - \psi_{\Delta t,h}\|_{0,-\frac{1}{2},\Gamma,*} + \frac{1}{\Delta t} \|\partial_t \phi - \partial_t \psi_{\Delta t,h}\|_{0,-\frac{1}{2},\Gamma,*} \right\}.$$

If in addition $\phi \in H^s_\sigma(\mathbb{R}^+, H^m(\Gamma))$ for $s > 1$ and $m > -\frac{1}{2}$, one obtains convergence rates.

For the proof, we assume for simplicity that $\partial_t f = (\partial_t f)_{\Delta t,h}$. Then from the weak coercivtiy of the bilinear form b, and adding a 0, we have

$$\|\phi - \phi_{\Delta t,h}\|^2_{0,-\frac{1}{2},\Gamma,*} \lesssim b(\phi_{\Delta t,h} - \phi, \phi_{\Delta t,h} - \psi_{\Delta t,h}) + b(\phi - \psi_{\Delta t,h}, \phi_{\Delta t,h} - \psi_{\Delta t,h})$$

for all test functions $\psi_{\Delta t,h}$. Using the Galerkin orthogonality

$$b(\phi_{\Delta t,h} - \phi, \phi_{\Delta t,h} - \psi_{\Delta t,h}) = 0\,,$$

the first term vanishes. For the second, we use the continuity of the duality pairing, the mapping properties of $\mathscr{V}$ and an inverse estimate in t:

$$\begin{aligned} b(\phi - \psi_{\Delta t,h}, \phi_{\Delta t,h} - \psi_{\Delta t,h}) &\leq \|\mathscr{V} \frac{\partial}{\partial t}(\phi - \psi_{\Delta t,h})\|_{-1,\frac{1}{2},\Gamma} \|\phi_{\Delta t,h} - \psi_{\Delta t,h}\|_{1,-\frac{1}{2},\Gamma,*} \\ &\lesssim \|\phi - \psi_{\Delta t,h}\|_{1,-\frac{1}{2},\Gamma,*} \|\phi_{\Delta t,h} - \psi_{\Delta t,h}\|_{1,-\frac{1}{2},\Gamma,*} \\ &\lesssim \frac{1}{\Delta t} \|\phi_{\Delta t,h} - \psi_{\Delta t,h}\|_{0,-\frac{1}{2},\Gamma,*} \|\phi - \psi_{\Delta t,h}\|_{1,-\frac{1}{2},\Gamma,*} \end{aligned}$$

Combining this with a triangle inequality, we obtain the claimed a priori bound:

$$\begin{aligned}\|\phi - \phi_{\Delta t,h}\|_{0,-\frac{1}{2},\Gamma,*} &\lesssim \|\phi - \psi_{\Delta t,h}\|_{0,-\frac{1}{2},\Gamma,*} + \|\phi_{\Delta t,h} - \psi_{\Delta t,h}\|_{0,-\frac{1}{2},\Gamma,*} \\ &\lesssim \left(1 + \frac{1}{\Delta t}\right) \|\phi - \psi_{\Delta t,h}\|_{1,-\frac{1}{2},\Gamma,*} \,.\end{aligned}$$

For the acoustic problem, we introduce the norm

$$|||p, \varphi||| = \left(||p||^2_{0,0,\Gamma} + ||\varphi||^2_{0,\frac{1}{2},\Gamma} + ||\partial_t \varphi||^2_{0,0,\Gamma}\right)^{\frac{1}{2}} .$$

Theorem 13.3 ([193]) *Assume (for simplicity) that* $\frac{1}{\alpha} \in L^\infty(\Gamma)$. *For the solutions* $\Phi = (p, \varphi) \in H^1_\sigma(\mathbb{R}^+, \widetilde{H}^{\frac{1}{2}}(\Gamma)) \times H^1_\sigma(\mathbb{R}^+, L^2(\Gamma))$ *of* (13.3) *and* $\Phi_{\Delta t,h} = (p_{\Delta t,h}, \varphi_{\Delta t,h}) \in V^{\tilde{p},\tilde{q}}_{\Delta t,h} \times V^{p,q}_{\Delta t,h}$ *of its discretisation there holds:*

$$\begin{aligned}&|||p - p_{\Delta t,h}, \varphi - \varphi_{\Delta t,h}||| \lesssim ||F_{\Delta t,h} - F||_{0,0,\Gamma} + ||G_{\Delta t,h} - G||_{0,0,\Gamma} \\ &+ \max\left(\frac{1}{\Delta t}, \frac{1}{\sqrt{h}}\right) \inf_{(q_{\Delta t,h}, \psi_{\Delta t,h}) \in V^{\tilde{p},\tilde{q}}_{\Delta t,h} \times V^{p,q}_{\Delta t,h}} \left(||p - q_{\Delta t,h}||_{1,0,\Gamma} + ||\varphi - \psi_{\Delta t,h}||_{1,\frac{1}{2},\Gamma}\right) .\end{aligned}$$

As for the Dirichlet problem, better estimates are obtained under smoothness assumptions, $\varphi \in H^{s_1}_\sigma(\mathbb{R}^+, H^{m_1}(\Gamma))$, $p \in H^{s_2}_\sigma(\mathbb{R}^+, H^{m_2}(\Gamma))$, [193].

Computable error indicators are a key ingredient to design adaptive mesh refinements. For the time-dependent boundary element methods efficient and reliable such estimates of residual type have been obtained in [194], see also [197] and [358] for alternative error indicators.

Using ideas going back to Carstensen [74] and Carstensen and Stephan [92] for the boundary element method for elliptic problems (see Section 10.1 and 10.2), we obtain an a posteriori error estimate with residual error estimator for the Galerkin solution to the Dirichlet problem in [194].

Theorem 13.4 *Let* $\phi, \phi_{\Delta t,h} \in H^1_\sigma(\mathbb{R}^+, H^{-\frac{1}{2}}(\Gamma))$ *the solutions to* (13.6) *resp.* (13.8). *Assume that* $R = \partial_t f - V\partial_t \phi_{\Delta t,h} \in H^0_\sigma(\mathbb{R}^+, H^1(\Gamma))$. *Then*

$$\begin{aligned}\|\phi - \phi_{\Delta t,h}\|^2_{0,-\frac{1}{2},\Gamma,*} &\lesssim \|R\|_{0,1,\Gamma}\big(\Delta t \|\partial_t R\|_{0,0,\Gamma} + \|h \cdot \nabla R\|_{0,0,\Gamma}\big) \\ &\lesssim \max\{\Delta t, h\}(\|\partial_t R\|_{0,0,\Gamma} + \|\nabla R\|_{0,0,\Gamma})^2\end{aligned}$$

Proof We first note that for all $\psi_{\Delta t,h} \in V^{p,q}_{\Delta t,h}$

$$\begin{aligned}\|\phi - \phi_{\Delta t,h}\|^2_{0,-\frac{1}{2},\Gamma,*} &\lesssim b(\phi - \phi_{\Delta t,h}, \phi - \phi_{\Delta t,h})\\ &= \int_{\mathbb{R}^+}\int_\Gamma \partial_t f(\phi - \phi_{\Delta t,h})\, ds_x\, d_\sigma t - b(\phi_{\Delta t,h}, \phi - \phi_{\Delta t,h})\\ &= \int_{\mathbb{R}^+}\int_\Gamma \partial_t f(\phi - \psi_{\Delta t,h})\, ds_x\, d_\sigma t - b(\phi_{\Delta t,h}, \phi - \psi_{\Delta t,h})\\ &= \int_{\mathbb{R}^+}\int_\Gamma (\partial_t f - V\partial_t \phi_{\Delta t,h})(\phi - \psi_{\Delta t,h})\, ds_x\, d_\sigma t\ .\end{aligned}$$

The last term may be estimated by:

$$\begin{aligned}&\int_{\mathbb{R}^+}\int_\Gamma (\partial_t f - V\dot{\phi}_{\Delta t,h})(\phi - \psi_{\Delta t,h})\, ds_x\, d_\sigma t\\ &\leq \|R\|_{0,\frac{1}{2},\Gamma}\|\phi - \psi_{\Delta t,h}\|_{0,-\frac{1}{2},\Gamma,*}\ .\end{aligned}$$

We use $\psi_{\Delta t,h} = \phi_{\Delta t,h}$ together with the interpolation inequality

$$\|R\|^2_{0,\frac{1}{2},\Gamma} \leq \|R\|_{0,0,\Gamma}\|R\|_{0,1,\Gamma}\ .$$

As the residual is perpendicular to $V^{p,q}_{\Delta t,h}$,

$$\begin{aligned}\|R\|^2_{0,0,\Gamma} = \langle R, R\rangle &= \langle R, R - \widetilde{\psi}_{\Delta t,h}\rangle\\ &\leq \|R\|_{0,0,\Gamma}\|R - \widetilde{\psi}_{\Delta t,h}\|_{0,0,\Gamma}\end{aligned}$$

for all $\widetilde{\psi}_{\Delta t,h} \in V^{p,q}_{\Delta t,h}$, we obtain

$$\|R\|_{0,0,\Gamma} \leq \inf\{\|R - \widetilde{\psi}_{\Delta t,h}\|_{0,0,\Gamma} : \widetilde{\psi}_{\Delta t,h} \in V^{p,q}_{\Delta t,h}\}\ .$$

Choosing $\widetilde{\psi}_{\Delta t,h} = \widetilde{\Pi}_{\Delta t,h}R$, based on the interpolation operator defined earlier, we obtain

$$\|R\|_{0,0,\Gamma} \lesssim \Delta t\|\partial_t R\|_{0,0,\Gamma} + \|h\cdot\nabla R\|_{0,0,\Gamma}\ .$$

The theorem follows. □

The result for the single layer potential generalizes to a theorem without any assumptions on the underlying meshes.

Theorem 13.5 *Let $\phi \in H^0_\sigma(\mathbb{R}^+, H^{-\frac{1}{2}}(\Gamma))$ be the solution to (13.6), and let $\phi_{h,\Delta t} \in H^0_\sigma(\mathbb{R}^+, H^{-\frac{1}{2}}(\Gamma))$ such that $\mathscr{R} = \partial_t f - \mathscr{V}\partial_t\phi_{h,\Delta t} \in H^0_\sigma(\mathbb{R}^+, H^1(\Gamma))$. Then*

$$\|\phi - \phi_{h,\Delta t}\|^2_{0,-\frac{1}{2},\Gamma,*} \lesssim \sum_{i,\Delta}\max\{\Delta t, h_\Delta\}\ \|\mathscr{R}\|^2_{0,1,[t_i,t_{i+1})\times\Delta}\ .$$

Because of the different norms in the upper and lower bounds for b, the a posteriori estimate only satisfies a weak variant of efficiency: For $\varepsilon \in (0, 1)$:

$$\max\{\Delta t, h\}^{-\frac{1-\varepsilon}{2}} \, \|\phi - \phi_{\Delta t,h}\|_{0,-\frac{1}{2},\Gamma} \lesssim \|R\|_{0,1-\varepsilon,\Gamma} = \|\mathcal{V}(\dot{\phi} - \dot{\phi}_{\Delta t,h})\|_{0,1-\varepsilon,\Gamma}$$
$$\lesssim \|\phi - \phi_{\Delta t,h}\|_{2,-\varepsilon,\Gamma} \le \|\phi - \phi_{\Delta t,h}\|_{2,0,\Gamma} \,.$$

A proof of the sharp estimate, $\varepsilon = 0$, would require sharp mapping properties of the layer potentials outside the energy spaces.

One then uses the mapping properties of V together with approximation properties of the finite element spaces to recover the same spatial Sobolev index $-\frac{1}{2}$ in the upper and lower estimates.

Theorem 13.6 ([194]) *Assume that the* $R \in H^0([0, T], H^1(\Gamma))$ *and that the ansatz functions* $V^{p,q}_{\Delta t,h} \subseteq H^2([0, T], H^0(\Gamma))$ *satisfy*

$$\inf_{\psi_{h\Delta t} \in V^{p,q}_{\Delta t,h}} \|\phi - \psi_{h\Delta t}\|_{2,0,\Gamma,*} \simeq \max\{\Delta t, h\}^{\beta} \tag{13.9}$$

for some $\beta > 0$. *Then for all* $\varepsilon \in (0, 1)$

$$\|R\|_{0,1-\varepsilon,\Gamma} \lesssim \max\{h^{-\frac{1}{2}}, (\Delta t)^{-\frac{1}{2}}\} \|\phi - \phi_{h\Delta t}\|_{2,-1/2,\Gamma,*}.$$

Remark 13.1 The hypothesis (13.9) can be verified using the singular expansion of the solution ϕ at the edges and corners [194].

For the acoustic problem, a simple error estimate reads as follows:

Theorem 13.7 ([194]) *Let* $(\varphi, p), (\varphi_{\Delta t,h}, p_{\Delta t,h}) \in H^1_0([0, T], H^{\frac{1}{2}}(\Gamma)) \times H^1([0, T], L^2(\Gamma))$ *be the solutions to* (13.3) *and its discretized variant, and assume that*

$$R_1 = F - \alpha\dot{\varphi}_{\Delta t,h} + 2K' p_{\Delta t,h} - 2W\varphi_{\Delta t,h} \in L^2([0, T], L^2(\Gamma)) \,,$$
$$R_2 = G + \alpha^{-1} p_{\Delta t,h} + 2S\dot{p}_{\Delta t,h} - 2K\dot{\varphi}_{\Delta t,h} \in L^2([0, T], L^2(\Gamma)) \,.$$

Then

$$|||p - p_{\Delta t,h}, \varphi - \varphi_{\Delta t,h}||| \lesssim \|R_1\|_{0,0,\Gamma} + \|R_2\|_{0,0,\Gamma} \,.$$

In [195] the Neumann problem is solved with a double layer potential ansatz leading to the hyper singular integral equation and corresponding a priori and a posteriori error estimates are given.

13.2.1 Adaptive Mesh Refinements

Space-time adaptive methods are still in their infancy. As a test case in [194] we concentrate on time-independent geometric singularities of the solution, e.g. in the horn geometry between the tyre and the street. In this case we expect to have time-independent meshes, refined near the singularities, which do not require an update of the Galerkin matrices in every time step.

From the discrete solution $\dot{\varphi}_{\Delta t,h}$ of the Dirichlet problem (13.8) and $\dot{f}$ we determine in every triangle $\triangle$ the time integrated local error indicator

$$\eta_\triangle^2 = \int_0^T \int_\triangle [h\nabla_\Gamma(\dot{f} - V\dot{\varphi}_{\Delta t,h})]^2 \,,$$

where the time integral is approximated by a Riemann sum.

The error indicators $\eta_\triangle$ lead to an adaptive algorithm, based on the 4 steps

$$\textbf{SOLVE} \longrightarrow \textbf{ESTIMATE} \longrightarrow \textbf{MARK} \longrightarrow \textbf{REFINE}.$$

Adaptive Algorithm [194]:
Input: Mesh $\mathscr{T} = \mathscr{T}_0$, refinement parameter $\theta \in (0, 1)$, tolerance $\epsilon > 0$, data f.

 (i) Solve $V\dot{\varphi}_{\Delta t,h} = \dot{f}$ on $\mathscr{T}$.
 (ii) Compute the error indicators $\eta(\triangle)$ in each triangle $\triangle \in \mathscr{T}$.
 (iii) Find $\eta_{max} = max_\triangle \eta(\triangle)$.
 (iv) Stop if $\sum_i \eta^2(\triangle_i) < \epsilon^2$.
 (v) Mark all $\triangle \in \mathscr{T}$ with $\eta(\triangle_i) > \theta\eta_{max}$.
 (vi) Refine each marked triangle into 4 new triangles to obtain a new mesh $\mathscr{T}$ (and project the new nodes onto the sphere). Choose Δt such that $\frac{\Delta t}{\Delta x} \leq 1$ for all traingles.
 (vii) Go to 1.

Output: Approximation of $\dot{\varphi}$.

Example We consider the single layer integral equation $\mathscr{V}\phi = f$ on the square screen $\Gamma = [-0.5, 0.5]^2 \times \{0\}$ with right hand side $f(t, x, y, z) = \sin(t)^5 x^2$ for times $[0, 2.5]$. Using a discretization by linear ansatz and test functions in space and time, we compare the error of a uniform discetization to the error of an adaptive series of meshes, steered by the residual error estimate. The time step is $\Delta t = 0.1$, and the uniform meshes consist of 18, 288, 648, 1352, and 6050 triangles, while the adaptive refinements correspond to 36, 74, 164, 370, 784, 1676, 3485, and 7432 triangles. Figure 13.1 shows the convergence in of the indicators and the energy error, for both the uniform and adaptive series of meshes.

Figure 13.2 shows some representative adaptive meshes, where the color scale highlights the residual-based indicator values for each element. Mesh refinements

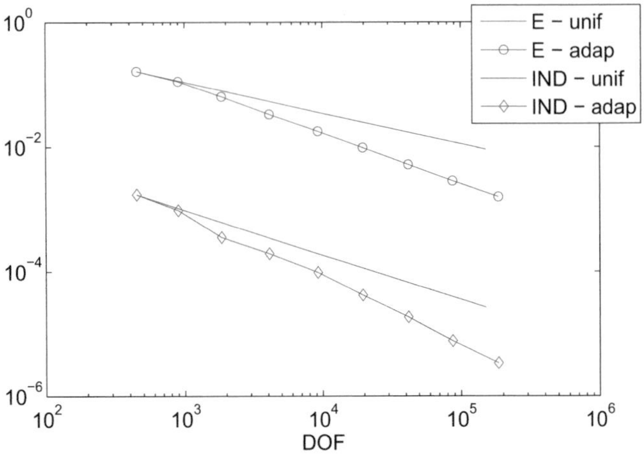

Fig. 13.1 Energy error and residual error indicators for Dirichlet problem on $\Gamma = [-0.5, 0.5]^2 \times \{0\}$ [194]

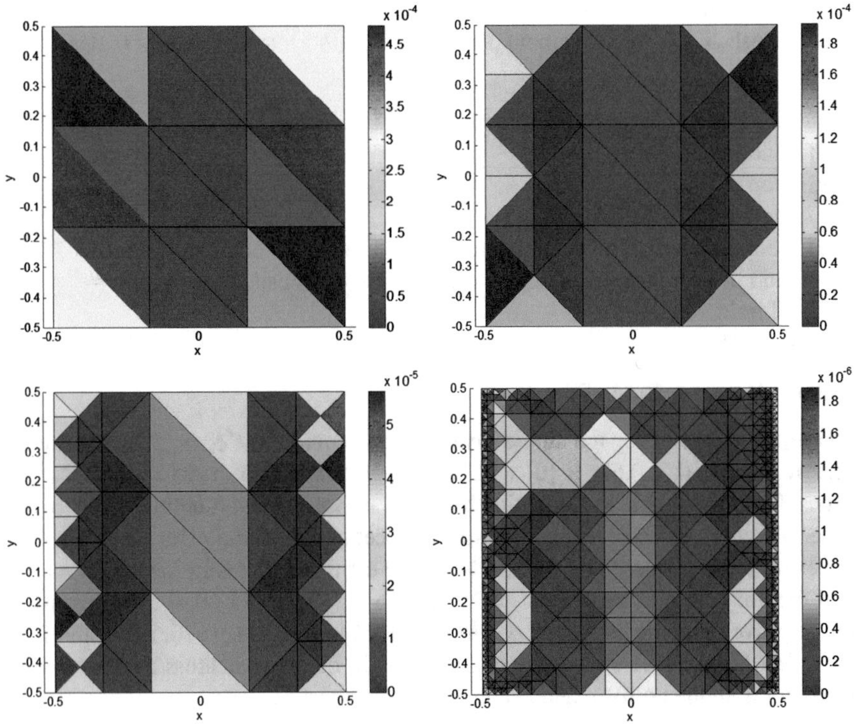

Fig. 13.2 Meshes 1, 2, 3 and 6 generated by adaptive refinements [194]

concentrate at the left and right edges, where the right hand side is steep, and to a lesser extent also at the top and bottom edges.

13.3 Time Domain BEM for Contact Problems

The previous section considered the use of time-independent adapted (graded) meshes to resolve the geometric singularities of the solution to the wave equation near edges and corners. In this section we discuss a class of problems where the singularities are moving in space and time, motivating future developments of space-time adaptive boundary element procedures. For the contact problems under consideration, the free contact boundary is changing with time, and so is the location of its associated edge and corner singularities.

In [192] we propose a time domain boundary element method for a dynamical contact problem for the scalar wave equation and provide a priori error estimates in the case of a flat contact area. For the (scalar) displacement $w : \Omega \mapsto \mathbb{R}$ contact conditions along a crack G with a non-penetrable material in a Lipschitz domain $\Omega \subset \mathbb{R}^n$ are described in terms of the traction $-\mu \frac{\partial w}{\partial \nu}|_G$ (with normal ν) and prescribed forces h:

$$\begin{cases} w|_G \geq 0\,, \; -\mu \frac{\partial w}{\partial \nu}|_G \geq h \\ w|_G > 0 \Rightarrow \mu \frac{\partial w}{\partial \nu} = h. \end{cases}$$

The full system of equations for the dynamical contact problem is given by:

$$\begin{cases} \frac{\partial^2 w}{\partial t^2} = c_s^2 \Delta w & \text{for } (t,x) \in \mathbb{R} \times \Omega \\ w = 0 & \text{on } \Gamma \backslash G \; (\Gamma = \partial \Omega) \\ w \geq 0\,, \; -mu \frac{\partial w}{\partial \nu} \geq 0 & \text{on } G \\ (-\mu \frac{\partial w}{\partial \nu} - h) w = 0 & \text{on } G \\ w = 0 & \text{for } (t,x) \in (-\infty, 0) \times \Omega\,. \end{cases} \tag{13.10}$$

In [192] we reduce (13.10) to an equivalent variational inequality (in a weak form) for the trace $u_\sigma = w_\sigma|_\Gamma$, $w_\sigma := e^{-\sigma t} w$, $h_\sigma := e^{-\sigma t} h$ (for some $\sigma > 0$) on the boundary: Find $u_\sigma \in H_\sigma^{1/2}(\mathbb{R}^+, \tilde{H}^{1/2}(G))$ such that $u_\sigma \geq 0$ and for all $v \in H_\sigma^{1/2}(\mathbb{R}^+, \tilde{H}^{1/2}(G))$ with $v \geq 0$:

$$\langle p_Q S_\sigma u_\sigma, v - u_\sigma \rangle \geq \langle h_\sigma, v - u_\sigma \rangle\,. \tag{13.11}$$

Here p_Q is the restriction to $Q = \mathbb{R} \times G$ and S_σ is the Dirichlet-to-Neumann operator, defined as

$$S_\sigma w_\sigma|_\Gamma := -\mu \frac{\partial w}{\partial \nu}|_\Gamma\,.$$

The discretized variational inequality then reads:

Find $u_{\Delta t,h} \in K^+_{\Delta t,h}$ such that

$$\langle p_Q S_\sigma u_{\Delta t,h}, v_{\Delta t,h} - u_{\Delta t,h}\rangle \geq \langle h, v_{\Delta t,h} - u_{\Delta t,h}\rangle \tag{13.12}$$

for all $v_{\Delta t,h} \in \tilde{K}^+_{\Delta t,h}$. Here, $\tilde{K}^+_{\Delta t,h} \subset \tilde{V}^{p,q}_{\Delta t,h}$ is the convex subset of nonnegative piecewise polynomials.

There holds the following a priori error estimate:

Theorem 13.8 *Let $h \in H^{\frac{3}{2}}_\sigma(\mathbb{R}^+, H^{-\frac{1}{2}}(G))$ and let $u \in H^{\frac{1}{2}}_\sigma(\mathbb{R}^+, \tilde{H}^{\frac{1}{2}}(G))^+$, $u_{\Delta t,h} \in \tilde{K}^+_{\Delta t,h} \subset H^{\frac{1}{2}}_\sigma(\mathbb{R}^+, \tilde{H}^{\frac{1}{2}}(G))^+$ be the solutions of (13.11), resp. (13.12). Then the following estimate holds:*

$$\|u_\sigma - u_{\Delta t,h}\|^2_{-\frac{1}{2},\frac{1}{2},\sigma,*} \lesssim_\sigma \tag{13.13}$$

$$\inf_{\phi_{\Delta t,h}\in\tilde{K}^+_{\Delta t,h}} (\|h - p_Q S_\sigma u_\sigma\|_{\frac{1}{2},-\frac{1}{2},\sigma}\|u_\sigma - \phi_{\Delta t,h}\|_{-\frac{1}{2},\frac{1}{2},\sigma,*} + \|u_\sigma - \phi_{\Delta t,h}\|^2_{\frac{1}{2},\frac{1}{2},\sigma,*}).$$

To assure conservation of energy in the numerical approximation, it proves useful to impose the constraints on the displacement only indirectly. We therefore reformulate the variational inequality as an equivalent mixed system. The Lagrange multiplier λ in this formulation also provides a measure to which extent the variational inequality is not an equality; physically, it indicates the contact area and the contact forces within the computational domain.

The variational inequality (13.11) is equivalent to the mixed formulation:

Find $(u, \lambda) \in X = H^{1/2}_\sigma(\mathbb{R}^+, \tilde{H}^{1/2}(G)) \times H^{1/2}_\sigma(\mathbb{R}^+, H^{-1/2}(G)), \lambda \geq 0$, such that

$$\begin{cases} \langle S_\sigma u, v\rangle - \langle \lambda, v\rangle = \langle h, v\rangle \\ \langle u, \mu - \lambda\rangle \geq 0 \end{cases} \tag{13.14}$$

holds for all $(v, \mu) \in X, \mu \geq 0$.

The corresponding discrete formulation with different meshes for the displacement and the Lagrange multiplier reads as follows:

Find $(u_{\Delta t,h}, \lambda_{\Delta t_2,h_2}) \in \tilde{V}^{1,1}_{t,h} \times (V^{0,0}_{t_2,h_2})^+$ such that

$$\begin{cases} \langle S_\sigma u_{\Delta t,h}, v_{\Delta t,h}\rangle - \langle \lambda_{\Delta t_2,h_2,v_{\Delta t,h}}\rangle = \langle h, v_{\Delta t,h}\rangle \\ \langle u_{\Delta t,h}, \mu_{\Delta t_2,h_2} - \lambda_{\Delta t_2,h_2}\rangle \geq 0 \end{cases} \tag{13.15}$$

holds for all $(v_{\Delta t,h}, \mu_{\Delta t_2,h_2}) \in \tilde{V}^{1,1}_{t,h} \times (V^{0,0}_{t_2,h_2})^+$.

Theorem 13.9 ([192]) *Let $C > 0$ sufficiently small, and $\frac{\max\{h_1,\Delta t_1\}}{\min\{h_2,\Delta t_2\}} < C$. Then the discrete mixed formulation* (13.15) *admits a unique solution, and the following a* *priori estimates hold:*

$$\|\lambda-\lambda_{\Delta t_2,h_2}\|_{0,-\frac{1}{2},\sigma} \lesssim \inf_{\tilde{\lambda}_{\Delta t_2,h_2}} \|\lambda-\tilde{\lambda}_{\Delta t_2,h_2}\|_{0,-\frac{1}{2},\sigma} + (\Delta t_1)^{-\frac{1}{2}} \|u-u_{\Delta t_1,h_1}\|_{-\frac{1}{2},\frac{1}{2},\sigma,*}, \tag{13.16}$$

$$\begin{aligned}\|u-u_{\Delta t_1,h_1}\|_{-\frac{1}{2},\frac{1}{2},\sigma,*} \lesssim_\sigma & \inf_{v_{\Delta t_1,h_1}} \|u-v_{\Delta t_1,h_1}\|_{\frac{1}{2},\frac{1}{2},\sigma,*} \\ & + \inf_{\tilde{\lambda}_{\Delta t_2,h_2}} \left\{\|\tilde{\lambda}_{\Delta t_2,h_2}-\lambda\|_{\frac{1}{2},-\frac{1}{2},\sigma} + \|\tilde{\lambda}_{\Delta t_2,h_2}-\lambda_{\Delta t_2,h_2}\|_{\frac{1}{2},-\frac{1}{2},\sigma}\right\}.\end{aligned} \tag{13.17}$$

A crucial ingredient in the proof is the inf-sup condition in space-time, which holds for $\frac{\max\{h_1,\Delta t_1\}}{\min\{h_2,\Delta t_2\}}$ sufficiently small [192]: There exists $\alpha > 0$ such that for all $\lambda_{\Delta t_2,h_2}$:

$$\sup_{\mu_{\Delta t_1,h_1}} \frac{\langle \mu_{\Delta t_1,h_1}, \lambda_{\Delta t_2,h_2}\rangle}{\|\mu_{\Delta t_1,h_1}\|_{0,\frac{1}{2},\sigma,*}} \geq \alpha \|\lambda_{\Delta t_2,h_2}\|_{0,-\frac{1}{2},\sigma}.$$

Numerically, system (13.15) is solved with an Uzawa algorithm, and the Dirichlet-to-Neumann operator is computed in terms of the retarded boundary layer potentials as $W + (K' - 1/2)V^{-1}(K - 1/2)$. See [192] for details of the discretization.

We now consider the discretization of the dynamical contact problem (13.10), (13.11) for a flat contact area. No exact solutions are known, so that we compare the numerical approximations to a reference solutions on an appropriate finer space-time mesh.

Example We choose $\Gamma = [-2, 2]^2 \times \{0\}$ with contact area $G = [-1, 1]^2 \times \{0\}$ for times $[0, 5]$, with the CFL-ratio $\frac{\Delta t}{h} = 0.7$. The right hand side of the contact problem (13.15) is given by

$$h(t,x) = e^{-2t} t^4 \cos(2\pi x)\cos(2\pi y)\chi_{[-0.25,0.25]}(x)\chi_{[-0.25,0.25]}(y).$$

The numerical solutions are compared to a reference solution on a mesh with 12800 triangles, and we use $\Delta t = 0.075$.

In this example, contact takes place from time $t = 4.25$ on. Figure 13.3 considers the relative error to the reference solution in $L^2([0, T] \times G)$. The numerical approximations converge at a rate of $\alpha = 0.8$ with increasing degrees of freedom. Algorithmically, the computational cost of the nonlinear solver is dominated by the cost of computing the matrix entries.

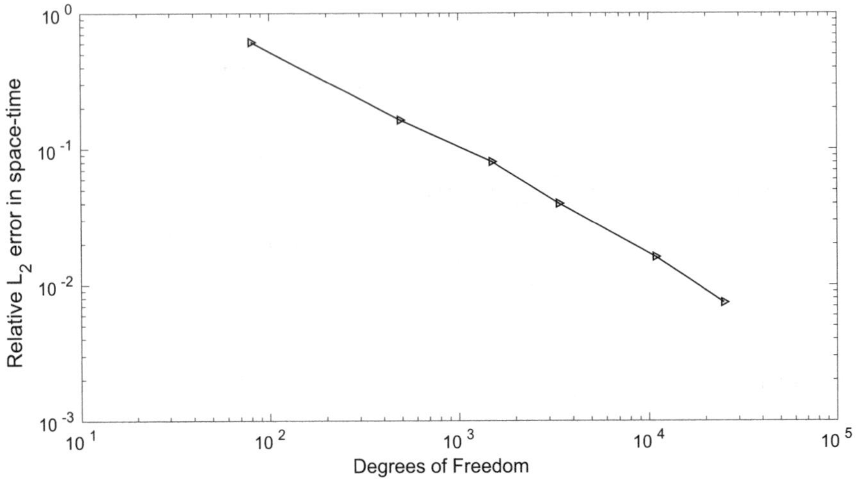

Fig. 13.3 Relative $L^2([0,T]\times\Gamma)$-error vs. degrees of freedom of the solutions to the contact problem for fixed $\frac{\Delta t}{h}$ [192]

13.4 Algorithmic Aspects of Time Domain BEM

13.4.1 MOT Algorithm

The Galerkin discretization in space and time leads to a block-lower-triangular system of equations, which can be solved by blockwise forward substitution. For ease of presentation we consider the Galerkin scheme for the Dirichlet problem (13.6) which can be rewritten for piecewise constant test and trial functions in space and time as:

Find $\phi_{\Delta t,h}\in V^{0,0}_{\Delta t,h}$ such that for all $\psi\in V^{0,0}_{\Delta t,h}$

$$\int_0^\infty\int_\Gamma (V\phi_{\Delta t,h}(t,x))\partial_t\psi(t,x)ds_x\,d_\sigma t\;=\int_0^\infty\int_\Gamma (f_{\Delta t,h}(t,x))\partial_t\psi(t,x)ds_x\,d_\sigma t\,. \tag{13.18}$$

This yields an algebraic system of the form

$$\sum_{m=1}^{n} V^{n-m}b^m = 2(f^{n-1}-f^n)$$

in time step $n=1,2,3,\ldots$. It can be solved by forward substitution, giving rise to the *marching in on time* (MOT) scheme

$$V^0b^n = 2(f^{n-1}-f^n)-\sum_{m=1}^{n-1} V^{n-m}b^m\,.$$

This can be seen as follows: Setting

$$\varphi_{\Delta t,h}(x,t) = \sum_{m=1}^{N_t} \sum_{i=1}^{N_s} b_i^m \beta^{m,0}(t)\varphi_i^0(x)$$

and

$$\psi(x,t) = \beta^{n,0}(t)\varphi_j^0(x)$$

one computes

$$\int_0^\infty \beta^{m,0}(t-|x-y|)\dot{\beta}^{n,0}(t)dt = H(t_{n-m}-|x-y|) - H(t_{n-m-1}-|x-y|) - H(t_{n-m+1}-|x-y|) + H(t_{n-m}-|x-y|)$$

with the Heaviside funcion

$$H(t_l-|x-y|) = \begin{cases} 1 & |x-y| \le t_l \\ 0 & \text{elsewhere} \end{cases}.$$

Therefore the left hand side in (13.18) becomes

$$\sum_{m=1}^{N_t}\sum_{i=1}^{N_s} b_i^m \Big[\iint_{\Gamma\times\Gamma\cap E_{n-m-1}} \frac{\varphi_i^0(y)\varphi_j^0(x)}{4\pi|x-y|} ds_y ds_x - \iint_{\Gamma\times\Gamma\cap E_{n-m}} \frac{\varphi_i^0(y)\varphi_j^0(x)}{4\pi|x-y|} ds_y ds_x \Big]$$

with

$$E_l := \{(x,y)\in\Gamma\times\Gamma : t_l \le |x-y| \le t_{l+1}\}.$$

Similarly, setting

$$f_{\Delta t,h}(x,t) = \sum_{m=1}^{N_t}\sum_{i=1}^{N_s} f_i^m \beta^{m,0}(t)\varphi_i^0(x)$$

yields for the right hand side in (13.18)

$$\sum_{i=1}^{N_s}[f_i^{n-1} - f_i^n]\int_\Gamma \varphi_i^0(x)\varphi_j^0(x)ds_x,$$

where

$$f_i^{n-1} = \sum_{m=1}^{N_t} f_i^m \beta^{m,0}(t_{n-1}),$$

because

$$\int_0^\infty \beta^{m,0}(t)\dot{\beta}^{n,0}(t)dt = \beta^{m,0}(t_{n-1}) - \beta^{m,0}(t_n).$$

The above fully discrete systems involve the computation of a series of matrices, that (if $\alpha_\infty = 0$) are sparsely populated, because the Dirac-delta fundamental solution restricts the number of interacting elements per time step. Note that the computation of each matrix only depends on the time difference. Furthermore, for bounded surfaces Γ the matrices V^{n-m} vanish whenever the time difference $l := n - m$ satisfies $l > \left[\frac{\operatorname{diam}\Gamma}{\Delta t}\right]$, i.e. the light cone has traveled through the entire surface Γ.

The most time consuming part in the MOT algorithm is the matrix computation, even though the resulting matrices are sparse. An efficient hp-composite Gauss-quadrature allows to compute the entries in V^l [402, 403].

13.4.2 An hp-Composite Quadrature of Matrix Elements

The most time consuming part in the MOT algorithm is the matrix computation, even though the resulting matrices are sparse in each time step. An efficient hp-composite Gauss-quadrature allows to compute the entries in V^l, and similarly for the other layer operators [331, 402, 403].

Recall the form of the matrix entries of V^l in $\mathbb{R}^3$ as an example:

$$\frac{1}{2\pi}\iiint_{\mathbb{R}^+\times\Gamma\times\Gamma} \frac{\varphi_i^p(y)\partial_t\beta^{n,q}(t-|x-y|)}{|x-y|}\varphi_j^p(x)\beta^{m,q}(t)\,ds_y\,ds_x\,d_\sigma t\,.$$

First, the time integrals are evaluated analytically and result in an integration domain

$$E = \{(x,y)\in\Gamma\times\Gamma : r_{\min} \le |x-y| \le r_{\max}\}$$

of the form of a light cone, $r_{\min}$ and $r_{\max}$ depending on t_m and t_n. It remains to evaluate terms like

$$G_{ij}^\nu = \iint_E k_\nu(x-y)\varphi_i^p(y)\varphi_j^p(x)\,ds_y\,ds_x\,, \tag{13.19}$$

where $k_\nu(x-y) = |x-y|^\nu$ denotes a weakly singular kernel function. Our numerical quadrature separates the outer spatial integration from the singular inner one. Define the domain of influence of $x \in \mathbb{R}^3$ by

$$E(x) := B_{r_{\max}}(x)\setminus B_{r_{\min}}(x) = \left\{y\in\mathbb{R}^3 : r_{\min}\le|x-y|\le r_{\max}\right\}$$

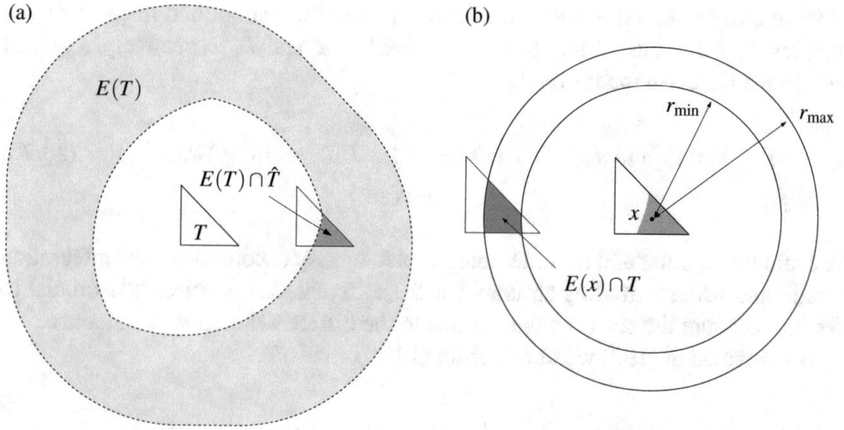

Fig. 13.4 Domains of influence and the illumination of test and trial element $\widehat{T}$ and T during the evaluation of the inner and outer integral [190]. (**a**) Outer integral: Domain of influence of triangle $\hat{T}$ intersected with triangle T. (**b**) Inner integral: Domain of influence $E(x)$ of point $x \in E(T) \cap \hat{T}$

as in Fig. 13.4b. Figure 13.4a similarly sketches the domain of influence of a triangle T,

$$E(T) := \bigcup_{x \in T} E(x) = \{y \in \mathbb{R}^3 : r_{\min} \leq |x - y| \leq r_{\max}\,,\ x \in T\}\,.$$

Defining $E(T_j, T_i) := E \cap (T_j \times T_i)$, we rewrite (13.19) as

$$\begin{aligned} G^{\nu}_{ij} &= \sum_{\substack{T_{i'} \subset \operatorname{supp} \varphi_i \\ T_{j'} \subset \operatorname{supp} \varphi_j}} \iint\limits_{E(T_{j'}, T_{i'})} k_\nu(x - y) \varphi^p_i(y) \varphi^p_j(x)\, ds_y\, ds_x \\ &= \sum_{\substack{T_{i'} \subset \operatorname{supp} \varphi_i \\ T_{j'} \subset \operatorname{supp} \varphi_j}} \int\limits_{T_{j'} \cap E(T_{i'})} \varphi^p_j(x) P^p_{i,i'}(x)\, ds_x\,, \end{aligned}$$

with a retarded potential $P_{i,i'}$ given by

$$P_{i,i'}(x) := \int\limits_{E(x) \cap T_{i'}} k_\nu(x - y) \varphi^p_i(y)\, ds_y\,.$$

To simplify notation, we explain the quadrature for a simplified integral. Given triangles T, $\hat{T}$ and basis functions φ, $\widehat{\varphi}$ defined on T and $\widehat{T}$, respectively, a typical entry in the Galerkin matrix reads

$$\int\limits_{E(T)\cap\widehat{T}} P\varphi(x)\widehat{\varphi}(x)\,ds_x\,,\quad P\varphi(x) := \int\limits_{E(x)\cap T} k_\nu(x-y)\varphi(y)\,ds_y\,. \tag{13.20}$$

We evaluate the outer and the inner integral step by step decomposing the integration domain and using a grading strategy for the different singularities. It is crucial to take into account the cut-off behavior due to the different domains of influence.

As explained in [190] we obtain from (13.20)

$$P\varphi(x) = \sum_{l=1}^{n_d} \int\limits_{\hat{D}_l} (d^2+r^2)^{\frac{\nu}{2}}\varphi(r,\theta)r\,dr\,d\theta\,,$$

where $d > 0$ and φ is sufficiently regular. For each of the domains $\hat{D}_l$ (see Fig. 13.5), we can write the integral as

$$I^{(\hat{D}_l)}f := \int_{\theta_1}^{\theta_2}\int_{r_1(\theta)}^{r_2(\theta)} f(r,\theta)\,dr\,d\theta\,,\quad f(r,\theta) := (d^2+r^2)^{\frac{\nu}{2}}\varphi(r,\theta)r. \tag{13.21}$$

To introduce our quadrature method, denote by $Q_n^{[a,b]}f := \sum_{i=1}^n w_i f(x_i)$ the Gauss-Legendre quadrature rule with n quadrature points to evaluate $\int_a^b f\,dx$. Given a subdivision of $[a,b]$ into m subintervals I_j, a variable order composite Gauss rule with degree vector $\mathbf{n} = (n_1,\ldots,n_m)$ is defined by $Q_{n,m,\sigma}f := \sum_{j=1}^m Q_{n_j}^{I_j}f$. We use a geometric subdivision of $[a,b]$ with m levels and grading parameter $\sigma \in (0,1)$: $[a,b] = \bigcup_{j=1}^m I_j$, where for $j = 1,\ldots,m$ we let $I_j := [x_{j-1},x_j]$, $x_0 := a$, $x_j := a+(b-a)\sigma^{m-j}$. For $n_r = (n_1^{(r)},\ldots,n_m^{(r)})$, $m_r \geq 1$ and $\sigma_r \in (0,1]$, the integral (13.21) is then computed as

$$Q^{\hat{D}_l}f := Q_{n_\theta}^{[\theta_1,\theta_2]}(Q_{n_r,m_r,\sigma_r}^{[r_1(\theta),r_2(\theta)]}f).$$

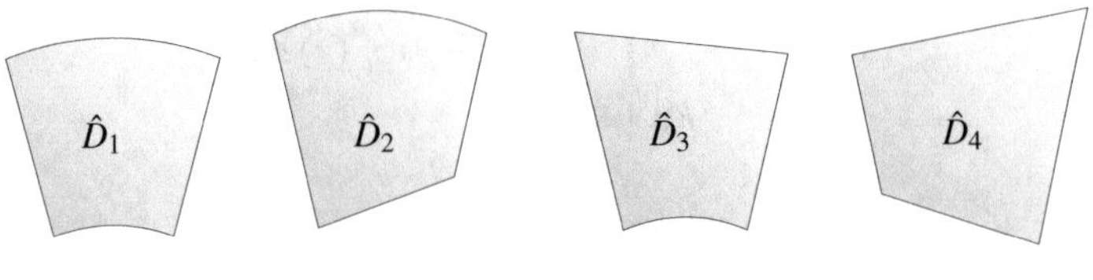

Fig. 13.5 Generic integration domains [190]

An error analysis for the evaluation of (13.21) is given in [331] by showing that the integrand belongs to the countably normed, weighted space $B^0_\beta(T)$ of Babuska [19].

Definition 13.2 (Countably normed space $B^l_\beta(T)$) We say $u \in B^l_\beta(T)$ with respect to a weight function $\Phi_{\beta,\alpha,l}$, if $u \in H^{l-1}(T)$ and if

$$\|\Phi_{\beta,\alpha,l} D^\alpha u\|_{L^2(\Omega)} \leq C d^{|\alpha|-l}(|\alpha| - l)!$$

for $|\alpha| = l, l+1, \dots$. Here the constants $C > 0$ and $d \geq 1$ are independent of $|\alpha|$.

If the number of angular quadrature points, n_θ, is chosen proportional to m_r, we obtain the following theorem on the accuracy of the quadrature in our TDBEM:

Theorem 13.10 ([331]) *Given a function $f \in B^0_\beta(T)$ with a weight function $\Phi_{\beta,\alpha,0}(r) = r^{|\alpha|+\beta}$, then there holds for $\hat{D}_l$:*

$$|I^{(\hat{D}_l)} f - Q^{(\hat{D}_l)} f| \leq C e^{-b\sqrt[3]{N}}$$

for $l = 1, \dots, 4$. Here N denotes the total number of quadrature points and C and b are positive constants independently of N, but depending on the grading factor σ_r, the number of levels m_r and on f.

13.5 Screen Problems and Graded Meshes

For solutions to elliptic equations in a polyhedral domain, the asymptotic behavior near the edges and corners, as well as its numerical approximation has been studied in Sect. 4.3 and Chap. 7.

In the case of the wave equation in domains with conical or wedge singularities, a similar asymptotic behavior has been obtained by Kokotov, Neittaanmäki and Plamenevskii since the late 1990's [268, 269]. Their results imply that at a fixed time t, the solution to the wave equation admits an explicit singular expansion with the same exponents as for elliptic equations.

The realistic scattering and diffraction of waves in $\mathbb{R}^3$ is crucially affected by geometric singularities of the scatterer, with significant new challenges for both the singular and numerical analysis. The article [191] studies the solution of the wave equation in the most singular case, outside a screen Γ in $\mathbb{R}^3$ or, equivalently, for an opening crack. From the singular expansion one obtains optimal convergence rates for piecewise polynomial approximations on graded meshes.

The computations below are conducted on graded meshes on the square $[-1, 1]^2$, respectively on the circular screen $\{(x, y, 0) : \sqrt{x^2 + y^2} \leq 1\}$. To define β-graded meshes on the square, due to symmetry, it suffices to consider a β-graded mesh on $[-1, 0]$. We define $y_k = x_k = -1 + (\frac{k}{N_l})^\beta$ for $k = 1, \dots, N_l$ and for a constant

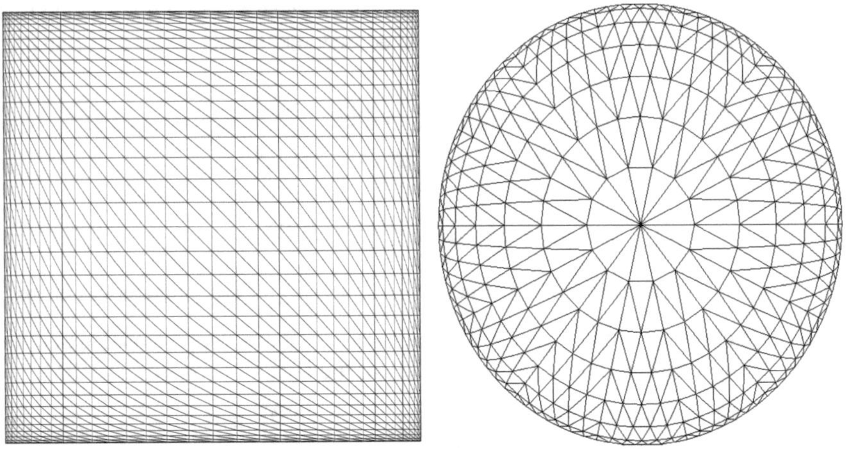

Fig. 13.6 β-graded meshes for square and circular screens, with $\beta = 2$ [191]

$\beta \geq 0$. The nodes of the β-graded mesh on the square are therefore (x_k, y_l), $k, l = 1, \ldots, N_l$. We note that for $\beta = 1$ we would have a uniform mesh.

In a general convex, polyhedral geometry graded meshes are locally modeled on this example. In particular, on the circular screen of radius 1, for $\beta = 1$ we take a uniform mesh with nodes on concentric circles of radius $r_k = 1 - \frac{k}{N_l}$ for $k = 0, \ldots, N_l - 1$. For the β-graded mesh, the radii are moved to $r_k = 1 - (\frac{k}{N_l})^\beta$ for $k = 0, \ldots, N_l - 1$. While the triangles become increasingly flat near the boundary, their total number remains proportional to N_l^2.

Examples of the resulting 2-graded meshes on the square and the circular screens are depicted in Fig. 13.6.

For these meshes one shows:

Theorem 13.11 ([191]) *Let* $\varepsilon > 0$.

a) Let ψ *be the solution to the hypersingular integral equation* $W\psi = g$ *and* $\psi^\beta_{h,\Delta t}$ *the best approximation in the norm of* $H^r_\sigma(\mathbb{R}^+, \widetilde{H}^{\frac{1}{2}-s}(\Gamma))$ *to* ψ *in* $\widetilde{V}^{p,1}_{\Delta t,h}$ *on a* β*-graded spatial mesh with* $\Delta t \lesssim h^\beta$. *Then*
$\|\psi - \psi^\beta_{h,\Delta t}\|_{r,\frac{1}{2}-s,\Gamma,*} \leq C_{\beta,\varepsilon} h^{\min\{\beta(\frac{1}{2}+s),\frac{3}{2}+s\}-\varepsilon}$*, where* $s \in [0, \frac{1}{2}]$ *and* $r \in [0, p)$.

b) Let ϕ *be the solution to the single layer integral equation* $V\phi = f$ *and* $\phi^\beta_{h,\Delta t}$ *the best approximation in the norm of* $H^r_\sigma(\mathbb{R}^+, \widetilde{H}^{-\frac{1}{2}}(\Gamma))$ *to* ϕ *in* $V^{p,0}_{\Delta t,h}$ *on a* β*-graded spatial mesh with* $\Delta t \lesssim h^\beta$. *Then* $\|\phi - \phi^\beta_{h,\Delta t}\|_{r,-\frac{1}{2},\Gamma,*} \leq C_{\beta,\varepsilon} h^{\min\{\frac{\beta}{2},\frac{3}{2}\}-\varepsilon}$*, where* $r \in [0, p+1)$.

Note that the energy norm associated to the weak form of the single layer integral equation is weaker than the norm of $H^1_\sigma(\mathbb{R}^+, H^{-\frac{1}{2}}(\Gamma))$ and stronger than the norm of $H^0_\sigma(\mathbb{R}^+, H^{-\frac{1}{2}}(\Gamma))$, according to the coercivity and continuity properties of V on screen. Similarly, for the weak form of the hypersingular integral equation, the energy norm is weaker than the norm of $H^1_\sigma(\mathbb{R}^+, H^{\frac{1}{2}}(\Gamma))$ and stronger than the norm of $H^0_\sigma(\mathbb{R}^+, H^{\frac{1}{2}}(\Gamma))$.

Together with the a priori estimates for the time domain boundary element methods on screens in this chapter, the theorem implies convergence rates for the Galerkin approximations, which recover those for smooth solutions (up to an arbitrarily small $\varepsilon > 0$) provided the grading parameter β is chosen sufficiently large.

The crucial ingredient in the proof of Theorem 13.11 is a precise description of the corner and edge singularities of the solution. In analogy with the work of Plamenevskii and coauthors, the asymptotic expansion of the solution u to the wave equation, respectively its normal derivative $\partial_\nu u$, near the corner of a polygonal screen in $\mathbb{R}^3$ in the time domain is as follows:

$$u(t,x)|_+ = \psi_0(t,r,\theta) + \chi(r)r^\gamma\alpha(t,\theta) + \tilde{\chi}(\theta)b_1(t,r)(\sin(\theta))^{\frac{1}{2}}$$
$$+ \tilde{\chi}(\tfrac{\pi}{2}-\theta)b_2(t,r)(\cos(\theta))^{\frac{1}{2}}\,,$$
$$\partial_\nu u(t,x)|_+ = \phi_0(t,r,\theta) + \chi(r)r^{\gamma-1}\alpha(t,\theta) + \tilde{\chi}(\theta)b_1(t,r)r^{-1}(\sin(\theta))^{-\frac{1}{2}}$$
$$+ \tilde{\chi}(\tfrac{\pi}{2}-\theta)b_2(t,r)r^{-1}(\cos(\theta))^{-\frac{1}{2}}\,.$$

The remainders ψ_0 and ϕ_0 are less singular, and γ is the singular exponent of the corner singularity known from the elliptic case. In particular, for the square screen $\gamma \simeq 0.2966$. For a circular screen, only the edge singularity with singular exponent $\frac{1}{2}$ (u), respectively $-\frac{1}{2}$ ($\partial_\nu u$), is present.

For the algorithmic details of the numerical experiments we refer to [191].

Example 13.1 We compute the solution to the integral equation $V\phi = f$ on $\mathbb{R}^+_t \times \Gamma$ with the circular screen $\Gamma = \{(x, y, 0) : 0 \leq \sqrt{x^2+y^2} \leq 1\}$ depicted in Fig. 13.6. We use constant test and ansatz functions in space and time. The right hand side is given by $f(t,x) = \cos(|k|t - k\cdot x)\exp -1/(10t^2))$, where $k = (0.2, 0.2, 0.2)$. The time discretization errors are negligibly small in this numerical experiment, when the time step is chosen to be $\Delta t = 0.005$. We compute the solution up to $T = 1$. The finest graded mesh consists of 2662 triangles, and we use the solution on this mesh as reference solution using the same $\Delta t = 0.005$.

We consider the error compared to the benchmark solution on the 2-graded mesh. For the error as a function of the degrees of freedom, Fig. 13.7 shows convergence in the energy norm with a rate -0.52 on the 2-graded mesh, respectively -0.26 on the uniform mesh. The error therefore behaves in agreement with the approximation properties proportional to $\sim h$ (equivalently, $\sim DOF^{-\frac{1}{2}}$) on the 2-graded mesh, while the convergence is $\sim h^{1/2}$ ($\sim DOF^{-1/4}$) on a uniform mesh.

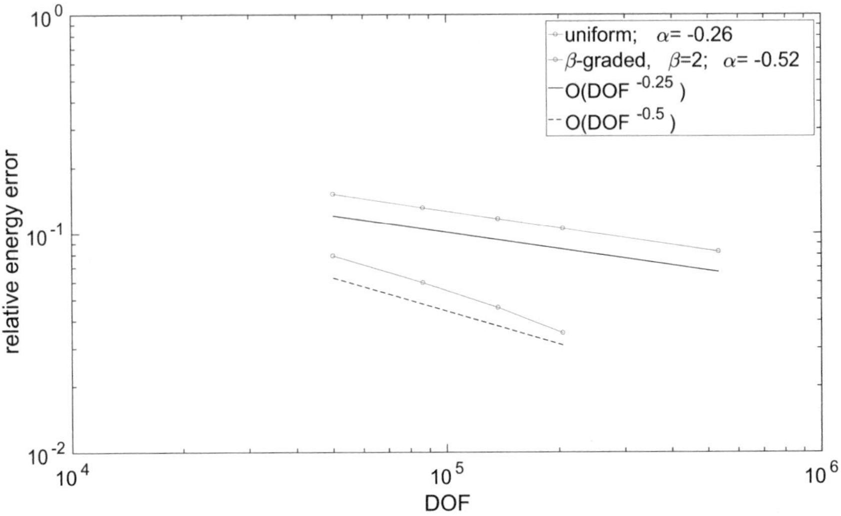

Fig. 13.7 Energy error for single layer equation on circular screen, Example 13.1 [191]

Example 13.2 We compute the solution to the integral equation $W\phi = g$ on $\mathbb{R}_t^+ \times \Gamma$ with the square screen $\Gamma = [-1, 1]^2 \times \{0\}$. We use linear ansatz and test functions in space, linear ansatz and constant test functions in time. The right hand side is given by

$$g(t, x) = (-\frac{3}{4} + \cos(\frac{\pi}{2}(4-t)) + \frac{\pi}{2}\sin(\frac{\pi}{2}(4-t)) - \frac{1}{4}(\cos(\pi(4-t)) + \pi\sin(\pi(4-t)))) \times [H(4-t) - H(-t)],$$

where H is the Heaviside function, and $\Delta t = 0.01$, $T = 4$. The finest graded mesh consists of 2312 triangles, and we use the solution on this mesh as reference solution using the same $\Delta t = 0.01$.

Figure 13.8 shows the error in both the energy and $L_2([0, T], L_2(\Gamma))$ norms with respect to the benchmark solution. The convergence rate in terms of the degrees of freedom on the 2-graded mesh is -0.51 in energy and -1.05 in L_2. On the uniform mesh the rate is -0.26 in energy and -0.50 in L_2. The rates on the 2-graded meshes are in close agreement with a convergence proportional to $\sim h$ (equivalently, $\sim DOF^{-1/2}$) predicted by the approximation properties in the energy norm, and $\sim h^{1/2}$ ($\sim DOF^{-1/4}$) on uniform meshes. Also in L_2 norm, the convergence corresponds to the expected rates: Approximately $\sim h^2$ (equivalently, $\sim DOF^{-1}$) on 2-graded meshes, $\sim h$ (equivalently, $\sim DOF^{-1/2}$) on uniform meshes. In all cases the convergence is twice as fast on the 2-graded compared to the uniform meshes.

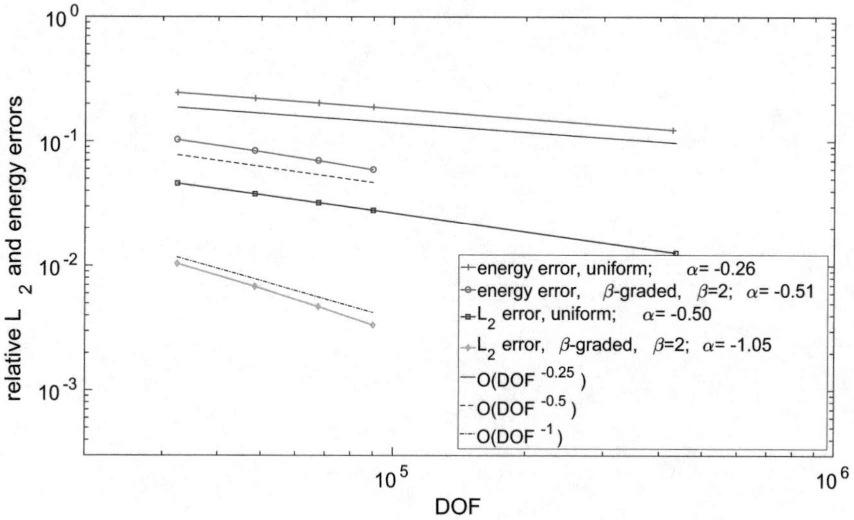

Fig. 13.8 $L_2([0,T], L_2(\Gamma))$ and energy error for hypersingular equation on square screen, Example 13.2 [191]

Appendix A
Linear Operator Theory

Here we recall some concepts from linear operator theory, in particular compact operators and Fredholm operators. It is assumed that the reader is familiar with basics of linear functional analysis like norm, metric, completeness, Banach space, Hilbert space etc. Some theorems are proved here, some are only stated.

Definition A.1 Let E, F be vector spaces (linear spaces) over a field Δ, a linear operator from E to F is a mapping $A : E \to F$ such that

$$A(\lambda x + \mu y) = \lambda Ax + \mu Ay \quad \forall \lambda, \mu \in \Delta;\ x, y \in E\,.$$

In the following, E, F are linear normed spaces.

Theorem A.1 *Let A be a linear operator from E into F. Then A is bounded, if and only if A is continuous, written $A \in \mathscr{B}(E, F)$.*

Definition A.2 The operator norm $\|A\| = \|A\|_{E\to F}$ of $A \in \mathscr{B}(E, F)$ is defined by

$$\|A\| := \sup_{\substack{x\in E \\ \|x\|\le 1}} \|Ax\|_F = \sup_{x\in E\setminus\{0\}} \frac{\|Ax\|_F}{\|x\|_E}$$

Definition A.3 An operator A is said to be continuously invertible if A^{-1} exists and is continuous.

Theorem A.2 (Banach's Theorem) *Let $A \in \mathscr{B}(X, Y)$ and X, Y be Banach spaces. Assume that A is injective and surjective (onto). Then A is continuously invertible.*

© Springer International Publishing AG, part of Springer Nature 2018

J. Gwinner, E. P. Stephan, *Advanced Boundary Element Methods*,
Springer Series in Computational Mathematics 52,
https://doi.org/10.1007/978-3-319-92001-6

Let E' denote the dual space of E. Let M be an arbitrary nonvoid subset of E. Then the set

$$\{x' \in E' : \langle x', x\rangle = 0 \ \forall x \in M\}$$

is the annihilator of M, written $M^\perp$. Similarly, $N^\perp \subset E$ for $N \subset E'$.

Let $A \in \mathscr{B}(E, F)$. Then the adjoint A' is in $\mathscr{B}(F', E')$. Moreover immediately

$$\ker A = (\mathrm{im}\ A')^\perp, \ \ker A' = (\mathrm{im}\ A)^\perp . \tag{A.1}$$

Also im $A \subset (\ker A')^\perp$ is obvious; however, equality holds only with closedness. More precisely, due to Banach:

Theorem A.3 (Closed Range Theorem) *Let $A \in \mathscr{B}(X, Y)$ and X, Y be Banach spaces. Then the following assertions are equivalent:*

(i) im A is closed in Y,
(ii) im A' is closed in X',
(iii) im $A = (\ker A')^\perp$,
(iv) im $A' = (\ker A)^\perp$.

A class of continuous linear operators that have closed range are Fredholm operators, see below Definition A.8.

Definition A.4 Let X, Y be Banach spaces. A sequence of linear operators $\{A_n : X \to Y\}$ is called strongly convergent (or pointwise convergent) to an operator $A : X \to Y$, written "$A_n \to A$", if for all $x \in X$, $\lim_{n\to\infty} \|A_n x - Ax\|_Y = 0$.

Definition A.5 Let X, Y be Banach spaces. A sequence of linear operators $\{A_n : X \to Y\}$ is called convergent in the norm, written "$A_n \Rightarrow A$", if $\lim\limits_{n\to\infty} \|A_n - A\| = 0$.

Theorem A.4 (Theorem of Banach-Steinhaus) *The sequence $\{A_n : X \to Y\}$ of continuous linear operators converges pointwise to a continuous linear operator $A : X \to Y$ if and only if*

(i) The sequence of operator norms $\|A_n\|$ is bounded.
(ii) The sequence $\{A_n x\}$ converges for all $x \in M$ where M is a dense subset of X.

Definition A.6 Let X, Y be Banach spaces and $A \in \mathscr{B}(X, Y)$. Then $A : X \to Y$ is said to be compact if the image of the closed unit ball $B_1^X = \{x \in X : \|x\| \le 1\}$ under A is relatively compact in Y.

Remark A.1 A set $K \subset Y$ is relatively compact in Y if for every $\varepsilon > 0$ there is a finite number of elements in K, say $y_1, \dots, y_m \in K$, such that the ε-balls around y_j $(j = 1, \dots, m)$ cover K, $K \subset \cup_{j=1}^m B_\varepsilon^Y(y_j)$. Thus the y_j are a finite ε-net in K. Furthermore A is compact if and only if, for every bounded sequence $\{x_n\}$ in X, the sequence $\{Ax_n\}$ has a convergent subsequence in Y.

Lemma A.1

(i) *Let $A : X \longrightarrow Y$ be compact and $B : Y \longrightarrow Z$ bounded. Then, the operator $AB : X \longrightarrow Z$ is compact, too.*
(ii) *Let again $A : X \longrightarrow Y$ be compact and let $B_n \longrightarrow 0$ be strongly convergent. Then, $B_n A \longrightarrow 0$ is strongly convergent, too.*

Proof (i) Exercise. (ii) Suppose the assertion does not hold. Then there exists a sequence $\{x_n\} \subset X$ satisfying $\|x_n\|_X = 1\,\forall n$ and $\|B_n A x_n\|_Z \geq \alpha > 0\,\forall n$. With A compact there further exists a subsequence $\{x'_n\}$ with $Ax'_n \longrightarrow y$ in Y. Thus,

$$0 < \alpha \leq \|B_n A x'_n\| \leq \|B_n y\| + \|B_n(Ax'_n - y)\|$$

But $\|B_n y\| \mapsto 0$ and $\|B_n(Ax'_n - y)\| \leq M\|Ax'_n - y\| \mapsto 0$ which is a contradiction. □

Definition A.7 Let X, Y be Banach spaces and $\{A_n\}$ a sequence of linear operators. Then $\{A_n : X \to Y\}$ is called collectively compact, if $K := \cup_{n\in\mathbb{N}} A_n(B_1^X(0))$ is relatively compact in Y.

Lemma A.2 *Assume that the sequence of the linear operators $A_n : X \to Y$ converges pointwise to A. Then there holds:*

(i) *If $K \subset X$ is relatively compact in X, then $A_n \Rightarrow A$ uniformly in K.*
(ii) *Let $B : W \to X$ (with another Banach space W) be a compact linear operator and assume that $B_n \Rightarrow B$. Then $A_n B_n \Rightarrow AB$.*
(iii) *Let $\{B_n : W \to X\}$ be collectively compact and $A = 0$. Then $A_n B_n \Rightarrow 0$.*

Proof

1. Assume that $\{x_1, \ldots, x_m\} \subset K$ is a ε-net for a given (but arbitrary) $\varepsilon > 0$, i.e. $K \subset \cup_{j=1}^m B_\varepsilon^X(x_j)$. Choose a $n_0(\varepsilon) \in \mathbb{N}$ such that for all $n > n_0(\varepsilon)$, $j = 1, \ldots, m$, $\|A_n x_j - Ax_j\|_Y < \varepsilon$. Let $x \in K$ arbitrary. Then there exists a $j \in \{1, \ldots, m\}$ with $\|x - x_j\|_X \leq \varepsilon$ and thus:

$$\begin{aligned}\|A_n x - Ax\| &\leq \|(A_n - A)(x - x_j)\| + \|(A_n - A)x_j\| \\ &\leq \|(A_n - A)\|\;\|(x - x_j)\| + \varepsilon \\ &\leq (1 + \|A_n - A\|)\cdot\varepsilon\end{aligned}$$

   By the Theorem of Banach-Steinhaus (see A.4) we have $\|A_n - A\| \leq C\,\forall n \in \mathbb{N}$ and therefore $\|A_n x - Ax\| \leq (1 + C)\cdot\varepsilon \to 0$ uniformly for all $x \in K$.
2. Since B is compact, $\{Bw | w \in W, \|w\| \leq 1\} \subset X$ is relatively compact in X. Therefore by 1., for any $\varepsilon > 0$ there exists $n_0 \in \mathbb{N}$ such that $\|(A_n - A)Bw\| \leq \varepsilon$ for $n \geq n_0$ and $\|w\| \leq 1$. Moreover since $B_n \Rightarrow B$, $\|(B_n - B)w\| \leq \varepsilon$ for $n \geq n_0$

and $\|w\| \leq 1$. Therefore for $\|w\| \leq 1$,

$$\begin{aligned}\|A_nB_nw - ABw\| &\leq \|A_n(B_n - B)w\| + \|(A_n - A)Bw\| \\ &\leq \|A_n\| \cdot \|(B_n - B)w\| + \varepsilon \\ &\leq (1 + C)\varepsilon \quad \text{for } \|w\| \leq 1.\end{aligned}$$

where $\|A_n\| \leq C$ by Banach-Steinhaus Theorem. Hence $A_nB_n \Rightarrow AB$.

3. From 1. we have that A_n converges uniformly to $0 = A$ on $K := \cup_n B_n(B_1^W(0))$, i.e. $\forall \varepsilon > 0\ \exists n_0 : \|A_nx\| \leq \varepsilon\ \forall n \geq n_0\ \forall x \in K$. Therefore $\|A_nB_nw\| \leq \varepsilon$ for $\forall n \geq n_0$, and $\|w\| \leq 1$ and so

$$\|A_nB_n\|_{W \to Y} := \sup_{\|w\| \leq 1} \|A_nB_nw\| \longrightarrow 0 \text{ for } n \to \infty.$$

□

Definition A.8 Let X, Y be Banach spaces and let $A \in \mathscr{B}(X, Y)$. Then A is called a Fredholm operator, written $A \in \mathscr{F}(X, Y)$, if A enjoys the following properties:

(I) The kernel ker A has finite dimension,
(II) im A is closed in Y,
(III) the range im A has finite codimension: codim im A = dim $(Y/\text{im } A) < \infty$.

The number

$$\text{ind } (A) := \dim \ker A - \text{codim im } A$$

is called the (Fredholm) index of A.

Remark A.2 By the closed range theorem, here Theorem A.3, the operator equation $Au = f$ with $A \in \mathscr{F}(X, Y)$ is solvable, if and only if, $f \in (\ker A')^{\perp}$.

Theorem A.5 *Let $A \in \mathscr{F}(X, Y)$ and X, Y be Banach spaces. Then the adjoint A' is in $\mathscr{F}(X, Y)$, too. Moreover,*

$$\textit{dim ker } A' = \textit{codim im } A \textit{ and codim im } A' = \textit{dim ker } A\,,$$

hence

$$\textit{ind } A' = -\textit{ind } A\,. \tag{A.2}$$

Proof By A.1, ker $A' = (\text{im } A)^{\perp} \cong (Y/\text{im } A)' = (\text{coker } A)'$. This shows the first formula in the theorem. To prove the second formula, by the closedness of im A, we can apply Theorem A.3, hence im $A' = (\ker A)^{\perp}$. This gives $(\ker A)' \cong X'/(\ker A)^{\perp} = X'/\text{im } A' = \text{coker } A'$, hence the second formula, (A.2), and also $A' \in \mathscr{F}(X, Y)$. □

Without proof we list from [364, 435] the following.

Theorem A.6 *Let $A \in \mathscr{B}(X, Y)$ and X, Y be Banach spaces. Then A is a Fredholm operator, if and only if, there exist $Q_1, Q_2 \in \mathscr{B}(Y, X)$ such that*

$$Q_1 A = I - C_1 \text{ in } X \text{ and } A Q_2 = I - C_2 \text{ in } Y \tag{A.3}$$

with compact operators C_1, C_2.

Let $A \in \mathscr{F}(X, Y)$, $B \in \mathscr{F}(Y, Z)$ and X, Y, Z Banach spaces. Then $B \circ A \in \mathscr{F}(X, Z)$ and

$$\mathit{ind}\,(B \circ A) = \mathit{ind}\, B + \mathit{ind}\, A\,. \tag{A.4}$$

For a Fredholm operator A and a compact operator C, the sum $A+C$ is a Fredholm operator and

$$\mathit{ind}\,(A + C) = \mathit{ind}\, A\,. \tag{A.5}$$

The set of Fredholm operators is an open subset in the space of bounded linear operators and the index is a continuous function.

To conclude Appendix A we recall from [259] Fredholm's alternative for a sesquilinear form a in a Hilbert space H under a Gårding inequality,

$$\Re\Big\{a(v, v) + (Cv, v)_H\Big\} \geq \alpha_0 \,\|v\|_H^2\,, \quad v \in H \tag{A.6}$$

for a constant $\alpha_0 > 0$ and a compact operator C from H into H.

Theorem A.7 *Suppose the continuous sesquilinear form $a : H \times H \to \mathbb{C}$ satisfies Gårding inequality (A.6). Then for the variational equation*

$$\text{Find } u \in H \text{ such that } a(u, v) = \ell(v)\,, \forall v \in H \tag{A.7}$$

there holds the alternative:

Either

(A.7) has exactly one solution $u \in H$ for every given $\ell \in H^*$

or

The homogeneous problem,

$$\text{Find } u_0 \in H \text{ such that } a(u_0, v) = 0\,, \forall v \in H \tag{A.8}$$

and its adjoint problem,

$$\text{Find } v_0 \in H \text{ such that } a(u, v_0) = 0\,, \forall u \in H \tag{A.9}$$

have finite dimensional kernels of the same dimension $k > 0$. *The nonhomogeneous problem (A.7) and its adjoint,*

$$\text{Find } w \in H \text{ such that } \overline{a(v, w)} = \ell^*(v)\,, \forall v \in H$$

have solutions iff the orthogonality conditions

$$\ell(v_{0(j)}) = 0, \text{ respectively, } \ell^*(u_{0(j)}) = 0 \text{ for } j = 1, \ldots, k$$

hold where $\{u_{0(j)}\}_{j=1}^k$ *spans the eigenspace of (A.8) and* $\{v_{0(j)}\}_{j=1}^k$ *spans the eigenspace of (A.9), respectively.*

Rewrite the variational equation (A.7) as

$$a(u, v) = (jAu, v)_H = (f, v)_H = \ell(v)\,,$$

where $f \in H$ represents $\ell \in H^*$ due to the Riesz representation theorem. Then the maps $jA : H \to H, A : H \to H^*$ are linear and bounded. Moreover from Fredholm's alternative above, Theorem A.7 we derive the following

Remark A.3 Suppose the Gårding inequality (A.6). Then the linear operator A has closed range, moreover A is a Fredholm operator with index zero.

Appendix B
Pseudodifferential Operators

Here we recall some concepts from Fourier transform and the theory of pseudodifferential operators. These are especially used in Chap. 4. For further reading see e.g. [253, 376, 415].

For $1 \leq p < \infty$ the space $L^p(E)$ consists of all measurable functions $f : E \to \mathbb{R}\ (\mathbb{C})$ with norm

$$\|f\|_{L^p} := \left(\int_E |f(x)|^p \, dx \right)^{1/p} < \infty,$$

whereas $L^\infty(E)$ has norm $\|f\|_{L^\infty} := \operatorname{ess\,sup}_{x \in E} |f(x)| < \infty$.

Definition B.1 Let $u \in L^1(\mathbb{R}^n)$. Then the Fourier transform of u is given by

$$\hat{u}(\xi) = (\mathscr{F}u)(\xi) = \int_{\mathbb{R}^n} e^{-ix\xi} u(x)\, dx \tag{B.1}$$

with its inverse

$$\tilde{u}(\xi) = \left(\tilde{\mathscr{F}}u\right)(\xi) = (2\pi)^{-n} \int_{\mathbb{R}^n} e^{ix\xi} u(x)\, dx. \tag{B.2}$$

Clearly $\hat{u} \in L^\infty(\mathbb{R}^n)$, $\| \hat{u} \|_{L^\infty} \leq \| u \|_{L^1}$.

Definition B.2 (Rapidly Decreasing Functions) $\varphi \in \mathscr{S}(\mathbb{R}^n) \iff \varphi \in C^\infty(\mathbb{R}^n, \mathbb{C})$, $\sup_{x \in \mathbb{R}^n} | < x >^\alpha D^\beta \varphi(x) | < \infty\ \forall \alpha, \beta \in \mathbb{N}_0^n$, with $< x > := 1 + |x|$.

Proposition B.1 $u \in \mathscr{S}(\mathbb{R}^n) \Longrightarrow \hat{u},\ \tilde{u} \in \mathscr{S}(\mathbb{R}^n)$.

© Springer International Publishing AG, part of Springer Nature 2018

J. Gwinner, E. P. Stephan, *Advanced Boundary Element Methods*,
Springer Series in Computational Mathematics 52,
https://doi.org/10.1007/978-3-319-92001-6

Proof $u \in \mathscr{S}(\mathbb{R}^n) \Longrightarrow x^\alpha u \in \mathscr{S}(\mathbb{R}^n) \quad \forall \alpha \in \mathbb{N}_0^n; \quad D^\beta u \in \mathscr{S}(\mathbb{R}^n) \quad \forall \beta \in \mathbb{N}_0^n$. Therefore we have $D^\beta (x^\alpha u) \in \mathscr{S}(\mathbb{R}^n) \subset L^1(\mathbb{R}^n)$ and consequently $\mathscr{F} D^\beta (x^\alpha u)$ is absolutely convergent. So we can interchange differentiation and integration in (B.1). With $\partial_j := \frac{\partial}{\partial x_j}$ and $D_j := \frac{1}{i}\partial_j$ we have

$$D_\xi^\alpha e^{-ix\xi} = (-i)^{|\alpha|} (-ix)^\alpha \, e^{-ix\xi} = (-1)^{|\alpha|} \, x^\alpha \, e^{-ix\xi}$$

and thus $\xi^\beta e^{-ix\xi} = (-1)^\beta \, D_x^\beta \, e^{-ix}$. It follows that

$$D_\xi^\alpha \hat{u}(\xi) = \int_{\mathbb{R}^n} (-x)^\alpha u(x) e^{-ix\xi} \, dx = \mathscr{F}\left((-x)^\alpha u\right)(\xi) \tag{B.3}$$

and

$$\begin{aligned} \xi^\beta \hat{u}(\xi) &= (-i)^\beta \int_{\mathbb{R}^n} u(x) \, D_x^\beta \, e^{-ix\xi} \, dx \\ &= \int_{\mathbb{R}^n} D_x^\beta \, u(x) e^{-ix\xi} \, dx = \mathscr{F}\left(D^\beta u\right)(\xi). \end{aligned} \tag{B.4}$$

Applying (B.3) and (B.4) yields

$$\xi^\beta D^\alpha \hat{u}(\xi) = \mathscr{F}\left(D^\beta (-x)^\alpha u\right)(\xi) \in L^\infty\left(\mathbb{R}^n\right) \quad \forall \alpha, \beta \in \mathbb{N}_0^n.$$

Hence $\hat{u} \in \mathscr{S}(\mathbb{R}^n)$. □

Proposition B.2 *The map $\mathscr{F} : \mathscr{S}(\mathbb{R}^n) \to \mathscr{S}(\mathbb{R}^n)$ is an isomorphism with $\mathscr{F}\widetilde{\mathscr{F}} = \widetilde{\mathscr{F}}\mathscr{F} = 1$. Moreover, for $u \in \mathscr{S}(\mathbb{R}^n)$ there holds:*

$$u(x) = (2\pi)^{-n} \int_{\mathbb{R}^n} \hat{u}(\xi) e^{ix\xi} \, d\xi. \tag{B.5}$$

Remark B.1 By Proposition B.1 we have $u = \widetilde{\mathscr{F}}(\mathscr{F}(u))$ for all $u \in \mathscr{S}(\mathbb{R}^n)$. Moreover, $\mathscr{F}$ and $\widetilde{\mathscr{F}}$ are related to each other via $(\mathscr{F} u)(x) = (2\pi)^n \left(\widetilde{\mathscr{F}} u\right)(-x)$.

It follows that

$$\mathscr{F}\mathscr{F} u(x) = (2\pi)^n u(-x). \tag{B.6}$$

Remark B.2 With equation (B.5) we write the differential operator $p(x, D) = \sum_{|\alpha|=0}^{N} a_\alpha(x) D_x^\alpha$ as

$$p(x, D) u(x) = (2\pi)^{-n} \int_{\mathbb{R}^n} e^{ix\xi} p(x, \xi) \hat{u}(\xi) \, d\xi.$$

Definition B.3 $p(x, \xi) = \sum_{|\alpha|=0}^{N} a_\alpha(x)\,\xi^\alpha$ is called the symbol of $p(x, D)$.

Definition B.4 A tempered distribution T on $\mathbb{R}^n$ is a continuous linear functional $T : \mathscr{S}(\mathbb{R}^n) \to \mathbb{C}$. The linear space of tempered distributions is denoted by $\mathscr{S}'(\mathbb{R}^n)$.

Next we extend $\mathscr{F}$ to a map on tempered distributions:

Find $\mathscr{F}^+ : \mathscr{S}(\mathbb{R}^n) \to \mathscr{S}(\mathbb{R}^n)$ such that

$$< \mathscr{F} f, \varphi > = < f, \mathscr{F}^+ \varphi > \quad \forall f \in \mathscr{S}'\left(\mathbb{R}^n\right), \varphi \in \mathscr{S}\left(\mathbb{R}^n\right),$$

$< f, \varphi > := \int\limits_{\mathbb{R}^n} f(x)\varphi(x)\,dx \quad \forall \varphi \in \mathscr{S}(\mathbb{R}^n)$ and $|< f, \varphi >| \le c$ bounded with $c \in \mathbb{R}$.

Then we define

$$\mathscr{F}\,|_{\mathscr{S}'(\mathbb{R}^n)} := (\mathscr{F}^+\,|_{\mathscr{S}(\mathbb{R}^n)})'.$$

If $f, \varphi \in \mathscr{S}(\mathbb{R}^n) \subset \mathscr{S}'(\mathbb{R}^n)$, then by Fubini's theorem

$$\begin{aligned} < \mathscr{F} f, \varphi > &= \int\limits_{\mathbb{R}^n}\int\limits_{\mathbb{R}^n} f(x)\, e^{-ix\xi}\, dx\, \varphi(\xi)\, d\xi \\ &= \int\limits_{\mathbb{R}^n} f(x) \int\limits_{\mathbb{R}^n} e^{-ix\xi} \varphi(\xi)\, d\xi\, dx = < f, \mathscr{F}\varphi > . \end{aligned}$$

So we get that $\mathscr{F}^+ = \mathscr{F}$ on $\mathscr{S}(\mathbb{R}^n)$.

Definition B.5 If $f \in \mathscr{S}'(\mathbb{R}^n)$, then $\mathscr{F} f = \hat{f} \in \mathscr{S}'(\mathbb{R}^n)$ is defined by

$$< \mathscr{F} f, \varphi > = < f, \mathscr{F}\varphi > \quad \forall \varphi \in \mathscr{S}\left(\mathbb{R}^n\right). \tag{B.7}$$

(B.7) is equivalent to

$$(\mathscr{F} f, \varphi) = (2\pi)^n \left(f, \widetilde{\mathscr{F}}\varphi\right) \quad \forall \varphi \in \mathscr{S}\left(\mathbb{R}^n\right) \tag{B.8}$$

with $(f, \varphi) := < f, \bar{\varphi} >$.

Remark B.3 Also on $\mathscr{S}'(\mathbb{R}^n)$ there holds $\mathscr{F}\widetilde{\mathscr{F}} = \widetilde{\mathscr{F}}\mathscr{F} = 1$.
Furthermore, $\mathscr{F}\,|_{\mathscr{S}'(\mathbb{R}^n)}$ is continuous extension of $\mathscr{F}\,|_{\mathscr{S}(\mathbb{R}^n)}$.
For $u \in \mathscr{S}'(\mathbb{R}^n)$ there holds

$$D^\alpha u = \mathscr{F}^{-1}\left(\xi^\alpha \hat{u}\right), \tag{B.9}$$

$$(-x)^\beta u = \mathscr{F}^{-1}\left(D^\beta \hat{u}\right). \tag{B.10}$$

For all $u \in \mathscr{S}(\mathbb{R}^n)$ there holds Parseval's equality:

$$\| \hat{u} \|^2_{L^2(\mathbb{R}^n)} = \left(\mathscr{F}u, \underbrace{\mathscr{F}u}_{=:\varphi} \right) = (2\pi)^n \left(u, \widetilde{\mathscr{F}}\varphi\right) = (2\pi)^n \| u \|^2_{L^2(\mathbb{R}^n)} .$$

Thus $u \mapsto (2\pi)^{\frac{-n}{2}} \hat{u}$ is an *isometry* on $L^2(\mathbb{R}^n)$. There holds

$$\left((2\pi)^{\frac{-n}{2}} \mathscr{F} \right)^{-1} = (2\pi)^{\frac{n}{2}} \widetilde{\mathscr{F}} = \left((2\pi)^{\frac{-n}{2}} \mathscr{F} \right)^{*},$$

where the adjoint A^* of a given operator A is defined by

$$(Af, \varphi) = \left(f, A^*\varphi\right) \quad \forall \varphi \in \mathscr{S}\left(\mathbb{R}^n\right), \; f \in \mathscr{S}'\left(\mathbb{R}^n\right).$$

Definition B.6 For all u, v in $\mathscr{S}(\mathbb{R}^n)$ the convolution is defined by $(u * v)(x) := \int_{\mathbb{R}^n} u(x-y)\, v(y)\, dy$.

The Fourier transform satisfies:

$$\left(\hat{u} * \hat{v}\right)(\xi) = \int_{\mathbb{R}^n} \hat{u}(\xi - \eta)\, \hat{v}(\eta)\, d\eta = < \hat{u}(\xi - \cdot), \hat{v} >$$
$$= < \mathscr{F}\hat{u}(\xi - \cdot), v > = (2\pi)^n \widehat{u \cdot v}(\xi).$$

The last equation follows from

$$\hat{u}(\xi - \eta) = \int_{\mathbb{R}^n} u(x)\, e^{-ix(\xi-\eta)}\, dx = (2\pi)^n \left(\widetilde{\mathscr{F}}_{x \mapsto \eta} e^{-ix\xi} u(x) \right)(\eta)$$

Thus

$$\left(\mathscr{F}_{\eta \mapsto x} \hat{u}(\xi - \eta)\right)(x) = (2\pi)^n e^{-ix\xi} u(x)$$

and

$$< \mathscr{F}\hat{u}(\xi - \cdot), v > = (2\pi)^n \int_{\mathbb{R}^n} e^{-ix\xi} u(x)\, v(x)\, dx = (2\pi)^n \widehat{u \cdot v}(\xi).$$

The following formulae are valid for $u, v \in \mathscr{S}(\mathbb{R}^n)$:

$$\hat{u} * \hat{v} = (2\pi)^n \widehat{u \cdot v} \tag{B.11}$$
$$\widehat{u * v} = \hat{u} \cdot \hat{v}. \tag{B.12}$$

We prove (B.12): $\widehat{\tilde{u} \cdot \tilde{v}} = \hat{\tilde{u}} * \hat{\tilde{v}}\, (2\pi)^{-n} = u * v\, (2\pi)^{-n}$.

Hence by Remark B.1

$$\widehat{u * v}\,(x) = (2\pi)^n \left(\mathscr{F}\widehat{\tilde{u}\cdot\tilde{v}}\right)(x) = (2\pi)^{2n}\,(\tilde{u}\cdot\tilde{v})\,(-x)$$
$$= (2\pi)^{2n}\,\tilde{u}\,(-x)\,\tilde{v}\,(-x) = (2\pi)^{2n}\,(2\pi)^{-n}\,\hat{u}\,(x)\,(2\pi)^{-n}\,\hat{v}\,(x) = \left(\hat{u}\cdot\hat{v}\right)(x)$$

which completes the proof. □

Next we give a brief introduction in pseudodifferential operators and symbol-classes. We define for $s \in \mathbb{R}$

$$||\phi||^2_{H^s(\mathbb{R}^n)} := (2\pi)^{-n} \int_{\mathbb{R}^n} (1 + |\xi|^2)^s |\widehat{\phi}(\xi)|^2\, d\xi \quad \phi \in \mathscr{S}(\mathbb{R}^n) \tag{B.13}$$

In the following we also use the abbreviation $<\xi>^{2s} := (1 + |\xi|^2)^s$.

We remark

$$||\phi||^2_{H^0(\mathbb{R}^n)} \equiv ||\phi||^2_{L^2(\mathbb{R}^n)}$$

and introduce the Sobolev space

$$H^s(\mathbb{R}^n) := \{\varphi \in \mathscr{S}'(\mathbb{R}^n) | \widehat{\varphi} \in L^2_{loc}(\mathbb{R}^n),\, ||\varphi||_{H^s(\mathbb{R}^n)} < \infty\}$$
$$= \text{completion of } \mathscr{S}'(\mathbb{R}^n) \text{ with } \; ||\cdot||_{H^s(\mathbb{R}^n)}$$

This gives rise to the Bessel potential operator $\Lambda^s : H^t \to H^{t-s} \quad \forall t, s \in \mathbb{R}$.

The differentiation $D^\alpha : H^s \longmapsto H^{s-|\alpha|}$ is continuous and for all $0 \le s \le t$,

$$\mathscr{S} \subset H^t \subset H^s \subset H^0 = L^2 \subset H^{-s} \subset \mathscr{S}'\,.$$

Now we introduce symbol classes. Let $p(x, D) = \sum_{|\alpha| \le k} a_\alpha(x) D^\alpha$. Then we have

$$p(x, D)u(x) = (2\pi)^{-n} \int_{\mathbb{R}^n} e^{ix\xi} p(x, \xi)\widehat{u}(\xi) d\xi\,, \forall u \in \mathscr{S}(\mathbb{R}^n)\,. \tag{B.14}$$

By the property of the Fourier transform,

$$\mathscr{F}^{-1}\widehat{D^\alpha u} = \mathscr{F}^{-1}(\xi^\alpha \widehat{u})\,,$$

hence

$$p(x, D)u(x) = \sum a_\alpha(x) \mathscr{F}^{-1}\widehat{D^\alpha u} = \sum a_\alpha(x)\mathscr{F}^{-1}(\xi^\alpha\widehat{u}) = \mathscr{F}^{-1} p(x, \xi)\widehat{u}\,.$$

Example B.1

(i) $p(\xi) =<\xi>^s$ $(= (1+|\xi|^2)^{\frac{s}{2}})$. This gives $p(D) = (1-\Delta)^s$, since $-\widehat{\Delta u} = -\widehat{\frac{d^2}{dx^2}u} = -(i\,\xi)^2\widehat{u}(\xi) = \xi^2\widehat{u}(\xi)$.

(ii) $p(\xi) = e^{ia\cdot\xi}$ with $a \in \mathbb{R}$ fixed gives $p(D)u(x) = u(x+a)$.

(iii) For $n = 1$, define using the Poisson kernel,

$$K_t f(x) := \frac{1}{\pi}\int_{-\infty}^{\infty} \frac{tf(y)}{t^2+(x-y)^2}\,dy\,.$$

Then $K_t = p(x, D)$ with $p(x,\xi) = e^{-t|\xi|}$.

Next we write (B.14) for some special functions $p(x,\xi)$.

$p(x,\xi) = p(x)$ gives a multiplication operator that maps $C_0^\infty \to C^\infty$

$p(x,\xi) = p(\xi)$ gives all convolution operators $\mathscr{S} \to \mathscr{S}'$, provided $p \in \mathscr{S}'$, (respectively $L^2 \to L^2 \Leftrightarrow p \in L^\infty$).

In the general case, formal calculus gives

$$\begin{aligned} p(x,D)u(x) &= (2\pi)^{-n}\int e^{ix\cdot\xi}p(x,\xi)\Big(\int e^{-iy\cdot\xi}u(y)\,dy\Big)\,d\xi \\ &= (2\pi)^{-n}\iint e^{-i(x-y)\cdot\xi}p(x,\xi)u(y)\,dy\,d\xi \\ &= \int \check{p}(x, x-y)u(y)\,dy\,, \end{aligned}$$

where $\check{p}(x,z) = (2\pi)^{-n}\int e^{iz\cdot\xi}p(x,\xi)\,d\xi$ or equivalently,

$$p(x,\xi) = \mathscr{F}_{z\to\xi}(\check{p}(x,z))(\xi) \tag{B.15}$$

Hence $p(x, D)$ has a kernel $k(x, y) = \check{p}(x, x-y)$ or equivalently, $\check{p}(x,z) = k(x, x-z)$, and thus

$$\begin{aligned} p(x,\xi) &= \int e^{-iz\cdot\xi}k(x, x-z)\,dz \\ &= \mathscr{F}k(x, x-\cdot)(\xi) \quad \text{if } k(x, x-\cdot) \in \mathscr{S}'\,. \end{aligned}$$

Let us introduce the following notions.

(i) Let Ω open in $\mathbb{R}^n$, fix $m, \rho, \delta \in \mathbb{R}$ with $\rho \le 1$, $\delta \ge 0$. Then define

$$\begin{aligned} S^m_{\rho,\delta}(\Omega) :=&\{p \in C^\infty(\Omega\times\mathbb{R}^n)|\forall K \subset\subset \Omega\ \forall\alpha,\beta \in \mathbb{N}_0^n\ \exists\, C_{K,\alpha,\beta} : \\ &|D_x^\alpha D_\xi^\beta p(x,\xi)| \le C_{K,\alpha,\beta} <\xi>^{n-\rho|\beta|+\delta|\alpha|}, \forall x \in K, \xi \in \mathbb{R}^n\} \end{aligned} \tag{B.16}$$

In the following we often have $\rho = 1, \delta = 0$.

(ii) Classical symbols:
$p \in S^m(\Omega) :\Leftrightarrow p \in S^m_{1,0}$ and
$\exists$ sequence $(p_{m-j})_{j\in\mathbb{N}_0} \subset C^\infty(\Omega \times \mathbb{R}^n)$ with

$$p_{m-j}(x, r\xi) = r^{m-j} p_{m-j}(x, \xi)\,, \forall\, |\xi| \geq 1, r \geq 1$$

(positively homogenous of degree $m - j$ for $|\xi| \geq 1, r \geq 1$) that decays faster than any power,

$$p - \sum_{j=0}^{N} p_{m-j} \in S^{m-N-1}_{1,0}(\Omega) \qquad (\Leftrightarrow\ p \sim \sum_{j\geq 0} p_{m-j})$$

(iii)

$$S^{-\infty} := \bigcap_m S^m = \bigcap_m S^m_{\rho,\delta} \quad \text{(independent of } \rho, \delta)$$

Note that when $v(x) \longmapsto \widehat{v}(\xi)$, the asymptotic behaviour of v for small x corresponds to the asymptotic behaviour of $\widehat{v}$ for large ξ.

Remember

$$(v * u)(x) = \int v(x - y)u(y)\,dy \qquad \widehat{v * u}(\xi) = \widehat{v}(\xi)\widehat{u}(\xi)$$

and consider the following

Example B.2

$$\Delta u = 0 \text{ in } \Omega = \mathbb{R}^+_2\,, u = g \text{ on } \partial\Omega = \mathbb{R}_1$$

Then

$$u(x) = -\frac{1}{2\pi}\int_{\mathbb{R}} \ln|x - y|\phi(y)\,dy$$

satisfies $\Delta u = 0$ in Ω. We have the convolution $u(x) = \int v(x - y)\,\phi(y)\,dy$ with the simple kernel $v(x) = \ln|x|$. Hence $\widehat{g}(\xi) = \widehat{v}(\xi)\widehat{\phi}(\xi)$ and thus

$$\phi(x) = \mathscr{F}^{-1}_{\xi\to x}\widehat{\phi}(\xi) = \mathscr{F}^{-1}_{\xi\to x}\frac{\widehat{g}}{\widehat{v}}(\xi)\,.$$

More generally let

$$u = A\phi\,,\ A\phi(x) = (2\pi)^{-n}\int e^{ix\cdot\xi}a(x, \xi)\widehat{\phi}(\xi)\,d\xi$$

Here the singularity of the kernel $k(x, x-y)$ of A when $x \to y$ is determined by the behaviour of $a(x,\xi)$ for $|\xi| \to \infty$.

Example B.3

(i) Let $p(x,\xi) = \sum_{|\alpha|\le k} a_\alpha(x)\xi^\alpha$, then $p(x,\xi) \in S^k$. Here we have $p_{k-j}(x,\xi) = \sum_{|\alpha|=k-j} a_\alpha(x)\xi^\alpha$. Note that we can write $D_\xi^\beta \xi^\alpha = C_{\alpha\,\beta}\xi^{\alpha-\beta}$, $\forall\, \beta \le \alpha$ with some constants $C_{\alpha\,\beta}$.

(ii) Let $q(\xi) \in C^\infty(\mathbb{R}^n \setminus \{0\})$ be homogeneous of degree m,

$$q(r\xi) = r^m q(\xi), \quad \forall r > 0, \xi \ne 0.$$

Let $\chi \in C_0^\infty(\mathbb{R}^n)$ with $\chi \equiv 1$ in a neighborhood of 0 and $\operatorname{supp}\chi \subset\subset B_1(0)$. Then put $p(x,\xi) = p(\xi) := (1-\chi(\xi))q(\xi)$. Hence there holds $p \in S^m_{1,0}(\mathbb{R}^n)$ (even $\in S^m(\mathbb{R}^n)$). Note that we can modify symbols for small $|\xi|$, since we are interested in the behaviour for large $|\xi|$.

(iii) Let $p(\xi) =< \xi >^{-2}= (1+|\xi|^2)^{-1}$. Then $p \in S^{-2} \subset S^{-2}_{1,0}$. Indeed, $p \in C^\infty$, $D^\alpha p(\xi) = (1+|\xi|^2)^{-1-|\alpha|} \cdot h^{|\alpha|}(\xi)$ with an appropriate polynomial h. Hence $|D^\alpha p(\xi)| \le C < \xi >^{-2-|\alpha|}$ and $p \in S^{-2}_{1,0}$ follows. Moreover, we use the asymptotic expansion

$$\frac{1}{1+|\xi|^2} = |\xi|^{-2}\frac{1}{1+|\xi|^{-2}} = -\sum_{k=1}^{\infty}(-1)^k|\xi|^{-2k} \; (|\xi| > 1)$$

and put

$$p_{-2k}(\xi) := (1-\chi(\xi))(-1)^{k+1}|\xi|^{-2k}$$

Then as seen above in (2.) $p_{-2k} \in S^{-2k}_{1,0}$ and hence

$$p(\xi) - \sum_{k=1}^{N} p_{-2k}(\xi) \in S^{-2N-2}_{1,0} \subset S^{-2N-1}_{1,0} \quad (N \ge 1)$$

(iv) Let $p(\xi) =< \xi >^s$. Then $p \in S^s \subset S^s_{1,0}$, $\forall s \in \mathbb{R}$

Lemma B.1 $D^\alpha(\frac{1}{u}) = \frac{1}{u} \sum_{k\le|\alpha|} C_{\alpha_1,\ldots,\alpha_k} \frac{D^{\alpha_1}u}{u} \cdot \ldots \cdot \frac{D^{\alpha_k}u}{u}$

Theorem B.1 *Let* $p \in S^m_{1,0}(\Omega)$ *and*

$$|\frac{1}{p(x,\xi)}| \le c < \xi >^{-m} \quad \textit{for } |\xi| \ge 1 \qquad (\Longleftrightarrow \textit{elliptic})$$

Then $\frac{1-\chi}{p} \in S^{-m}_{1,0}(\Omega)$.

Proof We have

$$|\frac{D_x^{\alpha_\varkappa} D_\xi^{\beta_\lambda} p(x,\xi)}{p(x,\xi)}| \le C < \xi >^{m-|\beta_\lambda|} < \xi >^{-m} = C < \xi >^{-|\beta_\lambda|}$$

and hence by the lemma above

$$|D_x^\alpha D_\xi^\beta \frac{1}{p(x,\xi)}| \le |\frac{1}{p(x,\xi)}| \sum_{\substack{\sum \alpha_\kappa = \alpha \\ \sum \beta_\lambda = \beta}} C_{\alpha\beta} \prod |\frac{D_x^{\alpha_\varkappa} D_\xi^{\beta_\lambda} p}{p}|$$
$$\le C < \xi >^{-m} < \xi >^{-|\beta|} .$$

□

Theorem B.2 *Let* $p \in S^m_{1,0}(\Omega)$, $q \in S^{m'}_{1,0}(\Omega)$ *. Then*

$$D_x^\alpha D_\xi^\beta p \in S_{1,0}^{m-|\beta|}(\Omega) \quad \textit{and } p \cdot q \in S_{1,0}^{m+m'}(\Omega) .$$

Proof Use Leibniz rule

$$D_x^\alpha D_\xi^\beta (p \cdot q)(x,\xi) = \sum_{\substack{\alpha'+\alpha''=\alpha \\ \beta'+\beta''=\beta}} \binom{\alpha}{\alpha'} \binom{\beta}{\beta'} D_x^{\alpha'} D_\xi^{\beta'} p(x,\xi)\, D_x^{\alpha''} D_\xi^{\beta''} q(x,\xi)$$

and estimate the partial derivatives of p by $< \xi >^{m-|\beta'|}$, respectively the partial derivatives of q by $< \xi >^{m'-|\beta''|}$ modulo some positive constant factor, what leads to the upper bound $< \xi >^{m+m'-|\beta|}$ modulo a positive constant.

□

We define a pseudodifferential operator of class $S^m_{1,0}$

$$p(x,D) \in OP\, S^m_{1,0}(\Omega) \quad :\Leftrightarrow \quad p(x,\xi) \in S^m_{1,0}(\Omega)$$

with

$$p(x,D)u(x) = (2\pi)^{-n} \int_{\mathbb{R}^n} e^{ix\cdot\xi} p(x,\xi) \widehat{u}(\xi)\, d\xi , \qquad \forall u \in C_0^\infty(\mathbb{R}^n) \tag{B.17}$$

Theorem B.3 *Let* $m \in \mathbb{R}$, , $p(x,D) \in OP\, S^m_{1,0}(\Omega)$. *There holds*

$$p(x,D) : C_0^\infty(\Omega) \longmapsto C^\infty(\Omega) \quad \textit{continuous, linear.}$$

Proof Now $u \in C_0^\infty(\mathbb{R}^n)$ implies $\widehat{u} \in \mathscr{S}(\mathbb{R}^n)$. Hence $p(x,\xi)\widehat{u}(\xi)$ still decays fast and the integral converges absolutely. Hence interchanging differentiation and integration yields with (B.17)

$$|D_x^{\alpha}(p(x,D)u(x))| \le c \sum_{\alpha'+\alpha''=\alpha} \int <\xi>^m <\xi>^{|\alpha''|} |\widehat{u}(\xi)|\, d\xi < \infty \quad \forall \alpha$$

Thus $p(x,D)u \in C^{\infty}$. □

Exercise: Show the mapping $p(x,D) \in OP\ S^m_{1,0} : H^s(\mathbb{R}^m) \longmapsto H^{s-m}(\mathbb{R}^m)$ is continuous, that is, there exists $C > 0$ such that $\|p(x,D)u\|_{H^{s-m}} \le C\,\|u\|_{H^s}$.

Next we consider the relation between strong ellipticity of a pseudodifferential operator and Gårding's inequality. As shown in Sect. 4.2 with the example of the single layer operator, considering integral operators as pseudodifferential operators allows to deduced the mapping properties of boundary integral operators by examining the symbols of the pseudodifferential operators. On the other hand, Garding's inequality for integal equations is the key property to guarantee convergence of Galerkin's method, see Theorem 6.1, Theorem 6.11. Now, Garding's inequality follows from the definition of uniform strong ellipticity of pseudodifferential operators, see Theorem 6.2.7 in [259].

Definition B.7 A system of pseudodifferential operators $A_{jk} \in OP\ S^{s_j+t_k}_{1,0}(\Omega)$ is called uniformly strongly elliptic if for the principal part matrix $a^0(\ x\ ;\ \xi) = ((a^{jk0}_{s_j+t_k}(x;\xi)))_{p\times p}$ there exist a C^{∞} -matrix valued function $\Theta(x) = ((\Theta_{jk}(x)))_{p\times p}$ and a constant $\gamma_0 > 0$ such that

$$\Re \zeta^T \Theta(x) a^0(x,\xi)\overline{\zeta} \ge \gamma_0 |\zeta|^2$$

for all $x \in \Omega, \zeta \in \mathbb{C}^p$ and $\xi \in \mathbb{R}^n$ with $|\xi| = 1$.

A uniformly strongly elliptic system of pseudodifferential operators satisfies a Gårding inequality, see [259, Theorem 6.2.7.] . In the following we present and prove the corresponding result for a single pseudodifferential operator:

Theorem B.4 (Gårding Inequality) *Let $p(x,\xi) \in OP\ S^m_{1,0}(\Omega)$ be strongly elliptic , i.e. $\forall K \subset\subset \Omega$ let there exist positive constants C_K, R_K such that there holds*

$$\Re p(x,\xi) \ge C_K <\xi>^m \ \forall x \in K, |\xi| \ge R_K$$

Then $\forall K \subset\subset \Omega$ and $\forall s \in \mathbb{R}$ there exist constants γ_K, $C_{K,s}$ such that

$$\Re(p(x,D)u,u) \ge \gamma_K ||u||^2_{H^{m/2}(\Omega)} - C_{K,s}||u||^s_{H^s(\Omega)}$$

$\forall u \in C_0^{\infty}(\Omega)$.

Lemma B.2 *Let* $p \in S^0_{1,0}(\Omega), \Re p(x,\xi) \geq C > 0\ \forall x, \xi$ *($|\xi|$ sufficiently large) then there exist* $B \in OP\ S^m_{1,0}(\Omega),\ K \in OP\ S^{-\infty}(\Omega)$ *such that*

$$\Re p(x, D) = B^* B + K.$$

Proof Setting $q(x, D) := \Lambda^{-\frac{m}{2}} p \Lambda^{-\frac{m}{2}}$ we have for $u \in C^\infty_0$ $(pu, u) = (q\Lambda^{\frac{m}{2}} u, \Lambda^{\frac{m}{2}} u)$ and $||u||^2_{H^{1/2}} \sim ||\Lambda^{\frac{m}{2}} u||^2_{L^2}$. Therefore it suffices to show for $m = 0$. Now we use the above lemma for $p_0(x,\xi) := \Re p(x,\xi) - c'$, with 0<c'<c. Hence $p_0(x,\xi) \geq c - c' > 0$. Then there exists $b \in OP\ S^0_{1,0}$ such that $\Re p_0(x, D) - B^* B =: S \in OP\ S^\infty$ such that

$$\Re(p(x, D)u, u) - c'||u||^2_{L^2} = ||Bu||^2_{L^2} + \Re(Su, u)$$

and this yields

$$\Re(p(x, D)u, u) \geq c''||u||^2_{L^2} + \Re(Su, u).$$

□

As a consequence, any strongly elliptic pseudodifferential operator defines a Fredholm operator of index zero since for the corresponding bilinear form one may apply the classical Fredholm alternative (A.3).

Example B.4 Writing the single layer potential as

$$V\psi(x) = (2\pi)^{1-n} \int_{\mathbb{R}^{n-1}} e^{ix'\xi'} \frac{\widehat{u}(\xi')}{|\xi'|} d\xi'$$

gives

$$\int_\Gamma (V\psi(x')\psi(x')dx' = \Re(\widehat{V\psi}(\xi'), \widehat{\psi}(\xi')) = \Re(2\pi)^{1-n} \int_{\mathbb{R}^{n-1}} \frac{1}{|\xi'|} \widehat{\psi}(\xi')\overline{\widehat{\psi}}(\xi')d\xi'$$
$$\geq \gamma||\psi||^2_{H^{-1/2}(\Gamma)} - \text{compact perturbation}.$$

Example B.5 The single layer potential in linear elasticity with fundamental solution

$$E(x, y) = 1/|x - y|I + \kappa(x - y)(x - y)^T$$

where $\kappa = \frac{\lambda+\mu}{\lambda+3\mu}$ has principal symbol

$$\sigma_0(V)(\xi) = \frac{\lambda + 3\mu}{2\mu(\lambda + 2\mu)} \frac{1}{|\xi|^3} \begin{pmatrix} |\xi|^2 + \kappa\xi_2^2 & -\kappa\xi_1\xi_2 & 0 \\ -\kappa\xi_1\xi_2 & |\xi|^2 + \kappa\xi_1^2 & 0 \\ 0 & 0 & |\xi|^2 \end{pmatrix}$$

The corresponding hypersingular operator has principal symbol

$$\sigma_0(W)(\xi) = \frac{-\mu^2}{|\xi|} \begin{pmatrix} |\xi|^2 + \epsilon\xi_1^2 & \epsilon\xi_1\xi_2 & 0 \\ \epsilon\xi_1\xi_2 & |\xi|^2 + \epsilon\xi_2^2 & 0 \\ 0 & 0 & (1+\epsilon)|\xi|^2 \end{pmatrix}$$

with $-1/2 < \epsilon := \lambda(\lambda + 2\mu)^{-1} < 1$ see [130, 396].

Appendix C
Convex and Nonsmooth Analysis, Variational Inequalities

C.1 Convex Optimization, Lagrange Multipliers

By this section of Appendix C we invite the reader to get acquainted with some fundamental concepts, methods, and results of convex optimization that are necessary for the proper understanding of the mathematical and numerical treatment of inequality constrained problems that occur in the Signorini boundary value problem and in further nonsmooth boundary value problems, see Chap. 5, and in contact problems, see Chap. 11 and also Sect. 12.5

Based on the monograph [55] of Blum and Oettli, we start with convex quadratic optimization in finite dimensions. Already at this level we encounter different formulations, namely a "primal" and a "mixed" formulation with signed Lagrange multipliers that are associated to inequality constraints. In fact, the existence of such Lagrange multipliers can be derived from the celebrated duality theory of linear optimization ("linear programming") without any further assumptions. Moreover, a solution in convex quadratic optimization is characterized by a "linear complementarity problem" and by a variational inequality (VI) of a special structure.

Then we proceed to convex variational problems in Hilbert space. As a straightforward extension of the finite dimensional case, we characterize solutions by variational inequalities with symmetric bilinear forms. Also guided by the finite dimensional case, we readily introduce the Lagrange function for convex cone constraints. However, the existence of Lagrange multipliers is more involved than in the finite dimensional case. First we construct the Lagrange multiplier in the space dual to the solution space of the primal variable, which is a Sobolev space of negative order in application to contact problems. Then in the subsequent subsection we follow [219] and present mixed formulations with Lagrange multipliers that live in the Hilbert space of constraints, which is the more regular L^2 function space on the contact boundary part in the application to unilateral contact problems. To this end we provide an extension of the famous Brezzi splitting theorem that originally covers sad-

© Springer International Publishing AG, part of Springer Nature 2018

J. Gwinner, E. P. Stephan, *Advanced Boundary Element Methods*,
Springer Series in Computational Mathematics 52,
https://doi.org/10.1007/978-3-319-92001-6

dle point problems with equality constraints, only, to a class of nonsmooth inequality constrained variational problems. Under the celebrated Babuška-Brezzi condition we obtain independent Lagrange multipliers in the ordering cone of the inequality constraints and in the subdifferential of the convex nonsmooth sublinear functional.

C.1.1 Convex Quadratic Optimization in Finite Dimensions

For given data $b \in \mathbb{R}^n$, $d \in \mathbb{R}^m$, $C \in \mathbb{R}^{m \times n}$, $A \in \mathbb{R}^{n \times n}$, where A is a symmetric positive semidefinite matrix, shortly $A = A^T \geq 0$, we consider the convex finite dimensional quadratic optimization problem with linear inequality constraints ("quadratic program")

$$(\mathrm{QP}) \begin{cases} \text{minimize } f(x) = \frac{1}{2} x^T A x - b^T x \\ \text{subject to } x \geq 0, \ Cx \leq d\,. \end{cases}$$

Put in another way, among all feasible solutions x to (QP), that is, $x \in \mathbb{R}^n_+$, shortly $x \geq 0$, that satisfy the constraints $(Cx)_j \leq d_j$ $(\forall j = 1, \ldots, m)$ we are looking for that feasible $\hat{x}$ that minimizes the objective function f.

The symmetry requirement $A = A^T$ is not essential, since we can replace the matrix A by its symmetric part $\dfrac{1}{2}(A + A^T)$ in the objective function f. In the formal discussion to follow, considering only signed variables x_i for $i = 1, \ldots, n$ does not lead to a loss of generality either, since for a free variable x_i we can use its decomposition $x_i = x_i^+ - x_i^-$, $x_i^+ \geq 0$, $x_i^- \geq 0$. Also an equality constraint $c_j^T x = d_j$ can be rewritten as

$$\begin{Bmatrix} c_j^T x \leq & d_j \\ -c_j^T x \leq & -d_j \end{Bmatrix}.$$

Of course, these two latter trivial reformulations are not appropriate in numerical computation, but are convenient here to reduce the discussion of constrained optimization problems to the standard form (QP) given above.

Now we take (QP) as primal optimization problem ("primal program") and proceed to its mixed formulation via the Lagrange function

$$L(x, y) = f(x) + y^T (Cx - d)\,.$$

In view of the sign conditions and the inequality constraints, the Lagrange function is considered only for $x \geq 0$, $y \geq 0$, since we have for any $x \geq 0$,

$$\sup_{y \geq 0} L(x, y) = \begin{cases} f(x) & \text{if } x \text{ is feasible;} \\ +\infty & \text{otherwise.} \end{cases}$$

This gives

$$\inf_{x\geq 0}\sup_{y\geq 0} L(x,y) = \inf(\mathrm{QP})\,,$$

where inf (QP) denotes the optimal value of (QP). Therefore in the sense of convex duality theory, the dual optimization problem ("dual quadratic program") to (QP) reads

$$(\mathrm{DQP})\ \begin{cases}\text{maximize } \inf_{x\geq 0} L(x,y)\\ \text{subject to } y\geq 0\,.\end{cases}$$

Obviously, $\inf\sup L \geq \sup\inf L$ is trivial. But in finite dimensions, without further assumptions, we have even the "duality equality" $\inf(\mathrm{QP}) = \sup(\mathrm{DQP})$; moreover, the dual problem attains an optimal solution, what is nothing else than a Lagrange multiplier to the inequality constrained optimization problem (QP):

Theorem C.1 *If (QP) has an optimal solution $\hat{x}$, then there exists a Lagrange multiplier $\hat{y}\geq 0$ such that $(\hat{x},\hat{y})$ is a saddle point of L on $\mathbb{R}^n_+\times\mathbb{R}^m_+$, that is, we have*

$$(SP)\quad L(\hat{x},y)\leq L(\hat{x},\hat{y})\leq L(x,\hat{y});\quad \forall x\in\mathbb{R}^n_+,\, y\in\mathbb{R}^m_+\,.$$

Proof To prove (SP) it is enough to establish the Karush–Kuhn–Tucker conditions, which read for the linear constraints in (QP) here

$$(\mathrm{KKT})\quad f(\hat{x})\leq f(x)+\hat{y}^T(Cx-d)\ \ \forall x\in\mathbb{R}^n_+\,.$$

Indeed, in view of feasibility, $C\hat{x}-d\leq 0$, (KKT) implies the equality

$$(*)\quad \hat{y}^T(C\hat{x}-d)=0\,.$$

Hence, the right hand side of (SP) follows from (KKT) directly, whereas the left hand side of (SP) is equivalent to (*) and the feasibility of $\hat{x}$.

Therefore it remains to show the existence of $\hat{y}\in\mathbb{R}^m_+$ that satisfies (KKT). Here we rely on the duality theorem of finite dimensional linear optimization ("linear programming") and first show the following

Proposition C.1 *Let $\hat{x}$ be an optimal solution to (QP). Then $\hat{x}$ is an optimal solution to the linear program*

$$(LP)\begin{cases}\textit{minimize } (A\hat{x}-b)^T x =: c^T x\\ \textit{subject } x\geq 0,\, Cx\leq d\,.\end{cases}$$

Proof of the Proposition Since the constraints of (QP) and (LP) are the same, it is enough to give the following contradiction argument. Suppose there exists $\tilde{x}\in\mathbb{R}^n_+$

such that $c^T\tilde{x} < c^T\hat{x}$ and $C\tilde{x} \le d$. Then consider $x_t = \hat{x} + t(\tilde{x} - \hat{x})$, where $0 < t < 1$; x_t is feasible for (QP). By $\nabla f(\hat{x})^T(\tilde{x} - \hat{x}) = c^T(\tilde{x} - \hat{x}) < 0$, for small enough $t > 0$, we arrive at $f(x_t) < f(\hat{x})$ contradicting the optimality of $\hat{x}$. □

Proof of the theorem continued. The dual linear optimization problem ("dual program") to (LP) reads

$$(DLP)\begin{cases}\text{maximize } -d^T y\\ \text{subject } y \ge 0,\, C^T y + c \ge 0\,;\end{cases}$$

this can be seen by means of the associated Lagrange function

$$l(x, y) = c^T x + y^T(Cx - d) = -d^T y + x^T(c + C^T y)$$

on $\mathbb{R}^n_+ \times \mathbb{R}^m_+$ and by the relation

$$\inf_{x\ge 0} l(x, y) = \begin{cases}-d^T y & \text{if } C^T y + c \ge 0\,;\\ -\infty & \text{otherwise.}\end{cases}$$

In virtue of the duality theorem of linear programming, see [55, 140], there exists $\hat{y} \in \mathbb{R}^m_+$ such that

$$\text{(i) } C^T\hat{y} \ge -c\,,$$
$$\text{(ii) } c^T\hat{x} = -d^T\hat{y}\,.$$

Then multiplying (i) by arbitrary $x \ge 0$ gives

$$x^T C^T\hat{y} \ge -x^T A\hat{x} + b^T x\,,$$

hence by (ii)

$$(Cx - d)^T\hat{y} \ge \hat{x}^T A(\hat{x} - x) + b^T x - b^T\hat{x}\,.$$

Thus we obtain

$$\begin{aligned} f(x) + (Cx - d)^T\hat{y} &\ge \hat{x}^T A(\hat{x} - x) + \tfrac{1}{2}x^T Ax - b^T\hat{x}\\ &= \tfrac{1}{2}\hat{x}^T A\hat{x} + \tfrac{1}{2}[(\hat{x} - x)^T A(\hat{x} - x)] - b^T\hat{x}\,,\end{aligned}$$

and since A is positive semidefinite,

$$f(x) + (Cx - d)^T\hat{y} \ge \frac{1}{2}\hat{x}^T A\hat{x} - b^T\hat{x}\,,$$

what is the claimed (KKT) inequality. □

We remark that the saddle point inequalities (SP) are clearly also sufficient for the optimality of $\hat{x}$.

We can characterize the optimality of $\hat{x}$ in another way using slack variables. Define the primal slack variable

$$v = d - Cx \in \mathbb{R}^m ,$$

then feasibility is equivalent to $v \geq 0$ and (*) reads $\hat{v}^T \hat{y} = 0$ with $\hat{v} = d - C\hat{x}$. Likewise define the dual slack variable

$$u = c + C^T y = A\hat{x} - b + C^T y \in \mathbb{R}^n .$$

Then for $y \geq 0$, feasibility in (DLP) is equivalent to $u \geq 0$ and with $\hat{u} = c + C^T \hat{y} = A\hat{x} - b + C^T \hat{y}$, we conclude from (ii) and (*) that

$$(**) \quad \hat{x}^T \hat{u} = 0 .$$

Since $\hat{v}_j \geq 0, \hat{y}_i \geq 0, \hat{u}_i \geq 0, \hat{x}_i \geq 0$, (*) means $\hat{v}_j = 0$ or $\hat{y}_j = 0$ and (**) means $\hat{u}_i = 0$ or $\hat{x}_i = 0$. In this sense $\hat{v}$ and $\hat{y}$, respectively $\hat{u}$ and $\hat{x}$ are “complementary variables”.

Altogether we obtain the following

Corollary C.1 *$\hat{x}$ is an optimal solution to (QP), if and only if $(\hat{x}, \hat{y}, \hat{u}, \hat{v}) \in \mathbb{R}^{n+m} \times \mathbb{R}^{n+m}$ satisfies*

$$\begin{pmatrix} u \\ v \end{pmatrix} = \begin{pmatrix} A & C^T \\ -C & 0 \end{pmatrix} \begin{pmatrix} x \\ y \end{pmatrix} + \begin{pmatrix} -b \\ d \end{pmatrix}$$

$$\begin{pmatrix} u \\ v \end{pmatrix} \geq 0 \quad \begin{pmatrix} x \\ y \end{pmatrix} \geq 0 \quad \begin{pmatrix} u \\ v \end{pmatrix}^T \begin{pmatrix} x \\ y \end{pmatrix} = 0 .$$

The above system of linear equations and sign inequalities can be considered as a “mixed formulation” of the convex quadratic optimization problem (QP). It leads to the

Definition Let $F : \mathbb{R}^N \to \mathbb{R}^N$ be given. Then the “complementarity problem” consists in finding $\hat{z} \in \mathbb{R}^N_+$ such that $F(\hat{z}) \in \mathbb{R}^N_+$ and $\hat{z}^T F(\hat{z}) = 0$ hold. We have a “linear complementarity problem” (LCP), if $F(z) = Bz - a$ is affine-linear for some $B \in \mathbb{R}^{N \times N}, a \in \mathbb{R}^N$.

Thus the solution of (QP) can be characterized as the solution of a special (LCP), where the matrix B has the special saddle point structure

$$B = \begin{pmatrix} A & C^T \\ -C & 0 \end{pmatrix} .$$

Furthermore a solution $\hat{z}$ to the complementarity problem can be characterized by the following "variational inequality":

$$\hat{z} \in \mathbb{R}^N_+ , \quad F(\hat{z})^T (z - \hat{z}) \geq 0 \quad \forall z \in \mathbb{R}^N_+ .$$

Indeed, the direct implication being obvious, only the reverse implication needs an argument; for that choose $z = \frac{1}{2}\hat{z}$ and $z = 2\hat{z}$. In the case of a linear complementarity problem, the variational inequality reads

$$\hat{z} \in \mathbb{R}^N_+ , \quad (B\hat{z})^T (z - \hat{z}) \geq a^T (z - \hat{z}) \quad \forall z \in \mathbb{R}^N_+ .$$

Remark Also in the case of a linear complementarity problem, the solution generally depends *nonlinear* on the data, e.g. on the datum a!

An unessential extension of the problem is obtained by a simple translation: Let $c \in \mathbb{R}^N$ be given; find $z \in \mathbb{R}^N$, such that $z \geq c$, $F(z) \geq 0$, $(z - c)^T F(z) = 0$.

An essential extension of the problem is obtained as follows. Instead of R^N_+, consider an arbitrary convex cone K (that is, $K + K \subseteq K$, $\mathbb{R}_+ K \subseteq K$) in $\mathbb{R}^N$, not necessarily polyhedric, define the positive polar cone $K^+ = \{u \in \mathbb{R}^N | u^T x \geq 0, \forall x \in K\}$. Then the complementarity problem consists in finding $\hat{z} \in \mathbb{R}^N$ such that $\hat{z} \in K$, $F(\hat{z}) \in K^+$, $\hat{z}^T F(\hat{z}) = 0$. Again, this can be characterized by a variational inequality. In the case of a linear complementarity problem with F affine-linear as above, this variational inequality reads

$$\hat{z} \in K , \quad (B\hat{z})^T (z - \hat{z}) \geq a^T (z - \hat{z}) \quad \forall z \in K .$$

For more information on linear complementarity problems and variational inequalities in finite dimensions we refer to the monographs of Cottle, Pang, and Stone [139] and of Facchinei and Pang [168, 169], respectively.

C.1.2 Convex Quadratic Optimization in Hilbert Spaces

Let V be a real Hilbert space (may be also a reflexive Banach space) and Z another real Hilbert space with its dual Z'. Let $A \in \mathscr{L}(V, V')$ with $A = A'$, $A \geq 0$ (i.e. $\langle Av, v\rangle \geq 0, \forall v \in V$), further $B \in \mathscr{L}(V, Z')$ and $f \in V'$, $g \in Z'$ fixed elements. We also need the adjoint $B' \in \mathscr{L}(Z, V')$. Moreover let an order $\leq$ be defined in Z via a convex closed cone $P \subset Z$ via $z \geq 0$ iff $z \in P$. Also $\zeta \in Z' \leq 0$ iff ζ lies in the negative dual cone $P^- = \{ \zeta \in Z' : \zeta(p) \leq 0, \forall p \in P\}$. With these given data, similar to (QP) in C.1.1, we consider the convex quadratic optimization problem

$$(CP) \begin{cases} \text{minimize } f(v) = \frac{1}{2}\langle Av, v\rangle - \langle f, v\rangle \\ \text{subject to } Bv \leq g . \end{cases}$$

This gives rise to the bilinear form $a(u, v) := \langle Au, v\rangle$ and the convex closed sets,

$$K(g) := \{v \in V \mid Bv \leq g\}$$

which is a translate of the convex closed cone (with vertex at zero)

$$K := \{v \in V \mid Bv \leq 0\}.$$

As in C.1.1 a solution u of (CP) is characterized by a variational inequality, here

$$u \in K(g),\ a(u, v-u) \geq \langle f, v-u\rangle\,, \forall v \in K(g)\,. \tag{C.1}$$

Analogously to C.1.1 , we introduce the Lagrangian

$$L(v, p) := f(v) + \langle p, Bv - g\rangle_{Z\times Z'} = f(v) + \langle B'p, v\rangle_{V'\times V} - g(p), \quad v \in V, p \in P\,,$$

to arrive at saddle points and to mixed formulations.

We can drop the requirement that $A = A'$ and now start from the primal VI (C.1). However, the existence of Lagrange multipliers in the cone P in the infinite dimensional space Z is more involved than in finite dimensions. As the recent paper [219] shows, this can be accomplished by an extension of the Brezzi splitting theorem under the Babuška-Brezzi condition. We postpone a sketch of this approach to Lagrange multipliers to the next subsection.

Before that we describe here first an easier approach under the assumption that there exists a preimage w of g under B, thus $Bw = g$. This allows to to work with the duality on $V \times V'$ and to obtain the following characterization via multipliers in the negative dual cone K^- to K.

Proposition C.2 *$u \in K(g)$ solves the VI (C.1), iff there exists $\lambda \in V'$ such that $(u, \lambda) \in K(g) \times K^-$ solves the mixed system*

$$(MP) \quad \begin{cases} a(u, v) + \langle \lambda, v\rangle = \langle f, v\rangle \\ \langle \kappa - \lambda, u - w\rangle \leq 0\,, \end{cases}$$

for all $v \in V, \kappa \in K^-$. Then there holds the complementarity condition

$$(*) \quad \langle \lambda, u - w\rangle = 0\,.$$

Proof Let $u \in K(g)$ solve the VI (C.1). Define $\lambda \in V'$ by $\lambda(v) = f(v) - a(u, v)$. Then $(MP)_1$ holds. Further, for any $v \in K$, $\tilde{v} := v + u$ lies in $K(g)$ and hence

$$-\lambda(v) = a(u, \tilde{v} - u) - f(\tilde{v} - u) \geq 0\,.$$

Thus $\lambda \in K^-$. Since $w \in K(g)$, $u - w \in K$,

$$\langle \kappa - \lambda, u - w\rangle = \langle \kappa, u - w\rangle + [a(u, u - w) - f(u - w)] \leq 0$$

for any $\kappa \in K^-$ and therefore (MP) holds.

The complementarity condition $(*)$ follows from $(MP)_2$ by the choice $\mu = 2\lambda$, $\mu = 0$.

Vice versa, let $v \in K(g)$, hence $v - w \in K$. This implies by the complementarity condition $(*)$

$$\langle \lambda, v - u\rangle = \langle \lambda, v - w\rangle - \langle \lambda, u - w\rangle \leq 0 .$$

Hence we arrive at $a(u, v - u) = (f - \lambda)(v - u) \geq f(v - u)$. □

From the proof above, it follows that $u \in K(g)$ solves the VI (C.1), iff there exists λ such that $[u, \lambda] \in K(g) \times K^-$ solves $(MP)_1$ and $(*)$ holds. Therefore the above mixed form does not depend on the chosen preimage w. Indeed, let $Bw_i = g$ $(i = 1, 2)$. Then $u \pm (w_1 - w_2) \in K(g)$ and thus by the VI (C.1), $\lambda(w_1 - w_2) = 0$.

The mixed formulation above applies to unilateral contact problems with Signorini condition on some boundary part Γ_c in appropriate function spaces, where for a boundary variable u the linear map $u \mapsto Bu$ is the restriction to the boundary part Γ_c; see Sect. 11.4.1.

With friction problems we encounter nonsmooth optimization problems of the form

$$(NOP) \quad \text{minimize } f(v) = \frac{1}{2}\langle Av, v\rangle - \langle f, v\rangle + \varphi(v), \;\; v \in V ,$$

where φ is convex, even positively homogeneous, hence sublinear on V, but not differentiable in the classic sense. A prominent example is

$$\varphi(v) = \int_{\Gamma_c} g|v| \, ds \quad (g \in L^\infty(\Gamma_c), g > 0) .$$

An optimal solution of (NOP) is characterized as solution to the so-called variational inequality of the second kind:

$$u \in V, \; \langle Au, v - u\rangle + \varphi(v) - \varphi(u) \geq f(v - u), \;\; \forall v \in V .$$

Here one can obtain by (L^1, L^∞) duality and density the useful duality formula

$$\varphi(v) = \int_{\Gamma_c} g|v| \, ds = \sup\{\int_{\Gamma_c} gv\mu \, ds \mid \mu \in C(\Gamma), |\mu| \leq 1\} .$$

C.1.3 *Lagrange Multipliers for Some Inequality Constrained Variational Inequalities*

In this subsection we deal with a canonical class of inequality constrained variational inequalities of the second kind, where the sum of a bilinear form and a sublinear functional and further a linear functional as right hand side occur and where the constraints are defined by linear inequalties with respect to a closed convex ordering cone. More precisely, let V, Z be real reflexive Banach spaces with (topological) dual spaces V', Z'. Let $P \subset Z$ be a closed convex cone with vertex at zero. Let $A \in \mathcal{L}(V, V'), B \in \mathcal{L}(V, Z')$ be continuous linear operators that give rise to the continuous bilinear forms $a : V \times V \to \mathbb{R},\ b : V \times Z \to \mathbb{R}$ via $a(v, w) = \langle Av, w\rangle_{V' \times V},\ b(v, z) = \langle Bv, z\rangle_{Z' \times Z}$. We use the null space $W := \ker B$ of B and its polar W° contained in V'. Further let $\varphi : V \to \mathbb{R}$ be sublinear, thus there holds the representation formula

$$\varphi(v) = \max_{\sigma \in S} \langle \sigma, v\rangle, \quad \forall v \in V, \tag{C.2}$$

where $S \subset V'$ is weak* compact and coincides with the convex subdifferential $\partial\varphi(0) = \{\xi \in V' | \langle \xi, \cdot\rangle \le \varphi\}$. In other words, φ is the support function [252] of S. Finally let $f \in V', g \in Z'$ be fixed. Then introduce the feasible set

$$K(g) = \{v \in V\ :\ b(v, p) \le \langle g, p\rangle,\ \forall p \in P\}$$

and pose the variational inequality in its primal form: Find $u \in V$ that satisfies

$$(VI) \quad u \in K(g),\ a(u, v - u) + \varphi(v) - \varphi(u) \ge \langle f, v - u\rangle,\ \forall v \in K(g)\,.$$

Our goal in this subsection is to arrive at the following mixed form with Lagrange multipliers $q \in P$ and $\tau \in S$:

$$(MF) \begin{cases} (MF-1)\ a(u, v) + b(v, q) + \langle \tau, v\rangle = \langle f, v\rangle\,, & \forall v \in V\,, \\ (MF-2)\ b(u, p - q) + \langle u, \sigma - \tau\rangle \le \langle g, p - q\rangle\,, & \forall [p, \sigma] \in P \times S\,. \end{cases}$$

To achieve this goal we use the famous Brezzi lemma which characterizes that B', the adjoint operator of B, is isomorph, i.e. is bijective with continuous inverse, by the celebrated Babuška-Brezzi condition (BB). More precisely, there holds

Lemma C.1 *The following assertions are equivalent.*

(i) There exists a number $\beta > 0$ *such that*

$$(BB) \qquad \sup_{v \in V, v \neq 0} \frac{b(v, z)}{\|v\|_V} \ge \beta\, \|z\|_Z\,, \qquad \forall z \in Z\,,$$

(ii) $B' : Z \to W^\circ$ *is isomorph with*

$$\|B'z\|_{V'} \geq \beta \, \|z\|_Z \,, \qquad \forall z \in Z \tag{C.3}$$

for some $\beta > 0$.

For the proof of the Brezzi lemma we can e.g. refer to [60, Theorem 3.6, Lemma 4.2].

Now we focus to the homogeneous case, where $g = 0$ with feasible set $K =: K(0)$, since the proof of this case is simpler and nearer to the linear functional analytical proof of the classic case of equality constrained variational problems than the proof for general g.

Theorem C.2 *The two problems* (VI) *and* (MF) *are related as follows. If* $[u, q, \lambda] \in V \times Z \times V'$ *solves* (MF) *(with* $g = 0$*), then* u *lies in* K *and solves* (VI)*. Vice versa, let* $u \in K$ *solve* (VI)*, then there exist* $q \in P$ *and* $\tau \in S$ *such that* $[u, q, \tau]$ *solves* (MF) *(with* $g = 0$*), provided* (BB) *holds for some* $\beta > 0$.

Proof We give a sketch of the proof divided in several steps.

I. Since P is a cone, we can choose $p = 1/2\, q$ and $p = 2q$ in $(MF - 2)$. Moreover we use (C.2). Thus we first observe that $(MF - 2)$ with $g = 0$ splits equivalently into the statements

$$(MF - 3) \qquad \begin{cases} b(u, p) \leq 0 \,, \forall p \in P \,, \\ b(u, q) = 0 \,, \\ \varphi(u) = \langle \tau, u \rangle \,. \end{cases}$$

II. Let $[u, q, \tau] \in V \times Z \times V'$ solve (MF) with $g = 0$. Then from $(MF - 3)_1$ it is immediate that $u \in K$.

To show that u solves (VI), let $v \in K$ be arbitrary. Then $b(v, q) \leq 0$ and from $(MF - 3)_2$, $b(v - u, q) \leq 0$. Hence from $(MF - 3)_3$, (C.2), and $(MF - 1)$,

$$\begin{aligned} & a(u, v - u) + \varphi(v) - \varphi(u) \\ & \geq a(u, v - u) + \langle \tau, v - u \rangle \\ & = -b(v - u, q) + \langle f, v - u \rangle \\ & \geq \langle f, v - u \rangle \,. \end{aligned}$$

III. The proof of the second part of the theorem runs in 5 steps.

1. Let $u \in K$ solve (VI). Since K is a cone, we can choose $v = 2u$ and $v = 1/2\, u$. This gives

$$a(u, u) + \varphi(u) = \langle f, u \rangle \,, \tag{C.4}$$

hence by addition,

$$a(u,v)+\varphi(v)\geq\langle f,v\rangle\,,\forall v\in K\,. \tag{C.5}$$

Note that (C.4) and (C.5) are equivalent to (VI).

2. By (C.2), (C.5) means: $\forall v \in K\ \exists\sigma \in S$ such that $a(u,v)+\langle\sigma,v\rangle \geq \langle f,v\rangle$. Since S is convex and weak* compact, it can be shown that there exists some $\tau\in S$ such that

$$a(u,v)+\langle\tau,v\rangle\geq\langle f,v\rangle\,,\forall v\in K\,. \tag{C.6}$$

3. By construction, $W=\ker B\subset K$. Hence (C.6) implies

$$a(u,w)+\langle\tau,w\rangle=\langle f,w\rangle\,,\forall w\in W\,,$$

or $f-Au-\tau\in W^{\circ}$. In virtue of the (BB) condition, Lemma C.1 applies and entails the existence of $q\in Z$ such that $B'q=f-Au-\tau$ or

$$a(u,v)+b(v,q)+\langle\tau,v\rangle=\langle f,v\rangle\,,\forall v\in V\,.$$

Thus we obtain $(MF-1)$.

4. We claim that $q\in P$. Indeed, (C.6) gives by definition of q,

$$\langle B'q,v\rangle=\langle Bv,q\rangle\leq 0\,,\forall v\in K\,.$$

This means $Bv\in P^{-}\Rightarrow Bv\in Q^{-}$, where $P^{-}=\{\zeta\in Z'|\langle\zeta,p\rangle\leq 0,\ \forall p\in P\}$ is the negative dual cone to P and $Q:=\mathbb{R}_{+}q\subset Z$. In virtue of the (BB) condition, Lemma C.1 applies and hence $B:(W^{\circ})'\to Z'$ is isomorph, in particular is onto. Therefore the implication above gives $P^{-}\subset Q^{-}$, what results by the bipolar theorem in $P^{--}=P\supset Q^{--}=Q$. This proves the claim.

5. To prove $(MF-2)$, we show $(MF-3)$. By feasibility of $u\in K$, $(MF-3_1)$ is obvious. Since $\tau\in S=\partial\varphi(0)$, $\varphi(u)\geq\langle\tau,u\rangle$. From (C.4) and (C.6), we get

$$\langle f,v\rangle=a(u,u)+\varphi(u)\geq a(u,u)+\langle\tau,u\rangle\geq\langle f,v\rangle\,,$$

hence $(MF-3)_3$, and also by definition of q, $\langle B'q,u\rangle=b(u,q)=0$, thus finally $(MF-3)_2$. □

For a more detailed proof and for the proof of the general case of arbitrary g, moreover for further references see [219].

Here let us first consider the special case $\varphi=0, S=\{0\}$. Our aim is to derive from the present mixed form (MF) the mixed form (MP) of the previous subsection.

The present mixed form becomes then with some preimage $w = B^{-1}g$

$$(MF)_0 \begin{cases} a(u,v) + \langle B'q, v\rangle = \langle f, v\rangle\,, & \forall v \in V\,,\\ \langle u, B'p - B'q\rangle \le \langle w, B'p - B'q\rangle\,, & \forall p \in P\,, \end{cases}$$

where the multiplier q exists in P. Note that $\lambda := B'q \in K^-$, the latter inequality $(MF)_{0-2}$ extends to the closure of $B'P$, what coincides with $[B^{-1}(P^-)]^- = K^-$. Hence we arrive at the mixed form (MP).

To conclude this subsection, we want to bring the present mixed form (MF) in relation to the mixed form used in BEM solution of frictional unilateral contact problems in [33, 37], see Sect. 11.4.1. To this end, we proceed as in the special case above and obtain from (MF) with again $Bw = g$ the pair $[\lambda, \tau] \in K^- \times S$ that together with $u \in V$ solves the mixed system

$$\begin{cases} a(u,v) + \langle \lambda, v\rangle + \langle \tau, v\rangle = \langle f, v\rangle, & \forall v \in V\,,\\ \langle \kappa - \lambda, u\rangle + \langle \sigma - \tau, u\rangle \le \langle \kappa - \lambda, w\rangle, & \forall [\kappa, \sigma] \in K^- \times S\,. \end{cases}$$

Note that $K^- + S$ is convex and closed in V'. Thus using the indicator function χ_K of K ($\chi_K(v) = 0$ iff $v \in K, = +\infty$ elsewhere),

$$\begin{aligned} K^- + S &= \partial\chi_K(0) + \partial\varphi(0) = \partial(\chi_K + \varphi)(0)\\ &= \{\mu | \langle \mu, \cdot\rangle \le \chi_K + \varphi\}\\ &= \{\mu | \langle \mu, v\rangle \le \varphi(v), \forall v \in V \text{ with } Bv \in P^-\} =: M \end{aligned}$$

what is the analog to the set of multipliers in [33, 37].

On the other hand, for any $\mu \in M$ - in the case of a general reflexive Banach space V in virtue of Troyanski's renorming theorem an equivalent norm can be introduced so that V and V' are locally uniformly convex, and thus also strictly convex - the constrained best approximation problem

$$\begin{aligned} &\text{minimize } \|\kappa\|^2 + \|\sigma\|^2,\ \kappa \in K^-, \sigma \in S\\ &\text{subject to } \kappa + \sigma = \mu \end{aligned}$$

admits unique solutions $\mu_- \in K^-, \mu_S \in S$ with $\mu = \mu_- + \mu_S$.

Therefore we arrive at the multiplier $\nu := \lambda + \tau \in M$ that together with $u \in V$ solves the somewhat condensed mixed system

$$(MF)_c \begin{cases} a(u,v) + \langle \nu, v\rangle = \langle f, v\rangle\,, & \forall v \in V\,,\\ \langle \mu - \nu, u\rangle \le \langle \mu_- - \nu_-, w\rangle, & \forall \mu \in M\,, \end{cases}$$

what corresponds to the mixed form in [33, 37].

C.2 Nonsmooth Analysis

With nonmonotone contact problems we encounter locally Lipschitz functions that are not necessarily convex or smooth in the sense of classical differentiability. Therefore in this section we draw some basics from Clarke's monograph [109] on nonsmooth analysis. We collect some fundamental concepts of the Clarke generalized differential calculus, in particular introduce his generalized directional derivative along with its basic properties. Following [332, 335] we also provide regularization techniques of nondifferentiable optimization to smooth locally Lipschitz functions that are minima or maxima of smooth functions. These regularization techniques are needed in addition for the numerical treatment of nonmonotone contact problems, see Sect. 11.5.

C.2.1 Nonsmooth Analysis of Locally Lipschitz Functions

Throughout this subsection, let X denote a (real) Banach space. Let $f : X \to \mathbb{R}$ be Lipschitz of rank K near a given point $x \in X$; that is, for some $\varepsilon > 0$, we have

$$|f(y) - f(z)| \leq K \, \|y - z\|; \; \forall y, z \in B(x, \varepsilon) .$$

Definition C.1

$$f^0(x; v) := \limsup\{\frac{f(y + tv) - f(y)}{t} \mid y \in X, y \to x; t > 0, t \to 0\}$$

is called the **generalized directional derivative** of f in the direction v.

Note that this definition does not presuppose the existence of a limit and that it differs from the common definition of the directional derivative (or Gâteaux derivative, which is continuous in v) in that the base point y in the difference quotient varies. Also note that in general $f^0(x; \cdot)$ is not linear. The utility of this definition is seen from the properties listed below.

Proposition C.3 *Let f be Lipschitz of rank K near x. Then:*

(i) The function $v \mapsto f^0(x; v) \in \mathbb{R}$ is sublinear, hence convex, and satisfies $|f^0(x;v)| \leq K \, \|v\|$ for all $v \in X$;
(ii) The function $(z, w) \mapsto f^0(z; w)$ is upper semicontinuous at (x, v); the function $w \mapsto f^0(x; w)$ is Lipschitz of rank K on X;
(iii) There holds $f^0(x; -v) = (-f)^0(x; v)$ for $v \in X$.

Definition C.2 The **generalized gradient** of the function f at x, denoted by (simply) $\partial f(x)$, is the unique nonempty weak* compact convex subset of the dual space X', whose support function is $f^0(x; .)$.

Thus

$$\xi \in \partial f(x) \Leftrightarrow f^0(x,v) \geq \langle \xi, v\rangle,\ \forall v \in X\,,$$
$$f^0(x;v) = \max\{\langle \xi, v\rangle\ :\ \xi \in \partial f(x)\},\ \forall v \in X\,.$$

A function $f : X \to \mathbb{R}$ which is continously differentiable near a point x is locally Lipschitz near x by the mean value theorem. Also a function $f : X \to \mathbb{R}$ which is convex and lower semicontinuous is locally Lipschitz on all of X. In either case, ∂f reduces to the familiar concept of the derivative, respectively of that of the subdifferential of convex analysis:

Theorem C.3 *If $f : X \to \mathbb{R}$ is continously differentiable near x, then $\partial f(x) = \{f'(x)\}$. If $f : X \to \mathbb{R}$ is convex and lower semicontinuous on X, then for any $x \in X$,*

$$\partial f(x) = \{\xi \in X^*\ :\ \langle \xi, y - x\rangle \leq f(y) - f(x),\ \forall y \in X\}\,.$$

On the other hand, let f be Lipschitz near x and suppose that $\partial f(x)$ is a singleton $\{\xi\}$, then f is Gâteaux differentiable with $f'(x) = \xi$.

Definition C.3 Let $f : X \to \mathbb{R}$ be locally Lipschitz near x. Then f is called **regular** at x, if $f^0(x;v)$ coincides with the classical directional derivative $f'(x,v)$ for all $v \in X$.

There is a calculus of generalized gradiens including a sum rule, mean value theorem, and chain rule; see [109] for details. Here we only provide an important formula of nonsmooth analysis ('Danskin's formula', see [109, (2.3.12)]) that chararacterizes the generalized directional derivative of max functions.

Let I be a finite index set and let $\{f_i : i \in I\}$ be a finite collection of functions that are Lipschitz near x. Then the function f defined by

$$f(x) := \max_{i \in I} f_i(x)$$

is Lipschitz near x as well. Let $I(x) := \{i \in I : f_i(x) = f(x)\}$ and "co" denote the convex hull.

Theorem C.4 *There holds*

$$\partial f(x) \subset co\,\{\partial f_i(x) : i \in I(x)\}.$$

If f_i is regular at x for each $i \in I(x)$, then equality holds and f is regular at x.

C.2.2 Regularization of Nonsmooth Functions

In this subsection we follow [332, 335] and present a unified approach to regularization of nonsmooth functions with focus to locally Lipschitz functions that are minima or maxima of smooth functions.

According to Bertsekas [49] the maximum function $f : \mathbb{R}^n \to \mathbb{R}$,

$$f(x) = \max\{g_1(x), g_2(x), \dots, g_m(x)\} \tag{C.7}$$

of m continuously differentiable functions g_i can be expressed by means of the plus function $\mathrm{p}(x) = x^+ = \max(x, 0)$, $x \in \mathbb{R}$ as

$$f(x) = g_1(x) + \mathrm{p}\left[g_2(x) - g_1(x) + \ldots + \mathrm{p}\left[g_m(x) - g_{m-1}(x)\right]\right]. \tag{C.8}$$

Replacing now the plus function by an approximation $P(\varepsilon, \dot{})$, the smoothing function $S : \mathbb{R}^n \times \mathbb{R}_{++} \to \mathbb{R}$ is given by

$$S(x, \varepsilon) := g_1(x) + P\left[\varepsilon, g_2(x) - g_1(x) + \ldots + P\left[\varepsilon, g_m(x) - g_{m-1}(x)\right]\right] \tag{C.9}$$

as suggested by Chen et al. in [100]. The advantage of this procedure lies in the use of one single regularization parameter ε to smooth an eventually larger number of kinks. Here, $P : \mathbb{R}_{++} \times \mathbb{R} \to \mathbb{R}$ is the smoothing function via convolution for the plus function p defined by

$$P(\varepsilon, t) = \int_{-\infty}^{\frac{t}{\varepsilon}} (t - \varepsilon s)\rho(s)\, ds.$$

We restrict $\rho : \mathbb{R} \to \mathbb{R}_+$ to be a density function of finite absolute mean; that is

$$k := \int_{\mathbb{R}} |s|\rho(s)\, ds < \infty.$$

The major properties of S, see [346], that follow from the properties of the function P, see [169, section 11.8.2], are collected in the following lemma.

Lemma C.2

(i) For any $\varepsilon > 0$ and for all $x \in \mathbb{R}^n$,

$$|S(x, \varepsilon) - f(x)| \le (m - 1)k\varepsilon. \tag{C.10}$$

(*ii*) *The function S is continuously differentiable on* $\mathbb{R}^n \times \mathbb{R}_{++}$ *and for any* $x \in \mathbb{R}^n$ *and* $\varepsilon > 0$ *there exist* $\Lambda_i \geq 0$ *such that* $\sum_{i=1}^{m} \Lambda_i = 1$ *and*

$$\nabla_x S(x, \varepsilon) = \sum_{i=1}^{m} \Lambda_i \nabla g_i(x). \tag{C.11}$$

Moreover,

$$co\{\xi \in \mathbb{R}^n \, : \, \xi = \lim_{k\to\infty} \nabla_x S(x_k, \varepsilon_k), \; x_k \to x, \; \varepsilon_k \to 0^+\} \subseteq \partial f(x), \tag{C.12}$$

where "co" denotes the convex hull and $\partial f(x)$ *is the Clarke subdifferential.*

We recall that the Clarke subdifferential of a locally Lipschitz function f at a point $x \in \mathbb{R}^n$ can be characterized by

$$\partial f(x) = co\,\{\xi \in \mathbb{R}^n \, : \, \xi = \lim_{k\to\infty} \nabla f(x_k), \; x_k \to x, \; f \text{ is differentiable at } x_k\},$$

since in finite dimensional case, according to Rademacher's theorem, f is differentiable almost everywhere.

The maximum function given by (C.7) is clearly locally Lipschitz continuous and by Theorem C.4, the Clarke subdifferential can be written as

$$\partial f(x) = co\{\nabla g_i(x) \, : \, i \in I(x)\}$$

with

$$I(x) := \{i \, : \, f(x) = g_i(x)\}.$$

In particular, if $x \in \mathbb{R}^n$ is a point such that $f(x) = g_i(x)$ then $\partial f(x) = \{\nabla g_i(x)\}$. For such a point $x \in \mathbb{R}^n$ we show later on that

$$\lim_{z\to x, \varepsilon\to 0^+} \nabla_x S(z, \varepsilon) = \nabla g_i(x).$$

Note that the set on the left-hand side in (C.12) goes back to [353]. In [99], this set is denoted by $G_S(x)$ and is called there the subdifferential associated with the smoothing function. The inclusion (C.12) shows in fact that $G_S(x) \subseteq \partial f(x)$. Moreover, according to the part (*b*) of Corollary 8.47 in [353], $\partial f(x) \subseteq G_S(x)$. Thus, $\partial f(x) = G_S(x)$.

Remark C.1 Note also that S is a smoothing approximation of f in the sense that

$$\lim_{z\to x, \varepsilon\to 0^+} S(z, \varepsilon) = f(x) \quad \forall x \in \mathbb{R}^n. \tag{C.13}$$

This is immediate from (C.10).

Remark C.2 The regularization procedure (C.9) can be also applied to a minimum function by

$$\min\{g_1(x), g_2(x), \ldots, g_m(x)\} = -\max\{-g_1(x), -g_2(x), \ldots, -g_m(x)\} \approx -S(x, \varepsilon).$$

Denote now

$$S_i = g_i - g_{i-1} + P\left[\varepsilon, g_{i+1} - g_i + P\left[\varepsilon, g_{i+2} - g_{i+1} + \ldots + P\left[\varepsilon, g_m - g_{m-1}\right]\right]\right].$$

This function should approximate

$$g_i - g_{i-1} + \mathrm{p}\left[g_{i+1} - g_i + \mathrm{p}\left[g_{i+2} - g_{i+1} + \ldots + \mathrm{p}\left[g_m - g_{m-1}\right]\right]\right]$$

$$\overset{(C.8)}{=} \max\{g_i - g_{i-1}, g_{i+1} - g_{i-1}, \ldots, g_m - g_{i-1}\} =: T_{i-1}.$$

Lemma C.3 *It holds*

$$\lim_{z \to x, \varepsilon \to 0+} P(\varepsilon, S_i(z, \varepsilon)) = \mathrm{p}(T_{i-1}(x)). \tag{C.14}$$

Proof First, for any $\varepsilon_0 > 0$ there exists $\delta_0 > 0$ such that

$$|P(\varepsilon, z) - \mathrm{p}(T_{i-1}(x))| < \varepsilon_0 \tag{C.15}$$

for any $z \in B_{\delta_0}(T_{i-1}(x))$ and $\varepsilon \in (0, \delta_0)$. Next, since S_i is a smoothing approximation of T_{i-1} in the sense of (C.13), there exists $\bar{\delta}_0 > 0$ such that

$$|S_i(z, \varepsilon) - T_{i-1}(x)| < \delta_0 \tag{C.16}$$

for any $z \in B_{\bar{\delta}_0}(x)$ and $\varepsilon \in (0, \bar{\delta}_0)$. Combining (C.15) and (C.16), it follows that

$$|P(\varepsilon, S_i(z, \varepsilon)) - \mathrm{p}(T_{i-1}(x))| < \varepsilon_0$$

holds for any $\varepsilon < \min\{\delta_0, \bar{\delta}_0\}$ and any $z \in B_{\bar{\delta}_0}(x)$. Thus, the assertion of the lemma is proved. □

Since the nonsmooth functions that occur in the nonmonotone contact problems can be reformulated by using the plus function, all our regularizations are based in fact on a class of smoothing approximations for the plus function. Some examples from [168] and the references therein are in order:

$$P_1(\varepsilon, t) = \int_{-\infty}^{\frac{t}{\varepsilon}} (t - \varepsilon s)\, \rho_1(s)\, ds = t + \varepsilon\, ln(1 + e^{-\frac{t}{\varepsilon}}) = \varepsilon\, ln(1 + e^{\frac{t}{\varepsilon}}), \tag{C.17}$$

where $\rho_1(s) = \dfrac{e^{-s}}{(1 + e^{-s})^2}$

$$P_2(\varepsilon, t) = \int_{-\infty}^{\frac{t}{\varepsilon}} (t - \varepsilon s)\, \rho_2(s)\, ds = \frac{\sqrt{t^2 + 4\varepsilon^2} + t}{2}, \tag{C.18}$$

where $\rho_2(s) = \dfrac{2}{(s^2 + 4)^{3/2}}$

$$P_3(\varepsilon, t) = \int_{-\infty}^{\frac{t}{\varepsilon}} (t - \varepsilon s)\, \rho_3(s)\, ds = \begin{cases} 0 & \text{if } t < -\frac{\varepsilon}{2} \\ \frac{1}{2\varepsilon}(t + \frac{\varepsilon}{2})^2 & \text{if } -\frac{\varepsilon}{2} \le t \le \frac{\varepsilon}{2} \\ t & \text{if } t > \frac{\varepsilon}{2}, \end{cases} \tag{C.19}$$

where $\rho_3(s) = \begin{cases} 1 \text{ if } -\frac{1}{2} \le s \le \frac{1}{2} \\ 0 \text{ otherwise.} \end{cases}$

$$P_4(\varepsilon, t) = \int_{-\infty}^{\frac{t}{\varepsilon}} (t - \varepsilon s)\, \rho_4(s)\, ds = \begin{cases} 0 & \text{if } t < 0 \\ \frac{t^2}{2\varepsilon} & \text{if } 0 \le t \le \varepsilon \\ t - \frac{\varepsilon}{2} & \text{if } t > \varepsilon, \end{cases} \tag{C.20}$$

where $\rho_4(s) = \begin{cases} 1 \text{ if } 0 \le s \le 1 \\ 0 \text{ otherwise.} \end{cases}$

In the following we need

$$A_i = \{x \in \mathbb{R}^n \,:\, g_i(x) > g_j(x),\ \forall j = 1, \ldots, m,\ j \neq i\} \quad \text{for all } i = 1, \ldots, m$$

and compute

$$P_t(\varepsilon, t) = \int_{-\infty}^{\frac{t}{\varepsilon}} \rho(s)\, ds. \tag{C.21}$$

Lemma C.4 *The following properties hold:*

a) If $x \in A_i$, $i = 1, 2, \ldots, m - 1$, then

$$\lim_{z \to x, \varepsilon \to 0^+} P_t(\varepsilon, S_{i+1}(z, \varepsilon)) = 0. \tag{C.22}$$

b) if $x \in A_i$, $i = 2, 3, \ldots, m$, then

$$\lim_{z \to x, \varepsilon \to 0^+} P_t(\varepsilon, S_j(z, \varepsilon)) = 1 \quad \textit{for all } j = 2, 3, \ldots, i. \tag{C.23}$$

Proof

a) Let $i \in \{1, 2, \ldots, m - 1\}$ and $x \in A_i$, i.e., $g_i(x) > g_j(x)$ for all $j = 1, \ldots, m$, $j \neq i$, and S_{i+1} be a smoothing approximation of T_i defined as above by

$$T_i = \max\{g_{i+1} - g_i, g_{i+2} - g_i, \ldots, g_m - g_i\}.$$

Clearly, $T_i(x) < 0$. Since by (C.16) in the proof of Lemma C.3

$$S_{i+1}(z, \varepsilon) \to T_i(x) \quad \text{as } z \to x \text{ and } \varepsilon \to 0^+ \tag{C.24}$$

and due to $T_i(x) < 0$, it follows from (C.21) that

$$P_t(\varepsilon, S_{i+1}(z, \varepsilon)) = \int_{-\infty}^{\frac{S_{i+1}(z,\varepsilon)}{\varepsilon}} \rho(s)\, ds \to 0 \quad \text{as } z \to x, \varepsilon \to 0^+$$

and (C.22) is verified.

b) Let now $x \in A_i$, $i \in \{2, 3, \dots, m\}$. We first prove the statement of the lemma for $j = i$. By the representation

$$S_i(z, \varepsilon) = g_i(z) - g_{i-1}(z) + P(\varepsilon, S_{i+1}(z, \varepsilon))$$

and using (C.14) from Lemma C.3, it follows that

$$S_i(z, \varepsilon) \to g_i(x) - g_{i-1}(x) \quad \text{as } z \to x \text{ and } \varepsilon \to 0^+. \tag{C.25}$$

Hence, since $g_i(x) - g_{i-1}(x) > 0$, we have

$$P_t(\varepsilon, S_i(z, \varepsilon)) = \int_{-\infty}^{\frac{S_i(z,\varepsilon)}{\varepsilon}} \rho(s)\, ds \to 1 \quad \text{as } z \to x, \varepsilon \to 0^+ \tag{C.26}$$

and therefore (C.23) is verified for $j = i$. Thus, we completely proved the lemma in the case $m = 2$. The remaining case can be based on an induction argument, see [335]. □

Now we are ready to show that the gradient of the given function g_i on A_i can be approximated by the gradients of the smoothing function.

Theorem C.5 *For any $x \in A_i$, $i = 1, 2, \dots, m$,*

$$\lim_{z \to x, \varepsilon \to 0^+} \nabla_x S(z, \varepsilon) = \nabla g_i(x).$$

Proof From (C.9), by direct differentiation with respect to x, it follows that

$$\begin{aligned} \nabla_x S(z, \varepsilon) &= \Big(1 - P_t(\varepsilon, S_2(z, \varepsilon))\Big)\nabla g_1(z) \\ &\quad + \sum_{i=2}^{m-1} \Big(1 - P_t(\varepsilon, S_{i+1}(z, \varepsilon))\Big) \prod_{j=2}^{i} P_t(\varepsilon, S_j(z, \varepsilon))\nabla g_i(z) \\ &\quad + \prod_{i=2}^{m} P_t(\varepsilon, S_i(z, \varepsilon))\nabla g_m(z). \end{aligned}$$

We shall distinguish the following three cases.

1) First, we take $x \in A_1$. From Lemma C.4 a)

$$\lim_{z \to x, \varepsilon \to 0^+} P_t(\varepsilon, S_2(z, \varepsilon)) = 0$$

and, consequently, the following relations hold as $z \to x$ and $\varepsilon \to 0^+$:

$$\Lambda_1 := 1 - P_t(\varepsilon, S_2(z, \varepsilon)) \to 1,$$

$$\Lambda_i := \Big(1 - P_t(\varepsilon, S_{i+1}(z, \varepsilon))\Big) \prod_{j=2}^{i} P_t(\varepsilon, S_j(z, \varepsilon)) \to 0, \quad i=2, \dots, m-1 \ (m \geq 3)$$

and

$$\Lambda_m := \prod_{j=2}^{m} P_t(\varepsilon, S_j(z, \varepsilon)) \to 0.$$

Hence, if $x \in A_1$ then $\lim_{z \to x, \varepsilon \to 0^+} \nabla_x S(z, \varepsilon) = \nabla g_1(x)$.

2) Let now $x \in A_i$ for some $i \in \{2, 3, \dots, m-1\}$, $m \geq 3$. By (C.22) and (C.23), it follows immediately that

$$\Lambda_i \to 1 \quad \text{as } z \to x, \ \varepsilon \to 0^+.$$

Further, we shall show that for any k, $k \in \{1, 2, \dots, m\}$, $k \neq i$, it holds for any $z \to x$ and $\varepsilon \to 0^+$ that $\Lambda_k \to 0$.
Indeed, the relation (C.23) implies

$$\lim_{z \to x, \varepsilon \to 0^+} P_t(\varepsilon, S_{k+1}(z, \varepsilon)) = 1 \quad \forall k = 1, \dots, i-1.$$

Therefore,

$$\Lambda_1 = 1 - P_t(\varepsilon, S_2(z, \varepsilon)) \to 0$$

and

$$\Lambda_k = (1 - P_t(\varepsilon, S_{k+1}(z, \varepsilon))) \prod_{j=2}^{k} P_t(\varepsilon, S_j(z, \varepsilon)) \to 0 \quad \forall k = 2, \dots, i-1 \tag{C.27}$$

as $z \to x$ and $\varepsilon \to 0^+$. Altogether, $\Lambda_k \to 0$ for all $k = 1, \dots, i-1$.
Let now $k \in \{i+1, i+2, \dots, m-1\}$. According to (C.22), the $(i+1)-$ multiplier $P_t(\varepsilon, S_{i+1}(z, \varepsilon))$ in (C.27) goes to zero and consequently, $\Lambda_k \to 0$.

Further, since $(i+1) \in \{3, 4, \ldots, m\}$ and $P_t(\varepsilon, S_{i+1}(z, \varepsilon))$ goes to zero, it follows that

$$\Lambda_m = \prod_{j=2}^{m} P_t(\varepsilon, S_j(z, \varepsilon)) \to 0 \quad \text{as } z \to x,\ \varepsilon \to 0^+.$$

In this way, we have proved that $\Lambda_k \to 0$ for every $k = 1, \ldots, m$, $k \neq i$, and therefore, if $x \in A_i$ then $\lim\limits_{z \to x, \varepsilon \to 0^+} \nabla_x S(z, \varepsilon) \to \nabla g_i(x)$.

3) Finally, let $x \in A_m$. From Lemma C.4 b),

$$\lim_{z \to x, \varepsilon \to 0^+} P_t(\varepsilon, S_i(z, \varepsilon)) = 1 \quad i = 2, \ldots, m.$$

Hence,

$$\Lambda_1 = 1 - P_t(\varepsilon, S_2(z, \varepsilon)) \to 0 \quad \text{and} \quad \Lambda_m \to 1.$$

Clearly, we can also write

$$P_t(\varepsilon, S_{i+1}(z, \varepsilon)) \to 1 \quad \forall i = 2, \ldots, m-1$$

and consequently,

$$\Lambda_i \to 0 \quad \text{for all } i = 2, \ldots, m-1.$$

Therefore, we have proved that if $x \in A_m$ then $\lim\limits_{z \to x, \varepsilon \to 0^+} \nabla_x S(z, \varepsilon) \to \nabla g_m(x)$.

Collecting all cases, the proof of the theorem is complete. □

Remark C.3 Note that if $x \in \mathbb{R}^n$ is a point such that $g_i(x) = g_j(x)$ for some i and j, $i \neq j$, then for any sequences $\{x_k\} \subset \mathbb{R}^n$, $\{\varepsilon_k\} \subset \mathbb{R}_{++}$ such that $x_k \to x$ and $\varepsilon_k \to 0^+$ we have

$$\lim_{k \to \infty} \nabla_x S(x_k, \varepsilon_k) \in \partial f(x).$$

C.3 Existence and Approximation Results for Variational Inequalities

C.3.1 Existence Results for Linear VIs

Let $(V, \|.\|, \langle ., . \rangle)$ be a real Hilbert space. Let $\lambda \in V^*$ be a continuous linear form, $K \subset V$ a nonvoid closed, convex set, and $\beta : V \times V \to \mathbb{R}$ be a continuous bilinear form, not necessarily symmetric. With these data given we consider the subsequent variational inequality (P): Find $\hat{u} \in K$ such that

$$\beta(\hat{u}, v - \hat{u}) \geq \lambda(v - \hat{u}) \quad \forall v \in K .$$

We require that β is positive semidefinite, i.e. $\beta(v, v) \geq 0$ for all $v \in V$. Hence the closed set

$$\mathscr{N} := \{u \in V : \beta(u, u) = 0\}$$

is a (generally nontrivial) subspace, as it is seen as follows. Clearly, $\mathbb{R}\mathscr{N} \subseteq \mathscr{N}$. The symmetric bilinear form

$$\beta^{\text{symm}}(u, v) := \frac{1}{2}\{\beta(u, v) + \beta(v, u)\}$$

satisfies the Schwarz inequality. Therefore for any $u, v \in \mathscr{N}$

$$0 \leq \beta(u + v, u + v) = 2\, \beta^{\text{symm}}(u, v) \leq 0 ,$$

hence $u + v \in \mathscr{N}$. This also shows that

$$\mathscr{N} = \{u \in V : \beta^{\text{symm}}(u, .) \equiv 0\} .$$

Although the solution of (P) generally depends nonlinearly on the datum λ, the solution set of (P) is convex. This is an easy consequence of the following useful characterization.

Lemma C.5 *Let $\hat{u} \in K$. Then $\hat{u}$ solves (P), if and only if*

$$\beta(v, \hat{u} - v) \leq \lambda(\hat{u} - v) \quad \forall v \in K .$$

Proof To show the "$\leq$" inequality, use positive semidefiniteness of β and obtain

$$\beta(v, \hat{u} - v) \leq -\beta(\hat{u}, v - \hat{u}) \leq -\lambda(v - \hat{u}) \quad \forall v \in K .$$

To show conversely (P), for any $v \in K$ take $w_t := \hat{u} + t(v - \hat{u})$, $t \in (0, 1)$. Then $w_t \in K$ and

$$\beta(w_t, \hat{u} - w_t) \leq \lambda(\hat{u} - w_t) ,$$

hence

$$\beta(w_t, v - \hat{u}) \geq \lambda(v - \hat{u}) .$$

Letting $t \to 0$, the inequality of (P) follows. □

To obtain existence results one needs further assumptions. In accordance to the Signorini problem in Sect. 5.1 , we assume that β is semicoercive in the sense that β should satisfy a Gårding inequality:

$$(G) \quad \beta(v,v) + \langle Cv, v\rangle \ \geq c\,\|v\|^2 \quad \text{for all } v \in V$$

with some real number $c > 0$ and a compact linear operator $C : V \to V^*$. If (G) holds with $C = 0$, then β is usually termed coercive or elliptic. In the coercive case, the Lions - Stampacchia theorem that extends the Lax - Milgram lemma guarantees unique solvability of (P) for each $\lambda \in V^*$:

Theorem C.6 (Lions - Stampacchia Theorem) *Let $\beta : V \times V \to \mathbb{R}$ be a continuous elliptic bilinear form on the Hilbert space V. Moreover, let $K \neq \emptyset$, convex, closed $\subset V$, $\lambda \in V^*$. Then the variational inequality (P) has a unique solution $\hat{u}$. Moreover, the mapping $\lambda \mapsto \hat{u}$ is Lipschitz continuous.*

Proof We give a sketch of the proof divided in three steps.

1. Let u_i be solutions to the data λ_i. Then choose $v = u_2$, respectively $v = u_1$ in (P), sum up and obtain $\beta(u_1 - u_2, u_1 - u_2) \leq (\lambda_1 - \lambda_2)(u_1 - u_2)$. Since β is elliptic, $c\,\|u_1 - u_2\|^2 \leq \|\lambda_1 - \lambda_2\|_{V^*}\|u_1 - u_2\|$, what shows Lipschitz continuity and uniqueness.
2. Existence in the case of symmetric β
   Method: Minimize "energy" $J(v) = \dfrac{1}{2}\beta(v,v) - \lambda(v)$, since minimization problem on K is equivalent to (P) in the symmetric case.
   Consider minimizing sequence $\{u_n\}$; this is a Cauchy sequence, what can be seen by the parallelogram rule. Then $u_n \to \hat{u} \in K$, $J(u_n) \to J(\hat{u})$, since J is continuous.
3. Existence in the general case.
   Let in addition σ a symmetric elliptic bilinear form, e. g. $\sigma(v,w) = \langle v, w\rangle$ or $\sigma(v,w) = \beta^{\text{symm}}(v,w) = \frac{1}{2}\Big(\beta(v,w) + \beta(w,v)\Big)$.
   For fixed $u \in K$, $\rho > 0$ $\exists^1 w \in K$ (according to the symmetric case above) such that

$$\sigma(w, v-w) \geq \sigma(u, v-w) - \rho[\beta(u, v-w) - \lambda(v-w)] \ \forall v \in K\,.$$

   Hence $u \mapsto w = S(u)$ gives a mapping S: $K \to K$. Clearly $\hat{u}$ solves (P), if and only if $\hat{u} = S(\hat{u})$. Choose now $\rho > 0$ sufficiently small, such that S is a contraction that gives the fixed point $\hat{u}$.
   For details see e.g. the monograph of Kinderlehrer and Stampacchia [267, theorem II.2.19]. □

Coercivity is also necessary for well-posedness; this is clarified in the following

Proposition C.4 *Let $A : H \to H$ be a linear continuous operator. Suppose that the bilinear form*

$$\alpha(x, y) = < Ax, y >$$

is symmetric and positive semidefinite. If A is bijective, then α is coercive.

Proof By Banach's inverse mapping theorem, A^{-1} is continuous. Then for any fixed $x \in H$ with $\|x\| = 1$ we have

$$\|A^{-1}\| \sup_{\|y\|=1} |\langle Ax, y\rangle| \geq \left\| A^{-1} \frac{Ax}{\|Ax\|} \right\| \left| \left\langle Ax, \frac{Ax}{\|Ax\|} \right\rangle \right| = \frac{\|x\|}{\|Ax\|} \cdot \frac{\|Ax\|^2}{\|Ax\|} = 1$$

and hence

$$\inf_{||x||=1} \sup_{||y||=1} |\langle Ax, y\rangle| \geq \frac{1}{||A^{-1}||}.$$

Thus by the Cauchy-Schwarz inequality applied to the positive semidefinite and symmetric bilinear form a,

$$\begin{aligned} \alpha(x, x) &\geq \sup_{y \neq 0} \frac{|\alpha(x, y)|}{\alpha(y, y)} \\ &\geq \frac{1}{\|A\|} [\sup_{y \neq 0} \frac{|\alpha(x, y)|}{\|y\|}]^2 \\ &\geq \frac{1}{\|A\| \ \|A^{-1}\|^2} \|x\|^2 \end{aligned}$$

what proves the asssertion. □

Thus for the general semicoercive bilinear form β under study, we need extra conditions for the specific $\lambda \in V^*$ to yield existence of solutions to (P). Referring to [28, 201] a sufficient condition for solvability is the recession condition $\mathscr{C} = -\mathscr{C}$, where the convex cone $\mathscr{C}$ is given by

$$\mathscr{C} := \{w \in \text{ac } K \cap \mathscr{N} : \beta(v, w) \leq \lambda(w) \, \forall v \in K\}$$

and with some fixed $k_0 \in K$

$$\text{ac } K := \bigcap_{t>0} t(K - k_0)$$

denotes the *asymptotic cone* or the *recession cone* of K. A stronger condition is that there exists some $v_0 \in K$ such that

$$\lambda(w) < \beta(v_0, w) \quad \forall w \in \text{ac } K \cap \mathcal{N} \setminus \{0\},$$

since this latter condition obviously implies that $\mathcal{C} = \{0\}$. In the case $0 \in K$, this latter condition simplifies to

$$\lambda(w) < 0 \quad \forall w \in \text{ac } K \cap \mathcal{N} \setminus \{0\},$$

which can already be found with Fichera [179] and Stampacchia [388].

In the Signorini problem discussed in Sect. 5.1, see (5.6), K is already a convex cone (with vertex at zero) and the set $K \cap \mathcal{N}$ coincides with the set of constant functions that are nonpositive on Γ_S, thus nonpositive throughout $\mathbb{R}^d$. Therefore the recession condition of Fichera–Stampacchia is here simply

$$\ell(\underline{1}) = \int_{\Gamma_N} g\, ds + \int_{\Gamma_S} h\, ds > 0\,. \tag{C.28}$$

This latter condition also guarantees the uniqueness of the solution of the Neumann–Signorini problem, where Γ_D may be empty.

C.3.2 Approximation of Linear VIs

In this subsection we present an approximation result, which is based on [213], for linear variational inequalities in Hilbert space. So we have the same setting as in the previous section and consider the problem (P), but now for simplicity K is assumed to be a nonvoid closed, convex cone (with vertex at zero).

To describe the approximation of our variational problem (P) we suppose that we are given a positive parameter h converging to 0 and a family $\{V^h\}_{h>0}$ of closed finite dimensional subspaces contained in V. In addition we have a family $\{K^h\}_{h>0}$ of closed convex nonempty cones of V^h. These sets K^h should approximate the given set K. However, piecewise polynomial interpolation - except piecewise linear interpolation - does not preserve order, thus generally K^h cannot assumed to be contained in K. To cope with this difficulty of nonconforming approximation we follow the discretization theory of Glowinski [199, Chapter 1], which refines the set convergence notion due to Mosco [309] (see also [6] for definition and further study) and independently to Stummel, see [414] and introduce the following two hypotheses (H1) and (H2):

(H1) If for some sequence $\{h_j\}_{j\in\mathbb{N}}$ with $h_j \to 0$, $v^{h_j} \in K^{h_j}$ $(j \in \mathbb{N})$ and v^{h_j} converges weakly to $v \in V$ $(j \to \infty)$, then $v \in K$.

(H2) There exist a subset $M \subset V$ such that $\overline{M} = K$ and mappings $r^h : M \to V^h$ with the property that, for each $v \in M$, $r^h v \to v$ $(h \to 0)$ and $r^h v \in K^h$ for all $h \le h_0(v)$ for some $h_0(v) > 0$.

Thus we approximate the problem (P) by the following variational inequality (P^h): Find $u^h \in K^h$ such that

$$\beta(u^h, v^h - u^h) \ge \lambda(v^h - u^h) \quad \forall v^h \in K^h .$$

By the existence theory in the infinite dimensional case, also solutions u^h to these finite dimensional problems exist.

Note that in most computations, however, it will be necessary to replace also β and λ by some approximations β^h and λ^h , defined by a numerical integration rule which is used in the finite element, respectively boundary element discretization. Since there is nothing new compared to the case of linear elliptic boundary value problems and variational equalities, we do not discuss this aspect here.

Now we can state and prove our basic convergence result.

Theorem C.7 *Let β , λ, K, and $\{K^h\}_h$ satisfy the conditions (G), (H1) and (H2). If the solution $\hat{u}$ of (P) is unique, then $\lim_{h\to 0} \|u^h - \hat{u}\| = 0$ holds.*

Proof We divide the proof in five parts. We first show a priori estimates for $\{u^h\}_h$, before we can establish the convergence results.

1) $|\,.\,|$ *– estimate for* $\{u^h\}$.
Fix $w_0 \in M$, let $w^h := r^h w_0 \in K^h$ for $0 < h = h_0 := h_0(w_0)$. Then we have $\lim \|w^h - w_0\| = 0$, and with u^h, a solution of (P^h)

$$|u^h|^2 = \beta(u^h, u^h) \le c_0 + c_1 \|u^h\| + \lambda(u^h) \tag{C.29}$$

$$\le c_0 + c_2 \|u^h\| . \tag{C.30}$$

Here and in the following $c_0, c_1, c_2, \ldots$ are generic positive constants. Moreover, by positive semidefiniteness,

$$\beta(w^h, u^h) - \lambda(u^h) \le \beta(w^h, w^h) - \lambda(w^h) \le c_3 . \tag{C.31}$$

2) *Norm-boundedness of* $\{u^h\}$.
Here we modify a contradiction argument, which in the existence theory of semicoercive variational inequalities goes back to Fichera [179] and Stampacchia [388]. We assume there exists a subsequence $\{u_\ell\}_{\ell\in\mathbb{N}} := \{u^{h_\ell}\}$ such that $\|u_\ell\| \to +\infty$ $(\ell \to \infty)$. With $y_\ell := \|u_\ell\|^{-1} u_\ell$ in the Hilbert space V, we can extract a subsequence, again denoted by $\{y_\ell\}$, that converges weakly to some $y \in V$. In virtue of (C.30) , we get

$$|y_\ell|^2 \, \|u_\ell\| \le c_4 .$$

Thus we have $|y_\ell| \to 0$. [Assume not. Then for a subsequence $|y_{\ell_k}| \geq c_5 > 0$ and hence

$$|y_{\ell_k}| \leq \frac{c_4}{c_5 \|u_{\ell_k}\|},$$

what by $\|u_{\ell_k}\| \to +\infty$ leads to a contradiction.]
Since $|\,.\,|$ is continuous and sublinear, hence weakly sequentially lower semicontinuous, we obtain $y \in \mathcal{N}$. Since $\{u_\ell\}$ belongs to the cone K^{h_ℓ}, (H1) implies that $y \in K$, too.
We claim that $y = 0$. From (C.31) we obtain

$$\beta(w_\ell, y_\ell) - \lambda(y_\ell) \leq \frac{c_3}{\|u_\ell\|},$$

hence

$$\beta(w_0, y) \leq \lambda(y) \quad \forall w_0 \in M, \tag{C.32}$$

which extends to $\overline{M} = K$ by continuity. Moreover, for the solution $\hat{u}$ we have by the characterization lemma C.5

$$\beta(u, \hat{u}) - \beta(u, u) \leq -\lambda(u - \hat{u}) \quad \forall u \in K. \tag{C.33}$$

From (C.32) and (C.33) it follows for any $t > 0$

$$\beta(u, \hat{u} + tu) - \beta(u, u) \leq \lambda(\hat{u} + tu) - \lambda(u) \quad \forall u \in K.$$

Hence by the characterization lemma C.5, $\hat{u} + ty$ solves (P), and by uniqueness, $y = 0$ follows.

Now we use (G). By compactness of C, for some subsequence $\lim_{k\to\infty} \|Cy_{\ell_k}\| = 0$ and

$$c\|y_{\ell_k}\|^2 \leq \beta(y_{\ell_k}, y_{\ell_k}) + \langle Cy_{\ell_k}, y_{\ell_k}\rangle,$$

hence $y_{\ell_k} \to 0$. However, $\|y_{\ell_k}\| = 1$, and a contradiction is reached proving the boundedness of $\{u^h\}$.

3) *Any weak limit point u^* of $\{u^h\}$ solves (P).*
By the preceding step, there exists a subsequence, again denoted by $\{u_\ell\}$ such that $u_\ell \rightharpoonup u^*$. By (H1), u^* belongs to K. To show that u^* solves (P), take $v \in M$ arbitrarily. Then $v_\ell := r^{h_\ell} v$ converges strongly to v, and for $h_\ell \leq h_0(v)$ we have

$$\beta(u_\ell, v_\ell - u_\ell) \geq \lambda(v_\ell - u_\ell).$$

Since β is positive semidefinite,

$$\beta(v_\ell, u_\ell - v_\ell) \leq \lambda(u_\ell - v_\ell).$$

Hence in the limit

$$\beta(v, u^* - v) \le \lambda(u^* - v) .$$

This inequality extends by continuity to $\overline{M} = K$. Finally by the characterization lemma C.5, we conclude for any $v \in K$

$$\beta(u^*, v - u^*) \ge \lambda(v - u^*) .$$

4) *Convergence with respect to* $| \, . \, |$.
Here we use an argument due to Glowinski [199, Chapter 1]. Since the solution $\hat{u}$ of (P) is unique, the entire family $\{u^h\}$ converges weakly to $\hat{u}$. Now take $v \in M$ arbitrarily. Then $v^h := r^h v$ converges strongly to v , and for $h_\ell \le h_0(v)$ we have

$$\begin{aligned}\beta(u^h - \hat{u}, u^h - \hat{u}) &= \beta(u^h, v^h - \hat{u}) - \beta(u^h, v^h - u^h) - \beta(\hat{u}, u^h - \hat{u}) \\ &\le c_6 \, \|v^h - \hat{u}\| + \lambda(u^h - v^h) - \beta(\hat{u}, u^h - \hat{u}) .\end{aligned}$$

Hence in the limit, for any $v \in M$,

$$0 \le \limsup_{h \to 0} |u^h - \hat{u}|^2 \le c_6 \, \|v - \hat{u}\| + \lambda(\hat{u} - v) .$$

The obtained inequality extends to K by density and continuity. Finally, the choice $v = \hat{u}$ leads to the desired $|.|$ – convergence.
5) *Convergence with respect to* $\|.\|$.
Assume there exists a sequence $\{u_\ell\}$ such that u_ℓ is a solution to (P^{h_ℓ}) and $\|u_\ell - \hat{u}\| \ge \delta > 0$. By part (2), $\|u_\ell - \hat{u}\|$ is bounded and therefore we can extract a subsequence, again denoted by $\{u_\ell\}$ such that $u_\ell - \hat{u}$ converges weakly to some $w \in V$. By part (3), $\hat{u} + w$ solves (P), hence by uniqueness $w = 0_V$. Now we again use (G). By compactness of C, we can extract a subsequence, again denoted by $\{u_\ell\}$ such that $C\, u_\ell$ converges strongly to $C\, \hat{u}$ and moreover by part (4), $|u_\ell - \hat{u}| \to 0$ $(\ell \to \infty)$. Therefore by (G), $\|u_\ell - \hat{u}\| \to 0$ $(\ell \to \infty)$, and a contradiction is reached. □

For the more general approximation of general convex closed sets (instead of cones) we refer to [213].

C.3.3 Pseudomonotone VIs—Existence Result

The Lions-Stampacchia theorem was substantially extended by Brézis to a very large class of (non-linear) operators, called *pseudomonotone* operators in [64, Theorem 24], see also [438, section 27.2]. With the symbol $\rightharpoonup$ denoting weak convergence on V, $T : V \to V^*$ is called *pseudomonotone*, if it is bounded and if for any sequence $\{u_n\}_{n\in\mathbb{N}}$ in V,

$$u_n \rightharpoonup u \text{ and } \liminf_{n\to\infty} \langle T(u_n), u - u_n \rangle \geq 0,$$

imply

$$\langle T(u), v - u \rangle \geq \limsup_{n\to\infty} \langle T(u_n), v - u_n \rangle, \ \forall v \in V.$$

Such a pseudomonotone operator $T : K \subset V \to V^*$ as defined above gives rise to the bifunction $\psi : K \times K \to \mathbb{R}$ via $\psi(u, v) := \langle T(u), v - u \rangle$. Then ψ is pseudomonotone (PM) in the sense that for any sequence $\{u_n\}$ in K,

$$u_n \rightharpoonup u \quad \text{and} \quad \liminf_{n\to\infty} \psi(u_n, u) \geq 0$$

imply that for any $v \in K$ there holds

$$\psi(u, v) \geq \limsup_{n\to\infty} \psi(u_n, v).$$

A simple example of a pseudomonotone bifunction (not represented by an operator) is $\psi(u, v) = g(v) - g(u)$, where g is a weakly lower semicontinuous function.

Let T be weakly continuous on subsets $\mathrm{F} \cap K$ of K, where F is a finite dimensional subspace of V. Then the function $\psi(\cdot, v)$ becomes upper semicontinuous on each finite dimensional part $\mathrm{F} \cap K$ of K. Here, we assume only that $\psi(u, u) \geq 0$ and $\psi(u, \cdot)$ is convex for any $u \in K$; thus we do not require that $\psi(u, \cdot)$ is linear-affine. This is a suitable extension for the treatment of hemivariational inequalities to follow. In this setting we have the following existence result from [212, Theorem 3].

Theorem C.8 *Let K be a closed convex nonvoid subset of a reflexive Banach space V. Let the bifunction $\psi : K \times K \to \mathbb{R}$ be pseudomonotone with $\psi(\cdot, v)$ upper semicontinuous on each finite dimensional part of K, $\psi(u, u) \geq 0$ and $\psi(u, \cdot)$ convex for any $u \in K$. Suppose that for some $u_0 \in K$, ψ satisfies the coercivity condition*

$$(CC) \quad \frac{\psi(u, u_0)}{\|u - u_0\|} \to -\infty \quad \text{as } u \in K, \ \|u\| \to \infty.$$

Then for any $f \in V^*$ the variational inequality $VI(\psi, f, K)$ admits a solution, i.e. there exists $u \in K$ such that

$$\psi(u, v) \geq \langle f, v - u \rangle, \ \forall v \in K. \tag{C.34}$$

C.3.4 Mosco Convergence, Approximation of Pseudomonotone VIs

In this subsection we present an approximation procedure for pseudomonotone variational inequalities, where the given data (ψ, f, K) of the variational inequality are approximated by bifunctions ψ_t, linear continuous functionals f_t and closed convex sets K_t, respectively, indexed by a directed set T. While K is contained in a general reflexive Banach space V, K_t is a subset of a subspace V_t of V. For the approximation of K by K_t we employ Mosco convergence, since we do not assume that K_t is a subset of K. We provide a general approximation result, which with finite-dimensional subspaces V_t of V can be considered as an abstract convergence result for the Galerkin method for the solution of $VI(\psi, f, K)$. Our approximation result includes also the existence of solutions to the approximate $VI(\psi_t, f_t, K_t)$ under an appropriate coerciveness condition.

We assume the following hypotheses:

(H1) If $\{v_{t'}\}_{t' \in T'}$ weakly converges to v in V, $v_{t'} \in K_{t'}$ $(t' \in T')$ for a subnet $\{K_{t'}\}_{t' \in T'}$ of the net $\{K_t\}_{t \in T}$, then $v \in K$.

(H2) For any $v \in K$ and any $t \in T$ there exists $v_t \in K_t$ such that $v_t \to v$ (strongly) in V.

(H3) ψ_t is pseudomonotone for any $t \in T$.

(H4) $f_t \to f$ in V^*.

(H5) For any nets $\{u_t\}$ and $\{v_t\}$ such that $u_t \in K_t$, $v_t \in K_t$, $u_t \rightharpoonup u$, and $v_t \to v$ in V it follows that

$$\liminf_{t \in T} \psi_t(u_t, v_t) \le \psi(u, v)\,.$$

(H6) The family $\{-\psi_t\}$ is uniformly bounded from below in the sense that there exist constants $c > 0$, d, $d_0 \in \mathbb{R}$ and $\alpha > 1$ (independent of $t \in T$) such that for some $w_t \in K_t$ with $w_t \to w$ there holds

$$-\psi_t(u_t, w_t) \ge c\|u_t\|_V^\alpha + d\|u_t\|_V + d_0, \quad \forall u_t \in K_t,\ \forall t \in T\,.$$

Remark C.4 The hypotheses (H1) and (H2) describe the Mosco convergence [6] of the family $\{K_t\}$ to K.

Remark C.5 Without loss of generality we can assume that $0 \in K \cap \{\cap_{t \in T} K_t\}$. Indeed, since K is nonvoid, by (H2) for any $w \in K$ there exist $w_t \in K_t$ such that $w_t \to w$. Then we consider the transformations $v \in K \mapsto v - w \in \tilde{K} := K - w$; $v_t \in K_t \mapsto v_t - w_t \in \tilde{K}_t := K_t - w_t$. Thus, $\tilde{K}_t$ Mosco converges to $\tilde{K}$ and the hypotheses (H3), (H5), (H6) hold for the transformed bifunctions $\tilde{\psi}$, $\tilde{\psi}_t$ as well.

Under these hypotheses we have the following basic convergence result.

Theorem C.9 (General Approximation Result) *Under conditions* (H1)–[(H6), *there exist solutions u_t to the approximate problem $VI(\psi_t, f_t, K_t)$ and the family $\{u_t\}$ is uniformly bounded in V. Moreover, there exists a subnet of $\{u_t\}$ that* *converges weakly in V to a solution of the problem $VI(\psi, f, K)$. Furthermore, any weak accumulation point of $\{u_t\}$ is a solution to the problem $VI(\psi, f, K)$.*

Proof Using (H3) and (H6), the existence of a solution u_t to $VI(\psi_t, f_t, K_t)$ follows from Theorem C.8. Inserting $v_t = 0$ in $VI(\psi_t, f_t, K_t)$ and using (H6) and (H4) we obtain

$$c\|u_t\|_V^\alpha + d\|u_t\|_V + d_0 \le -\psi_t(u_t, 0) \le \|f_t\|_{V^*}\|u_t\| \le C\,\|f\|_{V^*}\,\|u_t\|_V,$$

what proves the norm boundedness of $\{u_t\}$. So we can extract a subnet of $\{u_t\}$ denoted by $\{u_{t'}\}_{t'\in T'}$ such that $u_{t'}$ converges weakly to u in V. By (H1), $u \in K$. Now, take an arbitrary $v \in K$. By (H2), there exists a net $\{v_t\}$ such that $v_t \in K_t$ and $v_t \to v$ in V. By (H4) and (H5), we get from $VI(\psi_t, f_t, K_t)$ that for any $v \in K$

$$\psi(u, v) \ge \liminf_{t'\in T'} \psi_{t'}(u_{t'}, v_{t'}) \ge \lim_{t'\in T'} \langle f_{t'}, v_{t'} - u_{t'}\rangle = \langle f, v - u\rangle$$

and consequently u is a solution to $VI(\psi, f, K)$. At the same time we have proved that any weak accumulation point of $\{u_t\}$ is a solution to $VI(\psi, f, K)$. This should be understood in the sense that every weak limit of any subnet of $\{u_t\}$ is a solution to $VI(\psi, f, K)$. □

Remark C.6 Without the coercivity hypothesis (H6) we get a stability result in the sense of Painlevé-Kuratowski set convergence that guarantees the inclusion

$$\limsup_{t\in T} \mathscr{S}(\psi_t, f_t, K_t) \subset \mathscr{S}(\psi, f, K)\,.$$

Here, the set $\mathscr{S}(\psi, f, K)$, depending on ψ, f and K, consists of all functions $u \in K$ satisfying the variational inequality $VI(\psi; f; K)$.

C.3.5 A Hemivariational Inequality as a Pseudomonotone VI

Let V be the classical Sobolev space $H^1(\Omega; \mathbb{R}^d)$, where $\Omega \subset \mathbb{R}^d$ with $d = 2, 3$ is a bounded domain with Lipschitz boundary $\partial\Omega$, and let $K \subseteq V$ be a nonempty closed, convex set specified later. Further, let the boundary $\partial\Omega = \bar{\Gamma}_D \cup \bar{\Gamma}_c \cup \bar{\Gamma}_F$ be composed of three mutually disjoint parts: a Dirichlet boundary Γ_D, a contact boundary Γ_c and a part Γ_F, where given external forces are applied. We also assume that the measure of Γ_D and Γ_c is strictly positive.

With γ we denote the trace operator from V into $L^2(\Gamma_c; \mathbb{R}^d)$, which is a linear continuous mapping. Hence, there exists a constant c_0 depending on Ω and Γ_c such that

$$\|\gamma \mathbf{v}\|_{L^2(\Gamma_c;\mathbb{R}^d)} \le c_0\|\mathbf{v}\|_V, \quad \forall \mathbf{v} \in V. \tag{C.35}$$

Moreover, by the trace theorem [277, Theorem 6.10.5], γ is compact.

We introduce the linear elastic operator $A : V \to V^*$,

$$\langle A\mathbf{u}, \mathbf{v}\rangle = \int_\Omega \varepsilon(\mathbf{u}) : \sigma(\mathbf{v})\, dx, \tag{C.36}$$

where $\varepsilon(\mathbf{u}) = \frac{1}{2}(\nabla\mathbf{u} + (\nabla\mathbf{u})^T)$ is the linearized strain tensor and $\sigma(\mathbf{v}) = C : \varepsilon(\mathbf{v})$ is the stress tensor. Here, C is the elasticity tensor with symmetric positive L^∞ coefficients. Hence, the linear elastic operator $A : V \to V^*$ is continuous, symmetric and due to the Korn's inequality coercive, i.e. there exists a constant $c_K > 0$ such that

$$\langle A\mathbf{v}, \mathbf{v}\rangle \geq c_K \, \|\mathbf{v}\|_V^2, \quad \forall \mathbf{v} \in V\,. \tag{C.37}$$

We define the linear form $f : V \to \mathbb{R}$ by

$$\langle f, \mathbf{v}\rangle = \int_\Omega \mathbf{f}_0^T \mathbf{v}\, dx + \int_{\Gamma_F} \mathbf{f}_1^T \mathbf{v}\, ds,$$

where $\mathbf{f}_0 \in L^2(\Omega; \mathbb{R}^d)$ are the prescribed body forces and $\mathbf{f}_1 \in L^2(\Gamma_F; \mathbb{R}^d)$ are the prescribed surface tractions on Γ_F.

In what follows we consider a function $j : \Gamma_c \times \mathbb{R}^d \to \mathbb{R}$ such that $j(\cdot, \xi) : \Gamma_c \to \mathbb{R}$ is measurable on Γ_c for all $\xi \in \mathbb{R}^d$ and $j(s, \cdot) : \mathbb{R}^d \to \mathbb{R}$ is locally Lipschitz on $\mathbb{R}^d$ for almost all (a.a.) $s \in \Gamma_c$. Moreover, $j^0(s, \cdot\, ;\, \cdot)$ stands for the generalized Clarke directional derivative [109] of $j(s, \cdot)$, as used in Sect. 5.3 and analyzed in Sect. C.2.1 above. With this data we consider the following hemivariational inequality: Find $u \in K$ such that

$$\langle A\mathbf{u}, \mathbf{v}-\mathbf{u}\rangle + \int_{\Gamma_c} j^0(s, \gamma\mathbf{u}(s); \gamma\mathbf{v}(s) - \gamma\mathbf{u}(s))\, ds \geq \langle f, \mathbf{v}-\mathbf{u}\rangle, \quad \forall \mathbf{v} \in K. \tag{C.38}$$

We denote by $\partial j(s, \xi) := \partial j(s, \cdot)(\xi)$ the Clarke generalized subdifferential of $j(s, \cdot)$ at the point ξ. We assume that there exist positive constants c_1 and c_2 such that for a.a. $s \in \Gamma_c$, all $\xi \in \mathbb{R}^d$ and for all $\eta \in \partial j(s, \xi)$ the following inequalities hold

(i) $|\eta| \leq c_1(1 + |\xi|)$;
(ii) $\eta^T \xi \geq -c_2|\xi|$.

This growth condition assures that the integral in (C.38) is well defined. Indeed, it follows from (i) and (ii) that for a.a. $s \in \Gamma_c$

$$\left|j^0(s, \xi; \varsigma)\right| = \left|\max_{\eta \in \partial j(s,\xi)} \eta^T \varsigma\right| \leq \max_{\eta \in \partial j(s,\xi)} |\eta|\, |\varsigma| \leq c_1(1 + |\xi|)|\varsigma|, \quad \forall \xi, \varsigma \in \mathbb{R}^d \tag{C.39}$$

and

$$j^0(s, \xi; -\xi) = \max_{\eta \in \partial j(s,\xi)} \eta^T(-\xi) \leq c_2|\xi|, \quad \forall \xi \in \mathbb{R}^d. \tag{C.40}$$

The existence of a solution u to problem (P) can be derived from Theorem C.8. To this end we define the functional $\varphi : V \times V \to \mathbb{R}$ by

$$\varphi(\mathbf{u}, \mathbf{v}) = \int_{\Gamma_c} j^0(s, \gamma\mathbf{u}(s); \gamma\mathbf{v}(s) - \gamma\mathbf{u}(s))\, ds, \quad \forall \mathbf{u}, \mathbf{v} \in V. \tag{C.41}$$

Lemma C.6 ([220]) *The functional φ is pseudomonotone and satisfies*

$$\varphi(\mathbf{u}, 0) \leq c_3\|\mathbf{u}\|_V, \quad \forall \mathbf{u} \in V \tag{C.42}$$

for some positive constant c_3.

Proof Let $\{\mathbf{u}_m\}$ be a sequence in V such that $\mathbf{u}_m \rightharpoonup \mathbf{u}$ in V as $m \to \infty$. Since γ is compact, it follows for a subsequence of $\{\gamma\mathbf{u}_m\}$, which we denote again by $\{\gamma\mathbf{u}_m\}$, that

$$\gamma\mathbf{u}_m \to \gamma\mathbf{u} \text{ in } L^2(\Gamma_c; \mathbb{R}^d) \text{ as } m \to \infty. \tag{C.43}$$

Now, we fix $\mathbf{v} \in V$ and show that

$$\limsup_{m\to\infty} \varphi(\mathbf{u}_m, \mathbf{v}) \leq \varphi(\mathbf{u}, \mathbf{v}). \tag{C.44}$$

We first observe that by (C.43) there exists a subsequence of $\{\gamma\mathbf{u}_m\}$, which we denote again by $\{\gamma\mathbf{u}_m\}$, such that

$$\gamma\mathbf{u}_m(s) \to \gamma\mathbf{u}(s) \quad \text{for a.a. } s \in \Gamma_c \tag{C.45}$$

and

$$|\gamma\mathbf{u}_m(s)| \leq \kappa_0(s) \quad \text{for some nonnegative function } \kappa_0 \in L^2(\Gamma_c). \tag{C.46}$$

Using (C.39) and (C.46), it follows that

$$\begin{aligned} j^0(s, \gamma\mathbf{u}_m(s); \gamma\mathbf{v}(s) - \gamma\mathbf{u}_m(s)) &\leq c_1(1 + |\gamma\mathbf{u}_m(s)|)|\gamma\mathbf{v}(s) - \gamma\mathbf{u}_m(s)| \\ &\leq c_1(1 + \kappa_0(s))\big(|\gamma\mathbf{v}(s)| + \kappa_0(s)\big) \in L^1(\Gamma_c). \end{aligned}$$

From (C.45) and the upper semicontinuity of $j^0(s;\cdot,\cdot)$, we conclude by applying the Fatou lemma that

$$\begin{aligned}\limsup_{m\to\infty}\varphi(\mathbf{u}_m,\mathbf{v}) &= \limsup_{m\to\infty}\int_{\Gamma_c} j^0(s,\gamma\mathbf{u}_m(s);\gamma\mathbf{v}(s)-\gamma\mathbf{u}_m(s))\,ds\\ &\le \int_{\Gamma_c}\limsup_{m\to\infty} j^0(s,\gamma\mathbf{u}_m(s);\gamma\mathbf{v}(s)-\gamma\mathbf{u}_m(s))\,ds\\ &\le \int_{\Gamma_c} j^0(s,\gamma\mathbf{u}(s);\gamma\mathbf{v}(s)-\gamma\mathbf{u}(s))\,ds = \varphi(\mathbf{u},\mathbf{v}) \qquad (C.47)\end{aligned}$$

and thus, (C.44) is shown. Hence, the functional φ is pseudomonotone.

Furthermore, by (C.40) for any $\mathbf{u}\in V$ we can estimate

$$\begin{aligned}\varphi(\mathbf{u},0) &= \int_{\Gamma_c} j^0(s,\gamma\mathbf{u}(s);-\gamma\mathbf{u}(s))\,ds \le c_2\int_{\Gamma_c}|\gamma\mathbf{u}(s)|\,ds\\ &\le c_2((\text{meas}\,(\Gamma_c))^{1/2}\|\gamma\mathbf{u}\|_{L^2(\Gamma_c;\mathbb{R}^d)} \overset{(C.35)}{\le} c_2((\text{meas}\,(\Gamma_c))^{1/2}c_0\|\mathbf{u}\|_V,\end{aligned}$$

which implies (C.42). The proof of the lemma is thus complete. □

Since summation preserves pseudomonotonicity, see [216], the bifunction

$$\psi(\mathbf{u},\mathbf{v}) := \langle A\mathbf{u},\mathbf{v}-\mathbf{u}\rangle + \varphi(\mathbf{u},\mathbf{v})$$

is pseudomonotone and satisfies the assumptions of Theorem C.8; in particular the coercivity condition (CC) holds, since by (C.37) and (C.42),

$$-\psi(\mathbf{u},0) \ge c_K\|\mathbf{u}\|_V^2 - c_3\|\mathbf{u}\|_V\,.$$

We point out that uniqueness of the solution u to the hemivariational inequality (C.38) can be ensured for a large enough Korn constant c_K, see [332] for a proof of such an uniqueness result. We also refer to [326] for a similar uniqueness result to related nonconvex nonsmooth optimization problems.

Appendix D
Some Implementations for BEM

D.1 Symm's Equation on an Interval

$$Vu(x) := -\frac{1}{\pi}\int_{\Gamma} \ln|x-y|u(y)ds_y = f(x) \quad \text{for } x \in \Gamma = (-1,1). \tag{D.1}$$

On a uniform mesh with meshsize h on Γ, $x_j = -1+jh, \quad h = \frac{2}{n}, \quad j = 0,\dots,n$, we take the space $\overline{V}_h$ of piecewise constant functions and perform the h-version of the Galerkin scheme for (D.1):

Find $u_h \in \overline{V}_h$, such that

$$a(u_h, v_h) := \langle Vu_h, v_h\rangle = \langle f, v_h\rangle \quad \forall v_h \in \overline{V}_h. \tag{D.2}$$

With the auxiliary function $F(x) = x^2 \ln|x|$ the Galerkin element a_{ij} becomes:

$$\begin{aligned} -\frac{1}{\pi}\int_{x_i}^{x_{i+1}}\int_{x_j}^{x_{j+1}} \ln|x-y|\,dy\,dx = \frac{1}{2\pi}\Big(&F(x_{i+1}-x_{j+1}) - F(x_{i+1}-x_j) \\ &-F(x_i - x_{j+1}) + F(x_i - x_j) \\ &+3(x_{i+1}-x_i)(x_{j+1}-x_j)\Big). \end{aligned} \tag{D.3}$$

Exercise: Write a program,which implements the Galerkin scheme (D.2).

(i) Compute (D.2) for $f = 1$ and $f = x$. Note $u = \frac{1}{\ln 2}\frac{1}{\sqrt{1-x^2}}$ for $f = 1$.
(ii) Plot the solution for $n = 4$, $n = 8$ and $n = 16$.
(iii) Write for the program *unilap2.f90* a subroutine, which computes the energy norm of the solution of the Galerkin equations. Compute for $f = 1$ the error

© Springer International Publishing AG, part of Springer Nature 2018
J. Gwinner, E. P. Stephan, *Advanced Boundary Element Methods*,
Springer Series in Computational Mathematics 52,
https://doi.org/10.1007/978-3-319-92001-6

in the energy norm:

$$\|u - u_h\|_V = \sqrt{\|u\|_V^2 - \|u_h\|_V^2} = \sqrt{\frac{\pi}{\ln 2} - \|u_h\|_V^2}.$$

Compute for different n the errors in the energy norm and plot them in a double logarithmic scale.

D.2 The Dirichlet Problem in 2D

Now we consider the integral equation

$$\begin{aligned} Vu(x) &= (I+K)g(x), \qquad \text{for } x \in \Gamma \\ Vu(x) &:= -\frac{1}{\pi}\int_\Gamma \ln|x-y|u(y)\,ds_y \\ Ku(x) &:= -\frac{1}{\pi}\int_\Gamma \frac{\partial}{\partial n_y}\ln|x-y|u(y)\,ds_y = -\frac{1}{\pi}\int_\Gamma \frac{n_y(y-x)}{|x-y|^2}u(y)\,ds_y. \end{aligned} \tag{D.4}$$

As geometry we take the L-shape Γ with vertices $(0,0)$, $(0,0.5)$, $(-0.5,0.5)$, $(-0.5,-0.5)$, $(0.5,-0.5)$, $(0.5,0)$. We use a uniform mesh with length h and define there the space $\overline{V}_h$ of piecewise constant functions. Then the h-Version of the Galerkin BEM for the equation (D.4) reads: Find $u_h \in \overline{V}_h$, such that

$$a(u_h, v_h) := \langle Vu_h, v_h\rangle = \langle (I+K)g, v_h\rangle \quad \forall v_h \in \overline{V}_h.$$

Write a program, which implements this method. Use routines of `maiprogs`.
To describe the mesh use the following data structure

```
  integer:: ng
integer,parameter :: ngmax=2048
real(kind=dp):: rx(0:1,0:ngmax-1)
real(kind=dp):: rh(0:1,0:ngmax-1)
real(kind=dp):: rn(0:1,0:ngmax-1)
```

Here `ngmax` denotes the maximal number of elements and `ng` denotes the actual amount. `rx(0,i)` and `rx(1,i)` are the x- and y- components of a vertex of the element with the number `i`. `rh(.,i)` points from a vertex to the next vertex and `rn(.,i)` is the direction of the exterior normal of this element.

(i) Create a mesh generator, which creates for an arbitrary number of elements a uniform mesh.
(ii) Compute the Galerkin matrix for this data structure. Use the routine lapintegmd of liblap2.f90.

(iii) Compute the right hand side with Gaussian quadrature using lapid und lapkspot.
(iv) Test the last routine with $g \equiv 1$. (There holds $(I + K)1 \equiv 0$, why?)
(v) Solve the linear system with Gauss elimination.

The above mentioned subroutines can be downloaded from the home page of M. Maischak, http://people.brunel.ac.uk/~mastmmm/. Use

$$\underline{\langle Kg, v_h\rangle} = -\frac{1}{\pi}\int\limits_\Gamma v_h(x)\left[\int\limits_\Gamma g(y)\frac{\mathbf{n}_y\cdot(\mathbf{y}-\mathbf{x})}{|\mathbf{x}-\mathbf{y}|^2}ds_y\right]ds_x$$

$$\overset{\text{Fubini}}{=} -\frac{1}{\pi}\int\limits_\Gamma g(y)\left[\int\limits_\Gamma \frac{\mathbf{n}_y\cdot(\mathbf{y}-\mathbf{x})}{|\mathbf{x}-\mathbf{y}|^2}v_h(x)ds_x\right]ds_y = \underline{\langle g, K'v_h\rangle}.$$

Let $x \in \Gamma_i$, $y \in \Gamma_j$. If Γ_i and Γ_j are on the same edge then $\mathbf{n}_x \cdot (\mathbf{y}-\mathbf{x}) = 0$ and $\langle g, K'v_h\rangle = 0$. Otherwise we compute with the parametrisation of the vector $\mathbf{y}-\mathbf{x}$

$$\frac{1}{\pi}\int\limits_{\Gamma_i}\frac{\mathbf{n}\cdot(\mathbf{y}-\mathbf{x})}{|\mathbf{x}-\mathbf{y}|^2}ds_y = \frac{1}{\pi}\int\limits_{-1}^{1}\frac{\mathbf{n}\cdot(\mathbf{a}t+\mathbf{b})}{|\mathbf{a}t+\mathbf{b}|^2}dt$$

$$= \frac{1}{\pi}\left(\mathbf{n}\cdot\mathbf{a}\int\limits_{-1}^{1}\frac{t}{\mathbf{a}^2t^2+2\mathbf{a}\cdot\mathbf{b}t+\mathbf{b}^2}dt + \mathbf{n}\cdot\mathbf{b}\int\limits_{-1}^{1}\frac{1}{\mathbf{a}^2t^2+2\mathbf{a}\cdot\mathbf{b}t+\mathbf{b}^2}dt\right).$$

For the determination of these integrals, let $g_k^n(\alpha,\beta,\gamma) := \int\limits_{-1}^{1} t^k(\alpha t^2+\beta t+\gamma)^n dt$ with $\alpha = \mathbf{a}^2$, $\beta = 2\mathbf{a}\cdot\mathbf{b}$, $\gamma = \mathbf{b}^2$. Now with $\Delta = 4\alpha\gamma - \beta^2$:

$$g_0^{-1} = \int\limits_{-1}^{1}\frac{1}{\alpha t^2+\beta t+\gamma}dt = \frac{2}{\sqrt{\Delta}}\left(\arctan\frac{2\alpha+\beta}{\sqrt{\Delta}} - \arctan\frac{-2\alpha+\beta}{\sqrt{\Delta}}\right)$$

and

$$g_1^{-1} = \int\limits_{-1}^{1}\frac{t}{\alpha t^2+\beta t+\gamma}dt = \frac{2\alpha t+\beta}{\Delta(\alpha t^2+\beta t+\gamma)} + \frac{2\alpha}{\Delta}\int\limits_{-1}^{1}\frac{dt}{\alpha t^2+\beta t+\gamma}.$$

D.3 Symm's Equation on a Surface Piece

We can consider on a plane surface piece Γ

$$\frac{1}{4\pi}\int_\Gamma \frac{\psi(y)}{|x-y|}\,d\sigma_y = f(x)\,, \quad x \in \Gamma\,. \tag{D.5}$$

Decomposing Γ into rectangles $R_i \in \mathcal{T}_h$, we can choose our basis functions $\varphi_j(x)$ to be 1 only on one element,

$$\psi_h(x) = \sum_{i=1}^N \psi_j \varphi_j(x)\,, \qquad \varphi_j(x) = \begin{cases} 1\,, & x \in R_j\,, \\ 0\,, & \text{else.} \end{cases}$$

We get the Galerkin scheme: Find $\psi_h \in \tilde{H}^{-\frac{1}{2}}(\Gamma_1)$ such that, for $k = 1, \ldots, N$,

$$\int_\Gamma \frac{1}{4\pi}\int_\Gamma \frac{\psi_h(y)}{|x-y|}\,d\sigma_y \varphi_k(x)\,d\sigma_x = \sum_{j=1}^N \psi_j \underbrace{\frac{1}{4\pi}\int_\Gamma\int_\Gamma \frac{\varphi_j(y)\varphi_k(x)}{|y-x|}\,d\sigma_y\,d\sigma_x}_{=:I}$$

$$= \int_\Gamma f(x)\varphi_k(x)\,d\sigma_x \tag{D.6}$$

By definition of φ_j and φ_k, I can be written as

$$I = \frac{1}{4\pi}\int_{R_k}\int_{R_j} \frac{1}{|x-y|} d\sigma_y d\sigma_x = \frac{1}{4\pi}\int_{R_k}\int_{R_j} \frac{1}{\sqrt{(x_1-y_1)^2+(x_2-y_2)^2}} d\sigma_y d\sigma_x.$$

The inner integral can be calculated by transforming R_j to the reference square $\tilde{R} = [0,1]^2$:

$$\begin{aligned} R_j &:= \left\{(\xi,\eta) : \; x_j \le \xi \le x_j + h_x\,, \; y_j \le \eta \le y_j + h_y\right\} \\ \xi &= x_j + h_x u\,, \quad 0 \le u \le 1 \\ \eta &= y_j + h_y v\,, \quad 0 \le v \le 1\,. \end{aligned}$$

With fixed points $x = (a,b)$ and $y = (\xi,\eta)$, we get for the inner integral

$$\int_{R_j} \frac{1}{|x-y|}\,d\sigma_y =$$

$$= \int_{R_j} \frac{d\xi\,d\eta}{\sqrt{(\xi-a)^2+(\eta-b)^2}} = \int_{\tilde{R}} \frac{h_x\,du\,h_y\,dv}{\left((x_j+h_x u-a)^2 + (y_j+h_y v-b)^2\right)^{\frac{1}{2}}}$$

$$
\begin{aligned}
=&\alpha\left(\sinh^{-1}\left(\frac{h_y-\beta}{|\alpha|}\right)+\sinh^{-1}\left(\frac{\beta}{|\alpha|}\right)\right)+\beta\left(\sinh^{-1}\left(\frac{h_x-\alpha}{|\beta|}\right)+\sinh^{-1}\left(\frac{\alpha}{|\beta|}\right)\right)\\
&+(h_x-\alpha)\left(\sinh^{-1}\left(\frac{h_y-\beta}{|h_x-\alpha|}\right)+\sinh^{-1}\left(\frac{\beta}{|h_x-\alpha|}\right)\right)\\
&+(h_y-\beta)\left(\sinh^{-1}\left(\frac{h_x-\alpha}{|h_y-\beta|}\right)+\sinh^{-1}\left(\frac{\alpha}{|h_y-\beta|}\right)\right), \alpha=a-x_j,\ \beta=b-y_j.
\end{aligned}
$$

The outer integral can be approximated, e.g. by a 4-point quadrature formula that is exact for polynomials of degree ≤ 2: Let the quadrature nodes $\hat{x}_1, \hat{x}_2, \hat{x}_3, \hat{x}_4$ be given by $\hat{x}_i = (0.5 \pm \frac{\sqrt{3}}{6}, 0.5 \pm \frac{\sqrt{3}}{6}), i = 1, 2, 3, 4$ on $\tilde{R}$, then on the reference square, for the polynomial P there holds

$$
\int_{\tilde{R}} P(u,v)\,du\,dv = \frac{1}{4}\sum_{i=1}^{4} P(\hat{x}_i)\,.
$$

Note that the outer integration can also be performed analytically. This is implemented in the software package *maiprogs*, see [291].

For decomposing Γ we consider 4 different methods, firstly a uniform mesh with axis-parallel rectangles, secondly a graded mesh described by a tensor product mesh based on a 1-d graded mesh with grading constant β. And finally two adaptive strategies based on a two-level error estimator, one where we split each appropriate element into four equal sized elements and another one where we split the element horizontally or vertically into two parts or into four equal parts depending on the composition of the local error indicator (see Fig. D.1).

For the hierarchical error estimator we decompose every brick function ϕ_i^h associated with the element i and element size h into a set of three jump functions $\beta_{i,j}$ by uniformly refining the element i into four equal sized sub elements. Then there holds for $1 \leq j \leq 3$

$$
\beta_{i,j} = \sum_{l=1}^{4} c_{l,j}\phi_{i,l}^{h/2} = \phi_i^h + \sum_{l=2}^{4} \tilde{c}_{l,j}\phi_{i,l}^{h/2},
$$

where $\phi_{i,l}^{h/2}$ is the brick function on the sub element l to the element i. Further we define with the energynorm $\|\cdot\|_V$

$$
\begin{aligned}
\vartheta_{i,j} :=& \frac{\left|\left\langle V\psi_N - f, \beta_{i,j}\right\rangle\right|}{\|\beta_{i,j}\|_V} = \frac{\left|\left\langle V\psi_N - f, \phi_i^h + \sum_{l=2}^{4}\tilde{c}_{l,j}\phi_{i,l}^{h/2}\right\rangle\right|}{\|\beta_{i,j}\|_V}\\
=& \frac{\left|\left\langle V\psi_N - f, \sum_{l=2}^{4}\tilde{c}_{l,j}\phi_{i,l}^{h/2}\right\rangle\right|}{\|\beta_{i,j}\|_V}
\end{aligned}
$$

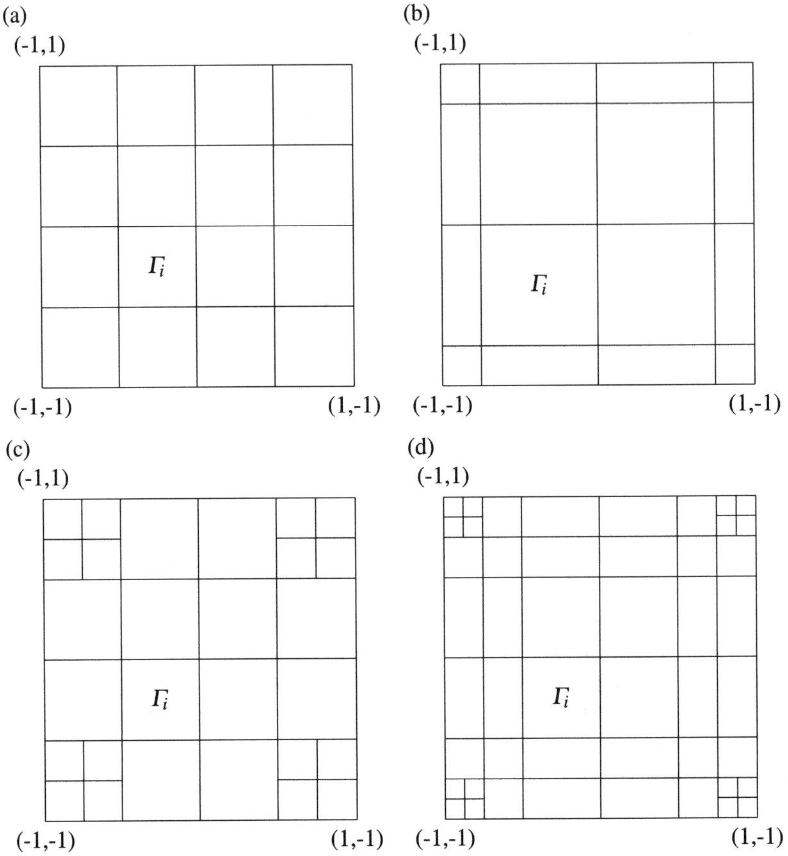

Fig. D.1 Different boundary decomposition techniques . (**a**) Uniform mesh decomposition. (**b**) Graded mesh decomposition. (**c**) Adaptive mesh decomposition strategies 1. (**d**) Adaptive mesh decomposition strategies 2

$$= \frac{\left|\left\langle V\sum_{i=1}^{N}\alpha_i\phi_i^h(x), \sum_{l=2}^{4}\tilde{c}_{l,j}\phi_{i,l}^{h/2}\right\rangle - \left\langle f, \sum_{l=2}^{4}\tilde{c}_{l,j}\phi_{i,l}^{h/2}\right\rangle\right|}{\|\beta_{i,j}\|_V}$$

Note that $\vartheta_{i,j}$ can be implemented efficiently when using the linearity of the scalar product and the reuse of old values. When making use of the Galerkin orthogonality as above we are able to reduce the computation time for $\vartheta_{i,j}$ by $\frac{1}{4}$. The local error indicator is now defined by

$$\vartheta_i := \sqrt{\vartheta_{i,1}^2 + \vartheta_{i,2}^2 + \vartheta_{i,3}^2}$$

For the adaptive strategy 2 we save in an additional vector if there holds $\vartheta_{i,2} \geq 1.5\vartheta_{i,1}$, $\vartheta_{i,1} \geq 1.5\vartheta_{i,2}$ or neither. If the element has been marked for refinement and the first condition is true split the element vertically into 2 equal sized rectangles, if the second condition is true then split horizontally into 2 equal sized rectangles and else into 4 equal sized rectangles. If the saturation assumption holds one can prove the efficiency and reliability of the error indicator $\eta = \|\vartheta\|_2$.

The numerical experiments were carried out by Lothar Banz on the Laptop Fujitsu Siemens Amilo M1439G with MatLab R2007. For solving the discrete linear system a CG algorithm is applied.

As we can see from Fig. D.3 the condition number of Galerkin matrix with an underlying uniform mesh behaves like $O(\sqrt{N})$.The condition numbers for the different mesh strategies are growing much faster than for the uniform mesh. For the graded meshes there holds the greater β is the faster the condition number grows.

The solution is obviously singular at the boundary of the boundary-domain Γ with strong singularities in the edges. The uniform mesh does not take the singular behavior into account which yields a lower convergence rate than for the different strategies. If we apply a graded mesh as described earlier we can improve the convergence rate in the energy norm of 0.25 for the uniform mesh to 0.73 for the graded mesh with $\beta = 4$ (see Fig. D.2). The more we take the singularity into account the better is the convergence rate. The local error is a product of the local

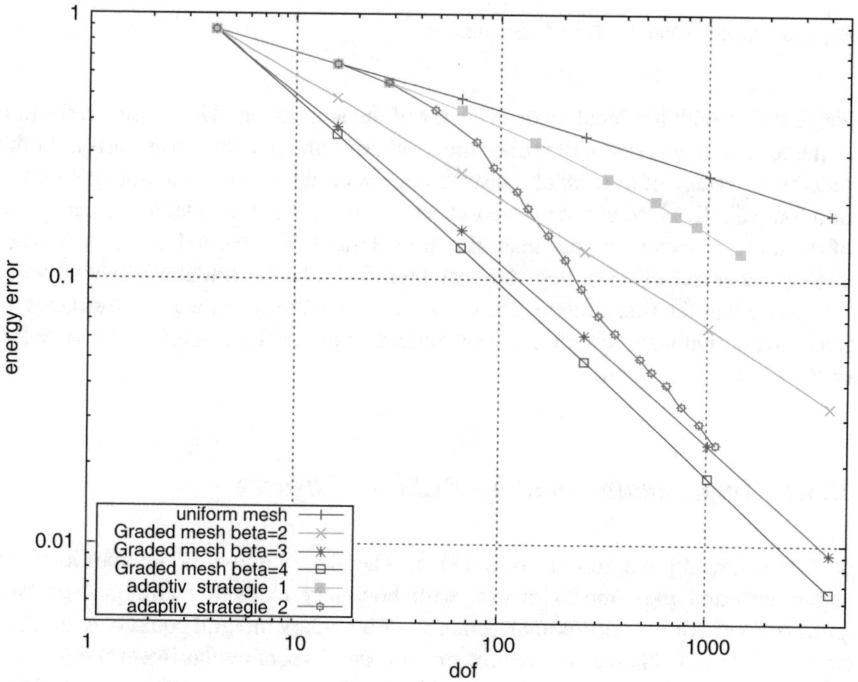

Fig. D.2 Energy error for different meshes

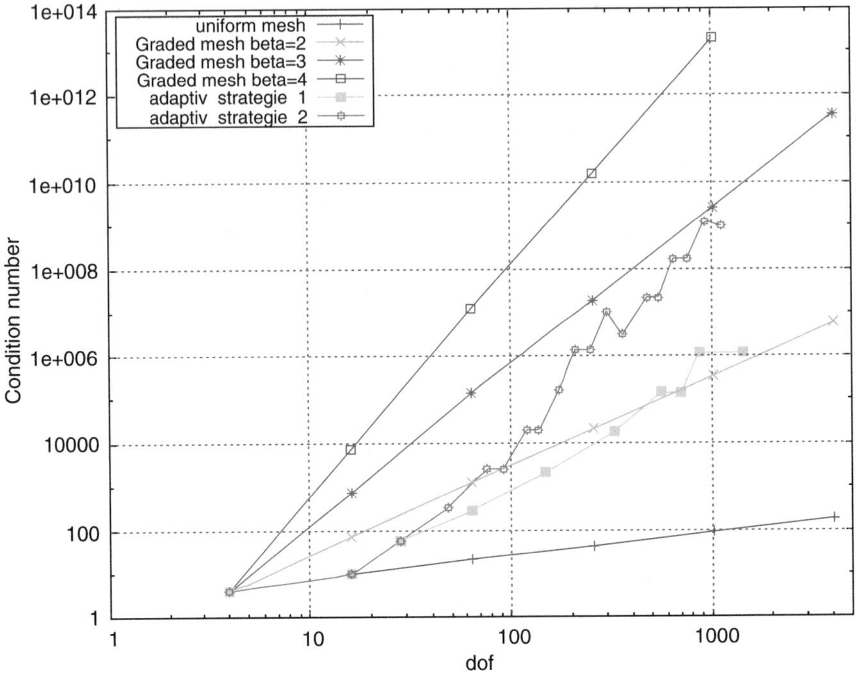

Fig. D.3 Condition number for different meshes

element size with the local error-behavior of the solution ψ. Therefore a reduction of the local element size will reduce the local error and thus the global error. As the reduction strategy of the first adaptive strategy is limited by no reduction or splitting into four equally sized elements we expect a convergence rate which is greater than of the uniform mesh and less than of a graded mesh. The second adaptive strategy has a broader reduction strategy and can therefore take the singular behavior better into account as the first strategy. However it is still worse than the graded strategy as it has no continuously, systematic, slow reduction of the elements close to the center of the boundary-domain.

D.3.1 Implementation of hp-BEM on Surfaces

In the following we report from [310]. The combination of geometric mesh refinement and h-p approximation with boundary element techniques gives a powerful tool for the approximate solution of boundary integral equations. In [235] an h-p Galerkin scheme for weakly singular and hypersingular integral equations on plane screens in $\mathbb{R}^3$ was analyzed and in [251] exponential convergence could be proved.

Although the singular integrals for plane surfaces in [235] can be evaluated analytically the assembly of the Galerkin matrix is extremely expensive. This becomes even worse if curved surfaces are considered and the entries of the Galerkin matrix have to be computed by a numerical quadrature rule. Here, an application of the *h*-*p* boundary element method has the advantage that the Galerkin error decays exponentially fast with the size of the Galerkin matrix, i.e. the number of Galerkin entries is kept low.

In this subsection we focus on the weakly singular integral equation on an open surface Γ, which corresponds to the direct single layer potential formulation of the Dirichlet problem for the homogeneous Laplace equation in $\mathbb{R}^3 \setminus \Gamma$. Our aim is now to define a quadrature rule which approximates the Galerkin entries exponentially fast with the number n of kernel evaluations. By increasing n at each *h*-*p* refinement step we may, hence, expect to preserve the exponential convergence of the Galerkin scheme while keeping the computational costs low.

The quadrature rules which we use are basically applications of Schwab's [372] graded quadrature rules for singular integrals to the inner *and* outer integrals in our Galerkin matrix. Schwab's rule can be applied directly to assemble collocation matrices or the inner integrals of the Galerkin entries. Based on the *h*-*p* approximation results in [17] and the interpolation property of e.g. Gaussian quadrature formulae exponential convergence could be proved [372]. For the outer integrals we need a similar rule which is designed to approximate the singularities of the single layer potential.

Let $G \subset \mathbb{R}^3$ be an open curved surface with parameter region $\Gamma = [-1, 1]^2$ and parameter function $\gamma : \Gamma \to G$. We assume that G satisfies a Lipschitz condition and that $\gamma(\partial\Gamma) = \partial G$. Let $V : \widetilde{H}^{-1/2}(G) \to H^{1/2}(G)$ be the single layer potential operator defined as

$$V\psi(x) = \frac{1}{4\pi}\int_\Gamma \frac{\psi(y)}{\|x-y\|}\, d\sigma_y\,.$$

Let $\widetilde{X}$ be a finite dimensional space of piecewise polynomial functions over Γ and let $X = \{\phi \circ \gamma^{-1};\, \phi \in \widetilde{X}\} \subset \widetilde{H}^{-1/2}(G)$. Let ψ_i, ψ_j be two basis functions in $\widetilde{X}$ and let Γ_i, Γ_j be elements in Γ with $\text{supp}(\psi_i) \subset \overline{\Gamma}_i$ and $\text{supp}(\psi_j) \subset \overline{\Gamma}_j$. Now, the entries of the Galerkin matrix and the right hand side vector are

$$\langle V\psi_j\circ\gamma^{-1}, \psi_i\circ\gamma^{-1}\rangle = \frac{1}{4\pi}\int_{\Gamma_i} \psi_i(x)\left(\int_{\Gamma_j} \frac{\psi_j(x)}{\|\gamma(x)-\gamma(y)\|} J(y)dy\right) J(x)dx \tag{D.7}$$

$$\langle f, \psi_i \circ \gamma^{-1}\rangle = \int_{\Gamma_i} \psi_i(x) f(\gamma(x)) J(x)dx \tag{D.8}$$

where $J(x) = \|\frac{\partial\gamma(x)}{\partial x_1} \times \frac{\partial\gamma(x)}{\partial x_2}\|$ and $\|.\|$ denotes the Euklidean norm in $\mathbb{R}^3$.

It could be shown in [251] for plane surfaces Γ that the *h*-*p* version of the boundary element method converges exponentially fast, see also Sect. 8.2. The

h-p meshes are geometrically graded towards $\partial\Gamma$ and the polynomial degrees of the test and trial functions in $x \in \Gamma$ are small if x is close to $\partial\Gamma$ and are increased perpendicular to $\partial\Gamma$ (for details see [235]). For plane surfaces, however, the integrals (D.7) can be evaluated analytically and the computational cost for the assembly of the Galerkin matrix grows only algebraically with the number N of h-p refinement steps, i.e. like $O(N^\alpha)$. We show that there is a quadrature rule which approximates the singular integrals (D.7) exponentially fast (with N) and which needs $O(N^\alpha)$ kernel evaluations ($\alpha \in \mathbb{N}$ fixed). Furthermore, we give (numerical) evidence that the h-p Galerkin method applied to (D.5) in combination with this quadrature rule leads to exponential convergence of the approximate solutions. To approximate both integrals we have to deal with point and edge singularites. The kernel $|\gamma(x) - \gamma(y)|^{-1}$ of the inner integral has obviously a point singularity at $y = x$ whereas the single layer potential has singular behaviour at $\gamma(\partial\,\mathrm{supp}(\psi))$.

For point singularities Schwab suggested the following rule for the approximation of the integral $\int_{\Gamma_0} \psi_0(x)\,dx$ where $\Gamma_0 = (0,1)^2$ and ψ_0 is singular at the origin: Given a fixed parameter $\sigma_1 \in (0,1)$ and an integer n one considers geometric subdivisions of Γ_0 into smaller rectangles $R_{l,k}$. We define $z_0 = 0$, $z_k = \sigma_1^{n-k}$, $1 \le k \le n$, and

$$\begin{aligned} R_{1,k} &= (z_{k-1}, z_k) \times (0, z_{k-1}) \quad \text{for } 2 \le k \le n\,, \\ R_{2,k} &= (z_{k-1}, z_k) \times (z_{k-1}, z_k) \quad \text{for } 1 \le k \le n\,, \\ R_{3,k} &= (0, z_{k-1}) \times (z_{k-1}, z_k) \quad \text{for } 2 \le k \le n\,. \end{aligned}$$

For fixed $\varrho_1, \varrho_2 \in \mathbb{N}_0$ let $Q_{l,k}$ denote the tensor product of the $(k+\varrho_1)$-point Gaussian quadrature rule in x_1-direction and the $(k+\varrho_2)$-point Gaussian quadrature rule in x_2-direction, scaled to $R_{l,k}$. Hence

$$Q_{l,k}\psi_0 \approx \int_{R_{l,k}} \psi_0(x)\,dx\,.$$

The composite quadrature rule $Q_n^{(1)}$ is now defined as

$$Q_n^{(1)}\psi_0 = Q_{2,1}\psi_0 + \sum_{k=2}^{n}\sum_{l=1}^{3} Q_{l,k}\psi_0 \approx \int_{\Gamma_0} \psi_0(x)\,dx\,.$$

The quadrature points and the subdivision of Γ_0 for $\sigma_1 = 0.4$, $n = 4$ and $\varrho_1 = \varrho_2 = 0$ are shown in Fig. D.4a.

Remark D.1 When approximating the Galerkin entries (D.7) we will choose $\varrho_k = p_k$ where p_k is the polynomial degree of ψ_j (or ψ_i) in x_k-direction ($k = 1, 2$).

For corner-edge singularities we consider again the reference element Γ_0. Let ϕ_0 have corner singularities at the origin $(0,0)$ and at the point $(x_1, x_2) = (0,1)$ and an edge singularity at $x_1 \equiv 0$. We use a geometric subdivision of Γ_0 towards the corners $(0,0)$ and $(0,1)$ and towards the corresponding edge with grading parameter

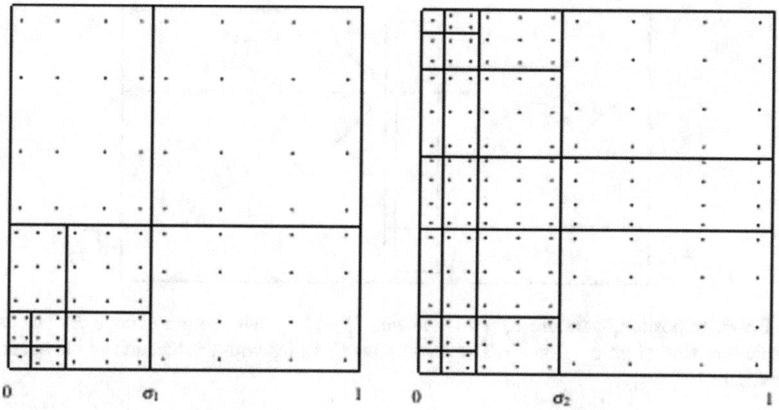

(a) $Q_n^{(1)}$ with $n = 4$ (b) $Q_n^{(2)}$ with $n = 4$

Fig. D.4 Subdivision of $[0, 1]^2$ and quadrature points for (**a**) and (**b**), where $\sigma_1 = \sigma_2 = 0.4$ and $\varrho_1 = \varrho_2 = 0$ [310]

$\sigma_2 \in (0, 1/2)$. This defines a quadrature rule $Q_n^{(2)}$. Hence, we have

$$Q_n^{(2)} \phi_0 \approx \int_{\Gamma_0} \phi_0(x)\, dx\,.$$

For an example see Fig. D.4b.

If the function ϕ_0 is singular at all four edges of Γ_0 we use a quadrature rule $Q_n^{(3)}$ with geometrical grading towards all the edges of Γ_0 and with grading parameter $\sigma_3 \in (0, 1/2)$.

To approximate the Galerkin entries (1.3) we have to deal with three critical cases where the kernel becomes singular:

i. Γ_i and Γ_j have a common node
ii. Γ_i and Γ_j have a common edge
iii. $\Gamma_i = \Gamma_j$

In the first case we use affine images of the quadrature rule $Q_n^{(1)}$ on Γ_i and Γ_j with grading towards the common node.

We define the integer $m = \lfloor n^{4/3} \rfloor$.

In the second case we use the affine image of $Q_m^{(2)}$ on Γ_i with grading towards the common edge and the common nodes. Let x_k denote the quadrature points of $Q_m^{(2)}$ on Γ_i. For any of these points x_k we consider the straight line L_k which contains x_k and which is orthogonal to the common edge E of Γ_i and Γ_j. The line L_k divides Γ_j into two rectangles $\Gamma_{j,k}^1$ and $\Gamma_{j,k}^2$. On each of these rectangles we apply the quadrature rule $Q_n^{(1)}$ with grading towards the point $L_k \cap E$ (see Fig. D.5).

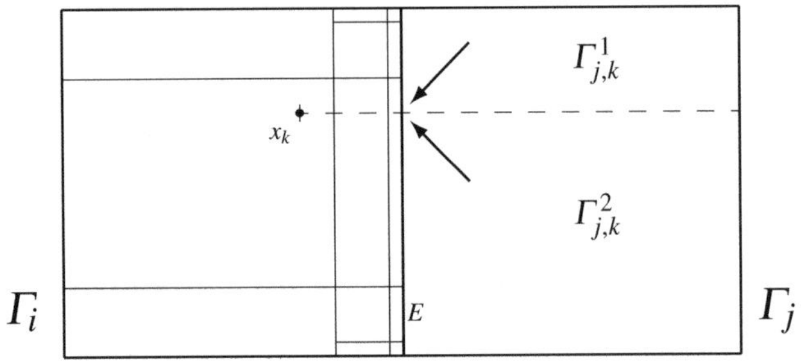

Fig. D.5 Composite quadrature for two elements Γ_i and Γ_j with common edge E. The arrows indicate direction of grading on Γ_j. The grading on Γ_j varies with the location of the quadrature points x_k in Γ_i [310]

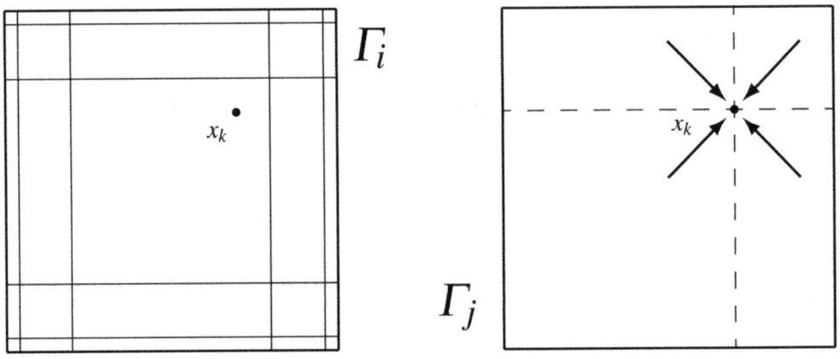

Fig. D.6 Composite quadrature for $\Gamma_i = \Gamma_j$. The grading for the inner quadrature (over Γ_j) varies with the location of the outer quadrature points $x_k \in \Gamma_i$ [310]

In the third case we use the affine image of $Q_m^{(3)}$ on Γ_i. For any quadrature point x_k which belongs to $Q_m^{(3)}$ we divide Γ_j into four rectangles with common node x_k and use the affine image of $Q_n^{(1)}$ on each of these rectangles (with grading towards x_k). See Fig. (D.6).

Next we prove exponential convergence of the quadrature rule introduced above for a simple example of two square elements in the (x_1, x_2)-plane with a common node. In the proof we restrict ourselves to piecewise constant test and trial functions, i.e. we have $\varrho_1 = \varrho_2 = 0$. Numerical results for higher polynomial degrees, for the case of two elements Γ_i and Γ_j with common edge and for the case $\Gamma_i = \Gamma_j$ are included in [310]: There the experimental results indicate exponential convergence in this case.

As a simple example we consider the elements $\Gamma_1 = (0,1)^2$ and $\Gamma_2 = (-1,0)^2$ and the parameter function $\gamma : [-1,1]^2 \to G$ defined as $\gamma(x_1,x_2) = (x_1,x_2,0)$. Let $\psi_1, \psi_2 \in \widetilde{H}^{-1/2}(G)$ be defined as

$$\psi_j(x) = \begin{cases} 1 & \text{if } x \in \Gamma_j \times \{0\} \\ 0 & \text{if } x \in G \setminus (\Gamma_j \times \{0\}) \end{cases} \qquad (j = 1, 2)\,.$$

Hence,

$$V(\psi_1, \psi_2) := \frac{1}{4\pi} \int_{\Gamma_2} \int_{\Gamma_1} \frac{1}{|x-y|}\, dy\, dx\,. \tag{D.9}$$

For $0 < \sigma < 1$ and $n \in \mathbb{N}$ define $Q_1 = Q_n^{(1)}$ and let Q_2 be the affine image of $Q_n^{(1)}$ on Γ_2 with grading towards the origin. For $i \in \{1, 2\}$ let $x_k^{(j)}$ and $w_k^{(j)}$ be the knots and weights of the rule Q_j, i.e.

$$Q_j g = \sum_{k=1}^{M} w_k^{(j)} g(x_k^{(j)})$$

where $M = 1 + 3\sum_{i=2}^{n} i^2 = n^3 + \frac{3}{2}n^2 + \frac{1}{2}n - 2$.

The following result states exponential convergence of the composite quadrature rule $Q_2 Q_1$ applied to $|x-y|^{-1}$.

Theorem D.1 ([310]) *For any* $0 < \sigma < 1$ *there are constants* $c_1, c_2 > 0$ *such that*

$$\left| V(\psi_1, \psi_2) - \sum_{k=1}^{M} \sum_{j=1}^{M} w_k^{(1)} w_j^{(2)} |x_k^{(1)} - x_j^{(2)}|^{-1} \right| \le c_1\, e^{-c_2 n} \tag{D.10}$$

for all $n \in \mathbb{N}$.

To prove the theorem we need the theory of countably normed spaces and the result from [372] on the exponential convergence of Q_j:

For a domain $A \subset \mathbb{R}^2 \setminus \{0\}$ and a parameter $0 < \beta < 1$ let $H_\beta^k(A)$ be the closure of $\mathscr{C}^\infty(\overline{A})$ with respect to the weighted Sobolev norm

$$\|g\|^2_{H_\beta^k(A)} = \sum_{j=0}^{k} \sum_{|\alpha|=j} \int_A |D^\alpha g(x)|^2 \phi^2_{\beta+j}(x)\, dx$$

where the weight function ϕ_s is defined as $\phi_s(x) = |x|^s$ $(s \in \mathbb{R})$.

The countably normed space $\mathscr{B}_\beta(A)$ is defined as the subspace of all functions g in $L^1(A) \cap \bigcap_{k=0}^{\infty} H_\beta^k(A)$ whose derivatives satisfy the growth condition

$$\left(\int_A |D^\alpha g(x)|^2 \phi^2_{\beta+j}(x)\, dx \right)^{1/2} \le D_g\, (d_g)^j\, j! \tag{D.11}$$

for all $\alpha \in \mathbb{N}_0^2$ with $|\alpha| = j$. The constants $d_g \ge 1$ and $D_g > 0$ depend on A and g but not on j.

As an example consider the family of functions

$$k_x(y) = \frac{1}{|x-y|} \in \mathscr{B}_\beta(\Gamma_1) \text{ for } 0 < \beta < 1 \text{ uniformly for } x \in \Gamma_2 .$$

The following result was proved in [372]:

Lemma D.1 *Let $g \in \mathscr{B}_\beta(\Gamma_1)$ with $\beta > 0$ sufficiently small. Then, for any $0 < \sigma_1 < 1$ there exist constants $b_1, b_2 > 0$ independent of n such that*

$$\left| \int_{\Gamma_1} g(y)\, dy - Q_n^{(1)} g \right| \le b_1 e^{-b_2 n} \tag{D.12}$$

where the constants b_1 and b_2 depend only on $\sigma_1, \beta, d_g, D_g$ and Γ_1.

We are now in the position to prove Theorem D.1.

Proof The triangle inequality yields

$$4\pi \left| V(\psi_1, \psi_2) - \sum_{k=1}^{M} \sum_{j=1}^{M} w_k^{(1)} w_j^{(2)} |x_k^{(1)} - x_j^{(2)}|^{-1} \right| \le$$

$$\le \underbrace{\left| \int_{\Gamma_1} \int_{\Gamma_2} \frac{1}{|x-y|}\, dy\, dx - Q_2^{(x)} \int_{\Gamma_1} \frac{1}{|x-y|}\, dy \right|}_{=\epsilon_1}$$

$$+ \underbrace{\left| Q_2^{(x)} \int_{\Gamma_1} \frac{1}{|x-y|}\, dy - Q_2^{(x)} Q_1^{(y)} \frac{1}{|x-y|} \right|}_{=\epsilon_2} .$$

We will estimate ϵ_2 and ϵ_1 seperately. By definition of Q_i and k_x we have

$$\epsilon_2 = \sum_{j=1}^{M} w_j^{(2)} \left| \int_{\Gamma_1} k_{x_j^{(2)}}(y)\, dy - Q_n^{(1)} k_{x_j^{(2)}} \right| .$$

From Lemma D.1 it follows that

$$\epsilon_2 \le \sum_{j=1}^{M} w_j^{(2)} b_1 e^{-b_2 n}$$

where b_1 and b_2 are independent of j. Hence,

$$\epsilon_2 \le |\Gamma_2|\, b_1 e^{-b_2 n} \le b_1 e^{-b_2 n} . \tag{D.13}$$

To estimate ϵ_1 we have to show that $V\psi_1(x) \in \mathscr{B}_\beta(\Gamma_2)$ for all $0 < \beta < 1$. For plane rectangular elements the single layer potential can be calculated analytically [235]. For $x \in \Gamma_2$ we have:

$$\begin{aligned} V\psi_1(x) &= \left. \left. (y_1 - x_1)\,\text{arsinh}\frac{y_2 - x_2}{|y_1 - x_1|} + (y_2 - x_2)\,\text{arsinh}\frac{y_1 - x_1}{|y_2 - x_2|} \right|_{y_1=0}^{1} \right|_{y_2=0}^{1} \\ &= -x_1 \ln(-x_2 + |x|) - x_2 \ln(-x_1 + |x|) + g_1(x) \\ &= -(\cos\theta + \sin\theta)\, r \ln r + g_2(r, \theta) \end{aligned}$$

where (r, θ) are the usual polar co-ordinates and g_1, g_2 are analytic in Γ_2.

With an alternative formulation of the growth condition in polar co-ordinates [19] it can be shown easily that $V\psi_1 \in \mathscr{B}_\beta(\Gamma_2)$ for all $0 < \beta < 1$. Hence, from Lemma D.1 it follows that

$$\epsilon_1 \le b_3 e^{-b_4 n}. \tag{D.14}$$

From (D.13) and (D.14) we conclude (D.10) with $c_1 = (b_1 + b_3)/(4\pi)$ and $c_2 = \min\{b_2, b_4\}$. □

For further reading see [104].

A comprehensive list of numerical experiments can be found e.g. in the *Book of Numerical Experiments – BONE* which can be downloaded from the home page of M. Maischak, http://people.brunel.ac.uk/mastmmm/.

References

1. H. Ammari, A. Buffa, J.-C. Nédélec, A justification of eddy currents model for the Maxwell equations. SIAM J. Appl. Math. **60**, 1805–1823 (2000)
2. P.M. Anselone, *Collectively Compact Operator Approximation Theory and Applications to Integral Equations* (Prentice-Hall, Englewood Cliffs, NJ, 1971)
3. D.N. Arnold, W.L. Wendland, On the asymptotic convergence of collocation methods. Math. Comput. **41**, 349–381 (1983)
4. K.E. Atkinson, A discrete Galerkin method for first kind integral equations with a logarithmic kernel. J. Integr. Equ. Appl. **1**, 343–363 (1988)
5. K.E. Atkinson, I.H. Sloan, The numerical solution of first-kind logarithmic-kernel integral equations on smooth open arcs. Math. Comput. **56**, 119–139 (1991)
6. H. Attouch, *Variational Convergence for Functions and Operators*. Applicable Mathematics Series (Pitman (Advanced Publishing Program), Boston, MA, 1984)
7. J.-P. Aubin, *Approximation of Elliptic Boundary-Value Problems*. Pure and Applied Mathematics, Vol. XXVI (Wiley-Interscience, New York-London-Sydney, 1972)
8. M. Aurada, M. Feischl, Th. Führer, M. Karkulik, J.M. Melenk, D. Praetorius, Classical FEM-BEM coupling methods: nonlinearities, well-posedness, and adaptivity. Comput. Mech. **51**, 399–419 (2014)
9. M. Aurada, M. Feischl, Th. Führer, M. Karkulik, J.M. Melenk, D. Praetorius, Local inverse estimates for non-local boundary integral operators. Math. Comput. **86**, 2651–2686 (2017)
10. M. Aurada, M. Feischl, Th. Führer, M. Karkulik, D. Praetorius, Efficiency and optimality of some weighted-residual error estimator for adaptive 2D boundary element methods. Comput. Methods Appl. Math. **13**, 305–332 (2013)
11. M. Aurada, M. Feischl, Th. Führer, M. Karkulik, D. Praetorius, Energy norm based error estimators for adaptive BEM for hypersingular integral equations. Appl. Numer. Math. **95**, 15–35 (2015)
12. M. Aurada, M. Feischl, M. Karkulik, D. Praetorius, A posteriori error estimates for the Johnson-Nédélec FEM-BEM coupling. Eng. Anal. Bound. Elem., 255–266 (2012)
13. M. Aurada, M. Feischl, D. Praetorius, Convergence of some adaptive FEM-BEM coupling for elliptic but possibly nonlinear interface problems. ESAIM Math. Model. Numer. Anal. **46**, 1147–1173 (2012)
14. I. Babuška, A.K. Aziz, *Survey Lectures on the Mathematical Foundations of the Finite Element Method*. The Mathematical Foundations of the Finite Element Method with Applications

© Springer International Publishing AG, part of Springer Nature 2018

J. Gwinner, E. P. Stephan, *Advanced Boundary Element Methods*,
Springer Series in Computational Mathematics 52,
https://doi.org/10.1007/978-3-319-92001-6

to Partial Differential Equations (Proc. Sympos., Univ. Maryland, Baltimore, MD, 1972). Academic Press, New York, 1972, With the collaboration of G. Fix and R. B. Kellogg, pp. 1–359
15. I. Babuška, A. Craig, J. Mandel, J. Pitkäranta, Efficient preconditioning for the p-version finite element method in two dimensions. SIAM J. Numer. Anal. **28**, 624–661 (1991)
16. I. Babuška, M.R. Dorr, Error estimates for the combined h and p versions of the finite element method. Numer. Math. **37**, 257–277 (1981)
17. I. Babuška, B.Q. Guo, The h-p version of the finite element method, part 1: The basic approximation results. Comput. Mech. **1**, 21–41 (1986)
18. I. Babuška, B.Q. Guo, The h-p version of the finite element method, part 2: General results and applications. Comput. Mech. **1**, 203–220 (1986)
19. I. Babuška, B.Q. Guo, Regularity of the solution of elliptic problems with piecewise analytic data. I. Boundary value problems for linear elliptic equation of second order. SIAM J. Math. Anal. **19**, 172–203 (1988)
20. I. Babuška, B.Q. Guo, E.P. Stephan, The h-p version of the boundary element method with geometric mesh on polygonal domains. Comput. Methods Appl. Mech. Eng. **80**, 319–325 (1990)
21. I. Babuška, B.Q. Guo, E.P. Stephan, On the exponential convergence of the h-p version for boundary element Galerkin methods on polygons. Math. Methods Appl. Sci. **12**, 413–427 (1990)
22. I. Babuška, B.Q. Guo, M. Suri, Implementation of nonhomogeneous Dirichlet boundary conditions in the p-version of the finite element method. Impact Comput. Sci. Eng. **1**, 36–63 (1989)
23. I. Babuška, M. Suri, The h-p version of the finite element method with quasi-uniform meshes. RAIRO Modél. Math. Anal. Numér. **21**, 199–238 (1987)
24. I. Babuška, M. Suri, The optimal convergence rate of the p-version of the finite element method. SIAM J. Numer. Anal. **24**, 750–776 (1987)
25. I. Babuška, M. Suri, The treatment of nonhomogeneous Dirichlet boundary conditions by the p-version of the finite element method. Numer. Math. **55**, 97–121 (1989)
26. I. Babuška, M. Vogelius, Feedback and adaptive finite element solution of one-dimensional boundary value problems. Numer. Math. **44**, 75–102 (1984)
27. M. Bach, S.A. Nazarov, W.L. Wendland, *Stable Propagation of a Mode-1 Planar Crack in an Isotropic Elastic Space. Comparison of the Irwin and the Griffith Approaches*. Current Problems of Analysis and Mathematical Physics (Taormina, Italy, 1998), Aracne, Rome, 2000, pp. 167–189
28. C. Baiocchi, A. Capelo, *Variational and Quasivariational Inequalities - Applications to Free Boundary Problems* (Wiley, New York, 1984)
29. A. Bamberger, T. Ha-Duong, Formulation variationnelle espace-temps pour le calcul par potentiel retardé de la diffraction d'une onde acoustique. I. Math. Methods Appl. Sci. **8**, 405–435 (1986)
30. L. Banjai, S. Sauter, Rapid solution of the wave equation in unbounded domains. SIAM J. Numer. Anal. **47**, 227–249 (2008/09)
31. L. Banjai, M. Schanz, *Wave Propagation Problems Treated with Convolution Quadrature and BEM*. Fast Boundary Element Methods in Engineering and Industrial Applications. Lect. Notes Appl. Comput. Mech., vol. 63 (Springer, Heidelberg, 2012), pp. 145–184
32. R.E. Bank, R.K. Smith, A posteriori error estimates based on hierarchical bases. SIAM J. Numer. Anal. **30**, 921–935 (1993)
33. L. Banz, H. Gimperlein, A. Issaoui, E.P. Stephan, Stabilized mixed hp-BEM for frictional contact problems in linear elasticity. Numer. Math. **135**, 217–263 (2017)
34. L. Banz, H. Gimperlein, Z. Nezhi, E.P. Stephan, Time domain BEM for sound radiation of tires. Comput. Mech. **58**, 45–57 (2016)
35. L. Banz, G. Milicic, N. Ovcharova, Improved stabilization technique for frictional contact problems solved with hp-bem. Comput. Math. Appl. Mech. Eng. (2018, to appear)

36. L. Banz, A. Schröder, *Biorthogonal Basis Functions in hp-adaptive fem for Elliptic Obstacle Problems*, 2015, pp. 1721–1742
37. L. Banz, E.P. Stephan, On hp-adaptive BEM for frictional contact problems in linear elasticity. Comput. Math. Appl. **69**, 559–581 (2015)
38. L. Banz, E.P. Stephan, Comparison of mixed hp-bem (stabilized and non-stabilized) for frictional contact problems. J. Comput. Appl. Math. **295**, 92–102 (2016)
39. H.J.C. Barbosa, T.J.R. Hughes, Circumventing the Babuška-Brezzi condition in mixed finite element approximations of elliptic variational inequalities. Comput. Methods Appl. Mech. Eng. **97**, 193–210 (1992)
40. R. Beck, P. Deuflhard, R. Hiptmair, R.H.W. Hoppe, B. Wohlmuth, Adaptive multilevel methods for edge element discretizations of Maxwell's equations. Surveys Math. Indust. **8**, 271–312 (1999)
41. A. Bendali, Numerical analysis of the exterior boundary value problem for the time-harmonic Maxwell equations by a boundary finite element method. I. The continuous problem. Math. Comput. **43**, 29–46 (1984)
42. A. Bendali, Numerical analysis of the exterior boundary value problem for the time-harmonic Maxwell equations by a boundary finite element method. II. The discrete problem. Math. Comput. **43**, 47–68 (1984)
43. C.A. Berenstein, R. Gay, *Complex Variables*. Graduate Texts in Mathematics, vol. 125 (Springer, New York, 1991)
44. J. Bergh, J. Löfström, *Interpolation Spaces. An Introduction*. Grundlehren der Mathematischen Wissenschaften, No. 223 (Springer, Berlin-New York, 1976)
45. C. Bernardi, Y. Maday, A.T. Patera, *Domain Decomposition by the Mortar Element Method*. Asymptotic and Numerical Methods for Partial Differential Equations with Critical Parameters (Beaune, 1992), NATO Adv. Sci. Inst. Ser. C Math. Phys. Sci., vol. 384 (Kluwer Acad. Publ., Dordrecht, 1993), pp. 269–286
46. Ch. Bernardi, Y. Maday, Polynomial interpolation results in Sobolev spaces. J. Comput. Appl. Math. **43**, 53–80 (1992)
47. Ch. Bernardi, Y. Maday, *Spectral Methods*. Handbook of Numerical Analysis, Vol. V. Handb. Numer. Anal. (North-Holland, Amsterdam, 1997), pp. 209–485
48. Ch. Bernardi, Y. Maday, *Spectral, Spectral Element and Mortar Element Methods*. Theory and Numerics of Differential Equations (Durham, 2000) (Springer, Berlin, 2001), pp. 1–57
49. D.P. Bertsekas, Nondifferentiable optimization via approximation. Math. Program. Stud. **3**, 1–25 (1975)
50. A. Bespalov, N. Heuer, The p-version of the boundary element method for a three-dimensional crack problem. J. Integr. Equ. Appl. **17**, 243–258 (2005)
51. A. Bespalov, N. Heuer, The p-version of the boundary element method for weakly singular operators on piecewise plane open surfaces. Numer. Math. **106**, 69–97 (2007)
52. A. Bespalov, N. Heuer, The hp-version of the boundary element method with quasi-uniform meshes in three dimensions. M2AN Math. Model. Numer. Anal. **42**, 821–849 (2008)
53. A. Bespalov, S. Nicaise, A priori error analysis of the BEM with graded meshes for the electric field integral equation on polyhedral surfaces. Comput. Math. Appl. **71**, 1636–1644 (2016)
54. J. Bielak, R.C. MacCamy, An exterior interface problem in two-dimensional elastodynamics. Quart. Appl. Math. **41**, 143–159 (1983/84)
55. E. Blum, W. Oettli, *Mathematische Optimierung* (Springer, Berlin-New York, 1975)
56. D. Boffi, F. Brezzi, M. Fortin, *Mixed Finite Element Methods and Applications*. Springer Series in Computational Mathematics, vol. 44 (Springer, Heidelberg, 2013)
57. D. Boffi, M. Costabel, M. Dauge, L. Demkowicz, R. Hiptmair, Discrete compactness for the p-version of discrete differential forms. SIAM J. Numer. Anal. **49**, 135–158 (2011)
58. A. Bossavit, The computation of eddy-currents, in dimension 3, by using mixed finite elements and boundary elements in association. Math. Comput. Modell. **15**, 33–42 (1991)
59. D. Braess, A posteriori error estimators for obstacle problems—another look. Numer. Math. **101**, 415–421 (2005)

60. D. Braess, *Finite Elements*, 3rd edn. (Cambridge University Press, Cambridge, 2007). Theory, Fast Solvers, and Applications in Elasticity Theory. Translated from the German by L.L. Schumaker
61. J.H. Bramble, A.H. Schatz, Higher order local accuracy by averaging in the finite element method. Math. Comput. **31**, 94–111 (1977)
62. J.H. Bramble, R.D. Lazarov, J.E. Pasciak, A least-squares approach based on a discrete minus one inner product for first order systems. Math. Comput. **66**, 935–955 (1997)
63. S.C. Brenner, L.R. Scott, *The Mathematical Theory of Finite Element Methods*, 3rd edn. Texts in Applied Mathematics, vol. 15 (Springer, New York, 2008)
64. H. Brezis, Èquations et inéquations non linéaires dans les espaces vectoriels en dualité. Ann. Inst. Fourier (Grenoble) **18**, 115–175 (1968)
65. F. Brezzi, M. Fortin, *Mixed and Hybrid Finite Element Methods*. Springer Series in Computational Mathematics, vol. 15 (Springer, New York, 1991)
66. U. Brink, E.P. Stephan, Adaptive coupling of boundary elements and mixed finite elements for incompressible elasticity. Numer. Methods Partial Differ. Equ. **17**, 79–92 (2001)
67. A. Buffa, *Trace Theorems on Non-smooth Boundaries for Functional Spaces Related to Maxwell Equations: An Overview*. Computational Electromagnetics (Kiel, 2001). Lect. Notes Comput. Sci. Eng., vol. 28 (Springer, Berlin, 2003), pp. 23–34
68. A. Buffa, S.H. Christiansen, The electric field integral equation on Lipschitz screens: definitions and numerical approximation. Numer. Math. **94**, 229–267 (2003)
69. A. Buffa, P. Ciarlet, Jr., On traces for functional spaces related to Maxwell's equations. I. An integration by parts formula in Lipschitz polyhedra. Math. Methods Appl. Sci. **24**, 9–30 (2001)
70. A. Buffa, P. Ciarlet, Jr., On traces for functional spaces related to Maxwell's equations. II. Hodge decompositions on the boundary of Lipschitz polyhedra and applications. Math. Methods Appl. Sci. **24**, 31–48 (2001)
71. A. Buffa, M. Costabel, D. Sheen, On traces for $\mathbf{H}(\mathbf{curl}, \Omega)$ in Lipschitz domains. J. Math. Anal. Appl. **276**, 845–867 (2002)
72. A. Buffa, R. Hiptmair, *Galerkin Boundary Element Methods for Electromagnetic Scattering*. Topics in Computational Wave Propagation. Lect. Notes Comput. Sci. Eng., vol. 31 (Springer, Berlin, 2003), pp. 83–124
73. C. Carstensen, Interface problem in holonomic elastoplasticity. Math. Methods Appl. Sci. **16**, 819–835 (1993)
74. C. Carstensen, Efficiency of a posteriori BEM-error estimates for first-kind integral equations on quasi-uniform meshes. Math. Comput. **65**, 69–84 (1996)
75. C. Carstensen, A posteriori error estimate for the symmetric coupling of finite elements and boundary elements. Computing **57**, 301–322 (1996)
76. C. Carstensen, An a posteriori error estimate for a first-kind integral equation. Math. Comput. **66**, 139–155 (1997)
77. C. Carstensen, S. Bartels, Each averaging technique yields reliable a posteriori error control in FEM on unstructured grids. I. Low order conforming, nonconforming, and mixed FEM. Math. Comput. **71**, 945–969 (2002)
78. C. Carstensen, B. Faermann, Mathematical foundation of a posteriori error estimates and adaptive mesh-refining algorithms for boundary integral equations of the first kind. Eng. Anal. Bound. Elem. **25**, 497–509 (2001)
79. C. Carstensen, M. Feischl, M. Page, D. Praetorius, Axioms of adaptivity. Comput. Math. Appl. **67**, 1195–1253 (2016)
80. C. Carstensen, S.A. Funken, Coupling of nonconforming finite elements and boundary elements. I. A priori estimates. Computing **62**, 229–241 (1999)
81. C. Carstensen, S.A. Funken, Coupling of mixed finite elements and boundary elements. IMA J. Numer. Anal. **20**, 461–480 (2000)
82. C. Carstensen, S.A. Funken, E.P. Stephan, A posteriori error estimates for hp-boundary element methods. Appl. Anal. **61**, 233–253 (1996)

83. C. Carstensen, S.A. Funken, E.P. Stephan, On the adaptive coupling of FEM and BEM in 2-d-elasticity. Numer. Math. **77**, 187–221 (1997)
84. C. Carstensen, D. Gallistl, J. Gedicke, Justification of the saturation assumption. Numer. Math. **134**, 1–25 (2016)
85. C. Carstensen, J. Gwinner, FEM and BEM coupling for a nonlinear transmission problem with Signorini contact. SIAM J. Numer. Anal. **34**, 1845–1864 (1997)
86. C. Carstensen, M. Maischak, D. Praetorius, E.P. Stephan, Residual-based a posteriori error estimate for hypersingular equation on surfaces. Numer. Math. **97**, 397–425 (2004)
87. C. Carstensen, M. Maischak, E.P. Stephan, A posteriori error estimate and h-adaptive algorithm on surfaces for Symm's integral equation. Numer. Math. **90**, 197–213 (2001)
88. C. Carstensen, D. Praetorius, Averaging techniques for the effective numerical solution of Symm's integral equation of the first kind. SIAM J. Sci. Comput. **27**, 1226–1260 (2006)
89. C. Carstensen, D. Praetorius, Convergence of adaptive boundary element methods. J. Integr. Equ. Appl. **24**, 1–23 (2012)
90. C. Carstensen, O. Scherf, P. Wriggers, Adaptive finite elements for elastic bodies in contact. SIAM J. Sci. Comput. **20**, 1605–1626 (1999)
91. C. Carstensen, E.P. Stephan, Adaptive coupling of boundary elements and finite elements. RAIRO Modél. Math. Anal. Numér. **29**, 779–817 (1995)
92. C. Carstensen, E.P. Stephan, A posteriori error estimates for boundary element methods. Math. Comput. **64**, 483–500 (1995)
93. C. Carstensen, E.P. Stephan, Adaptive boundary element methods for some first kind integral equations. SIAM J. Numer. Anal. **33**, 2166–2183 (1996)
94. J. Chabassier, A. Chaigne, P. Joly, Time domain simulation of a piano. Part 1: Model description. ESAIM Math. Model. Numer. Anal. **48**, 1241–1278 (2014)
95. G.A. Chandler, I.G. Graham, Product integration-collocation methods for noncompact integral operator equations. Math. Comput. **50**(181), 125–138 (1988)
96. G.A. Chandler, I.H. Sloan, Spline qualocation methods for boundary integral equations. Numer. Math. **58**, 537–567 (1990)
97. D. Chen, V.A. Menegatto, X. Sun, A necessary and sufficient condition for strictly positive definite functions on spheres. Proc. Am. Math. Soc. **131**, 2733–2740 (2003)
98. G. Chen, J. Zhou, *Boundary Element Methods with Applications to Nonlinear Problems*, 2nd edn. Atlantis Studies in Mathematics for Engineering and Science, vol. 7b (Atlantis Press, Paris; World Scientific Publishing, Hackensack, NJ, 2010)
99. X. Chen, Smoothing methods for nonsmooth, nonconvex minimization. Math. Program. **134**, 71–99 (2012)
100. X. Chen, L. Qi, D. Sun, Global and superlinear convergence of the smoothing Newton method and its application to general box constrained variational inequalities. Math. Comput. **67**, 519–540 (1998)
101. A. Chernov, *Nonconforming boundary elements and finite elements for interface and contact problems with friction – hp-version for mortar, penalty and Nitsche's methods*, Ph.D. thesis, Leibniz Universität Hannover, 2006
102. A. Chernov, M. Maischak, E.P. Stephan, A priori error estimates for hp penalty BEM for contact problems in elasticity. Comput. Methods Appl. Mech. Eng. **196**, 3871–3880 (2007)
103. A. Chernov, M. Maischak, E.P. Stephan, hp-mortar boundary element method for two-body contact problems with friction. Math. Methods Appl. Sci. **31**, 2029–2054 (2008)
104. A. Chernov, T. von Petersdorff, Ch. Schwab, Exponential convergence of hp quadrature for integral operators with Gevrey kernels. ESAIM Math. Model. Numer. Anal. **45**, 387–422 (2011)
105. P.G. Ciarlet, *Basic Error Estimates for Elliptic Problems*. Handbook of Numerical Analysis, Vol. II (North-Holland, Amsterdam, 1991), pp. 17–351
106. P.G. Ciarlet, *The Finite Element Method for Elliptic Problems*. Classics in Applied Mathematics, vol. 40 (Society for Industrial and Applied Mathematics (SIAM), Philadelphia, PA, 2002)

107. X. Claeys, R. Hiptmair, Integral equations on multi-screens. Integr. Equ. Oper. Theory **77**, 167–197 (2013)
108. X. Claeys, R. Hiptmair, Integral equations for electromagnetic scattering at multi-screens. Integr. Equ. Oper. Theory **84**, 33–68 (2016)
109. F.H. Clarke, *Optimization and Nonsmooth Analysis*, 2nd edn. Classics in Applied Mathematics, vol. 5 (Society for Industrial and Applied Mathematics (SIAM), Philadelphia, PA, 1990)
110. Ph. Clément, Approximation by finite element functions using local regularization. RAIRO Analyse Numérique **9**, 77–84 (1975)
111. A.R. Conn, N.I.M. Gould, P.L. Toint, *Trust-Region Methods*. MPS/SIAM Series on Optimization (Society for Industrial and Applied Mathematics (SIAM); Mathematical Programming Society (MPS), Philadelphia, PA, 2000)
112. M. Costabel, *Principles of Boundary Element Methods*. Finite Elements in Physics (Lausanne, 1986) (North-Holland, Amsterdam, 1987), pp. 243–274
113. M. Costabel, *Symmetric Methods for the Coupling of Finite Elements and Boundary Elements (Invited Contribution)*. Boundary Elements IX, Vol. 1 (Stuttgart, 1987) (Comput. Mech., Southampton, 1987), pp. 411–420
114. M. Costabel, Boundary integral operators on Lipschitz domains: elementary results. SIAM J. Math. Anal. **19**, 613–626 (1988)
115. M. Costabel, *A Symmetric Method for the Coupling of Finite Elements and Boundary Elements*. The Mathematics of Finite Elements and Applications, VI (Uxbridge, 1987) (Academic Press, London, 1988), pp. 281–288
116. M. Costabel, *Randelemente* (Vorlesungsmanuskript, TU Darmstadt, 1989)
117. M. Costabel, *Time-Dependent Problems with a Boundary Integral Equation Method*. Encyclopedia of Computational Mechanics, 2004
118. M. Costabel, *Some Historical Remarks on the Positivity of Boundary Integral Operators*. Boundary Element Analysis. Lect. Notes Appl. Comput. Mech., vol. 29 (Springer, Berlin, 2007), pp. 1–27
119. M. Costabel, M. Dauge, S. Nicaise, Analytic regularity for linear elliptic systems in polygons and polyhedra. Math. Models Methods Appl. Sci. **22**, 1250015, 63 (2012)
120. M. Costabel, V.J. Ervin, E.P. Stephan, Symmetric coupling of finite elements and boundary elements for a parabolic-elliptic interface problem. Quart. Appl. Math. **48**, 265–279 (1990)
121. M. Costabel, V.J. Ervin, E.P. Stephan, Experimental convergence rates for various couplings of boundary and finite elements. Math. Comput. Model. **15**, 93–102 (1991)
122. M. Costabel, V.J. Ervin, E.P. Stephan, Quadrature and collocation methods for the double layer potential on polygons. Z. Anal. Anwendungen **12**, 699–707 (1993)
123. M. Costabel, A. McIntosh, On Bogovskiĭ and regularized Poincaré integral operators for de Rham complexes on Lipschitz domains. Math. Z. **265**, 297–320 (2010)
124. M. Costabel, F. Penzel, R. Schneider, Convergence of boundary element collocation methods for Dirichlet and Neumann screen problems in $\mathbf{R}^3$. Appl. Anal. **49**, 101–117 (1993)
125. M. Costabel, E. Stephan, W.L. Wendland, *Zur Randintegralmethode für das erste Fundamentalproblem der ebenen Elastizitätstheorie auf Polygongebieten*. Recent Trends in Mathematics (Reinhardsbrunn, 1982). Teubner-Texte zur Math., vol. 50 (Teubner, Leipzig, 1982), pp. 56–68
126. M. Costabel, E.P. Stephan, Curvature terms in the asymptotic expansions for solutions of boundary integral equations on curved polygons. J. Integr. Equ. **5**, 353–371 (1983)
127. M. Costabel, E.P. Stephan, The normal derivative of the double layer potential on polygons and Galerkin approximation. Appl. Anal. **16**, 205–228 (1983)
128. M. Costabel, E.P. Stephan, *Boundary Integral Equation for Mixed Boundary Value Problem in Polygonal Domains and Galerkin Approximation*. Mathematical Models and Methods in Mechanics. Banach Center Publications, vol. 15 (PWN - Polish Scientific Publishers, 1985), pp. 175–251
129. M. Costabel, E.P. Stephan, A direct boundary integral equation method for transmission problems. J. Math. Anal. Appl. **106**, 367–413 (1985)

130. M. Costabel, E.P. Stephan, An improved boundary element Galerkin method for three-dimensional crack problems. Integr. Equ. Oper. Theory **10**, 467–504 (1987)
131. M. Costabel, E.P. Stephan, On the convergence of collocation methods for boundary integral equations on polygons. Math. Comput. **49**, 461–478 (1987)
132. M. Costabel, E.P. Stephan, *Coupling of Finite Elements and Boundary Elements for Inhomogeneous Transmission Problems in* $\mathbf{R}^3$. The Mathematics of Finite Elements and Applications, VI (Uxbridge, 1987) (Academic Press, London, 1988), pp. 289–296
133. M. Costabel, E.P. Stephan, *Coupling of Finite Elements and Boundary Elements for Transmission Problems of Elastic Waves in* $\mathbf{R}^3$. Advanced Boundary Element Methods (San Antonio, TX, 1987) (Springer, Berlin, 1988), pp. 117–124
134. M. Costabel, E.P. Stephan, Duality estimates for the numerical solution of integral equations. Numer. Math. **54**, 339–353 (1988)
135. M. Costabel, E.P. Stephan, Strongly elliptic boundary integral equations for electromagnetic transmission problems. Proc. Roy. Soc. Edinburgh Sect. A **109**, 271–296 (1988)
136. M. Costabel, E.P. Stephan, Coupling of finite and boundary element methods for an elastoplastic interface problem. SIAM J. Numer. Anal. **27**, 1212–1226 (1990)
137. M. Costabel, E.P. Stephan, Integral equations for transmission problems in linear elasticity. J. Integr. Equ. Appl. **2**, 211–223 (1990)
138. M. Costabel, E.P. Stephan, W.L. Wendland, On boundary integral equations of the first kind for the bi-Laplacian in a polygonal plane domain. Ann. Scuola Norm. Sup. Pisa Cl. Sci. (4) **10**, 197–241 (1983)
139. R.W. Cottle, J.S. Pang, R.E. Stone, *The Linear Complementarity Problem* (Academic Press, Boston, 1992)
140. G.B. Dantzig, *Linear Programming and Extensions*, corrected edn. Princeton Landmarks in Mathematics (Princeton University Press, Princeton, NJ, 1998)
141. M. Dauge, *Elliptic Boundary Value Problems on Corner Domains. Smoothness and Asymptotics of Solutions*. Lecture Notes in Mathematics, vol. 1341 (Springer, Berlin, 1988)
142. R.A. DeVore, G.G. Lorentz, *Constructive Approximation*. Grundlehren der Mathematischen Wissenschaften, vol. 303 (Springer, Berlin, 1993)
143. W. Dörfler, A convergent adaptive algorithm for Poisson's equation. SIAM J. Numer. Anal. **33**, 1106–1124 (1996)
144. W. Dörfler, R.H. Nochetto, Small data oscillation implies the saturation assumption. Numer. Math. **91**, 1–12 (2002)
145. C. Eck, *Existence and regularity of solutions to contact problems with friction (german)*, Ph.D. thesis, Universität Stuttgart, 1996
146. C. Eck, J. Jarušek, M. Krbec, *Unilateral Contact Problems*. Pure and Applied Mathematics, vol. 270 (Chapman & Hall/CRC, Boca Raton, FL, 2005)
147. C. Eck, S.A. Nazarov, W.L. Wendland, Asymptotic analysis for a mixed boundary-value contact problem. Arch. Ration. Mech. Anal. **156**, 275–316 (2001)
148. C. Eck, H. Schulz, O. Steinbach, W.L. Wendland, *An Adaptive Boundary Element Method for Contact Problems*. Error Controlled Adaptive Finite Elements in Solid Mechanics (Wiley, 2003), pp. 181–209
149. C. Eck, O. Steinbach, W.L. Wendland, A symmetric boundary element method for contact problems with friction. Math. Comput. Simul. **50**, 43–61 (1999)
150. C. Eck, W.L. Wendland, A residual-based error estimator for BEM-discretizations of contact problems. Numer. Math. **95**, 253–282 (2003)
151. I. Ekeland, R. Témam, *Convex Analysis and Variational Problems*. Classics in Applied Mathematics, vol. 28 (Society for Industrial and Applied Mathematics (SIAM), Philadelphia, PA, 1999)
152. J. Elschner, The double layer potential operator over polyhedral domains. I. Solvability in weighted Sobolev spaces. Appl. Anal. **45**, 117–134 (1992)
153. J. Elschner, The double layer potential operator over polyhedral domains. II. Spline Galerkin methods. Math. Methods Appl. Sci. **15**, 23–37 (1992)

154. J. Elschner, The h-p-version of spline approximation methods for Mellin convolution equations. J. Integr. Equ. Appl. **5**, 47–73 (1993)
155. J. Elschner, I.G. Graham, An optimal order collocation method for first kind boundary integral equations on polygons. Numer. Math. **70**, 1–31 (1995)
156. J. Elschner, I.G. Graham, Quadrature methods for Symm's integral equation on polygons. IMA J. Numer. Anal. **17**, 643–664 (1997)
157. J. Elschner, I.G. Graham, Numerical methods for integral equations of Mellin type. J. Comput. Appl. Math. **125**(1–2), 423–437 (2000)
158. J. Elschner, Y. Jeon, I.H. Sloan, E.P. Stephan, The collocation method for mixed boundary value problems on domains with curved polygonal boundaries. Numer. Math. **76**, 355–381 (1997)
159. J. Elschner, E.P. Stephan, A discrete collocation method for Symm's integral equation on curves with corners. J. Comput. Appl. Math. **75**, 131–146 (1996)
160. H. Engels, *Numerical Quadrature and Cubature*. Computational Mathematics and Applications (Academic Press, London-New York, 1980)
161. C. Erath, S. Ferraz-Leite, S. Funken, D. Praetorius, Energy norm based a posteriori error estimation for boundary element methods in two dimensions. Appl. Numer. Math. **59**, 2713–2734 (2009)
162. C. Erath, S. Funken, P. Goldenits, D. Praetorius, Simple error estimators for the Galerkin BEM for some hypersingular integral equation in 2D. Appl. Anal. **92**, 1194–1216 (2013)
163. E. Ernst, M. Théra, A converse to the Lions-Stampacchia theorem. ESAIM Control Optim. Calc. Var. **15**, 810–817 (2009)
164. V.J. Ervin, N. Heuer, An adaptive boundary element method for the exterior Stokes problem in three dimensions. IMA J. Numer. Anal. **26**, 297–325 (2006)
165. V.J. Ervin, N. Heuer, E.P. Stephan, On the h-p version of the boundary element method for Symm's integral equation on polygons. Comput. Methods Appl. Mech. Eng. **110**, 25–38 (1993)
166. V.J. Ervin, E.P. Stephan, Collocation with Chebyshev polynomials for a hypersingular integral equation on an interval. J. Comput. Appl. Math. **43**, 221–229 (1992)
167. G.I. Eskin, *Boundary Value Problems for Elliptic Pseudodifferential Equations*. Translations of Mathematical Monographs, vol. 52 (American Mathematical Society, Providence, RI, 1981)
168. F. Facchinei, J.-S. Pang, *Finite-Dimensional Variational Inequalities and Complementarity Problems*, Vol. I. Springer Series in Operations Research (Springer, New York, 2003)
169. F. Facchinei, J.-S. Pang, *Finite-Dimensional Variational Inequalities and Complementarity Problems*, Vol. II. Springer Series in Operations Research (Springer, New York, 2003)
170. B. Faermann, Local a-posteriori error indicators for the Galerkin discretization of boundary integral equations. Numer. Math. **79**, 43–76 (1998)
171. B. Faermann, Localization of the Aronszajn-Slobodeckij norm and application to adaptive boundary element methods. I. The two-dimensional case. IMA J. Numer. Anal. **20**, 203–234 (2000)
172. B. Faermann, Localization of the Aronszajn-Slobodeckij norm and application to adaptive boundary element methods. II. The three-dimensional case. Numer. Math. **92**, 467–499 (2002)
173. R.S. Falk, Error estimates for the approximation of a class of variational inequalities. Math. Comput. **28**, 963–971 (1974)
174. M. Feischl, Th. Führer, N. Heuer, M. Karkulik, D. Praetorius, Adaptive boundary element methods. Arch. Comput. Methods Eng. **22**, 309–389 (2015)
175. M. Feischl, Th. Führer, M. Karkulik, J.M. Melenk, D. Praetorius, Quasi-optimal convergence rates for adaptive boundary element methods with data approximation, part I: Weakly-singular integral equation. Calcolo **51**, 531–562 (2014)
176. M. Feischl, Th. Führer, G. Mitscha-Eibl, D. Praetorius, E.P. Stephan, Convergence of adaptive BEM and adaptive FEM-BEM coupling for estimators without h-weighting factor. Comput. Methods Appl. Math. **14**, 485–508 (2014)

177. M. Feischl, M. Karkulik, J.M. Melenk, D. Praetorius, Quasi-optimal convergence rate for an adaptive boundary element method. SIAM J. Numer. Anal. **51**, 1327–1348 (2013)
178. S. Ferraz-Leite, D. Praetorius, Simple a posteriori error estimators for the h-version of the boundary element method. Computing **83**, 135–162 (2008)
179. G. Fichera, *Boundary Value Problems of Elasticity with Unilateral Constraints*. Handbuch der Physik, Festkörpermechanik, vol. VIa/2, ed. by S. Flügge (Springer, Berlin, 1972), pp. 391–424
180. G.J. Fix, S. Gulati, G.I. Wakoff, On the use of singular functions with finite element approximations. J. Comput. Phys. **13**, 209–228 (1973)
181. Th. Führer, *Zur Kopplung von Finiten Elementen und Randelementen*, Ph.D. thesis, TU Wien, 2014
182. A. Gachechiladze, R. Gachechiladze, J. Gwinner, D. Natroshvili, A boundary variational inequality approach to unilateral contact problems with friction for micropolar hemitropic solids. Math. Methods Appl. Sci. **33**, 2145–2161 (2010)
183. A. Gachechiladze, R. Gachechiladze, J. Gwinner, D. Natroshvili, Contact problems with friction for hemitropic solids: boundary variational inequality approach. Appl. Anal. **90**, 279–303 (2011)
184. R. Gachechiladze, J. Gwinner, D. Natroshvili, A boundary variational inequality approach to unilateral contact with hemitropic materials. Mem. Differ. Equ. Math. Phys. **39**, 69–103 (2006)
185. D. Gaier, Integralgleichungen erster Art und konforme Abbildung. Math. Z. **147**, 113–129 (1976)
186. G.N. Gatica, G.C. Hsiao, F.-J. Sayas, Relaxing the hypotheses of Bielak-MacCamy's BEM-FEM coupling. Numer. Math. **120**, 465–487 (2012)
187. G.N. Gatica, N. Heuer, E.P. Stephan, An implicit-explicit residual error estimator for the coupling of dual-mixed finite elements and boundary elements in elastostatics. Math. Methods Appl. Sci. **24**, 179–191 (2001)
188. G.N. Gatica, M. Maischak, E.P. Stephan, Numerical analysis of a transmission problem with Signorini contact using mixed-FEM and BEM. ESAIM Math. Model. Numer. Anal. **45**, 779–802 (2011)
189. H. Gimperlein, M. Maischak, E. Schrohe, E.P. Stephan, Adaptive FE-BE coupling for strongly nonlinear transmission problems with Coulomb friction. Numer. Math. **117**, 307–332 (2011)
190. H. Gimperlein, M. Maischak, E.P. Stephan, Adaptive time domain boundary element methods with engineering applications. J. Integr. Equ. Appl. **29**, 75–105 (2017)
191. H. Gimperlein, F. Meyer, C. Özdemir, D. Stark, E.P. Stephan, Boundary elements with mesh refinements for the wave equation. Numer. Math. (2018)
192. H. Gimperlein, F. Meyer, C. Özdemir, D. Stark, E.P. Stephan, Time domain boundary elements for dynamic contact problems. Comput. Methods Appl. Mech. Eng. **333**, 147–175 (2018)
193. H. Gimperlein, Z. Nezhi, E.P. Stephan, A priori error estimates for a time-dependent boundary element method for the acoustic wave equation in a half-space. Math. Methods Appl. Sci. **40**, 448–462 (2017)
194. H. Gimperlein, C. Özdemir, D. Stark, E.P. Stephan, A residual a posteriori error estimate for the time-domain boundary element method. Preprint 2017
195. H. Gimperlein, C. Özdemir, E.P. Stephan, Time domain boundary element methods for the Neumann problem: error estimates and acoustics problems. J. Comput. Math. **36**, 70–89 (2018)
196. V. Girault, P.-A. Raviart, *Finite Element Methods for Navier-Stokes Equations*. Springer Series in Computational Mathematics, vol. 5 (Springer, Berlin, 1986). Theory and Algorithms
197. M. Gläfke, *Adaptive methods for time domain boundary integral equations*, Ph.D. thesis, Brunel University, 2012
198. M. Gläfke, M. Maischak, E.P. Stephan, Coupling of FEM and BEM for a transmission problem with nonlinear interface conditions. Hierarchical and residual error indicators. Appl. Numer. Math. **62**, 736–753 (2012)

199. R. Glowinski, *Numerical Methods for Nonlinear Variational Problems*. Scientific Computation (Springer, Berlin, 2008)
200. R. Glowinski, J.-L. Lions, R. Trémolières, *Numerical Analysis of Variational Inequalities*. Studies in Mathematics and Its Applications, vol. 8 (North-Holland Publishing, Amsterdam-New York, 1981)
201. D. Goeleven, *Noncoercive Variational Problems and Related Results*. Pitman Research Notes in Mathematics Series, vol. 357 (Longman, Harlow, 1996)
202. I.C. Gohberg, I.A. Feldman, *Convolution Equations and Projection Methods for Their Solution* (American Mathematical Society, Providence, RI, 1974). Translations of Mathematical Monographs, Vol. 41
203. M.O. González, *Classical Complex Analysis*. Monographs and Textbooks in Pure and Applied Mathematics, vol. 151 (Marcel Dekker, New York, 1992)
204. P. Grisvard, *Elliptic Problems in Nonsmooth Domains*. Monographs and Studies in Mathematics, vol. 24 (Pitman (Advanced Publishing Program), Boston, MA, 1985). Reprint 2011
205. H. Guediri, On a boundary variational inequality of the second kind modelling a friction problem. Math. Methods Appl. Sci. **25**, 93–114 (2002)
206. N.M. Günter, *Potential Theory and Its Applications to Basic Problems of Mathematical Physics* (Frederick Ungar Publishing, New York, 1967)
207. B.Q. Guo, *The h-p Version of the Finite Element Method for Solving Boundary Value Problems in Polyhedral Domains*. Boundary Value Problems and Integral Equations in Nonsmooth Domains (Luminy, 1993). Lecture Notes in Pure and Appl. Math., vol. 167 (Dekker, New York, 1995), pp. 101–120
208. B.Q. Guo, N. Heuer, The optimal rate of convergence of the p-version of the boundary element method in two dimensions. Numer. Math. **98**, 499–538 (2004)
209. B.Q. Guo, N. Heuer, The optimal convergence of the h-p version of the boundary element method with quasiuniform meshes for elliptic problems on polygonal domains. Adv. Comput. Math. **24**, 353–374 (2006)
210. B.Q. Guo, N. Heuer, E.P. Stephan, The h-p version of the boundary element method for transmission problems with piecewise analytic data. SIAM J. Numer. Anal. **33**, 789–808 (1996)
211. B.Q. Guo, T. von Petersdorff, E.P. Stephan, *An hp Version of the Boundary Element Method for Plane Mixed Boundary Value Problems*. Advances in Boundary Elements. Comput. Mech., vol. 1 (Springer, Berlin, 1989), pp. 95–103
212. J. Gwinner, On fixed points and variational inequalities—a circular tour. Nonlinear Anal. **5**, 565–583 (1981)
213. J. Gwinner, Discretization of semicoercive variational inequalities. Aequationes Math. **42**, 72–79 (1991)
214. J. Gwinner, *Boundary Element Convergence for Unilateral Harmonic Problems*. Progress in Partial Differential Equations: Calculus of Variations, Applications (Pont-à-Mousson, 1991). Pitman Res. Notes Math. Ser., vol. 267 (Longman Sci. Tech., Harlow, 1992), pp. 200–213
215. J. Gwinner, A discretization theory for monotone semicoercive problems and finite element convergence for p-harmonic Signorini problems. Z. Angew. Math. Mech. **74**, 417–427 (1994)
216. J. Gwinner, A note on pseudomonotone functions, regularization, and relaxed coerciveness. Nonlinear Anal. **30**, 4217–4227 (1997)
217. J. Gwinner, On the p-version approximation in the boundary element method for a variational inequality of the second kind modelling unilateral contact and given friction. Appl. Numer. Math. **59**, 2774–2784 (2009)
218. J. Gwinner, hp-FEM convergence for unilateral contact problems with Tresca friction in plane linear elastostatics. J. Comput. Appl. Math. **254**, 175–184 (2013)
219. J. Gwinner, Lagrange multipliers and mixed formulations for some inequality constrained variational inequalities and some nonsmooth unilateral problems. Optimization **66**, 1323–1336 (2017)
220. J. Gwinner, N. Ovcharova, From solvability and approximation of variational inequalities to solution of nondifferentiable optimization problems in contact mechanics. Optimization **64**, 1683–1702 (2015)

221. J. Gwinner, E.P. Stephan, *Boundary Element Convergence for a Variational Inequality of the Second Kind*. Parametric Optimization and Related Topics, III (Güstrow, 1991). Approx. Optim., vol. 3 (Lang, Frankfurt am Main, 1993), pp. 227–241
222. J. Gwinner, E.P. Stephan, A boundary element procedure for contact problems in plane linear elastostatics. RAIRO M2AN **27**, 457–480 (1993)
223. T. Ha-Duong, *On Retarded Potential Boundary Integral Equations and Their Discretisation*. Topics in Computational Wave Propagation. Lect. Notes Comput. Sci. Eng., vol. 31 (Springer, Berlin, 2003), pp. 301–336
224. T. Ha-Duong, B. Ludwig, I. Terrasse, A Galerkin BEM for transient acoustic scattering by an absorbing obstacle. Int. J. Numer. Methods Eng. **57**, 1845–1882 (2003)
225. W. Hackbusch, *Integral Equations*. International Series of Numerical Mathematics, vol. 120 (Birkhäuser Verlag, Basel, 1995)
226. G. Hämmerlin, K.-H. Hoffmann, *Numerische Mathematik*, 4th edn. (Springer-Lehrbuch, Springer, Berlin, 1994)
227. H. Han, A direct boundary element method for Signorini problems. Math. Comput. **55**, 115–128 (1990)
228. H. Han, A new class of variational formulations for the coupling of finite and boundary element methods. J. Comput. Math. **8**, 223–232 (1990)
229. F. Hartmann, The Somigliana identity on piecewise smooth surfaces. J. Elasticity **11**, 403–423 (1981)
230. T. Hartmann, E.P. Stephan, Rates of convergence for collocation with Jacobi polynomials for the airfoil equation. J. Comput. Appl. Math. **51**, 179–191 (1994)
231. T. Hartmann, E.P. Stephan, *A Discrete Collocation Method for a Hypersingular Integral Equation on Curves with Corners*. Contemporary Computational Mathematics - a Celebration of the 80th Birthday of Ian Sloan (Springer, 2018)
232. J. Haslinger, C.C. Baniotopoulos, P.D. Panagiotopoulos, A boundary multivalued integral "equation" approach to the semipermeability problem. Appl. Math. **38**, 39–60 (1993)
233. N. Heuer, Additive Schwarz method for the p-version of the boundary element method for the single layer potential operator on a plane screen. Numer. Math. **88**, 485–511 (2001)
234. N. Heuer, An hp-adaptive refinement strategy for hypersingular operators on surfaces. Numer. Methods Partial Differ. Equ. **18**, 396–419 (2002)
235. N. Heuer, M. Maischak, E.P. Stephan, Exponential convergence of the hp-version for the boundary element method on open surfaces. Numer. Math. **83**, 641–666 (1999)
236. N. Heuer, M. Maischak, E.P. Stephan, Preconditioned minimum residual iteration for the h-p version of the coupled FEM/BEM with quasi-uniform meshes. Numer. Linear Algebra Appl. **6**, 435–456 (1999)
237. N. Heuer, S. Meddahi, F.-J. Sayas, Symmetric coupling of LDG-FEM and DG-BEM. J. Sci. Comput. **68**, 303–325 (2016)
238. N. Heuer, M.E. Mellado, E.P. Stephan, hp-adaptive two-level methods for boundary integral equations on curves. Computing **67**, 305–334 (2001)
239. N. Heuer, M.E. Mellado, E.P. Stephan, A p-adaptive algorithm for the BEM with the hypersingular operator on the plane screen. Int. J. Numer. Methods Eng. **53**, 85–104 (2002)
240. N. Heuer, E.P. Stephan, The hp-version of the boundary element method on polygons. J. Integr. Equ. Appl. **8**, 173–212 (1996)
241. N. Heuer, E.P. Stephan, Boundary integral operators in countably normed spaces. Math. Nachr. **191**, 123–151 (1998)
242. N. Heuer, E.P. Stephan, *The Poincaré-Steklov Operator Within Countably Normed Spaces*. Mathematical Aspects of Boundary Element Methods (Palaiseau, 1998). Res. Notes Math., vol. 414 (Chapman & Hall/CRC, Boca Raton, FL, 2000), pp. 152–164
243. N. Heuer, T. Tran, Radial basis functions for the solution of hypersingular operators on open surfaces. Comput. Math. Appl. **63**, 1504–1518 (2012)
244. S. Hildebrandt, E. Wienholtz, Constructive proofs of representation theorems in separable Hilbert space. Commun. Pure Appl. Math. **17**, 369–373 (1964)
245. E. Hille, *Analytic Function Theory*, Vol. II (Ginn and Co., Boston-New York-Toronto, 1962)

246. R. Hiptmair, Symmetric coupling for eddy current problems. SIAM J. Numer. Anal. **40**, 41–65 (2002)
247. R. Hiptmair, Coupling of finite elements and boundary elements in electromagnetic scattering. SIAM J. Numer. Anal. **41**, 919–944 (2003)
248. R. Hiptmair, C. Schwab, Natural boundary element methods for the electric field integral equation on polyhedra. SIAM J. Numer. Anal. **40**, 66–86 (2002)
249. I. Hlaváček, J. Haslinger, J. Nečas, J. Lovíšek, *Solution of Variational Inequalities in Mechanics*. Applied Mathematical Sciences, vol. 66 (Springer, New York, 1988)
250. H. Holm, M. Maischak, E.P. Stephan, The hp-version of the boundary element method for Helmholtz screen problems. Computing **57**, 105–134 (1996)
251. H. Holm, M. Maischak, E.P. Stephan, Exponential convergence of the h-p version BEM for mixed boundary value problems on polyhedrons. Math. Methods Appl. Sci. **31**, 2069–2093 (2008)
252. L. Hörmander, Sur la fonction d'appui des ensembles convexes dans un espace localement convexe. Ark. Mat. **3**, 181–186 (1955)
253. L. Hörmander, *Linear Partial Differential Operators* (Springer, Berlin-New York, 1976)
254. P. Houston, E. Süli, A note on the design of hp-adaptive finite element methods for elliptic partial differential equations. Comput. Methods Appl. Mech. Eng. **194**, 229–243 (2005)
255. G.C. Hsiao, O. Steinbach, W.L. Wendland, Domain decomposition methods via boundary integral equations. J. Comput. Appl. Math. **125**, 521–537 (2000)
256. G.C. Hsiao, E.P. Stephan, W.L. Wendland, On the Dirichlet problem in elasticity for a domain exterior to an arc. J. Comput. Appl. Math. **34**, 1–19 (1991)
257. G.C. Hsiao, W.L. Wendland, A finite element method for some integral equations of the first kind. J. Math. Anal. Appl. **58**, 449–481 (1977)
258. G.C. Hsiao, W.L. Wendland, The Aubin-Nitsche lemma for integral equations. J. Integr. Equ. **3**, 299–315 (1981)
259. G.C. Hsiao, W.L. Wendland, *Boundary Integral Equations*. Applied Mathematical Sciences, vol. 164 (Springer, Berlin, 2008)
260. G.C. Hsiao, W.L. Wendland, *Boundary Element Methods: Foundation and Error Analysis*. Encyclopedia of Computational Mechanics, vol. 1, ed. by E. Stein et al. (Wiley, Chichester, 2004), pp. 339–373
261. J. Jarušek, Contact problems with bounded friction coercive case. Czechoslov. Math. J. **33**(108), 237–261 (1983)
262. Y. Jeon, I.H. Sloan, E.P. Stephan, J. Elschner, Discrete qualocation methods for logarithmic-kernel integral equations on a piecewise smooth boundary. Adv. Comput. Math. **7**, 547–571 (1997)
263. C. Johnson, J.C. Nédélec, On the coupling of boundary integral and finite element methods. Math. Comput. **35**, 1063–1079 (1980)
264. M. Karkulik, *Zur Konvergenz und Quasioptimalität adaptiver Randelementmethoden*, Ph.D. thesis, TU Wien, 2012
265. M. Karkulik, J.M. Melenk, Local high-order regularization and applications to hp-methods. Comput. Math. Appl. **70**, 1606–1639 (2015)
266. N. Kikuchi, J.T. Oden, *Contact Problems in Elasticity: A Study of Variational Inequalities and Finite Element Methods*. SIAM Studies in Applied Mathematics, vol. 8 (Society for Industrial and Applied Mathematics (SIAM), Philadelphia, PA, 1988)
267. D. Kinderlehrer, G. Stampacchia, *An Introduction to Variational Inequalities and Their Applications*. Classics in Applied Mathematics, vol. 31 (Society for Industrial and Applied Mathematics (SIAM), Philadelphia, PA, 2000)
268. A.Yu. Kokotov, P. Neittaanmäki, B.A. Plamenevskii, The Neumann problem for the wave equation in a cone. J. Math. Sci. (New York) **102**, 4400–4428 (2000)
269. A.Yu. Kokotov, P. Neittaanmäki, B.A. Plamenevskii, Diffraction on a cone: the asymptotic of solutions near the vertex.. J. Math. Sci. (New York) **109**, 1894–1910 (2002)
270. V.A. Kondratiev, Boundary value problems for elliptic equations in domains with conical and angular points. Trans. Moscow Math. Soc. **16**, 227–313 (1967)

271. V.A. Kozlov, V.G. Maz'ya, J. Rossmann, *Elliptic Boundary Value Problems in Domains with Point Singularities*. Mathematical Surveys and Monographs, vol. 52 (American Mathematical Society, Providence, RI, 1997)
272. J. Král, *Integral Operators in Potential Theory*. Lecture Notes in Mathematics, vol. 823 (Springer, Berlin, 1980)
273. M.A. Krasnosel'skiĭ, G.M. Vaĭnikko, P.P. Zabreĭko, Ya.B. Rutitskii, V.Ya. Stetsenko, *Approximate Solution of Operator Equations* (Wolters-Noordhoff Publishing, Groningen, 1972)
274. A. Krebs, M. Maischak, E.P. Stephan, Adaptive FEM-BEM coupling with a Schur complement error indicator. Appl. Numer. Math. **60**, 798–808 (2010)
275. A. Krebs, E.P. Stephan, A p-version finite element method for nonlinear elliptic variational inequalities in 2D. Numer. Math. **105**, 457–480 (2007)
276. R. Kress, *Linear Integral Equations*, 3rd edn. Applied Mathematical Sciences, vol. 82 (Springer, New York, 2014)
277. A. Kufner, O. John, S. Fučík, *Function Spaces* (Noordhoff International Publishing, Leyden; Academia, Prague, 1977)
278. V.D. Kupradze, T.G. Gegelia, M.O. Basheleĭshvili, T.V. Burchuladze, *Three-Dimensional Problems of the Mathematical Theory of Elasticity and Thermoelasticity* (North-Holland, Amsterdam-New York, 1979)
279. U. Lamp, T. Schleicher, E.P. Stephan, W.L. Wendland, Galerkin collocation for an improved boundary element method for a plane mixed boundary value problem. Computing **33**, 269–296 (1984)
280. U. Langer, O. Steinbach, Boundary element tearing and interconnecting methods. Computing **71**, 205–228 (2003)
281. F. Leydecker, *hp-version of the boundary element method for electromagnetic problems : error analysis, adaptivity, preconditioners*, Ph.D. thesis, Leibniz Universität Hannover, 2006
282. F. Leydecker, M. Maischak, E.P. Stephan, M. Teltscher, Adaptive FE-BE coupling for an electromagnetic problem in $\mathbb{R}^3$—a residual error estimator. Math. Methods Appl. Sci. **33**, 2162–2186 (2010)
283. F. Leydecker, M. Maischak, E.P. Stephan, M. Teltscher, A p-hierarchical error estimator for a fe-be coupling formulation applied to electromagnetic scattering problems in $\mathbb{R}^3$. Appl. Anal. **91**, 277–293 (2012)
284. J.-L. Lions, E. Magenes, *Non-homogeneous Boundary Value Problems and Applications. Vol. I.* Grundlehren der mathematischen Wissenschaften, vol. 181 (Springer, New York-Heidelberg, 1972). Translated from the French by P. Kenneth
285. R.C. MacCamy, E.P. Stephan, *A Simple Layer Potential Method for Three-Dimensional Eddy Current Problems*. Ordinary and Partial Differential Equations. Lecture Notes in Math., vol. 1964 (Springer, Berlin, 1982), pp. 477–484
286. R.C. MacCamy, E.P. Stephan, A boundary element method for an exterior problem for three-dimensional Max-well's equations. Appl. Anal. **16**, 141–163 (1983)
287. R.C. MacCamy, E.P. Stephan, Solution procedures for three-dimensional eddy current problems. J. Math. Anal. Appl. **101**, 348–379 (1984)
288. R.C. MacCamy, E.P. Stephan, A skin effect approximation for eddy current problems. Arch. Ration. Mech. Anal. **90**, 87–98 (1985)
289. R.C. MacCamy, M. Suri, A time-dependent interface problem for two-dimensional eddy currents. Quart. Appl. Math. **44**, 675–690 (1987)
290. M. Maischak, *hp-methoden für randintegralgleichungen bei 3d-problemen, theorie und implementierung*, Ph.D. thesis, Leibniz Universität Hannover, 1996
291. M. Maischak, *The analytical computation of the Galerkin elements for the Laplace, Lamé and Helmholtz equation in 3d-BEM, MAIPROGS documentation*, 2000
292. M. Maischak, *Fem/bem Methods for Signorini-Type Problems: Error Analysis, Adaptivity, Preconditioners*. Habilitationsschrift (Leibniz Universität Hannover, 2004)
293. M. Maischak, *Manual of the Software Package Maiprogs*, version 3.7.1 edn., 2012
294. M. Maischak, P. Mund, E.P. Stephan, Adaptive multilevel BEM for acoustic scattering. Comput. Methods Appl. Mech. Eng. **150**, 351–367 (1997)

295. M. Maischak, E.P. Stephan, The hp-version of the boundary element method in R^3: the basic approximation results. Math. Methods Appl. Sci. **20**, 461–476 (1997)
296. M. Maischak, E.P. Stephan, A least squares coupling method with finite elements and boundary elements for transmission problems. Comput. Math. Appl. **48**, 995–1016 (2004)
297. M. Maischak, E.P. Stephan, Adaptive hp-versions of BEM for Signorini problems. Appl. Numer. Math. **54**, 425–449 (2005)
298. M. Maischak, E.P. Stephan, A FEM-BEM coupling method for a nonlinear transmission problem modelling Coulomb friction contact. Comput. Methods Appl. Mech. Eng. **194**, 453–466 (2005)
299. M. Maischak, E.P. Stephan, Adaptive hp-versions of boundary element methods for elastic contact problems. Comput. Mech. **39**, 597–607 (2007)
300. M. Maischak, E.P. Stephan, *The hp-Version of the Boundary Element Method for the Lamé Equation in 3D*. Boundary Element Analysis. Lect. Notes Appl. Comput. Mech., vol. 29 (Springer, Berlin, 2007), pp. 97–112
301. V.G. Maz'ya, *Boundary Integral Equations*. Analysis, IV, Encyclopaedia Math. Sci., vol. 27 (Springer, Berlin, 1991), pp. 127–222
302. V.G. Maz'ya, B.A. Plamenevskiĭ, The coefficients in the asymptotic expansion of the solutions of elliptic boundary value problems near an edge. Dokl. Akad. Nauk SSSR **229**, 33–36 (1976)
303. V.G. Maz'ya, B.A. Plamenevskiĭ, The coefficients in the asymptotics of solutions of elliptic boundary value problems with conical points. Math. Nachr. **76**, 29–60 (1977)
304. W. McLean, *Strongly Elliptic Systems and Boundary Integral Equations* (Cambridge University Press, Cambridge, 2000)
305. W. McLean, S. Prössdorf, W.L. Wendland, A fully-discrete trigonometric collocation method. J. Integr. Equ. Appl. **5**, 103–129 (1993)
306. S. Meddahi, J. Valdés, O. Menéndez, P. Pérez, On the coupling of boundary integral and mixed finite element methods. J. Comput. Appl. Math. **69**, 113–124 (1996)
307. M. Mitrea, M. Wright, Boundary value problems for the Stokes system in arbitrary Lipschitz domains. Astérisque **344**, viii+241 (2012)
308. P. Morin, K.G. Siebert, A. Veeser, A basic convergence result for conforming adaptive finite elements. Math. Models Methods Appl. Sci. **18**, 707–737 (2008)
309. U. Mosco, Convergence of convex sets and of solutions of variational inequalities. Adv. Math. **3**, 510–585 (1969)
310. P. Mund, *On the Implementation of the h-p Boundary Element Method on Curved Surfaces*. Boundary Elements: Implementation and the Analysis of Advanced Algorithms. Notes on Numerical Fluid Mechanics, vol. 54 (Vieweg, Braunschweig, 1996), pp. 182–193
311. P. Mund, E.P. Stephan, Adaptive coupling and fast solution of FEM-BEM equations for parabolic-elliptic interface problems. Math. Methods Appl. Sci. **20**, 403–423 (1997)
312. P. Mund, E.P. Stephan, An adaptive two-level method for the coupling of nonlinear FEM-BEM equations. SIAM J. Numer. Anal. **36**, 1001–1021 (1999)
313. P. Mund, E.P. Stephan, An adaptive two-level method for hypersingular integral equations in $\mathbf{R}^3$. In: Proceedings of the 1999 International Conference on Computational Techniques and Applications (Canberra), vol. 42, 2000, pp. C1019–C1033
314. P. Mund, E.P. Stephan, J. Weisse, Two-level methods for the single layer potential in $\mathbf{R}^3$. Computing **60**, 243–266 (1998)
315. F.J. Narcowich, J.D. Ward, Scattered data interpolation on spheres: error estimates and locally supported basis functions. SIAM J. Math. Anal. **33**, 1393–1410 (2002)
316. S.A. Nazarov, Asymptotic behavior of the solution of the Dirichlet problem in an angular domain with a periodically changing boundary Mat. Zametki **49**, 86–96 (1991)
317. S.A. Nazarov, B.A. Plamenevskiĭ, *Elliptic Problems in Domains with Piecewise Smooth Boundaries*. De Gruyter Expositions in Mathematics, vol. 13 (Walter de Gruyter & Co., Berlin, 1994)
318. M. Neamtu, L.L. Schumaker, On the approximation order of splines on spherical triangulations. Adv. Comput. Math. **21**, 3–20 (2004)

319. J. Nečas, J. Jarušek, J. Haslinger, On the solution of the variational inequality to the Signorini problem with small friction. Boll. Un. Mat. Ital. B (5) **17**, 796–811 (1980)
320. J.C. Nédélec, *Approximation des équations intégrales en mécanique et en physique* (Ecole Polytechnique, Palaiseau, 1977)
321. J.C. Nédélec, Mixed finite elements in $\mathbf{R}^3$. Numer. Math. **35**, 315–341 (1980)
322. J.C. Nédélec, Integral equations with nonintegrable kernels. Integr. Equ. Oper. Theory **5**, 562–572 (1982)
323. J.C. Nédélec, *Acoustic and Electromagnetic Equations. Integral Representations for Harmonic Problems*. Applied Mathematical Sciences, vol. 144 (Springer, New York, 2001)
324. J.C. Nédélec, J. Planchard, Une méthode variationnelle d' éléments finis pour la résolution numérique d'un problème extérieur dans R^3. Rev. Franc. Automat. Informat. Rech. Operat. Sér. Rouge **7**, 105–129 (1973)
325. J.C. Nédélec, Computation of eddy currents on a surface in $\mathbf{R}^3$ by finite element methods. SIAM J. Numer. Anal. **15**, 580–594 (1978)
326. L. Nesemann, E.P. Stephan, Numerical solution of an adhesion problem with FEM and BEM. Appl. Numer. Math. **62**, 606–619 (2012)
327. J. Nečas, *Direct Methods in the Theory of Elliptic Equations*. Springer Monographs in Mathematics (Springer, Heidelberg, 2012). Translated from the 1967 French original by G. Tronel and A. Kufner, and a contribution by Ch. G. Simader
328. J. Nitsche, Ein Kriterium für die Quasi-Optimalität des Ritzschen Verfahrens. Numer. Math. **11**, 346–348 (1968)
329. M. Ochmann, Closed form solutions for the acoustical impulse response over a masslike or an absorbing plane. J. Acoust. Soc. Am. **129**, 3502–3512 (2011)
330. G. Of, O. Steinbach, Is the one-equation coupling of finite and boundary element methods always stable?. ZAMM Z. Angew. Math. Mech. **93**, 476–484 (2013)
331. E. Ostermann, *Numerical methods for space-time variational formulations of retarded potential boundary integral equations*, Ph.D. thesis, Leibniz Universität Hannover, 2009
332. N. Ovcharova, *Regularization Methods and Finite Element Approximation of Hemivariational Inequalities with Applications to Nonmonotone Contact Problems*, Ph.D. thesis, Universität der Bundeswehr München, 2012
333. N. Ovcharova, On the coupling of regularization techniques and the boundary element method for a hemivariational inequality modelling a delamination problem. Math. Methods Appl. Sci. **40**, 60–77 (2017)
334. N. Ovcharova, L. Banz, Coupling regularization and adaptive *hp*-BEM for the solution of a delamination problem. Numer. Math. **137**, 303–337 (2017)
335. N. Ovcharova, J. Gwinner, A study of regularization techniques of nondifferentiable optimization in view of application to hemivariational inequalities. J. Optim. Theory Appl. **162**, 754–778 (2014)
336. P.D. Panagiotopoulos, J. Haslinger, On the dual reciprocal variational approach to the Signorini-Fichera problem. Convex and nonconvex generalization. Z. Angew. Math. Mech. **72**, 497–506 (1992)
337. L.F. Pavarino, Additive Schwarz methods for the p-version finite element method. Numer. Math. **66**, 493–515 (1994)
338. T.D. Pham, T. Tran, Strongly elliptic pseudodifferential equations on the sphere with radial basis functions. Numer. Math. **128**, 589–614 (2014)
339. T.D. Pham, T. Tran, A. Chernov, Pseudodifferential equations on the sphere with spherical splines. Math. Models Methods Appl. Sci. **21**, 1933–1959 (2011)
340. F.V. Postell, E.P. Stephan, On the h-, p- and h-p versions of the boundary element method—numerical results. Comput. Methods Appl. Mech. Eng. **83**, 69–89 (1990)
341. R.A. Prato Torres, E.P. Stephan, F. Leydecker, A FE/BE coupling for the 3D time-dependent eddy current problem. Part I: a priori error estimates. Computing **88**, 131–154 (2010)
342. R.A. Prato Torres, E.P. Stephan, F. Leydecker, A FE/BE coupling for the 3D time-dependent eddy current problem. Part II: a posteriori error estimates and adaptive computations. Computing **88**, 155–172 (2010)

343. S. Prössdorf, *Linear Integral Equations*. Analysis, IV, Encyclopaedia Math. Sci., vol. 27 (Springer, Berlin, 1991), pp. 1–125
344. S. Prössdorf, A. Rathsfeld, Quadrature and collocation methods for singular integral equations on curves with corners. Z. Anal. Anwendungen **8**, 197–220 (1989)
345. S. Prössdorf, B. Silbermann, *Numerical Analysis for Integral and Related Operator Equations*. Operator Theory: Advances and Applications, vol. 52 (Birkhäuser Verlag, Basel, 1991)
346. L. Qi, D. Sun, Smoothing functions and smoothing Newton method for complementarity and variational inequality problems. J. Optim. Theory Appl. **113**, 121–147 (2002)
347. J. Radon, Über die Randwertaufgabe beim logarithmischen Potential. Math.–Nat. Kl. Abt. IIa **128**, 1123–1167 (1919)
348. E. Rank, A posteriori error estimates and adaptive refinement for some boundary integral element method. In: Proceedings Int. Conf. on Accuracy Estimates and Adaptive Refinements in FE Computations (ARFEC, Lisbon, 1984), pp. 55–64
349. R. Rannacher, W.L. Wendland, On the order of pointwise convergence of some boundary element methods. I. Operators of negative and zero order. RAIRO Modél. Math. Anal. Numér. **19**, 65–87 (1985)
350. R. Rannacher, W.L. Wendland, Pointwise convergence of some boundary element methods. II. RAIRO Modél. Math. Anal. Numér. **22**, 343–362 (1988)
351. S. Rempel, G. Schmidt, Eigenvalues for spherical domains with corners via boundary integral equations. Integr. Equ. Oper. Theory **14**, 229–250 (1991)
352. J.E. Roberts, J.-M. Thomas, *Mixed and Hybrid Methods*. Handbook of Numerical Analysis, Vol. II. Handb. Numer. Anal., II (North-Holland, Amsterdam, 1991), pp. 523–639
353. R.T. Rockafellar, R.J.-B. Wets, *Variational Analysis*. Grundlehren der Mathematischen Wissenschaften, vol. 317 (Springer, Berlin, 1998)
354. E.B. Saff, A.B.J. Kuijlaars, Distributing many points on a sphere. Math. Intell. **19**, 5–11 (1997)
355. J. Saranen, Local error estimates for some Petrov-Galerkin methods applied to strongly elliptic equations on curves. Math. Comput. **48**, 485–502 (1987)
356. J. Saranen, G. Vainikko, *Periodic Integral and Pseudodifferential Equations with Numerical Approximation*. Springer Monographs in Mathematics (Springer, Berlin, 2002)
357. J. Saranen, W.L. Wendland, On the asymptotic convergence of collocation methods with spline functions of even degree. Math. Comput. **45**, 91–108 (1985)
358. S. Sauter, A. Veit, Retarded boundary integral equations on the sphere: exact and numerical solution. IMA J. Numer. Anal. **34**, 675–699 (2014)
359. S.A. Sauter, Ch. Schwab, *Boundary Element Methods*. Springer Series in Computational Mathematics, vol. 39 (Springer, Berlin, 2011)
360. F.-J. Sayas, The validity of Johnson-Nédélec's BEM-FEM coupling on polygonal interfaces. SIAM J. Numer. Anal. **47**, 3451–3463 (2009)
361. F.-J. Sayas, *Retarded Potentials and Time Domain Boundary Integral Equations. A Road Map*. Springer Series in Computational Mathematics, vol. 50 (Springer, Cham, 2016)
362. A.H. Schatz, V. Thomée, W.L. Wendland, *Mathematical Theory of Finite and Boundary Element Methods*. DMV Seminar, vol. 15 (Birkhäuser Verlag, Basel, 1990)
363. A.H. Schatz, L.B. Wahlbin, Interior maximum norm estimates for finite element methods. Math. Comput. **31**, 414–442 (1977)
364. M. Schechter, *Principles of Functional Analysis* (Academic Press, New York-London, 1971,1973)
365. G. Schmidt, On spline collocation methods for boundary integral equations in the plane. Math. Methods Appl. Sci. **7**, 74–89 (1985)
366. G. Schmidt, Boundary element discretization of Poincaré-Steklov operators. Numer. Math. **69**, 83–101 (1994)
367. H. Schmitz, K. Volk, W.L. Wendland, Three-dimensional singularities of elastic fields near vertices. Numer. Methods Partial Differ. Equ. **9**, 323–337 (1993)
368. D. Schötzau, Ch. Schwab, Exponential convergence for hp-version and spectral finite element methods for elliptic problems in polyhedra. Math. Models Methods Appl. Sci. **25**, 1617–1661 (2015)

369. H. Schulz, O. Steinbach, A new a posteriori error estimator in adaptive direct boundary element methods: the Dirichlet problem. Calcolo **37**, 79–96 (2000)
370. H. Schulz, O. Steinbach, *A New a Posteriori Error Estimator in Adaptive Direct Boundary Element Methods. The Neumann Problem.* Multifield Problems (Springer, Berlin, 2000), pp. 201–208
371. H. Schulz, O. Steinbach, W.L. Wendland, *On Adaptivity in Boundary Element Methods.* Aspects of the Boundary Element Method (Kluwer, 2001). In: IUTAM/IACM/IABEM Symposium on Advanced Mathematical and Computational Mechanics, pp. 315–325
372. Ch. Schwab, Variable order composite quadrature of singular and nearly singular integrals. Computing **53**, 173–194 (1994)
373. Ch. Schwab, *p- and hp-Finite Element Methods.Theory and Applications in Solid and Fluid Mechanics*. Numerical Mathematics and Scientific Computation (The Clarendon Press, Oxford University Press, New York, 1998)
374. Ch. Schwab, M. Suri, The optimal p-version approximation of singularities on polyhedra in the boundary element method. SIAM J. Numer. Anal. **33**, 729–759 (1996)
375. Ch. Schwab, W.L. Wendland, On the extraction technique in boundary integral equations. Math. Comput. **68**, 91–122 (1999)
376. R. Seeley, *Topics in Pseudo-Differential Operators*. Pseudo-Diff. Operators (C.I.M.E., Stresa, 1968; Edizioni Cremonese, Rome, 1969), pp. 167–305
377. K.G. Siebert, A convergence proof for adaptive finite elements without lower bound. IMA J. Numer. Anal. **31**, 947–970 (2011)
378. I.H. Sloan, A quadrature-based approach to improving the collocation method. Numer. Math. **54**, 41–56 (1988)
379. I.H. Sloan, *Error Analysis of Boundary Integral Methods*. Acta Numerica, 1992 (Cambridge Univ. Press, Cambridge, 1992), pp. 287–339
380. I.H. Sloan, A. Spence, The Galerkin method for integral equations of the first kind with logarithmic kernel: theory. IMA J. Numer. Anal. **8**, 105–122 (1988)
381. I.H. Sloan, E.P. Stephan, Collocation with Chebyshev polynomials for Symm's integral equation on an interval. J. Aust. Math. Soc. Ser. B **34**, 199–211 (1992)
382. I.H. Sloan, T. Tran, The tolerant qualocation method for variable-coefficient elliptic equations on curves. J. Integr. Equ. Appl. **13**, 73–98 (2001)
383. I.H. Sloan, W.L. Wendland, Commutator properties for periodic splines. J. Approx. Theory **97**, 254–281 (1999)
384. I.H. Sloan, W.L. Wendland, Spline qualocation methods for variable-coefficient elliptic equations on curves. Numer. Math. **83**, 497–533 (1999)
385. I.H. Sloan, W.L. Wendland, Qualocation methods for elliptic boundary integral equations. Numer. Math. **79**, 451–483 (1998)
386. W. Spann, On the boundary element method for the Signorini problem of the Laplacian. Numer. Math. **65**, 337–356 (1993)
387. W. Spann, Error estimates for the boundary element approximation of a contact problem in elasticity. Math. Methods Appl. Sci. **20**, 205–217 (1997)
388. G. Stampacchia, *Variational Inequalities*. Theory and Applications of Monotone Operators, ed. by A. Ghizetti (Edizione Odersi, Gubbio, 1969), pp. 101–192
389. O. Steinbach, Adaptive finite element-boundary element solution of boundary value problems. J. Comput. Appl. Math. **106**, 307–316 (1999)
390. O. Steinbach, Adaptive boundary element methods based on computational schemes for Sobolev norms. SIAM J. Sci. Comput. **22**, 604–616 (2000)
391. O. Steinbach, *Numerische Näherungsverfahren für elliptische Randwertprobleme* (Springer, 2003)
392. O. Steinbach, A note on the stable one-equation coupling of finite and boundary elements. SIAM J. Numer. Anal. **49**, 1521–1531 (2011)
393. O. Steinbach, Boundary element methods for variational inequalities. Numer. Math. **126**, 173–197 (2014)

394. O. Steinbach, W.L. Wendland, On C. Neumann's method for second-order elliptic systems in domains with non-smooth boundaries. J. Math. Anal. Appl. **262**, 733–748 (2001)
395. E.P. Stephan, *Solution Procedures for Interface Problems in Acoustics and Electromagnetics*. Theoretical Acoustics and Numerical Techniques. CISM Courses and Lectures, vol. 277 (Springer, Vienna, 1983), pp. 291–348
396. E.P. Stephan, A boundary integral equation method for three-dimensional crack problems in elasticity. Math. Methods Appl. Sci. **8**, 609–623 (1986)
397. E.P. Stephan, Boundary integral equations for mixed boundary value problems in $\mathbf{R}^3$. Math. Nachr. **134**, 21–53 (1987)
398. E.P. Stephan, Boundary integral equations for screen problems in $\mathbf{R}^3$. Integr. Equ. Oper. Theory **10**, 236–257 (1987)
399. E.P. Stephan, Coupling of finite elements and boundary elements for some nonlinear interface problems. Comput. Methods Appl. Mech. Eng. **101**, 61–72 (1992)
400. E.P. Stephan, The h-p boundary element method for solving 2- and 3-dimensional problems. Comput. Methods Appl. Mech. Eng. **133**, 183–208 (1996)
401. E.P. Stephan, M. Maischak, A posteriori error estimates for fem-bem couplings of three-dimensional electromagnetic problems. Comput. Methods Appl. Mech. Eng. **194**, 441–452 (2005)
402. E.P. Stephan, M. Maischak, E. Ostermann, Transient boundary element method and numerical evaluation of retarded potentials. In: Computational Science - ICCS 2008, 8th International Conference, Kraków, Poland, June 23–25, 2008, Proceedings, Part II, 2008, pp. 321–330
403. E.P. Stephan, M. Maischak, E. Ostermann, TD-BEM for sound radiation in three dimensions and the numerical evaluation of retarded potentials. In: International Conference on Acoustics, NAG/DAGA, 2009
404. E.P. Stephan, M. Suri, On the convergence of the p-version of the boundary element Galerkin method. Math. Comput. **52**, 31–48 (1989)
405. E.P. Stephan, M. Suri, The h-p version of the boundary element method on polygonal domains with quasiuniform meshes. RAIRO Modél. Math. Anal. Numér. **25**, 783–807 (1991)
406. E.P. Stephan, M.T. Teltscher, Collocation with trigonometric polynomials for integral equations to the mixed boundary value problem. Numerische Mathematik (2018, to appear)
407. E.P. Stephan, T. Tran, Localization and post processing for the Galerkin boundary element method applied to three-dimensional screen problems. J. Integr. Equ. Appl. **8**, 457–481 (1996)
408. E.P. Stephan, W.L. Wendland, *Remarks to Galerkin and Least Squares Methods with Finite Elements for General Elliptic Problems*. Ordinary and Partial Differential Equations (Proc. Fourth Conf., Univ. Dundee, Dundee, 1976). Lecture Notes in Math., vol. 564 (Springer, Berlin, 1976), pp. 461–471
409. E.P. Stephan, W.L. Wendland, *Mathematische Grundlagen der finiten Element-Methoden*. Methoden und Verfahren der Mathematischen Physik [Methods and Procedures in Mathematical Physics], vol. 23 (Verlag Peter D. Lang, Frankfurt, 1982)
410. E.P. Stephan, W.L. Wendland, An augmented Galerkin procedure for the boundary integral method applied to two-dimensional screen and crack problems. Appl. Anal. **18**, 183–219 (1984)
411. E.P. Stephan, W.L. Wendland, An augmented Galerkin procedure for the boundary integral method applied to mixed boundary value problems. Appl. Numer. Math. **1**, 121–143 (1985)
412. G. Strang, G.J. Fix, *An Analysis of the Finite Element Method* (Prentice-Hall, Englewood Cliffs, NJ, 1973)
413. J.A. Stratton, *Electromagnetic Rheory* (Wiley, 2007)
414. F. Stummel, Perturbation theory for Sobolev spaces. Proc. Roy. Soc. Edinburgh Sect. A **73**, 5–49 (1975)
415. M.E. Taylor, *Pseudodifferential Operators*. Princeton Mathematical Series, vol. 34 (Princeton University Press, Princeton, NJ, 1981)
416. T. Tran, The K-operator and the Galerkin method for strongly elliptic equations on smooth curves: local estimates. Math. Comput. **64**, 501–513 (1995)

417. T. Tran, Local error estimates for the Galerkin method applied to strongly elliptic integral equations on open curves. SIAM J. Numer. Anal. **33**, 1484–1493 (1996)
418. T. Tran, Q.T. Le Gia, I.H. Sloan, E.P. Stephan, Boundary integral equations on the sphere with radial basis functions: error analysis. Appl. Numer. Math. **59**, 2857–2871 (2009)
419. T. Tran, I.H. Sloan, Tolerant qualocation—a qualocation method for boundary integral equations with reduced regularity requirement. J. Integr. Equ. Appl. **10**, 85–115 (1998)
420. T. Tran, E.P. Stephan, Additive Schwarz methods for the h-version boundary element method. Appl. Anal. **60**, 63–84 (1996)
421. G. Verchota, Layer potentials and regularity for the Dirichlet problem for Laplace's equation in Lipschitz domains. J. Funct. Anal. **59**, 572–611 (1984)
422. T. von Petersdorff, Boundary integral equations for mixed Dirichlet, Neumann and transmission problems. Math. Methods Appl. Sci. **11**, 185–213 (1989)
423. T. von Petersdorff, *Randwertprobleme der Elastizitätstheorie für Polyeder - Singularitäten und Approximation mit Randelementmethoden*, Ph.D. thesis, TU Darmstadt, 1989
424. T. von Petersdorff, E.P. Stephan, *A Direct Boundary Element Method for Interface Crack Problems*. Computational Mechanics '88 (Springer, 1988). In: Proceedings Atlanta, Vol 1, pp. 11.vii.1–11.vii.5
425. T. von Petersdorff, E.P. Stephan, Decompositions in edge and corner singularities for the solution of the Dirichlet problem of the Laplacian in a polyhedron. Math. Nachr. **149**, 71–103 (1990)
426. T. von Petersdorff, E.P. Stephan, Regularity of mixed boundary value problems in $\mathbf{R}^3$ and boundary element methods on graded meshes. Math. Methods Appl. Sci. **12**, 229–249 (1990)
427. H. Wendland, *Scattered Data Approximation*. Cambridge Monographs on Applied and Computational Mathematics, vol. 17 (Cambridge University Press, Cambridge, 2005)
428. W.L. Wendland, *Asymptotic Convergence of Boundary Element Methods; Integral Equation Methods for Mixed bvp's*. Preprint 611 (TU Darmstadt, Fachbereich Mathematik, 1981)
429. W.L. Wendland, *Boundary Element Methods and Their Asymptotic Convergence*. Theoretical Acoustics and Numerical Techniques. CISM Courses and Lectures, vol. 277 (Springer, Vienna, 1983), pp. 135–216
430. W.L. Wendland, *Randelementmethoden – eine Einführung* (Vorlesungsmanuskript, TU Darmstadt, 1985)
431. W.L. Wendland, Qualocation, the new variety of boundary element methods. Wiss. Z. Tech. Univ. Karl-Marx-Stadt **31**, 276–284 (1989)
432. W.L. Wendland, E.P. Stephan, A hypersingular boundary integral method for two-dimensional screen and crack problems. Arch. Ration. Mech. Anal. **112**, 363–390 (1990)
433. W.L. Wendland, E.P. Stephan, G.C. Hsiao, On the integral equation method for the plane mixed boundary value problem of the Laplacian. Math. Methods Appl. Sci. **1**, 265–321 (1979)
434. W.L. Wendland, D.H. Yu, Adaptive boundary element methods for strongly elliptic integral equations. Numer. Math. **53**, 539–558 (1988)
435. J. Wloka, *Partial Differential Equations* (Cambridge University Press, Cambridge, 1987). Translated from the German by C. B. Thomas and M. J. Thomas
436. H. Yserentant, On the multilevel splitting of finite element spaces. Numer. Math. **49**, 379–412 (1986)
437. E. Zeidler, *Nonlinear Functional Analysis and Its Applications. IV* (Springer, New York, 1988)
438. E. Zeidler, *Nonlinear Functional Analysis and Its Applications. II/B* (Springer, New York, 1990)
439. O.C. Zienkiewicz, D.W. Kelly, P. Bettess, *Marriage à la Mode—The Best of Both Worlds (Finite Elements and Boundary Integrals)*. Energy Methods in Finite Element Analysis (Wiley, Chichester, 1979), pp. 81–107

Index

© Springer International Publishing AG, part of Springer Nature 2018

J. Gwinner, E. P. Stephan, *Advanced Boundary Element Methods*,
Springer Series in Computational Mathematics 52,
https://doi.org/10.1007/978-3-319-92001-6

Printed in the United States
By Bookmasters